# UNIVERSE

# UNIVERSE

## FIFTH EDITION

**William J. Kaufmann III**

*Late of San Diego State University*

**Roger A. Freedman**

*University of California, Santa Barbara*

W. H. FREEMAN AND COMPANY

NEW YORK

Dedicated to
Lee Johnson Kaufmann
and
Caroline Robillard-Freedman,
Strong survivors

ACQUISITIONS EDITOR: Holly Hodder

DEVELOPMENTAL EDITOR: John Haber

PROJECT EDITOR: Diane Cimino Maass

DIRECTOR OF MEDIA: Patrick Shriner

COVER AND TEXT DESIGNER: Vicki Tomaselli

ILLUSTRATION COORDINATOR: Bill Page

ILLUSTRATIONS: Fine Line Illustrations

PRODUCTION COORDINATOR: Paul W. Rohloff

COMPOSITION: Sheridan Sellers and Bonnie Stuart
W.H. Freeman and Company Electronic Publishing Center

MANUFACTURING: RR Donnelley and Sons

MARKETING MANAGER: Kimberly Manzi

. . .

Copyright © 1985, 1988, 1991, 1994, 1999 by W. H. Freeman and Company
Library of Congress Cataloging-in-Publication Data
Kaufmann, William J.
Universe/William J. Kaufmann III, Roger A. Freedman. — 5th ed.
p. cm.
Includes bibliographical references and index.
ISBN 0-7167-3495-8
1. Astronomy.   2. Cosmology.   I. Freedman, Roger A. II. Title.
QB43.2.K38   1998
520—dc21                                                    98-19098
CIP

Printed in the United States of America
First printing 1998

# Contents Overview

# Contents

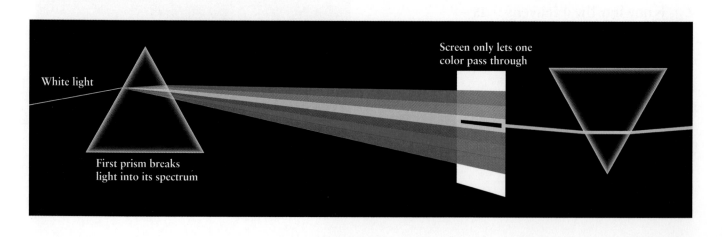

White light

First prism breaks light into its spectrum

Screen only lets one color pass through

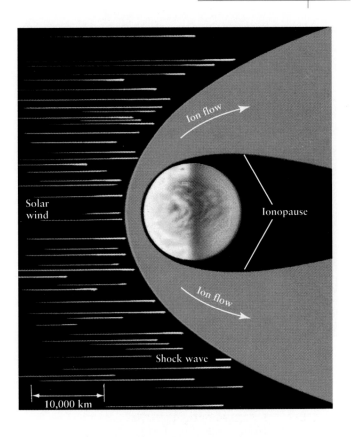

Solar wind

Ion flow

Ionopause

Ion flow

Shock wave

10,000 km

## PART IV    Galaxies and Cosmology    611

# To The Instructor

I have been a loyal user of Bill Kaufmann's *Universe* in my own classes at the University of California, Santa Barbara. Since the very first edition, it has taught my students the breadth of contemporary astronomy and the nature of scientific inquiry. Its spectacular images and lucid text have opened their eyes to the beauty of the sky and the thrill of scientific discovery. No other astronomy text I've come across explains so clearly and engagingly *how* science is made.

I was honored to be given the opportunity to revise *Universe* after Bill Kaufmann's untimely passing in 1994. The challenge for me was to make this edition even easier for students to learn from and for instructors to teach from. I've endeavored to bring to this book the insights I've gained from fifteen years of teaching astronomy. Above all, I've tried to retain the strengths that have made *Universe* a standard in astronomy education while enhancing its particular focus on *how scientists work and think*.

## Our developing picture of the universe guides the features of this book

By studying astronomy, students should learn not only *what* we know about our fascinating physical universe but also *how* we know it. This emphasis is at the heart of both the text's organization and its features. Together, they emphasize the scientific way of knowing, based on observation and continued questioning.

*Universe* still retains the classic features that Bill Kaufmann gave it:

- **A flexible four-part sequence** and optional box essays make it easy to structure a course as an instructor sees fit.

- **A narrative of scientific discovery** underscores each chapter.

- **Extensive review aids** include a large number of end-of-chapter questions of different levels of difficulty.

But along with these time-honored features, this fifth edition incorporates several innovations designed to help students meet the challenges of learning astronomy:

- **More help with mathematics and special topics** are found throughout the text and in an expanded program of optional boxes: "Tools of the Astronomer's Trade" and "Looking Deeper into Astronomy." Students will better understand how science builds models of nature.

- **NEW analogies** will help students use those models for themselves. I have added a third and entirely new kind of box, "The Heavens on the Earth," and have highlighted useful analogies within the text. These features will show students how the principles of astronomy can help them understand their own experience.

- **NEW questions and cautions** will help students learn that science depends on thinking critically about what we know. The chapter-opening questions will stimulate their curiosity, and cautions will help readers avoid common misconceptions and conceptual pitfalls.

- **More guest essays** will stimulate student curiosity. Readers share the excitement of unresolved puzzles and new answers.

- **Along with the latest discoveries and images,** I have added new information about how observations are made and labeled telescopic images with the wavelength of electromagnetic radiation used to make them. Students will learn more about how science is actually done.

- **NEW links to electronic media and Observing Tips and Tools** give students the tools to make their *own* discoveries. With the text and its accompanying CD-ROM, students can participate actively in the scientific process.

The rest of this preface describes the features of this edition in more detail. I'll begin with the features that have made this text so successful over four editions.

# Continuing Features

Because this book is designed for both one- and two-term courses, it is more comprehensive than its sister text, *Discovering the Universe*. To make its coverage accessible to a wide range of students, *Universe* has a flexible, modular structure that lets instructors teach the topics in the order desired.

## This book is organized for instructor flexibility

The chapters are organized into four parts, in a scheme that stresses how our understanding of the universe has developed. Part One introduces the foundations of astronomy: naked-eye observations; basic tools, such as Kepler's laws; the properties of light; and the design of telescopes. Part Two moves outward through the solar system, starting with its formation. Part Three introduces stellar astronomy with the star most familiar to us, our Sun, and subsequent chapters mirror the life history of a star. Part Four describes galaxies, cosmology, and current thinking about extraterrestrial intelligence.

Instructors may cover more or less of each part as time, preference, and the preparation of their students permit. An instructor who wants to teach a course that emphasizes the stars or that presents stellar astronomy before the solar system—as I often do in the introductory course at the University of California, Santa Barbara—has the freedom to do so. For example, a course could begin with the introductory chapters on gravitation and light (Chapters 4 and 5), then proceed directly to Chapter 18, which introduces the Sun and stars, without any loss of continuity.

*New to this edition,* I have enhanced this structure so that each part will orient students better and permit greater instructor flexibility:

- **A brief overview** and a dramatic image now open each part.

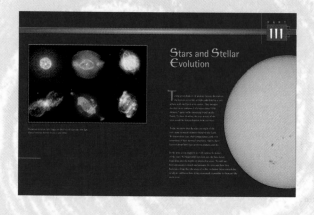

- **Links to other chapters** now appear in blue underlined type. These will help students see the connection between different chapters. They will also make it easier for students to locate material they need to review.

a nearly perfect blackbody with an average temperature of about 5800 K. (Blackbody spectra are described in Sections 5-3 and 5-4; in particular, see Figure 5-10). However, superimposed on this spectrum are many dark absorption lines, shown in Figure 5-11. Hence, the net spectrum of the photosphere is an *absorption line spectrum*, which we introduced in Figure 5-6. We see an absorption line spectrum whenever we view a hot, glowing object through a relatively cool gas. In

## Chapters emphasize scientific discovery

*Universe* does not rely solely on its Earth-outward sequence to show how our knowledge of the universe has grown. In each chapter, the first section draws on historic questions and exciting observations. These, in turn, raise new questions that help draw the reader toward the outer limits of our understanding.

For example, each of the chapters on the planets begins with the view of the planet from the Earth. Each chapter then explores the planet's atmosphere and surface before describing current models of the interior. It concludes with prospects for new discoveries by future spacecraft missions.

*New to this edition,* I have done more to place theory in the context of observations:

- **Einstein's theories of relativity** now appear in Chapter 24 along with black holes, rather than in Part One. Students appreciate the theories more when they can see immediately how they are applied.

- **Enhanced accounts of the history of astronomy** include Chapter 4 on the Copernican revolution. I also work history more fully into the text, as in Chapter 23 on neutron stars.

## Review aids help students get the most out of this book

My students have always appreciated the many features of *Universe* that help them review what they've learned. As in past editions, full-sentence section headings again tell the evolving story of astronomy, and highlighted tables of key data again introduce each planet in our solar system.

Chapters again end with summaries of Key Ideas and with extensive questions (with selected answers at the back of the book). These questions incorporate Problem-Solving Tips and Tools as one more step in building new skills.

*New to this edition,* I have added a student-oriented introduction to prepare students for their textbook, as well as further end-of-chapter tools:

- **How to Use This Textbook** follows this preface. A major stumbling block for many students is that they don't know how best to study astronomy. To this end, I tell students how to get the most out of the features of this book. I've included hints for students who may never have taken a college science course before.

- **Key Words** lists now include a page references to a term's first appearance in the chapter, so that students can more easily test their understanding.

- **Observing Tips and Tools** provide students with practical hints to help them with the Observing Projects.

### Observing Projects

Observing tips and tools:

If you do not have access to a telescope, you can learn a lot by observing the Moon through binoculars. Note that the Moon will appear right-side-up through binoculars, but inverted through a telescope;

# New to This Edition

Now I'll describe new features that build on Bill Kaufmann's belief in letting students share in how astronomy is done. I'll also list some of the many changes to both illustrations and text that keep this edition current.

## Optional boxes help with mathematics and with challenging concepts

As in past editions, boxes set aside the most technical material. Any or all may simply be omitted without loss of continuity, so that instructors can tailor their courses to an appropriate level of difficulty. Boxes on important mathematical formulas always include worked examples to prepare students better for the many end-of-chapter questions.

*New to this edition,* I have expanded the box program and enhanced the presentation of fundamental equations to make varied depths of coverage that much easier:

- **Tools of the Astronomer's Trade** boxes explain how to use essential mathematics, such as the small-angle formula (Box 1-2), the radiation laws for hot, glowing bodies (Box 5-2), and the formula for determining the size of a star (Box 19-5).

- **Looking Deeper into Astronomy** boxes offer interesting extensions, such as why some planets have atmospheres while others do not (Box 7-2), how astronomers discover comets (Box 17-2), and how a black hole might act as a time machine (Box 24-4).

- **Fundamental equations** within the text now have helpful titles and a recap of the meaning of algebraic symbols just below.

**box 1-2** | Tools of the Astronomer's Trade

### The Small-Angle Formula

You can estimate the angular sizes of objects in the sky with your hand and fingers (Figure 1-11). Using rather more sophisticated equipment, astronomers can measure angular sizes to a fraction of a second of arc. Keep in mind, however, that *angular* size is not the same as *actual* size. As an example, if you extend your arm while looking at a full moon, you can completely cover the Moon with yo... ...your perspective, your th... (that is, it subtends a lar... actual size of your thum... less than the actual diam...

The small-angle formula

$$D = \frac{\alpha d}{206{,}265}$$

$D =$ linear size of an object
$\alpha =$ angular size of the object, in arcsec
$d =$ distance to the object

**box 24-4** | Looking Deeper into Astronomy

### Wormholes and Time Machines

General relativity is so rich, complex, and fascinating—and so may be the geometry of black holes. For example, in the 1930s, Einstein and his colleague, Nathan Rosen, discovered that a black hole could possibly connect our universe with a second domain of space and time that is separate from ours. The first diagram shows this connection, called an **Einstein-Rosen bridge.** You could think of the upper surface as our universe and the lower surface as a "parallel universe." Alternatively, the upper and lower surfaces could be different regions of our own universe. In that case, an Einstein-Rosen bridge would connect our universe with itself, forming a **wormhole**, shown in the...

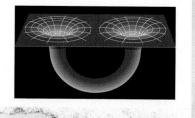

## NEW features show science as a process of questioning

Bill Kaufmann introduced Observing Projects and chapter reading lists to *Universe.* He knew that students learn best about the scientific process when they participate in it for themselves. I have tried to apply that same philosophy to pedagogical aids within each chapter.

*New to this edition,* several features are designed to get students thinking about astronomy and about their own preconceptions:

- **A set of questions** begins each chapter, with one for each section of the chapter. These are questions that students may already be asking themselves, and they will also serve students as the first of each chapter's many review aids.

*In this chapter you will find the answers to the following questions:*

13-1 Can you stand on the surface of Jupiter, as you can on the Earth or the Moon?

13-2 What is going on in Jupiter's Great Red Spot?

13-3 What is the origin of Jupiter's multicolored clouds?

13-4 What happened to the comet that crashed into Jupiter?

13-5 How do astronomers know about Jupiter's interior?

13-6 What is the origin of Jupiter's strong magnetic field?

13-7 Where do you find the highest temperatures in the solar system?

- **Caution! paragraphs** directly address student misconceptions. This new feature helps students learn ideas that may conflict with their "common sense." For example, many people think that astronauts in orbit feel weightless because they are beyond the pull of the Earth's gravity. They cannot understand the correct explanation until they understand why their wrong ideas are indeed wrong!

CAUTION! An astronaut on board an orbiting spacecraft feels "weightless," but this is *not* because she is "beyond the pull of gravity." The astronaut is herself an independent satellite of the Earth, and the Earth's gravitational pull is what holds her in orbit. She feels "weightless" because she and her spacecraft are falling *together* around the Earth, so there is nothing pushing her against any of the spacecraft walls. You feel the same "weightless" sensation

## NEW analogies help students draw connections

Many concepts in astronomy are utterly foreign to my students. To help them understand these concepts, I find it very useful to use analogies that draw connections between astronomy and everyday life. When Bill Kaufmann described a car's motion as it drives onto sand, my students grasped refraction immediately. Other analogies in past editions introduced my students to angular momentum and the expanding universe.

*New to this edition,* I have highlighted useful analogies and added many new ones, often with new artist renderings:

- **The Heavens on the Earth,** a third and entirely new kind of box, emphasizes the connection between astronomy and the underlying laws of physics as they apply on the Earth. Examples of celestial concepts applied to everyday life include why cans of diet soda float in water while cans of regular soda sink (Box 7-1), why sunsets are red and milk is white (Box 15-1), and why a bicycle pump gets warm when you inflate a tire (Box 21-1).

**Box 15-1** | The Heavens on the Earth

### *Light Scattering*

The scattering of light by particles is used by scientists to study Saturn's rings, but it also explains a number of phenomena that you can see here on the Earth.

The light that comes from the daytime sky is sunlight that has been scattered by the molecules that make up our atmosphere (see part a of the accompanying figure). Very small particles like molecules are quite effective at scattering blue, short-wavelength light but less effective at scattering long-wavelength red light. That's why the sky

as of blue light, and the scattered light will appear white. This explains the white color of clouds, fog, and haze, in which the scattering particles are ice crystals or water droplets. Whole milk looks white because of light scattering from tiny fat globules; nonfat milk has only a very few of these globules and thus has a slight bluish cast.

Light scattering also has many applications to astronomy outside the solar system. In Chapter 20 we will see how light scattering explains the colors of certain clouds, or nebulae, in interstellar space.

- **Analogy icons** call out useful discussions throughout the text. New analogies include a comparison of finding our place in the Galaxy to the plight of a motorist lost on a foggy night (Chapter 25).

ANALOGY Herschel and Kapteyn faced much the same dilemma as a lost motorist on a foggy night. Unable to see more than a city block in any direction, he would have a hard time deciding what part of town he was in or even if he was in a small town or a large city. If the fog layer were relatively shallow, however, our hapless motorist

## More essays introduce the personal views of leading astronomers

Each part again contains guest essays, in which leading astronomers and a well-known debunker of pseudoscience share their experiences of how science works.

Three of the essays in *Universe* –by Sandra M. Faber, Alan Dressler, and Stephen W. Hawking–have fired the interest and imagination of students.

## GEOFF MARCY

### Alien Planets

When Geoff Marcy sat down to write this essay, eight planets were known to orbit other stars. He leads the team that has found six. He is an astronomer at San Francisco State University and the University of California at Berkeley.

Dr. Marcy became interested in astronomy at age 14, when his parents bought him a small reflecting telescope. Since receiving his Ph.D. from the University of California, Santa Cruz, he has studied stars similar to our Sun. He helped show that magnetic regions on their surfaces cause dark "star spots" and stellar flares. He also showed that brown dwarfs—stars too little to burn hydrogen—rarely orbit other stars. His work may soon reveal whether a planetary system like our own is common or a quirk of the cosmos.

*Four essays new to this edition* give the text a depth, quality, and intimate view of science that could not otherwise have been possible, "Why Astrology Is Not Science," by James "The Amazing" Randi; "Alien Planets," by Geoff Marcy; "Discovering a Comet," by Alan Hale; and "Searching for Neutrinos Beyond the Textbooks," by John N. Bahcall.

## Updated color illustrations are an essential part of this book

*Universe* was the first full-color astronomy textbook. In this edition, I have continued this tradition with a wealth of new astronomical images. I chose these not merely for aesthetics, but with the goal of helping students see the universe with the eye of a scientist. Many new explanatory diagrams help students visualize important concepts.

*New to this edition,* the selection of illustrations has been substantially updated to reflect the latest developments in astronomy and to show how astronomy is practiced today:

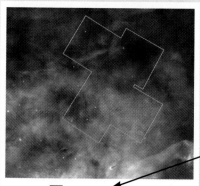

a   R I V U X G

*f*igure 20-22

**The Core of the Orion Nebula (a)** This Hubble Space Telescope view at visible wavelengths shows the inner regions of the Orion Nebula (compare Figure 20-1*b*). At the lower left are the four massive stars, called the Trapezium,

- **Labels on all telescopic images** note the wavelength used to make them.

- **New images** from planetary probes and the Hubble Space Telescope (HST) accompany many other new photographs and diagrams.

Chapters on the solar system contain new images of the Antarctic ozone hole (Chapter 8), a *Clementine* map of the Moon's composition (Chapter 9), images from *Mars Pathfinder* and *Mars Global Surveyor* (Chapter 12) and *Galileo* (Chapters 13 and 14), HST observations of the outer planets (Chapters 15 and 16), and close-ups of the asteroids Ida and Mathilde (Chapter 17).

Dramatic new images can also be found in the chapters on stars and galaxies. Examples include a *SOHO* image of a solar coronal ejection (Chapter 18) and HST images of star formation (Chapter 20), planetary nebulae (Chapter 22), protogalaxies (Chapter 26), and quasar host galaxies (Chapter 27).

New and improved explanatory diagrams can also be found throughout the book. Among them are illustrations that help explain the apparent motions of the stars (Chapter 2) and Moon (Chapter 3), the origin of tides (Chapter 9), Mercury's spin-orbit coupling (Chapter 10), differential rotation (Chapter 13), interstellar reddening (Chapter 20), how spiral arms induce star formation (Chapter 25), superluminal motion (Chapter 27), and the isotropy problem in cosmology (Chapter 29).

## A thoroughly up-to-date text shows science in the making

I have endeavored to make this book easier to read without lowering its level in any way. I have checked the *entire* text word by word and revised as needed for clarity. I took special care to smooth the transitions between sections of a chapter and to give students a sense of what each section is about.

I also include the latest information as this edition goes to press. Currency is essential not just for scientific accuracy, but also to show students how science is made and remade. Besides the dramatic new color images I have mentioned, I also include the latest discoveries, such as evidence for ice at the lunar poles (Chapter 9) and the suspicion that it may also exist at the poles of Mercury (Chapter 10), the detection of extrasolar planets and brown dwarfs (Chapter 7), discoveries from *Mars Pathfinder* and *Mars Global Surveyor* (Chapter 12), evidence for massive neutrinos (Chapter 18), the first view of an isolated neutron star (Chapter 23), confirmation that a supermassive black hole lies at the center of our Galaxy (Chapter 25), and the discovery of a galaxy with a redshift greater than $z = 5$ (Chapter 28).

*New to this edition,* I have devoted a full chapter to the search for extraterrestrial intelligence and used the World Wide Web to help students stay current:

- **Chapter 30, "The Search for Extraterrestrial Intelligence,"** is now also a theme of relevant preceding chapters.

- **Where to Learn More** lists at the end of each chapter include not only up-to-date books and magazine articles, but also web sites where students can find the most current information. This textbook is now planned as a media edition from the very first.

**Where to Learn More**

*Books and magazine articles*

Beatty, J. K. "Galileo: An Image Gallery." *Sky & Telescope,* November 1996 and March 1997. These articles include many *Galileo* images with explanatory text.

Beebe, R. *Jupiter: The Giant Planet.* Smithsonian Institution Press, 1997. This book, written for a popular audience by a leading astronomer who specializes in observations of Jupiter and its satellites, includes some of the early results

**W** *World Wide Web*

By far the best place to find the latest information about the Galilean satellites is the *Galileo* mission web site (**http://www.jpl.nasa.gov/galileo/**). Two other good starting places are the Jupiter page of "Views of the Solar System" (**http://www.hawastsoc.org/solar/eng/jupiter.htm**) and "The Nine Planets" page for Io (**http://www.seds.org/nineplanets/nineplanets/io.html**), both of which have links to information about all of Jupiter's satellites.

Many spacecraft images of the Galilean satellites can

## Supplements

In addition to a web site and CD-ROM, a constellation of supplements complements this text. The package now includes planetarium software and a free special issue of *Scientific American* with every student copy.

## A NEW web site and CD-ROM make this textbook truly a media edition

To keep the book itself up to date as new discoveries are made, a **web site** accompanies this textbook: **http://www.whfreeman.com/astronomy**

This site is continually updated with the latest astronomical findings. It includes questions for students and is linked to specific sections of the text.

Every copy of this textbook also comes with the *Universe 5.0* **CD-ROM**. They work together, because they were planned together from the first. This free CD-ROM, for both Macintosh and Windows computers, offers a vast array of study and learning features that will strongly appeal to today's students.

The CD-ROM includes a wealth of videos, animations, simulations, drill questions, interactive exercises, a pop-up glossary, and a rich source of other material, *all* explicitly linked to sections of the text. Students will view the surface of Mars in virtual reality, right along with *Mars Pathfinder*. They will benefit from Expanded Toolboxes and a calculator tailored to key formulas in their textbook. Many of the images and animations are particularly suitable for use in lectures.

Last, but not least, the CD-ROM includes a special student edition of *Starry Night*™, the critically acclaimed planetarium software from Sienna Software. I've used planetarium programs for student assignments for many years, and my students find them to be excellent aids to visualizing the workings of the universe. *Starry Night*™ is unquestionably the best program available for this purpose and for use in lectures.

## A comprehensive set of print supplements enhances the text

The revised *Instructor's Manual and Resource Guide,* prepared by George A. Carlson of Citrus College, contains chapter synopses and outlines, hints for teaching and discussion, a current list of teaching resources that includes audiovisual materials, as well as readings, and fully worked out solutions to all computational problems in the book.

*Overhead Transparencies* and *Slides* contain a selection of 100 color drawings from the text. They benefit from this edition's extra care to make diagrams vivid and three-dimensional.

T. Alan Clark and William J. F. Wilson of the University of Calgary have revised and updated the printed *Test Bank* and the *Computerized Test Bank,* available in Macintosh and IBM formats. It now includes some 2000 multiple-choice questions, indexed by chapter and topic. The *Test Bank* also contains chapter outlines, which are convenient for class distribution or as a basis for lectures.

*New to this edition,* with every copy of this textbook, students will receive a special issue of *Scientific American.* **"Magnificent Cosmos"** describes the latest discoveries in astronomy. Its fourteen articles have all been updated by their authors from their original appearance in *Scientific American.*

For more information about these supplements, contact your W. H. Freeman and Company representative: **http://www.whfreeman.com/College/FindRep.html**.

## Acknowledgments

I would like to thank my colleagues who carefully scrutinized the manuscript of this edition. This is a stronger and better textbook due to their conscientious efforts:

Robert R. J. Antonucci, *University of California, Santa Barbara*

Omer Blaes, *University of California, Santa Barbara*

George L. Cassiday, *University of Utah*

John J. Cowan, *University of Oklahoma*

John E. Crawford, *McGill University*

Robert A. Egler, *North Carolina State University*

Steven Giddings, *University of California, Santa Barbara*

Carl Gwinn, *University of California, Santa Barbara*

Paul A. Heckert, *Western Carolina University*

Scott B. Johnson, *Idaho State University*

Stanton J. Peale, *University of California, Santa Barbara*

Robert L. Pompi, *State University of New York, Binghamton*

Charles W. Rogers, *Southwestern Oklahoma State University*

Alan F. Sill, *Texas Tech University*

Caroline Simpson, *Florida International University*

My heartfelt thanks also go out to the following instructors, who reviewed my proposal for this edition or examined the fourth edition with particular care with an eye to the plans for revision. Their suggestions were tremendously helpful:

Bel Campbell, *University of New Mexico*

George A. Carlson, *Citrus College*

John J. Cowan, *University of Oklahoma*

Robert Dick, *Carleton University*

James N. Douglas, *University of Texas, Austin*

Debra Meloy Elmegreen, *Vassar College*

Juhan Frank, *Louisiana State University*

Buford M. Guy, Cleveland State

Robert O'Dell, *Rice University*

Robert L. Pompi, *State University of New York, Binghamton*

Thomas Tsung, *Grossmont College*

George Wegner, *Dartmouth College*

Louis Winkler, *The University of Pennsylvania*

Robert L. Zimmerman, *University of Oregon*

I would also like to thank the many people whose advice on four editions has had an ongoing influence:

Robert Allen, *University of Wisconsin, La Crosse;* Robert R. J. Antonucci, *University of California, Santa Barbara;* Alice L. Argon, *Harvard-Smithsonian Center for Astrophysics;* David Van Blerkom, *University of Massachusetts;* John M. Burns, *Mount San Antonio College;* Bruce W. Carney, *University of North Carolina;* Bradley W. Carroll, *Weber State University;* Roger B. Culver, *Colorado State University;* James N. Douglas, *University of Texas at Austin;* Robert J. Dukes, Jr., *College of Charleston;* David S. Evans, *University of Texas at Austin;* George W. Ficken, Jr., *Cleveland State University;* Andrew Fraknoi, *Astronomical Society of the Pacific;* Owen Gingerich, *Harvard University;* Paul F. Goldsmith, *University of Massachusetts;* J. Richard Gott III, *Princeton University;* Austin F. Gulliver, *Brandon University;* Bruce Hanna, *Old Dominion University;* Paul Hodge, *University of Washington;* Douglas P. Hube, *University of Alberta;* Icko Iben, Jr., *Pennsylvania State University;* John K. Lawrence, *California State University, Northridge;* Laurence A. Marschall, *Gettysburg College;* Dimitri Mihalas, *University of Illinois;* L. D. Opplinger, *Western Michigan University;* John R. Percy, *University of Toronto;* Carlton Pryor, *Rutgers University;* James

L. Regas, *California State University, Chico;* Terry Retting, *University of Notre Dame;* Tina Riedinger, *University of Tennessee;* James A. Roberts, *University of North Texas;* Kenneth S. Rumstay, *Valdosta State College;* Richard Saenz, *California Polytechnic State University;* Thomas F. Scanlon, *Grossmont College;* Richard L. Sears, *University of Michigan;* Isaac Shlosman, *University of Kentucky;* Michael L. Sitko, *University of Cincinnati;* David B. Slavsky, *Loyola University of Chicago;* Joseph S. Tenn, *Sonoma State University;* Gordon B. Thomson, *Rutgers University;* Charles R. Tolbert, *University of Virginia;* Virginia Trimble, *University of California, Irvine;* Bruce A. Twarog, *University of Kansas;* Donat G. Wentzel, *University of Maryland;* Nicholas Wheeler, *Reed College;* and Raymond E. White, *University of Arizona.*

I'm also particularly grateful to James Randi, Geoff Marcy, Alan Hale, and John Bahcall, whose new essays for this edition greatly enhance its coverage. Alan Dressler was kind enough, too, to revise his essay in light of our latest understanding of the universe's large-scale structure.

Many others have participated in the preparation of this book, and I thank them for their efforts. Foremost among them are my good friends Holly Hodder, the acquisitions editor whose faith in my abilities made this edition possible, and John Haber, the development editor who made this a better and stronger book in every way. Special thanks go to Liz Widdicombe, president; Louise B. Ketz, copy editor; Diane Cimino Maass, project editor; Paul W. Rohloff, production coordinator; Sheridan Sellers and Bonnie Stuart, compositors; Larry Marcus, photo editor; Vicki Tomaselli, the designer of this attractive new edition; Patrick Shriner, director of media; Matthew Fitzpatrick, associate editor; and Kimberly Manzi, marketing manager.

Although we have made a valiant effort to make this edition error free, some mistakes may have crept in. I would appreciate hearing from anyone who finds an error or wishes to comment on the text. You may write or e-mail me.

Roger A. Freedman
College of Creative Studies
University of California, Santa Barbara
Santa Barbara, CA 93106
e-mail:airboy@physics.ucsb.edu

# How to Use this Textbook

I f you're like most students just opening this textbook, you're enrolled in one of the few science courses you'll take in college. As you study astronomy, you'll probably do relatively little reading compared to a literature or history course—at least in terms of the number of pages. But your readings will be packed with information, much of it new to you and (I hope) exciting. You can't read your textbook like a novel and expect to learn much from it.

If you're a little worried, don't be. I wrote this book with you in mind. In this section I'll suggest how this textbook can help you succeed in astronomy.

## Read Before a Lecture

You'll get the most out of your astronomy course if you read each chapter *before* hearing a lecture about its subject matter. That way, many of the topics will already be clear in your mind, and you'll understand the lecture better. You'll be able to spend more of your listening and note-taking time on the more challenging ideas presented in the lecture.

## Take Notes as You Read and Make Use of Office Hours

Have paper handy as you read, and write down the key points of each section so that you can review them later. If any parts of the section don't seem clear on first reading, make a note of them, too, including the page numbers.

Once you've gone through a chapter, reread it with special emphasis on the ideas that gave you trouble the first time. If you're still unsure about these after the lecture, consult your instructor, either during office hours or after class. Bring your notes with you, so your instructor can see which concepts are giving you trouble. Once your instructor has helped clarify things for you, revise your notes so you'll remember your newfound insights. You'll end up with a chapter summary in your own words. This will be a tremendous help when studying for exams!

## Use the Special Features of This Book

In writing this textbook, we've included several features that will make it easier for you to learn and appreciate astronomy. Look for these as you read through each chapter:

- **Chapter-opening questions:** Each chapter opens with a series of questions. Use these questions as a guide to help you in your reading. You'll find the answer to each question by reading the corresponding section of the chapter.

- **Key Words and Glossary:** When we introduce a new term that has a special importance in astronomy, it appears in **boldface type**. A list of Key Words also appears at the end of each chapter, along with a page number telling you where the word first appears. Be sure that you understanding the meaning of each one and can explain them *in your own words*. Don't just memorize the definitions! The Glossary, which you'll find near the back of the book, has brief definitions of all the key words in the book.

- **Links to other sections:** A number of topics appear in different part of the textbook. For example, the laws that govern the orbits of the planets (discussed in Chapter 4) also apply to binary star systems (Chapter 19) and to the motion of our entire solar system within the Milky Way (Chapter 25). References to topics from a previous chapter appears in blue underlined type. These links can help you find and review these important ideas.

- **Caution! paragraphs:** You may think that the phases of the Moon are caused by the Earth's shadow falling on the Moon and that the Earth is closer to the Sun in summer than in winter. In fact, these "commonsense" ideas are incorrect! (You'll learn the answers in Chapters 3 and 4.) Paragraphs signaled by the caution icon will alert you to such potential pitfalls.

- **Analogy paragraphs:** To learn astronomical ideas, it can be helpful to relate them to your experiences on the Earth. Throughout the book you'll find paragraphs labeled with the analogy icon. They relate the motions of the planets to children on a merry-go-round; the bending of light in a telescope lens to a car driving on sand; Einstein's theories to what happens inside a bicycle tire; and more.

- **Wavelength tabs:** When astronomers observe the heavens, they do not use visible light alone. Many of the images in this textbook were made with special telescopes that are sensitive to nonvisible forms of light. Understanding different forms of light, as well as what astronomers learn from them, is an important part of your astronomy education. To help you with this, we've included

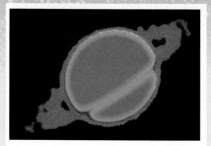

R I V U X G  This VLA image shows radio emission from Saturn at a wavelength of 2 cm. This emission is caused by electrically charged particles moving within Saturn's strong magnetic field and acting as tiny radio transmitters. In this

wavelength tabs with the telescope images in this textbook. Each tab has the letters R I V U X G; the highlighted letter shows whether the image was made with **R**adio waves, **I**nfrared radiation, **V**isible light, **U**ltraviolet light, **X** rays, or **G**amma rays. (We discuss the different kinds of nonvisible light in Chapter 5.) For example, the image of Saturn shown here was made with a radio telescope (*R*).

- **Supplementary topics in numbered boxes:** There are three different kinds of boxes, each with a different purpose. *Tools of the Astronomer's Trade* boxes offer help with mathematical formulas and solving problems. *Looking Deeper into Astronomy* boxes take you above and beyond what appears in the main text. The *Heavens on the Earth* boxes shows how the same principles that astronomers use to describe the heavens also apply here on the Earth.

- **Worked examples:** It's guaranteed that you'll be doing some math problems as part of your astronomy course. To help you learn how, look carefully at examples in the Tools of the Astronomer's Trade boxes. Don't just read them; work the examples out on a piece of paper. Your goal is to be able to do these calculations yourself just as you'll be asked to do on exams.

- **Chapter summaries:** At the end of each chapter you'll find a section called Key Ideas, a brief summary in outline format. Make sure you fully understand each description in the summary. If something's unclear, you'll need to go back and review the text. To get the most out of the summaries, use them in conjunction with the notes you make yourself while reading.

- **Questions and projects:** At the end of each chapter you will find a list of questions and Observing Projects, which your instructor may assign as homework. You will also find Problem-Solving Tips and Tools with more advanced questions. These suggest where you can look for guidance in the textbook. You'll also find Observing Tips and Tools with several of the observing projects. Near the end of the textbook you'll find answers to selected questions.

- **Where to learn more:** There is no way that a single textbook can describe all of astronomy. The Where to Learn More section at the end of each chapter lists other books, magazine articles, and World Wide Web sites that you may want to investigate. It is worth remembering that anybody can post anything on the web. As a result, you can easily find crackpot theories that are not at all supported by observation. The web sites listed are to help you locate *solid* science on the web. You'll find even more useful links at the web site for this book: **http://www.whfreeman.com/astronomy**

- **Guest essays:** At various places throughout this textbook, you'll find essays by scientists at the cutting edge of astronomy, including Geoff Marcy on the discovery of alien planets and Stephen W. Hawking on the edge of space-time. By reading these, you'll get a sense of their excitement about astronomy. We hope that you catch this excitement!

Useful appendices summarize some important data and list formulas you'll need to use time and time again. The index is an often-neglected part of a textbook, but if you don't know where to find out about a particular topic, first look in the index.

## Try Astronomy for Yourself with Your Star Charts and CD-ROM

At the back of this book, you'll find star charts for the four seasons of the year. These charts can get you started with your own observations of the universe. Hold the chart overhead in the same orientation as the compass points, with southern horizon toward the south and western horizon toward the west. (To save strain on your arms, you may want to cut these pages out of the book.)

In addition, your copy of this textbook includes a CD-ROM. This disc, which works on both Macintosh and Windows computers, contains a wealth of images, movies, and animations and even a planetarium program, all keyed to chapters in this book. This CD-ROM can bring astronomy alive for you! Many students also complain of difficulty using a calculator to solve problems, especially when dealing with scientific notation. I've made that easier for you, with the CD-ROM's built-in calculator, tailored to key formulas in astronomy. The CD-ROM also has interactive exercises to help you with the questions in your textbook and your homework.

## Make Use of Other Students

Many students find it useful to form study groups for astronomy. You can hash out challenging topics with each other and have a good time while you're doing it. But make sure that you write up your homework by yourself, because the penalties for copying or plagiarizing the work of other students can be severe in the extreme.

Some students also find individual assistance useful. If you think a tutor will be helpful, link up with one *early*. Getting a tutor late in the course, in the belief that you'll be able to catch up with what you missed earlier, is almost always a pointless exercise.

## The Most Important Advice of All

I haven't mentioned the most important thing you should do when studying astronomy: *have fun!* Of all the different kinds of scientists, astronomers are among the most excited about what they do and what they study. Let some of that excitement about the universe rub off on you, and you'll have a great time with this course and with this textbook.

In preparing this edition of *Universe*, I tried very hard to make it the kind of textbook that a student like you would find useful. I'm very interested in your comments and opinions! Please feel free to write me or send me e-mail, and I will respond personally.

Best wishes for success in your studies!

Roger A. Freedman
College of Creative Studies
University of California, Santa Barbara
Santa Barbara, CA 93106
e-mail:airboy@physics.ucsb.edu

# UNIVERSE

Comet Hale–Bopp on a spring evening in 1997. (Courtesy of Johnny Horne)

# Introducing Astronomy

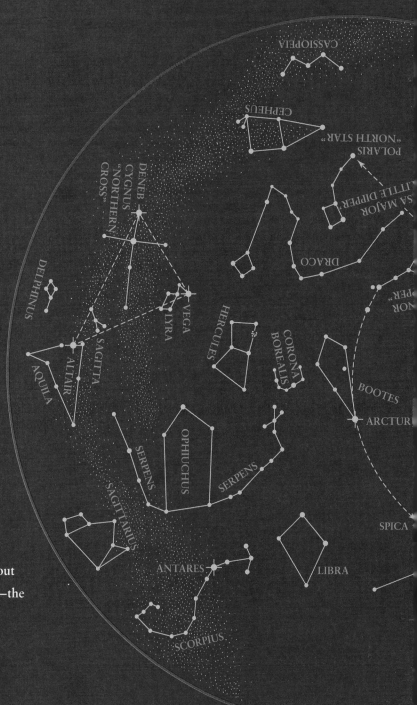

**f**or as long as there have been humans on Earth, we have gazed into the night sky and wondered: How far away is the Moon? What are planets and comets? What are the stars, and why do they shine?

Today we know the answers to these ancient questions. But for every question that astronomy answers, a host of new questions appear. If planets are other worlds that orbit the Sun, where did they come from? If stars are objects like the Sun but more distant, how did they form and how do they evolve? If we live in a universe of stars, how will the universe change over billions of years?

Astronomy is the process of asking and answering questions such as these. In the next six chapters we will see how astronomers work. We will learn about the tools—including the laws of physics—that they use in their quest. And we will see what we have learned about the heavens using the most common tool of astronomy—the naked eye.

# Astronomy and the Universe

**Stars and Interstellar Gas**

R I **V** U X G

New stars are forming in the clouds of interstellar gas and dust shown in this photograph. The gas clouds glow because of the radiation emitted by newborn stars. The denser clouds of interstellar dust block light; they appear as dark regions silhouetted against glowing background gas. The Horsehead Nebula is in the lower left; the Orion Nebula is toward the upper right. Both nebulae are about 1500 light-years from Earth. (Royal Observatory, Edinburgh)

*In this chapter you will find the answers to the following questions:*

*The splendor of the star-filled night sky is one of the truly remarkable experiences of life. From a dark site far from city lights, we see thousands of stars scattered from horizon to horizon. The delicate glow of the Milky Way traces a luminous path across the sky, and the entire spectacle swings slowly overhead from east to west as the night progresses. No planetarium show, no artist's brush, no poet's words can truly capture the beauty of this breathtaking panorama (Figure 1-1).*

*For thousands of years people have looked up at the heavens and contemplated the universe. Like our ancestors, we find our thoughts turning to profound questions as we gaze at the stars. How was the universe created? Where did the Earth, Moon, and Sun come from? What are the planets and stars made of? And how do we fit in? What is our place in the cosmic scope of space and time?*

*To wonder about the universe is a particularly human endeavor. Our curiosity, our desire to explore and discover, and most importantly our ability to reason about what we have discovered are qualities that distinguish us from other animals. The study of the stars transcends all boundaries of culture, geography, and politics. In a literal sense, astronomy is a universal subject—its subject is the entire universe.*

## 1-1 Astronomers use the laws of physics and construct testable theories and models to understand the universe

Astronomy has a rich heritage that dates back to the myths and legends of antiquity. Centuries ago, the heavens were thought to be populated with demons and heroes, gods and goddesses. Astronomical phenomena were explained as the result of supernatural forces and divine intervention.

The course of civilization was greatly affected by one realization: *The universe is comprehensible.* This first awareness is one of the great gifts to come to us from ancient Greece. Greek astronomers discovered that by observing the heavens and carefully reasoning about what they saw, they could learn something about how the universe operates. For example, as we shall see in Chapter 3, they measured the size of Earth and understood and predicted eclipses. Modern science is a direct descendant of the intellectual endeavors of these ancient Greek pioneers.

Like art, music, or any other human creative activity, science makes use of intuition and experience. But the approach used by scientists to explore physical reality differs from other forms of intellectual endeavor in that it is based fundamentally on *observation, logic,* and *skepticism.* This approach, called the **scientific method,** requires that our ideas about the world around us be consistent with what we actually observe. In practice, the scientific method goes something like this: A scientist trying to understand some phenomenon proposes a **hypothesis,** which is a collection of ideas that seems to explain the phenomenon. It is in develop-

ing such hypotheses that scientists are at their most creative, imaginative, and intuitive. But their hypotheses must always be developed in agreement with existing observations and experiments, because a discrepancy with what is observed implies that the hypothesis is wrong. (The exception is if the scientist thinks that the existing results are wrong and can give compelling evidence to show that they are wrong.) The scientist then considers how the implications of the hypothesis might lead to predictions that can be tested. Only after a hypothesis has accurately forecast the results of new experiments or observations does the scientist feel confident that the hypothesis is on firm ground.

Scientists describe reality in terms of **models,** which are hypotheses that have withstood observational or experimental tests. A model tells us about the properties and behavior of some object or phenomenon. A familiar example is a model of the atom, which scientists picture as electrons orbiting a central nucleus. Another example, which we will encounter in Chapter 18, is a model of the Sun that tells us about physical conditions (for example, temperature, pressure, density) within the interior of the Sun. A well-developed model makes use of mathematics to make detailed predictions; for example, a successful model of the Sun's interior should describe in detail how the temperature changes as you look deeper into the Sun. For this reason, mathematics is one of the most important tools used by scientists.

A body of related hypotheses can be pieced together into a self-consistent description of nature called a **theory.** An example from Chapter 24 is Einstein's general theory of relativity, which describes the effect of gravity on space and time. Without models and theories there is no understanding and no science, only collections of facts.

In everyday language the word "theory" is often used to mean an idea that looks good on paper, but has little to do with reality. In science, however, a good theory is one that explains reality very well. An excellent example is the general theory of relativity, which does a superb job of describing a host of gravitational effects. It predicted, for example, that light rays can be bent by gravity; careful observation shows that this is just what happens and that the amount of bending is just what the theory predicts. Indeed, the general theory of relativity has so far agreed in detail with every experimental and observational test it has undergone.

Skepticism is an essential part of the scientific method. New hypotheses have to be able to withstand the close scrutiny of other scientists. The more radical the hypothesis, the more skepticism and critical evaluation it will receive from the scientific community, because the general rule in science is that extraordinary claims require extraordinary evidence. (That is why scientists as a rule do not accept claims that people have been abducted by aliens and taken aboard UFOs. The evidence presented for these claims is flimsy, secondhand, and unverifiable.) At the same time, scientists must be open-minded. They must be willing to discard long-held ideas if they fail to agree with new observations and experiments, provided the new data has survived evaluation. (If an alien spacecraft were truly to land on Earth, scientists would be the first to accept that aliens existed—provided they could take a careful look at the spacecraft and its occupants.) That is why scientific knowledge is always provisional. As we go through this book, we will encounter many instances where new observations have transformed our understanding of Earth, the planets, the Sun and stars, and indeed the very structure of the universe. We will place great emphasis on understanding how these observations are made, because they are at the heart of the scientific method.

Theories that accurately describe the workings of physical reality have a significant effect on civilization. For example, based in part on observations of how the planets orbit the Sun, the seventeenth-century scientist Isaac Newton deduced a set of fundamental mathematical laws that describe how *all* objects move. These laws, which are described in Chapter 4, work equally well on Earth as in the most distant corner of the universe and represent our first complete, coherent description of the behavior of the physical universe. **Newtonian mechanics** had an immediate practical application in the construction of machines, buildings, and bridges. It is no coincidence that the Industrial Revolution followed hard on the heels of these theoretical and mathematical advances inspired by astronomy.

Newtonian mechanics and other proven physical principles are collectively referred to as the **laws of physics**. Astronomers use these laws to interpret their observations and to understand the phenomena of the universe. Light and its relationship to matter are of particular importance, because the only information that we can gather about distant stars and galaxies is in the light that we receive from them. Using the

**ƒigure 1-1** R I **V** U X G

**The Starry Sky** The star-filled sky is a beautiful and inspiring sight. This photograph, taken from northern Mexico, shows Halley's Comet and a portion of the Milky Way (upper right). To get a good view of the heavens, you must be far from city lights. (Courtesy of D. L. Mammana)

physical laws that describe how objects absorb and emit light, astronomers have been able to analyze starlight to learn what stars are made of, to deduce the sizes of stars, and even to measure a star's temperature. (We will see how this is done in Chapter 19.) The analysis of light even allows astronomers to investigate objects so remote that their light takes billions of years to reach us on Earth.

An important part of science is the development of new tools for research and new techniques of observation. As an example, until recently everything we knew about the distant universe was based on visible light. Astronomers would peer through telescopes to observe and analyze visible starlight. By the end of the nineteenth century, however, scientists had begun discovering forms of light invisible to the human eye: X rays, gamma rays, radio waves, microwaves, and ultraviolet and infrared radiation.

As we will see in Chapter 6, in recent years astronomers have constructed telescopes that can detect such nonvisible forms of light (Figure 1-2). Whether located on Earth or in orbit above the obscuring effects of the atmosphere, these

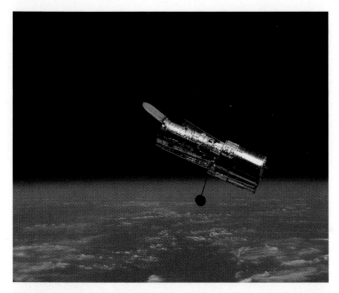

**figure 1-2**

**A Telescope in Space** Because it orbits high above Earth in the vacuum of space, the Hubble Space Telescope (HST) can detect not only visible light but also ultraviolet and near-infrared light. These forms of nonvisible light are absorbed by our atmosphere and hence are difficult or impossible to detect with a telescope on the surface of Earth. Since it was first placed in orbit in 1990, HST has helped to revolutionize astronomers' understanding of the universe. It is shown here after being serviced by the crew of the space shuttle *Discovery* in 1997. The Hubble Space Telescope is large enough—13.1 meters (43 feet) end-to-end and 4.3 meters (14 feet) in diameter at its widest point—to easily be seen with the naked eye from the Earth's surface. (NASA)

instruments give us views of the universe vastly different from anything our eyes can see. These new views are crucial to our understanding not only of familiar objects like the Sun and stars, but also such exotic objects as neutron stars, pulsars, quasars, and black holes. Aided by high-technology telescopes, today's astronomers carry on the program of careful observation and logical analysis begun thousands of years ago by their ancient Greek predecessors.

## 1-2 By exploring the planets, astronomers uncover clues about the formation of the solar system

In this book, we take three major steps out into the universe: from the Earth to the planets, on to the stars, and finally to the galaxies. The star we call the Sun and all the celestial bodies that orbit the Sun (including Earth, the other eight planets, and all their various moons) make up the **solar system.** We explore the solar system in Chapters 7 through 17, beginning with the Earth and the Moon and then on to the

other planets, moving outward toward the frigid depths of space where comets spend most of their time.

Since the 1960s a series of unmanned spacecraft have been used to visit and explore all the planets except Pluto (Figure 1-3). Using the remote "eyes" of such spacecraft, we have flown over Mercury's cratered surface, peered beneath Venus's poisonous cloud cover, and discovered enormous canyons and extinct volcanoes on Mars. We have found active volcanoes on a moon of Jupiter, seen the rings of Saturn and Uranus up close, and looked down on the active atmosphere of Neptune. Along with rocks brought back by the Apollo astronauts from the Moon (the only world beyond Earth yet visited by humans), this new information has revolutionized theories about the creation and evolution of the solar system. We have come to realize that many of the planets and their satellites were shaped by collisions with other objects. Craters on the Moon and on many other worlds are the relics of innumerable impacts by interplanetary rocks. More important, the Moon may itself be the result of a catastrophic collision between the Earth and a planet-sized object shortly after the solar system was formed. Such a collision could have torn sufficient material from the primordial Earth to create the Moon.

**figure 1-3** R I **V** U X G

**Saturn and Its Satellites** This montage shows the planet Saturn along with several of its moons. The individual images that were combined to make this montage were made by the *Voyager 1* spacecraft as it flew past Saturn in December 1980. Space probes to the planets have provided us with a wealth of information about other worlds. This new knowledge gives us important insights into the formation and evolution of the solar system, as well as an appreciation of the variety found in nature. (NASA)

**figure 1-5**

**A Thermonuclear Explosion** Understanding the Sun's source of energy has given humanity the ability to build thermonuclear weapons. The hydrogen bomb and the thermonuclear reactions at the Sun's center both operate under the same basic physical principle—the conversion of matter into energy by way of nuclear reactions. This thermonuclear detonation on October 31, 1952, had an energy output, or "yield," equivalent to 10.4 million tons of TNT. This is more than 500 times the output of the bombs exploded over Hiroshima and Nagasaki in 1945, but a mere ten-billionth of the amount of energy released by the Sun in one second. (Defense Nuclear Agency)

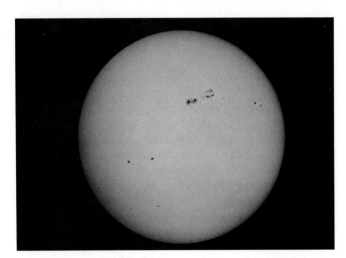

**figure 1-4** R I V U X G

**Our Star, the Sun** The Sun is a typical star. Its diameter is about 1.39 million kilometers (roughly a million miles), and its surface temperature is about 5500°C (10,000°F). The Sun draws its energy from thermonuclear reactions occurring at its center, where the temperature is about 15 million degrees Celsius. (Celestron International)

In Chapter 20 we look deep into space to find star clusters and clouds of glowing gas, called **nebulae** (singular **nebula**), scattered across the sky. These beautiful objects can tell us much about the lives of stars. Stars are born within huge clouds of interstellar gas and dust such as the Orion Nebula, shown in Figure 1-6. After millions or billions of years, stars die. As discussed in Chapter 22, some end their lives with a spectacular detonation called a **supernova** (plural **supernovae**) that blows the star apart. The Crab Nebula seen in Figure 1-7 is a striking example of a remnant left behind by a supernova.

Dying stars can produce some of the strangest objects in the sky. In Chapter 23 we will see that some dead stars become **pulsars,** which emit pulses of radio waves, or **X–ray bursters,** which emit powerful bursts of X rays. In Chapter 24 we encounter massive, almost inconceivably dense objects called **black holes,** which are surrounded by gravity so powerful that nothing—not even light—can escape. A number of black holes have been discovered only recently by Earth-orbiting telescopes that detect nonvisible light, such as the X rays emitted by gases falling toward a black hole.

During their death throes, stars return the gas of which they are made to interstellar space. (Figure 1-7 shows these

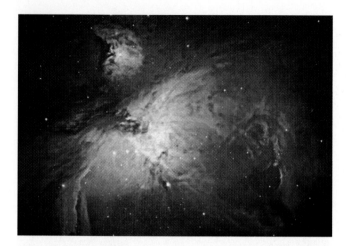

**Figure 1-6** R I V U X G

**The Orion Nebula** This beautiful nebula is a fine example of a stellar "nursery" where stars are born. Intense ultraviolet light from the newly formed stars causes the surrounding gases to glow. Many of the stars embedded in this nebula are less than a million years old. The Orion Nebula is about 1500 light-years from Earth, and the distance across the nebula is about 5 light-years. (Anglo-Australian Observatory)

**Figure 1-7** R I V U X G

**The Crab Nebula** This nebula is a fine example of a supernova remnant. A dying star exploded, leaving behind this beautiful funeral shroud of glowing gases blasted violently into space. A thousand years after the explosion these gases are still moving outward at about 1000 km/s (roughly 2 million miles per hour). The Crab Nebula is 6500 light-years from Earth, and the distance across the nebula is about 10 light-years. (Lick Observatory)

expelled gases expanding away from the site of a supernova explosion.) This gas contains heavy elements created by thermonuclear reactions in the interiors of the stars. Interstellar space thus becomes enriched with newly manufactured atoms and molecules. The Sun and its planets were formed from interstellar material enriched in heavy elements by preceding generations of stars. Indeed, the iron and nickel that make up the body of Earth, as well as the carbon in our bodies and the oxygen we breathe, was created deep inside ancient stars. By studying stars and their evolution, we are really studying our own origins.

---

### 1-4    By observing galaxies, astronomers learn about the creation and fate of the universe

Stars are not spread uniformly across the universe but are grouped together in huge assemblages called **galaxies.** Galaxies come in a wide range of shapes and sizes. A typical galaxy, like our own Milky Way, contains several hundred billion stars. As described in Chapter 26, some galaxies are much smaller, containing only a few million stars. Others are monstrosities that devour neighboring galaxies in a process called "galactic cannibalism."

Our Milky Way Galaxy has arching spiral arms like those of the galaxy shown in Figure 1-8. These arms are particular-

ly active sites of star formation. As we explore our Galaxy in Chapter 25, we find that there is a mysterious object at its center with a mass millions times greater than that of our Sun. Many astronomers suspect that this curious object is an enormous black hole.

Some of the most intriguing galaxies appear to be in the throes of violent convulsions and are rapidly expelling matter. The centers of these strange galaxies, which may harbor even more massive black holes, are often powerful sources of X rays and radio waves.

Even more awesome sources of energy are found still deeper in space. At distances so great that their light takes billions of years to reach Earth, we find the mysterious **quasars.** Although quasars look like stars (Figure 1-9), they are the most distant and most luminous objects in the sky. A typical quasar shines with the brilliance of a hundred galaxies. As explained in Chapter 27, recent observations imply that quasars draw their energy from material falling into enormous black holes.

The motions of clusters of galaxies reveal that we live in an expanding universe. Extrapolating into the past, we learn that the universe must have been born from an incredibly dense state (perhaps infinitely dense) roughly 15 billion years ago.

A variety of evidence indicates that the universe began with a cosmic explosion, known as the **Big Bang,** that occurred throughout all space at the beginning of time. Events shortly after the Big Bang dictated the present nature of the universe. Chapters 28 and 29 describe the significant progress that astronomers are making in understanding these

ƒigure 1-8  R I **V** U X G

**A Galaxy** This spectacular galaxy, called M100, contains several hundred billion stars. The galaxy's spiral arms are outlined by many nebulae, which are sites of active star formation. M100 has a diameter of about 120,000 light-years and lies about 56 million light-years from Earth. (Anglo-Australian Observatory)

ƒigure 1-9  R I **V** U X G

**A Quasar** The two bright starlike objects in this image look almost identical at first glance but are actually dramatically different. The object on the left is indeed a star that lies a few hundred light-years from Earth. But the "star" on the right is actually a quasar located about 9 billion light-years away. To appear so bright even though they are so distant, quasars like this one must be the most luminous objects ever seen. The other objects in this image are galaxies like that in Figure 1-8. (Charles Steidel, California Institute of Technology/ NASA)

cosmic events. This knowledge may even reveal the origin of some of the most basic properties of physical reality. In addition, the motions of distant clusters of galaxies may answer questions about the ultimate fate of the universe. Will it continue expanding forever, or will it someday stop and collapse back in on itself?

The work of unraveling the deepest mysteries of the universe requires specialized tools, including telescopes and spacecraft. Box 1-1 describes how astronomers use a no less important tool, the computer, to analyze data and test new theories. But for many purposes the most useful device for studying the universe is the human brain itself. Our goal in this book is to help you use *your* brain to share in the excitement of scientific discovery. In the remainder of this chapter we introduce some of the key concepts and mathematics that we will use in subsequent chapters and that you will use throughout your own study of astronomy.

## 1-5  Astronomers use angles to denote the apparent sizes and positions of objects in the sky

An important part of astronomy is keeping track of the positions and apparent sizes of objects in the sky. To this end, astronomers use angles and a system of angular measure. An **angle** is the opening between two lines that meet at a point. **Angular measure** describes the size of an angle exactly. The basic unit of angular measure is the **degree**, designated by the symbol °. A full circle is divided into 360°, and a right

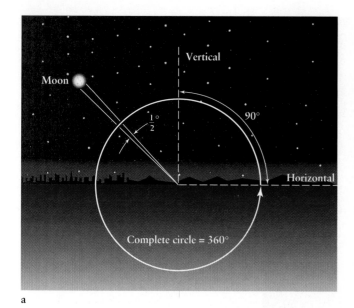

a

b

## figure 1-10

**Angles and Angular Measure** (a) Angles are measured in degrees (°). There are 360° in a complete circle and 90° in a right angle. For example, the angle between the vertical direction (directly above you) and the horizontal direction (towards the horizon) is 90°. The angular diameter of the Moon in the sky is about ½°. (b) The Big Dipper is an easily recognized grouping of seven bright stars. The angular distance between the two pointer stars at the front of the Big Dipper is about 5°, roughly ten times the angular diameter of the Moon.

angle measures 90° (see Figure 1-10a). As shown in Figure 1-10b, the angle between the two "pointer stars" in the Big Dipper is about 5°. (These two stars "point" to Polaris, the North Star, as described in Chapter 2.)

Astronomers also use angular measure to describe the apparent size of a celestial object. For example, the angle covered by the diameter of the full moon is about ½° (Figure 1-10a). We therefore say that the **angular diameter** (or **angular size**) of the Moon is ½°. Alternatively, astronomers say that the Moon **subtends** an angle of ½°. Ten full moons could fit side by side between the two pointer stars in the Big Dipper.

The adult human hand held at arm's length provides a means of estimating angles, as shown in Figure 1-11. For example, your fist covers an angle of 10°, whereas the tip of your finger is about 1° wide. Various segments of your index

## figure 1-11

**Estimating Angles with the Human Hand** Various parts of the adult human hand extended to arm's length can be used to estimate angular distances and sizes in the sky.

## box 1-1 | Looking Deeper into Astronomy

### *Computers in Astronomical Research*

If you ask the average person what astronomers do on the job, the response will likely be, "They look through telescopes." Telescopes for visible light are indeed important tools of astronomy, and they make the universe accessible to countless amateur astronomers. Hardly any professional astronomers actually look through telescopes, however. Instead, images are recorded electronically using much the same technology as is found inside a video camera. (In Chapter 6 we will see the advantages of this technology over the human eye.) Other telescopes are sensitive to invisible forms of light such as X rays or radio waves; there is no way that one could "look through" such telescopes. Furthermore, many astronomers seldom, if ever, even go near a telescope. These researchers analyze new data or develop theories to explain the data.

The only piece of hardware that is used on a daily basis by *all* astronomers is not the telescope but rather the computer. A number of computers of different kinds can be found at any modern observatory. Some are used to control the orientation of the telescope and to keep it pointed at the object under study; other computers may regulate the shape of the telescope optics to keep the image in the sharpest possible focus; and ordinary personal computers help astronomers keep in contact with their colleagues around the world through electronic mail. Perhaps most important, a computer records the data gathered by the telescope's electronic detector and saves it (on magnetic media such as a hard disk) for later study.

Once the stored data are returned to the astronomer's office, which may be thousands of miles from the telescope, other computers are used for data analysis. Several decades ago photographic film was used to record almost all telescopic images, and extracting all of the information from such photographs was often a slow and difficult process. But when images are stored in electronic form, computers can digest those images directly and process them quickly to bring out the most subtle details. Many of the astronomical images in this book were processed in just this way. Software for this purpose is even available for personal computers and is used by many amateur astronomers who have electronic detectors on their telescopes.

Besides their role in collecting and analyzing data, computers have also changed the character of much astronomical research in a fundamental way. Consider the plight of an astronomer who wants to understand why the galaxy shown in Figure 1-8 has such a beautiful pattern of spiral arms. The answer must lie in the laws of physics that describe the behavior of the stars and nebulae that constitute the galaxy. Although these laws are known, the problem seems completely intractable, because the galaxy contains an astronomically large number of stars. An equally large number of intertwined equations would have to be solved to describe the motions of these stars, and a scientist with pencil and paper will not live long enough to solve them. But such problems have become tractable in recent years with the development of high-speed supercomputers.

Supercomputers are a valuable research tool because the laws of physics can be written in a way that they can be solved by a supercomputer. In their basic form, the laws of physics are valid at every point in space and at every moment of time. But scientists seldom need to apply these laws everywhere at all times. Instead, they use points in space separated by distances that are small compared to the size of the object under study. They also use time intervals that are short compared to the duration of the process they want to examine. By programming a supercomputer with the laws of physics expressed only at these selected points and time intervals, scientists can reduce a complicated problem to a form that the machine can handle.

For example, to analyze the spiral galaxy in Figure 1-8, you start with all of the stars at certain positions and moving with certain speeds. Knowing the gravitational influences that each star feels from all of the others, you can compute how each star will move over a brief time interval. (In a spiral galaxy, individual stars may take hundreds of millions of years to complete one orbit

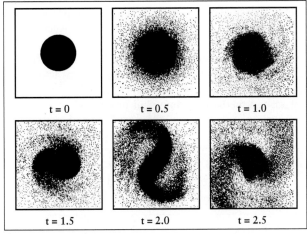

t = 0     t = 0.5     t = 1.0

t = 1.5     t = 2.0     t = 2.5

(After F. Hohl et al.)

*(continued on the following page)*

**box 1-1** *(continued)*

around the galaxy's center, so a "brief" time interval may be a hundred thousand years or so.) Now you can repeat the entire calculation using the data for the stars in their new positions. Step by step, you compute the motions of all of the stars within the galaxy. Modern supercomputers are ideal for performing these repetitive computations.

The accompanying figure shows the result of just such a calculation using 100,000 simulated stars. The individual pictures are like frames from a movie, showing how the simulated galaxy evolves. Note that a spiral structure forms naturally, although it subsequently fades away. In Chapters 25 and 26 we will discuss our present-day understanding of

galaxies, much of which is based on such supercomputer simulations.

Using supercomputers, astronomers can simulate phenomena that might otherwise never be observed. For example, later in this book we will see how the Moon might have been created by a Mars-sized object striking Earth. We will also see what happens when gas is captured and swallowed by a black hole. We will watch as two galaxies, consisting of thousands of stars and gas clouds, collide and interact with each other, a process that in reality takes millions of years. These are just a few examples of the essential role of the computer in modern astronomy.

---

finger extended to arm's length can be similarly used to estimate angles a few degrees across.

To talk about smaller angles, we subdivide the degree into 60 **arcminutes** (also called minutes of arc), which is commonly abbreviated as 60 arcmin or 60′. An arcminute is further subdivided into 60 **arcseconds** (or seconds of arc), usually written as 60 arcsec or 60″. Thus,

$$1° = 60 \text{ arcmin} = 60′$$

$$1′ = 60 \text{ arcsec} = 60″$$

For example, on January 1, 1998, the planet Saturn had an angular diameter of 17.9 arcsec as viewed from Earth. That is a convenient, precise statement of how big the planet appeared in Earth's sky on that date. (Because this angular diameter is so small, to the naked eye Saturn appears simply as a point of light. To see any detail on Saturn, such as the planet's rings, requires a telescope.)

If we know the angular size of an object as well as the distance to that object, we can determine the actual linear size of the object (measured in kilometers or miles, for example). We describe a method for doing this in Box 1-2.

| 1-6 | Powers-of-ten notation is a useful shorthand system of writing numbers |

Astronomy is a subject of extremes. Astronomers investigate the largest structures in the universe, including galaxies and clusters of galaxies, but they must also study atoms and atomic nuclei, among the smallest objects in the universe, in

order to explain why stars shine. They also study conditions ranging from the incredibly hot and dense centers of stars to the frigid near-vacuum of interstellar space. To describe such a wide range of phenomena, we need an equally wide range of both large and small numbers.

Astronomers avoid such confusing terms as "a million billion billion" by using a standard shorthand system called **powers-of-ten notation**. All the cumbersome zeros that accompany a large number are consolidated into one term consisting of 10 followed by an **exponent**, which is written as a superscript. The exponent indicates how many zeros you would need to write out the long form of the number. Thus,

$$10^0 = 1$$
$$10^1 = 10$$
$$10^2 = 100$$
$$10^3 = 1000$$
$$10^4 = 10,000$$

and so forth. The exponent also tells you how many tens must be multiplied together to give the desired number, which is why the exponent is also called the **power of ten**. For example, ten thousand can be written as $10^4$ ("ten to the fourth" or "ten to the fourth power") because $10^4 = 10 \times 10 \times 10 \times 10 = 10,000$.

With this notation, numbers are written as a figure between one and ten multiplied by the appropriate power of ten. The approximate distance between Earth and the Sun, for example, can be written as $1.5 \times 10^8$ kilometers (or $1.5 \times 10^8$ km for short). Once you get used to it, this is more convenient than writing "150,000,000 kilometers" or "one hundred and fifty million kilometers." (The same number could also be

# box 1-2 | Tools of the Astronomer's Trade

## *The Small-Angle Formula*

You can estimate the angular sizes of objects in the sky with your hand and fingers (Figure 1-11). Using rather more sophisticated equipment, astronomers can measure angular sizes to a fraction of a second of arc. Keep in mind, however, that *angular* size is not the same as *actual* size. As an example, if you extend your arm while looking at a full moon, you can completely cover the Moon with your thumb. That's because from your perspective, your thumb has a larger angular size (that is, it subtends a larger angle) than the Moon. But the actual size of your thumb (about 2 centimeters) is much less than the actual diameter of the Moon (more than 3000 kilometers).

The figure below shows how the angular size of an object is related to its linear size. Part **a** of the figure shows that for a given angular size, the more distant the object, the larger its actual size. For example, your thumb held at arm's length just covers the full Moon; the angular size of your thumb and the Moon are about the same, but the Moon is much farther away and is far larger in linear size. Part **b** shows that for a given linear size, the angular size decreases the further away the object. This is why a car looks smaller and smaller as it drives away from you. We can put these

relationships together into a single mathematical expression called the **small-angle formula**. Suppose that an object subtends an angle α (the Greek letter alpha) and is at a distance *d* from the observer, as in part **c** of the diagram. If the angle α is small, as is almost always the case for objects in the sky, the small-angle formula tells us that the linear size (*D*) of the object is given by the following expression:

**The small-angle formula**

$$D = \frac{\alpha d}{206,265}$$

$D$ = linear size of an object
$\alpha$ = angular size of the object, in arcsec
$d$ = distance to the object

The number 206,265 is required in the formula; mathematically, it is equal to the number of arcseconds in a complete circle (that is, 360°) divided by the number $2\pi$ (the ratio of the circumference of a circle to that circle's radius).

**EXAMPLE:** On December 31, 1996, Jupiter was at a distance of 909 million kilometers from Earth. Jupiter's angular diameter on that date was 32.4 seconds of arc. Using the small-angle formula, we can calculate the planet's diameter as follows:

$$D = \frac{32.4 \times 909,000,000 \text{ km}}{206,265} = 143,000 \text{ km}$$

This answer agrees very well with an equatorial diameter of 142,984 km determined from spacecraft flybys.

**EXAMPLE:** Under excellent conditions, a telescope on Earth can see details with an angular size as small as 1 arcsec. We can use the small-angle formula to determine the greatest distance at which you could see details as small as 1.7 m (the height of a typical person) under these conditions. We rewrite the formula to solve for the distance *d*:

$$d = \frac{206,265D}{\alpha} = \frac{206,265 \times 1.7 \text{ m}}{1} = 350,000 \text{ m} = 350 \text{ km}$$

This is much less than the distance to the Moon—384,000 km. Thus, even the best telescope on Earth could not be used to see an astronaut walking on the surface of the Moon.

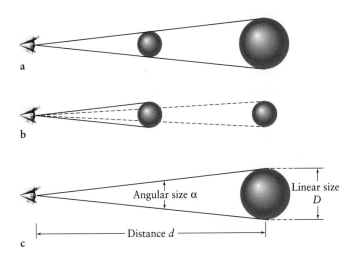

a

b

Angular size α — Linear size D — Distance *d*

c

(a) Two objects that have the same angular size may have different linear sizes if they are at different distances from the observer. (b) For an object of a given linear size, the angular size is larger the closer the object is to the observer. (c) The small-angle formula relates the linear size of an object to its angular size and its distance from the observer.

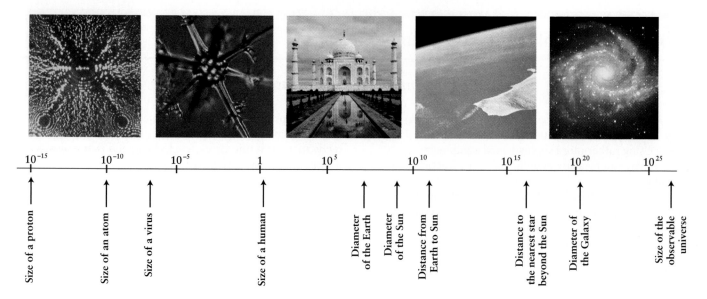

## figure 1-12

**Examples of Powers-of-Ten Notation** The scale gives the sizes of objects in meters, ranging from subatomic particles at the left to the entire observable universe on the right. The photograph at the left shows tungsten atoms, $10^{-10}$ meter in diameter. Second from left is the crystalline skeleton of a diatom (a single-celled organism) $10^{-4}$ meter (0.1 millimeter) in size. At the center is the Taj Mahal, within reach of our unaided senses. On the right, looking across the Indian Ocean toward the South Pole, we see the curvature of the Earth, $10^7$ meters in diameter. At the far right is a galaxy, $10^{21}$ meters (100,000 light-years) in diameter. (Courtesy of Scientific American Books; NASA; AAT)

written as $15 \times 10^7$ or $0.15 \times 10^9$, but the preferred form is *always* to have the first figure be between 1 and 10.)

Note that most electronic calculators use a shorthand for powers-of-ten notation. To enter the number $1.5 \times 10^8$, you first enter 1.5, then press a key labeled "EXP" or "EE," then enter the exponent 8. (The EXP or EE key takes care of the "$\times 10$" part of the expression.) The number will then appear on your calculator's display as "1.5 E 8," "1.5 8," or some variation of this; usually the "$\times 10$" is not displayed as such. There are some variations from one kind of calculator to another, so you should spend a few minutes reading over your calculator's instruction manual to make sure you know the correct procedure for working with numbers in powers-of-ten notation. You will be using this notation continually in your study of astronomy, so this is time well spent.

**CAUTION** Confusion can result from the way that calculators display powers-of-ten notation. Since $1.5 \times 10^8$ is displayed as "1.5 8" or "1.5 E 8," it is not uncommon to think that $1.5 \times 10^8$ is the same as $1.5^8$. That is not correct, however; $1.5^8$ is equal to 1.5 multiplied by itself eight times, or 25.63, which is not even close to $150,000,000 = 1.5 \times 10^8$. Another, not uncommon mistake is to write $1.5 \times 10^8$ as $15^8$. If you are inclined to do this, perhaps you are thinking that you can multiply 1.5 by 10, then tack on the exponent later. This also does not work;

$15^8$ is equal to 15 multiplied by itself eight times, or 2,562,890,625, which again is nowhere near $1.5 \times 10^8$. Reading over the manual for your calculator will help you to avoid these common errors.

Powers-of-ten notation can be applied to numbers that are less than one by using a minus sign in front of the exponent. A negative exponent tells you to divide by the appropriate number of tens, for example, $10^{-1} = \frac{1}{10}$. The location of the decimal point is as follows:

$$10^0 = 1$$
$$10^{-1} = 0.1$$
$$10^{-2} = 0.01$$
$$10^{-3} = 0.001$$
$$10^{-4} = 0.0001$$

and so forth.

The easiest way to understand negative exponents is to realize that they tell you how many tenths must be multiplied together to give the desired number. For example, one ten–thousandth can be written as $10^{-4}$ (ten to the minus four) because $10^{-4} = \frac{1}{10} \times \frac{1}{10} \times \frac{1}{10} \times \frac{1}{10} = 0.0001$. Expressed with this notation, the diameter of a hydrogen atom is $1.1 \times 10^{-10}$ meter, or $1.1 \times 10^{-10}$ m. That is much more convenient than saying "0.00000000011 meters" or "110 trillionths of a meter."

## $box$ 1-3 | Tools of the Astronomer's Trade

### *Arithmetic with Exponents*

Using powers-of-ten notation, it is easy to multiply numbers. For example, suppose you want to multiply 100 by 1000. If you use ordinary notation, you have to write a lot of zeros:

$$100 \times 1000 = 100{,}000 \text{ (one hundred thousand)}$$

By converting these numbers to powers-of-ten notation, this same multiplication can be written more compactly as

$$10^2 \times 10^3 = 10^5$$

Because $2 + 3 = 5$, we are led to the following general rule for *multiplying* numbers expressed in terms of powers of ten—simply *add* the exponents.

**EXAMPLE:**

$$10^4 \times 10^3 = 10^{4+3} = 10^7$$

To *divide* numbers expressed in terms of powers of ten, remember that $10^{-1} = \frac{1}{10}$, $10^{-2} = \frac{1}{100}$, and so on. The general rule for any exponent $n$ is

$$10^{-n} = \frac{1}{10^n}$$

In other words, dividing by $10^n$ is the same as multiplying by $10^{-n}$. To carry out a division, you first transform it into multiplication by changing the sign of the exponent, and then carry out the multiplication by adding the exponents.

**EXAMPLE:**

$$\frac{10^4}{10^6} = 10^4 \times 10^{-6} = 10^{4 + (-6)} = 10^{4-6} = 10^{-2}$$

Usually a computation involves numbers like $3.0 \times 10^{10}$, that is, an ordinary number multiplied by a factor of 10 with an exponent. In such cases, to perform multiplication or division, you can treat the numbers separately from the factors of $10^n$.

**EXAMPLE:** We can redo the first numerical example from Box 1-2 in a straightforward manner by using exponents:

$$D = 32.4 \times \frac{909{,}000{,}000 \text{ km}}{206{,}265} = \frac{3.24 \times 10 \times 9.09 \times 10^8}{2.06265 \times 10^5} \text{ km}$$

$$= \frac{3.24 \times 9.09 \times 10^{1+8-5}}{2.06265} \text{ km} = 14.3 \times 10^4 \text{ km}$$

$$= 1.43 \times 10 \times 10^4 \text{ km} = 1.43 \times 10^5 \text{ km}$$

---

Using powers-of-ten notation, one can write familiar numerical terms as follows:

$$\text{one hundred} = 100 = 10^2$$
$$\text{one thousand} = 1000 = 10^3$$
$$\text{one million} = 1{,}000{,}000 = 10^6$$
$$\text{one billion} = 1{,}000{,}000{,}000 = 10^9$$
$$\text{one trillion} = 1{,}000{,}000{,}000{,}000 = 10^{12}$$

and also

$$\text{one-hundredth} = 0.01 = 10^{-2}$$
$$\text{one-thousandth} = 0.001 = 10^{-3}$$
$$\text{one-millionth} = 0.000001 = 10^{-6}$$
$$\text{one-billionth} = 0.000000001 = 10^{-9}$$
$$\text{one-trillionth} = 0.000000000001 = 10^{-12}$$

Because powers-of-ten notation bypasses all the awkward zeros, we can conveniently describe the size of objects as small as atoms or as big as galaxies (Figure 1-12). Powers-of-ten notation also makes it easy to multiply and divide numbers, as described in Box 1-3.

### 1-7 | Astronomical distances are often measured in astronomical units, parsecs, or light-years

Astronomers use many of the same units of measurement as do other scientists. They often measure lengths in meters (abbreviated m), masses in kilograms (kg), and time in seconds (s). (We discuss these units of measurement, as well as techniques for converting between different sets of units, in Box 1-4.) Like other scientists, astronomers often find it useful to combine these units with powers of ten and create new units using prefixes. As an example, the number 1000 (= $10^3$) is represented by the prefix "kilo," and so a distance of 1000

## box 1-4 | Tools of the Astronomer's Trade

### *Units of Length, Time, and Mass*

To understand and appreciate the universe, we need to describe phenomena not only on the large scales of galaxies but also on the submicroscopic scale of the atom. Astronomers generally use units that are best suited to the topic at hand. For example, interstellar distances are conveniently expressed in either light–years or parsecs, whereas the diameters of the planets are more comfortably presented in kilometers.

Most scientists prefer to use a version of the metric system called the International System of Units, abbreviated **SI** (after the French name Système International). In SI units, length is measured in meters (m), time is measured in seconds (s), and mass (a measure of the amount of material in an object) is measured in kilograms (kg). How are these basic units related to other measures?

When discussing objects on a human scale, sizes and distances are usually expressed in millimeters (mm), centimeters (cm), and kilometers (km). These units of length are related to the meter as follows:

$$1 \text{ millimeter} = 0.001 \text{ m}$$

$$1 \text{ centimeter} = 0.01 \text{ m}$$

$$1 \text{ kilometer} = 1000 \text{ m}$$

The English system of inches (in.), feet (ft), and miles (mi) is actually based on the SI system—the inch is defined to be exactly 2.54 cm. A useful set of conversions is

$$1 \text{ in.} = 2.54 \text{ cm}$$

$$1 \text{ ft} = 0.3048 \text{ m}$$

$$1 \text{ mi} = 1.609 \text{ km}$$

Each of these equalities can also be written as a fraction equal to 1. For example, you can write

$$\frac{0.3048 \text{ m}}{1 \text{ ft}} = 1$$

These fractions are useful for converting a quantity from one set of units to another. For example, the *Saturn V* rocket used to send astronauts to the Moon stands about 363 feet tall. How can we convert this height to meters? The trick is to remember that a quantity does not change if you multiply it by 1. Expressing the number 1 by the fraction (0.3048 m/1 ft), we can write the height of the rocket as

$$363 \text{ ft} \times 1 = 363 \text{ ft} \times \frac{0.3048 \text{ m}}{1 \text{ ft}} = 111 \frac{\text{ft} \times \text{m}}{\text{ft}} = 111 \text{ m}$$

The unwanted units of feet cancel, just like ordinary numbers, giving you an answer in meters. You can make a quick check of this result by noting that a meter is larger than a foot. Therefore, the number of meters in the height should be less than the number of feet, as indeed it is.

**EXAMPLE:** The diameter of Mars is 6794 km. Let's try expressing this in miles.

 You can get into trouble if you are careless in applying the trick of taking the number whose units are to be converted and multiplying it by 1. For example, if we multiply the diameter by 1 expressed as (1.609 km)/(1 mi), we get

meters is the same as 1 kilometer (1 km). Here are some of the most common prefixes, with examples of how they are used:

$$\text{one-billionth meter} = 10^{-9} \text{ m} = 1 \text{ nanometer}$$

$$\text{one-millionth second} = 10^{-6} \text{ s} = 1 \text{ microsecond}$$

$$\text{one-thousandth arcsecond} = 10^{-3} \text{ arcsec} = 1 \text{ milliarcsecond}$$

$$\text{one-hundredth meter} = 10^{-2} \text{ m} = 1 \text{ centimeter}$$

$$\text{one thousand meters} = 10^{3} \text{ m} = 1 \text{ kilometer}$$

$$\text{one million tons} = 10^{6} \text{ T} = 1 \text{ megaton}$$

In principle, we could express all sizes and distances in astronomy using units based on the meter. Indeed, we will use kilometers to give the diameters of craters on the Moon, as well as the diameter of the Moon itself. But, while a kilometer (roughly equal to three-fifths of a mile) is an easy distance for humans to visualize, a megameter ($10^{6}$ m) is not. For this reason, astronomers have devised units of measure that are more appropriate for the tremendous distances between the planets and the far greater distances between the stars.

When discussing distances across the solar system, astronomers use a unit of length called the **astronomical unit**

$$6794 \text{ km} \times 1 = 6794 \text{ km} \times \frac{1.609 \text{ km}}{1 \text{ mi}} = 10{,}930 \frac{\text{km}^2}{\text{mi}}$$

This cannot be right, because the unwanted units of km did not cancel. Furthermore, a mile is larger than a kilometer, so the diameter expressed in miles should be a smaller number than when expressed in kilometers.

The correct approach is to write the number 1 so that the units will cancel. Because the number we are starting with is in kilometers, we must write the number 1 with kilometers in the denominator ("downstairs" in the fraction). Thus, we express 1 as (1 mi)/(1.609 km):

$$6794 \text{ km} \times 1 = 6794 \text{ km} \times \frac{1 \text{ mi}}{1.609 \text{ km}} = 4222 \frac{\text{km} \times \text{mi}}{\text{km}}$$
$$= 4222 \text{ mi}$$

Now the units of km cancel as they should, and the distance in miles is a smaller number than in kilometers (as it must be).

When discussing very small distances such as the size of an atom, astronomers often use the micrometer (μm) or the nanometer (nm). These are related to the meter as follows:

$$1 \text{ micrometer} = 1 \text{ μm} = 10^{-6} \text{ m}$$
$$1 \text{ nanometer} = 1 \text{ nm} = 10^{-9} \text{ m}$$

Thus, 1 μm = $10^3$ nm. (Note that the micrometer is often, but erroneously, called the micron.)

The basic unit of time is the second (s). It is related to other units of time as follows:

$$1 \text{ minute (min)} = 60 \text{ s}$$
$$1 \text{ hour (h)} = 3600 \text{ s}$$
$$1 \text{ day (d)} = 86{,}400 \text{ s}$$
$$1 \text{ year (yr)} = 3.156 \times 10^7 \text{ s}$$

In the SI system, speed is properly measured in meters per second (m/s). Quite commonly, however, speed is also expressed in km/s and mi/h:

$$1 \text{ km/s} = 10^3 \text{ m/s}$$
$$1 \text{ km/s} = 2237 \text{ mi/h}$$
$$1 \text{ mi/h} = 0.447 \text{ m/s}$$
$$1 \text{ mi/h} = 1.47 \text{ ft/s}$$

In addition to using kilograms, astronomers sometimes express mass in grams (g) and in solar masses ($M_\odot$), where the subscript $\odot$ is the symbol denoting the Sun. It is especially convenient to use solar masses when discussing the masses of stars and galaxies. These units are related to each other as follows:

$$1 \text{ kg} = 1000 \text{ g}$$
$$1 \text{ M}_\odot = 1.99 \times 10^{30} \text{ kg}$$

CAUTION! You may be wondering why we have not given a conversion between kilograms and pounds. The reason is that these units do not refer to the same physical quantity! A kilogram is a unit of *mass*, which is a measure of the amount of material in an object. By contrast, a pound is a unit of *weight*, which tells you how strongly gravity pulls on that object's material. Consider a person who weighs 110 pounds on Earth, corresponding to a mass of 50 kg. Gravity is only about one-sixth as strong on the Moon as it is on Earth, so this person would weigh only one-sixth as much, or about 18 pounds, on the Moon. But that person's mass of 50 kg is the same on the Moon; wherever you go in the universe, you take all of your material along with you. We will explore the relationship between mass and weight in Chapter 4.

(abbreviated AU). This is the average distance between Earth and the Sun:

$$1 \text{ AU} = 1.496 \times 10^8 \text{ km} = 93 \text{ million miles}$$

Thus the distance between the Sun and Jupiter can be conveniently stated as 5.2 AU.

To talk about distances to the stars, astronomers have two different units of length. The **light-year** (abbreviated ly) is the distance that light travels in one year. This is a useful concept because the speed of light in empty space always has the same value, $3.00 \times 10^5$ km/s (kilometers per second) or

$1.86 \times 10^5$ mi/s (miles per second). In terms of kilometers or astronomical units, one light-year is given by

$$1 \text{ ly} = 9.46 \times 10^{12} \text{ km} = 63{,}240 \text{ AU}$$

This distance is roughly equal to 6 trillion miles.

CAUTION! Keep in mind that despite its name, the light-year is a unit of distance and not a unit of time. As an example, Proxima Centauri, the nearest star other than the Sun, is a distance of 4.3 light-years from Earth. This means that light takes 4.3 years to travel to us from Proxima Centauri.

Physicists often measure interstellar distances in light-years, because the speed of light is one of nature's most important numbers. But many astronomers prefer to use another unit of length, the **parsec** (abbreviated pc), because its definition is closely related to a method of measuring distances to the stars (you might want to glance ahead to Figure 19-2). Imagine taking a journey far into space, beyond the orbits of the outer planets. As you look back toward the Sun, Earth's orbit subtends a smaller angle in the sky the farther you are from the Sun. The distance at which 1 AU subtends an angle of 1 arcsec, as shown in Figure 1-13, is defined as one parsec:

$$1 \text{ pc} = 3.09 \times 10^{13} \text{ km} = 3.26 \text{ ly}$$

The distance to Proxima Centauri can be stated as 1.3 pc as well as 4.3 ly. Whether you choose to use parsecs or light-years is a matter of personal taste.

For even greater distances, astronomers commonly use **kiloparsecs** and **megaparsecs** (abbreviated kpc and Mpc). As we saw before, these prefixes simply mean "thousand" and "million," respectively:

$$1 \text{ kiloparsec} = 1 \text{ kpc} = 1000 \text{ pc} = 10^3 \text{ pc}$$

$$1 \text{ megaparsec} = 1 \text{ Mpc} = 1,000,000 \text{ pc} = 10^6 \text{ pc}$$

For example, the distance from Earth to the center of our Milky Way Galaxy is about 8 kpc, and the galaxy shown in Figure 1-8 is about 17 Mpc away.

Some astronomers prefer to talk about thousands or millions of light-years rather than kiloparsecs and megaparsecs. Once again, the choice is a matter of personal taste. As a general rule, astronomers use whatever yardsticks seem best suited for the issue at hand and do not restrict themselves to one system of measurement. For example, an astronomer might say that the supergiant star Antares has a diameter of 860 million kilometers and is located at a distance of 150 parsecs from Earth.

## 1-8 Astronomy is an adventure of the human mind

An underlying theme of this book is that the universe is rational. It is not a hodgepodge of unrelated things behaving in unpredictable ways. Rather, we find strong evidence that fundamental laws of physics govern the nature of the universe and the behavior of everything in it. These unifying concepts enable us to explore realms far removed from our earthly experience. Thus, a scientist can do experiments in a laboratory to determine the properties of light or the behavior of atoms and then use this knowledge to investigate the structure of the universe.

The discovery of fundamental laws of nature has had a profound influence on humanity. These laws have led to an immense number of practical applications that have fundamentally transformed commerce, medicine, entertainment, transportation, and other aspects of our lives. In particular, space technology has given us instant contact with any point on the globe through communication satellites, accurate weather forecasts from meteorological satellites, precise navigation to any point on Earth using signals from the satellites of the Global Positioning System (GPS), and long-term monitoring of Earth's climate and environment from orbiting spacecraft (Figure 1-14).

As important as the applications of science are, the pursuit of scientific knowledge for its own sake is no less important. We are fortunate to live in an age where this pursuit is in full flower. Just as explorers such as Columbus and Magellan discovered the true size and nature of our planet in the fifteenth and sixteenth centuries, astronomers of the twentieth and twenty-first centuries are exploring the universe to an extent that is unparalleled in human history. Indeed, even the voyages into space imagined by such great science fiction writers as Jules Verne and H. G. Wells pale in comparison to the reality of today. In recent years humans have walked on the Moon, sent robot spacecraft to dig into

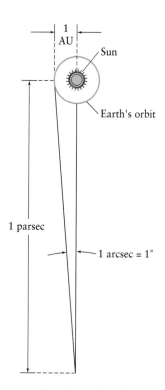

**Figure 1-13**

**A Parsec** The parsec, a unit of length commonly used by astronomers, is equal to 3.26 light-years. The parsec is defined as the distance at which 1 AU perpendicular to the observer's line of sight subtends an angle of 1 arcsec.

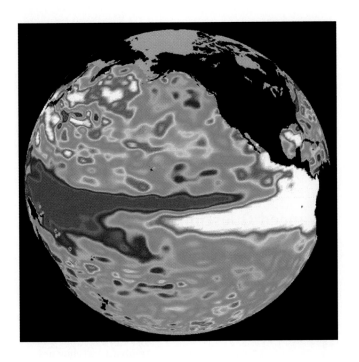

**figure 1-14**

**The Earth's Oceans from Space** As it orbits Earth, the unmanned *TOPEX/Poseidon* spacecraft (a joint project of the United States and France) measures variations in the height of the oceans. Areas of warm water bulge upward by several centimeters, so these measurements keep track of changes in ocean temperature. In this image, areas of high water level and thus high temperature have been colored white and red, while areas where the water level and temperature are both low are shown in blue and purple. While the temperature variations are no more than a few degrees above or below normal, they have a direct and profound effect on Earth's global climate. By monitoring these variations, scientists can give early warning of the potential for floods, droughts, and other forms of severe weather. The savings in lives and property that result from these early warnings more than pay for the cost of the spacecraft. (NASA)

the Martian soil and explore the satellites of Jupiter, and used the most powerful telescopes ever built to probe the limits of the observable universe. Never before has so much been revealed in so short a time.

As you proceed through this book, you will learn about the tools that scientists use to explore the natural world, as well as what they observe with these tools. But most importantly, you will see how astronomers build from their observations an understanding of the universe in which we live. This search for understanding is what makes science more than merely a collection of data and elevates it to one of the great adventures of the human mind. It is an adventure that will continue as long as there are mysteries left in the universe—an adventure which we hope you will come to appreciate and share.

## Key Words

*Terms preceded by an asterisk (\*) are discussed in the Boxes.*

angle, p. 9

angular diameter (angular size), p. 10

angular measure, p. 9

arcminute (′, minute of arc), p. 12

arcsecond (″, second of arc), p. 12

astronomical unit (AU), p. 16

Big Bang, p. 8

black hole, p. 7

degree (°), p. 9

exponent, p. 12

galaxy, p. 8

hypothesis, p. 4

kiloparsec (kpc), p. 18

laws of physics, p. 5

light-year (ly), p. 17

megaparsec (Mpc), p. 18

model, p. 4

nebula (*plural* nebulae), p. 7

Newtonian mechanics, p. 5

parsec (pc), p. 18

power of ten, p. 12

powers-of-ten notation, p. 12

pulsar, p. 7

quasar, p. 8

scientific method, p. 4

\* SI units, p. 16

\* small-angle formula, p. 13

solar system, p. 6

subtend (an angle), p. 10

supernova (*plural* supernovae), p. 7

theory, p. 4

X–ray burster, p. 7

## Key Ideas

• **Astronomy, Science, and the Nature of the Universe:** The universe is comprehensible. The scientific method is a procedure for formulating hypotheses about the universe. These are tested by observation or experimentation in order to build consistent models or theories that accurately describe phenomena in nature.

• **Observations of the heavens** have led to discovery of some of the fundamental laws of physics. The laws of physics are in turn used by astronomers to interpret their observations.

• **The Solar System:** Exploration of the planets provides information about the origin and evolution of the solar system, as well as the history and resources of Earth.

• **Stars and Nebulae:** Study of the stars and nebulae provides information about the origin and history of the Sun and the solar system.

• **Galaxies:** Observations of galaxies provide information about the origin and history of the universe.

• **Angular Measure:** Astronomers use angles to denote the positions and sizes of objects in the sky. The size of an angle is measured in degrees, arcminutes, and arcseconds.

• **Powers-of-Ten Notation:** The powers-of-ten notation system is a convenient shorthand for writing numbers. It allows very large and very small numbers to be expressed in a compact form.

• **Units of Distance:** A variety of distance units are used by astronomers. These include the astronomical unit (the average distance from Earth to Sun), the light-year (the distance that light travels in one year), and the parsec.

## Review Questions

1. What are degrees, arcminutes, and arcseconds used for? What are the relationships between these units of measure?

2. With the aid of a diagram, explain what it means to say that the Moon subtends an angle of ½°.

3. How many arcseconds equal 1°?

4. What is an exponent? How are exponents used in powers-of-ten notation?

5. What are the advantages of using powers-of-ten notation?

6. How is an astronomical unit (AU) defined? Give an example in which this unit of measure would be convenient to use.

7. What is the advantage to the astronomer of using the light-year as a unit of distance?

8. What is a parsec? How is it related to a kiloparsec and a megaparsec?

9. Write the following numbers using powers-of-ten notation: (a) one hundred million, (b) eighty thousand, (c) seven one-thousandths, (d) thirty-five billion, (e) your age in years, (f) the number of pages in this book.

10. Give the word or phrase that corresponds to the following standard abbreviations: (a) km, (b) cm, (c) s, (d) km/s, (e) mi/h, (f) m, (g) m/s, (h) h, (i) yr, (j) g, (k) kg. Which of these are units of speed? (*Hint:* You may have to refer to a dictionary. All of these abbreviations should be part of your working vocabulary.)

## Advanced Questions

*Questions preceded by an asterisk (\*) involve topics discussed in the Boxes.*

### Problem-solving tips and tools:

The size of an astronomical object is related to the angle it subtends by the small-angle formula given in Box 1-2. Box 1-4 illustrates the method for converting from one unit of measure to another. An object traveling at speed $v$ for a time $t$ covers a distance $d$ given by $d = vt$; for example, a car traveling at 90 km/h ($v$) for 3 hours ($t$) covers a distance $d = (90 \text{ km/h})(3 \text{ h}) = 270$ km. Similarly, the time $t$ required to cover a given distance $d$ at speed $v$ is $t = d/v$; for example, if $d = 270$ km and $v = 90$ km/h, $t = (270 \text{ km})/(90 \text{ km/h}) = 3$ hours.

11. The bright star Procyon is 3.50 pc from Earth. (a) What is the distance to Procyon in kilometers? Use powers-of-ten notation. (b) How long does it take for light emanating from Procyon to reach Earth? Give your answer in years.

12. The speed of light is $3.00 \times 10^8$ m/s. How long does it take light to travel from the Sun to Earth? Give your answer in seconds, using powers-of-ten notation.

13. A hydrogen atom has a radius of about $5 \times 10^{-9}$ cm. The radius of the observable universe is about 15 billion light-years. How many times larger than a hydrogen atom is the universe? Use powers-of-ten notation.

14. The diameter of the Sun is $1.4 \times 10^{11}$ cm, and the distance to the nearest star, Proxima Centauri, is 4.3 ly. Suppose you want to build an exact scale model of the Sun and Proxima Centauri, and you are using a basketball 30 cm in diameter to represent the Sun. In your scale model, how far away would Proxima Centauri be from the Sun? Give your answer in kilometers, using powers-of-ten notation.

15. How many Suns would it take, laid side by side, to reach the nearest star? Use powers–of–ten notation. (*Hint:* See the preceding question.)

16. When *Voyager 2* sent back pictures of Neptune during its historic flyby of that planet in 1989, the spacecraft's

radio signals traveled for 4 hours to reach Earth. How far away was the spacecraft? Give your answer in kilometers, using powers-of-ten notation.

**17.** The age of the universe is about 15 billion years. What is this age in seconds? Use powers-of-ten notation.

**18.** The Sun's mass is $1.99 \times 10^{30}$ kg, three-quarters of which is hydrogen. The mass of a hydrogen atom is $1.67 \times 10^{-27}$ kg. How many hydrogen atoms does the Sun contain? Use powers-of-ten notation.

**\*19.** On January 1, 1998, the planet Uranus was at a distance of 20.721 AU from Earth. The diameter of Uranus is 51,118 km. What was the angular size of Uranus as seen from Earth on January 1, 1998? Give your answer in arcminutes.

**\*20.** The average distance to the Moon is 384,000 km, and the Moon subtends an angle of ½°. Use this information to calculate the diameter of the Moon in kilometers.

**\*21.** At what distance would a person have to hold a nickel (which has a diameter of about 2.0 cm) in order for the nickel to subtend an angle of **(a)** 1°? **(b)** 1 arcmin? **(c)** 1 arcsec? Give your answers in meters.

**\*22.** A person with good vision can see details that subtend an angle of as small as 1 arcminute. If two dark lines on an eye chart are 2 millimeters apart, how far can such a person be from the chart and still be able to tell that there are two distinct lines? Give your answer in meters.

**\*23.** Suppose your telescope can give you a clear view of objects and features that subtend angles of at least 2 arcsec. What is the diameter of the smallest crater you can see on the Moon?

**\*24.** Explain where the number 206,265 in the small-angle formula comes from.

## Discussion Questions

**25.** How do astronomical observations differ from those of other sciences?

**26.** Scientists assume that "reality is rational." Discuss what this means and the thinking behind it.

## Observing Projects

**27.** Look up at the sky on a clear, cloud-free night. Is the Moon in the sky? If so, does it interfere with your ability to see the fainter stars? Why do you suppose astronomers prefer to schedule their observations on nights when the Moon is not in the sky?

**28.** On a dark, clear, moonless night, can you see the Milky Way from where you live ? If so, briefly describe its appearance. If not, what seems to be interfering with your ability to see the Milky Way?

**29.** Look up at the sky on a clear, cloud-free night and note the positions of a few prominent stars relative to such reference markers as rooftops, telephone poles, and treetops. Also note the location from where you make your observations. A few hours later, return to that location and again note the positions of the same bright stars that you observed earlier. How have their positions changed? From these changes, can you deduce the general direction in which the stars appear to be moving?

## Where to Learn More

*Books*

Asimov, I. *The Measure of the Universe*. Harper & Row, 1983. An extensive journey through the universe that uses powers of ten.

Audouze, J., Israël, G., and Falque, J.-C., eds. *The Cambridge Atlas of Astronomy*. Cambridge University Press, 1994. A remarkably complete description of modern astronomy for the general reader and amateur astronomer, with more than a thousand photographs and illustrations.

Hartmann, W., and Miller, R. *Cycles of Fire*. Workman, 1987. A colorful album of paintings by two noted astronomical artists (one of them a professional astronomer), accompanied by an excellent nontechnical tour of stars, nebulae, and galaxies.

Henbest, N., and Marten, M. *The New Astronomy*. Cambridge University Press, 1996. A very accessible description of the most recent astronomical discoveries, with a very extensive collection of images of objects from the planets to the most remote galaxies.

Morrison, P., and others. *Powers of Ten*. Scientific American Library, 1995. A tour of the universe where each step corresponds to a power of ten.

*Magazines*

The two most popular monthly magazines covering the latest developments in astronomy for a general audience are *Sky & Telescope*, published by Sky Publishing, and *Astronomy*, published by Kalmbach Publishing. These two excellent magazines also publish, on an annual basis, guides to help beginners get started in amateur astronomy. They include information about how to buy and use binoculars and telescopes and how to find objects in the sky. *SkyWatch* is available from Sky Publishing, and *Explore the Universe* from Kalmbach Publishing.

[W] *World Wide Web*

The World Wide Web contains a tremendous and ever-growing amount of information about astronomy. The sites listed below provide good starting points for exploring the astronomical riches available on the web. Other web sites can be found on the CD-ROM that accompanies this book. In later chapters we will point out additional web sites where you can find information about specific astronomical topics. Note that web site addresses can change; you may need to use a search engine to locate the sites listed below.

The Space Telescope Science Institute site (**http://www. stsci.edu/**) has a wealth of beautiful pictures from the Hubble Space Telescope, as well as explanations of what the pictures tell us about objects in the heavens.

The National Aeronautics and Space Administration (NASA) home page (**http://www.nasa.gov/**) will lead you to literally hundreds of web sites that describe NASA's past, present, and future missions to explore space. The archives of images and videos are especially extensive.

If you are looking for information about a specific topic in astronomy, a good place to begin your search is **http://www. yahoo.com/Science/**, which is a large list of science-related web sites arranged by topic. For example, try searching this site for information about the *TOPEX/ Poseidon* spacecraft, which made the measurements depicted in Figure 1-14.

Another useful site is AstroWeb, a search engine used by many professional astronomers (available at **http://www. stsci.edu/astroweb/astronomy.html** or **http://fits.cv.nrao. edu/www/astronomy.html**).

Every day of the year, you'll find a different astronomical image at Astronomy Picture of the Day (**http://antwrp. gsfc.nasa.gov/apod/**), along with an explanation and links to related web sites. There is also a searchable archive of all past "Pictures of the Day." If you're looking for a picture of a particular astronomical subject, especially ones that have been in the news, this is an excellent place to start.

*Sky & Telescope* and *Astronomy* magazines are represented on the web at **http://www.skypub.com/** and **http://www.astronomy.com/**, respectively. These sites provide up-to-the-minute information about what to see in the nighttime sky, as well as excerpts from the magazines and links to other astronomy web sites.

# SANDRA M. FABER

## Why Astronomy?

Sandra M. Faber, professor of astronomy at the University of California, Santa Cruz, and astronomer at Lick Observatory, was intrigued by the origins of the universe when she entered Swarthmore College. She completed her doctorate in astronomy at Harvard University but did her dissertation at the Carnegie Institution's Department of Terrestrial Magnetism, where she was influenced by the distinguished astronomer Vera Rubin. Dr. Faber chaired the now legendary group of astronomers called the Seven Samurai, who surveyed the nearest 400 elliptical galaxies and discovered a new mass concentration, the Great Attractor. She is the recipient of many honors and awards for her research, including the Bok Prize from Harvard University in 1978. She was elected to the National Academy of Sciences in 1985 and in 1986 won the coveted Dannie Heineman Prize from the American Astronomical Society, given in recognition of a sustained body of especially influential astronomical research. She is also on the Board of Trustees of the Carnegie Institution.

As you study astronomy, it is appropriate for you to ask, "Why am I studying this subject? What good is it for human beings in general and for me in particular?" Astronomy, it must be admitted, does not offer the same practical benefits as other sciences. So how can astronomy be important to your life?

On the most basic level, I like to think of astronomy as providing the introductory chapters for the ultimate textbook on human history. Ordinary texts start with recorded history, going back some 3000 years. For events before that, we consult archeologists and anthropologists, who tell us about the early history of our species. For knowledge of the time before that, we consult paleontologists, biologists, and geologists about the origin and evolution of life and the evolution of our planet, altogether going back some five billion years. Astronomy tells us about the vast stretch of time before the origin of the Earth, the ten billion years or so that saw the formation of the Sun and solar system, the Milky Way Galaxy, and the origin of the universe in the Big Bang. Knowledge of astronomy is essential for a well-educated person's view of history.

Astronomy challenges our belief system and impels us to put our "philosophical house" in order. Take the origin of the world, for example. According to Genesis in the Bible, the world and all in it were created in six days by the hand of God. However, the ancient Egyptians believed that the Earth arose spontaneously from the infinite waters of the eternal universe, Nun, and Alaskan legends taught that the world was created by the conscious imaginings of a creator deity named Father Raven.

The modern astronomical story of the creation of the Earth differs from all these in that it claims to be buttressed by physics and by observational fact. The Sun, it is asserted, formed via gravitational collapse from a dense cloud of interstellar dust and gas about five billion years ago. The planets coalesced at the same time as condensates within the swirling solar nebula. This process took not weeks or days but several hundred thousand years. Confidence in this theory stems from our looking out into the Galaxy and seeing young stars actually form in this way.

At issue here, really, is the deep question of how we are to gain information about the nature of the physical world—whether by revelation and intuition or by logic and observation. Where science stops and faith begins is a thorny issue for all human beings, but particularly so for astronomers—and for astronomy students.

Astronomy cultivates our notions about cosmic time and cosmic evolution. It is all too easy, given the short span of human life, to overlook the fact that the universe is a dynamic place. When I first entered astronomy, I found myself handicapped by a conservative mind-set that assumed, subtly, that celestial objects were unchanging. This might seem strange for someone whose avowed interest was in learning how galaxies formed. Such are the vagaries of the human mind! Fortunately, I lost this bias as I grew older, mainly, I think, by observing evolution all around me. Many things that had seemed immutable to me as a child—social structures, customs, even the physical environment—turned out not to be so. With that realization, the veil fell from my eyes, and I was able to accept cosmic evolution as a core concept in my thinking about the universe.

Following closely on the concept of evolution is the concept of fragility—if something can change, it might even actually disappear someday. The most obvious example is the limited lifetime of our Sun. In another five billion years or so, the Sun will enter its death throes, during which it will swell up and brighten to 1000 times its present luminosity and, in the process, incinerate the Earth.

Five billion years is far enough in the future that neither you nor I feel any personal responsibility for preparing the human race to meet this challenge. However, there are many other cosmic catastrophes that will befall us in the meantime. The Earth will be hit by sizable pieces of space junk; eroded impact craters show this happens all the time—every few million years or so. Debris thrown up into the atmosphere from these events could be so extensive as to totally alter the climate for an extended period of time, possibly leading to mass extinctions. There will be enormous volcanic eruptions that will make Mount St. Helens look like a small firecracker. Another Ice Age, which will totally disrupt modern agriculture, is virtually certain within the next 20,000 years, unless we cook the Earth ourselves first by burning too much fossil fuel.

Closer to the human sphere, such common notions as the inevitability of human progress, the desirability of endless economic growth, the ability of the Earth to support its growing human population, are all based on limited experience and may not—and probably *will* not—prove viable in the long run. Consequently, we must rethink who we are as a species and what is our proper activity on Earth. Contemplating cosmic time puts us into the appropriate frame of mind for grappling with these issues. These are problems that are very long range and involve the whole human race yet are totally under human control and are vital to our long-term survival and well-being.

Astronomy is essential to developing a perspective on human existence and its relation to the cosmos. What question could be grander and at the same time more relevant to us? We can hardly contemplate human existence in general without taking at least a brief look at our own lives.

Earth is a mote in the vast cosmic sea, and our lives, in the evolutionary scheme of the universe, would seem at face value to be insignificant. Should that thought terrify us, depriving our lives of any meaning? Should we seek to provide meaning by introducing an intelligent creator? Or should we be fully comforted merely to know that this great universe has given birth to us in an entirely natural, perhaps even inevitable, chain of events?

Not long ago I visited the Southern Hemisphere for the first time to observe the sky from Chile. At that latitude, the center of the Milky Way passes overhead, where it makes a grand show, not the miserable, fuzzy patch that is visible, low on the horizon, from North America. Stepping out on the observatory catwalk one morning before dawn, I saw for the first time the galactic center in all its glory soaring overhead. At that instant, my perceptions underwent a profound alteration. One moment I was standing in an earthbound "room" with the Galaxy painted on the "ceiling." The next instant, the walls of this earthly room fell away, and I was floating freely in outer space, viewing the center of the Galaxy from an outpost on its periphery. The sense of depth was breathtaking: Sparkling star clusters and dense dust clouds were etched against the blazing inner bulge of the Galaxy many thousands of light-years distant. I saw the Galaxy as an immense, three-dimensional object in space for the first time. It became "real." At that moment, my sense of place in the universe underwent a fundamental and irreversible change. I remained a citizen of the Earth, but I also became a citizen of the Galaxy.

Many astronomers believe that the ultimate, proper concept of "home" for the human race is our universe. It seems more and more likely that there are a very large number of other universes. Virtually none of them would appear to resemble ours; they may have different numbers of space and time dimensions—perhaps even no space or time—and very different physical laws. The vast majority probably are incapable of harboring intelligent life as we know it.

The parallel with Earth is striking. Among the solar system planets, only Earth can support human life. Among the great number of planets that likely exist in our Galaxy, only a small fraction are apt to be such that we could call them home. The fraction of hospitable universes is likely to be smaller still. Our universe would seem, then, to be the ultimate instance of "home": a sanctuary in a vast sea of inhospitable universes.

I began this essay by talking about history and ended with questions that border on the ethical and religious. Astronomy is like that: It offers a modern-day version of Genesis—and of Apocalypse, too. Astronomy stimulates and challenges us on our deepest, most human levels. I hope that, during this course, you will be able to take time out to contemplate the broader implications of what you are studying. This is one of the rare opportunities in life to think about who you are and where you and the human race are going. Don't miss it.

# Knowing the Heavens

**Circumpolar Star Trails** This long exposure, taken from Australia's Siding Spring Mountain and aimed at the south celestial pole, shows the rotation of the sky. The building in the foreground houses the Anglo-Australian Telescope, one of the largest telescopes in the southern hemisphere. During the exposure, someone carrying a flashlight walked along the catwalk on the outside of the telescope dome. Another flashlight made the wavy trail at ground level. (Anglo-Australian Observatory)

*In this chapter you will find the answers to the following questions:*

2-1 What role did astronomy play in ancient civilizations?

2-2 Are the stars that make up a constellation actually close to each other?

2-3 Are the same stars visible every night of the year? What is so special about the North Star?

2-4 Are the same stars visible from any location on Earth?

2-5 What causes the seasons? Why are they opposite in the northern and southern hemispheres?

2-6 Has the same star always been the North Star?

2-7 Can we use the rising and setting of the Sun as the basis of our system of keeping time?

2-8 Why are there leap years?

The telescope is the symbol of modern astronomy. It allows us to see weather systems on other planets and to monitor the births and deaths of distant stars. Using telescopes, astronomers can peer at galaxies so remote that their light has taken billions of years to reach us. But using telescopes to study the nature of distant objects is a relatively recent innovation. For most of human history, astronomy meant the study of the positions of objects in the sky as observed with the naked eye. This part of astronomy is known today as **positional astronomy.** You do this sort of astronomy when you notice where the Moon is in the sky, or when you trace the pattern of the Big Dipper.

Positional astronomy cannot tell us what the Sun is made of or how far away the stars are. We need telescopes, and their attendant equipment, to answer those questions. But, as you will learn in this chapter, naked-eye positional astronomy is nonetheless tremendously important and useful. Before astronomers can train a telescope on a distant star, they must know where in the sky that star can be found. The cycle of day and night and the annual rhythm of the seasons are directly related to the apparent motions of the Sun across the sky. By studying positional astronomy, you will come to understand how the Earth moves through space. In this way you will get a sense of our true place in the cosmos.

## 2-1 Positional astronomy had an important place in ancient civilizations

Four to five thousand years ago, the inhabitants of the British Isles erected stone structures, such as Stonehenge, that suggest a preoccupation with the motions of the sky. Alignments of these stones appear to show where the Sun rose and set at key times during the year. A similar structure in the New World is the Medicine Wheel, a circular ring of stones constructed by the Plains Indians atop a windswept plateau in Wyoming. Certain stones in this ring mark the rising points of the Sun and certain bright stars on the first day of summer.

Aztec, Mayan, and Incan architects in Central and South America also designed buildings with astronomical orientations. At the ruined city of Tiahuanaco in Bolivia, the walls of the Temple of the Sun were aligned north-south and east-west with an accuracy of better than one degree. The great Egyptian pyramids, built around 3000 B.C., are likewise oriented north-south and east-west with remarkable precision.

In addition to temples and tombs, some ancient buildings appear to have been dedicated expressly to astronomy. One of the best examples was built nearly a thousand years ago in the Mayan city of Chichén Itzá, on the Yucatán Peninsula (Figure 2-1). The Caracol's cylindrical tower contains windows aligned with the northernmost and southernmost rising and setting points of both the Sun and the planet Venus. A similar four-story adobe building, probably constructed during the fourteenth century, is located at the Casa Grande site in Arizona.

These structures bear witness to an awareness of naked-eye astronomy by the peoples of many cultures. Many of the concepts of modern positional astronomy come to us from these ancients, including the idea of dividing the sky into constellations.

### figure 2-1

**The Caracol at Chichén Itzá** This ancient Mayan observatory in the Yucatán was built around A.D. 1000. Its architecture is based on alignments with important celestial events. Mayan astronomers developed a very accurate calendar and measured the motions of celestial bodies with great precision. Special significance was associated with the planet Venus, which inspired sacrificial rites and other ceremonies. (Courtesy of E. C. Krupp)

## 2-2 Eighty-eight constellations cover the entire sky

Looking at the sky on a clear, dark night, you might think that you can see millions of stars. Actually, the unaided human eye can detect only about 6000 stars. Because half of the sky is below the horizon at any one time, you can see only roughly 3000 stars at most. When ancient peoples looked at these thousands of stars, they imagined that groupings of stars traced out pictures in the sky. Astronomers still refer to these groupings, called **constellations** (from the Latin for "group of stars").

You may already be familiar with some of these pictures or patterns in the sky, such as the Big Dipper, which is actually part of the large constellation Ursa Major (the Great Bear). Many constellations, such as Orion in Figure 2-2, have names derived from the myths and legends of antiquity.

a R I **V** U X G

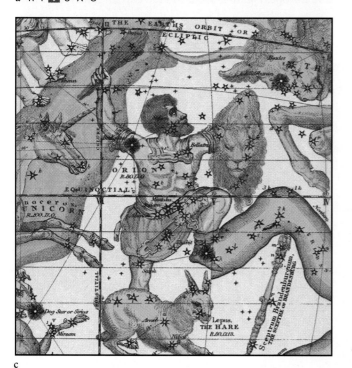

c

b

**figure 2-2** R I **V** U X G

**Three Views of Orion** Orion is a prominent winter constellation. From North America, Orion is easily seen high above the southern horizon from December through March. **(a)** This four-minute time-exposure photograph of Orion shows that stars come in different colors, some red and others blue. You can see these colors with the naked eye; they are even more pronounced as seen through binoculars or a telescope (R. C. Mitchell, Central Washington University). **(b)** A portion of a modern star atlas shows the names of some of the brightest stars in Orion, along with the distances in light-years (ly) to those stars. The lines show the borders between Orion and its neighboring constellations (labeled in capitals), as defined by modern astronomers. The entire sky is divided up into a total of 88 constellations of different shapes and sizes. **(c)** This fanciful drawing from a star atlas published in 1835 shows Orion the Hunter as well as other celestial creatures. (Courtesy of Janus Publications)

Although some star groupings vaguely resemble the figures they are supposed to represent (see Figure 2-2c), most do not.

The term "constellation" has a broader definition in present-day astronomy. On modern star charts, the entire sky is divided into 88 regions, each of which is called a constellation. For example, the constellation Orion is now defined to be an irregular patch of sky whose borders are shown in Figure 2-2b. When astronomers refer to nebula M42 in Orion, they mean that as seen from Earth this nebula appears to be within Orion's patch of sky. Some constellations cover large areas of the sky (Ursa Major being one of the biggest) and others very small areas (Crux, the Southern Cross, being the smallest). But because the modern constellations cover the entire sky, *every* star lies in one constellation or another.

CAUTION! When you look at a constellation's star pattern, it is tempting to conclude that you are seeing a group of stars that are all relatively close to each other. In fact, most of these stars are nowhere near each other. As an example, Figure 2-2b shows the distances in light-years to four stars in Orion. (Chapter 19 discusses how astronomers determine the distances to the stars.) Although Bellatrix (Arabic for "the Amazon") and Mintaka ("the belt") appear to be close to each other, Mintaka is actually about a thousand light-years further away than is Bellatrix. They merely appear to be close to each other, because they are in nearly the same directions as seen from Earth. The same illusion often appears when you see an airliner's lights at night. It is very difficult to tell how far away a single bright light is, which is why you can mistake an airliner a few kilometers away for a star trillions of times more distant.

The star names shown in Figure 2-2b are from the Arabic. For example, Betelgeuse means "armpit," which makes sense when you look at the star atlas drawing in Figure 2-2c. Other types of names are also used for stars. For example, Betelgeuse is also known as α Orionis (α is the Greek letter alpha) and as HD 39801. The various systems used for naming stars are discussed in Box 2-1.

## 2-3 The appearance of the sky changes during the course of the night and from one night to the next

Go outdoors soon after dark, find a spot away from bright lights, and note the patterns of stars in the sky. Do the same a few hours later. You will find that the entire pattern of stars (including the Moon, if it is visible) has shifted its position. New constellations will have risen above the eastern horizon, and some will have disappeared below the western horizon. If you look again before dawn, you will see that the stars that were just rising in the east when the night began are now low in the western sky. This daily motion, or **diurnal motion,** of the stars is apparent in time-exposure photographs (Figure 2-3). We can understand this apparent motion of the sky by considering the rotation of the Earth.

At any given moment, it is daytime on the half of the Earth illuminated by the Sun and nighttime on the other half. The Earth rotates from west to east, making one complete rotation every 24 hours, which is why there is a daily cycle of day and night. Because of this rotation, stars appear to us to rise in the east and set in the west, as do the Sun and Moon. Figure 2-4 shows two views of the Earth as seen from a point above the North Pole. At the instant shown in Figure 2-4a, it is day in Asia but night in most of North America and Europe. Figure 2-4b shows the Earth 4 hours later. Four

**figure 2-3**

**Diurnal Motion of the Night Sky** This photograph is a time exposure that recorded the motion of the stars across the night sky over the Kitt Peak National Observatory near Tucson, Arizona. Because the Earth rotates from west to east, the stars appear to be moving from east to west and thus appear as long trails in this photograph. (NOAO)

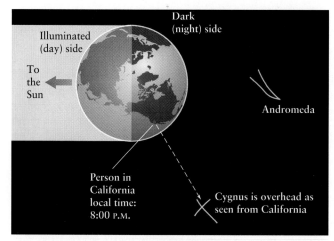

**a** Earth as seen from above the north pole

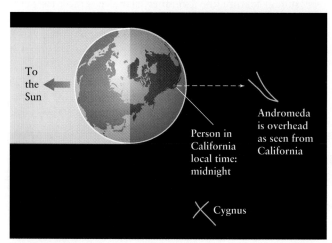

**b** 4 hours later

## figure 2-4

**Why Diurnal Motion Happens** The diurnal (daily) motion of the stars, the Sun, and the Moon is a consequence of the Earth's rotation. **(a)** This drawing shows the Earth as seen from a vantage point above the North Pole. It is day on the side of the Earth that is illuminated by the Sun, and night on the dark, unilluminated side. At the time shown, the local time in California is 8:00 P.M. and a person in California sees the constellation Cygnus directly overhead. **(b)** Four hours later, the Earth has rotated to the east by one-sixth of a complete rotation; from the perspective of a person on Earth, the entire sky appears to have rotated to the west by one-sixth of a complete rotation. It is now midnight in California, and the constellation directly over California is Andromeda.

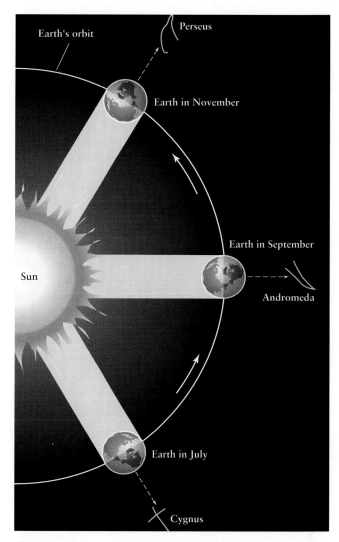

## figure 2-5

**Why the Night Sky Changes During the Year** The particular stars that you see in the night sky are different at different times during the year. This is a consequence of the Earth's orbital motion around the Sun; as we orbit the Sun, the nighttime side of the Earth gradually turns toward different parts of the sky. This figure shows which constellation is overhead at midnight local time—when the Sun is on the opposite side of the Earth from your location—at different times during the year for observers at midnorthern latitudes (including the United States). If you want to view the constellation Andromeda, the best time of the year to do it is in late September when Andromeda is nearly overhead at midnight.

hours is one-sixth of a complete 24-hour day, so the Earth has made one-sixth of a rotation between Figures 2-4a and 2-4b. Europe is now in the illuminated half of the Earth (the Sun has risen in Europe), while Alaska has moved from the illuminated to the dark half of the Earth (the Sun has set in Alaska). As seen from California, in Figure 2-4a the time is 8:00 P.M. and the constellation Cygnus (the Swan) is directly overhead. Four hours later, the constellation over California

is Andromeda (named for a mythological princess). Because the Earth is rotating from west to east, it appears to us on Earth that the entire sky is rotating around us in the opposite direction, from east to west.

In addition to the diurnal motion of the sky, the constellations visible in the night sky also change slowly over the course of a year. This happens because the Earth orbits, or revolves around, the Sun (Figure 2-5). Over the course of a year, the Earth makes one complete orbit, and the darkened, nighttime side of the Earth is gradually turned toward different parts of the heavens. For example, as seen from the northern hemisphere, at midnight in late July the constellation Cygnus is close to overhead; at midnight in late September the constellation Andromeda is close to overhead; and at midnight in late November the constellation Perseus (commemorating a mythological hero) is close to overhead. If you follow a particular star on successive evenings, you will find that it rises approximately 4 minutes earlier each night, or 2 hours earlier each month.

Constellations can help you find your way around the sky. For example, if you live in the northern hemisphere, you can use the Big Dipper in Ursa Major to find the north direction by drawing a straight line through the two stars at the front of the Big Dipper's bowl (Figure 2-6). The first moderately bright star you come to is Polaris, also called the North Star because it is located almost directly over the Earth's north pole. If you draw a line from Polaris straight down to the horizon, you will find the north direction.

By drawing a line through the two stars at the rear of the bowl of the Big Dipper, you can find Leo (the Lion). As shown in Figure 2-6, that line points toward Regulus, the brightest star in the "sickle" tracing the lion's mane. By following the handle of the Big Dipper, you can locate the bright reddish star Arcturus in Boötes (the Shepherd) and the prominent bluish star Spica in Virgo (the Virgin). The saying, "Follow the arc to Arcturus and speed to Spica," may help you remember these stars, which are conspicuous in the evening sky during the spring and summer.

During winter in the northern hemisphere, you can see some of the brightest stars in the sky. Many of them are in the vicinity of the "winter triangle," which connects bright stars in the constellations of Orion (the Hunter), Canis Major (the Large Dog), and Canis Minor (the Small Dog), as shown in Figure 2-7. The "winter triangle" is high above the southern horizon during the middle of winter at midnight.

A similar feature, the "summer triangle," graces the summer sky in the northern hemisphere. This triangle connects the brightest stars in Lyra (the Harp), Cygnus (the Swan),

---

## *b*ox 2-1 | Looking Deeper into Astronomy

### *Star Names and Catalogs*

The customs for naming stars have changed over the centuries. Many of the brightest stars in the sky have Arabic names that were assigned in medieval times, when astronomy was widely studied among Islamic nations. The illustration shows the Big Dipper with the Arabic names for its seven brightest stars. The North Star received its formal name, Polaris, later, when Latin was the language used by European astronomers.

As you might suspect, memorizing exotic star names is an unwelcome burden for many astronomers. A simpler system, invented by Johann Bayer in 1603, uses the constellation names and the 24 lowercase letters of the Greek alphabet:

| | | |
|---|---|---|
| α alpha | ι iota | ρ rho |
| β beta | κ kappa | σ sigma |
| γ gamma | λ lambda | τ tau |
| δ delta | μ mu | υ upsilon |
| ε epsilon | ν nu | φ phi |
| ζ zeta | ξ xi | χ chi |
| η eta | ο omicron | ψ psi |
| θ theta | π pi | ω omega |

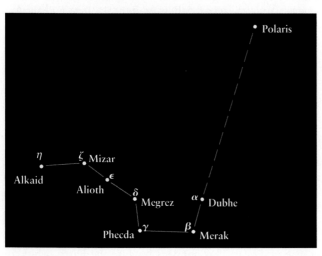

In constructing a star name, a Greek letter is used with the Latin possessive form of the name of the constellation in which the star is located. In most cases, the brightest star in the constellation is α, the second brightest is β, the third is γ, and so on. For example, the brightest star in Libra (the Scales) is called α Librae, or α Lib for short. This name is more informative, if less melodious, than its Arabic name, Zubenelgenubi. The 12 constellations of the zodiac with their Latin possessives are as follows:

**figure 2-6**

**The Big Dipper as a Guide** The North Star can be seen by any observer in the northern hemisphere on any night of the year. This star chart shows how the Big Dipper can be used to point out the North Star as well as the brightest stars in three other constellations. The chart is drawn to show the sky at around 10 P.M. (standard time) on March 1, 9 P.M. on April 1, or 8 P.M. on May 1. Because of the Earth's rotation and orbital motion around the Sun, the view will be different on different dates and at different times, but the relative positions of the stars will be the same. The angular distance from Polaris to Spica is about 101°, so this chart covers a large fraction of the sky.

| Constellation | Possessive |
| --- | --- |
| Aries | Arietis |
| Taurus | Tauri |
| Gemini | Geminorum |
| Cancer | Cancri |
| Leo | Leonis |
| Virgo | Virginis |
| Libra | Librae |
| Scorpius | Scorpii |
| Sagittarius | Sagittarii |
| Capricornus | Capricornii |
| Aquarius | Aquarii |
| Pisces | Piscium |

In Bayer's system, only the brightest two dozen stars in a constellation can be named. However, astronomers are often interested in very faint stars, many of which are too dim to be seen with the naked eye. Astronomers refer to these fainter stars by using designations from standard star catalogs.

One of the first major star catalogs, *Bonner Durch-musterung* (Bonn Comprehensive Survey), was produced in Germany in the mid-1800s by F. W. Argelander of the Bonn Observatory. This catalog lists the positions of 324,188 stars, each of which is designated by a BD number. For example, the star BD+5° 1668 is a dim star in the constellation Monoceros (the Unicorn). Another frequently used catalog is the *Henry Draper Catalogue*, which was compiled in the United States between 1911 and 1915. (It is named after a physician and avid amateur astronomer whose widow financed the project.) The 225,300 stars in this catalog are listed by their HD numbers. For example, HD 87901 is α Leonis (also called Regulus), the brightest star in Leo (the Lion).

The most extensive catalog of stars yet compiled is the *Hubble Space Telescope Guide Star Catalog*. This is an immense list of 15,169,873 stars, all too faint to be seen by the naked eye, whose positions are used to help in pointing the Hubble Space Telescope (Figure 1-2). Stars from this catalog have GSC numbers, such as the star GSC 1234 1132 in the constellation Taurus. Unlike other star catalogs, this one is only available in electronic form (on two CD-ROMs); if it were to be printed, it would require hundreds of thousands of pages.

There are many star catalogs, including some that list only stars of specific types. As a result, a single star may have a plethora of different names. For example, the bright star Vega in the constellation Lyra is also known as α Lyrae, BD +38° 3238, HD 172167, 3 Lyrae, HR 7001, GC 25466, SAO 67174, and ADS 11510. In this book we use the Bayer system to refer to stars in most cases and thus will refer to this star simply as "Vega (α Lyrae)."

**CAUTION!** A number of unscrupulous commercial firms offer to name a star for you for a fee. The money that they charge you for this "service" is real, but the star names are not; none of these names are recognized by astronomers. If you want to use astronomy to commemorate your name or the name of a friend or relative, consider making a donation to your local planetarium or science museum. The money will be put to much better use!

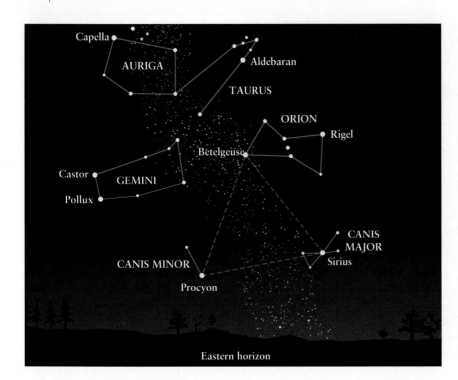

**Figure 2-7**

**The "Winter Triangle"** This star chart shows the eastern sky as it appears on a winter evening in the northern hemisphere (around 10 P.M. on December 1, around 9 P.M. on January 1, or around 8 P.M. on February 1). Three of the brightest stars in the sky make up the "winter triangle," which is about 26° on a side. In addition to the constellations involved in the triangle, the prominent constellations Gemini (the Twins), Auriga (the Charioteer), and Taurus (the Bull) are also shown.

and Aquila (the Eagle) (Figure 2-8). A conspicuous portion of the Milky Way forms a beautiful background for these constellations, which are nearly overhead during the middle of summer at midnight.

A set of selected star charts for the evening hours of all twelve months of the year is included at the end of this book. You may find stargazing an enjoyable experience,

and these star charts will help you identify many well-known constellations.

Note that all of the star charts in this section and at the end of this book are drawn for an observer in the northern hemisphere. If you live in the southern hemisphere, you can see constellations that are not visible from the northern hemisphere, and vice versa. In the next section we will see why this is so.

**Figure 2-8**

**The "Summer Triangle"** This star chart shows the eastern sky as it appears in the evening during the summer in the northern hemisphere (around 11 P.M.—daylight savings time—on June 1, around 10 P.M. on July 1, and around 9 P.M. on August 1). The angular distance from Deneb to Altair is about 38°. In addition to the three constellations in the triangle, the faint constellations of Sagitta (the Arrow) and Delphinus (the Dolphin) are shown.

## 2-4 It is convenient to imagine that the stars are located on a celestial sphere

Many ancient societies believed that all the stars are the same distance from the Earth. They imagined the stars to be bits of fire imbedded into the inner surface of an immense hollow sphere, called the **celestial sphere,** with the Earth at its center. In this picture of the universe, the Earth was fixed and did not rotate. Instead, the entire celestial sphere rotated once a day around the Earth from east to west, thereby causing the diurnal motion of the sky. The picture of a rotating celestial sphere fit well with naked-eye observations, and for its time was a useful model of how the universe works. (We discussed the role of models in science in Section 1-1).

We now know that this simple model of the universe is not correct. Diurnal motion is due to the rotation of the Earth, not the rest of the universe. Furthermore, as we saw in the discussion of the constellations, it is impossible to tell with the naked eye how far away the stars are. Indeed, the brightest stars that you can see with the naked eye range from 10 to more than 1000 light-years away (see Figure 2-2), and telescopes allow us to see objects at distances of billions of light-years. Thus, astronomers now recognize that the celestial sphere is an *imaginary* object that has no basis in physical reality. Nonetheless, the celestial sphere model remains a useful tool of positional astronomy. If we imagine, as did the ancients, that the Earth is stationary and that the celestial sphere rotates around us, it is relatively easy to specify the directions to different objects in the sky and to visualize the motions of these objects.

Figure 2-9 depicts the celestial sphere, with the Earth at its center. (A more correct drawing would show the celestial sphere as being millions of times larger than the Earth.) If we project the Earth's equator out into space, we obtain the **celestial equator.** The celestial equator divides the sky into northern and southern hemispheres, just as the Earth's equator divides the Earth into two hemispheres.

If we project the Earth's north and south poles into space, we obtain the **north celestial pole** and the **south celestial pole.** The two celestial poles are where the Earth's axis of rotation (extended out into space) intersects the celestial sphere (see Figure 2-9). The star Polaris is less than 1° away from the north celestial pole, which is why it is called the North Star or the Pole Star.

Using the celestial equator and poles, we can define a coordinate system to specify the positions of stars on the celestial sphere. The most commonly used coordinate system uses two angles, *right ascension* and *declination*, that are quite like longitude and latitude on Earth, as described in Box 2-2. These coordinates tell us in what direction we should look to see the star. To locate the star's true position in three-dimensional space, we must also know the distance to the star. In Chapter 19 we will see how such distances are measured.

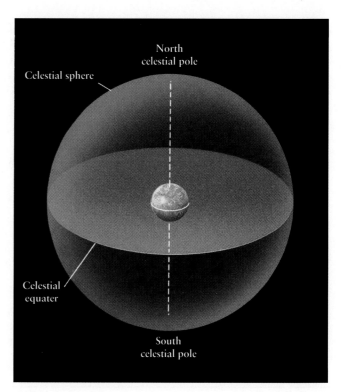

### ƒigure 2-9

**The Celestial Sphere** The celestial sphere is the apparent sphere of the sky. The view in this figure is from the outside of the sphere. Because the Earth is at the center of the celestial sphere, however, our view is always of the *inside* of the sphere. The celestial equator and poles are the projections of the Earth's equator and axis of rotation out into space. The celestial poles are therefore located directly over the Earth's poles.

The point in the sky directly overhead an observer anywhere on Earth is called that observer's **zenith.** The zenith and celestial sphere are shown in Figure 2-10 for an observer located at 35° north latitude (that is, at a location on the Earth's surface 35° north of the equator). Because the zenith is shown at the top of Figure 2-10, the Earth and the celestial sphere appear "tipped" compared to Figure 2-9. At any time, an observer can see only half of the celestial sphere; the other half is below the horizon, hidden by the body of the Earth. The hidden half of the celestial sphere is shaded dark in Figure 2-10.

For an observer anywhere in the northern hemisphere, including the observer in Figure 2-10, the north celestial pole is always above the horizon. As the Earth turns from west to east—or, equivalently, as the celestial sphere apparently turns from east to west—stars near the north celestial pole revolve around the pole, never rising or setting. This is why Polaris can be seen from North America at any time of night on any night of the year. Similarly, stars near the south celestial pole revolve around that pole but always remain below the horizon of an observer in the northern hemisphere. Hence, these

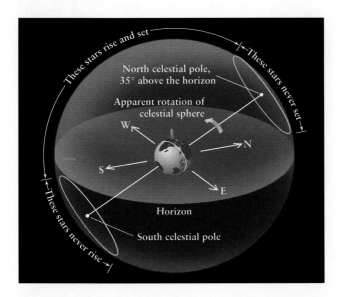

## figure 2-10

**The View from 35° North Latitude**
As viewed from North America, the north celestial pole is always above the horizon. To an observer at 35° north latitude (roughly the latitude of Los Angeles, Albuquerque, Memphis, and Atlanta), the north celestial pole is 35° above the horizon. Stars within 35° of the *north* celestial pole are always above the horizon on any night of the year. These stars trace out circles in the sky around the north celestial pole during the course of the night. Stars within 35° of the *south* celestial pole are always below the horizon and can never be seen from this latitude. Stars that lie between these two extremes rise in the east and set in the west.

---

## box 2-2 | Looking Deeper into Astronomy

## *Celestial Coordinates*

To denote the positions of objects in the sky, astronomers use a system based on *right ascension* and *declination*. Declination corresponds to latitude. As shown in the illustration, the **declination** of an object is its angular distance north or south of the celestial equator, measured along a circle passing through both celestial poles. It is measured in degrees, arcminutes, and arcseconds (see Section 1-5).

Right ascension corresponds to longitude. Astronomers measure right ascension from a specific point, called the vernal equinox, on the celestial equator. This point is one of two locations where the Sun crosses the celestial equator during its apparent annual motion, as discussed in Section 2-5. In the Earth's northern hemisphere, spring officially begins when the Sun reaches the vernal equinox in late March. The **right ascension** of an object in the sky is the angular distance from the vernal equinox eastward along the celestial equator to the circle used in measuring its declination (see illustration). Astronomers do not measure this angular distance in degrees; instead, they use time units (hours, minutes, and seconds) corresponding to the time required for the celestial sphere to rotate through this angle. For example, suppose there is a star at your zenith right now with right ascension $6^h 0^m 0^s$. Two hours and 30 minutes from now, there will be a different object at your zenith with right ascension $8^h 30^m 0^s$.

Right ascension and declination are very useful because they tell an astronomer precisely where in the sky an object is located. For example, the coordinates of the bright star Rigel (β Orionis) for the year 2000 are R.A. = $5^h 14^m 32.2^s$, Decl. = $-8° 12' 06''$. (R.A. and Decl. are abbreviations for right ascension and declination.) A minus sign on the declination indicates that the star is south of the celestial equator;

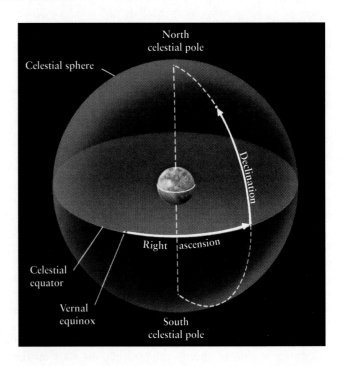

a plus sign (or no sign at all) indicates that an object is north of the celestial equator. As discussed in Box 2-4, right ascension is particularly useful for determining when a particular object can best be observed.

It is important to state the year for which a star's right ascension and declination are valid. This is so because of precession, as discussed in Section 2-6.

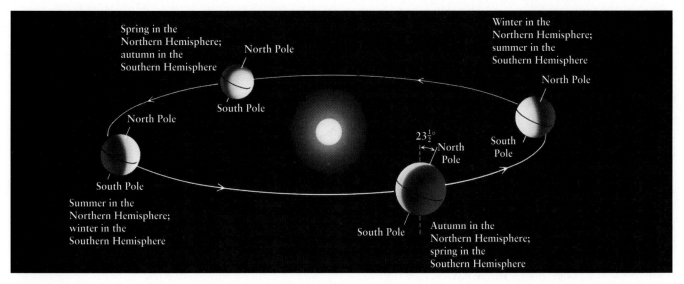

**figure 2-11**

**The Seasons** The Earth's axis of rotation is inclined 23½° away from the perpendicular to the plane of the Earth's orbit. The Earth maintains this orientation (with its north pole aimed at the north celestial pole, near the star Polaris) throughout the year as it orbits the Sun. Consequently, the amount of solar illumination and the number of daylight hours at any location on Earth vary in a regular pattern throughout the year.

stars can never be seen by this observer (Figure 2-10). Stars at intermediate positions on the celestial sphere rise in the east and set in the west, as in Figure 2-3.

For an observer at 35° south latitude (the latitude of Sydney, Australia), the roles of the north and south celestial poles are reversed: Objects close to the *south* celestial pole revolve around that pole and never rise or set. (The photograph that opens this chapter, taken in Australia, shows the circular trails of stars around the south celestial pole.) Stars close to the *north* celestial pole are always below the horizon and can never be seen by an observer in the southern hemisphere. Hence, Australian astronomers never see the North Star, but are able to see other stars that are forever hidden from North American observers.

## 2-5 The seasons are caused by the tilt of Earth's axis of rotation

In addition to rotating on its axis every 24 hours, the Earth revolves around the Sun in about 365¼ days. This orbital motion, along with the tilt of the Earth's axis of rotation, gives rise to the cycle of seasons. It also explains why the seasons are opposite in the northern and southern hemispheres. For example, February is midwinter in North America but midsummer in Australia.

As shown in Figure 2-11, the Earth's axis of rotation is not perpendicular to the plane of the Earth's orbit. Instead, it is tilted about 23½° away from the perpendicular. As it orbits the Sun, the Earth maintains this tilt, with the Earth's north pole pointing toward the north celestial pole. (This stability is a hallmark of all rotating objects. A top will not fall over as long as it is spinning, and the rotating wheels of a motorcycle help to keep the rider upright.)

During part of the year, when the Earth is in the part of its orbit shown on the left side of Figure 2-11, the northern hemisphere is tilted toward the Sun. As the Earth spins on its axis, a point in the northern hemisphere spends more than 12 hours in the sunlight, so the days there are long and the nights are short. Hence, it is summer in the northern hemisphere. The summer is hot not only because of the extended daylight hours but also because the Sun is high in the northern hemisphere's sky. As a result, sunlight strikes the ground at a nearly perpendicular angle that heats the ground efficiently (Figure 2-12a). During this same time of year in the southern hemisphere, the days are short and the nights are long, because a point in this hemisphere spends fewer than 12 hours a day in the sunlight. The Sun is low in the sky, so sunlight strikes the surface at a grazing angle that causes little heating (Figure 2-12b) and it is winter in the southern hemisphere.

Half a year later, the Earth is in the part of its orbit shown on the right side of Figure 2-11. Now the situation is reversed, with winter in the northern hemisphere (which is now tilted away from the Sun) and summer in the southern hemisphere. During spring and autumn, the two hemispheres receive roughly equal amounts of illumination from the Sun, and daytime and nighttime are of equal length everywhere on Earth.

a          b

### figure 2-12

**Solar Energy in Summer and Winter**
At different times of the year, sunlight strikes the ground at different angles. **(a)** When it is summer at your location, the Sun is high in the sky. The energy in a shaft of light falls onto a small, nearly circular area. This concentrated solar energy heats the ground effectively and makes the days warm. The days are also longest in summer, which further increases the heating. **(b)** In winter the Sun is low in the sky even at midday. The same shaft of light in (a) now falls on a larger, oval area. Because the sunlight is less concentrated and the days are also their shortest in duration, little heating of the ground takes place. This accounts for the low temperatures in winter.

*CAUTION!* It is a common misconception that the Earth is closer to the Sun in summer and farther away in winter. In fact, the Earth is closest to the Sun in January, when it is winter in the northern hemisphere, and farthest away in July. In addition, if the seasons were caused by variations in the Earth-Sun distance, the seasons would be the same in both hemispheres! The fact is that the distance from the Earth to the Sun varies only slightly (about 3%) over the course of a year. This small variation has little influence on the cycle of the seasons.

As the Earth orbits around the Sun, the Sun's position (as seen from the Earth) gradually shifts with respect to the background stars. As shown in Figure 2-13, the Sun therefore appears to trace out a circular path, called the **ecliptic,** on the celestial sphere. (This word suggests that the path traced out by the Sun has something to do with eclipses. We discuss the connection in Chapter 3.) Because there are 365¼ days in a year and 360° in a circle, the Sun appears to move along the ecliptic at a rate of about 1° per day. This motion is from west to east, that is, in the direction opposite to the apparent motion of the celestial sphere.

*ANALOGY* Envision the celestial sphere as a merry-go-round rotating clockwise, and the Sun as a restless child who is walking slowly around the merry-go-round's rim in the counterclockwise direction. During the time it takes the child to make a round trip, the merry-go-round rotates 365¼ times.

The ecliptic can also be thought of as the projection onto the celestial sphere of the plane of the Earth's orbit. This is *not* the same as the plane of the Earth's equator, thanks to the 23½° tilt of the Earth's rotation axis shown in Figure 2-11. Instead, the ecliptic and the celestial equator are inclined to each other by that same 23½° angle.

The ecliptic and the celestial equator intersect at only two points, which are exactly opposite each other on the celestial sphere. Each point is called an **equinox** (from the Latin for "equal night"), because when the Sun appears at either of these points, day and night are each about 12 hours long at

all locations on Earth. The term "equinox" is also used to refer to the date on which the Sun passes through one of these special points on the ecliptic.

On about March 21 of each year, the Sun crosses northward across the celestial equator at the **vernal equinox.** This marks the beginning of spring in the northern hemisphere

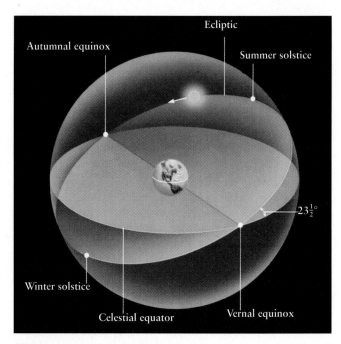

### figure 2-13

**The Ecliptic, Equinoxes, and Solstices** The ecliptic is the apparent annual path of the Sun as projected onto the celestial sphere. Inclined to the celestial equator by 23½° because of the tilt of the Earth's axis of rotation, the ecliptic intersects the celestial equator at two points, called equinoxes. The northernmost point on the ecliptic is the summer solstice, and the corresponding southernmost point is the winter solstice. The Sun is shown in its approximate position for August 1.

("vernal" is from the Latin for "spring"). On about September 22 the Sun moves southward across the celestial equator at the **autumnal equinox,** marking the moment when fall begins in the northern hemisphere. Since the northern and southern hemispheres experience opposite seasons at any given time, for Australians and South Africans the vernal equinox actually marks the beginning of autumn. The names of the equinoxes come from a time when virtually all astronomers lived north of the equator.

Between the vernal and autumnal equinoxes lie two other significant locations along the ecliptic. The point on the ecliptic farthest north of the celestial equator is called the **summer solstice** ("solstice" is Latin for "solar standstill"), and it is at the summer solstice that the Sun stops moving northward on the celestial sphere. At this point, the Sun is as far north of the celestial equator as it can get. It marks the location of the Sun at the moment summer begins in the northern hemisphere (about June 21). At the beginning of the northern hemisphere's winter, the Sun is farthest south of the celestial equator, at a point called the **winter solstice** (about December 21).

Because the Sun's position on the celestial sphere varies slowly over the course of a year, its daily path across the sky (due to the Earth's rotation) also varies with the seasons, as shown in Figure 2-14. On the first day of spring or the first day of fall, when the Sun is at one of the equinoxes, the Sun rises directly in the east and sets directly in the west. Day and night are of equal duration.

During summer in the northern hemisphere, when the northern hemisphere is tilted toward the Sun, the Sun rises in the northeast and sets in the northwest. The Sun is at its northernmost position at the summer solstice, giving the northern hemisphere the greatest number of daylight hours. There are even certain locations on Earth (north of the Arctic Circle, as discussed in Box 2-3) where the Sun does not set at all during summer nights.

When the northern hemisphere is tilted away from the Sun and it is winter in the northern hemisphere, the Sun rises in the southeast. Daylight lasts for fewer than 12 hours as the Sun skims low over the southern horizon and sets in the southwest. Northern hemisphere nights are longest when the Sun is at the winter solstice. In fact, north of the Arctic Circle a winter night lasts 24 hours, as described in Box 2-3.

---

| 2-6 | The Moon helps to cause precession, a slow, conical motion of Earth's axis of rotation |

The Moon is by far the brightest and most obvious naked-eye object in the nighttime sky. Like the Sun, the Moon slowly changes its position relative to the background stars; unlike the Sun, the Moon makes a complete trip around the celestial sphere in only about four weeks, or about a month. (The word "month" comes from the same Old English root as the word "moon.") Ancient astronomers realized that this

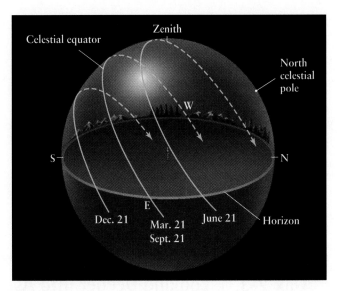

**figure 2-14**

**The Sun's Daily Path Across the Sky** On the first day of spring and the first day of fall, the Sun rises precisely in the east and sets precisely in the west. During summer in the northern hemisphere, the Sun rises in the northeast and sets in the northwest. The Sun reaches its northernmost position at the summer solstice. In winter in the northern hemisphere, the Sun rises in the southeast and sets in the southwest. It reaches its southernmost position at the winter solstice.

motion occurs because the Moon orbits the Earth in roughly four weeks. In one hour the Moon moves on the celestial sphere by about ½°, or roughly its own angular size.

The Moon's path on the celestial sphere is never far from the Sun's path (that is, the ecliptic). This is because the plane of the Moon's orbit around the Earth is inclined only slightly from the plane of the Earth's orbit around the Sun. The Moon's path varies somewhat from one month to the next, but always remains within a band called the **zodiac** that extends about 8° on either side of the ecliptic. Twelve famous constellations (listed in Box 2-1) lie along the zodiac. The Moon is generally found in one of these 12 constellations. As it moves along its orbit, the Moon appears north of the celestial equator for about two weeks and then south of the celestial equator for about the next two weeks. We will describe more about the Moon's motion, as well as why the Moon goes through phases, in Chapter 3.

The Moon not only moves around the Earth but, in concert with the Sun, also helps to change the Earth's rotation. This is because both the Sun and the Moon exert a gravitational pull on the Earth. We will discuss gravity in greater detail in Chapter 4; for now, it is sufficient to realize that gravity is a universal attraction of matter for other matter.

The gravitational pull of the Sun and the Moon affects the Earth's rotation because the Earth is slightly fatter across the equator than it is from pole to pole: The equatorial diameter of the Earth is 43 kilometers (27 miles) larger than the diameter measured from pole to pole. The Earth is therefore said to have an "equatorial bulge." Because of the gravitational pull of the Moon and the Sun on this bulge, the orientation of the Earth's axis of rotation gradually changes.

The Earth behaves somewhat like a spinning top, as illustrated in Figure 2-15. If the top were not spinning, gravity would pull the top over on its side. When the top is spinning, gravity causes the top's axis of rotation to trace out a circle, producing a motion called **precession.**

As the Sun and Moon move along the zodiac, each spends half its time north of the Earth's equatorial bulge and half its time south of it. The gravitational pull of the Sun and Moon tugging on the equatorial bulge tries to pull the Earth's axis of rotation perpendicular to the plane of the ecliptic. But because the Earth is spinning, the combined actions of gravity and rotation cause Earth's axis to trace out a circle in the sky, much like what happens to the toy top. As the axis precesses, it remains tilted about 23½° to the perpendicular.

As the Earth's axis of rotation slowly changes its orientation, the north and south celestial poles—which are the projections of that axis onto the celestial sphere—change their positions relative to the stars. At present the north celestial pole lies within 1° of the star Polaris, which is why Polaris is the North Star. But in 3000 B.C., the north celestial pole was closest to the star Thuban in the constellation of Draco (the Dragon), and thus that star and not Polaris was the North

---

## box 2-3 | Looking Deeper into Astronomy

### *Tropics and Circles*

The 23½° inclination of the ecliptic to the celestial equator produces some noteworthy events at particular locations on the Earth. For example, on any date during the year there is a band of locations encircling the Earth where the Sun appears directly overhead at high noon. The southernmost band where this happens is called the **Tropic of Capricorn,** which circles the Earth at a latitude of 23½° south of the equator. As shown on the left in the illustration, on the day of the winter solstice the Sun is at the zenith at high noon all along the Tropic of Capricorn.

The corresponding northern location is called the **Tropic of Cancer,** 23½° north of the equator. On the day of the summer solstice, when the Sun has reached its northernmost declination, the Sun passes directly overhead at high noon at all locations along this tropic. This is shown on the right in the illustration.

There are also regions on Earth near the two poles where the Sun spends many consecutive days either above or below the horizon. For example, at the north pole, the Sun rises on the day of the vernal equinox, stays above the horizon for six months, and then sets on the day of the autumnal equinox. During the next six months, the north pole is subjected to continuous night because the Sun is too far south to be seen.

North of the **Arctic Circle** you can see the Sun for 24 continuous hours on at least one day of the year. The Arctic Circle lies 23½° south of the north pole (that is, at a latitude of 90° − 23½° = 66½° N).

The corresponding region around the south pole is bounded by the **Antarctic Circle,** as shown in the illustration. At the time of the winter solstice, explorers south of the Antarctic Circle enjoy the "midnight sun," while there is continuous night north of the Arctic Circle.

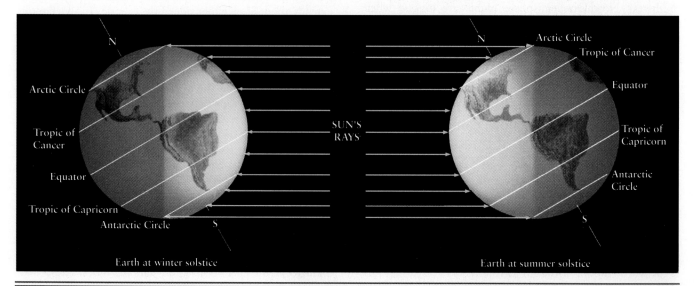

Earth at winter solstice                              Earth at summer solstice

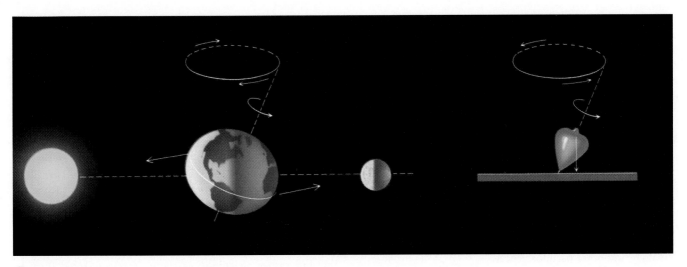

## figure 2-15

**Precession** The gravitational pull of the Moon and the Sun on the Earth's equatorial bulge causes the Earth to precess. As the Earth precesses, its axis of rotation slowly traces out a circle in the sky, like the shifting axis of a spinning top.

Star. In A.D. 14,000, the North Star will be the bright star Vega in Lyra (the Harp). It takes 26,000 years for the north celestial pole to complete one full precessional circle around the sky (Figure 2-16). The south celestial pole executes a similar circle in the southern sky.

Precession also causes the Earth's equatorial plane to change its orientation. Because this plane defines the location of the celestial equator in the sky, the celestial equator precesses as well. The intersections of the celestial equator and the ecliptic define the equinoxes, as seen in Figure 2-13, so these key locations in the sky also shift slowly from year to year. For this reason, the precession of the Earth is also called the **precession of the equinoxes.** The first person to detect the precession of the equinoxes was the Greek astronomer

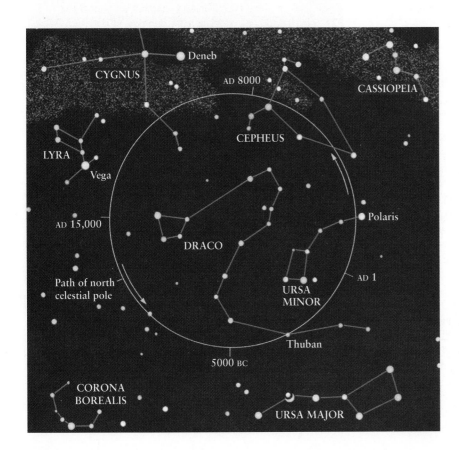

## figure 2-16

**Precession and the Path of the North Celestial Pole** As the Earth precesses, the north celestial pole slowly traces out a circle among the northern constellations. At present, the north celestial pole is near the moderately bright star Polaris, which serves as the North Star. Twelve thousand years from now the bright star Vega will be the North Star.

Hipparchus in the second century B.C., who compared his own observations with those of Babylonian astronomers three centuries earlier. Today, the vernal equinox is located in the constellation Pisces (the Fishes). Two thousand years ago, it was in Aries (the Ram). Around the year A.D. 2600, the vernal equinox will move into Aquarius (the Water Bearer).

CAUTION, Astrological terms like the "Age of Aquarius" involve boundaries in the sky that are not recognized by astronomers and are generally not even related to the positions of the constellations. Indeed, astrology is *not* a science at all, but merely a collection of superstitions and hokum that tries to use some of the terminology of astronomy but which rejects the logical thinking that is at the heart of science. James Randi's essay "Why Astrology Is Not Science" at the end of this chapter has more to say about astrology and other pseudosciences.

As discussed in Box 2-2, the astronomer's system of locating heavenly bodies by their right ascension and declination is tied to the positions of the celestial equator and the vernal equinox. Because of precession, these positions are changing, and thus the coordinates of stars in the sky are also constantly changing. These changes are very small and gradual, but they add up over the years. To cope with this difficulty, astronomers always make note of the date (called the **epoch**) for which a particular set of coordinates is precisely correct. Consequently, star catalogues and star charts are periodically updated. Most current catalogs and star charts are prepared for the epoch 2000. The coordinates in these reference books, which are precise for January 1, 2000, will require very little correction over the next few decades.

---

| 2-7 | Keeping track of time is traditionally a responsibility of astronomers |

Since the dawn of civilization, people have needed accurate timekeeping. Ancient Egyptians wanted to know when the Nile would flood, farmers needed to know when to plant crops, and priests had to schedule religious observances. Even today, we have many reasons to keep the calendar consistent with the cycle of the seasons. For example, many of our holiday traditions involve seasonal observances.

Astronomers have traditionally been responsible for telling time. We want the system of timekeeping used in everyday life to reflect the position of the Sun in the sky. The Sun's position determines whether we are awake or asleep and whether it is time for breakfast or dinner.

Thousands of years ago, the sundial was invented to keep track of **apparent solar time**. To obtain more formal measurements astronomers use the **meridian**, which is a north-south circle on the celestial sphere that passes through the zenith (the point directly overhead) and both celestial poles, as shown in Figure 2-17. *Local noon* is defined to be when the Sun crosses the **upper meridian**, which is the half of the meridian above the horizon. At *local midnight*, the Sun cross-

es the **lower meridian**, the half of the meridian below the horizon; this crossing cannot be observed directly. The crossing of the meridian by any object in the sky is called a **meridian transit**. If the crossing occurs above the horizon, it is an *upper* meridian transit. An **apparent solar day** is formally defined as the interval between two successive upper meridian transits of the Sun as observed from any fixed spot on the Earth. Stated less formally, an apparent solar day is the time from one local noon to the next local noon, or from when the Sun is highest in the sky to when it is again highest in the sky.

Unfortunately, the Sun is not a good timekeeper. The length of an apparent solar day (as measured by a device such as an hourglass) varies from one time of year to another. There are two main reasons why this is so, both having to do with the way in which the Earth orbits the Sun.

The first reason is that the Earth's orbit is not a perfect circle but rather an ellipse, as shown in exaggerated form in Figure 2-18a. As we will learn in Chapter 4, the Earth moves more rapidly along its orbit when it is near the Sun than when it is farther away. Hence as seen from the Earth, the Sun appears to move eastward along the ecliptic at a speed that varies during the year. The Sun appears to move more than 1° per day along the ecliptic in January, when the Earth is nearest the Sun, and less than 1° per day in July, when the Earth is farthest from the Sun. By itself, this effect would cause the apparent solar day to be longer in January than in July.

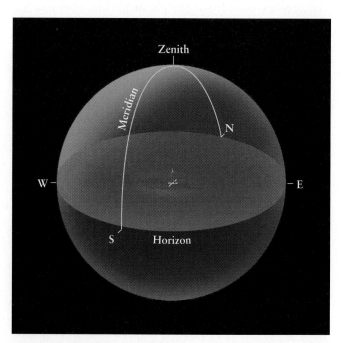

**ƒigure 2-17**

**The Meridian** The meridian is a circle that passes through the observer's zenith (the point directly overhead) and the north and south points on the observer's horizon. The passing of celestial objects across the meridian can be used to measure time. The part of the meridian above the horizon is the upper meridian, and the part below the horizon is the lower meridian.

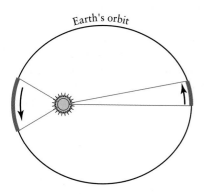

a A month's motion of the Earth along its orbit

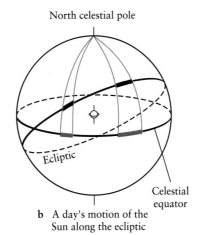

b A day's motion of the Sun along the ecliptic

figure 2-18

**Why the Sun Is a Poor Timekeeper** There are two main reasons that the Sun is a poor timekeeper. **(a)** The Earth's speed along its orbit varies during the year; it moves fastest when closest to the Sun in January and slowest when furthest from the Sun in July. Hence, the apparent speed of the Sun along the ecliptic is not constant. **(b)** Because of the tilt of the Earth's rotation axis, the ecliptic is inclined with respect to the celestial equator. Therefore, the projection of the Sun's daily progress along the ecliptic onto the celestial equator (shown in blue) varies during the year. Hence the Sun's net daily eastward progress against the stars is not constant.

The second reason why the Sun is not a good timekeeper is the 23½° angle between the ecliptic and the celestial equator (see Figure 2-13). As shown in Figure 2-18b, this causes a significant part of the Sun's apparent motion when near the equinoxes to be in a north-south direction. The net daily eastward progress in the sky is then somewhat foreshortened. At the summer and winter solstices, by contrast, the Sun's motion is parallel to the celestial equator. Thus, there is no comparable foreshortening around the beginning of summer or winter. This effect by itself makes the apparent solar day longer in March and September than in June or December. When we combine these effects with those due to the Earth's noncircular orbit, we find that the length of the apparent solar day varies in a complicated fashion over the course of a year.

To avoid these difficulties, astronomers invented an imaginary object called the **mean sun** that moves along the celestial *equator* at a uniform rate. (In science and mathematics, "mean" is a synonym for "average.") The mean sun is sometimes slightly ahead of the real Sun in the sky, sometimes behind. As a result, mean solar time and apparent solar time can differ by as much as a quarter of an hour at certain times of the year.

Because the mean sun moves at a constant rate, it serves as a fine timekeeper. A **mean solar day** is the interval between successive upper meridian transits of the mean sun. It is exactly 24 hours long, the average length of an apparent solar day. One 24-hour day as measured by your alarm clock or wristwatch is a mean solar day. In practice, the United States standard for time is an ultraprecise atomic clock located at the National Institute for Standards and Technology in Boulder, Colorado. But even this clock is adjusted so that its readings are in agreement with mean solar time.

**Time zones** were invented for convenience in commerce, transportation, and communication. In a time zone, all clocks and watches are set to the mean solar time for a meridian of longitude that runs approximately through the center of the zone. Time zones around the world are generally centered on meridians of longitude at 15° intervals. In most cases, going from one time zone to the next requires you to change the

time on your wristwatch by exactly one hour. The time zones for most of North America are shown in Figure 2-19.

In order to coordinate their observations with colleagues elsewhere around the globe, astronomers often keep track of time using Coordinated Universal Time, somewhat confusingly abbreviated UTC or UT. This is the time in a zone that includes Greenwich, England, a seaport just outside of London where the first internationally accepted time standard was kept. In former times UTC was known as Greenwich Mean Time. In North America, Eastern Standard Time (EST) is 5 hours different from UTC; 9:00 A.M. EST is 14:00 UTC. Coordinated Universal Time is also used by

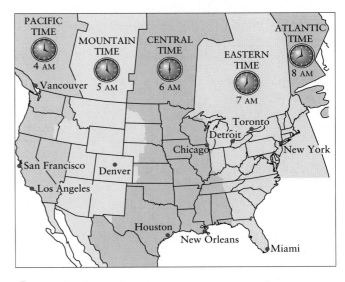

figure 2-19

**Time Zones in North America** For convenience, the Earth is divided into 24 time zones, generally centered on 15° intervals of longitude around the globe. Consequently, there are four time zones across the continental United States, making for a 3-hour time difference between New York and California.

aviators and sailors, who can easily travel from one time zone to another.

Although it is natural to want our clocks and method of timekeeping to be related to the Sun, astronomers often use a system that is related to the stars. **Sidereal time,** which is based on the apparent motion of the stars rather than that of the Sun, is useful when aiming a telescope. Most observatories are therefore equipped with a sidereal clock, as discussed in Box 2-4.

---

| 2-8 | Astronomical observations led to the development of the modern calendar |
|---|---|

Just as the day is a natural unit of time based on the Earth's rotation, the year is a natural unit of time based on the Earth's revolution about the Sun. Unfortunately, nature has

not arranged things for our convenience. The year does not divide into exactly 365 whole days. Ancient astronomers realized that the length of a year is approximately 365¼ days, so the Roman emperor Julius Caesar established the system of "leap years" to account for this extra quarter of a day. By adding an extra day to the calendar every four years, he hoped to ensure that seasonal astronomical events, such as the beginning of spring, would occur on the same date year after year.

Caesar's system would have been perfect if the year were exactly 365¼ days long and if there were no precession. Unfortunately, this is not the case. To be more accurate, astronomers now use several different types of years. For example, the **sidereal year** is defined to be the time required for the Sun to return to the same position with respect to the stars. It is equal to 365.2564 mean solar days, or $365^d\ 6^h\ 9^m\ 10^s$.

The sidereal year is the true orbital period of Earth around the Sun, but it is not the year on which we base our calendar. Like Caesar, most people want annual events to fall

---

| ƀox 2-4 | Tools of the Astronomer's Trade |
|---|---|

## *Sidereal Time*

If you want to observe a particular object in the heavens, the ideal time to do so is when the object is high in the sky, on or close to the upper meridian. This minimizes the distorting effects of the Earth's atmosphere, which increase as you view closer to the horizon. For astronomers

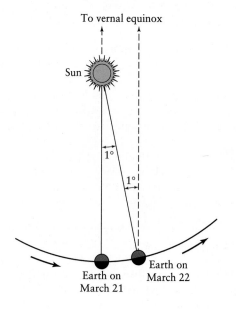

To vernal equinox

Sun

1°

1°

Earth on
March 21

Earth on
March 22

who study the Sun, this means making observations at local noon, which is not too different from noon as determined using mean solar time. For astronomers who observe stars or galaxies, however, the optimum time to observe depends on the particular object to be studied. The problem is this: Given the location of a given object on the celestial sphere, when will that object be on the upper meridian?

To answer this question, astronomers use sidereal time rather than solar time. It is different from the time on your wristwatch. In fact, a sidereal clock and an ordinary clock even tick at different rates, because they are based on different astronomical objects. Ordinary clocks are related to the position of the Sun, while sidereal clocks are based on the position of the vernal equinox, the location from which right ascension is measured (see Box 2-2 for a discussion of right ascension).

Regardless of where the Sun is, midnight sidereal time at your location is defined to be when the vernal equinox crosses your upper meridian. (Like solar time, sidereal time depends on where you are on Earth.) A **sidereal day** is the time between two successive upper meridian passages of the vernal equinox. In contrast, an apparent solar day is the time between two successive upper meridian crossings of the Sun. These two kinds of day are not equal for reasons shown in the illustration. Because the Earth moves

on the same date each year. For example, we want the first day of spring to occur on March 21. But spring begins when the Sun is at the vernal equinox, and the vernal equinox moves slowly against the background stars because of precession. Therefore, to set up a calendar we make use of the **tropical year,** which is equal to the time needed for the Sun to return to the vernal equinox. This period is equal to 365.2422 mean solar days, or $365^d 5^h 48^m 46^s$. Because of precession, the tropical year is 20 minutes and 24 seconds shorter than the sidereal year.

Caesar's assumption that the tropical year equals 365¼ years was off by 11 minutes and 14 seconds. This tiny error adds up to about three days every four centuries. Although Caesar's astronomical advisers were aware of the discrepancy, they felt that it was too small to matter. However, by the sixteenth century the first day of spring was occurring on March 11.

The Roman Catholic Church became concerned because Easter kept shifting to progressively earlier dates. To straighten things out, Pope Gregory XIII instituted a calendar reform in 1582. He began by dropping ten days (October 4, 1582, was followed by October 15, 1582), which brought the first day of spring back to March 21. Next, he modified Caesar's system of leap years.

Caesar had added February 29 to every calendar year that is evenly divisible by four. Thus, for example, 1988, 1992, and 1996 were all leap years with 366 days. But we have seen that this system produces an error of about three days every four centuries. To solve the problem, Pope Gregory decreed that only the century years evenly divisible by 400 should be leap years. For example, the years 1700, 1800, and 1900 (which would have been leap years according to Caesar) were not leap years in the improved Gregorian system, but the year 2000, which can be divided evenly by 400, is a leap year.

We use the Gregorian system today. It assumes that the year is 365.2425 mean solar days long, which is very close to the true length of the tropical year. In fact, the error is only one day in every 3300 years. That won't cause any problems for a long time.

---

along its orbit around the Sun, the Earth must rotate through nearly 361° to get from one local noon to the next. This extra 1° of rotation corresponds to 4 minutes of time, which is the amount by which a solar day exceeds a sidereal day. To be precise:

$$1 \text{ sidereal day} = 23^h 56^m 4.091^s$$

where the hours, minutes, and seconds are in mean solar time.

One day according to your wristwatch is one mean solar day, which is exactly 24 hours of solar time long. A **sidereal clock,** which measures sidereal time, measures sidereal hours, minutes, and seconds, where one sidereal day is divided into 24 sidereal hours. This explains why a sidereal clock ticks at a slightly different rate than your wristwatch. As a result, at some times of the year a sidereal clock will show a very different time than an ordinary clock. (On local noon on March 21, when the Sun is at the vernal equinox, a sidereal clock will say that it is midnight. Do you see why?)

We can now answer the question in the opening paragraph. The vernal equinox, whose celestial coordinates are R.A. = $0^h 0^m 0^s$, Decl. = 0° 0′ 0″, crosses the upper meridian at midnight sidereal time (0:00). The autumnal equinox, which is on the opposite side of the celestial sphere at R.A. = $12^h 0^m 0^s$, Decl. = 0° 0′ 0″, crosses the upper meridian 12 sidereal hours later at noon sidereal time (12:00). Comparing the right ascensions of these points with the sidereal times of their upper meridian passage, we conclude that *any* object crosses the upper meridian when the sidereal time is equal to the object's right ascension. That is why astronomers measure right ascension in units of time rather than degrees, and why measurements of right ascension are always measured in sidereal hours, minutes, and seconds.

**EXAMPLE:** Suppose you want to observe the bright star Regulus in the constellation of Leo (the Lion). What is the best time to do this? From reference books, such as the *Astronomical Almanac,* you find that the epoch 2000 coordinates of this star are R.A. = $10^h 08^m 22.2^s$, Decl. = +11° 58′ 02″. Regulus will therefore be on the upper meridian and thus ideally suited for observation, when the sidereal time at your location is equal to its right ascension, or about 10:08. As this example shows, it is useful to keep track of sidereal time, which is why most observatories are equipped with a side-real clock.

While we have been emphasizing the usefulness of sidereal time, mean solar time is still the best method of timekeeping for most earthbound purposes. All time measurements in this book are expressed in mean solar time unless otherwise stated.

## Key Words

*Terms preceded by an asterisk (\*) are discussed in the Boxes.*

*Antarctic Circle, p. 38
apparent solar day, p. 40
apparent solar time, p. 40
*Arctic Circle, p. 38
autumnal equinox, p. 37
celestial equator, p. 33
celestial sphere, p. 33
constellation, p. 27
*declination, p. 34
diurnal motion, p. 28
ecliptic, p. 36
epoch, p. 40

equinox, p. 36
lower meridian, p. 40
mean solar day, p. 41
mean sun, p. 41
meridian, p. 40
meridian transit, p. 40
north celestial pole, p. 33
positional astronomy, p. 26
precession, p. 38
precession of the equinoxes, p. 39
*right ascension, p. 34
*sidereal clock, p. 43
*sidereal day, p. 42

sidereal time, p. 42
sidereal year, p. 42
south celestial pole, p. 33
summer solstice, p. 37
time zone, p. 41
tropical year, p. 43
*Tropic of Cancer, p. 38
*Tropic of Capricorn, p. 38
upper meridian, p. 40
vernal equinox, p. 36
winter solstice, p. 37
zenith, p. 33
zodiac, p. 37

## Key Ideas

*Ideas preceded by an asterisk (\*) are discussed in the Boxes.*

**Constellations and the Celestial Sphere:** It is convenient to imagine the stars fixed to the celestial sphere with the Earth at its center.

• The surface of the celestial sphere is divided into 88 regions called constellations.

**Diurnal (Daily) Motion of the Celestial Sphere:** The celestial sphere appears to rotate around the Earth once in each 24-hour period. In fact, it is actually the Earth that is rotating.

• The poles and equator of the celestial sphere are determined by extending the axis of rotation and the equatorial plane of the Earth out to the celestial sphere.

• *The positions of objects on the celestial sphere are described by specifying their right ascension (in time units) and declination (in angular measure).

**Seasons and the Tilt of the Earth's Axis:** The Earth's axis of rotation is tilted at an angle of about 23½° from the perpendicular to the plane of the Earth's orbit.

• The seasons are caused by the tilt of the Earth's axis.

• Over the course of a year, the Sun appears to move around the celestial sphere along a path called the ecliptic. The ecliptic is inclined to the celestial equator by about 23½°.

• The ecliptic crosses the celestial equator at two points in the sky, the vernal and autumnal equinoxes. The northernmost point that the Sun reaches on the celestial sphere is the summer solstice, and the southernmost point is the winter solstice.

**Precession:** The orientation of the Earth's axis of rotation moves slowly, a phenomenon called precession.

• Precession is caused by the gravitational pull of the Sun and Moon on the Earth's equatorial bulge.

• Precession of the Earth's axis causes the positions of the equinoxes and celestial poles to shift slowly.

• *Because the system of right ascension and declination is tied to the position of the vernal equinox, the date (or epoch) of observation must be specified when giving the position of an object in the sky.

**Timekeeping:** Astronomers use several different means of keeping time.

• Apparent solar time is based on the apparent motion of the Sun across the celestial sphere, which varies over the course of the year.

• Mean solar time is based on the motion of an imaginary mean sun along the celestial equator, which produces a uniform mean solar day of 24 hours. Ordinary watches and clocks measure mean solar time.

• Sidereal time is based on the apparent motion of the celestial sphere.

**The Calendar:** The tropical year is the period between two passages of the Sun across the vernal equinox. Leap year corrections are needed because the tropical year is not exactly 365 days. The sidereal year is the actual orbital period of the Earth.

## Review Questions

**1.** How are constellations useful to astronomers? Are there any stars in the sky that are not part of a constellation?

**2.** Why can't a person in Antarctica use the Big Dipper to find the north direction?

**3.** What is the celestial sphere, and why is this ancient concept still useful today?

**4.** Imagine that someone suggests sending a spacecraft to land on the surface of the celestial sphere. How would you respond to such a suggestion?

**5.** What is the celestial equator, and how is it related to the Earth's equator? How are the north and south celestial poles related to the Earth's axis of rotation? Where on Earth would you have to be for the celestial equator to pass through your zenith?

**6.** Where on Earth do you have to be for the north celestial pole to appear on the horizon? Explain.

**7.** How many degrees is the angle from the horizon to the zenith?

**8.** Where on Earth do you have to be in order to see the Sun at the zenith? If you stay at that location for a full year, on how many days will the Sun pass through the zenith?

**9.** Where on Earth do you have to be in order to see the south celestial pole directly overhead? What is the maximum possible elevation of the Sun above the horizon at that location? On what date can this maximum elevation be observed?

**10.** Using a diagram, explain why the tilt of the Earth's axis relative to the Earth's orbit causes the seasons as we orbit the Sun.

**11.** What are the vernal and the autumnal equinoxes? What are the summer and winter solstices? How are these four points related to the ecliptic and the celestial equator?

**12.** At what point on the horizon does the vernal equinox rise? Where on the horizon does it set? (*Hint:* See Figure 2-14.)

**13.** How does the daily path of the Sun across the sky change with the seasons?

**14.** What causes precession of the equinoxes? How long does it take for the vernal equinox to move 1° along the ecliptic?

**15.** Why is the fictitious mean sun a better timekeeper than the actual Sun in the sky?

**16.** Why is the time given by a sundial not necessarily the same as the time on your wristwatch?

**17.** Why is it convenient to have the Earth divided into time zones?

**18.** What is the difference between the sidereal year and the tropical year? Why are calendars based on the tropical year?

**19.** When is the next leap year? Is 2000 a leap year? Will 2100 be a leap year?

## Advanced Questions

### Problem-solving tips and tools:

To help you visualize the heavens, it is worth taking the time to become familiar with various types of star charts. These include the simple star charts at the end of this book, the monthly star charts published in such magazines as *Sky & Telescope* and *Astronomy*, and the more detailed maps of the heavens found in star atlases (see under "Where to Learn More"). A number of very useful star atlas programs are available for personal computers. These allow you to view the sky on any date and at any time, as seen from any point on Earth, and to animate the sky to visualize its diurnal and annual motions. A superb example is the program *Starry Night*, a version of which is on the CD-ROM that accompanies this book. Shareware sky atlas programs can also be downloaded from the World Wide Web (for example, see **http://www.shareware.com/**). You may also find it useful to examine a planisphere, a device consisting of two rotatable disks. The bottom disk shows all the stars in the sky (for a particular latitude), and the top one is opaque with a transparent, oval window through which only some of the stars can be seen. By rotating the top disk, you can immediately see which constellations are above the horizon at any time of the year.

**20.** The time-exposure photograph that opens this chapter shows the trails made by individual stars as the celestial sphere appears to rotate around the Earth. **(a)** If you were standing in Australia next to the camera that took this picture and looking in the same direction that the camera was pointing, would you see the stars moving clockwise or counterclockwise? **(b)** In taking this photograph, for approximately what length of time was the camera shutter left open?

**21.** Figure 2-4 shows the situation on September 21, when Cygnus is highest in the sky at 8:00 P.M. local time and Andromeda is highest in the sky at midnight. But as Figure 2-5 shows, on July 21 Cygnus is highest in the sky at midnight. On July 21, at approximately what local time is Andromeda highest in the sky? Explain your reasoning.

**22.** Figure 2-5 shows which constellations are high in the sky (for observers in the northern hemisphere) in the months of July, September, and November. Based on this figure, would you be able to see Andromeda at midnight on March 21? Draw a picture to justify your answer.

**23.** (a) Redraw Figure 2-10 for an observer at the north pole. (*Hint:* The north celestial pole is directly overhead this observer.) (b) Redraw Figure 2-10 for an observer at the equator. (*Hint:* The celestial equator passes through this observer's zenith.) (c) Using Figure 2-10 and your drawings from (a) and (b), justify the following rule, long used by navigators: The latitude of an observer in the northern hemisphere is equal to the angle in the sky between that observer's horizon and the north celestial pole. (d) State the rule that corresponds to (c) for an observer in the southern hemisphere.

**24.** On December 1 at 10:00 P.M. you look toward the eastern horizon and see the bright star Procyon rising. At approximately what time will Procyon rise two weeks later, on December 15?

**25.** How would the sidereal and solar days change (a) if the Earth's rate of rotation increased, (b) if the Earth's rate of rotation decreased, and (c) if the Earth's rotation were retrograde (that is, if the Earth rotated about its axis opposite to the direction in which it revolves about the Sun)?

**\*26.** (a) What is the sidereal time when the vernal equinox rises? (b) On what date is the sidereal time nearly equal to the solar time? Explain.

**\*27.** Consult a star map of the southern hemisphere and determine which, if any, bright southern stars could someday become south celestial pole stars.

**28.** Are there stars in the sky that never set? Are there stars that never rise? Does your answer depend on the observer's location on Earth? Explain.

**29.** Using a star map, determine which bright stars, if any, could someday mark the location of the vernal equinox. Give the approximate years when this should happen.

**30.** Suppose that you live at a latitude of 40° N. What is the elevation of the Sun above the southern horizon at noon at the time of the winter solstice? Explain your reasoning.

**31.** The Great Pyramid at Giza has a tunnel that points toward the north celestial pole. At the time the pyramid was built, about 2600 B.C., toward which star did it point? Toward which star does this same tunnel point today? (See Figure 2-16.)

**32.** In the northern hemisphere, houses are designed to have "southern exposure," that is, with the largest windows on the southern side of the house. But in the southern hemisphere houses are designed to have "northern exposure." Why are houses designed this way, and why is there a difference between the hemispheres?

**\*33.** What is the right ascension of a star that is on the meridian at midnight at the time of the autumnal equinox? Explain.

**\*34.** The coordinates on the celestial sphere of the summer solstice are R.A. = $6^h$ $0^m$ $0^s$, Decl. = +23° 27′. What are the right ascension and declination of the winter solstice? Explain your answer.

**\*35.** At local noon on March 21, when the Sun is at the vernal equinox, a sidereal clock will say that it is midnight. Explain why.

**\*36.** Right ascension is measured in hours, minutes, and seconds. Because 24 hours of right ascension takes you all the way around the celestial equator, it follows that $24^h = 360°$. What is the angle in the sky (measured in degrees) between a star with R.A. = $6^h$ $0^m$ $0^s$, Decl. = 0° 0′ 0″ and a second star with R.A. = $11^h$ $20^m$ $0^s$, Decl. = 0° 0′ 0″? Explain your answer.

## Discussion Questions

**37.** Examine a list of the 88 constellations. Are there any constellations whose names obviously date from modern times? Where are these constellations located? Why do you suppose they do not have archaic names?

**38.** Why is it useful to astronomers to have telescopes in both the southern hemisphere and the northern hemisphere? (See Figure 2-10.)

**39.** Describe how the seasons would be different if the Earth's axis of rotation, rather than its present 23½° tilt, were tilted (a) by 0° and (b) by 90°.

**40.** In William Shakespeare's *Julius Caesar* (Act 3, Scene 1), Caesar says the following :

> But I am constant as the northern star.
> Of whose true-fix'd and resting quality
> There is no fellow in the firmament.

Translate Caesar's statement about the "northern star" into modern astronomical language. Is the northern star truly "constant"? Was the northern star the same in Shakespeare's day (1564–1616) as it is today? You can find the entire play on the World Wide Web (**http://the-tech.mit.edu/Shakespeare/works.html**).

## Observing Projects

**Observing tips and tools:**

Moonlight is so bright that it interferes with seeing the stars. For the best view of the constellations, do your observing when the Moon is below the horizon. You can find the times of moonrise and moonset in your local newspaper. Each monthly issue of the magazines *Sky & Telescope* and *Astronomy* includes much additional observing information.

**41.** On a clear, cloud-free night, use the star charts at the end of this book to see how many constellations of the zodiac you can identify. Which ones were easy to find? Which were difficult? Are the zodiacal constellations the most prominent ones in the sky?

42. Examine the star charts that are published monthly in such popular astronomy magazines as *Sky & Telescope* and *Astronomy*. How do they differ from the star charts at the back of this book? On a clear, cloud-free night, use one of these star charts to locate the celestial equator and the ecliptic. Note the inclination of the Milky Way to the ecliptic and celestial equator. The Milky Way traces out the plane of our galaxy. What do your observations tell you about the orientation of the Earth and its orbit relative to the galaxy's plane?

43. Suppose you wake up before dawn and want to see which constellations are in the sky. Explain how the star charts at the end of this book can be quite useful, even though chart times are given only for the evening hours. Which chart most closely depicts the sky at 4:00 A.M. tomorrow morning? Set your alarm clock for 4:00 A.M. to see if you are correct.

## Where to Learn More

*Books and magazine articles*

Allen, R. *Star Names: Their Lore and Meaning*. Dover, 1963 (originally published in 1899). A classic book detailing the origins of constellations and star names.

Kaler, J. *The Ever-Changing Sky*. Cambridge University Press, 1996. Written by a well-known expert on stars and their evolution, this is an in-depth but nonmathematical guide to all aspects of observing the night sky.

Lovi, G. "That Mysterious Zodiac," *Sky & Telescope*, April 1993. A short but illuminating article about the history of the zodiacal constellations.

MacRobert, A. "Understanding Celestial Coordinates." *Sky & Telescope*, September 1995. This brief, clear article gives a simple explanation of the sometimes confusing celestial coordinates right ascension and declination (described in Box 2-2).

Ottewell, G. *The Astronomical Calendar*. Published yearly by the Universal Workshop (Furman University, Greenville, South Carolina 29613; check out the web site at http://www.kalend.com/). This easy-to-use, atlas-sized book describes all the celestial events that will take place in the following year. The illustrations alone are more than worth the modest price of the book.

Pasachoff, J. M., and Menzel, D. H. *A Field Guide to the Stars and Planets*. Houghton Mifflin, 1992. A pocket-size guide with basic skywatching information as well as detailed sky charts.

Ridpath, I., and Tirion, W. *The Monthly Sky Guide*, 4th ed. Cambridge University Press, 1996. An excellent guide to the night sky for each month of the year, with large, clear sky charts and tips for naked-eye observing.

Staal, J. *The New Patterns in the Sky: Myths and Legends of the Stars*. McDonald & Woodward, 1996. This entertaining book describes the mythology of all 88 constellations, including much star lore from non-Western cultures.

Tirion, W. *Sky Atlas 2000.0*. Sky Publishing Corporation and Cambridge University Press, 1981. One of the finest star atlases available, this is an indispensable resource for any serious amateur astronomer.

### W *World Wide Web*

To learn about observing the night sky, two excellent places to start are the web sites for the magazines *Sky & Telescope* (http://www.skypub.com/) and *Astronomy* (http://www.astronomy.com/). An especially handy web page is This Week's Sky at a Glance (http://www.skypub.com/whatsup/saag.shtml), which gives viewing hints about current celestial events. You can determine the local time of sunrise and sunset at any location on the Earth using a handy web page at the U. S. Naval Observatory (http://tycho.usno.navy.mil/srss2.html). A web site at George Mason University has detailed maps of all the constellations, along with a list of interesting objects for binoculars and small telescopes (http://astro.gmu.edu/constellation/constellation.html). Observations of the sky are at the heart of the time-honored art of celestial navigation (http://peck.ipph.purdue.edu/al/space.html), which only recently has been replaced by electronic navigation using the Global Positioning System. You can learn about modern, high-tech methods of timekeeping at the web site for the National Institute for Standards and Technology (http://www.boulder.nist.gov/timefreq/).

# JAMES RANDI

## Why Astrology Is Not Science

James ("the Amazing") Randi works tirelessly to expose trickery, so that others can relish the greater wonder of science. As a magician, he has had his own TV show and an enormous public following. As a lecturer, he addresses teachers, students, and others worldwide. His newsletter and column for *The Skeptic* are key resources for educators. His many books include *Flim-Flam!, The Faith Healers,* and *The Mask of Nostradamus,* about a legendary con man with secrets of his own.

Mr. Randi helped found the Committee for the Scientific Investigation of Claims of the Paranormal, and his $10,000 prize for "the performance of any paranormal event . . . under proper observing conditions" has gone unclaimed for 25 years. An amateur archeologist and astronomer as well, he lives in Florida with several untalented parrots and the occasional visiting magus.

I'm involved in the strange business of telling folks what they should already know. I meet audiences who believe in all sorts of impossible things, often despite their education and intelligence. My job is to explain how science differs from the unproven, illogical assumptions of pseudoscience—and why it matters. Perhaps my best example is the difference between astronomy and astrology.

Both astrology and astronomy arose from the wonders of the night sky, from the stars to comets, planets, the Sun, and the Moon. Surely, humans have long reasoned, there must be some meaning in their motions. Surely the Moon's effect on tides hints at hidden "causes" for strange events. *Judiciary* (literally "judging") astrology therefore attempted to foretell the future—our earthly future. To serve it, *horary* (literally "hourly") astrology carefully tracked the heavens.

It is the latter that has become astronomy. Thanks to its process of careful measurement and testing, we now understand more about the true nature of the starry universe than astrologers could ever have imagined. With the birth of a new science, astronomers had a logical framework based on physical causes and systematic observations.

Astrology remains a popular delusion. Far too many believe today that patterns in the sky govern our lives. They accept the vague tendencies and portents of seers who cast horoscopes. They shouldn't. Just a glance at the tenets of astrology provides ample evidence of its absurdity.

An individual is said to be born under a sign. To the astrologer, the Sun was located "in" that sign at the moment of birth. (Stars are not seen in the daytime, but no matter—a calculation tells where the Sun is.) Each sign takes its name from a constellation, a totally imaginary figure invented for our convenience in referring to stars. Different cultures have different mythical figures up there, and so different schools of astrology assign different meanings to the signs they use.

In the spirit of equal-opportunity swindling, astrologers divide up the year fairly, ignoring variations in the size of constellations. Since Libra is tiny, while Virgo is huge, they chop some of the sky off Virgo and add it—along with bits of Scorpio—to bring Libra up to size. The Sun could well be declared "in" Libra when it is actually outside that constellation.

It gets worse. Science constantly challenges itself and changes. The rules of astrology could not, although they were made up thousands of years ago, and since then the "fixed" stars have moved. In particular, precession of the equinoxes has shifted objects in the sky relative to our calendar. The constellations have changed but astrology has not. If you were born August 7, you are said to be a Leo, but the Sun that day was really in the same part of the sky as the constellation Cancer.

With a theory like this to back it up, we should not be surprised at the bottom line: *a pseudoscience does not work.* Test after test has checked its predictions, and the result is always the same. One such investigator is Shawn Carlson of the University of California, San Diego. As he put it in *Nature* magazine, astrology is "a hopeless cause." Johannes Kepler, the pioneering astronomer, himself cast horoscopes, but they are little remembered today. Owen Gingerich, a historian of science at Harvard, puts it well: Kepler was the astrologer who destroyed astrology.

Astronomy works, and it works very well indeed. That isn't easy. Because we humans tend to find what we want in any body of data, it takes science's careful process of observation, creative insight, and critical thinking to understand and predict changes in nature. As I write, a transit of Ganymede is due next Thursday at 21:47:20. At exactly that time, the satellite of Jupiter will cross in front of its planet as seen from Earth, and yet most of us will never know it. Still other moons of Jupiter may hold fresh clues to the formation of our entire solar system and the conditions for life elsewhere.

For most people, astronomy has too little fantasy or money in it, and they will never experience the beauty in its predictions. The dedicated labors of generations of scientists have enabled us to perform a genuine wonder.

# Eclipses and the Motion of the Moon

A Solar Eclipse  On July 11, 1991, the Moon passed directly between the Sun and the Earth, and a total solar eclipse was visible from Hawaii, Mexico, and Central and South America. This multiple-exposure photograph, taken from La Paz, Baja California, shows the Sun at 5-minute intervals as it moved across the sky from left to right. More and more of the Sun became hidden from view as it passed behind the Moon. During the almost 7 minutes when the Sun's disk was completely hidden by the Moon, only the Sun's corona, or outer halo of gas, was visible; the unearthly pearlescent glow of the corona can be seen in the middle image. (Akira Fujii, Hiroyuki Tomioka, and Yonematsu Shiono)

*In this chapter you will find the answers to the following questions:*

3-1  Why does the Moon go through phases? Why do we always see the same face of the Moon?

3-2  What is the difference between a lunar eclipse and a solar eclipse?

3-3  How often do lunar eclipses happen? When one is taking place, where do you have to be to see it?

3-4  How often do solar eclipses happen? Why are they visible only from certain special locations on Earth?

3-5  How is it possible to predict eclipses?

3-6  How did ancient astronomers deduce the sizes of the Earth, the Moon, and the Sun?

*The Moon is by far the easiest astronomical object to find in the nighttime sky. It is also one of the most dynamic. Even to the naked eye, its appearance changes dramatically from night to night. We are all familiar with its changes in phase, from new moon to full moon and back again. The times when the Moon rises and sets also differ noticeably from one night to the next. If you look closely, you can even see the Moon's motion against the backdrop of stars. It shifts noticeably over the space of only an hour or two!*

*In Chapter 9 we discuss the Moon as a world in its own right. We will see why it is a dry, dead, airless place so different from our Earth. In this chapter, however, we are concerned with the motions of the Moon as they can be seen from Earth. We explore why the Moon goes through a regular cycle of phases. We describe how the Moon's orbit around Earth leads to* lunar eclipses, *in which the Earth comes between the Sun and the Moon, and* solar eclipses, *in which the Moon comes between the Sun and the Earth. A solar eclipse, like that in the photograph that opens this chapter, is both awe-inspiring to behold and of scientific importance, because it provides astronomers with information about the Sun's outermost layers.*

*Because the Moon is so prominent to the naked eye, it played an important role in the beginnings of astronomy more than 2000 years ago. From their naked-eye observations of the Moon, ancient Greek astronomers were able to determine the size and shape of the Earth, as well as other features of the solar system. Thus, the Moon—which has for millennia loomed large in the minds of poets, lovers, and dreamers—has also played a key role in the development of our modern picture of the universe.*

| 3-1 | The phases of the Moon are caused by the Moon's orbital motion |
|---|---|

As seen from Earth, both the Sun and the Moon appear to move relative to the background of stars. Both objects move from west to east on the celestial sphere, although at different rates. In one year the Sun appears to make a complete trip around the celestial sphere along the path we call the *ecliptic* (Section 2-4). By comparison, the Moon takes only about four weeks. In the past these similar motions led people to believe that both the Sun and the Moon orbit around the Earth. We now know that only the Moon orbits the Earth, while the Earth-Moon system as a whole (Figure 3-1) orbits the Sun. (In Chapter 4 we will discuss how this realization came to be.)

One key difference between the Sun and the Moon is in the nature of the light that we receive from them. The Sun emits its own light. So do the stars, which are objects like the Sun but much farther away, and so does an ordinary light-bulb. By contrast, the light that we see from the Moon is *reflected* light. This is sunlight that has struck the Moon's surface, bounced off, and ended up in our eyes here on Earth.

CAUTION! You probably associate *reflection* with shiny objects like a mirror or the surface of a lake, but in science the term refers to light bouncing off any object. You see most objects around you by reflected light. When you look at your hand, for example, you are seeing light from the Sun (or from a light fixture) that has been reflected from the skin of your hand and into your eye. In the same way, moonlight is really sunlight that has been reflected by the Moon's surface.

**ƒigure 3-1** R I **V** U X G

**The Earth and the Moon** The Moon orbits the Earth every 27.3 days, at an average distance of 384,400 kilometers (238,900 miles). This picture of the Earth and the Moon was taken in 1992 by the *Galileo* spacecraft on its way toward Jupiter. The Sun, which provides the illumination for both the Earth and the Moon, was far to the right and out of the camera's field of view when this photograph was taken. (NASA)

Figure 3-1 shows both the Moon and the Earth as seen from a spacecraft. When this photograph was taken, the Sun was far off to the right. Therefore, the right-hand hemispheres of both worlds were illuminated by the Sun, and the left-hand hemispheres were in darkness. Hence only the right-hand hemispheres are visible in the photograph. In the same way, when we view the Moon from the Earth we see only the half of the Moon that faces the Sun and is illuminated. However, not all of the illuminated half of the Moon is necessarily facing us. As the Moon moves around the Earth, from one night to the next we see different amounts of the illuminated half of the Moon. These different appearances of the Moon are called **lunar phases** (Figure 3-2).

Figure 3-3 shows the relationship between the position of the Moon in its orbit and the lunar phase visible from Earth. For example, when the Moon is at position A, it is in roughly the same direction in the sky as the Sun as seen from Earth. Hence the dark hemisphere of the Moon faces the Earth. This phase, in which the Moon is not visible, is called **new moon.**

As the Moon continues around its orbit from position A in Figure 3-3, more of the illuminated half of the Moon becomes exposed to our view. The result, shown at position B, is a phase called **waxing crescent moon** ("waxing" is a synonym for "increasing"). About a week after new moon, the Moon is at position C; we then see half of the Moon's illuminated hemisphere and half of the dark hemisphere. This phase is called **first quarter moon.** The name means that this phase is one-quarter of the way through the complete cycle of lunar phases. Note that a first quarter moon appears to be half illuminated, *not* one-quarter illuminated!

During the next week, the Moon reaches position D in Figure 3-3. Now, still more of the illuminated hemisphere can be seen from Earth, giving us the phase called **waxing gibbous moon** ("gibbous" is another word for "swollen"). When the Moon stands opposite the Sun in the sky (position E) we see the fully illuminated hemisphere. This phase is called **full moon.**

Over the following two weeks, we see less and less of the illuminated hemisphere as the Moon continues along its orbit. The moon is said to be *waning*, or decreasing in illumination as seen from Earth. The phases are called **waning gibbous moon** (position F), **last quarter moon** (position G), and **waning crescent moon** (position H).

The Moon takes about four weeks to complete one orbit around the Earth, so it likewise takes about four weeks for a complete cycle of phases from new moon to full moon and back to new moon. To help you better visualize lunar phases, Box 3-1 describes a way that you can simulate the cycle shown in Figure 3-3 using ordinary objects on Earth.

Although the phase of the Moon is continuously changing, one constant aspect of the Moon is that it always keeps essentially the same hemisphere facing the Earth. Hence you will always see the same craters and mountains on the Moon, no matter when you look at it; the only difference will be the angle at which these surface features are illuminated by the Sun. (You can verify this by carefully examining the photographs of the Moon in Figure 3-2.)

Why is it that we only ever see one face of the Moon? You might think that it is because the Moon does not rotate (unlike the Earth, which rotates around an axis that passes from its north pole to its south pole). To see that this cannot be the case, consider Figure 3-4. This figure shows the Earth and the orbiting Moon from a vantage point far above the Earth's north pole; two craters on the lunar surface have been colored, one in red and one in blue. If the Moon did not rotate on its axis, as in Figure 3-4a, at some times the red crater would be visible from Earth, while at other times the blue crater would be visible. Thus, we would see different parts of the lunar surface over time, which does not happen in reality.

## figure 3-2   R I **V** U X G

**The Moon's Phases** Half of the Moon (one hemisphere) is always illuminated by the Sun. As the Moon orbits the Earth, varying amounts of the illuminated half of the Moon are exposed to our Earth-based view. The "age" refers to the time that has elapsed since new moon phase. It takes about 29½ days for the Moon to go through all its phases. Note that as the Moon orbits the Earth, it always keeps the same side facing the Earth. Consequently, Earth-based observers always see the same face of the Moon, regardless of the Moon's phase. (Lick Observatory)

Waxing crescent
(age: 4 days)

First quarter
(age: 7 days)

Waxing gibbous
(age: 10 days)

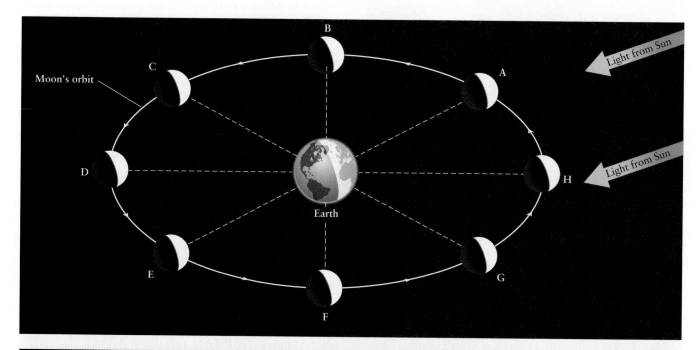

## figure 3-3

**Why the Moon Goes Through Phases** This figure shows the Moon at eight positions on its orbit, along with a picture of what the Moon looks like at each position as seen from Earth. The changes in phase occur because light from the Sun illuminates one half of the Moon, and as the Moon orbits the Earth we see varying amounts of the Moon's illuminated half.

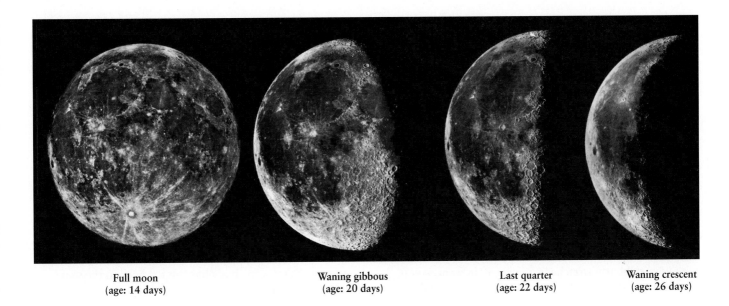

| Full moon (age: 14 days) | Waning gibbous (age: 20 days) | Last quarter (age: 22 days) | Waning crescent (age: 26 days) |

In fact, the Moon always keeps the same face toward us because it *is* rotating in a special way: It takes exactly as long to rotate on its axis as it does to make one orbit around the Earth. This situation is called **synchronous rotation.** As shown in Figure 3-4*b*, this keeps the crater shown in red always facing the Earth, so that we always see the same face

of the Moon. In Chapter 9 we will see why the Moon's rotation and orbital motion are in step with each other. Box 3-2 describes some additional details of the relationship between the Moon's orbit and its rotation.

An astronaut standing at the spot shown in red in Figure 3-4*b* would spend two weeks (half of a lunar orbit) in

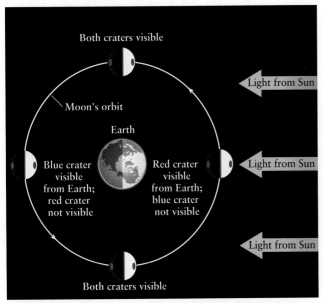

a  If the Moon did not rotate, we could see all sides of the Moon

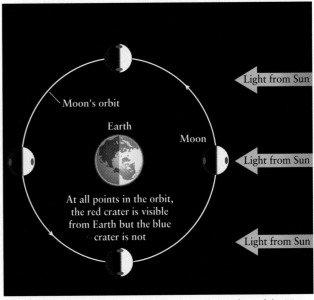

b  In fact the Moon does rotate and we see only one face of the Moon

## ƒigure 3-4

**The Moon's Rotation** These figures show the Moon's orbit as viewed from a point high above the Earth's north pole. The Moon is shown at four points in its orbit. **(a)** If the Moon did not rotate around its own north-south axis, then at various times during a lunar orbit the red crater would be visible from Earth while at other times the blue crater would be visible. Over a complete orbit, the entire surface of the Moon would be visible. **(b)** In reality, we only see one face of the Moon, because the Moon rotates on its axis and it makes one rotation in exactly the same time that it makes one orbit around the Earth.

## box 3-1 | The Heavens on the Earth

### *Phases and Shadows*

**F**igure 3-3 shows how the relative positions of the Earth, the Moon, and the Sun explain the phases of the Moon that we see. You can visualize lunar phases more clearly by doing a simple experiment here on Earth. All you need are a small round object, such as an orange or a baseball, and a bright source of light, such as a street lamp or the Sun.

In this experiment, you play the role of an observer on the Earth looking at the Moon, and the round object plays the role of the Moon. Hold the object in your right hand with your right arm stretched straight out in front of you. Now orient yourself and the object in your hand so that the object is directly between you and the light source, which plays the role of the Sun. In this orientation, the illuminated half of the object faces away from you, like the Moon when it is at position A in Figure 3-3. Because the illuminated half of the Moon then faces away from the Earth, the lunar phase is new moon. By analogy, the round object in your hand is in its "new" phase when you hold it between your eyes and the light source.

Now, slowly turn your body to the left so that the object in your hand "orbits" around you. As you do so, more and more of the illuminated side of the "moon" in your hand will become visible, and it will appear to go through the same cycle of phases—waxing crescent, first quarter, and waxing gibbous—as does the real Moon. When your body

has rotated through half a turn so that the light source is directly behind you, you will be looking face on at the illuminated side of the object in your hand. This corresponds to a full moon, which occurs when the Moon is at position E in its orbit as shown in Figure 3-3. (Make sure your body doesn't cast a shadow on the "moon" in your hand—that would correspond to a lunar eclipse!)

As you continue turning to the left, more of the unilluminated half of the object will become visible as its phase moves through waning gibbous, last quarter, and waning crescent. When your body has rotated back to the same orientation that you were in originally, the unilluminated half of your hand-held "moon" will again be facing toward you, and its phase will again be new. If you continued to rotate, the object in your hand would repeat the cycle of "phases," just as the Moon does as it orbits around the Earth.

The experiment works best when there is just one light source around. If there are several light sources, such as in a room with several lamps turned on, the different sources will create multiple shadows and it will be difficult to see the phases of your hand-held "moon." If you do the experiment outdoors using sunlight, you may find that it's best to perform it in the early morning or late afternoon when shadows are most pronounced and the Sun's rays are nearly horizontal.

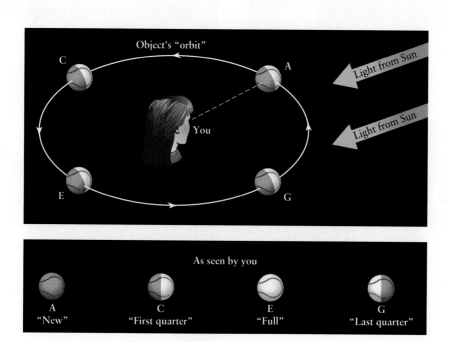

darkness, or lunar nighttime, and the next two weeks in sunlight, or lunar daytime. In other words, the Sun rises and sets as seen from the Moon. Hence, no part of the Moon is perpetually in darkness. There really is no "dark side of the Moon"; the side of the Moon that constantly faces away from the Earth is properly called the *far* side. For example, the blue crater on the far side of the Moon in Figure 3-4*b* is in sunlight for half of each lunar orbit.

The time for a complete lunar "day" is about four weeks, which is the time that it takes the Moon to rotate once on its axis. (Remember that it takes the same time for one complete lunar orbit.) The four weeks that the Moon takes to complete one cycle of its phases inspired our ancestors to invent the concept of a month. For historical reasons, none of which have much to do with the heavens, the calendar we use today has months of differing lengths. In contrast, astronomers find it useful to define two other types of months, depending on whether the Moon's motion is measured relative to the stars or to the Sun. Neither corresponds exactly to the familiar months of our usual calendar.

The **sidereal month** is the time it takes the Moon to complete one full orbit of the Earth, measured with respect to the stars. This true orbital period is equal to 27.3 days. The **synodic month**, or **lunar month**, is the time it takes the Moon to complete one cycle of phases (that is, from new moon to new moon or from full moon to full moon) and thus is measured with respect to the Sun (rather than the stars). The length of the "day" on the Moon is a synodic month, not a sidereal month.

The synodic month is longer than the sidereal month because the Earth is orbiting the Sun while the Moon goes through its phases. As Figure 3-5 shows, the Moon must travel *more* than 360° along its orbit to complete a cycle of phases (for example, from one new moon to the next). Because of this extra distance, the synodic month is equal

to about 29.53 days, about two days longer than the sidereal month.

Both the sidereal and synodic month vary somewhat from one orbit to another, the latter by as much as half a day. The reason is that the Sun's gravity sometimes causes the Moon to speed up or slow down slightly in its orbit, depending on the relative positions of the Sun, Moon, and Earth.

## 3-2 Eclipses occur only when the Sun and Moon are both on the line of nodes

From time to time the Sun, Earth, and Moon all happen to lie along a straight line. When this occurs, the shadow of the Earth can fall on the Moon or the shadow of the Moon can fall on the Earth. Such phenomena are called **eclipses**. They are perhaps the most dramatic astronomical events that can be seen with the naked eye.

A **lunar eclipse** occurs when the Moon passes through the Earth's shadow. This occurs when the Sun, Earth, and Moon are in a straight line, with the Earth between the Sun and Moon, so that the Moon is at full phase (position E in Figure 3-3). At this point in the Moon's orbit, the face of the Moon seen from Earth would normally be fully illuminated by the Sun; instead, it appears quite dim because the Earth casts a shadow on the Moon.

A **solar eclipse** occurs when the Earth passes through the Moon's shadow. As seen from Earth, the Moon moves in front of the Sun. Once again, this can happen only when the Sun, Moon, and Earth are in a straight line. However, for a solar eclipse to occur, the Moon must be between the Earth and the Sun. Therefore, a solar eclipse can occur only at new moon (position A in Figure 3-3).

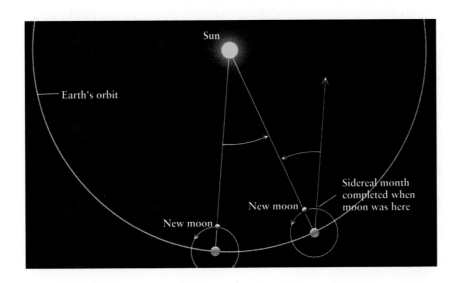

**Figure 3-5**

**The Sidereal and Synodic Months** The sidereal month is the time the Moon takes to complete one full revolution around the Earth with respect to the background stars. However, because the Earth is constantly moving along its orbit about the Sun, the Moon must travel through slightly more than 360° to get from one new moon to the next. Thus, the synodic month is longer than the sidereal month.

## box 3-2 | Looking Deeper into Astronomy

### *Some Details of the Moon's Orbit*

The Moon's orbit about the Earth is not a circle. Rather, it is an elongated curve called an *ellipse*, as sketched in figure *a*. (We discuss elliptical orbits in more detail in Chapter 4.) The Moon is said to be at *perigee* when it is nearest the Earth and at *apogee* when it is farthest from the Earth. The line connecting the points of perigee and apogee passes through the Earth and is called the **line of apsides** (figure *a*).

The average center-to-center distance from the Earth to the Moon is 384,400 km (238,900 mi). Because the orbit is not circular, the actual distance can vary by more than 50,000 km, from a minimum of 356,410 km (221,510 mi) at perigee to a maximum of 406,697 km (252,763 mi) at apogee. (The difference between the Earth-Moon distances at perigee and at apogee has been greatly exaggerated in figure *a*.) Consequently, the apparent angular size of the Moon as seen from the Earth varies over the course of a month. At perigee, the Moon has an angular diameter of 33′ 31″ (33 arcminutes and 31 arcseconds); at apogee, the angular diameter is only 29′ 22″. The average apparent size of the Moon is 31′ 5″, or just a bit more than ½°.

The average speed of the Moon along its orbit is 1.02 km/s (2280 mi/h). As seen from Earth, the Moon appears to move eastward through the constellations from one day to the next. The Moon's daily eastward progress averages 13.2° (which is 360° divided by the 27.3 days in the sidereal month). In one hour, the Moon moves slightly more than ½°, which is just a bit more than its own diameter. As a result of this motion, moonrise is about 50 minutes later from one day to the next.

The Moon's motion along its orbit is not completely uniform. It travels faster when it is near perigee and slower when near apogee, which means that the Moon's rotation is not in perfect lockstep with its orbital motion. As a result, the Moon appears to wobble back and forth around its rotation axis by as much as 7.9°. Furthermore, the Moon's rotation axis has a 7° tilt (in the same way that the rotation axis of the Earth has a 23½° tilt, shown in Figure 2-11). Because of this tilt, the Moon appears to nod its north pole toward and away from us as it goes around its orbit. Thanks to the apparent wobbling and nodding of the Moon, collectively called **libration**, at one time or another we can see more than half (about 59%) of the Moon's surface.

The mutual gravitational attraction between the Earth and the Moon keeps the Moon in orbit about the Earth. However, the Sun's gravitational pull is also constantly tugging on the Moon, affecting its orbit.

One effect of the Sun's gravitational pull is that the orientation of the Moon's orbital plane constantly changes. To simulate this change, hold your right arm straight in front of you with the palm of your hand down. Keeping your arm horizontal, tilt your hand upward a little. Keeping your arm stationary, move your tilted hand sideways. As your hand moves, the plane of the hand remains

---

**CAUTION** Both new moon and full moon occur at intervals of 29½ days. Hence, you might expect that there would be a solar eclipse every 29½ days, followed by a lunar eclipse about two weeks (half a lunar orbit) later. But in fact, there are only a few solar eclipses and lunar eclipses per year. Solar and lunar eclipses are so infrequent because the plane of the Moon's orbit and the plane of the Earth's orbit are not exactly aligned, as shown in Figure 3-6. The angle between the plane of the Earth's orbit and the plane of the Moon's orbit is about 5°. Because of this tilt, new moon and full moon usually occur when the Moon is either above or below the plane of the Earth's orbit. When the Moon is not in the plane of the Earth's orbit, the Sun, Moon, and Earth cannot align perfectly, and an eclipse cannot occur.

In order for the Sun, Earth, and Moon to be lined up for an eclipse, the Moon must lie in the same plane—the ecliptic plane—as the Earth's orbit around the Sun. Thus, whenever an eclipse occurs, the Moon appears from Earth to be on the ecliptic (which is how the ecliptic gets its name).

The plane of the Earth's orbit and the plane of the Moon's orbit intersect along a line called the **line of nodes.** The line of nodes passes through the Earth and is pointed in a particular direction in space. Eclipses can occur only if the line of nodes is pointed toward the Sun—that is, if the Sun lies on or near the line of nodes—and if, at the same time, the Moon lies on or very near the line of nodes. Only then do the Sun, Earth, and Moon lie in a straight enough line, as shown in Figure 3-7.

Anyone who wants to predict eclipses must know the orientation of the line of nodes. But the line of nodes is gradually shifting because of the gravitational pull of the Sun on the Moon, as described in Box 3-2. As a result, the line of nodes rotates slowly westward. Astronomers calculate such details to fix the dates and times of upcoming eclipses.

at the same angle above the horizontal, but the orientation of that plane changes. In the same way, the plane of the Moon's orbit maintains its 5° tilt to the plane of the ecliptic (see Figure 3-6), but the orientation of the orbit's plane changes. This causes the line of nodes, which is the line along which the plane of the Moon's orbit intersects the plane of the ecliptic, to move westward as sketched in figure *b*. This movement, called the **regression of the line of nodes**, is quite slow; it takes 18.61 years for the line of nodes to complete one full rotation.

The Sun's gravitational pull also changes the orientation of the Moon's elliptical orbit within its plane. To see what

this means, put this book on a horizontal table or desk (leave it open to this page). Keeping the book on the table, rotate it so that the text changes from right side up to upside down. The book is still in the same horizontal plane, but its orientation within that plane has changed. For the Moon, the orientation change means that the line of apsides, shown in figure *a*, moves slowly within the Moon's orbital plane. This effect, sketched in figure *c*, is called the **rotation of the Moon's orbit.** While the line of nodes is moving westward among the constellations, the Sun's gravity causes the line of apsides to move eastward. It takes 8.85 years for the line of apsides to complete one full rotation.

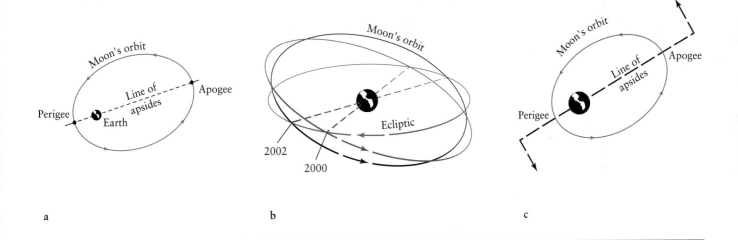

a      b      c

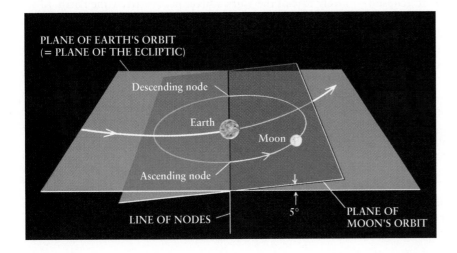

ƒigure 3-6

**The Inclination of the Moon's Orbit** The plane of the Moon's orbit (shown in red) is slightly tilted with respect to the plane of the Earth's orbit, also called the plane of the ecliptic (shown in blue). These two planes intersect along a line called the line of nodes.

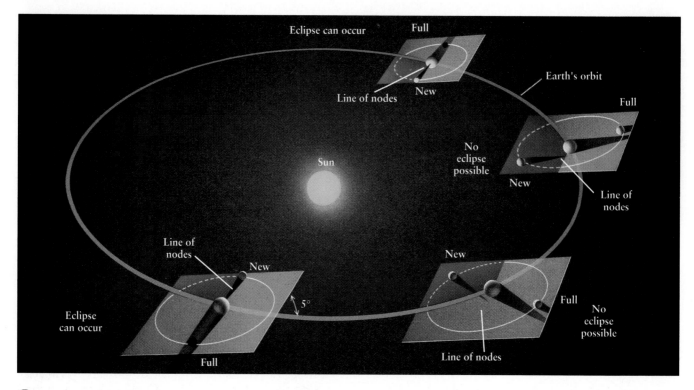

## figure 3-7

**Conditions for Eclipses** Eclipses can only take place if the Sun and Moon are very near or on the line of nodes. A solar eclipse occurs only if the Moon is very near the line of nodes at new moon; a lunar eclipse occurs only if the Moon is very near the line of nodes at full moon. Only under these circumstances can the Sun, Earth, and Moon all lie along a straight line. These conditions are met in two of the circumstances (upper right and lower left) shown here. No eclipses would be possible in the other two circumstances (shown to the right of the Sun and at the lower right), because the Moon is not near the line of nodes. In these situations, the shadow of the new Moon does not fall on the Earth and the Earth's shadow does not fall on the full Moon.

There are at least two—but never more than five—solar eclipses each year. The last year in which five solar eclipses occurred was 1935. The least number of eclipses possible (two solar, zero lunar) happened in 1969. Lunar eclipses occur just about as frequently as solar eclipses, but the maximum possible number of eclipses (lunar and solar combined) in a single year is seven.

## 3-3 Lunar eclipses can be either total, partial, or penumbral, depending on the alignment of the Sun, Earth, and Moon

There are three kinds of lunar eclipses, depending on exactly how the Moon travels through the Earth's shadow. As shown in Figure 3-8, the shadow of the Earth has two distinct parts. The **umbra** is the darkest part of the Earth's shadow, within which no portion of the Sun's surface can be seen. The **penumbra** is not quite as dark, because only part of the Sun's surface is covered by the Earth. Most people notice a lunar eclipse only if the Moon passes into the Earth's umbra, called the umbral phase of the eclipse. As this phase begins, a bite seems to be taken out of the Moon.

Figure 3-9 shows the different ways in which the Moon can pass into the Earth's shadow. If the Moon travels completely into the umbra, a **total lunar eclipse** occurs. If only part of the Moon passes through the umbra, we have a **partial lunar eclipse**. When the Moon passes through only the Earth's penumbra, we see a **penumbral eclipse**. During a penumbral eclipse, none of the lunar surface is completely shaded by the Earth, because only part of the Sun's light is blocked by the Earth. Because the Moon still looks full but only a little dimmer than usual, penumbral eclipses are easy to miss.

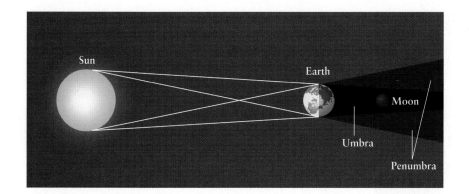

**figure 3-8**

**The Geometry of a Lunar Eclipse** People on the nighttime side of the Earth see a lunar eclipse when the Moon moves through the Earth's shadow. In the umbra, the darkest part of the shadow, the Sun is completely covered by the Earth. In the penumbra, only part of the Sun is covered by the Earth. For a total lunar eclipse to occur, the Moon must lie completely within the umbra (as shown here).

There are on average two or three lunar eclipses per year; about half of all lunar eclipses are penumbral. Table 3-1 lists all total and partial lunar eclipses from 1998 to 2004. (The only lunar eclipses in 1998 and 2002 are penumbral eclipses and are not listed in the table.)

If you were on the Moon during a total lunar eclipse, the Sun would be hidden behind the Earth. But some sunlight would be visible through the thin ring of atmosphere around the Earth, just as you can see sunlight through a person's hair if they stand with their head between your eyes and the Sun. As a result, a small amount of light reaches the Moon during a total lunar eclipse, and so the Moon does not completely disappear from the sky as seen from Earth. Most of the light that passes through the Earth's atmosphere is red, and thus the darkened Moon glows faintly in reddish hues (Figure 3-10).

Because lunar eclipses occur at full moon, and a full moon is directly opposite the Sun in the sky, a lunar eclipse can be seen at any place on Earth where the Sun is below the horizon (that is, where it is nighttime). The maximum duration of a lunar eclipse occurs when the Moon travels directly through the center of the umbra. The Moon's speed through the Earth's shadow is roughly 1 kilometer per second (2280 miles per hour), which means that totality (the period when the Moon is completely within the Earth's umbra) can last for as long as 1 hour and 42 minutes.

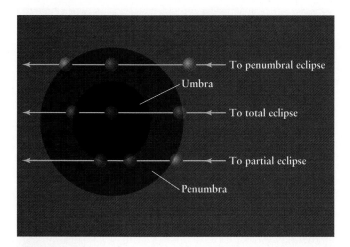

**figure 3-9**

**Various Types of Lunar Eclipse** This figure shows the Earth's umbra and penumbra at the distance of the Moon's orbit, as viewed from the Earth. A total lunar eclipse happens if the path that the Moon takes along its orbit passes completely within the umbra, as shown by the middle line. In a partial lunar eclipse (bottom line), only part of the Moon passes through the umbra, and in a penumbral lunar eclipse (top line), the Moon passes through only the penumbra.

| table 3-1 | Lunar Eclipses, 1998–2004 | |
|---|---|---|
| Date | Percentage eclipsed (100% = total) | Duration of of totality |
| 1999 July 28 | 40% | — |
| 2000 January 21 | 100% | 1 h 18 m |
| 2000 July 16 | 100% | 1 h 42 m |
| 2001 January 9 | 100% | 1 h 2 m |
| 2001 July 5 | 50% | — |
| 2003 May 16 | 100% | 52 m |
| 2003 November 9 | 100% | 24 m |
| 2004 May 4 | 100% | 1 h 16 m |
| 2004 October 28 | 100% | 1h 22m |

*Only partial and total lunar eclipses are given.*
*Penumbral eclipses are not included.*
*Dates are given in standard astronomical format—year, month, day.*

*Data from Fred Espenak, NASA/Goddard Space Flight Center*

**f igure 3-10** R I **V** U X G

**A Total Eclipse of the Moon** This photograph was taken by an amateur astronomer during the lunar eclipse of September 6, 1979. The only sunlight that reaches the Moon during a lunar eclipse is stray light that has been filtered through the Earth's atmosphere and deflected into the Earth's shadow. This light is quite red, which gives the Moon a distinctly reddish color during a total lunar eclipse. (Courtesy of M. Harms)

To see the remarkable spectacle of a total solar eclipse, you must be inside the darkest part of the Moon's shadow, also called the umbra, where the Moon completely blocks the Sun. Because the Sun and the Moon have nearly the same angular diameter, only the tip of the Moon's umbra reaches the Earth's surface, as shown in Figure 3-12. As the Earth rotates, the tip of the umbra traces an **eclipse path** across the Earth's surface. Only those locations within the eclipse path are treated to the spectacle of a total solar eclipse. Figure 3-13 shows the dark spot on the Earth's surface produced by the Moon's umbra.

Immediately surrounding the Moon's umbra is the region of partial shadow called the penumbra. As seen from this area, the Sun's surface appears only partially covered by the Moon. During a solar eclipse, the Moon's penumbra covers a large portion of the Earth's surface, and anyone standing inside the penumbra sees a **partial solar eclipse.**

The width of the eclipse path depends primarily on the Earth-Moon distance during totality. The eclipse path is widest if the Moon happens to be at **perigee,** the point in its orbit nearest the Earth (see Box 3-2). Although this nearness can produce an eclipse path up to 270 kilometers (170 miles) wide, it is usually much narrower.

Sometimes the Moon's umbra does not reach down to Earth's surface, resulting in a third type of solar eclipse, called an **annular eclipse.** This can happen when the Moon is at or near **apogee,** its farthest position from Earth, so its umbra falls short of the Earth and the Moon appears too small to cover the Sun completely. During an annular eclipse, a thin

---

| **3-4** | **Solar eclipses can be either total, partial, or annular, depending on the alignment of the Sun, Earth, and Moon** |
|---|---|

Because of a coincidence of nature, as seen from Earth the angular diameter of the Moon is almost exactly the same as the angular diameter of the Sun—about 0.5°. For this reason, the Moon "fits" over the Sun during a **total solar eclipse.** For a few minutes, the Moon blocks out the dazzling solar disk and not much else. The thin, hot gases that surround the Sun, called the **solar corona,** can be photographed and studied in detail during the brief period when an eclipse is total (Figure 3-11).

*CAUTION!* If you are fortunate enough to see a solar eclipse, keep in mind that the *only* time when it is safe to look at the Sun is during **totality,** when the solar disk is blocked by the Moon and only the solar corona is visible. Viewing this magnificent spectacle cannot harm you in any way. But if you look directly at the Sun when even a portion of its intensely brilliant disk is exposed, you can suffer permanent eye damage or even blindness. *Be careful!*

**f igure 3-11** R I **V** U X G

**A Total Eclipse of the Sun** During a total solar eclipse, the Moon completely covers the Sun's disk, allowing the solar corona to be seen. This halo of hot gases extends for hundreds of thousands of kilometers into space. This photograph was taken by an amateur astronomer during the solar eclipse of July 11, 1991, the same eclipse shown in the photograph that opens this chapter. (Courtesy of W. Dellinges)

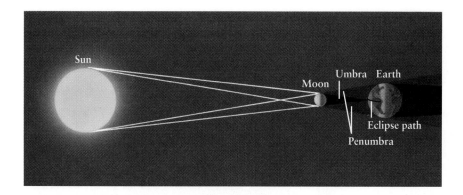

### figure 3-12

**The Geometry of a Total Solar Eclipse**
During a total solar eclipse, the tip of the Moon's umbra reaches the Earth's surface. As the Earth and Moon move along their orbits, this tip traces an eclipse path across the Earth's surface. People within the eclipse path see a total eclipse, whereas anyone within the penumbra sees only a partial eclipse of the Sun. (Compare this illustration with Figure 3-8, which shows the geometry of a total lunar eclipse.)

### figure 3-13   R I V U X G

**The Moon's Shadow on Earth** These three photographs were taken at one-hour intervals from an Earth-orbiting satellite during the total solar eclipse of July 11, 1991. The arrows indicate the Moon's umbra, which appears as a dark spot that moves across the Pacific (top picture), over the southern tip of Baja California (center), and down the isthmus of Panama (bottom). (M. Magisos and J. P. Riser, University of Arizona)

### figure 3-14   R I V U X G

**An Annular Eclipse of the Sun** This composite of six photographs taken at sunrise in Costa Rica shows the progress of an annular eclipse of the Sun on December 24, 1974. (Five photographs were made of the Sun, and one was made of the hills and sky.) Note that at mideclipse the limb, or outer edge, of the Sun is visible around the Moon. (Courtesy of D. di Cicco)

However, there is a complication. The line of nodes gradually shifts its position with respect to the background stars (see Box 3-2). It takes 346.6 days to go from one alignment of the line of nodes pointing toward the Sun to the next identical alignment. This period is called the **eclipse year.**

To predict when you will see another solar eclipse, you need to know how many whole lunar months equal some whole number of eclipse years. This information will tell you how long you will have to wait for the next virtually identical alignment of the Sun, the Moon, and the line of nodes. By trial and error, you find that 223 lunar months is the same length of time as 19 eclipse years, because

$$223 \times 29.53 \ = \ 19 \times 346.6 \ = \ 6585 \text{ days}$$

This calculation is accurate to within a few hours. A more accurate calculation gives an interval, called the **saros,** that is about one-third day longer, or 6585.3 days. Eclipses separated by the saros interval are said to form an *eclipse series.*

You might think that you and your neighbors would simply have to wait one full saros interval to go from one solar eclipse to the next. However, because of the extra one-third day, the Earth will have rotated by an extra 120° (one-third of a complete rotation) when the next solar eclipse of a particular series occurs. The eclipse path will thus be one-third of the way around the world from you. Therefore, you must wait three full saros intervals (54 years, 34 days) before the eclipse path comes back around to your part of the Earth. Figure 3-16 shows a series of six solar eclipse paths, each separated from the next by one saros interval.

There is evidence that ancient Babylonian astronomers knew about the saros interval. The discovery of the saros is more likely to have come from lunar eclipses than solar eclipses. If you are far from the eclipse path but still within the Moon's penumbra, there is a good chance that you could fail to notice a solar eclipse. Even if half the Sun is covered by the Moon, the remaining solar surface provides enough sunlight for the outdoor illumination not to be greatly diminished. In contrast, everyone on the nighttime side of the Earth can see an eclipse of the Moon unless clouds block the view.

| 3-6 | Ancient astronomers measured the size of the Earth and attempted to determine distances to the Sun and Moon |

The prediction of eclipses was not the only problem attacked by ancient astronomers. More than 2000 years ago, centuries before sailors of Columbus's era crossed the oceans, Greek astronomers were fully aware that the Earth is not flat. They had come to this conclusion using a combination of observation and logical deduction, much like modern scientists. The Greeks noted that during lunar eclipses, when the Moon passes through the Earth's shadow, the edge of the shadow is

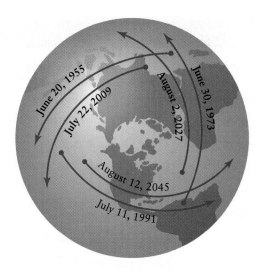

ƒigure 3-16

**An Eclipse Series** Eclipses separated by the saros interval (6565.3 days, or just over 18 calendar years) are said to form an "eclipse series," because their geometric details, such as the orientation of the line of nodes, are nearly identical. The six eclipse paths shown here are all members of the same series. The duration of totality is nearly the same for all members of a particular eclipse series.

always circular. Because a sphere is the only shape that always casts a circular shadow from any angle, they concluded that the Earth is spherical.

Around 200 B.C., the Greek astronomer Eratosthenes devised a way to measure the circumference of the spherical Earth. It was known that on the summer solstice (the first day of summer; see Section 2-5) in the town of Syene in Egypt, near present-day Aswan, the Sun shone directly down the vertical shafts of water wells. Hence, at local noon on that day the Sun was at the zenith (see Section 2-4) as seen from Syene. Eratosthenes knew that the Sun never appeared at the zenith at his home in the Egyptian city of Alexandria, which is on the Mediterranean Sea almost due north of Syene. Rather, on the summer solstice in Alexandria, the position of the Sun at local noon was about 7° south of the zenith (Figure 3-17). This angle is about one-fiftieth of a complete circle, so he concluded that the distance from Alexandria to Syene must be one-fiftieth of the Earth's circumference.

In Eratosthenes's day, the distance from Alexandria to Syene was said to be 5000 stades. Therefore, Eratosthenes found the Earth's circumference to be

$$50 \times 5000 \text{ stades} \ = \ 250,000 \text{ stades}$$

Unfortunately, no one today is sure of the exact length of the Greek unit called the stade. One guess is that the stade

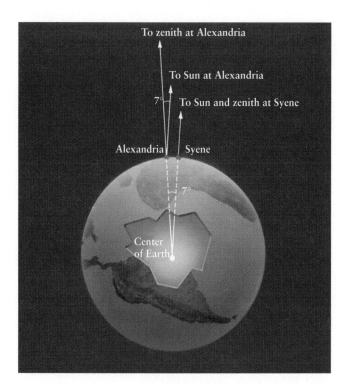

## figure 3-17

**Eratosthenes's Method of Determining the Diameter of the Earth** Around 200 B.C., Eratosthenes noticed that the Sun is about 7° south of the zenith at Alexandria when it is directly overhead at Syene. This angle is about one-fiftieth of a circle, so the distance between Alexandria and Syene must be about one-fiftieth of the Earth's circumference.

was about one-sixth of a kilometer, which would mean that Eratosthenes obtained a circumference for the Earth of about 42,000 kilometers. This is remarkably close to the modern value of 40,000 kilometers.

Eratosthenes was only one of several brilliant astronomers to emerge from the so-called Alexandrian school, which by his time had a distinguished tradition. One of the first Alexandrian astronomers, Aristarchus of Samos, had proposed a method of determining the relative distances to the Sun and Moon, perhaps as long ago as 280 B.C.

Aristarchus knew that the Sun, Moon, and Earth form a right triangle at the moment of first or last quarter moon, with the right angle at the location of the Moon (Figure 3-18).

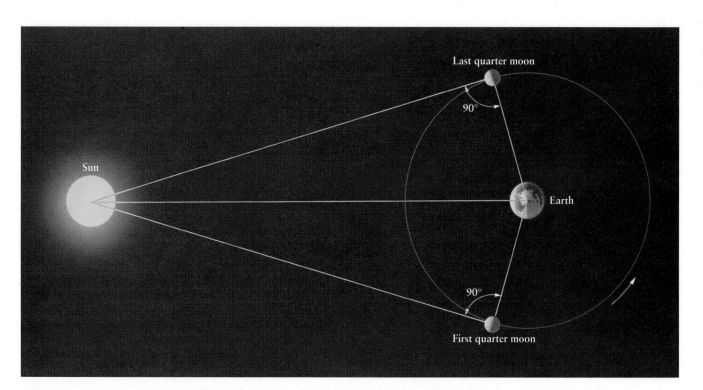

## figure 3-18

**Aristarchus's Method of Determining Distances to the Sun and Moon** Aristarchus knew that the Sun, Moon, and Earth form a right triangle at first and last quarter phases. Using geometrical arguments, he calculated the relative lengths of the sides of these triangles, thereby obtaining the distances to the Sun and Moon.

He estimated that, as seen from Earth, the angle between the Moon and the Sun at first and third quarters is 87°, or 3° less than a right angle. Using the rules of geometry, Aristarchus concluded that the Sun is about 20 times farther from us than is the Moon. We now know that Aristarchus erred in measuring angles and that the average distance to the Sun is about 390 times larger than the average distance to the Moon. It is nevertheless impressive that people were trying to measure distances across the solar system more than 2000 years ago.

Using lunar eclipses, Aristarchus also made an equally bold attempt to determine the relative sizes of the Earth, Moon, and Sun. From his observations of how long the Moon takes to move through the Earth's shadow, Aristarchus estimated the diameter of the Earth to be about three times larger than the diameter of the Moon. To determine the diameter of the Sun, Aristarchus simply pointed out that the Sun and the Moon have the same angular size in the sky. Therefore, their diameters must be in the same proportion as their distances. In other words, because Aristarchus believed the Sun to be 20 times farther from the Earth than the Moon, he concluded that the Sun must be 20 times larger than the Moon. Once Eratosthenes had measured the Earth's circumference, astronomers of the Alexandrian school

| Table 3-3 | A Comparison of Ancient and Modern Astronomical Measurements | |
|---|---|---|
| | Ancient measure (km) | Modern measure (km) |
| Earth's diameter | 13,000 | 12,756 |
| Moon's diameter | 4,300 | 3,476 |
| Sun's diameter | $9 \times 10^4$ | $1.39 \times 10^6$ |
| Earth-Moon distance | $4 \times 10^5$ | $3.84 \times 10^5$ |
| Earth-Sun distance | $10^7$ | $1.50 \times 10^8$ |

could estimate the diameters of the Sun and Moon as well as their distances from Earth.

Table 3-3 summarizes some ancient and modern measurements of the sizes of Earth, the Moon, and the Sun and the distances between them. Some of these ancient measurements are far from the modern values. Yet the achievements of our ancestors still stand as impressive applications of observation and reasoning and important steps toward the development of the scientific method.

## Key Words

*Terms preceded by an asterisk (\*) are discussed in the Boxes.*

annular eclipse, p. 60

apogee, p. 60

eclipse, p. 55

eclipse path, p. 60

eclipse year, p. 64

first quarter moon, p. 51

full moon, p. 51

last quarter moon, p. 51

*libration, p. 56

*line of apsides, p. 56

line of nodes, p. 56

lunar eclipse, p. 55

lunar month, p. 55

lunar phases, p. 51

new moon, p. 51

partial lunar eclipse, p. 58

partial solar eclipse, p. 60

penumbra (*plural* penumbrae), p. 58

penumbral eclipse, p. 58

perigee, p. 60

*regression of the line of nodes, p. 57

*rotation of the Moon's orbit, p. 57

saros, p. 64

sidereal month, p. 55

solar corona, p. 60

solar eclipse, p. 55

synchronous rotation, p. 53

synodic month, p. 55

totality, p. 60

total lunar eclipse, p. 58

total solar eclipse, p. 60

umbra (*plural* umbrae), p. 58

waning crescent moon, p. 51

waning gibbous moon, p. 51

waxing crescent moon, p. 51

waxing gibbous moon, p. 51

## Key Ideas

**Lunar Phases:** The phases of the Moon occur because light from the Moon is actually reflected sunlight. As the relative positions of the Earth, the Moon, and the Sun change, we see more or less of the illuminated half of the Moon.

**Length of the Month:** Two types of months are used in describing the motion of the Moon.

• With respect to the stars, the Moon completes one orbit around the Earth in a sidereal month, averaging 27.3 days.

• The Moon completes one cycle of phases (one orbit around the Earth with respect to the Sun) in a synodic month (also called a lunar month), averaging 29.5 days.

**The Moon's Orbit:** The plane of the Moon's orbit is tilted by about 5° from the plane of the Earth's orbit, or ecliptic.

• The line of nodes is the line where the planes of the Moon's orbit and the Earth's orbit intersect. The gravitational pull of the Sun gradually shifts the orientation of the line of nodes with respect to the stars.

**Conditions for Eclipses:** During a lunar eclipse, the Moon moves through the Earth's shadow. During a solar eclipse, the Earth passes through the Moon's shadow.

• Lunar eclipses occur at full moon, while solar eclipses occur at new moon.

• Either type of eclipse can only occur when the Sun and Moon are both on or very near the line of nodes. If this condition is not met, the Earth's shadow cannot fall on the Moon and the Moon's shadow cannot fall on the Earth.

**Umbra and Penumbra:** The shadow of an object has two parts: the umbra, within which the light source is completely blocked, and the penumbra, where the light source is only partially blocked.

**Lunar Eclipses:** Depending on the relative positions of the Sun, Moon, and Earth, lunar eclipses may be total (the Moon passes completely into the Earth's umbra), partial (only part of the Moon passes into the Earth's umbra), or penumbral (the Moon passes only into the Earth's penumbra).

**Solar Eclipses:** Solar eclipses may be total, partial, or annular.

• During a total solar eclipse, the Moon's umbra traces out an eclipse path over the Earth's surface as the Earth rotates. Observers outside the eclipse path but within the penumbra see only a partial solar eclipse.

• During an annular eclipse, the umbra falls short of the Earth, and the outer edge of the Sun's disk is visible around the Moon at mideclipse.

**Predicting Eclipses:** Using an interval called the saros, it is possible to group total solar eclipses into series and to predict the time of the next eclipse in each series.

**The Moon and Ancient Astronomers:** Ancient astronomers such as Aristarchus and Eratosthenes made great progress in determining the sizes and relative distances of the Earth, the Moon, and the Sun.

## Review Questions

1. Why does the Moon exhibit phases?

2. A common misconception about the Moon's phases is that they are caused by the Earth's shadow. Use Figure 3-3 to explain why this not correct.

3. Suppose it is the first day of spring in the northern hemisphere. What is the phase of the Moon if the Moon is located at (a) the vernal equinox; (b) the summer solstice; (c) the autumnal equinox; (d) the winter solstice? (*Hint:*

Make a drawing showing the relative positions of the Sun, Earth, and Moon. Compare to Figure 3-3.)

4. (a) If you lived on the Moon, would you see the Sun rise and set, or would it always be in the same place in the sky? Explain. (b) Would you see the Earth rise and set, or would it always be in the same place in the sky? Explain.

5. Is the far side of the Moon (the side that can never be seen from Earth) the same as the dark side of the Moon? Explain.

6. What is the difference between a sidereal month and a synodic month? Which is longer? Why?

7. What is the difference between the umbra and the penumbra of a shadow?

8. Why isn't there a lunar eclipse at every full moon and a solar eclipse at every new moon?

9. What is the line of nodes? Why is it important to the subject of eclipses?

10. What is a penumbral eclipse of the Moon? Why do you suppose that it is easy to overlook such an eclipse?

11. The maximum duration of totality of a lunar eclipse is 1 hour, 42 minutes. But none of the lunar eclipses listed in Table 3-1 last this long. Why is this?

12. If you were looking at the Earth from the side of the Moon that faces the Earth, what would you see (a) during a total lunar eclipse? (b) during a total solar eclipse?

13. Which type of eclipse—lunar or solar—do you think most people on Earth have seen? Why?

14. Can a total eclipse of the Sun be followed three months later by a lunar eclipse? Why or why not?

15. How is an annular eclipse of the Sun different from a total eclipse of the Sun? What causes this difference?

16. Can one ever observe an annular eclipse of the Moon? Why or why not?

17. What is the saros? How did ancient astronomers use it to predict eclipses?

18. How did Eratosthenes determine the size of the Earth?

19. How did Aristarchus try to estimate the distance from the Earth to the Sun and Moon?

## Advanced Questions

### Problem-solving tips and tools:

It is helpful to know that the saros interval of 6585.3 days equals 18 years and $11\frac{1}{3}$ days if the interval includes four leap years, but is 18 years and $10\frac{1}{3}$ days if it includes five leap years. The average angular speed of the Moon along its orbit (that is, how many degrees around its orbit the Moon travels per day) can be estimated by dividing 360° by the length of a sidereal month.

**20.** How many more sidereal months than synodic months are there in a year? Explain.

**21.** If the Moon revolved about the Earth in the same orbit but in the opposite direction, would the synodic month be longer or shorter than the sidereal month? Explain your reasoning.

**22.** A "blue moon" is the second full moon within the same calendar month. There is usually only one full moon within a calendar month, which is why the phrase "once in a blue moon" means "hardly ever." Why are blue moons so rare? Are there any months of the year in which it would be impossible to have two full moons? Explain your answer.

**23.** You are watching a lunar eclipse from some place on the Earth's night side. Will you see the Moon enter the Earth's shadow from the east or from the west? Explain your reasoning.

**24.** The total lunar eclipse of September 27, 1996, was visible from North America. Was it also visible from India, on the opposite side of the Earth? Explain your reasoning.

**25.** On February 26, 1998, residents of Central America were treated to a total solar eclipse. (**a**) When and over what part of the world will the next eclipse of that series occur? Explain. (**b**) When might you next expect an eclipse of that series to be visible from Central America? Explain.

**26.** Figure 3-13 shows that during a total solar eclipse, the Moon's umbra moves in a generally eastward direction across the Earth's surface. This is true for total solar eclipses in general. Use a drawing to explain why the motion is eastward, not westward.

**27.** Just as the distance from the Earth to the Moon varies somewhat as the Moon orbits the Earth, the distance from the Sun to the Earth changes as the Earth orbits the Sun. The Earth is closest to the Sun at its *perihelion;* it is farthest from the Sun at its *aphelion.* In order for a total solar eclipse to have the maximum duration of totality, should the Earth be at perihelion or aphelion? Assume that the Earth-Moon distance is the same in both situations. As part of your explanation, draw two pictures like Figure 3-12, one with the Earth relatively close to the Sun and one with the Earth relatively far from the Sun.

**28.** During an occultation, or "covering up," of Jupiter by the Moon, an astronomer notices that it takes the Moon's edge 90 seconds to cover Jupiter's disk completely. If the Moon's motion is assumed to be uniform and the occultation was "central" (that is, center over center), find the angular diameter of Jupiter. (*Hint:* Assume that Jupiter does not appear to move in the sky during this brief 90-second interval. You will need to convert the Moon's angular speed from degrees per day to arcseconds per second.)

## Discussion Questions

**29.** Describe the cycle of lunar phases that would be observed if the Moon moved about the Earth in an orbit perpendicular to the plane of the Earth's orbit. Would it be possible for both solar and lunar eclipses to occur under these circumstances? Explain your reasoning.

**30.** How would a lunar eclipse look if the Earth had no atmosphere? Explain your reasoning.

**31.** In his novel *King Solomon's Mines,* author H. Rider Haggard described a total solar eclipse that was reportedly seen in both South Africa and in the British Isles. Is such an eclipse possible? Why or why not?

**32.** Why do you suppose that total solar eclipse paths fall more frequently on oceans than on land? (You may find it useful to look at Figure 3-15.)

**33.** Examine Figure 3-15, which shows all of the total solar eclipses from 1997 to 2020. What are the chances that you might be able to travel to one of the eclipse paths? Do you think you might go through your entire life without ever seeing a total eclipse of the Sun?

## Observing Projects

**34.** Observe the Moon on each clear night over the course of a month. On each night, note the Moon's location among the constellations and record that location on a star chart that also shows the ecliptic. After a few weeks, your observations will begin to trace the Moon's orbit. Identify the orientation of the line of nodes by marking the points where the Moon's orbit and the ecliptic intersect. On what dates is the Sun near the nodes marked on your star chart? Compare these dates with the dates of the next solar and lunar eclipses.

**35.** It is quite possible that a lunar eclipse will occur while you are taking this course. Look up the date of the next lunar eclipse in Table 3-1, in the current issue of a reference such as the *Astronomical Almanac* or *Astronomical Phenomena,* or on the World Wide Web. Then make arrangements to observe the next lunar eclipse. You can observe the eclipse with the naked eye, but binoculars or a small telescope will enhance your viewing experience. If the eclipse is partial or total, note the times at which the Moon enters and exits the Earth's umbra. If the eclipse is penumbral, can you see any changes in the Moon's brightness as the eclipse progresses?

## Where to Learn More

*Books and magazine articles*

Di Cicco, D., "The Great Eclipse of 1991." *Sky & Telescope,* December 1991. A collection of fascinating observer reports and magnificent photographs from the July 11, 1991, total solar eclipse, which had an unusually long duration of totality. Other articles in this same issue describe some of the scientific results from this eclipse and some of the crazy behavior that the eclipse induced in skywatchers.

Harrington, P. S. *Eclipse!* Wiley, 1997. This practical guide to observing lunar and solar eclipses includes full details on all eclipses through 2017, with advice about where on Earth the best view of each eclipse can be expected.

Hoskin, M., ed. *The Cambridge Illustrated History of Astronomy.* Cambridge University Press, 1997. This very complete, lucidly written one-volume history traces the story of astronomical thought from prehistory to the present day, including attempts to predict eclipses.

Schaeffer, B. E. "Solar Eclipses That Changed the World." *Sky & Telescope*, May 1994. As this interesting article describes, religious and political leaders throughout history have used solar eclipses to amaze and terrify their followers.

W *World Wide Web*

The web sites for the magazines *Sky & Telescope* (**http://www.skypub.com/**) and *Astronomy* (**http://www.astronomy.com/**) have information about upcoming lunar and solar eclipses, as do the magazines themselves. For complete information about eclipses past and present, including information about where to go to see upcoming solar eclipses, see the web site maintained by Fred Espenak of NASA's Goddard Space Flight Center (**http://planets.gsfc.nasa.gov/eclipse/eclipse.html**). A number of other sites have information about the astronomical ideas and beliefs of ancient cultures. A useful and extensive list of these sites is maintained at the Max Planck Institute for Extraterrestrial Physics in Garching, Germany (**http://www.gamma.mpgarching.mpg.de/~hcs/history.html**).

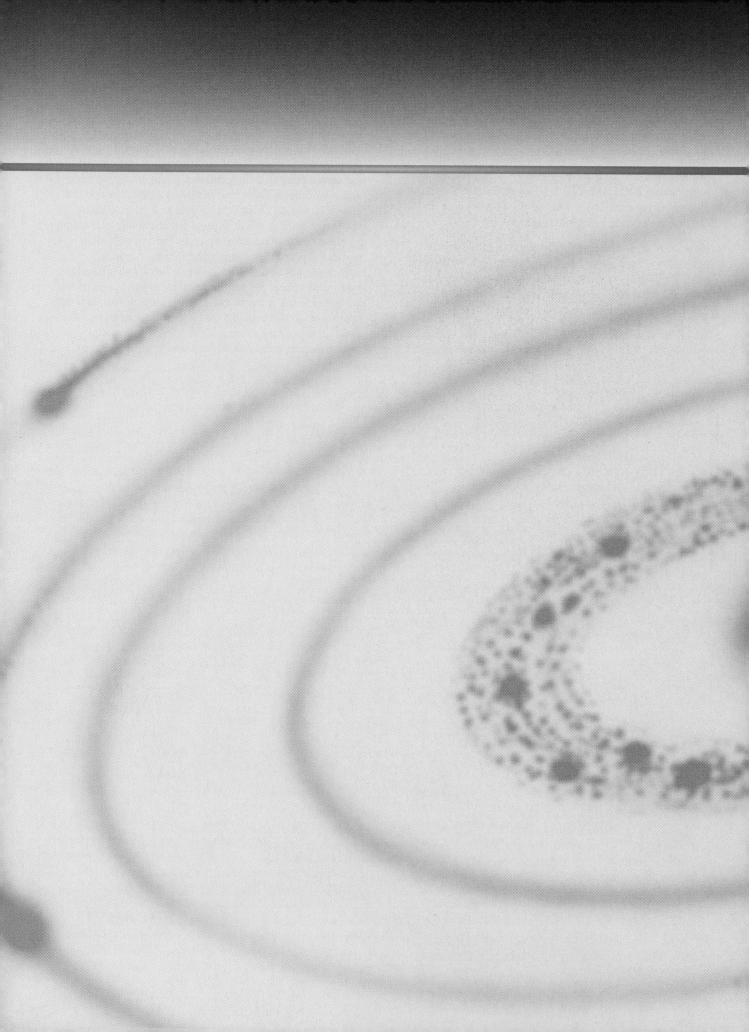

# Gravitation and the Motions of the Planets

**In this chapter you will find the answers to the following questions:**

4-1 How did ancient astronomers explain the motions of the planets?

4-2 Why did Copernicus think that the Earth and the other planets go around the Sun?

4-3 How did Tycho Brahe attempt to test the ideas of Copernicus?

4-4 What paths do the planets follow as they move around the Sun?

4-5 What did Galileo see in his telescope that confirmed that planets orbit the Sun?

4-6 Do the same fundamental laws of nature that apply on Earth also explain the motions of the planets?

4-7 Why don't the planets fall into the Sun?

**Apollo 11 Leaving the Moon** This historic photograph shows the lunar module *Eagle* in orbit around the Moon after completing the first successful manned lunar landing in July 1969. The photograph was taken from the command module *Columbia*, in which the astronauts returned to Earth. All of the orbital maneuvers to dock the *Eagle* and the *Columbia* and then to set course for Earth were based on Newtonian mechanics and Newton's law of gravity. Astronomers use these same physical principles to understand a wide range of phenomena, from the orbits of the planets to the rotation of an entire galaxy. (NASA)

The entire celestial sphere of stars appears to rotate around the Earth once a day, and the Sun and Moon both appear to move relative to the stars from one day to the next. In the preceding two chapters, we saw how to understand these apparent motions: They are the result of the Earth rotating on its axis, the Earth orbiting around the Sun, and the Moon orbiting the Earth.

But it is by no means obvious that the Earth rotates on its axis and moves around the Sun. Indeed, our daily experience strongly suggests that it does not. The steadiness of the ground could easily lead us to believe that the Earth is stationary. The daily rising and setting of the Sun, Moon, and stars might make us think that the entire cosmos revolves around the Earth.

In fact, until a few centuries ago it was almost universally believed that the Earth was a fixed and immovable object at the center of the universe. Since ancient times, scholars had attempted to understand the complicated ways in which the celestial objects they called planets move in the sky. But it was not until the seventeenth century that scientists discovered the true motions of the planets and showed that the Earth is itself one of several planets orbiting the Sun.

In this chapter we learn what led to this revolutionary change in the way that humans view the universe. It is a story of several remarkable individuals making painstaking observations, laborious calculations, and great intellectual leaps, aided by the courage to challenge accepted ideas. A final act of this drama was Isaac Newton's discovery of why the planets move the way that they do. He created a precise mathematical description of the force of gravity, which holds the planets in their orbits. In addition, Newton formulated the fundamental laws of physics. These laws govern the motions of objects here on Earth as well as the motions of the planets.

In Chapters 7 through 17 we will study the properties of the planets and other bodies that make up our solar system. For now, our focus is on understanding how these bodies move and on the forces that control their motions.

---

| 4-1 | Ancient astronomers invented geocentric cosmogonies to explain planetary motions |

Since the dawn of civilization, scholars have attempted to explain the nature of the universe. In their approach to the problem, the ancient Greeks were the first to use the principles that still guide scientists today. For example, around 550 B.C., Pythagoras and his followers put forth the idea that nature can be described with mathematics. About 200 years later, Aristotle asserted that the universe is governed by physical laws. In short, the Greeks were the first to believe that the universe can be described and understood logically.

As they attempted to describe the universe, most Greeks assumed that the Sun, the Moon, the stars, and the planets revolve about a stationary Earth. A theory of the Earth's place in the universe is called a **cosmogony**, and a theory in which the Earth is at the center of the universe is called a **geocentric cosmogony**. (By contrast, *cosmology* deals with the origin and evolution of the entire universe, not just that of the Earth and planets. We will study cosmology in Chapters 28 and 29.)

In the geocentric cosmogony of the ancient Greeks, the entire celestial sphere rotated around the stationary Earth once a day. The stars were fixed in their positions on the rotating celestial sphere, which explained the diurnal motion of the stars. The Sun and Moon both participated in this daily rotation of the sky, which explained their rising and setting motions. To explain why the Sun and Moon both move slowly with respect to the stars, the ancient Greeks imagined that both of these objects orbit around the Earth.

ANALOGY Picture two children (representing the Sun and the Moon) who walk at different speeds around a rotating merry-go-round (representing the celestial sphere) and its wooden horses (which represent the stars). The children rotate along with the merry-go-round, but also change their positions with respect to the wooden horses (Figure 4-1*a*). In the same way, the Greeks imagined that the Sun and Moon turned with the celestial sphere and yet changed their positions with respect to the stars (Figure 4-1*b*).

The ancient Greeks and other cultures of that time knew of five planets: Mercury, Venus, Mars, Jupiter, and Saturn. These planets are quite obvious because they are bright objects in the night sky. For example, when Venus is at its maximum brilliancy, it is 16 times brighter than the brightest star. (By contrast, Uranus, Neptune and Pluto are quite dim and were not discovered until after the invention of the telescope.)

Like the Sun and Moon, every planet rises in the east and sets in the west once a day. And like the Sun and Moon, from night to night the planets slowly move on the celestial sphere, that is, with respect to the background of stars. However, the character of this motion on the celestial sphere is quite different for the planets. Both the Sun and the Moon always move from west to east on the celestial sphere, that is, opposite the direction in which the celestial sphere appears to rotate. Furthermore, the Sun and the Moon each move at relatively constant speeds around the celestial sphere (the Moon's speed is faster than that of the Sun, because it travels all the way around the celestial sphere in about a month while the Sun takes an entire year). But each of the planets appears to wander back and forth on the celestial sphere, moving from west to east at some times but at other times from east to west, and as it wanders its speed on the celestial sphere can vary substantially. Indeed, the name *planet* is well-deserved; it comes from a Greek term meaning "wanderer." This distinctive behavior is depicted in Figure 4-2, which shows how Mars will move with respect to the background of stars during the year 2003. In our merry-go-round analogy, the Greeks imagined the planets as children walking around the rotating merry-go-round but who keep changing their minds about which direction to walk!

As seen from Earth, the planets wander primarily across the 12 constellations of the zodiac. (Figure 4-2 shows that during 2003, Mars moves through Capricorn, Aquarius, and Pisces.) As mentioned in Section 2-5, these constellations encircle the sky in a continuous band centered on the ecliptic, which is the apparent path of the Sun across the celestial sphere. If you observe a planet as it travels across the zodiac from night to night, you will find that the planet usually moves slowly eastward against the background stars. This eastward progress is called **direct motion.** Occasionally, however, the planet will seem to stop and then back up for several weeks or months. This occasional westward movement is called **retrograde motion.** Both direct and retrograde motions are much slower than the apparent daily rotation of the sky caused by the Earth's rotation. Hence they are best detected by mapping the position of a planet against the background stars from night to night over a long period. This leads to maps such as that shown in Figure 4-2. Mars undergoes retrograde motion about every 22½ months; all the other planets go through retrograde motion, but at different intervals.

Explaining the nonuniform motions of the five planets was one of the main challenges facing the astronomers of antiquity. The Greeks developed many theories to account for retrograde motion and the loops that the planets trace out against the background stars. One of the most successful and enduring models was originated by Hipparchus in the second century B.C. and expounded upon by Ptolemy, the last of the great Greek astronomers, during the second century A.D. It is usually called the **Ptolemaic system.** The basic concept is sketched in Figure 4-3*a*. Each planet is assumed to move in a small circle called an **epicycle,** whose center in turn moves

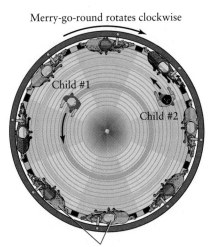

Merry-go-round rotates clockwise

Child #1

Child #2

Wooden horses fixed on merry-go-round

**a**   A rotating merry-go-round

Celestial sphere rotates to the west

Sun

Moon

Earth

Stars fixed on celestial sphere

**b**   The Greek geocentric cosmogony

ƒigure 4-1

**A Merry-go-round Analogy** **(a)** Imagine a merry-go-round that rotates clockwise as seen from above. Two children walk around the merry-go-round in a counterclockwise direction, each walking at a different speed from one of the merry-go-round's wooden horses to another. **(b)** The ancient Greeks pictured the universe as a celestial sphere that rotated to the west around a stationary Earth. In this picture, the stars rotate along with the celestial sphere just as the wooden horses rotate along with the merry-go-round in (a). The Sun and Moon are analogous to the two children; they both move eastward within the celestial sphere but at different speeds.

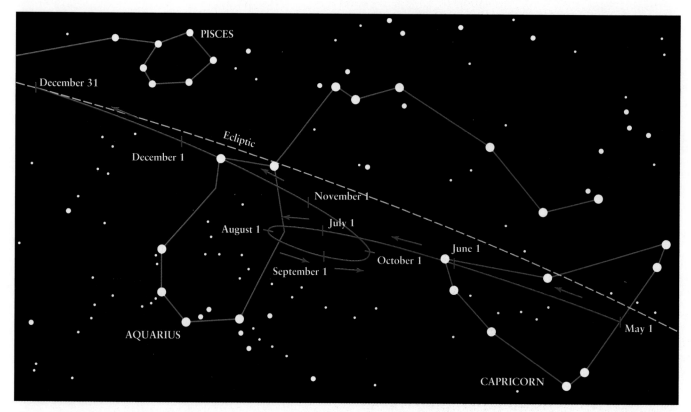

**figure 4-2**

**The Path of Mars in 2003** During the second half of 2003, Mars will move across the constellations Capricorn, Aquarius, and Pisces. Mars's motion will be direct (from west to east, or from right to left in this figure) most of the time but will be retrograde (from east to west, or from left to right in this figure) during August and September. Notice that the speed with which Mars moves relative to the stars is not constant. Mars moves farther from May 1 to June 1 than it does from June 1 to July 1, which means that it moves faster during May than during June.

in a larger circle, called a **deferent,** which is centered approximately on the Earth. Both the epicycle and deferent rotate in the same direction, shown as counterclockwise in Figure 4-3a.

As viewed from Earth, the epicycle moves eastward along the deferent. Most of the time the eastward motion of the planet on its epicycle adds to the eastward motion of the epicycle on the deferent (Figure 4-3b). Then the planet is seen to be in direct (eastward) motion against the background stars. However, when the planet is on the part of its epicycle nearest Earth, the motion of the planet along the epicycle is opposite to the motion of the epicycle along the deferent. The planet therefore appears to slow down and halt its usual eastward movement among the constellations, and actually goes backward in retrograde (westward) motion for a few weeks or months (Figure 4-3c). Thus, the concept of epicycles and deferents enabled Greek astronomers to explain the retrograde loops of the planets.

Using the wealth of astronomical data in the library at Alexandria, including records of planetary positions for hundreds of years, Ptolemy deduced the sizes and rotation rates

of the epicycles and deferents needed to reproduce the recorded paths of the planets. After years of tedious work, Ptolemy assembled his calculations into 13 volumes, collectively called the *Almagest*. His work was used to predict the positions and paths of the Sun, Moon, and planets with unprecedented accuracy. In fact, the *Almagest* was so successful that it became the astronomer's bible, and for more than 1000 years, the Ptolemaic system endured as a useful description of the workings of the heavens.

There were, however, certain aspects of the Ptolemaic system that were less than satisfactory. Perhaps its greatest problem was a philosophical one: It treated each planet independent of the others. There was no rule in the *Almagest* that related the size and rotation speed of one planet's epicycle and deferent to the corresponding sizes and speeds for other planets. This made the Ptolemaic cosmogony very unsatisfying to many astronomers of the Middle Ages. They felt that a correct model of the universe should be based on a simple set of underlying principles that applied to all of the planets.

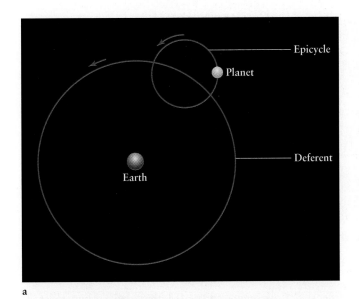

a

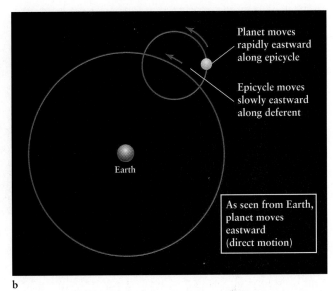

b

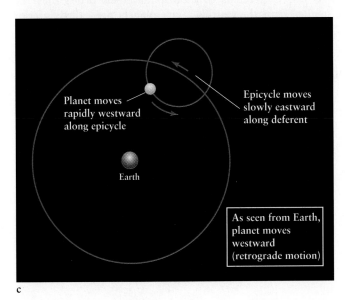

c

## figure 4-3

**A Geocentric Explanation of Retrograde Motion** **(a)** In the geocentric cosmogony of the ancient Greeks, each planet was imagined to move along an epicycle, which in turn moves along a deferent centered approximately on the Earth. The planet moves along the epicycle at a faster speed than the epicycle moves along the deferent. **(b)** At most times the eastward motion of the planet on the epicycle adds to the eastward motion of the epicycle on the deferent, and the planet moves eastward in direct motion as seen from Earth. **(c)** When the planet is on the inside of the deferent, its motion along the epicycle is westward. Because this motion is faster than the eastward motion of the epicycle on the deferent, the planet appears from Earth to be moving westward in retrograde motion.

The idea that simple, straightforward explanations of phenomena are most likely to be correct is called **Occam's razor,** after the fourteenth-century English philosopher who first expressed it. (The word "razor" refers to shaving extraneous details from an argument or explanation.) Although Occam's razor has no proof or verification, it appeals to the scientist's sense of beauty and elegance, and it has helped lead to the simple and powerful laws of nature that scientists use today.

Half a century after the death of William of Occam, the Polish astronomer Nicolaus Copernicus began to construct a Sun-centered system model of the universe. This model explained planetary motions in a more natural way than Ptolemaic cosmogony and helped lay the foundations of modern physical science.

| 4-2 | Nicolaus Copernicus devised the first comprehensive heliocentric cosmogony |
|---|---|

Imagine riding on a fast racehorse. As you pass a slowly walking pedestrian, he appears to move backward, even though he is traveling in the same direction as you and your horse.

In the third century B.C., this sort of observation inspired the Greek astronomer Aristarchus to suggest a straightforward explanation of retrograde motion. In Aristarchus's model, all the planets, including Earth, revolve about the Sun. Hence, this model is a **heliocentric** (Sun-centered) **cosmogony.** Different planets take different lengths of time to complete an

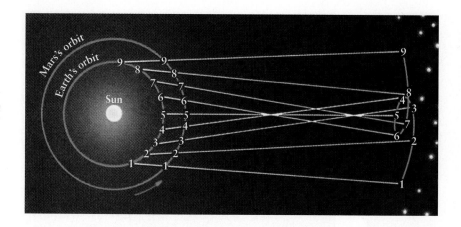

**figure 4-4**

**A Heliocentric Explanation of Retrograde Motion** In the heliocentric cosmogony of Aristarchus, the Earth and the other planets orbit the Sun. The Earth travels around the Sun more rapidly than Mars. Consequently, as the Earth overtakes and passes this slower-moving planet, Mars appears for a few months (from points 4 through 6) to fall behind and move backward with respect to the background of stars.

orbit, so from time to time one planet will overtake another, just as a fast-moving horse overtakes a person on foot. When the Earth overtakes Mars, for example, Mars appears to move backward in retrograde motion, as shown in Figure 4-4. Thus, in the heliocentric picture, the occasional retrograde motion of a planet is merely the result of our own motion.

As described in Section 3-6, Aristarchus demonstrated that the Sun is bigger than the Earth (see Table 3-3). This made it sensible to imagine the Earth orbiting the larger Sun. He also imagined that the Earth rotated on its axis once a day, thereby explaining the daily rising and setting of the Sun, Moon, and planets and the diurnal motions of the stars. To explain why the apparent motions of the planets never take them far from the ecliptic, Aristarchus proposed that the orbits of the Earth and all the planets must lie in nearly the same plane. (Recall from Section 2-5 that the ecliptic is the projection onto the celestial sphere of the plane of the Earth's orbit.)

This heliocentric cosmogony is conceptually much simpler than an Earth-centered system, such as that of Ptolemy, with all its "circles upon circles." In Aristarchus's day, however, the idea of an orbiting, rotating Earth seemed inconceivable, given the Earth's apparent stillness and immobility, and his model never found broad acceptance.

Almost 2000 years elapsed before someone had both the insight and determination to work out the details of a heliocentric cosmogony. That person was a Polish lawyer, physician, economist, canon of the church, and artist named Nicolaus Copernicus (Figure 4-5). Especially gifted in mathematics, and influenced by the ideas of Aristarchus, Copernicus turned his attention to astronomy in the early 1500s.

Copernicus realized that a heliocentric model had several advantages beyond providing a natural explanation of retrograde motion. In the Ptolemaic system, the arrangement of the planets—that is, which are close to the Earth and which are far away—was chosen in large part by guesswork. But using a heliocentric model, Copernicus could determine the arrangement of the planets without ambiguity. Because Mercury and Venus are always observed fairly near the Sun in

the sky, Copernicus concluded that their orbits must be smaller than the Earth's. Such planets are called **inferior planets** (Figure 4-6). The other visible planets—Mars, Jupiter, and Saturn—are sometimes seen on the side of the celestial sphere opposite the Sun, so that these planets appear high above the horizon at midnight (when the Sun is far below the horizon). When this happens, the Earth must lie between the Sun and these planets. Copernicus therefore concluded that the orbits of Mars, Jupiter, and Saturn must be larger than the Earth's orbit. Hence, these planets are called **superior planets.** Three additional planets (Uranus,

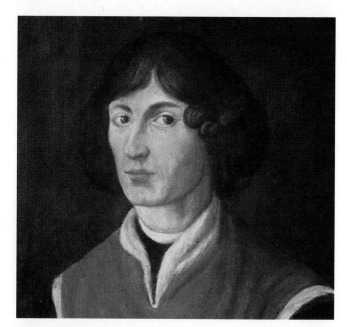

**figure 4-5**

**Nicolaus Copernicus (1473–1543)** Copernicus was the first person to work out the details of a heliocentric system in which the planets, including the Earth, orbit the Sun. (E. Lessing/Magnum)

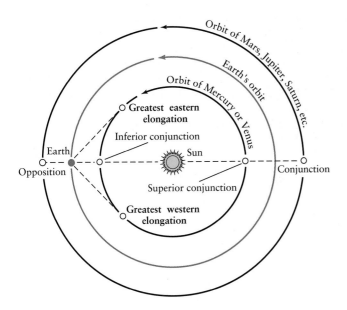

## figure 4-6

**Planetary Orbits and Configurations** The size of a planet's orbit and the planet's location on its orbit together determine when and where in the sky the planet can be seen from Earth. Planets whose orbits are smaller than the Earth's orbit (Mercury and Venus) are called *inferior planets*. Because these planets move in such tight orbits, they are never seen very far from the Sun. The greatest angle between the Sun and an inferior planet occurs when the planet is at the points called *greatest eastern elongation*, when the planet is visible after sunset, and *greatest western elongation*, when the planet is visible before sunrise. When an inferior planet is closest to us (at *inferior conjunction*) or farthest away (at *superior conjunction*), it is in the same direction of the sky as the Sun and is only above the horizon during the day. Planets whose orbits are larger than that of the Earth are called *superior planets*. When a superior planet is at *opposition*, it is in the part of the sky opposite the Sun and can be seen around midnight. When the planet is at *conjunction*, it is in the same part of the sky as the Sun; hence, the planet is above the horizon in the daytime and cannot be seen at night. In this figure you are looking down onto the plane of the Earth's orbit from a point far above the Earth's northern hemisphere. The orbits of all the planets lie in nearly the same plane. The Sun and the planets are not shown to scale.

Neptune, and Pluto), discovered after the telescope was invented (and after the death of Copernicus), can also be seen in the midnight sky. Thus, these are also superior planets with orbits larger than the Earth's.

The heliocentric model also explains why planets appear in different parts of the sky on different dates. Both of the inferior planets (Mercury and Venus) go through cycles: The planet is seen in the west after sunset for several weeks or months, then for several weeks or months in the east before sunrise, and then in the west after sunset again. Figure 4-6

shows the reason for this cycle. When Mercury or Venus is visible after sunset, it is near **greatest eastern elongation.** (In general, the angle between the Sun and a planet as viewed from Earth is called the planet's **elongation.**) The planet's position in the sky is as far east of the Sun as possible, so it appears above the western horizon after sunset (that is, to the east of the Sun) and is often called an "evening star." At **greatest western elongation,** Mercury or Venus is as far west of the Sun as it can possibly be and rises before the Sun, gracing the predawn sky as a "morning star" in the east. When Mercury or Venus is at **inferior conjunction,** it is between us and the Sun, and it is moving from the evening sky into the morning sky. At **superior conjunction,** when the planet is on the opposite side of the Sun, it is moving back into the evening sky.

A superior planet such as Mars, whose orbit is larger than the Earth's, is best seen in the night sky when it is at **opposition.** At this point the planet is in the part of the sky opposite the Sun and is highest in the sky at midnight. This is also when the planet appears brightest, because it is closest to us. But when a superior planet like Mars is located behind the Sun at **conjunction,** it is above the horizon during the daytime and so is not well placed for nighttime viewing.

The Ptolemaic cosmogony has no simple rules relating the motion of one planet to another. But Copernicus showed that there *are* such rules in a heliocentric model. In particular, he found a correspondence between the time a planet takes to complete one orbit—that is, its **period**—and the size of the orbit. Determining the period of a planet takes some care, because the Earth, from which we must make the observations, is also moving. Realizing this, Copernicus was careful to distinguish between two different periods of each planet. The **synodic period** is the time that elapses between two successive identical configurations, as seen from Earth—from one opposition to the next, for example, or from one conjunction to the next. The **sidereal period** is the true orbital period of a planet, the time it takes the planet to complete one full orbit of the Sun relative to the stars.

The synodic period of a planet can be determined by observing the sky, but the sidereal period must be calculated. Copernicus figured out how to do this (Box 4-1). The results for all of the planets are shown in Table 4-1.

To find a relationship between the sidereal period of a planet and the size of its orbit, Copernicus still had to determine the relative distances of the planets from the Sun. He devised a straightforward geometric method of determining the relative distances of the planets from the Sun using trigonometry. His answers turned out to be remarkably close to the modern values, as shown in Table 4-2.

The distances in Table 4-2 are given in terms of the astronomical unit, which is the average distance from Earth to Sun (Section 1-7). Copernicus did not know the precise value of this distance, so he could only determine the *relative* sizes of the orbits of the planets. One method used by modern astronomers to determine the astronomical unit is to measure the Earth-Venus distance very accurately using radar. At the same time they measure the angle in the sky between Venus

### box 4-1 | Tools of the Astronomer's Trade

## *Relating Synodic and Sidereal Periods*

We can derive a mathematical formula that relates a planet's sidereal period (the time required for the planet to complete one orbit) to its synodic period (the time between two successive identical configurations). To start with, let's consider an inferior planet (Mercury or Venus) orbiting the Sun as shown in the figure. Let $P$ be the planet's sidereal period, $S$ the planet's synodic period, and $E$ the sidereal period of Earth or sidereal year (see Section 2-7), which Copernicus knew to be nearly 365¼ days.

The rate at which the Earth moves around its orbit is the number of degrees around the orbit divided by the time to complete the orbit, or $360°/E$ (equal to a little less than 1° per day). Similarly, the rate at which the inferior planet moves along its orbit is $360°/P$.

During a given time interval, the angular distance that the Earth moves around its orbit is its rate, $(360°/E)$, multiplied by the length of the time interval. Thus, during a time $S$, or one synodic period of the inferior planet, the Earth covers an angular distance of $(360°/E)S$ around its orbit. In that same time, the inferior planet covers an angular distance of $(360°/P)S$. Note, however, that the inferior planet has gained one full lap on the Earth, and hence has covered 360° more than the Earth has. Thus, $(360°/P)S = (360°/E)S + 360°$. Dividing each term of this equation by $360° S$ gives

**For an inferior planet:**

$$\frac{1}{P} = \frac{1}{E} + \frac{1}{S}$$

$P$ = inferior planet's sidereal period
$E$ = Earth's sidereal period = 1 year
$S$ = inferior planet's synodic period

A similar analysis for a superior planet (for example, Mars, Jupiter, or Saturn) yields

**For a superior planet:**

$$\frac{1}{P} = \frac{1}{E} - \frac{1}{S}$$

$P$ = superior planet's sidereal period
$E$ = Earth's sidereal period = 1 year
$S$ = superior planet's synodic period

Using these formulas, we can calculate a planet's sidereal period $P$ from its synodic period $E$. Often astronomers express $P$, $E$, and $S$ in terms of years, by which they mean Earth years of approximately 365.26 days.

**EXAMPLE:** Jupiter has an observed synodic period $S$ of 398.9 days, or 1.092 years. Because Jupiter is a superior planet and because $E = 1$ year exactly, we determine Jupiter's sidereal period using the second of the two equations:

$$\frac{1}{P} = \frac{1}{1} - \frac{1}{1.092} = 0.08425,$$

so

$$P = \frac{1}{0.08425} = 11.87 \text{ years}$$

It takes 11.87 years for Jupiter to complete one full orbit of the Sun. The much shorter synodic period of 1.092 years is the time from one opposition to the next, or the time that elapses from when the Earth overtakes Jupiter to when it next overtakes Jupiter. This is much shorter than the sidereal period, because Jupiter moves quite slowly around its orbit. The Earth overtakes it a little less often than once per Earth orbit (that is, at intervals of a little bit more than a year).

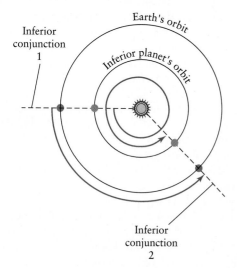

Earth's orbit

Inferior planet's orbit

Inferior conjunction 1

Inferior conjunction 2

| Table 4-1 | Synodic and Sidereal Periods of the Planets | |
|---|---|---|
| Planet | Synodic period | Sidereal period |
| Mercury | 116 days | 88 days |
| Venus | 584 days | 225 days |
| Earth | — | 1.0 year |
| Mars | 780 days | 1.9 years |
| Jupiter | 399 days | 11.9 years |
| Saturn | 378 days | 29.5 years |
| Uranus | 370 days | 84.0 years |
| Neptune | 368 days | 164.8 years |
| Pluto | 367 days | 248.5 years |

| Table 4-2 | Average Distances of the Planets from the Sun | |
|---|---|---|
| Planet | Copernican value | Modern value |
| Mercury | 0.38 AU* | 0.39 AU |
| Venus | 0.72 AU | 0.72 AU |
| Earth | 1.00 AU | 1.00 AU |
| Mars | 1.52 AU | 1.52 AU |
| Jupiter | 5.22 AU | 5.20 AU |
| Saturn | 9.07 AU | 9.54 AU |
| Uranus | — | 19.19 AU |
| Neptune | — | 30.06 AU |
| Pluto | — | 39.53 AU |

*1 AU = 1 astronomical unit = average distance of the Earth from the Sun

and the Sun, and then calculate the Earth-Sun distance using trigonometry. In this way the astronomical unit is found to be $1.496 \times 10^8$ km (92.96 million miles). Once the astronomical unit is known, the average distances from the Sun to each of the planets in kilometers can be determined from Table 4-2. A table of these distances is given in Appendix 2.

By comparing Tables 4-1 and 4-2, you can see the unifying relationship between planetary orbits in the Copernican cosmogony: The farther a planet is from the Sun, the longer it takes to travel around its orbit (that is, the longer its sidereal period). That is not simply because a planet must travel farther to complete a larger orbit; in addition, a planet moves more slowly in a larger orbit. For example, Mercury, with its small orbit, moves at an average speed of 47.9 km/s (107,000 mi/h). Saturn travels around its large orbit much more slowly, at an average speed of 9.64 km/s (21,600 mi/h). The older Ptolemaic model offers no such simple relations between the motions of different planets.

At first Copernicus assumed that Earth travels around the Sun along a circular path. He found that perfectly circular orbits could not accurately describe the paths of the other planets, so he had to add an epicycle to each planet. (This was *not* to explain retrograde motion, which Copernicus realized was because of the differences in orbital speeds of different planets, as shown in Figure 4-4. Rather, the small epicycles helped Copernicus account for slight variations in each planet's speed along its orbit.) Even though he clung to the old notion that orbits must be made up of circles, Copernicus had shown that a heliocentric cosmogony could explain the motions of the planets. He compiled his ideas and calculations into a book entitled *De revolutionibus orbium coelestium* (On the Revolutions of the Celestial Spheres), which was published in 1543, the year of his death.

Modern astronomers agree with Copernicus that the planets orbit the Sun. But for several decades after his death, most astronomers saw little reason to change their allegiance to the older geocentric cosmogony of Ptolemy. The predictions that the Copernican model makes for the apparent positions of the planets are, on average, no better or worse than those of the Ptolemaic model. The test of Occam's razor does not really favor either model, because both use a combination of circles to describe each planet's motion. More concrete evidence was needed to convince scholars to abandon the old, comforting idea of a stationary Earth at the center of the universe. This evidence was not long in coming, and it was literally an explosion in the celestial sphere.

## 4-3 Tycho Brahe made astronomical observations that disproved ancient ideas about the heavens

On November 11, 1572, a bright star suddenly appeared in the constellation of Cassiopeia. At first, it was even brighter than Venus, but then it began to grow dim. After 18 months, it faded from view. Modern astronomers recognize this event as a supernova explosion, the violent death of a certain type of star (see Chapter 22). In the sixteenth century, however, the vast majority of scholars held with the ancient teachings of Aristotle and Plato, who had argued that the heavens are permanent and unalterable. Consequently, the "new star" of 1572 could not really be a star at all, because the heavens do not change; it must instead be some sort of bright object quite near the Earth, perhaps not much farther away than the clouds overhead.

The 25-year-old Danish astronomer Tycho Brahe (1546–1601) realized that straightforward observations might reveal the distance to the new star. It is common experience

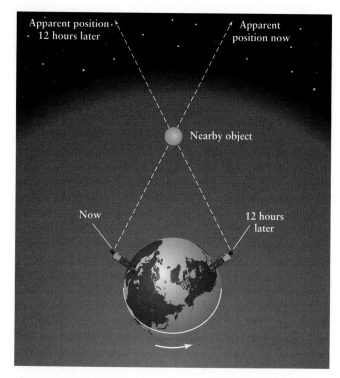

## figure 4-7

**The Parallax of a Nearby Object** Tycho Brahe argued that if an object is near the Earth, its position relative to the background stars should change over the course of a night. Tycho failed to measure such changes for a supernova in 1572 and a comet in 1577 and therefore concluded that these objects were far from the Earth.

that when you walk from one place to another, nearby objects appear to change position against the background of more distant objects. This phenomenon, whereby the apparent position of an object changes because of the motion of the observer, is called **parallax**. Tycho pointed out that if the new star was nearby, then its position should shift against the background stars over the course of a single night because Earth's rotation changes our viewpoint. This is shown in Figure 4-7. (Actually, Tycho believed that the heavens rotate about the Earth, as in the Ptolemaic cosmogony, but the net effect is the same.)

Tycho's careful observations failed to disclose any parallax. The farther away an object is, the less it appears to shift against the background as we change our viewpoint, and the smaller the parallax. Hence, the new star had to be quite far away, farther from Earth than anyone had imagined. Tycho also attempted to measure the parallax of a bright comet that appeared in 1577 and, again, found it too small to measure. Thus, the comet, too, had to be far beyond the Earth. With these observations, Tycho showed that the heavens are by no means pristine and unchanging, a discovery that flew in the face of nearly 2000 years of wisdom.

Tycho's observations of the supernova attracted the attention of the king of Denmark, who was so impressed that he financed the construction of a magnificent observatory for Tycho on the island of Hven, just off the Danish coast. Tycho named his palatial observatory Uraniborg ("sky castle"). The bequest for Uraniborg allowed Tycho to design and have built a set of astronomical instruments, like the wall quadrant shown in Figure 4-8, vastly superior in quality to any earlier instruments. With this state-of-the-art equipment, Tycho proceeded to measure the positions of stars and planets with unprecedented accuracy. In addition, he and his assistants were careful to make several observations of the same star or planet with different instruments, in order to identify any errors that might be caused by the instruments themselves. This painstaking approach, which revolutionized the practice of astronomy, is used by astronomers today.

## figure 4-8

**Tycho Brahe (1546–1601) at His Observatory** During the late sixteenth century, Tycho Brahe measured the positions of stars and planets with unprecedented accuracy. He is shown here with some of the measuring apparatus at Uraniborg, the grand observatory that he built under the patronage of Frederik II of Denmark. (This magnificent observatory lacked a telescope, which had not yet been invented.) His measurements, which were analyzed after Tycho's death by Johannes Kepler, were crucial to the development of astronomy in the seventeenth century. (Photo Researchers, Inc.)

**figure 4-9**
**Johannes Kepler (1571–1630)** By analyzing Tycho Brahe's detailed records of planetary positions, Kepler developed three general principles, called Kepler's laws, that describe how the planets move about the Sun. He was the first to realize that the orbits of the planets are ellipses and not circles. (E. Lessing/Magnum)

Using his instruments at Uraniborg, Tycho Brahe attempted to test Copernicus's ideas about the Earth going around the Sun. Tycho argued that if the Earth was in motion, then nearby stars should appear to shift their positions with respect to background stars as we orbit the Sun. Tycho failed to detect any such parallax, and he concluded that the Earth was at rest and the Copernican system was wrong.

On this point Tycho was in error, for nearby stars do in fact shift their positions as he had suggested. But even the nearest stars are so far away that the shifts in their positions are too small (less than an arcsecond) to be seen with the naked eye. Tycho would have needed a telescope to detect the parallax that he was looking for, but the telescope was not invented until after his death. Indeed, the first accurate determination of stellar parallax was not made until 1838. (We wll discuss the significance of these more modern parallax measurements in Chapter 19.)

Although he remained convinced that the Earth was at the center of the universe, Tycho nonetheless made a tremendous contribution toward putting the heliocentric model on a solid foundation. From 1576 to 1597, he used the instruments of Uraniborg to make comprehensive measurements of the positions of the planets with an accuracy of 1 arcmin. This is as well as can be done with the naked eye and was far superior to any earlier measurements. Within the reams of data that Tycho compiled lay the truth about the motions of the planets. The person who would extract this truth was a German mathematician who became Tycho's assistant in 1600, a year before the great astronomer's death. His name was Johannes Kepler (Figure 4-9).

## 4-4 Johannes Kepler proposed elliptical paths for the planets about the Sun

The task that Johannes Kepler took on at the beginning of the seventeenth century was to find a model of planetary motion that agreed completely with Tycho's extensive and very accurate observations of planetary positions. To do this, Kepler found that he had to break with an ancient prejudice about planetary motions. Astronomers had long assumed that heavenly objects move in circles, which were considered the most perfect and harmonious of all geometric shapes. They believed that if a perfect God resided in heaven along with the stars and planets, then the motions of these bodies must be perfect too. Against this context, Kepler dared to try to explain planetary motions with noncircular curves. In particular, he found that he had the best success with a particular kind of curve called an **ellipse**.

An ellipse can be drawn by using a loop of string, two thumbtacks, and a pencil, as shown in Figure 4-10*a*. Each thumbtack in the figure is at a **focus** (plural **foci**) of the ellipse; an ellipse has two foci. The longest diameter of an ellipse, called the **major axis**, passes through both foci. Half of that distance is called the **semimajor axis**, whose length is usually designated by the letter *a*. A circle is a special case of an ellipse in which the two foci are at the same point (this corresponds to using only a single thumbtack in Figure 4-10*a*).

By assuming that planetary orbits were ellipses, Kepler found, to his delight, that he could make his theoretical calculations match precisely to Tycho's observations. This important discovery, first published in 1609, is now called **Kepler's first law:**

*The orbit of a planet about the Sun is an ellipse with the Sun at one focus.*

There is no object at the other focus of a planet's elliptical orbit. This "empty focus" has geometrical significance, because it helps to define the shape of the ellipse, but plays no other role. The length of the semimajor axis of a planet's orbit is the average distance between the planet and the Sun.

Ellipses come in different shapes, depending on the elongation of the ellipse. The shape of an ellipse is described by its **eccentricity**, designated by the letter *e*. The value of *e* can range from 0 (a circle) to just under 1 (nearly a straight line). The greater the eccentricity, the more elongated the ellipse. A few examples of ellipses with different eccentricities are

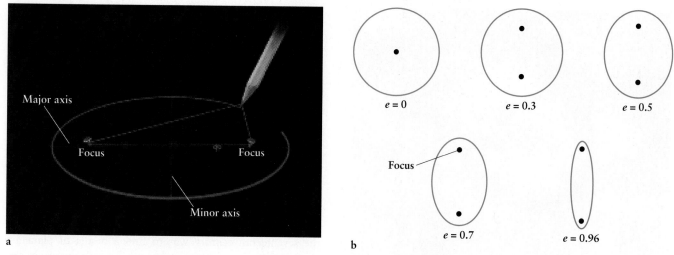

**Figure 4-10**

**Ellipses** **(a)** You can draw an ellipse with a pencil, a loop of string, and two thumbtacks. If the string is kept taut, the pencil traces out an ellipse. The thumbtacks are located at the two foci. The ellipse's major axis is the longest diameter across the ellipse. The minor axis is the perpendicular bisector of the major axis, that is, it is perpendicular to the major axis and passes through the center of the ellipse. **(b)** A series of ellipses with different eccentricities. Eccentricities range from $e = 0$ (a circle) to just under $e = 1$ (virtually a straight line).

shown in Figure 4-10b. Because a circle is a special case of an ellipse, it is possible to have an orbit that is a perfect circle. But all of the planets have orbits that are at least slightly elliptical, with eccentricities that range from 0.007 for Venus (the most nearly circular of any planetary orbit) to 0.248 for Pluto (the most elongated orbit).

Once he knew the shape of a planet's orbit, Kepler was ready to describe exactly *how* it moves on that orbit. As a planet travels in an elliptical orbit, its distance from the Sun varies. Kepler realized that the speed of a planet also varies along its orbit. A planet moves most rapidly when it is nearest the Sun, at a point on its orbit called **perihelion**. Conversely, a planet moves most slowly when it is farthest from the Sun, at a point called **aphelion** (Figure 4-11).

After much trial and error, Kepler found a way to describe just how a planet's speed varies as it moves along its orbit. This discovery, also published in 1609, is illustrated in Figure 4-11. Suppose that it takes 30 days for a planet to go from point A to point B. During that time, an imaginary line joining the Sun and the planet sweeps out a nearly triangular area. Kepler discovered that a line joining the Sun and the planet also sweeps out exactly the same area during any other 30-day interval. In other words, if the planet also takes 30 days to go from point C to point D, then the two shaded segments in Figure 4-11 are equal in area. **Kepler's second law** can be stated thus:

*A line joining a planet and the Sun sweeps out equal areas in equal intervals of time.*

This relationship is also called the **law of equal areas**. In the idealized case of a circular orbit, a planet would have to move at a constant speed around the orbit in order to satisfy Kepler's second law.

Kepler's second law describes how the speed of a given planet changes as it orbits the Sun. Kepler also deduced from Tycho's data a relationship that can be used to compare the motions of *different* planets. Published in 1618 and now called **Kepler's third law,** it states a relationship between the size of a planet's orbit and the time the planet takes to go once around the Sun:

*The square of the sidereal period of a planet is directly proportional to the cube of the semimajor axis of the orbit.*

Kepler's third law says that the larger a planet's orbit—that is, the larger the semimajor axis, or average distance from the planet to the Sun—the longer the sidereal period, which is the time it takes the planet to complete an orbit. From Kepler's third law one can show that the larger the semimajor axis, the slower the average speed at which the planet moves around its orbit. (By contrast, Kepler's *second* law describes how the speed of a given planet is sometimes faster and sometimes slower than its average speed.) This qualitative relationship between orbital size and orbital speed is just what Aristarchus and Copernicus used to explain retrograde motion, as described in Section 4-2. Kepler's great contribution was to make this relationship a quantitative one.

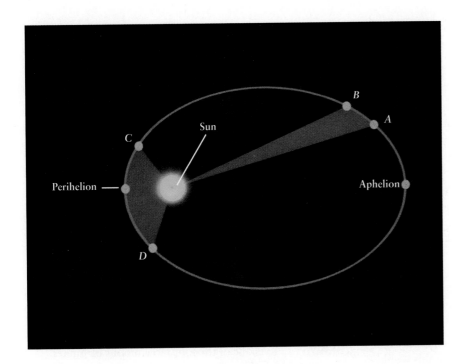

figure 4-11

**Kepler's First and Second Laws** According to Kepler's first law, a planet travels around the Sun along an elliptical orbit with the Sun at one focus. There is nothing at the other focus. According to his second law, a planet moves fastest when closest to the Sun (at perihelion) and slowest when farthest from the Sun (at aphelion). As it moves, an imaginary line joining the planet and the Sun sweeps out equal areas in equal intervals of time (from *A* to *B* or from *C* to *D*). Kepler found that by using these laws, his calculations gave a perfect fit to the apparent motions of the planets.

It is useful to restate Kepler's third law as an equation. If a planet's sidereal period $P$ is measured in years and the length of its semimajor axis $a$ is measured in astronomical units (AU), where 1 AU is the average distance from the Earth to the Sun (Section 1-7), then Kepler's third law is

**Kepler's third law**

$$P^2 = a^3$$

$P$ = planet's sidereal period, in years

$a$ = planet's semimajor axis, in AU

If you know either the sidereal period of a planet or the semimajor axis of its orbit, you can find the other one using this equation. Box 4-2 gives some examples.

We can verify Kepler's third law for all of the planets, including those that were discovered after Kepler's death, using data from Tables 4-1 and 4-2. If Kepler's third law is correct, for each planet the numerical values of $P^2$ and $a^3$ should be equal. This is indeed true to very high accuracy, as shown in Table 4-3.

Kepler's laws are a landmark in the history of astronomy. They made it possible to calculate the motions of the planets with far better accuracy than any geocentric model ever had, and they helped to justify the idea of a heliocentric cosmogony. Kepler's laws also pass the test of Occam's razor, for they are simpler in every way than the schemes of Ptolemy or Copernicus, both of which used a complicated combination of circles. But their significance goes beyond

understanding planetary orbits. Kepler's laws are also obeyed by spacecraft orbiting the Earth, by two stars revolving about each other in a binary star system, and even by galaxies in their orbits about each other. Throughout this book, we shall use Kepler's laws in a wide range of situations.

| Table 4-3 | A Demonstration of Kepler's Third Law | | |
|---|---|---|---|
| **Planet** | **Sidereal period** $P$ **(years)** | **Semimajor axis** $a$ **(AU)** | $P^2$ | $a^3$ |
| Mercury | 0.24 | 0.39 | 0.06 | 0.06 |
| Venus | 0.61 | 0.72 | 0.37 | 0.37 |
| Earth | 1.00 | 1.00 | 1.00 | 1.00 |
| Mars | 1.88 | 1.52 | 3.53 | 3.51 |
| Jupiter | 11.86 | 5.20 | 140.7 | 140.6 |
| Saturn | 29.46 | 9.54 | 867.9 | 868.3 |
| Uranus | 84.01 | 19.19 | 7,058 | 7,067 |
| Neptune | 64.79 | 30.06 | 27,160 | 27,160 |
| Pluto | 248.54 | 39.53 | 61,770 | 61,770 |

## box 4-2 | Tools of the Astronomer's Trade

# *Kepler's Third Law*

Kepler's third law comes in two different versions. The first version is

$$P^2 = a^3$$

This applies *only* to the special case of an object, like a planet, that orbits the Sun. In this form, $P$ is the sidereal period of the object, measured in years, and $a$ is the semimajor axis of the orbit, measured in astronomical units (AU).

The second version, called Newton's form of Kepler's third law (Section 4-7), is

$$P^2 = \left[ \frac{4\pi^2}{G(m_1 + m_2)} \right] a^3$$

This form applies to any situation where two bodies of masses $m_1$ and $m_2$ orbit each other. For example, this would be the equation to use for a moon orbiting a planet or a spacecraft orbiting the Earth. When using Newton's form of Kepler's third law, $P$ is normally expressed in seconds, $a$ in meters, and $m_1$ and $m_2$ in kilograms. The universal constant of gravitation, $G$, is then equal to $6.67 \times 10^{-11}$.

The following are a few examples of Kepler's third law used in different situations.

**EXAMPLE:** The average distance from Venus to the Sun—that is, the semimajor axis $a$ of Venus's orbit—is 0.72 AU. Because Venus orbits the Sun, we can use the first version of Kepler's third law. To determine the sidereal period of Venus, we first cube the semimajor axis (multiply it by itself twice):

$$a^3 = (0.72)^3 = 0.72 \times 0.72 \times 0.72 = 0.373$$

According to Kepler's third law this is also equal to $P^2$, the square of the sidereal period. So, to find $P$, we have to "undo" the square, that is, take the square root. Using a calculator, we find

$$P = \sqrt{P^2} = \sqrt{0.373} = 0.61$$

The sidereal period of Venus is 0.61 years, or a bit more than seven Earth months.

**EXAMPLE:** Consider a small asteroid (a rocky body a few tens of kilometers across) that takes eight years to complete one orbit around the Sun. Because the asteroid orbits the Sun, we can again use the first version of Kepler's third law. To determine the semimajor axis $a$ of the asteroid's orbit, we first square the period:

$$P^2 = 8^2 = 8 \times 8 = 64$$

From Kepler's third law, this is also equal to $a^3$. To determine $a$, we must take the *cube root* of $a^3$, that is, find the number whose cube is 64. If your calculator has a cube root function, denoted by the symbol $\sqrt[3]{\phantom{x}}$, you can use it to find that the cube root of 64 is 4: $\sqrt[3]{64} = 4$. Otherwise, you can determine by trial and error that the cube of 4 is 64:

$$4^3 = 4 \times 4 \times 4 = 64$$

Because the cube of 4 is 64, it follows that the cube root of 64 is 4 (taking the cube root "undoes" the cube). With either technique you find that the orbit of this asteroid has semimajor axis $a = 4$ AU. This is intermediate between the orbits of Mars and Jupiter. We will discuss asteroids in detail in Chapter 17.

**EXAMPLE:** Io is one of the four large moons of Jupiter discovered by Galileo. It orbits at a distance of 421,600 km from the center of Jupiter and has an orbital period of 1.77 days. Because this is not an orbit around the Sun, we must use Newton's form of Kepler's third law. We can use this form to determine the combined mass of Jupiter ($m_1$) and Io ($m_2$). We first rewrite the equation in the form

$$m_1 + m_2 = \frac{4\pi^2 a^3}{G P^2}$$

To solve this equation, we have to convert the distance $a$ from kilometers to meters and convert the period $P$ from days to seconds. There are 1000 meters in 1 kilometer and 86,400 seconds in 1 day, so:

$$a = (421{,}600 \text{ km}) \times \frac{1000 \text{ m}}{1 \text{ km}} = 4.216 \times 10^8 \text{ m}$$

$$P = (1.77 \text{ days}) \times \frac{86{,}400 \text{ s}}{1 \text{ day}} = 1.529 \times 10^5 \text{ s}$$

We can now put these values and the value of $G$ into the above equation:

$$m_1 + m_2 = \frac{4\pi^2 (4.216 \times 10^8)^3}{(6.67 \times 10^{-11})(1.529 \times 10^5)^2} = 1.90 \times 10^{27} \text{ kg}$$

Io is very much smaller than Jupiter, so its mass is only a small fraction of the mass of Jupiter. Thus, $m_1 + m_2$ is very nearly the mass of Jupiter alone. We conclude that Jupiter has a mass of $1.90 \times 10^{27}$ kg, or about 300 times the mass of Earth. In later chapters we will do calculations of this kind to find the masses of stars, black holes, and entire galaxies of stars.

## 4-5 Galileo's discoveries with a telescope strongly supported a heliocentric cosmogony

Astronomy was changed forever when Dutch opticians invented the telescope during the first decade of the seventeenth century, the same time that Kepler was making such rapid progress. The scholar who used this new tool to give convincing evidence that the planets orbit the Sun, not the Earth, was an Italian mathematician and physical scientist named Galileo Galilei (Figure 4-12).

While Galileo did not invent the telescope, he was the first to point one of these new devices toward the sky and to publish his observations. Beginning in 1610, he saw things that no one had ever dreamed of. He discovered mountains on the Moon, sunspots on the Sun, and the rings of Saturn, and he was the first to see that the Milky Way is not a featureless band of light but rather "a mass of innumerable stars."

One of Galileo's most important discoveries with the telescope was that Venus exhibits phases like those of the Moon (Figure 4-13). Galileo also noticed that the apparent size of Venus as seen through his telescope was related to the planet's phase. Venus appears smallest at gibbous phase and largest at crescent phase. There is also a correlation between the phases of Venus and the planet's angular distance from the Sun.

Figure 4-14 shows that these relationships are entirely compatible with a heliocentric model in which the Earth and Venus both go around the Sun. They are also completely *incompatible* with the Ptolemaic system, in which the Sun and Venus both orbit the Earth. To explain why Venus is never seen very far from the Sun, the Ptolemaic cosmogony had to assume that the deferents of Venus and of the Sun move together in lockstep, with the epicycle of Venus centered on a straight line between the Earth and the Sun (Figure 4-15). In this model, Venus was never on the opposite side of

## figure 4-12

**Galileo Galilei (1564–1642)** Galileo was one of the first people to use a telescope to observe the heavens. He discovered craters on the Moon, sunspots on the Sun, the phases of Venus, and four moons orbiting Jupiter. He also made the key observations that show that the Earth orbits the Sun, not vice versa. (Art Resource)

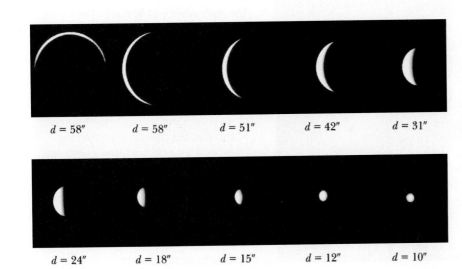

$d = 58''$  $d = 58''$  $d = 51''$  $d = 42''$  $d = 31''$

$d = 24''$  $d = 18''$  $d = 15''$  $d = 12''$  $d = 10''$

## figure 4-13   R I **V** U X G

**The Phases of Venus** This series of photographs shows how the appearance of Venus changes as it moves along its orbit. The number below each view is the angular diameter of the planet in arcseconds. Venus has the largest angular diameter when it is a crescent, and the smallest angular diameter when it is gibbous (nearly full). (New Mexico State University Observatory)

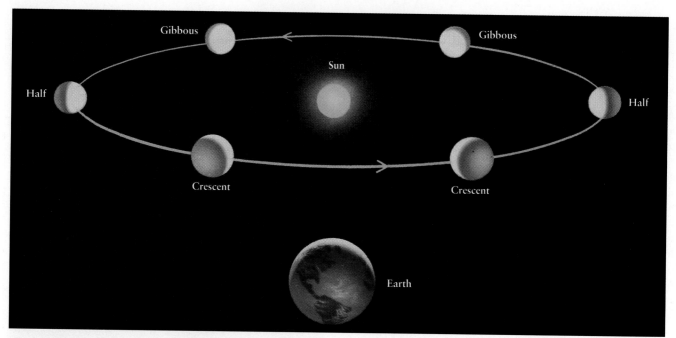

## figure 4-14

**The Changing Appearance of Venus Explained in a Heliocentric Model** In a heliocentric model, in which the Earth and Venus both orbit the Sun, the changing appearance of Venus is explained naturally. When Venus is on the opposite side of the Sun from the Earth, it appears gibbous (nearly fully illuminated) and of a small angular size (because it is farthest from Earth).

When Venus is on the same side of the Sun as we are, we see it in a crescent phase and, because it is now relatively close, with a larger angular size. When Venus is at the greatest angle to either side of the Sun (at maximum eastern or western elongation), it appears half-illuminated and of an intermediate angular size. (Compare Figure 4-13.)

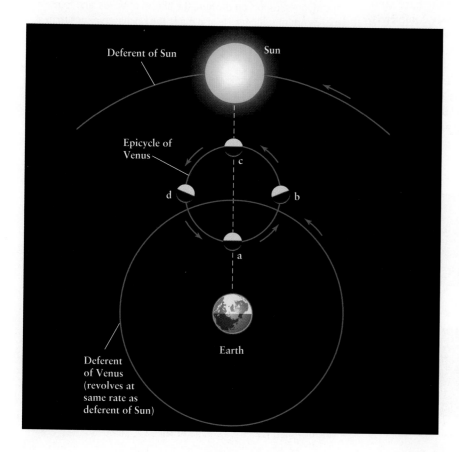

## figure 4-15

**The Appearance of Venus in the Ptolemaic Model** In the geocentric Ptolemaic model, the deferents of Venus and the Sun rotate together, with the epicycle of Venus centered on a line (shown dashed) connecting the Sun and the Earth. In this model Venus is never on the side of the Sun opposite the Earth, and so an Earth observer would never see Venus as more than half illuminated. (At positions *a* and *c*, Venus will appear in a "new" phase; at positions *b* and *d*, Venus will appear as a crescent. Compare with Figure 3-3, which shows the phases of the Moon.) Because Galileo in fact saw Venus in gibbous phases in which it was nearly fully illuminated, he concluded that the Ptolemaic cosmogony must be incorrect.

the Sun from the Earth, and so it could never have shown the gibbous phases that Galileo observed.

Galileo also found more unexpected evidence for the ideas of Copernicus. In 1610 Galileo discovered four moons, now called the Galilean satellites, orbiting Jupiter (Figure 4-16). He realized that they were orbiting Jupiter because they appeared to move back and forth from one side of the planet to the other. Confirming observations made by Jesuit observers in 1620 are shown in Figure 4-17. Astronomers soon realized that the larger the orbit of one of the moons around Jupiter, the slower that moon moves and the longer it takes that moon to travel around its orbit. These are the same relationships that Copernicus deduced for the motions of the planets around the Sun. Thus, the moons of Jupiter behave like a Copernican system in miniature.

Galileo's telescopic observations constituted the first fundamentally new astronomical data in almost 2000 years. Contradicting prevailing opinion, his discoveries strongly suggested a heliocentric structure of the universe. At the time, the Roman Catholic Church attacked Galileo's ideas, because they could not be reconciled with certain passages in the Bible or with the writings of Aristotle and Plato. Galileo was condemned to spend his latter years under house arrest "for vehement suspicion of heresy." Nevertheless, there was no turning back.

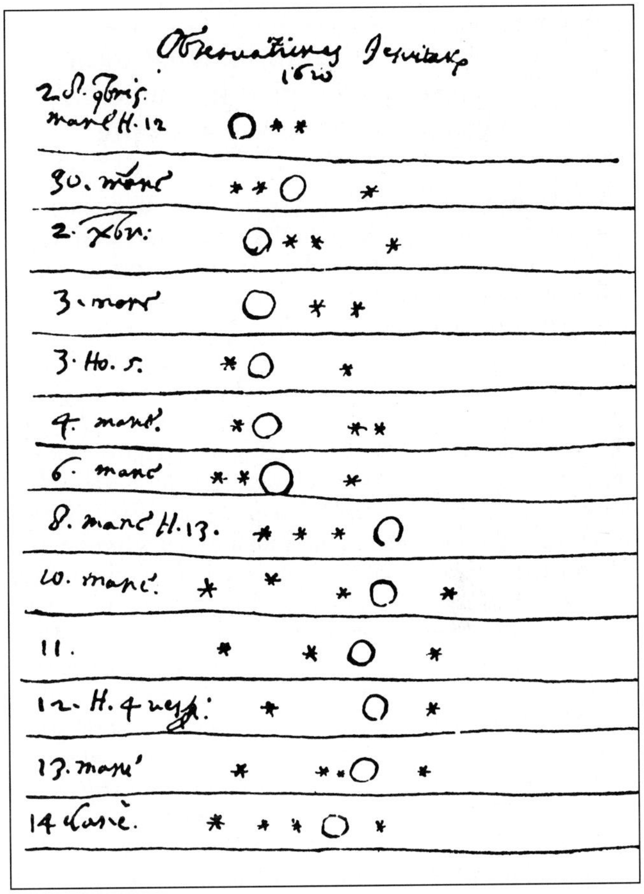

**figure 4-17**

**Early Observations of Jupiter's Moons** In 1610 Galileo discovered four "stars" that move back and forth across Jupiter from one night to the next. He concluded that these are four moons that orbit Jupiter, much as our Moon orbits the Earth. This drawing shows notations made by Jesuit observers on successive nights in 1620. The circle represents Jupiter and the stars its moons. (Yerkes Observatory)

**figure 4-16**  R I **V** U X G

**Jupiter and Its Largest Moons** This photograph, taken by an amateur astronomer with a small telescope, shows the four Galilean satellites alongside an overexposed image of Jupiter. Each satellite is bright enough to be seen with the unaided eye, were it not overwhelmed by the glare of Jupiter. (Courtesy of C. Holmes)

Galileo's observations showed convincingly that the Ptolemaic cosmogony was entirely wrong, and that a heliocentric cosmogony is the more correct one. Galileo did not, however, prove that the particular heliocentric model proposed by Copernicus is the correct one. As we have seen, Kepler's work showed that the orbits of the planets are actually ellipses, not combinations of circles, as in the Copernican model.

The revolutionary ideas of Galileo were soon to inspire the Englishman Isaac Newton, born on Christmas Day of 1642, a dozen years after the death of Kepler and the same year that Galileo died. While Galileo and Kepler revolutionized our understanding of planetary motions, Newton would revolutionize our concepts of the basic laws that govern all motions on Earth as well as in the heavens.

## 4-6 Isaac Newton formulated three laws that describe fundamental properties of physical reality

Until the mid-seventeenth century, virtually all attempts to describe the heavens mathematically were empirical, or based directly on data and observations. From Ptolemy to Kepler, astronomers would adjust their ideas and calculations by trial and error until they ended up with answers that agreed with observation.

Isaac Newton (Figure 4-18) introduced a new approach. He began with three quite general statements, now called **Newton's laws of motion.** These laws, deduced from experimental observation, apply to all forces and all bodies. Newton then showed that Kepler's three laws follow logically from these laws of motion and from a formula for the force of gravity that he derived from observation. In other words, Kepler's laws are not just an empirical description of the motions of the planets, but a direct consequence of the fundamental laws of physical matter. Using this deeper insight into the nature of motions in the heavens, Newton and his successors were able to describe accurately not just the orbits of the planets but also the orbits of the Moon and comets.

Newton's laws of motion describe objects on Earth as well as in the heavens. Thus, we can understand each of these laws by considering the motions of objects around us. We begin with **Newton's first law** or the **law of inertia:**

*A body remains at rest, or moves in a straight line at a constant speed, unless acted upon by a net outside force.*

By **force** we mean any push or pull that acts on the body. An *outside* force is one that is exerted on the body by something other than the body itself. The net, or total, outside force is the combined effect of all of the individual outside forces that act on the body

Right now, you are demonstrating the first part of Newton's first law. As you sit in your chair reading this, there are two outside forces acting on you: The force of gravity pulls you downward, and the chair pushes up on you. These two forces are of equal strength but of opposite direction, so their effects cancel—there is no *net* outside force. Hence, your body remains at rest as stated in Newton's first law. If you try to lift yourself out of your chair by grabbing your knees and pulling up, you will remain at rest because this force is not an outside force.

 The second part of Newton's first law, about objects in motion, may seem to go against common sense. If you want to make this book move across the floor in a straight line at a constant speed, you must continually push on it. You might therefore think that there *is* a net outside force, the force of your push. But another force also acts on the book—the force of friction as the book rubs across the floor. As you push the book across the floor, the force of your push exactly balances the force of friction, so

again there is no net outside force. The effect is to make the book move in a straight line at constant speed, just as Newton's first law says. If you stop pushing, there will be nothing to balance the effects of friction. Then there will be a net outside force and the book will slow to a stop.

Newton's first law tells us that a force must be acting on the planets. A planet moving through empty space encounters no friction, so it would tend to fly off into space along a straight line if there were no other outside force acting on it. Because this does not happen, Newton concluded that there must be a force that acts continuously on the planets to keep them in their elliptical orbits.

Newton's second law describes how the motion of an object changes if there is a net outside force acting on it. To appreciate Newton's second law, we must first understand three quantities that describe motion—speed, velocity, and acceleration.

Speed is a measure of how fast an object is moving. Speed and direction of motion together constitute an object's **veloc-**

## figure 4-18

**Isaac Newton (1642–1727)** Using mathematical techniques that he devised, Isaac Newton formulated the universal law of gravitation and demonstrated that the planets orbit the Sun according to simple mechanical rules. (National Portrait Gallery, London)

ity. Compared to a car driving north at 100 km/h (62 mi/h), a car driving east at 100 km/h has the same speed but a different velocity.

**Acceleration** is the rate at which velocity changes. Because velocity involves both speed and direction, acceleration can result from changes in either. Contrary to popular use of the term, acceleration does not simply mean speeding up. A car is accelerating if it is speeding up, but it is also accelerating if it is slowing down or turning (that is, changing the direction in which it is moving).

You can verify these statements about acceleration if you think about the sensations of riding in a car. If the car is moving with a constant velocity (in a straight line at a constant speed), you feel the same as if the car were not moving at all. But you can feel it when the car accelerates: You feel thrown back in your seat if the car speeds up, thrown forward if the car slows down, and thrown sideways if the car changes direction in a tight turn. In Box 4-3 we discuss the reasons for these sensations, along with other applications of Newton's laws to everyday life.

An apple falling from a tree is a good example of acceleration that involves only an increase in speed. Initially, at the moment the stem breaks, the apple's speed is zero. After 1 second, its downward speed is 9.8 meters per second, or

---

## box 4-3 | The Heavens on the Earth

### *Newton's Laws in Everyday Life*

In our study of astronomy, we use Newton's three laws of motion to help us understand the motions of objects in the heavens. But you can see applications of Newton's laws every day in the world around you. By considering these everyday applications, we can gain insight into how Newton's laws apply to celestial events that are far removed from ordinary human experience.

Newton's *first* law, or principle of inertia, says that an object at rest naturally tends to remain at rest and that an object in motion naturally tends to remain in motion. This explains the sensations that you feel when riding in an automobile. When you are waiting at a red light, your car and your body are both at rest. When the light turns green and you press on the gas pedal, the car accelerates forward but your body attempts to stay where it was. Hence, the seat of the accelerating car pushes forward into your body, and it feels as though you are being pushed back in your seat.

Once the car is up to cruising speed, your body wants to keep moving in a straight line at this cruising speed. If the car makes a sharp turn to the left, the right side of the car will move toward you. Thus, you will feel as though you are being thrown to the car's right side (the side on the outside of the turn). If you bring the car to a sudden stop by pressing on the brakes, your body will continue moving forward until the seat belt stops you. In this case, it feels as though you are being thrown toward the front of the car.

Newton's *second* law states that the net outside force on an object equals the product of the object's mass and its acceleration. You can accelerate a crumpled-up piece of paper to a pretty good speed by throwing it with a moderate force. But if you try to throw a backpack full of books by using the same force, the acceleration will be much less because the backpack has much more mass than the crumpled paper. Because of the reduced acceleration during the throw, the backpack will leave your hand moving at only a slow speed.

Automobile airbags are based on the relationship between force and acceleration. It takes a large force to bring a fast-moving object suddenly to rest because this requires a large acceleration. In a collision, the driver of a car not equipped with airbags is jerked to a sudden stop and the large forces that act can cause major injuries. But if the car has airbags that deploy in an accident, the driver's body will slow down more gradually as it contacts the airbag, and so the driver's acceleration will be less. (Remember that *acceleration* can refer to slowing down as well as to speeding up.) Hence, the force on the driver and the chance of injury will both be greatly reduced.

Newton's *third* law, the principle of action or reaction, explains how a car can accelerate at all. It is not correct to say that the engine pushes the car forward, because Newton's second law tells us that it takes a force acting from outside the car to make the car accelerate. Rather, the engine makes the wheels and tires turn, and the tires push backward on the ground. (You can see this backward force acting when a car drives through mud and sprays mud backward from the spinning tires.) From Newton's third law, the ground must exert an equally large forward force on the car, and this is the force that pushes the car forward.

You use the same principles when you walk: You push backward on the ground with your foot, and the ground pushes forward on you. Icy pavement or a freshly waxed floor have greatly reduced friction. In these situations, your feet and the surface under you can exert only weak forces on each other, and it is much harder to walk.

9.8 m/s (32 feet per second, or 32 ft/s). After 2 seconds, the apple's speed is twice this, or 19.6 m/s. After 3 seconds, the speed is 29.4 m/s. Because the apple's speed increases by 9.8 m/s for each second of free fall, the rate of acceleration is 9.8 meters per second per second, or 9.8 m/s$^2$ (32 ft/s$^2$). Thus, the Earth's gravity produces a constant acceleration of 9.8 m/s$^2$ downward, toward the center of the Earth.

A planet revolving about the Sun along a perfectly circular orbit is an example of acceleration that involves change of direction only. As the planet moves along its orbit, its speed remains constant. Nevertheless, the planet is continuously being accelerated because its direction of motion is continuously changing.

**Newton's second law** says that the acceleration of an object is proportional to the net outside force acting on the object. In other words, the harder you push on an object, the greater the resulting acceleration. This law can be succinctly stated as an equation. If a net outside force $F$ acts on an object of mass $m$, the object will experience an acceleration $a$ such that

**Newton's second law**

$$F = ma$$

$F$ = net outside force on an object
$m$ = mass of object
$a$ = acceleration of object

The **mass** of an object is a measure of the total amount of material in the object. It is usually expressed in kilograms (kg) or grams (g). For example, the mass of the Sun is $2 \times 10^{30}$ kg, the mass of a hydrogen atom is $1.7 \times 10^{-27}$ kg, and the mass of an average adult is 75 kg. The Sun, a hydrogen atom, and a person have these masses regardless of where they happen to be in the universe.

It is important not to confuse the concepts of mass and weight. **Weight** is the force of gravity that acts on a body and, like any force, is usually expressed in pounds or newtons (1 newton = 0.225 pound).

We can use Newton's second law to relate mass and weight. We have seen that the acceleration caused by the Earth's gravity is 9.8 m/s$^2$. When a 50-kg swimmer falls from a diving board, the only outside force acting on her as she falls is her weight. Thus, from Newton's second law ($F = ma$), her weight is equal to her mass multiplied by the acceleration due to gravity:

$$50 \text{ kg} \times 9.8 \text{ m/s}^2 = 490 \text{ newtons} = 110 \text{ pounds}$$

Note that this answer is correct only when the person is on Earth. She would weigh less on the Moon, where the pull of gravity is weaker, and more on Jupiter, where the gravitational pull is stronger. Floating deep in space, she would have no weight at all; she would be "weightless." Nevertheless, under all these circumstances, she would always have exactly the same mass, because mass is an inherent property of matter unaffected by details of the environment. Whenever we describe the properties of planets, stars, or galaxies, we speak of their masses, never of their weights.

We have seen that a planet is continually accelerating as it orbits the Sun. From Newton's second law, this means that there must be a net outside force that acts continually on each of the planets. As discussed in the next section, this force is the gravitational attraction of the Sun.

The last of Newton's general laws of motion, called **Newton's third law,** is the famous statement about action and reaction:

*Whenever one body exerts a force on a second body, the second body exerts an equal and opposite force on the first body.*

For example, if you weigh 110 pounds, when you are standing up you are pressing down on the floor with a force of 110 pounds. Newton's third law tells us that the floor is also pushing up against your feet with an equal force of 110 pounds.

Newton realized that because the Sun is exerting a force on each planet to keep it in orbit, each planet must also be exerting an equal and opposite force on the Sun. However, the planets are much less massive than the Sun (for example, the Earth has only $\frac{1}{300,000}$ of the Sun's mass). Therefore, although the Sun's force on a planet is the same as the planet's force on the Sun, the planet's much smaller mass gives it a much larger acceleration, according to Newton's second law. This is why the planets circle the Sun instead of vice versa. Newton's laws reveal the reason for our heliocentric solar system.

---

**4-7** | Newton's description of gravity accounts for Kepler's laws and explains the motions of the planets

Tie a ball to one end of a piece of string, hold the other end of the string in your hand, and whirl the ball around in a circle. As the ball "orbits" your hand, it is continuously accelerating because its velocity is changing. (Even if its speed is constant, its direction of motion is changing.) In accordance with Newton's second law, this can happen only if the ball is continuously acted on by an outside force, the pull of the string. The pull is directed along the string toward your hand. In the same way, Newton saw, the force that keeps a planet in orbit around the Sun is a pull that always acts toward the Sun. That pull is **gravity,** or **gravitational force.**

Newton's discovery about the forces that act on planets led him to suspect that the force of gravity pulling a falling apple straight down to the ground is fundamentally the same as the force on a planet that is always directed straight at the Sun. In other words, gravity is the force that shapes the orbits of the planets. What is more, he was able to determine how the force of gravity depends on the distance between the Sun and the planet. His result was a law of gravitation that could apply to the motion of distant planets as well as to the flight of a football on Earth. Using this law, Newton achieved the remarkable goal of deducing Kepler's laws from fundamental principles of nature.

To see how Newton reasoned, think again about a ball attached to a string. If you use a short string, so that the ball orbits in a small circle, and whirl the ball around your hand at a high speed, you will find that you have to pull fairly hard on the string (Figure 4-19*a*). But if you use a longer string, so that the ball moves in a larger orbit, and if you make the ball orbit your hand at a slow speed, you only have to exert a light tug on the string (Figure 4-19*b*). The orbits of the planets behave in the same way; the larger the size of the orbit, the slower the planet's speed. By analogy to the force of the string on the orbiting ball, Newton concluded that the force that attracts a planet toward the Sun must decrease with increasing distance between the Sun and the planet.

Using his own three laws and Kepler's three laws, Newton succeeded in formulating a general statement that describes the nature of the gravitational force. Newton's **universal law of gravitation** is as follows:

*Two bodies attract each other with a force that is directly proportional to the mass of each body and inversely proportional to the square of the distance between them.*

This law states that *any* two objects exert gravitational pulls on each other. Normally, you notice only the gravitational force that the Earth exerts on you, otherwise known as your weight. In fact, you feel gravitational attractions to *all* the objects around you. For example, this book is exerting a gravitational force on you as you read it. But because the force exerted on you by this book is proportional to the book's mass, which is very small compared to the Earth's mass, the force is too small to notice. (It can actually be measured with sensitive equipment.)

Consider two 1-kg objects separated by a distance of 1 meter. Newton's universal law of gravitation says that the force is directly proportional to the mass, so if we double the mass of one object to 2 kg, the force between the objects will double. If we double both masses so that we have two 2-kg objects separated by 1 meter, the force will be $2 \times 2 = 4$ times what it was originally (the force is directly proportional to the mass of *each* object). If we go back to two 1-kg masses, but double their separation to 2 meters, the force will be only one-quarter its original value. This is because the force is inversely proportional to the distance: If we double the distance, the force is multiplied by a factor of

$$\frac{1}{2^2} = \frac{1}{4}$$

Newton's universal law of gravitation can be stated more succinctly as an equation. If two objects have masses $m_1$ and

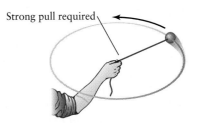

a   Ball moves at a high speed
in a small circle

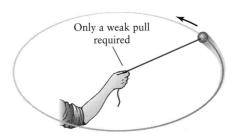

b   Ball moves at a low speed

## figure 4-19

**A Ball on a String** **(a)** To make a ball on a string move at high speed around a small circle, you have to exert a substantial pull on the string. Similarly, a planet that orbits close to the Sun moves at high speed and requires a substantial gravitational force from the Sun. **(b)** If you lengthen the string and make the same ball move at low speed around a large circle, much less pull is required. In the same way, a planet in a large orbit moves at low speed and requires less gravitational force to stay in orbit. This example shows that the Sun's gravitational pull must decrease with increased distance.

$m_2$ and are separated by a distance $r$, then the gravitational force $F$ between these two objects is given by the equation

**Newton's universal law of gravitation**

$$F = G \left( \frac{m_1 m_2}{r^2} \right)$$

$F$ = gravitational force between two objects
$m_1$ = mass of first object
$m_2$ = mass of second object
$r$ = distance between objects
$G$ = universal constant of gravitation

If the masses are measured in kilograms and the distance between them in meters, then the force is measured in newtons. In this formula, $G$ is a number called the **universal**

**constant of gravitation.** Laboratory experiments have yielded a value for $G$ of

$$G = 6.67 \times 10^{-11} \text{ newton m}^2/\text{kg}^2$$

We can use Newton's universal law of gravitation to calculate the force with which any two bodies attract each other. For example, to compute the gravitational force that the Sun exerts on the Earth, we substitute values for the Earth's mass ($m_1 = 5.98 \times 10^{24}$ kg), the Sun's mass ($m_2 = 1.99 \times 10^{30}$ kg), the distance between them ($r = 1$ AU $= 1.5 \times 10^{11}$ m), and the value of $G$ into Newton's equation. We get

$$F_{\text{Sun-Earth}} = 6.67 \times 10^{-11} \left[ \frac{5.98 \times 10^{24} \times 1.99 \times 10^{30}}{(1.50 \times 10^{11})^2} \right]$$

$$= 3.53 \times 10^{22} \text{ newtons}$$

If we calculate the force that the Earth exerts on the Sun, we get exactly the same result. (Mathematically, we just let $m_1$ be the Sun's mass and $m_2$ be the Earth's mass instead of the other way around. The product of the two numbers is the same, so the force is the same.) This is in accordance with Newton's third law: Any two objects exert *equal* gravitational forces on each other.

Because there is a gravitational force between any two objects, Newton concluded that gravity is also the force that keeps the Moon in orbit around the Earth. It is also the force that keeps artificial satellites in orbit. But if the force of gravity attracts two objects to each other, why don't satellites immediately fall to Earth? Why doesn't the Moon fall into the Earth? And, for that matter, why don't the planets fall into the Sun?

To see the answer, imagine (as Newton did) dropping a ball from a great height above the Earth's surface, as in Figure 4-20. After you drop the ball, it of course falls straight down (path A in Figure 4-20). But if you *throw* the ball horizontally, it travels some distance across the Earth's surface before hitting the ground (path B). If you throw the ball harder, it travels a greater distance (path C). If you could throw at just the right speed, the curvature of the ball's path will exactly match the curvature of the Earth's surface (path E). Although the Earth's gravity is making the ball fall, the Earth's surface is falling away under the ball at the same rate. Hence, the ball does not get any closer to the surface, and the ball is in circular orbit. So the ball in path E is in fact falling, but it is falling *around* the Earth rather than *toward* the Earth.

A spacecraft is launched into orbit in just this way—by throwing it fast enough. The thrust of a rocket is used to give the spacecraft the necessary orbital speed. Once the spacecraft is in orbit, the rocket turns off and the spacecraft falls continually around the Earth.

**CAUTION!** An astronaut on board an orbiting spacecraft feels "weightless," but this is *not* because she is "beyond the pull of gravity." The astronaut is herself an independent satellite of the Earth, and the Earth's gravitational pull is what holds her in orbit. She feels "weightless" because she and her spacecraft are falling *together* around

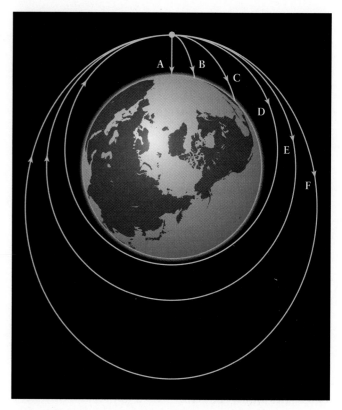

**f**igure 4-20

**An Explanation of Orbits** If a ball is dropped from a great height above the Earth's surface, it falls straight down (A). If the ball is thrown with some horizontal speed, it follows a curved path before hitting the ground (B, C). If thrown with just the right speed (E), the ball goes into circular orbit; the ball's path curves but it never gets any closer to the Earth's surface. If the ball is thrown with a speed that is slightly less (D) or slightly more (F) than the speed for a circular orbit, the ball's orbit is an ellipse.

the Earth, so there is nothing pushing her against any of the spacecraft walls. You feel the same "weightless" sensation whenever you are falling, such as when you jump off a diving board or ride the free-fall ride at an amusement park.

If the ball in Figure 4-20 is thrown with a slightly slower speed than that required for a circular orbit, its orbit will be an ellipse (path D). An elliptical orbit also results if instead the ball is thrown a bit too fast (path F). In this way, spacecraft can be placed into any desired orbit around the Earth by adjusting the thrust of the rockets.

Just as the ball in Figure 4-20 will not fall to Earth if given enough speed, the Moon does not fall to Earth and the planets do not fall into the Sun. The planets were given their initial speeds around the Sun by circumstances that prevailed when the solar system first formed 4.6 billion years ago. (We will discuss the formation of the planets in Chapter 7.) Figure 4-20 shows that a circular orbit is a very special case, so it is not surprising that none of the planets has an orbit that is precisely circular.

Using his three laws of motion and his law of gravity, Newton found that he could prove Kepler's three laws mathematically. Kepler's first law, concerning the elliptical shape of planetary orbits, proved to be a direct consequence of the $1/r^2$ factor in the law of universal gravitation. (Had the nature of gravity in our universe been different, so that this factor was given by a different function such as $1/r$ or $1/r^3$, elliptical orbits would not have been possible.) The law of equal areas, or Kepler's second law, is a consequence of the Sun's gravitational force on a planet being directed straight toward the Sun.

**ANALOGY** A good analogy for Kepler's second law is a twirling ice skater. If the skater pulls her arms straight in to her body, she spins faster; if she lets her arms extend away from her body, her rate of spin decreases. In the same way, a planet in an elliptical orbit travels at a higher speed when it moves closer to the Sun (toward perihelion) and travels at a lower speed when it moves away from the Sun (toward aphelion).

Newton also demonstrated that Kepler's third law follows logically from his law of gravity. Specifically, he proved that if two objects with masses $m_1$ and $m_2$ orbit each other, the period $P$ of their orbit and the semimajor axis $a$ of their orbit (that is, the average distance between the two objects) are related by the equation

$$P^2 = \left[ \frac{4\pi^2}{G(m_1 + m_2)} \right] a^3$$

This equation is called **Newton's form of Kepler's third law.** It is valid whenever two objects orbit each other because of their mutual gravitational attraction. As we discuss in Chapter 19, it is invaluable in the study of binary star systems, in which two stars orbit each other. If the orbital period $P$ and semimajor axis $a$ of the two stars in a binary system are known, an astronomer can use this formula to calculate the sum $m_1 + m_2$ of the masses of the two stars. As we describe in Chapters 21 and 22, knowing the mass of a star tells us a great deal about the way in which that star evolves. Box 4-2 gives an example of using Newton's form of Kepler's third law.

Newton also discovered new features of orbits around the Sun. For example, his equations soon led him to conclude that the orbit of an object around the Sun need not be an ellipse. It could be any one of a family of curves called conic sections.

A **conic section** is any curve that you get by cutting a cone with a plane, as shown in Figure 4-21. You can get circles and ellipses by slicing all the way through the cone. You can also get two open curves called **parabolas** and **hyperbolas.** If you were to throw the ball in Figure 4-20 with a fast enough speed, it would follow a parabolic or hyperbolic orbit and would fly off into space, never to return. Comets hurtling toward the Sun from the depths of space sometimes follow hyperbolic orbits.

Newton's ideas turned out to be applicable to an incredibly wide range of situations. The orbits of the planets and their satellites could now be calculated with unprecedented precision. Using his laws of motion, Newton himself proved that the Earth's axis of rotation must precess because of the gravitational pull of the Moon and the Sun on the Earth's equatorial bulge (see Figure 2-15). Similarly, all the details of the Moon's orbit (see Box 3-2) could be demonstrated mathematically with a body of knowledge built on Newton's work that is today called **Newtonian mechanics.**

Not only could Newtonian mechanics explain a variety of known phenomena in detail, but it could also predict new phenomena. For example, one of Newton's friends, Edmund Halley, was intrigued by three similar historical records of a comet that had been sighted at intervals of 76 years. Assuming these records to be accounts of the same comet, Halley used Newton's methods to work out the details of the comet's orbit and predicted its return in 1758. It was first sighted on Christmas night of 1757, a fitting memorial to Newton's birthday, and to this day the comet bears Halley's name (Figure 4-22).

Perhaps the most dramatic success of Newton's ideas was their role in the discovery of the eighth planet from the Sun. The seventh planet, Uranus, had been discovered accidentally by William Herschel in 1781 during a telescopic survey of the sky. Fifty years later, however, it was clear that Uranus was not following its predicted orbit. Two mathematicians,

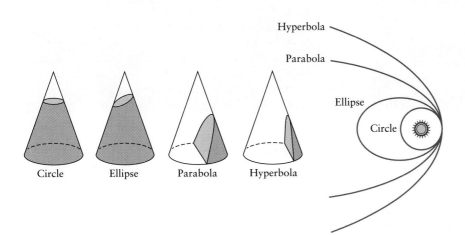

**Figure 4-21**

**Conic Sections** A conic section is any one of a family of curves obtained by slicing a cone with a plane. The orbit of one body about another can be any one of these curves: a circle, an ellipse, a parabola, or a hyperbola.

**figure 4-22** R I **V** U X G

**Halley's Comet** Halley's Comet orbits the Sun with an average period of about 76 years. During the twentieth century, the comet passed near the Sun in 1910 and again in 1986. It will next be prominent in the sky in 2061. This photograph shows how the comet looked in 1986. (Courtesy of J. Marling)

John Couch Adams in England and Urbain Le Verrier in France, independently calculated that the gravitational pull of a yet unknown, more distant planet could explain the deviations of Uranus from its orbit. They each predicted that the planet would be found at a certain location in the constellation of Aquarius. A brief telescopic search on September 23, 1846, revealed the planet Neptune within 1° of the calculated position. (Figure 4-23 shows photographs of both Uranus and Neptune.) Before it was sighted with a telescope, Neptune was actually discovered with pencil and paper.

Because it has been so successful in explaining and predicting many important phenomena, Newtonian mechanics has become the cornerstone of modern physical science. Even today, as we send astronauts into Earth orbit and spacecraft to the outer planets, Newton's equations are used to calculate the orbits and trajectories of these spacecraft.

In the twentieth century, scientists found that Newton's laws do not apply in all situations. A new theory called *quantum mechanics* had to be developed to explain the behavior of matter on the very smallest of scales, such as within the atom and within the atomic nucleus. Albert Einstein developed the *theory of relativity* to explain what happens at very high speeds approaching the speed of light and in places where gravitational forces are very strong. (We shall learn more about both of these theories in later chapters.) For many purposes in astronomy, however, Newton's laws are as useful today as when Newton formulated them more than three centuries ago.

**figure 4-23** R I **V** U X G

**Uranus and Neptune** The discovery of Neptune was a major triumph for Newtonian mechanics. In an effort to explain deviations in the predicted orbit of Uranus (shown on the left with three of its moons), astronomers predicted the existence of Neptune (shown on the right with one of its moons). Uranus and Neptune are nearly the same size; both have diameters about four times that of the Earth.

## Key Words

| | | |
|---|---|---|
| acceleration, p. 89 | greatest western elongation, p. 77 | Occam's razor, p. 75 |
| aphelion, p. 82 | heliocentric cosmogony, p. 75 | opposition, p. 77 |
| conic section, p. 93 | hyperbola, p. 93 | parabola, p. 93 |
| conjunction, p. 77 | inferior conjunction, p. 77 | parallax, p. 80 |
| cosmogony, p. 72 | inferior planet, p. 76 | perihelion, p. 82 |
| deferent, p. 74 | Kepler's first law, p. 81 | period (of a planet), p. 77 |
| direct motion, p. 73 | Kepler's second law, p. 82 | Ptolemaic system, p. 73 |
| eccentricity, p. 81 | Kepler's third law, p. 82 | retrograde motion, p. 73 |
| ellipse, p. 81 | law of equal areas, p. 82 | semimajor axis (of an ellipse), p. 81 |
| elongation, p. 77 | law of inertia, p. 88 | sidereal period, p. 77 |
| epicycle, p. 73 | major axis (of an ellipse), p. 81 | superior conjunction, p. 77 |
| focus (of an ellipse; *plural* foci), p. 81 | mass, p. 90 | superior planet, p. 76 |
| force, p. 88 | Newtonian mechanics, p. 93 | synodic period, p. 77 |
| geocentric cosmogony, p. 72 | Newton's first law of motion, p. 88 | universal constant of gravitation, p. 91 |
| gravitational force, p. 90 | Newton's form of Kepler's third law, p. 93 | universal law of gravitation, p. 91 |
| gravity, p. 90 | Newton's second law of motion, p. 90 | velocity, p. 88 |
| greatest eastern elongation, p. 77 | Newton's third law of motion, p. 90 | weight, p. 90 |

## Key Ideas

**Apparent Motions of the Planets:** Like the Sun and Moon, the planets move on the celestial sphere with respect to the background of stars. Most of the time a planet moves eastward in direct motion, in the same direction as the Sun and the Moon, but from time to time it moves westward in retrograde motion.

**Geocentric Cosmogony:** Ancient astronomers believed the Earth to be at the center of the universe. They invented a complex system of epicycles and deferents to explain the direct and retrograde motions of the planets on the celestial sphere.

**Heliocentric Cosmogony:** Copernicus's heliocentric (Sun-centered) theory simplified the general explanation of planetary motions.

• In a heliocentric system, the Earth is one of the planets orbiting the Sun.

• A planet undergoes retrograde motion as seen from Earth when the Earth and the planet pass each other.

• The sidereal period of a planet, its true orbital period, is measured with respect to the stars. Its synodic period is measured with respect to the Earth and the Sun (for example, from one opposition to the next).

**Evidence for the Heliocentric Model:** The invention of the telescope led Galileo to new discoveries that supported a heliocentric cosmogony. These included his observations of the phases of Venus and of the motions of four moons around Jupiter.

**Elliptical Orbits and Kepler's Laws:** Copernicus thought that the orbits of the planets were combinations of circles. Using data collected by Tycho Brahe, Kepler deduced three laws of planetary motion: (1) the orbits are in fact ellipses; (2) a planet's speed varies as it moves around its elliptical orbit; and (3) the orbital period of a planet is related to the size of its orbit.

**Newton's Laws of Motion:** Isaac Newton developed three principles, called the laws of motion, that apply to the motions of objects on Earth as well as in space. These are (1) the law of inertia, (2) the relationship between the net outside force on an object and the object's acceleration, and (3) the principle of action and reaction. These laws and Newton's universal law of gravitation can be used to deduce Kepler's laws. They lead to extremely accurate descriptions of planetary motions.

• The mass of an object is a measure of the amount of matter in the object. Its weight is a measure of the force with which the gravity of some other object pulls on it.

• In general, the path of one object about another, such as that of a planet or comet about the Sun, is one of the curves called conic sections: circle, ellipse, parabola, or hyperbola.

# Review Questions

**1.** What is an epicycle, and how is it important in Ptolemy's explanation of the retrograde motions of the planets?

**2.** What is the significance of Occam's razor as a tool for analyzing theories?

**3.** How did Copernicus explain the retrograde motion of the planets?

**4.** Which planets can never be seen at opposition? Which planets can never be seen at inferior conjunction? Explain your answers.

**5.** At what configuration (for example, superior conjunction, greatest eastern elongation, etc.) would it be best to observe Mercury or Venus with an Earth-based telescope? At what configuration would it be best to observe Mars, Jupiter, or Saturn? Explain your answers.

**6.** What is the difference between the synodic period and the sidereal period of a planet?

**7.** What observations did Tycho Brahe make in an attempt to test the heliocentric model? What were his results and why?

**8.** What observations did Galileo make that reinforced the heliocentric model? Why did these observations contradict the older model of Ptolemy?

**9.** What are Kepler's three laws? Why are they important?

**10.** A line joining the Sun and an asteroid is found to sweep out 4.9 AU² of space in the year 2001. How much area is swept out in 2002? Over a period of five years?

**11.** A comet with a period of 1000 years moves in a highly elongated orbit about the Sun. What is the comet's average distance from the Sun? What is the farthest it can get from the Sun? (*Hint:* The *closest* that a comet can get to the Sun is a very small distance indeed.)

**12.** The orbit of a spacecraft about the Sun has a perihelion distance of 0.5 AU and an aphelion distance of 3.5 AU. What is the semimajor axis of the spacecraft's orbit? What is its orbital period?

**13.** What are Newton's three laws? Give an everyday example of each law.

**14.** What is your weight in pounds and in newtons? What is your mass in kilograms?

**15.** How much force do you have to exert on a 1-kg brick to give it an acceleration of 4 m/s²? If you double this force, what is the brick's acceleration? Explain.

**16.** The mass of the Moon is $7.35 \times 10^{22}$ kg, while that of the Earth is $5.98 \times 10^{24}$ kg. The average distance from the center of the Moon to the center of the Earth is 384,400 km. What is the size of the gravitational force that the Earth and the Moon exert upon each other? How does this compare with the force between the Sun and the Earth calculated in the text?

**17.** Suppose that the Earth were moved to a distance of 5 AU from the Sun. How much stronger or weaker would the Sun's gravitational pull be on the Earth? Explain.

**18.** How far would you have to go from Earth to be completely beyond the pull of its gravity? Explain.

**19.** What is the difference between weight and mass?

**20.** What are conic sections, and in what way are they related to the orbits of planets in the solar system?

**21.** Why was the discovery of Neptune an important confirmation of Newton's universal law of gravitation?

# Advanced Questions

*Questions preceded by an asterisk (\*) involve topics discussed in the Boxes.*

### Problem-solving tips and tools:

Details about sidereal and synodic periods are supplied in Box 4-1. Data about the planets and their satellites are found in the appendixes at the back of this book. If you want to calculate the gravitational force that you feel on the surface of a planet, the distance *r* to use is the planet's radius (the distance between you and the center of the planet).

**\*22.** The synodic period of Venus (an inferior planet) is 583.92 days. Calculate its sidereal period.

**\*23.** Is it possible for a planet's sidereal period to equal its synodic period? Why or why not?

**24.** Look up the dates of the greatest eastern and western elongations of Mercury for any year you choose. (*Hint:* See Chapter 10.) Does it take longer to go from eastern to western elongation, or vice versa? Why do you suppose this is the case?

**25.** One trajectory that can be used to send spacecraft from the Earth to Venus is an ellipse that has its aphelion at the Earth and its perihelion at Venus. The spacecraft is launched from Earth and coasts along this ellipse until it reaches Venus, when a rocket is fired either to put the spacecraft into orbit around Venus or to cause it to land on Venus. (a) Find the semimajor axis of the ellipse. (*Hint:* Draw a picture showing this elliptical orbit along with the orbits of the Earth and Venus. The semimajor axis is half the length of the long axis of the ellipse. Treat the orbits of the Earth and Venus as circles.) (b) Calculate how long (in days) such a one-way trip to Venus would take.

**26.** A satellite is said to be in a "geosynchronous" orbit if it appears always to remain over the exact same spot on Earth. (a) What is the period of this orbit? (b) At what distance from the center of the Earth must such a satellite be placed into orbit? (c) Explain why the orbit must be in the plane of the Earth's equator.

27. Suppose that you traveled to a planet whose mass is the same as Earth's but whose diameter is three times as large. Would you weigh more or less on that planet than on Earth? By what factor?

28. If you landed on Mars, would you weigh more or less than you do on Earth? By what factor? See Appendix 2 for relevant data about Mars.

29. The mass of Saturn is approximately 100 times that of Earth, and the semimajor axis of Saturn's orbit is approximately 10 AU. To this approximation, how does the gravitational force that the Sun exerts on Saturn compare to the gravitational force that the Sun exerts on the Earth? How do the accelerations of Saturn and the Earth compare?

30. Suppose you have discovered an alien solar system in which a planet circles a star once every three years at an average distance of 9 AU. How does the mass of this star compare with that of our Sun? (Assume the planet's mass is very small.)

*31. In Box 4-2 we analyzed the orbit of Jupiter's moon Io. Look up information about the orbits of Jupiter's three other large moons (Europa, Ganymede, and Callisto) in the appendices. Demonstrate that these data are in agreement with Newton's form of Kepler's third law.

## Discussion Questions

32. Which planet would you expect to exhibit the greatest variation in apparent brightness as seen from Earth? Which planet would you expect to exhibit the greatest variation in angular diameter? Explain your answers.

33. Use two thumbtacks, a loop of string, and a pencil to draw several ellipses. Describe how the shape of an ellipse varies as the distance between the thumbtacks changes.

## Observing Projects

34. It is quite probable that within a few weeks of your reading this chapter one of the planets will be near opposition or greatest eastern elongation, making it readily visible in the evening sky. Select a planet that is at or near such a configuration by consulting a reference book, such as the current issue of the *Astronomical Almanac* or the pamphlet entitled *Astronomical Phenomena* (both published by the U.S. government), or by looking in magazines or on the World Wide Web (see below under "Where to Learn More"). At that configuration, would you expect the planet to be moving rapidly or slowly from night to night against the background stars? Verify your expectations by observing the planet once a week for a month, recording your observations on a star chart.

35. If Jupiter happens to be visible in the evening sky, observe the planet with a small telescope on five consecutive clear nights. Record the positions of the four Galilean satellites by making nightly drawings, just as the Jesuit priests did in 1620 (see Figure 4-17). From your drawings, can you tell which moon orbits closest to Jupiter and which orbits farthest? Was there a night when you could see only three of the moons? What do you suppose happened to the fourth moon on that night?

36. If Venus happens to be visible in the evening sky, observe the planet with a small telescope once a week for a month. On each night, make a drawing of the crescent that you see. From your drawings, can you determine if the planet is nearer or farther from the Earth than the Sun is? Do your drawings show any changes in the shape of the crescent from one week to the next? If so, can you deduce if Venus is coming toward us or moving away from us?

## Where to Learn More

*Books and magazine articles*

Cohen, I. B. "Newton's Discovery of Gravity." *Scientific American,* March 1981. An intriguing account of how Newton compared the real world with a mathematical model of it.

Caspar, M. *Kepler.* Dover, 1993. This biography describes how Kepler triumphed over physical handicaps and religious persecution to transform our view of the planets.

Drake, S. "Newton's Apple and Galileo's Dialogue." *Scientific American,* August 1980. This article shows how the work of Galileo influenced Newton's thinking.

————. *Galileo.* Oxford University Press, 1983. A slim, readable scientific biography, written by a distinguished historian of science.

Fauvel, J., et al. *Let Newton Be!* Oxford University Press, 1988. A lively collection of essays about Newton the man and about his contributions to astronomy and physics.

Gingerich, O. "How Galileo Changed the Rules of Science." *Sky & Telescope,* March 1993. This entertaining article describes the impact of Galileo's observations on science and the Roman Catholic Church.

————. *The Great Copernicus Chase, and Other Adventures in Astronomical History.* Sky Publishing/ Cambridge University Press, 1992. This collection of interesting essays covers a variety of topics from the period of Copernicus, Kepler, Galileo, and Newton.

Krupp, E. C. "Observing the Occasion." *Sky & Telescope,* December 1996. A short biography of the remarkably colorful life of Tycho Brahe, written on the occasion of his 450th birthday.

Thoren, V. E. *The Lord of Uraniborg: A Biography of Tycho Brahe.* Cambridge University Press, 1990. An in-depth look at Tycho's life and considerable accomplishments.

W *World Wide Web*

The web sites for the magazines *Sky & Telescope* (**http://www.skypub.com/**) and *Astronomy* (**http://www.astronomy.com/**) give up-to-date information about which planets are visible and where to look for them. (Detailed charts for the planets can be found in the magazines themselves.)

In addition to the planets, several spacecraft in orbit around Earth are visible to the naked eye. You can see the positions of these spacecraft, updated in real time, at a web site maintained by NASA's Marshall Space Flight Center (**http://liftoff.msfc.nasa.gov/RealTime/JTrack/Spacecraft.html**).

The Galileo Project web site at Rice University (**http://es.rice.edu/ES/humsoc/Galileo/**) has a wealth of information about Galileo's life and times. The site even includes directions for making a telescope like Galileo's using inexpensive materials. The Institute and Museum of the History of Science in Florence, Italy (**http://galileo.imss.firenze.it/**), has a multimedia exhibition on the life of Galileo, including images of his original telescope (and of the preserved middle finger of his right hand!). Biographies of Copernicus, Tycho, Kepler, and other figures of importance in the history of astronomy can also be found at this site.

Newton's laws and the law of universal gravitation are the principles by which spacecraft navigate through our solar system. An excellent on-line primer on spacecraft navigation is available from NASA's Jet Propulsion Laboratory (**http://www.jpl.nasa.gov/basics/**).

# The Nature of Light and Matter

Molecular
hydrogen

Neon

Lithium

Iron

Barium

Calcium

The Sun

Incandescent
lamp

Fluorescent
lamp

**Spectra and Spectral Lines** When an electric spark passes through a gas, atoms of the gas emit radiation at certain wavelengths. As a result, a spectrum of that gas contains bright spectral lines. Here the spectra of various elements are displayed. Each element has a unique pattern of spectral lines. Spectra of the Sun and two common household lamps are also shown. By analyzing the spectrum of light coming from a distant object in space, astronomers can determine such properties as the object's chemical composition, temperature, and motion through space. (Courtesy of Bausch and Lomb)

*In this chapter you will find the answers to the following questions:*

5-1 How fast does light travel? How can this speed be measured?

5-2 Why do we think light is a wave? What kind of wave is it?

5-3 How is the light from an ordinary light bulb different from the light emitted by a neon sign?

5-4 How can astronomers measure the temperatures of stars?

5-5 What is a photon? How does an understanding of photons help explain why ultraviolet light causes sunburns?

5-6 How can astronomers tell what distant celestial objects are made of?

5-7 What are atoms made of?

5-8 How does the structure of atoms explain what kind of light those atoms can emit or absorb?

5-9 How can we tell if a star or galaxy is approaching us or receding from us?

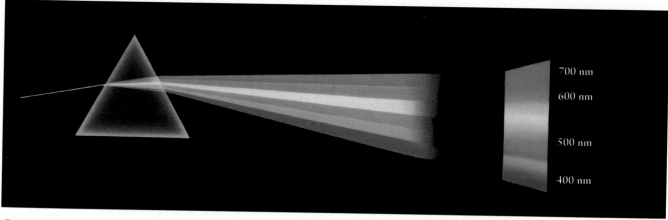

**figure 5-3**

**A Prism and a Spectrum** When a beam of white light passes through a glass prism, the light is broken into a rainbow-colored band called a spectrum. The numbers on the right side of the spectrum indicate the wavelengths of different colors of light.

in the form of waves rather than particles. Scientists now realize that light possesses both particlelike and wavelike characteristics.

The wavelike aspect of light was convincingly demonstrated around 1801 by the English experimenter Thomas Young. He passed a beam of light through two thin, parallel slits in an opaque screen, as shown in Figure 5-5a. On a white surface some distance beyond the slits, the light formed a pattern of alternating bright and dark bands. Young reasoned that if a beam of light was a stream of particles (as Newton had suggested), the two beams of light from

the slits should simply form bright images of the slits on the white surface. The pattern of bright and dark bands he observed is just what would be expected, however, if light had wavelike properties. An analogy with water waves demonstrates why.

ANALOGY Imagine ocean waves pounding against a reef or breakwater that has two openings (Figure 5-5b). A pattern of ripples is formed inside the barrier as the waves coming through the two openings interfere with each other. At certain points along the shore, crests arrive simul-

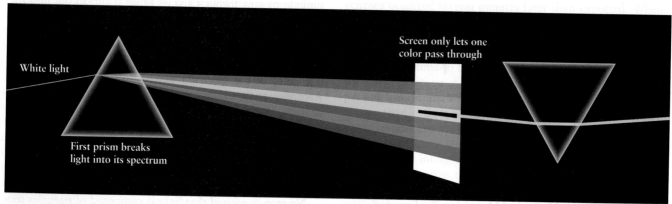

**figure 5-4**

**Newton's Experiment on the Nature of Light** In a crucial experiment, Newton took sunlight that had passed through a prism and sent it through a second prism. Between the two prisms was a screen with a hole in it that only allowed one color of the spectrum to pass through. This same color emerged from the second prism. Newton's experiment proved that prisms do not add color to light but merely bend different colors through different angles. It also proved that white light, such as sunlight, is actually a combination of all the colors that appear in its spectrum.

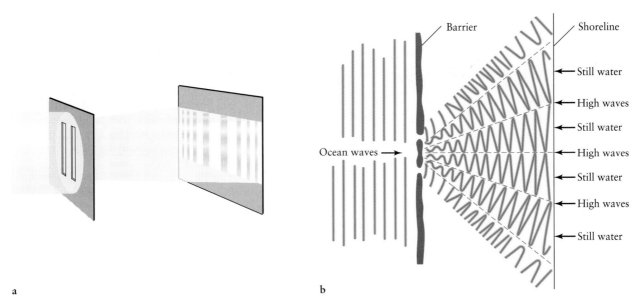

**figure 5-5**

**Young's Double-Slit Experiment** **(a)** In the modern laboratory, Thomas Young's classic double-slit experiment is most easily repeated by shining light from a laser onto two closely spaced parallel slits. Alternating dark and light bands appear on a screen beyond the slits. **(b)** The intensity of light on the screen in the double-slit experiment is analogous to the height of water waves that strike a shore after passing through a barrier with two openings. In certain locations, ripples from both openings reinforce each other to produce extra high waves. At intermediate locations along the shoreline, ripples and troughs cancel each other, producing still water.

taneously from the two openings, producing high waves. At intermediate points along the shore, a crest from one opening meets a trough from the other opening, thus canceling each other and producing bands of still water. A similar explanation accounts for the pattern of bright and dark bands that Young observed on the white surface behind the slits in his experiment.

The discovery of the wave nature of light posed some obvious questions. What exactly is "waving" in light? That is, what is it about light that goes up and down like water waves on the ocean? Because we can see light from the Sun, planets, and stars, light waves must be able to travel across empty space. Hence, whatever is "waving" cannot be any material substance. What, then, is it? The answer came from a seemingly unlikely source—a comprehensive theory that described electricity and magnetism. Numerous experiments during the first half of the nineteenth century demonstrated an intimate connection between electric and magnetic forces. A central idea to emerge from these experiments is the concept of a *field,* an immaterial yet measurable disturbance of any region of space in which electric or magnetic forces are felt. Thus, an electric charge is surrounded by an electric field, and a magnet is surrounded by a magnetic field. Experiments in the early 1800s demonstrated that moving electric charges produce a magnetic field; conversely, moving a magnet gives rise to an electric field.

In the 1860s, the Scottish mathematician and physicist James Clerk Maxwell succeeded in describing all the basic properties of electricity and magnetism in four equations. This mathematical achievement demonstrated that electric and magnetic forces are really two aspects of the same phenomenon, which we now call **electromagnetism.** By combining his four equations, Maxwell showed that electric and magnetic fields should travel through space in the form of waves at a speed of $3 \times 10^8$ m/s—exactly equal to the best available value for the speed of light. Maxwell's suggestion that these waves do exist and are observed as light was soon confirmed by experiments. Because of its electric and magnetic properties, light is also called **electromagnetic radiation.**

CAUTION You may associate the term *radiation* with radioactive materials like uranium, but this term refers to anything that radiates, or spreads away, from its source. For example, scientists sometimes refer to sound waves as "acoustic radiation." Radiation does not have to be related to radioactivity!

Electromagnetic radiation consists of perpendicular, oscillating electric and magnetic fields, as shown in Figure 5-6. The distance between two successive wave crests is called the **wavelength** of the light, usually designated by the Greek letter $\lambda$ (lambda). No matter what the wavelength, electromagnetic

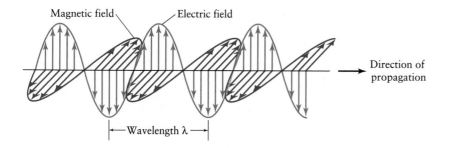

Magnetic field   Electric field

Direction of propagation

Wavelength λ

**ƒigure 5-6**

**Electromagnetic Radiation** All forms of light consist of oscillating electric and magnetic fields that move through space at a speed of $3 \times 10^8$ m/s. This figure shows a "snapshot" of these fields at one instant. The distance between two successive crests, called the wavelength of the light, is usually designated by the Greek letter λ (lambda).

radiation always travels at the same speed $c = 3 \times 10^8$ m/s in a vacuum.

More than a century elapsed between Newton's experiments with a prism and the confirmation of the wave nature of light. One reason for this delay is that the wavelength of **visible light** is extremely short—less than a thousandth of a millimeter—and thus is not easily detectable. To express such tiny distances conveniently, scientists use a unit of length called the *nanometer* (abbreviated nm), where 1 nm = $10^{-9}$ m. Experiments demonstrated that visible light has wavelengths covering the range from about 400 nm for violet light to about 700 nm for red light. Intermediate colors of the rainbow like yellow (550 nm) have intermediate wavelengths, as shown in Figure 5-7. (Some astronomers prefer to measure wavelengths in *angstroms*. One angstrom, abbreviated Å, is one-tenth of a nanometer: 1 Å = 0.1 nm = $10^{-10}$ m. In these units, the wavelengths of visible light extend from about 4000 Å to about 7000 Å. We will not use these units in this book, however.)

Maxwell's equations place no restrictions on the wavelength of electromagnetic radiation. Hence, electromagnetic waves could and should exist with wavelengths both longer and shorter than the 400–700 nm range of visible light. Consequently, researchers began to look for *invisible* forms of light. These are forms of electromagnetic radiation to which the cells of the human retina do not respond.

The first kind of invisible radiation to be discovered actually preceded Maxwell's work by more than a half century. Around 1800 the British astronomer William Herschel passed sunlight through a prism and held a thermometer just beyond the red end of the visible spectrum. The thermometer registered a temperature increase, indicating that it was being exposed to an invisible form of energy. This invisible energy, now called **infrared radiation,** was later realized to be electromagnetic radiation with wavelengths somewhat longer than those of visible light. In experiments with electric sparks in 1888, the German physicist Heinrich Hertz succeeded in producing electromagnetic radiation with even longer wavelengths of a few centimeters or more. These are now known as **radio waves.** In 1895 another German physicist, Wilhelm Röntgen, invented a machine that produces electromagnetic radiation with wavelengths shorter than 10 nm, now known as **X rays.** The X-ray machines in mod-

ern medical and dental offices are direct descendants of Röntgen's invention. Over the years radiation has been discovered with many other wavelengths.

Visible light occupies only a tiny fraction of the full range of possible wavelengths, collectively called the **electromagnetic spectrum.** As shown in Figure 5-7, the electromagnetic spectrum stretches from the longest-wavelength radio waves to the shortest-wavelength gamma rays.

On the long-wavelength side of the visible spectrum, infrared radiation covers the range from about 700 nm to 1 mm. Astronomers interested in infrared radiation often express wavelength in *micrometers* or *microns*, abbreviated μm, where 1 μm = $10^3$ m = $10^{-6}$ m. **Microwaves** have wavelengths from roughly 1 mm to 10 cm, while radio waves have even longer wavelengths.

At wavelengths shorter than those of visible light, **ultraviolet radiation** extends from about 400 nm down to 10 nm. Next are X rays, which have wavelengths between about 10 and 0.01 nm, and beyond them at even shorter wavelengths are **gamma rays.** It should be noted that these rough boundaries are simply arbitrary divisions in the electromagnetic spectrum.

Astronomers who work with radio telescopes often prefer to speak of *frequency* rather than wavelength. The **frequency** of a light wave is the number of wave crests that pass a given point in one second. Equivalently, it is the number of complete *cycles* of the wave that pass per second (a complete cycle is from one crest to the next). Frequency is usually denoted by the Greek letter ν (nu). The unit of frequency is the cycle per second, also called the hertz (abbreviated Hz) in honor of Heinrich Hertz, the physicist who first produced radio waves. For example, if 500 crests of a wave pass you in one second, the frequency of the wave is 500 cycles per second or 500 Hz.

In working with frequencies, it is often convenient to use the prefix *mega-* (meaning "million," or $10^6$, and abbreviated M) or *kilo-* (meaning "thousand," or $10^3$, and abbreviated k). For example, AM radio stations broadcast at frequencies between 535 and 1605 kHz, while FM radio stations broadcast at frequencies in the range from 88 to 108 MHz.

There is a simple relationship between the frequency and wavelength of an electromagnetic wave. Because light moves at a constant speed $c = 3 \times 10^8$ m/s, if the wavelength (dis-

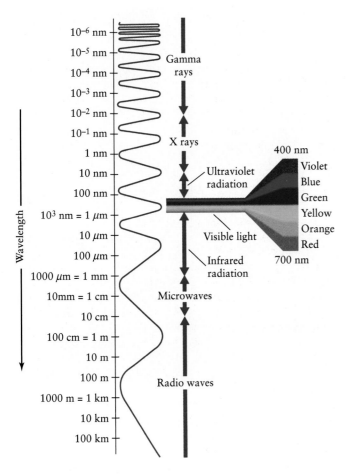

**Figure 5-7**

**The Electromagnetic Spectrum** The full array of all types of electromagnetic radiation is called the electromagnetic spectrum. It extends from the longest-wavelength radio waves to the shortest-wavelength gamma rays. Visible light occupies only a tiny portion of the full electromagnetic spectrum.

tance from one crest to the next) is made shorter, the frequency must increase (more of those closely spaced crests pass you each second). Mathematically, the frequency ν of light is related to its wavelength λ by

**Frequency and wavelength of an electromagnetic wave**

$$\nu = \frac{c}{\lambda}$$

ν = frequency of an electromagnetic wave (in Hz)
c = speed of light = $3 \times 10^8$ m/s
λ = wavelength of the wave (in meters)

For example, hydrogen atoms in space emit radio waves with a wavelength of 21.12 cm. To calculate the frequency of this radiation, we must first express the wavelength in

meters rather than centimeters: λ = 0.2112 m. Then we can use the above formula to find the frequency ν:

$$\nu = \frac{c}{\lambda} = \frac{3 \times 10^8 \text{ m/s}}{0.2112 \text{ m}} = 1.42 \times 10^9 \text{ Hz} = 1420 \text{ MHz}$$

Visible light has a much shorter wavelength and higher frequency that radio waves. You can use the above formula to show that for yellow-orange light of wavelength 600 nm, the frequency is $5 \times 10^{14}$ Hz or 500 *million* megahertz!

## 5-3 A dense object emits electromagnetic radiation according to its temperature

To learn about objects in the heavens, astronomers study the character of the electromagnetic radiation coming from those objects. Such studies can be very revealing because different kinds of electromagnetic radiation are typically produced in different ways. As an example, on Earth the most common way to generate radio waves is to make an electric current oscillate back and forth (as is done in the broadcast antenna of a radio station). By contrast, X rays for medical and dental purposes are usually produced by bombarding atoms with fast-moving particles extracted from within other atoms. Our own Sun emits radio waves from near its glowing surface and X rays from its corona (see Figure 3-11). Hence, these observations indicate the presence of electric currents near the Sun's surface, and of fast-moving particles in the Sun's outermost regions. (We will discuss the Sun at length in Chapter 18.)

The simplest and most common way to produce electromagnetic radiation, either on or off the Earth, is to heat an object. The hot filament of wire inside an ordinary light bulb emits white light, and a neon sign has a characteristic red glow because neon gas within the tube is heated by an electric current. In like fashion, almost all of the visible light that we receive from space comes from hot objects like the Sun and the stars. The kind and amount of light emitted by a hot object tells us not only how hot it is but also about other properties of the object.

We can tell whether the hot object is made of relatively dense or relatively thin material. Consider the difference between a light bulb and a neon sign. The dense, solid filament of a light bulb makes white light, which is a mixture of all different visible wavelengths, while the thin neon gas produces light of a rather definite red color and hence a rather definite wavelength. For now we will concentrate our attention on the light produced by dense objects. (We will return to gases in Section 5-6.) Even though the Sun and stars are gaseous, not solid, it turns out that they emit light with many of the same properties as a hot, glowing, solid object.

Imagine a welder or blacksmith heating a bar of iron. As the bar becomes hot, it begins to glow deep red, as shown in Figure 5-8a. (You can see this same glow from the coils of a toaster, or from an electric range turned on "high.") As the temperature rises farther, the bar begins to give off a brighter, reddish-orange light (Figure 5-8b). At still higher temperatures, it shines with a brilliant yellowish-white light (Figure 5-8c). If the bar could be prevented from melting and vaporizing, at extremely high temperatures it would emit a dazzling blue-white light.

As this example shows, the amount of energy emitted by the hot, dense object and the dominant wavelength of the emitted radiation both depend on the temperature of the object. The hotter the object, the more energy it emits and the shorter the wavelength at which most of the energy is emitted. Colder objects emit relatively little energy, and this emission is primarily at long wavelengths.

These observations explain why you can't see in the dark. The temperatures of people, animals, and furniture are rather less than even the iron bar in Figure 5-8a. So, while these objects emit radiation even in a darkened room, most of this emission is at wavelengths greater than those of red light, in the infrared part of the spectrum (see Figure 5-7). Your eye is not sensitive to infrared, and you thus cannot see ordinary objects in a darkened room. But you *can* detect this

**figure 5-9**  R **I** V U X G

**Infrared Radiation from Room-Temperature Objects** This image of a boy and his dog was made with a camera that is sensitive to infrared radiation. The different colors in the image represent regions of different temperature; white denotes the warmest areas that emit the most infrared light, while black areas (including the dog's cold nose) are at the lowest temperatures and emit the least radiation. (R. P. Clark and M. Goff)

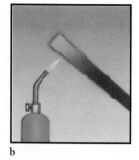

**figure 5-8**

**Heating a Bar of Iron** This sequence of drawings shows how the appearance of a heated bar of iron changes with temperature. As the temperature increases, the amount of energy radiated by the bar increases. The color of the bar also changes as the temperature goes up because the dominant wavelength of light emitted by the bar decreases.

radiation by using a camera that is sensitive to infrared light (Figure 5-9).

To better understand the relationship between the temperature of a dense object and the radiation it emits, it is helpful to know just what "temperature" means. The temperature of a substance is directly related to the average speed of the tiny **atoms** that make up the substance. (Typical atoms are about $10^{-10}$ m = 0.1 nm in diameter, or about $\frac{1}{5000}$ as large as a typical wavelength of visible light). If something is hot, its atoms are moving at high speeds; if a substance is cold, its atoms are moving slowly. In discussing temperature, scientists usually prefer the Kelvin temperature scale, on which temperature is measured in **kelvins** (K) upward from **absolute zero**. This is the coldest possible temperature, at which atoms move as slowly as possible (they can never quite stop completely). On the more familiar Celsius and Fahrenheit temperature scales, absolute zero (0 K) is –273°C and –460°F. Ordinary room temperature is 293 K, 20°C, or 68°F. The relationships between the Kelvin, Celsius, and Fahrenheit temperature scales are discussed in Box 5-1.

## box 5-1 | Tools of the Astronomer's Trade

### *Temperatures and Temperature Scales*

Three temperature scales are in common use. Throughout most of the world, temperatures are expressed in **degrees Celsius** (°C). The Celsius temperature scale is based on the behavior of water, which freezes at 0°C and boils at 100°C at sea level on Earth. This scale is named in honor of the Swedish astronomer Anders Celsius, who proposed it in 1742.

Astronomers usually prefer the Kelvin temperature scale. This is named after the nineteenth-century British physicist Lord Kelvin, who made many important contributions to our understanding of heat and temperature. Absolute zero, the temperature at which atomic motion is at the absolute minimum, is –273°C in the Celsius scale but 0 K in the Kelvin scale. Because it is impossible for atomic motion to be any less than the minimum, nothing can be colder than 0 K; hence, there are no negative temperatures on the Kelvin temperature scale. Note that we do *not* use degree (°) with the Kelvin temperature scale.

A temperature expressed in kelvins is always equal to the temperature in degrees Celsius plus 273. On the Kelvin scale, water freezes at 273 K and boils at 373 K. Water must be heated through a change of 100 K or 100°C to go from its freezing point to its boiling point. Thus, the "size" of a kelvin is the same as the "size" of a Celsius degree. When considering temperature changes, measurements in kelvins and Celsius degrees are the same.

The now-archaic Fahrenheit scale, which expresses temperature in **degrees Fahrenheit** (°F), is used only in the United States. When the German physicist Gabriel Fahrenheit introduced this scale in the early 1700s, he intended 0°F to represent the temperature of a healthy human body. On the Fahrenheit scale, water freezes at 32°F and boils at 212°F. There are 180 Fahrenheit degrees between the freezing and boiling points of water, so a degree Fahrenheit is only 100/180 = 5/9 as large as either a Celsius degree or a kelvin.

Two simple equations allow you to convert a temperature from the Celsius scale to the Fahrenheit scale and from Fahrenheit to Celsius:

$$T_F = \frac{9}{5} T_C + 32$$

$$T_C = \frac{5}{9} (T_F - 32)$$

$T_F$ = temperature in degrees Fahrenheit
$T_C$ = temperature in degrees Celsius

**EXAMPLE:** A typical room temperature is 68°F. We can convert this to the Celsius scale using the second equation:

$$T_C = \frac{5}{9} (68 - 32) = 20°C$$

To convert this to the Kelvin scale, we simply add 273 to the Celsius temperature. Thus,

$$68°F = 20°C = 293 K$$

The diagram displays the relationships between these three temperature scales.

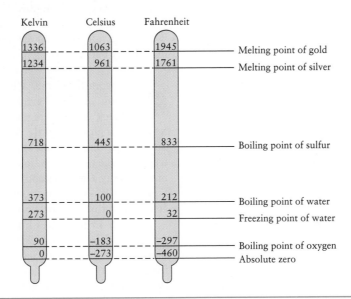

| Kelvin | Celsius | Fahrenheit | |
|---|---|---|---|
| 1336 | 1063 | 1945 | Melting point of gold |
| 1234 | 961 | 1761 | Melting point of silver |
| 718 | 445 | 833 | Boiling point of sulfur |
| 373 | 100 | 212 | Boiling point of water |
| 273 | 0 | 32 | Freezing point of water |
| 90 | –183 | –297 | Boiling point of oxygen |
| 0 | –273 | –460 | Absolute zero |

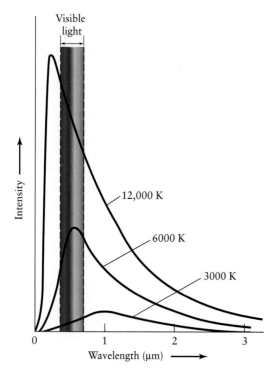

**Figure 5-10**

**Blackbody Curves** A blackbody curve shows the intensity of light at every wavelength that is emitted by a blackbody (an idealized case of a dense object) at a particular temperature. Three such curves are shown here. The higher the temperature, the shorter the wavelength of maximum emission $\lambda_{max}$ (where the curve peaks) and the greater the amount of light emitted at every wavelength. Note that all three of these blackbodies emit substantial amounts of light in the visible range (shown by the rainbow-colored band). Note also that the vertical scale has been compressed so that all three curves can be seen; the peak intensity for the 12,000-K curve is actually about 1,000 times greater than the peak intensity for the 3000-K curve. The intensity for a blackbody at room temperature (around 300 K) is so low at all wavelengths that the corresponding blackbody curve could not be seen on this graph; it would nearly coincide with the horizontal axis.

Figure 5-10 depicts quantitatively how the radiation from a dense object depends on its Kelvin temperature. Each curve in this figure shows the intensity of light emitted at each wavelength by a dense object at a given temperature. In other words, the curves show the *spectrum* of light emitted by such an object. At any temperature, a hot, dense object emits at all wavelengths, so its spectrum is a smooth, continuous curve with no gaps in it. The shape of the spectrum depends on temperature, however. An object at relatively low temperature (say, 3000 K) has a low curve, indicating

a low intensity of radiation. The **wavelength of maximum emission,** at which the curve has its peak and the emission of energy is strongest, is at a long wavelength. The higher the temperature, the higher the curve (indicating greater intensity) and the shorter the wavelength of maximum emission.

*CAUTION!* Figure 5-10 shows that for a dense object at a temperature of 3000 K, the wavelength of maximum emission is around 1 μm (1000 nm). Because this is an infrared wavelength well outside the visible range, you might think that you cannot see the radiation from an object at this temperature. In fact, the glow from such an object *is* visible; the curve shows that this object emits plenty of light within the visible range, as well as at even shorter wavelengths. The 3000-K curve is quite a bit higher at the red end of the visible spectrum than at the violet end, so a dense object at this temperature will appear red in color. Similarly, the 12,000-K curve has its wavelength of maximum emission in the ultraviolet part of the spectrum, at a wavelength shorter than visible light. But such a hot, dense object also emits copious amounts of visible light (much more than at 6000 K or 3000 K, for which the curves are lower) and thus will have a very visible glow. The curve for this temperature is higher for blue light than for red light, and so the color of a dense object at 12,000 K is a brilliant blue or blue-white. These conclusions agree with the color changes of a heated rod shown in Figure 5-8. The same principles apply to stars: A star that looks blue has a high surface temperature, while a red star has a relatively cool surface.

The curves in Figure 5-10 are drawn for an idealized type of dense object called a **blackbody**. A perfect blackbody does not reflect any light at all; instead, it absorbs all radiation falling on it. Because it reflects no electromagnetic radiation, the radiation that it does emit is entirely the result of its temperature. Ordinary objects, like tables, textbooks, and people, are not perfect blackbodies; they reflect light, which is why they are visible. A star such as the Sun, however, behaves very much like a perfect blackbody, because it absorbs almost completely any radiation falling on it from outside. The curves in Figure 5-10 are often called **blackbody curves.**

*CAUTION!* Despite its name, a blackbody does not necessarily *look* black. The Sun, for instance, does not look black because its temperature is high (around 5800 K) and so it glows brightly. But a room-temperature (around 300 K) blackbody would appear very black indeed. Even if it were as large as the Sun, it would emit only about $\frac{1}{100,000}$ as much energy. (Its blackbody curve is far too low to graph in Figure 5-10.) Furthermore, most of this radiation would be at wavelengths that are too long for our eyes to perceive.

Figure 5-11 illustrates how accurately a star like the Sun can be modeled as a blackbody. The graph shows how the intensity of sunlight varies with wavelength, as measured from above the Earth's atmosphere. (This is necessary because the Earth's atmosphere absorbs certain wavelengths.)

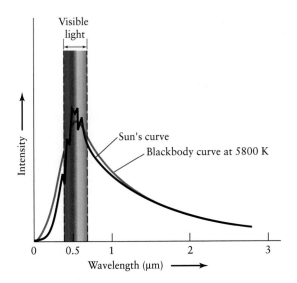

## figure 5-11

**The Sun as a Blackbody** This graph compares the intensity of sunlight over a wide range of wavelengths (shown as a black curve) with the intensity of radiation coming from a blackbody at a temperature of 5800 K (shown as a red curve). The measurements of the Sun's intensity were made above the Earth's atmosphere. The Sun mimics a blackbody remarkably well. Note that the Sun's spectrum peaks in the visible spectrum. This is not surprising, because the human eye evolved to take advantage of the most plentiful light available.

The peak of the curve is at a wavelength of about 0.5 μm or 500 nm, corresponding to yellow light. The blackbody curve for a temperature of 5800 K is also plotted in Figure 5-11. Note how closely the observed intensity curve for the Sun matches the blackbody curve. This is a strong indication that the temperature of the Sun's glowing surface is about 5800 K—a temperature that we can measure across a distance of 150 million kilometers! The close correlation between blackbody curves and the observed intensity curves for most stars is the paramount reason why astronomers are interested in the physics of blackbody radiation.

An important attribute of **blackbody radiation** is that it depends only on the temperature of the object emitting the radiation, not on the chemical composition of the object. The light emitted by molten gold at 2000 K is very nearly the same as that emitted by molten lead at 2000 K. Therefore, it might seem that analyzing the light from a star can tell astronomers the star's temperature but not what the star is made of. As Figure 5-11 shows, however, the intensity curve for the Sun (a typical star) is not precisely that of a blackbody. We will see later in this chapter that the differences between a star's spectrum and that of a blackbody can in fact be used to determine the chemical composition of the star.

## 5-4 Wien's law and the Stefan-Boltzmann law are useful tools for analyzing glowing objects like stars

The mathematical formula that describes the blackbody curves in Figure 5-10 is a rather complicated one. But there are two simpler formulas for blackbody radiation that prove to be very useful in many branches of astronomy. We will use both of these formulas many times, not only in our study of the stars, but also as we explore the planets (which are dense, relatively cool objects that emit infrared radiation). One of these formulas relates the temperature of a blackbody to its wavelength of maximum emission, and the other relates the temperature to the amount of energy that the blackbody emits.

Figure 5-10 shows that the higher the temperature ($T$) of a blackbody, the shorter its wavelength of maximum emission ($\lambda_{max}$). In 1893 the German physicist Wilhelm Wien used ideas about both heat and electromagnetism to make this relationship quantitative. The formula that he derived, which today is called **Wien's law,** is

**Wien's law for a blackbody**

$$\lambda_{max} = \frac{0.0029}{T}$$

$\lambda_{max}$ = wavelength of maximum emission of the object (in meters)

$T$ = temperature of object (in kelvins)

According to Wien's law, the wavelength of maximum emission of a blackbody is inversely proportional to its temperature in kelvins. In other words, if the temperature of the blackbody doubles, its wavelength of maximum emission is halved, and vice versa. For example, Figure 5-10 shows blackbody curves for temperatures of 3000 K, 6000 K, and 12,000 K. From Wien's law, a blackbody with a temperature of 6000 K has a wavelength of maximum emission $\lambda_{max}$ = $(0.0029)/(6000) = 4.8 \times 10^{-7}$ m, or 0.48 μm, in the visible part of the electromagnetic spectrum. At 12,000 K, or twice the temperature, the blackbody has a wavelength of maximum emission half as great, or $\lambda_{max} = 0.24$ nm; this is in the ultraviolet. At 3000 K, just half our original temperature, the value of $\lambda_{max}$ is twice the original value—0.96 μm, which is an infrared wavelength. You can see that these wavelengths agree with the peaks of the curves in Figure 5-10.

Wien's law is very useful for determining the surface temperatures of stars. It is not necessary to know how far away the star is, how large it is, or how much energy it radiates into space. All we need to know is the dominant wavelength of the star's electromagnetic radiation. Box 5-2 includes two additional examples of using Wien's law to relate the temperature $T$ of a star (which is an almost perfect blackbody) to the star's wavelength of maximum emission.

The other useful formula for the radiation from a blackbody involves the *total* amount of energy the blackbody

## box 5-2 | Tools of the Astronomer's Trade

### *Using the Laws of Blackbody Radiation*

The Sun and stars behave like nearly perfect black-bodies. Wien's law and the Stefan-Boltzmann law can therefore be used to relate a star's surface temperature to the energy flux and wavelength of maximum emission of the star's radiation. This is demonstrated in the following examples.

**EXAMPLE:** The surface temperature of the Sun can be determined using Wien's law. The Sun emits energy over a wide range of wavelengths, but the maximum intensity of sunlight is at a wavelength of roughly 500 nm = $5 \times 10^{-7}$ m. From Wien's law, we find the Sun's surface temperature $T_\odot$ to be

$$T_\odot = \frac{0.0029}{\lambda_{max}} = \frac{0.0029}{5 \times 10^{-7}} = 5800 \text{ K}$$

(The symbol $\odot$ is the standard astronomical symbol for the Sun.) This is about the same temperature as an iron welding arc.

**EXAMPLE:** We can also find the surface temperature of the Sun using the Stefan-Boltzmann law. Using detectors above the Earth's atmosphere, astronomers have measured the average flux of solar energy arriving at Earth; this value, called the **solar constant**, is equal to 1370 W/m². However, the quantity $F$ in the Stefan-Boltzmann law refers to the flux measured at the Sun's surface, not at the Earth.

To determine the value of $F$, we first imagine a huge sphere of radius 1 AU with the Sun at its center, as shown in the figure. Each square meter of that sphere received 1370 watts of power from the Sun, so the total energy radiated by the Sun per second is equal to the solar constant multiplied by the

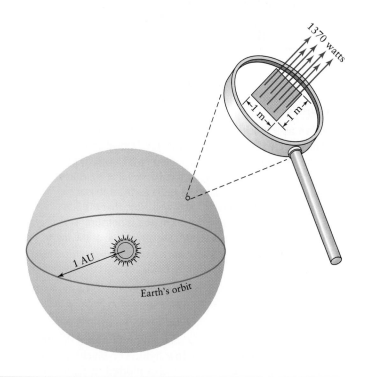

radiates at *all* wavelengths. (By contrast, the curves in Figure 5-10 show how much energy a blackbody radiates at each individual wavelength.)

Energy is usually measured in joules (J), named after the nineteenth-century English physicist James Joule. A **joule** is the amount of energy contained in the motion of a 2-kilogram mass moving at a speed of one meter per second. The joule is a convenient unit of energy because it is closely related to the familiar watt (W): 1 watt is 1 joule per second, or 1 W = 1 J/s. For example, the energy consumption of a 100-watt light bulb is 100 J/s. The energy content of food is also often measured in joules; in most of the world, diet soft drinks are labeled as "low joule" rather than "low calorie."

The amount of energy emitted by a blackbody depends both on its temperature and on its surface area. This makes

sense—a large burning log radiates much more heat than a burning match, even though the temperatures are the same. To consider the effects of temperature alone, it is convenient to look at the amount of energy emitted from each square meter of an object's surface in a second. This quantity is called the **energy flux** ($F$). Flux means "rate of flow," and thus $F$ is a measure of how much energy is flowing out of the object. It is measured in joules per square meter per second, usually written as J/m²–s or J m⁻² s⁻¹. (The superscript –1 means you are dividing by that quantity.) Alternatively, because 1 watt equals 1 joule per second, we can express flux in watts per square meter (W/m², or W m⁻²).

In 1879 the Austrian physicist Josef Stefan summarized the results of his experiments. The flux from a blackbody is proportional to the fourth power of the object's temperature

sphere's surface area. The result, called the **luminosity** of the Sun and denoted by the symbol $L_\odot$, is $L_\odot = 3.90 \times 10^{26}$ W. That is, in one second the Sun radiates $3.90 \times 10^{26}$ joules of energy into space.

Because we know the size of the Sun, we can compute the energy flux (energy emitted per square meter per second) at its surface. The radius of the Sun is $R_\odot = 6.96 \times 10^8$ m, and its surface area is $4\pi R_\odot^2$. Therefore, its energy flux is the luminosity (total energy emitted by the Sun per second) divided by the surface area (the number of square meters of surface):

$$F_\odot = \frac{L_\odot}{4\pi R_\odot^2} = \frac{3.90 \times 10^{26}\,\text{W}}{4\pi\,(6.96 \times 10^8\text{m})^2} = 6.41 \times 10^7\ \text{W/m}^2$$

Notice that the solar constant of 1370 W/m$^2$ is very much less than this; by the time the Sun's radiation reaches Earth, it is spread over a greatly increased area.

Once we have the Sun's flux $F_\odot$, we can use the Stefan-Boltzmann law to find the Sun's surface temperature $T_\odot$:

$$T_\odot^4 = \frac{F_\odot}{\sigma} = 1.13 \times 10^{15}\ \text{K}^4$$

Taking the fourth root (the square root of the square root) of this value, we find the surface temperature of the Sun to be $T_\odot = 5800$ K. This result agrees with the value we computed in the previous example using Wien's law.

**EXAMPLE:** Sirius, the brightest star in the night sky, has a surface temperature of about 10,000 K. We can use Wien's law to calculate the wavelength ($\lambda_{max}$) at which Sirius emits most intensely:

$$\lambda_{max} = \frac{0.0029}{T} = \frac{0.0029}{10,000} = 2.9 \times 10^{-7}\text{m} = 290\ \text{nm}$$

Sirius therefore emits light most intensely in the ultraviolet. In the visible part of the spectrum, it emits more blue light than red light (see the curve for 12,000 K in Figure 5-10), so Sirius has a distinct blue color.

**EXAMPLE:** We can use the Stefan-Boltzmann law to compare the energy flux from Sirius with that from the Sun. For the Sun, the Stefan-Boltzmann law is $F_\odot = \sigma T_\odot^4$, and for Sirius we can likewise write $F_* = \sigma T_*^4$, where the subscripts $\odot$ and $*$ refer to the Sun and Sirius, respectively. If we divide one equation by the other, the Stefan-Boltzmann constants cancel out, giving us

$$\frac{F_*}{F_\odot} = \frac{T_*^4}{T_\odot^4}$$

In the preceding examples, we found that the temperatures of Sirius and the Sun are 10,000 K and 5800 K, respectively. Thus, we get

$$\frac{F_*}{F_\odot} = \frac{(10,000)^4}{(5800)^4} = \left(\frac{10,000}{5800}\right)^4 = 8.8$$

In other words, each square meter of Sirius's surface emits 8.8 times more energy per second than a square meter of the Sun's surface.

(measured in kelvins). Five years after Stefan announced his law, another Austrian physicist, Ludwig Boltzmann, showed how it could be derived mathematically from basic assumptions about atoms and molecules. For this reason Stefan's law is commonly known as the **Stefan-Boltzmann law**. Written as an equation, the Stefan-Boltzmann law is

**Stefan-Boltzmann law for a blackbody**

$$F = \sigma T^4$$

$F$ = energy flux, in joules per square meter of surface per second

$\sigma$ = a constant

$T$ = object's temperature, in kelvins

The value of the constant $\sigma$ (the Greek letter sigma) is known from laboratory experiments to have the value

$$\sigma = 5.67 \times 10^{-8}\ \text{W m}^{-2}\ \text{K}^{-4}$$

The Stefan-Boltzmann law says that if you double the temperature of an object (for example, from 300 K to 600 K), then the energy emitted from the object's surface each second increases by a factor of $2^4 = 16$. If you increase the temperature by a factor of 10 (for example, from 300 to 3000 K), the rate of energy emission increases by a factor of $10^4 = 10,000$. Thus, a chunk of iron at room temperature (around 300 K) emits very little electromagnetic radiation (and essentially no visible light), but an iron bar heated to 3000 K glows quite intensely.

Box 5-2 shows how the Stefan-Boltzmann law can be used to calculate the surface temperature of the Sun and to compare the Sun to other stars.

## 5-5 Light has properties of both waves and particles

Because the light emitted by stars is so close to that emitted by a blackbody, it is useful to know as much about blackbody radiation as possible. Indeed, a major research topic in the late nineteenth century was to explain on theoretical grounds all of the characteristics of blackbody radiation. Young's 1801 experiment (Section 5-2) had firmly established that light is a wave, so various theories of blackbody radiation were developed based on a wave picture of light. But all such theories failed to explain the characteristic shapes of blackbody curves shown in Figure 5-10.

In 1900, however, the German physicist Max Planck discovered that he could derive a formula that correctly described blackbody curves, if he assumed that electromagnetic energy is emitted in discrete, particlelike packets. He further had to assume that the energy of each such packet, today called a **photon**, is related to the wavelength of light: The greater the wavelength, the lower the energy of a photon associated with that wavelength. Thus, a photon of red light (wavelength $\lambda = 700$ nm) has less energy than a photon of violet light ($\lambda = 400$ nm). In this picture, light has a dual personality; it behaves as a stream of particlelike photons, but each photon has wavelike properties. In this sense, the best answer to the question "Is light a wave or a stream of particles?" is "Yes!"

It was soon realized that Planck's photon hypothesis explains more than just the detailed shape of blackbody curves. For example, it explains why only ultraviolet light causes suntans and sunburns. The reason is that tanning or burning involves a chemical reaction in the skin. High-energy, short-wavelength ultraviolet photons can trigger these reactions, but the lower-energy, longer-wavelength photons of visible light cannot. Similarly, normal photographic film is sensitive to visible light but not to infrared light; a long-wavelength infrared photon does not have enough energy to cause the chemical change that occurs when film is exposed to the higher-energy photons of visible light.

Another phenomenon explained by the photon hypothesis is the **photoelectric effect**. In this effect, a metal plate is illuminated by a light beam. If ultraviolet light is used, tiny negatively charged particles called **electrons** are emitted from the metal plate. (We will see in Section 5-6 that the electron is one of the basic particles of the atom.) But if visible light is used, no matter how bright, no electrons are emitted. In 1905 the great German physicist Albert Einstein explained this by noting that a certain minimum amount of energy is required to remove an electron from the metal

plate. The energy of a short-wavelength ultraviolet photon is greater than this minimum value, so an electron that absorbs a photon of ultraviolet light will have enough energy to escape from the plate. But an electron that absorbs a photon of visible light, with its longer wavelength and lower energy, does not gain enough energy to escape and so remains within the metal. Einstein and Planck both won Nobel Prizes for their contributions to understanding the nature of light.

The relationship between the energy $E$ of a single photon and the wavelength of the electromagnetic radiation can be expressed in a simple equation:

**Energy of a photon (in terms of wavelength)**

$$E = \frac{hc}{\lambda}$$

$E$ = energy of a photon
$h$ = Planck's constant
$c$ = speed of light
$\lambda$ = wavelength of light

The value of the constant $h$ in this equation, now called *Planck's constant,* has been shown in laboratory experiments to be

$$h = 6.625 \times 10^{-34} \text{ J s}$$

Because this is a tiny value, a single photon carries a very small amount of energy. For example, a photon of red light with wavelength 633 nm has an energy of only $3.14 \times 10^{-19}$ J (Box 5-3). This is why we ordinarily do not notice that light comes in the form of photons; even a dim light source emits so many photons per second that it seems to be radiating a continuous stream of energy.

The energies of photons are sometimes expressed in terms of a small unit of energy called the **electron volt** (eV). One electron volt is equal to $1.602 \times 10^{-19}$ J, so a 633-nm photon has an energy of 1.96 eV. In terms of electron volts, Planck's constant may be written as

$$h = 4.135 \times 10^{-15} \text{ eV s}$$

Because the frequency $\nu$ of light is related to the wavelength $\lambda$ by $\nu = c/\lambda$, we can rewrite the equation for the energy of a photon as

**Energy of a photon (in terms of frequency)**

$$E = h\nu$$

$E$ = energy of a photon
$h$ = Planck's constant
$\nu$ = frequency of light

The two equations $E = hc/\lambda$ and $E = h\nu$ are together called **Planck's law.** Note that both equations express a rela-

## box 5-3 | The Heavens on the Earth

### *Photons at the Supermarket*

A beam of light can be regarded as a stream of tiny packets of energy called photons. The Planck relationships $E = hc/\lambda$ and $E = h\nu$ can be used to relate the energy $E$ carried by a photon to its wavelength $\lambda$ and frequency $\nu$.

As an example, the laser bar-code scanners used at retail stores and supermarkets emit orange-red light of wavelength 633 nm. To calculate the energy of a single photon of this light, we must first express the wavelength in meters. A nanometer (nm) is equal to $10^{-9}$ m, so the wavelength is

$$\lambda = (633 \text{ nm}) \left( \frac{10^{-9} \text{ m}}{1 \text{ nm}} \right)$$

$$= 633 \times 10^{-9} \text{ m}$$

$$= 6.33 \times 10^{-7} \text{ m}$$

Then, using the Planck formula $E = hc/\lambda$, we find that the photon energy is

$$E = \frac{hc}{\lambda} = \frac{(6.625 \times 10^{-34} \text{ J s})(3 \times 10^8 \text{ m s}^{-1})}{6.33 \times 10^{-7} \text{ m}} = 3.14 \times 10^{-19} \text{ J}$$

This is a very small amount of energy. The laser in a typical bar-code scanner emits $10^{-3}$ joule of light energy per second, so the number of photons emitted per second is

$$\frac{10^{-3} \text{ joule per second}}{3.14 \times 10^{-19} \text{ joule per photon}} = 3.2 \times 10^{15} \text{ photons per second}$$

This is such a large number that the laser beam seems like a continuous flow of energy rather than a stream of little energy packets.

---

tionship between a particlelike property of light (the energy $E$ of a photon) and a wavelike property (the wavelength $\lambda$ or the frequency $\nu$).

The photon picture of light is essential for understanding the detailed shapes of blackbody curves. As we will see, it also helps to explain how and why the spectra of stars differ from those of perfect blackbodies.

## 5-6    Each chemical element produces its own unique set of spectral lines

In 1814 the German master optician Joseph von Fraunhofer repeated Newton's classic experiment of shining a beam of sunlight through a prism (see Figure 5-3). But this time he subjected the resulting rainbow-colored spectrum to intense magnification. To his surprise, Fraunhofer discovered that the solar spectrum contains hundreds of fine, dark lines, now called **spectral lines.** By contrast, if the light from a perfect blackbody were sent through a prism, it would produce a smooth, continuous spectrum with no dark lines. Fraunhofer counted over 600 dark lines in the Sun's spectrum; today we know of more than 30,000. Hundreds of spectral lines are visible in the photograph of the Sun's spectrum shown in Figure 5-12.

Half a century later, chemists discovered that they could produce spectral lines in the laboratory and use these spectral lines to analyze what kinds of atoms different substances are made of. Chemists had long known that many substances emit distinctive colors when sprinkled into a flame. As an example, ordinary table salt sprinkled into the flame of a gas stove produces a characteristic orange-yellow glow. To facilitate study of these colors, around 1857 the German chemist Robert Bunsen invented a gas burner (today called a Bunsen burner) that produces a clean flame with no color of its own. Bunsen's colleague, Prussian-born physicist Gustav Kirchhoff, suggested that the colored light produced by substances in a flame might best be studied by passing the light through a prism (Figure 5-13). The two scientists promptly discovered that the spectrum from a flame consists of a pattern of thin, bright spectral lines against a dark background. The same kind of spectrum is produced by heated gases such as neon or argon.

Kirchhoff and Bunsen then found that each chemical element produces its own unique pattern of spectral lines. Thus was born in 1859 the technique of **spectral analysis,** the identification of chemical substances by their unique patterns of spectral lines.

A chemical **element** is a fundamental substance that cannot be broken down into more basic chemicals. Some examples are hydrogen, oxygen, carbon, iron, gold, and silver. After Kirchhoff and Bunsen had recorded the prominent

figure 5-12

**The Sun's Spectrum** Numerous dark spectral lines are seen in this photograph of the Sun's spectrum. The spectrum is spread out so much that it had to be cut into segments to fit on this page. (NOAO)

spectral lines of all the then-known elements, they soon began to discover other spectral lines in the spectra of vaporized mineral samples. In this way they discovered elements whose presence had never before been suspected. In 1860 Kirchhoff and Bunsen found a new line in the blue portion of the spectrum of a sample of mineral water. After isolating the previously unknown element responsible for making the line, they named it cesium (from the Latin *caesium,* "grayblue"). The next year, a new line in the red portion of the spectrum of a mineral sample led them to discover the element rubidium (Latin *rubidium,* "red").

Spectral analysis even allowed the discovery of new elements outside Earth. During the solar eclipse of 1868, astronomers found a new spectral line in light coming from the hot gases at the upper surface of the Sun while the main body of the Sun was hidden by the Moon. This line was attributed to a new element that was named helium (from the Greek *helios,* "sun"). Helium was not discovered on Earth until 1895, when it was found in gases obtained from a uranium mineral.

The spectrum of the Sun, with its dark spectral lines superimposed on a bright background (Figure 5-12), may seem to be unrelated to the spectra of bright lines against a dark background produced by substances in a flame (Figure 5-13). But by the early 1860s Kirchhoff's experiments had revealed a direct connection between these two types of spectra. His conclusions are summarized in three important statements about

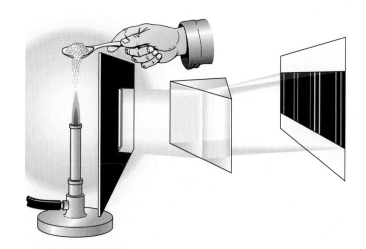

figure 5-13

**The Kirchhoff–Bunsen Experiment** In the mid-1850s, Gustav Kirchhoff and Robert Bunsen discovered that when a chemical substance is heated and vaporized, the spectrum of the emitted light exhibits a series of bright spectral lines. They also found that each chemical element produces its own characteristic pattern of spectral lines. (In an actual laboratory experiment, lenses would be needed to focus the image of the slit onto the screen.)

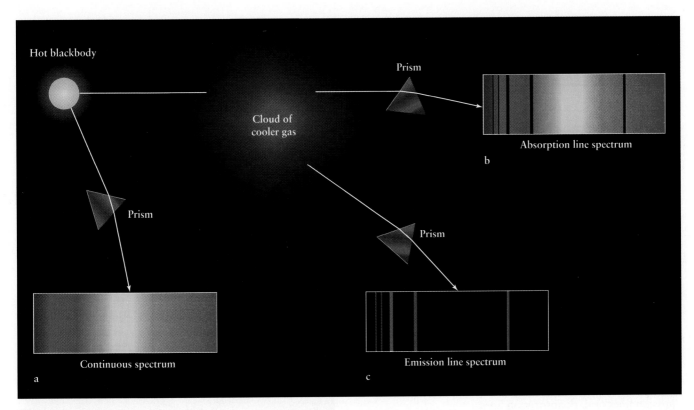

## figure 5-14

**Continuous, Absorption Line, and Emission Line Spectra** This schematic diagram summarizes how different types of spectra are produced. A hot, glowing object such as a blackbody emits a continuous spectrum of light (spectrum *a*). If this light is passed through a cloud of a cooler gas, the cloud selectively absorbs light of certain specific wavelengths. As a result of this absorption, the light from the hot, glowing object that passes directly through the cloud is depleted in these wavelengths, and the spectrum of this light has dark absorption lines (spectrum *b*). The cloud does not retain all the light energy that it absorbs but radiates it outward in all directions. If the light radiated from the gas is viewed against a cold, dark background, its spectrum contains bright emission lines (spectrum *c*). The wavelengths of the dark absorption lines in spectrum *b* are the same as those of the bright emission lines in spectrum *c*. The chemical composition of the cloud determines the specific wavelengths that the cloud both absorbs and emits.

spectra that are today called **Kirchhoff's laws.** These laws, which are illustrated in Figure 5-14, are as follows:

**Law 1** A hot opaque body, such as a perfect blackbody or a hot, dense gas produces a **continuous spectrum**—a complete rainbow of colors without any spectral lines.

**Law 2** A hot, transparent gas produces an **emission line spectrum**—a series of bright spectral lines against a dark background.

**Law 3** A cool, transparent gas in front of a source of a continuous spectrum produces an **absorption line spectrum**—a series of dark spectral lines among the colors of the continuous spectrum. Furthermore, the dark lines

in the absorption spectrum of a particular gas occur at exactly the same wavelengths as the bright lines in the emission spectrum of that same gas.

Whether an emission line spectrum or an absorption line spectrum is observed from a gas cloud depends on the relative temperatures of the gas cloud and its background. Absorption lines are seen if the background is hotter than the gas, and emission lines are seen if the background is cooler.

For example, if sodium is placed in the flame of a Bunsen burner in a darkened room, the flame will emit a characteristic orange-yellow glow. (This same glow is produced if we use ordinary table salt, which is a compound of sodium and chlorine.) When the light from the flame is passed

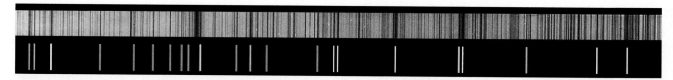

## figure 5-15

**Iron in the Sun** The upper part of this figure is a portion of the Sun's absorption line spectrum from 420 to 430 nm. Numerous dark spectral lines are visible. The lower part of the figure is a corresponding portion of the emission line spectrum of vaporized iron. Several bright spectral lines are seen against a black background. The iron lines coincide with some of the solar lines, which proves that there is some iron (albeit a very tiny amount) in the Sun's atmosphere. (Carnegie Observatories)

through a prism, it displays an emission line spectrum with two closely spaced spectral lines at wavelengths of 588.99 and 589.59 nm, in the orange-yellow part of the spectrum. We now turn on a light bulb whose filament is hotter than the flame and shine the bulb's white light through the flame. The spectrum of this light after it passes through the flame is the continuous spectrum from the light bulb, but with two closely spaced *dark* lines at 588.99 and 589.59 nm (produced by the sodium vapor within the flame). Thus, the chemical composition of the gas is revealed by either bright emission lines or dark absorption lines.

Kirchhoff's laws also mean that if a beam of white light is passed through a gas, the atoms of the gas somehow extract light of very specific wavelengths from the white light. Hence, an observer who looks straight through the gas at the white-light source (the blackbody in Figure 5-14) will receive light whose spectrum has dark absorption lines superimposed on the continuous spectrum of the white light. The gas atoms then radiate light of precisely these same wavelengths in all directions. An observer at an oblique angle (that is, one who is not sighting directly through the cloud toward the blackbody) will receive only this light radiated by the gas cloud; the spectrum of this light is bright emission lines on a dark background.

**Spectroscopy** is the systematic study of spectra and spectral lines. Spectral lines are tremendously important in astronomy, because they provide extremely reliable evidence about the chemical composition of distant objects. As an example, the spectrum of the Sun shown in Figure 5-12 is an absorption line spectrum. The continuous spectrum comes from the hot interior of the Sun, which acts like a blackbody. The dark absorption lines are caused by this light passing through a cooler gas; this gas is the atmosphere that surrounds the Sun. Therefore, by identifying the spectral lines present in the solar spectrum, we can determine the chemical composition of the Sun's atmosphere.

Figure 5-15 shows both a portion of the Sun's spectrum and the emission line spectrum of iron vapor over the same wavelength range. This pattern of bright spectral lines in the lower spectrum is iron's own distinctive "fingerprint," which no other substance can imitate. Because some absorption lines in the Sun's spectrum coincide with the iron lines, some vaporized iron must exist in the Sun's atmosphere.

Spectroscopy is also applied to the light emitted from gas clouds in space, such as the nebula NGC 2363 shown in Figure 5-16. Such glowing clouds have emission line spectra, because we see them against the black background of space. The dominant red color of NGC 2363 is due to an emission

## figure 5-16   R I **V** U X G

**The Nebula NGC 2363** This Hubble Space Telescope image shows the nebula NGC 2363, a glowing gas cloud within a galaxy located some 10 million light-years away in the constellation Camelopardis (the Camel). The hot stars within the nebula emit high-energy, ultraviolet photons, which are absorbed by the surrounding gas and heat the gas to high temperature. This heated gas produces light with an emission line spectrum. The particular wavelength of red light emitted by the nebula is 656 nm, characteristic of hydrogen gas. (L. Drissen, J.-R. Roy, and C. Robert/NASA)

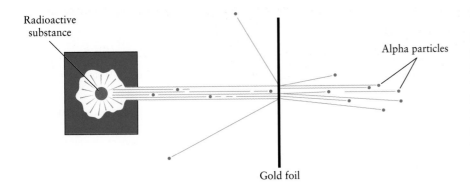

Radioactive substance

Alpha particles

Gold foil

**figure 5-17**

**Rutherford's Experiment** Alpha particles from a radioactive source are directed at a thin metal foil. Most alpha particles pass through the foil with very little deflection. Occasionally, however, an alpha particle will recoil, indicating that it has collided with the massive nucleus of an atom. This experiment provided the first evidence that the nuclei of atoms are relatively massive and compact.

line at a wavelength near 656 nm. This is one of the characteristic wavelengths emitted by hydrogen gas, so we can conclude that this nebula contains hydrogen. More detailed analyses of this kind show that hydrogen is the most common element in gaseous nebulae, and indeed in the universe as a whole.

What is truly remarkable about spectroscopy is that it can determine chemical composition at any distance. The 656-nm red light produced by a sample of heated hydrogen gas on Earth is the same as that observed coming from NGC 2363 in Figure 5-16, located about 10 million light-years away. By using the basic principles outlined by Kirchhoff, astronomers have the tools to make chemical assays of objects that are almost inconceivably distant.

To make full use of Kirchhoff's laws, it is helpful to understand why they work. Why does an atom absorb light of only particular wavelengths? And why does it then emit light of only these same wavelengths? Maxwell's theory of electromagnetism (Section 5-2) could not answer these questions. The answers did not come until early in the twentieth century, when scientists began to discover the structure and properties of atoms.

## 5-7   An atom consists of a small, dense nucleus surrounded by electrons

The first important clue about the internal structure of atoms came from an experiment conducted in 1910 by Ernest Rutherford, a gifted chemist and physicist from New Zealand. Rutherford and his colleagues at the University of Manchester in England had been investigating the recently discovered phenomenon of radioactivity. Certain radioactive elements, such as uranium and radium, were known to emit particles. One type of particle, an alpha particle, has about the same mass as a helium atom and is emitted from a radioactive substance with considerable speed.

In one series of experiments, Rutherford and his colleagues were using alpha particles as projectiles to probe the structure of solid matter. They directed a beam of these particles at a thin sheet of metal (Figure 5-17). Almost all the alpha parti-

cles passed through the metal sheet with little or no deflection from their straight-line paths. To the surprise of the experimenters, however, an occasional alpha particle bounced back from the metal sheet as though it had struck something quite dense. Rutherford later remarked, "It was almost as incredible as if you fired a fifteen-inch shell at a piece of tissue paper and it came back and hit you."

Rutherford concluded from this experiment that most of the mass of an atom is concentrated in a compact, massive lump of matter that occupies only a small part of the atom's volume. Most of the alpha particles pass freely through the nearly empty space that makes up most of the atom, but a few particles happen to strike the dense mass at the center of the atom and bounce back.

Rutherford proposed a new model for the structure of an atom, shown in Figure 5-18. According to this model, a massive, positively charged **nucleus** at the center of the atom is

**figure 5-18**

**Rutherford's Model of the Atom** Electrons orbit the atom's nucleus, which contains most of the atom's mass. The nucleus contains two types of particles, protons and neutrons.

## Box 5-4 | Looking Deeper into Astronomy

### *Atoms, the Periodic Table, and Isotopes*

Each different chemical element is made of a specific type of atom. For example, an atom of hydrogen has 1 proton in its nucleus, an oxygen atom has 8 protons in its nucleus, and so on. The number of protons in an atom's nucleus equals the **atomic number** for that particular element. A listing of the chemical elements is most conveniently displayed in the form of a **periodic table** (shown in the figure). Elements are arranged in the periodic table in order of increasing atomic number. With only a few exceptions, this sequence also corresponds to increasing average mass of the atoms of the elements. Thus, hydrogen (symbol H), with atomic number 1, is the lightest element. Iron (symbol Fe) has atomic number 26 and is a relatively heavy element. All the elements that appear in a single vertical column of the periodic table have similar chemical properties. For example, the elements listed in the far right column are all gases under the conditions of temperature and pressure found at the Earth's surface, and they are all very reluctant to react chemically with other elements.

In addition to nearly 100 naturally occurring elements, the periodic table also includes a number of artificially produced elements. All these elements are heavier than uranium (symbol U) and are highly radioactive, which means that they decay into lighter elements within a short time of being created in laboratory experiments.

The number of protons in the nucleus of an atom determines what element that atom is. Nevertheless, the same element may have different numbers of neutrons in its nucleus. Typically, there are more neutrons than protons in a nucleus, especially in the case of the heaviest elements. For example, consider oxygen, the eighth element on the periodic table. Its atomic number is 8, so every oxygen nucleus has exactly 8 protons, but it can have 8, 9, or 10 neutrons. Thus, there are three slightly different kinds of oxygen, called **isotopes.** The isotope with 8 neutrons is by far the most abundant variety. It is written as $^{16}O$, or oxygen-16. The rarer isotopes with 9 and 10 neutrons are designated as $^{17}O$ and $^{18}O$, respectively.

The superscript that precedes the chemical symbol for an element is equal to the total number of protons and neutrons in a nucleus of that particular isotope. For example, the common isotope of iron is $^{56}Fe$, or iron-56, which means that its nucleus contains a total of 56 protons and neutrons. From the periodic table, however, we see that the atomic number of iron is 26. This means that every iron atom has 26 protons in its nucleus. Therefore, the number of neutrons in an iron-56 nucleus is $56 - 26 = 30$.

It is extremely difficult to distinguish chemically between the various isotopes of a particular element. Ordinary chemical reactions involve only the electrons that orbit the atom, never the neutrons buried in its nucleus. But there are small differences in the wavelengths of the spectral lines for different isotopes of the same element. For example, the spectral line wavelengths of the

orbited by tiny, negatively charged electrons. Rutherford concluded that at least 99.98% of the mass of an atom must be concentrated in its nucleus, whose diameter is only about $10^{-14}$ m. (The diameter of a typical atom is $10^{-10}$ m.)

We know today that the nucleus of an atom contains two types of particles, **protons** and **neutrons.** A proton has a positive electric charge, equal and opposite to that of an electron. As its name suggests, a neutron has no electric charge—it is electrically neutral. As an example, an alpha particle (such as the one Rutherford's team used) is actually a nucleus of the helium atom, with two protons and two neutrons. Protons and neutrons are held together in a nucleus by the so-called strong nuclear force, whose great strength overcomes the electric repulsion between the positively charged protons. A proton and a neutron have almost the same mass, $1.7 \times 10^{-27}$ kg, and each has about 2000 times as much mass as an electron ($9.1 \times 10^{-31}$ kg). In an ordinary atom there are as many posi-

tive protons as there are negative electrons, so the atom has no net electric charge. Because the mass of the electron is so small, the mass of an atom is not much greater than the mass of its nucleus. That is why an alpha particle has nearly the same mass as an atom of helium.

Whereas the solar system is held together by gravitational forces, atoms are held together by electrical forces. Each proton carries a positive charge, and each electron carries an equal but opposite negative charge. The electric forces attracting the positively charged protons and the negatively charged electrons keep the atom from coming apart. Box 5-4 describes more about the connection between the structure of atoms and the chemical and physical properties of substances made of those atoms.

Rutherford's experiments clarified the structure of the atom, but they did not explain how these tiny particles within the atom give rise to spectral lines. The task of reconciling

hydrogen isotope $^2$H are about 0.03% greater than the wavelengths for the most common hydrogen isotope, $^1$H. Thus, different isotopes can be distinguished by careful spectroscopic analysis.

Isotopes are important in astronomy for a variety of reasons. By measuring the relative amounts of different isotopes of a given element in a geologic sample, the age of that sample can be determined. The mixture of isotopes left behind when a star explodes into a supernova (see Section 1-3) tells astronomers about the processes that led to the explosion. And astronomers can learn a great deal about the origin of the universe by comparing the amounts of different isotopes of hydrogen present in the universe today. We will discuss all these topics further in later chapters.

| 1 H | | | | | | | | | | | | | | | | | 2 He |
|---|---|---|---|---|---|---|---|---|---|---|---|---|---|---|---|---|---|
| 3 Li | 4 Be | | | | | | | | | | | 5 B | 6 C | 7 N | 8 O | 9 F | 10 Ne |
| 11 Na | 12 Mg | | | | | | | | | | | 13 Al | 14 Si | 15 P | 16 S | 17 Cl | 18 Ar |
| 19 K | 20 Ca | 21 Sc | 22 Ti | 23 V | 24 Cr | 25 Mn | 26 Fe | 27 Co | 28 Ni | 29 Cu | 30 Zn | 31 Ga | 32 Ge | 33 As | 34 Se | 35 Br | 36 Kr |
| 37 Rb | 38 Sr | 39 Y | 40 Zr | 41 Nb | 42 Mo | 43 Tc | 44 Ru | 45 Rh | 46 Pd | 47 Ag | 48 Cd | 49 In | 50 Sn | 51 Sb | 52 Te | 53 I | 54 Xe |
| 55 Cs | 56 Ba | 71 Lu | 72 Hf | 73 Ta | 74 W | 75 Re | 76 Os | 77 Ir | 78 Pt | 79 Au | 80 Hg | 81 Tl | 82 Pb | 83 Bi | 84 Po | 85 At | 86 Rn |
| 87 Fr | 88 Ra | 103 Lr | 104 Rf | 105 Db | 106 Sg | 107 Bh | 108 Hs | 109 Mt | 110 Uun | 111 Uuu | 112 Uub | | | | | | |

| 57 La | 58 Ce | 59 Pr | 60 Nd | 61 Pm | 62 Sm | 63 Eu | 64 Gd | 65 Tb | 66 Dy | 67 Ho | 68 Er | 69 Tm | 70 Yb |
|---|---|---|---|---|---|---|---|---|---|---|---|---|---|
| 89 Ac | 90 Th | 91 Pa | 92 U | 93 Np | 94 Pu | 95 Am | 96 Cm | 97 Bk | 98 Cf | 99 Es | 100 Fm | 101 Md | 102 No |

Rutherford's atomic model with Kirchhoff's laws of spectral analysis was undertaken by the young Danish physicist Niels Bohr, who joined Rutherford's group at Manchester in 1911.

## 5-8 Spectral lines are produced when an electron jumps from one energy level to another within an atom

Niels Bohr began his study of the connection between atomic spectra and atomic structure by trying to understand the structure of hydrogen, the simplest and lightest of the elements. (As mentioned in Section 5-6, hydrogen is also the most common element in the universe.) When he was done, he had not only found a way to explain this atom's spectrum but had also found a justification for Kirchhoff's laws in terms of atomic physics.

The most common type of hydrogen atom consists of a single electron and a single proton. Hydrogen has a simple visible-light spectrum consisting of a pattern of lines that begins at a wavelength of 656.3 nm and ends at 364.6 nm. The first spectral line is called $H_\alpha$ (H-alpha), the second spectral line is called $H_\beta$ (H-beta), the third is $H_\gamma$ (H-gamma), and so forth. The closer you get to the short-wavelength end of the spectrum at 364.6 nm, the more spectral lines you see.

The regularity in this spectral pattern was described mathematically in 1885 by Johann Jakob Balmer, a Swiss schoolteacher. The spectral lines of hydrogen at visible wavelengths are today called **Balmer lines,** and the entire pattern from $H_\alpha$ onward is called the **Balmer series.** More than two dozen Balmer lines are seen in the spectrum of the star shown in Figure 5-19. Stars in general have Balmer absorption lines in

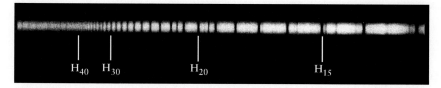

## figure 5-19

**Balmer Lines in the Spectrum of a Star** This portion of the spectrum of the star HD 193182 shows nearly two dozen Balmer lines, from $H_{13}$ through $H_{40}$ (numbers are used beyond the first few Balmer lines so you do not have to memorize the entire Greek alphabet). The series converges at 364.6 nm, just to the left of $H_{40}$. This star's spectrum also contains the first 12 Balmer lines ($H_\alpha$ through $H_{12}$), but they are located beyond the right edge of this photograph. (Carnegie Observatories)

their spectra, which shows that there is hydrogen in the atmospheres of stars.

Using trial and error, Balmer discovered a formula from which the wavelengths ($\lambda$) of hydrogen's spectral lines can be calculated. Balmer's formula is usually written

$$\frac{1}{\lambda} = R\left(\frac{1}{4} - \frac{1}{n^2}\right)$$

where $n$ can be any integer (whole number) greater than 2. Here $R$ is the *Rydberg constant* ($R = 1.097 \times 10^7$ m$^{-1}$), named in honor of the Swedish spectroscopist J. R. Rydberg. To get the wavelength of $H_\alpha$, you put $n = 3$ into Balmer's formula:

$$\frac{1}{\lambda_\alpha} = (1.097 \times 10^7 \text{ m}^{-1})\left(\frac{1}{4} - \frac{1}{3^2}\right) = 1.524 \times 10^6 \text{ m}^{-1}$$

so,

$$\lambda_\alpha = \frac{1}{1.524 \times 10^6 \text{ m}^{-1}} = 6.563 \times 10^{-7} \text{ m} = 656.3 \text{ nm}$$

To get $H_\beta$, use $n = 4$, and to get $H_\gamma$, use $n = 5$. If you use $n = \infty$ (the symbol $\infty$ stands for infinity), you get the short-wavelength end of the hydrogen spectrum at 364.6 nm. (Note that 1 divided by infinity equals zero.)

Bohr realized that to fully understand the structure of the hydrogen atom, he had to be able to derive Balmer's formula using the laws of physics. He first made the rather wild assumption that the electron in a hydrogen atom can only orbit the nucleus in certain specific orbits. (This was a significant break with the ideas of Newton, in whose mechanics any orbit should be possible.) As shown in Figure 5-20, it is customary to label the permissible **Bohr orbits** by the numbers $n = 1$, $n = 2$, $n = 3$, and so on. Although confined to one of these allowed orbits while circling the nucleus, an electron can jump from one Bohr orbit to another.

For an electron to jump from one Bohr orbit to another, the hydrogen atom must gain or lose a specific amount of energy. For the electron to go from an inner to an outer orbit, the atom must absorb energy; for the electron to go from an outer to an inner orbit, the atom must release

energy. As an example, Figure 5-21 shows an electron jumping between the $n = 2$ and $n = 3$ orbits of a hydrogen atom as it absorbs or emits an $H_\alpha$ photon.

The energy gained or released by the atom when the electron jumps from one orbit to another is the difference in energy between these two orbits. In other words, the amount of energy change in a jump from a low orbit to a high orbit (Figure 5-21a) is the same as in a jump from the high orbit back to the low one (Figure 5-21b). According to Planck and Einstein, the packet of energy gained or released is a photon with an energy given by $E = hc/\lambda$, where $h$ is Planck's con-

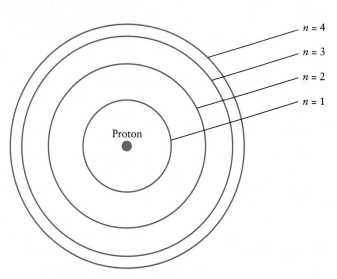

## figure 5-20

**The Bohr Model of the Hydrogen Atom** According to the Bohr model of the hydrogen atom, an electron circles the nucleus (a proton) only in allowed orbits $n = 1, 2, 3$, and so forth. The first four Bohr orbits are shown here. This figure is not shown to scale; compared to the $n = 1$ orbit, in the Bohr model the $n = 2$ orbit has 4 times the radius, the $n = 3$ orbit has 9 times the radius, and the $n = 4$ orbit has 16 times the radius.

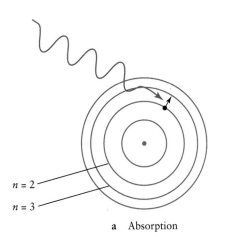
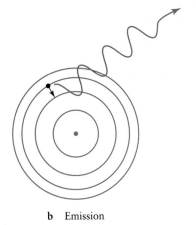

$n = 2$
$n = 3$

a   Absorption          b   Emission

Figure 5-21

**The Absorption and Emission of an H$_\alpha$ Photon** This schematic diagram, drawn according to the Bohr model of the atom, shows what happens when a hydrogen atom absorbs or emits a photon whose wavelength is 656.3 nm. **(a)** The photon is absorbed by the atom, causing the electron to jump from the $n = 2$ orbit up to $n = 3$. **(b)** The photon is emitted by the atom as the electron falls from the $n = 3$ orbit down to the $n = 2$ orbit.

stant, $c$ is the speed of light, and $\lambda$ is the photon's wavelength. It follows that the photons that an atom can emit and the photons that an atom can absorb have the same energy and, hence, the same wavelength. Thus, Bohr's picture explains Kirchhoff's observation that atoms emit and absorb the same wavelengths of light.

The Bohr picture also helps us visualize what happens to produce an emission line spectrum. When a gas is heated, the atoms of the gas move around rapidly and can collide forcefully with each other. These energetic collisions excite the electrons and kick them into high orbits. The electrons then cascade back down to the innermost possible orbit, emitting photons whose energies are equal to the energy differences between different Bohr orbits. In this fashion, a hot gas produces an emission line spectrum with a variety of different wavelengths.

To produce an absorption line spectrum, begin with a relatively cool gas in which most of the atoms have their electrons in inner, low-energy orbits. If a beam of light with a continuous spectrum is shone through the gas, most wavelengths will pass through undisturbed. Only those photons will be absorbed whose energies are just right to excite an electron to an allowed outer orbit. Hence, only certain wavelengths will be absorbed, and dark lines will appear in the spectrum at those wavelengths.

Using his picture of allowed orbits and the formula $E = hc/\lambda$, Bohr was able to prove mathematically that the wavelength $\lambda$ of the photon emitted or absorbed as an electron jumps between an inner orbit $N$ and an outer orbit $n$ is

**Bohr formula for hydrogen wavelengths**

$$\frac{1}{\lambda} = R\left(\frac{1}{N^2} - \frac{1}{n^2}\right)$$

$N$ = number of inner orbit
$n$ = number of outer orbit
$R$ = Rydberg constant = $1.097 \times 10^7$ m$^{-1}$
$\lambda$ = wavelength (in meters) of emitted or absorbed photon

In deriving this equation from the laws of physics. Bohr found that he could express $R$ entirely in terms of basic physical quantities, such as the speed of light, Planck's constant, and the mass and charge of the electron.

If Bohr let $N = 2$ in his formula, he got back the formula that Balmer discovered by trial and error. Hence, Bohr concluded that all the Balmer lines are produced by electron transitions between the second Bohr orbit ($N = 2$) and higher orbits ($n = 3, 4, 5,$ and so on).

Bohr's formula also correctly predicts the wavelengths of other series of spectral lines that occur at nonvisible wavelengths. The value $N = 1$ gives the **Lyman series**, which is entirely in the ultraviolet. All the spectral lines in this series involve electron transitions between the lowest Bohr orbit and all higher orbits ($n = 2, 3, 4,$ and so on). This pattern of spectral lines begins with L$_\alpha$ (Lyman alpha) at 122 nm and converges on L$_\infty$ at 91 nm. At infrared wavelengths is the **Paschen series**, for which $N = 3$. This series, which involves transitions between the third Bohr orbit and all higher orbits, begins with P$_\alpha$ (Paschen alpha) at 1875 nm and converges on P$_\infty$ at 821 nm. Additional series exist at still longer wavelengths. Figure 5-22 shows some examples of electron transitions in the Bohr atom for these series, along with the wavelength for each transition. These transitions can also be depicted in terms of an *energy-level diagram*, described in Box 5-5.

The shortest wavelength in the Lyman series (L$_\infty$), $\lambda = 91$ nm, has a special significance. If an electron in the lowest Bohr orbit ($n = 1$) absorbs a photon with this wavelength, it is lifted to the $n = \infty$ orbit. But this orbit is infinitely large, which means that the electron has actually escaped from the atom altogether. The process of removing an electron from an atom is called **ionization**. In general, a hydrogen atom will be ionized when it absorbs an ultraviolet photon with a wavelength of 91 nm or less. (According to the Planck formula $E = hc/\lambda$, shorter-wavelength photons have higher energy and so are even more capable of blasting an electron completely out of a hydrogen atom.) Ionization plays a key role in making the nebula in Figure 5-16 glow (see Box 5-5 for details).

## Atomic Energy-Level Diagrams

Today's view of the atom owes much to the Bohr model, but it is different in certain ways. The modern picture is based on **quantum mechanics,** a branch of physics dealing with photons and subatomic particles that was developed during the 1920s. As a result of this work, physicists no longer picture electrons as moving in specific orbits about the nucleus. Instead, electrons are now said to occupy certain **energy levels** in the atom.

An extremely useful way of displaying the structure of an atom is with an **energy-level diagram,** such as that shown for hydrogen in Figure a. The lowest energy level, called the **ground state,** corresponds to the $n = 1$ Bohr orbit. Higher energy levels, called **excited states,** correspond to successively larger Bohr orbits. An electron can jump from the ground state up to the $n = 2$ level only if the atom absorbs a Lyman-alpha photon with a wavelength of 122 nm. The energy of a photon is determined by the relationship $E = h\nu = hc/\lambda$ and is commonly expressed in electron volts (eV). As explained in Section 5-5, an electron volt is a tiny amount of energy ($1 \text{ eV} = 1.6 \times 10^{-19}$ J). The Lyman-alpha photon has an energy of 10.19 eV, so the energy level of $n = 2$ is shown in the figure as having an energy 10.19 eV above the energy of the ground state (conventionally assigned a value of 0 eV). Similarly, the $n = 3$ level is 12.07 eV above the ground state, and so forth. Electrons can make transitions to a higher energy level by absorption of a photon or in a collision between atoms; they can make transitions to lower energy levels by emission of a photon.

On the energy-level diagram for hydrogen, the $n = \infty$ level has an energy of 13.6 eV. If the electron is initially in the ground state and the atom absorbs a photon of any energy greater than 13.6 eV, the electron will be removed completely from the atom and the atom will be *ionized*. The gaseous nebula NGC 2363 shown in Figure 5-16 has hot stars in its neighborhood that produce copious amounts of ultraviolet photons with energies greater than 13.6 eV. These photons cause some of the hydrogen atoms in the nebula to become ionized. When the electrons recombine with the nuclei, they cascade down the energy levels to the ground state and emit visible light in the process.

The atoms of heavier elements have more complex energy-level diagrams. Figure b shows the energy-level diagram for sodium, along with the wavelengths of photons absorbed or emitted in some of sodium's major electron transitions. At visible wavelengths, the sodium spectrum is dominated by two strong lines, called the sodium D lines, at 588.99 and 589.59 nm. These two lines are strong because they correspond to two transitions that are the primary avenue through which electrons cascade from high orbits down to the ground state. Astronomers find energy-level diagrams useful in understanding the lines they observe in the spectra of stars and nebulae.

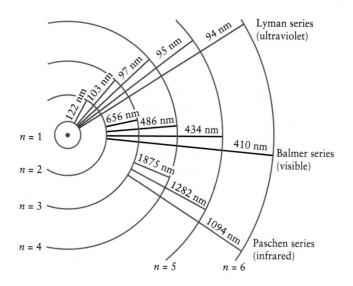

**figure 5-22**

**Electron Transitions in the Hydrogen Atom** This figure shows some of the electron transitions that can occur in the hydrogen atom. Transitions between the lowest orbit ($N = 1$) and higher orbits ($n = 2, 3, 4, \dots$) give rise to spectral lines in the ultraviolet, called the Lyman series. Transitions between the second orbit ($N = 2$) and higher orbits ($n = 3, 4, 5, \dots$) cause visible spectral lines in the Balmer series, and transitions between the third orbit ($N = 3$) and higher orbits ($n = 4, 5, 6, \dots$) create the Paschen series of spectral lines in the infrared. The same wavelengths occur in emission (the electron drops from a high orbit to a low one, emitting a photon) and in absorption (the electron absorbs a photon and jumps from a low orbit to a high one). The orbits are not shown to scale.

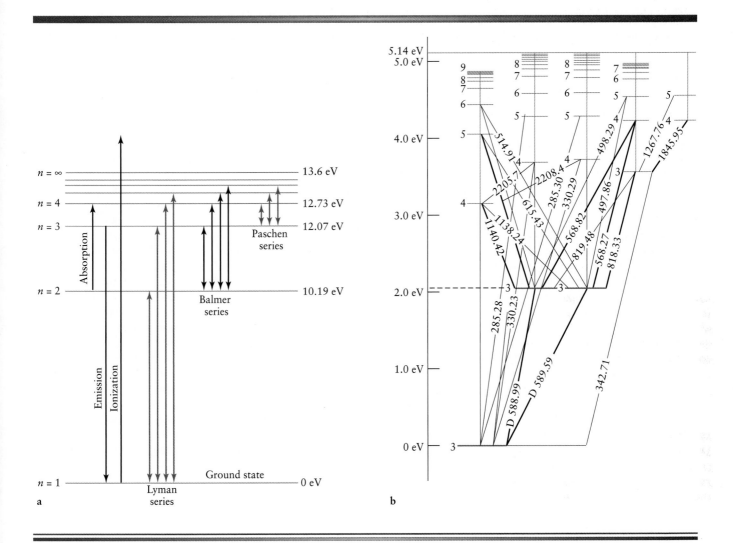

The same basic principles of the Bohr model for hydrogen also apply to the atoms of other elements. Electrons in each kind of atom can be only in certain orbits, so only photons of certain wavelengths can be emitted or absorbed. Each type of atom has a unique arrangement of electron orbits, so the pattern of spectral lines is likewise unique to that particular type of atom. These patterns are in general much more complicated than for the hydrogen atom, so there is no simple relationship analogous to the Bohr formula that applies to the spectra of all atoms. As an example, Box 5-5 shows the complex spectrum of sodium.

The Bohr picture explains the emission line spectra and absorption line spectra of gases. But what about the continuous spectra produced by dense objects like the filament of a light bulb or the coils of a toaster? These objects are made of atoms, so why don't they emit light with an emission line spectrum characteristic of the particular atoms of which

they are made? The reason is directly related to the difference between a gas on the one hand and a liquid or solid on the other. In a gas, atoms are widely separated and can emit photons without interference from other atoms. But in a liquid or a solid, atoms are so close that they almost touch, and thus these atoms interact strongly with each other. These interactions interfere with the process of emitting photons. As a result, the pattern of distinctive bright spectral lines that the atoms would emit in isolation becomes "smeared out" into a continuous spectrum.

ANALOGY Think of atoms like tuning forks. If you strike a single tuning fork, it produces a sound wave with a single clear frequency and wavelength, just as an isolated atom emits light of definite wavelengths. But if you shake a box packed full of tuning forks, you will hear a clanging noise that is a mixture of sounds of all different frequencies and wavelengths. This is directly analogous to the continuous spectrum emitted by a dense object with closely packed atoms.

With the work of such people as Planck, Einstein, Rutherford, and Bohr, the interchange between astronomy and physics came full circle. Modern physics was born when Newton set out to understand the motions of the planets. Two and a half centuries later, physicists in their laboratories probed the properties of light and the structure of atoms. Their labors had immediate applications in astronomy. Armed with this new understanding of light and matter, astronomers were able to probe in detail the chemical and physical properties of planets, stars, and galaxies.

By analyzing the spectra of starlight, we have been able to learn about the temperatures and chemical compositions of stars. In later chapters we see how spectra can also tell us how strong the pull of gravity is at a star's surface. And, as we see in the following section, the spectrum of a star can reveal something about the star's motion through space.

## 5-9  The wavelength of a spectral line is affected by the relative motion between the source and the observer

Christian Doppler, a professor of mathematics in Prague, pointed out in 1842 that the observed wavelength of light is affected by motion. As shown in Figure 5-23, light waves from an approaching light source are compressed. The circles represent the peaks of waves emitted from various positions as the source moves along. Each successive wave is emitted from a position slightly closer to you, so you see a shorter wavelength than you would if the source were stationary. All the lines in the spectrum of an approaching source are shifted toward the short-wavelength (blue) end of the spectrum. This phenomenon is called a **blueshift**.

Conversely, light waves from a receding source are stretched out, so that you see a longer wavelength than you would if the source were stationary. All the lines in the spectrum of a receding source are shifted toward the longer-wavelength (red) end of the spectrum, producing a **redshift**. In general, the effect of relative motion on wavelength is called the **Doppler effect**.

You have probably noticed a similar Doppler effect for sound waves. The sound of a police car's siren has a high pitch (that is, the sound waves have high frequency and short wavelength) when the police car is approaching. When the police car is moving away, the siren has a low pitch (the sound waves have low frequency and long wavelength).

Suppose that $\lambda_0$ is the wavelength of a particular spectral line from a light source that is not moving. It is the wavelength that you might look up in a reference book or determine in a laboratory experiment for this spectral line. If the source is moving, this particular spectral line is shifted to a different wavelength $\lambda$. The size of the wavelength shift is usually written as $\Delta\lambda$, where $\Delta\lambda = \lambda - \lambda_0$. Thus, $\Delta\lambda$ is the difference between the wavelength listed in reference books and the wavelength that you actually observe in the spectrum of a star or galaxy.

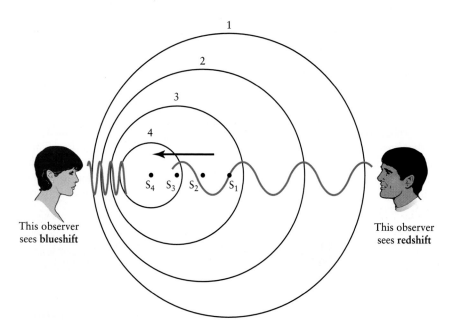

**This observer sees blueshift**

**This observer sees redshift**

**Figure 5-23**

**The Doppler Effect** This figure shows why wavelength is affected by motion between the light source and an observer. A source of light is moving toward the left. The four circles (numbered 1 through 4) indicate the location of the light waves that were emitted by the moving source when it was at points $S_1$ through $S_4$, respectively. Note that the waves are crowded together in front of the moving source but are spread out behind it. Consequently, wavelengths appear shortened (blueshifted) if the source is moving toward the observer. They appear lengthened (redshifted) if the source is moving away from the observer. Motion perpendicular to an observer's line of sight does not affect wavelength.

Doppler proved that the wavelength shift ($\Delta\lambda$) is governed by the following simple equation:

**Doppler shift equation**

$$\frac{\Delta\lambda}{\lambda_0} = \frac{v}{c}$$

$\Delta\lambda$ = wavelength shift
$\lambda_0$ = wavelength if source is not moving
$v$ = velocity of the source measured along the line of sight
$c$ = speed of light = $3 \times 10^5$ km/s

Box 5-6 includes two examples of using this formula. (A note of caution: This formula is valid only if the velocity $v$ is much less than the speed of light $c$. There are objects in the universe that move at speeds approaching $c$, and for these objects we must use a different formula for the Doppler shift. We will describe this in Chapter 27.)

The Doppler shift enables astronomers to determine the line-of-sight motion of astronomical objects. As we will see in the next chapter, astronomers use a device called a spectrograph, which they attach to a telescope, to obtain spectra of planets, stars, or galaxies. As an example, the spectrum of the bright star Vega has prominent Balmer absorption lines. Laboratory experiments have determined that the normal wavelength $\lambda_0$ of the $H_\alpha$ Balmer line is 656.285 nm. But in Vega's spectrum, $H_\alpha$ is located at $\lambda$ = 656.255 nm. The wavelength of $H_\alpha$ has thus been shortened, which means that the spectral line is blueshifted, and thus Vega is coming

---

## box 5-6 | Tools of the Astronomer's Trade

### *Applications of the Doppler Effect*

Doppler's formula relates the radial velocity of an astronomical object to the wavelength shift of its spectral lines. Here are two examples showing how this incredibly powerful formula can be used.

**EXAMPLE:** As measured in the laboratory, the prominent $H_\alpha$ spectrum line of hydrogen has a wavelength of $\lambda_0$ = 656.285 nm. But in the spectrum of the star Vega, this line has a wavelength $\lambda$ = 656.255 nm. The wavelength shift is

$$\lambda - \lambda_0 = 656.255 \text{ nm} - 656.285 \text{ nm} = -0.030 \text{ nm}$$

The negative value means that we see the light from Vega shifted to shorter wavelengths (that is, a blueshift). Using Doppler's formula, we can calculate the star's radial velocity:

$$v = c \frac{\Delta\lambda}{\lambda_0} = 3 \times 10^5 \text{ km/s} \left( \frac{-0.030 \text{ nm}}{656.285 \text{ nm}} \right) = -13.7 \text{ km/s}$$

The minus sign indicates that the star is coming toward us at 13.7 km/s.

By plotting the motions of different stars toward and away from us, astronomers have been able to learn how the Milky Way Galaxy (of which our Sun is a part) is rotating. From this knowledge, and aided by Newton's universal law of gravitation (Section 4-7), they have made the surprising discovery that the Milky Way contains roughly ten times more matter than had heretofore been thought. We will discuss the significance of this result in Chapter 25.

**EXAMPLE:** In the radio region of the electromagnetic spectrum, hydrogen atoms emit and absorb photons with a wavelength of 21.12 cm., giving rise to a spectral feature commonly called the *21-centimeter line*. The galaxy NGC 3840 in the constellation Leo (the Lion) is receding from us at a speed of 7370 km/s, or about 2.5% of the speed of light. Using the Doppler formula, we can predict the wavelength at which we expect to detect the 21-cm line from this galaxy. The wavelength shift is

$$\lambda = \lambda_0 \left( \frac{v}{c} \right) = 21.12 \text{ cm} \left( \frac{7370 \text{ km/s}}{3 \times 10^5 \text{ km/s}} \right) = 0.52 \text{ cm}$$

Therefore, the 21-cm line of hydrogen received from this galaxy will be detected at a wavelength of

$$\lambda = \lambda_0 + \Delta\lambda = 21.12 \text{ cm} + 0.52 \text{ cm} = 21.64 \text{ cm}$$

The 21-cm line has been redshifted to a longer wavelength because the galaxy is receding from us. In fact, as we will see in Chapter 26, most galaxies are receding from us. In Chapter 28, we will see how this implies that the universe is expanding outward from the Big Bang that took place some 15 billion years ago.

toward us. The arithmetic of calculating Vega's speed is given in Box 5-6 along with other examples.

 The redshifts and blueshifts of stars visible to the naked eye, or even through a small telescope, are quite small. As an example, the blueshift of the H$_\alpha$ line of Vega is only 0.030 nm. These tiny wavelength changes are too small to detect with the eye, which is why a spectrograph is required. So, if you see a star with a red color, it means that the star is really red; it does not mean that it is moving rapidly away from us.

Speed determined from the Doppler effect is called **radial velocity,** because v is the component of the star's motion parallel to our line of sight, or along the "radius" drawn from Earth to the star. Of course, a sizable fraction of a star's motion may be perpendicular to our line of sight. The speed of this transverse movement across the sky does not affect wavelengths if the speed is small compared with $c$.

The Doppler effect is an important tool in astronomy because it tells us how to calculate speeds from wavelength shifts, thereby uncovering basic information about the mo-tions of planets, stars, and galaxies. As we will see in later chapters, the rotation of Venus was deduced from the Doppler shift of radar waves reflected from its surface. The back-and-forth Doppler shifting of the spectral lines of certain stars has revealed that these stars are being orbited by unseen companions; in this way, astronomers have discovered planets around other stars and massive objects that may be black holes. Astronomers also use the Doppler effect along with Kepler's third law to measure the masses of galaxies. Near the end of this book, we will see that the redshifts of spectral lines of remote galaxies reveal that we live in an expanding universe. These are but a few examples of how Doppler's discovery has empowered astronomers in their quest to understand the universe.

In this chapter we had a glimpse of how much can be learned by analyzing light from the heavens. To analyze this light, however, it is first necessary that we collect as much of it as possible, because most light sources in space are very dim. Collecting the faint light from distant objects is a key purpose of telescopes. In the next chapter we will describe both how telescopes work and how they are used.

## Key Words

*Terms preceded by an asterisk (*) are discussed in the Boxes.*

absolute zero, p. 106

absorption line spectrum, p. 115

atom, p. 106

*atomic number, p. 118

Balmer line, p. 119

blackbody, p. 108

blackbody curves, p. 108

blueshift, p. 124

Bohr orbits, p. 120

continuous spectrum, p. 115

*degrees Celsius, p. 107

*degrees Fahrenheit, p. 107

Doppler effect, p. 124

electromagnetic radiation, p. 103

electromagnetic spectrum, p. 104

electromagnetism, p. 103

electron, p. 112

electron volt, p. 112

element, p. 113

emission line spectrum, p. 115

energy flux, p. 110

*energy level, p. 122

*energy-level diagram, p. 122

*excited state, p. 122

frequency, p. 104

gamma rays, p. 104

*ground state, p. 122

infrared radiation, p. 104

ionization, p. 121

*isotope, p. 118

joule, p. 110

kelvin, p. 106

Kirchhoff's laws, p. 115

*luminosity, p. 111

Lyman series, p. 121

microwaves, p. 104

neutron, p. 118

nucleus, p. 117

Paschen series, p. 121

*periodic table, p. 118

photoelectric effect, p. 112

photon, p. 112

Planck's law, p. 112

proton, p. 118

*quantum mechanics, p. 122

radial velocity, p. 126

radio waves, p. 104

redshift, p. 124

*solar constant, p. 110

spectral analysis, p. 113

spectral line, p. 113

spectroscopy, p. 116

spectrum (*plural* spectra), p. 101

Stefan-Boltzmann law, p. 111

ultraviolet radiation, p. 104

visible light, p. 104

wavelength, p. 103

wavelength of maximum emission, p. 108

Wien's law, p. 109

X rays, p. 104

# Key Ideas

**The Nature of Light:** Light is electromagnetic radiation. It has wavelike properties described by its wavelength $\lambda$ and frequency $\nu$, and travels through empty space at the constant speed $c = 3 \times 10^8$ m/s.

**Blackbody Radiation:** A blackbody is a hypothetical object that is a perfect absorber of electromagnetic radiation at all wavelengths. Stars closely approximate the behavior of blackbodies, as do other hot, dense bodies.

• The intensities of radiation emitted at various wavelengths by a blackbody at a given temperature are shown by a *blackbody curve*.

• *Wien's law* states that the dominant wavelength at which a blackbody emits electromagnetic radiation is inversely proportional to the Kelvin temperature of the object: $\lambda_{max}$ (in meters) = 0.0029/$T$.

• The *Stefan-Boltzmann law* states that a blackbody radiates electromagnetic waves with a total energy flux $F$ directly proportional to the fourth power of the Kelvin temperature $T$ of the object: $F = \sigma T^4$.

**Photons:** An explanation of blackbody curves led to the discovery that light has particlelike properties. The particles of light are called photons.

• *Planck's law* relates the energy $E$ of a photon to its frequency $\nu$ or wavelength $\lambda$: $E = h\nu = hc/\lambda$, where $h$ is Planck's constant.

**Kirchhoff's Laws:** Kirchhoff's three laws of spectral analysis describe conditions under which different kinds of spectra are produced.

• A hot, dense object such as a blackbody emits a continuous spectrum covering all wavelengths.

• A hot, transparent gas produces a spectrum that contains bright (emission) lines.

• A cool, transparent gas in front of a light source that itself has a continuous spectrum produces dark (absorption) lines in the continuous spectrum.

**Atomic Structure:** An atom has a small dense nucleus composed of protons and neutrons. The nucleus is surrounded by electrons that occupy only certain orbits or energy levels.

• When an electron jumps from one energy level to another, it emits or absorbs a photon of appropriate energy (and hence of a specific wavelength).

• The spectral lines of a particular element correspond to the various electron transitions between energy levels in atoms of that element.

• Bohr's model of the atom correctly predicts the wavelengths of hydrogen's spectral lines.

**The Doppler Shift:** The Doppler shift enables us to determine the radial velocity of a light source from the displacement of its spectral lines.

• The spectral lines of an approaching light source are shifted toward short wavelengths (a blueshift); the spectral lines of a receding light source are shifted toward long wavelengths (a redshift).

• The size of a wavelength shift is proportional to the radial velocity between the light source and the observer.

# Review Questions

1. Approximately how many times around the world could a beam of light travel in one second?

2. (a) Describe an experiment in which light behaves like a wave. (b) Describe an experiment in which light behaves like a particle.

3. What is meant by the frequency of light? How is frequency related to wavelength?

4. A light source emits infrared radiation at a wavelength of 825 nm. What is the frequency of this radiation?

5. Announcers at a certain FM radio station say that they are at "89.9 on your dial," meaning that they transmit at a frequency of 89.9 MHz. What is the wavelength of the radio waves emitted from this station?

6. Using Wien's law and the Stefan-Boltzmann law, explain the color and intensity changes that are observed as the temperature of a hot, glowing object increases.

7. The bright star Deneb in the constellation Cygnus (the Swan) has a surface temperature of 8700 K. What is the dominant wavelength ($\lambda_{max}$) of its energy emission? What color is this star?

8. The bright star Antares in the constellation Scorpio (the Scorpion) emits the greatest intensity of radiation at a wavelength ($\lambda_{max}$) of 1210 nm. What is the surface temperature of the star? What color is this star?

9. What is a blackbody? In what way is a blackbody black? If a blackbody is black, how can it emit light?

10. Explain why astronomers are interested in blackbody radiation.

11. How is the energy of a photon related to its wavelength? What kind of photons carry the most energy? What kind of photons carry the least energy?

12. How do we know that atoms have massive, compact nuclei?

13. Describe the spectrum of hydrogen at visible wavelengths and explain how Bohr's model of the atom accounts for the Balmer lines.

14. Why do different elements have different patterns of lines in their spectra?

15. What is the Doppler effect? Why is it important to astronomers?

## Advanced Questions

### Problem-solving tips and tools:

Formulas for converting between temperature scales are found in Box 5-1. Relationships between a star's radius, its luminosity, and its surface temperature are discussed in Box 5-2. A simple application of Planck's law to calculate the energy of a photon is given in Box 5-3.

16. What is the temperature of the Sun's surface in degrees Fahrenheit?

17. What wavelength of electromagnetic radiation is emitted with greatest intensity by this book? To what region of the electromagnetic spectrum does this wavelength correspond?

18. Your body temperature is 98.6°F. What kind of radiation do you predominantly emit? At what wavelength do you emit the most radiation?

19. The bright star Altair in the constellation Aquila (the Eagle) has a surface temperature of 7400 K. Compared to the Sun, how much more energy is emitted each second from each square meter of Altair's surface?

20. Imagine a star the same size as the Sun but with a surface temperature one-half that of the Sun. At what wavelength would that star emit most of its radiation? How many times dimmer than the Sun would that star be?

21. Black holes (discussed in Chapter 24) do not themselves emit light. But it is possible to detect radiation from material falling toward a black hole. Calculations suggest that as this matter falls, it is compressed and heated to temperatures around $10^6$ K. Calculate the wavelength of maximum emission for this temperature. In what part of the electromagnetic spectrum does this wavelength lie?

\*22. The bright star Sirius in the constellation of Canis Major (the Large Dog) has a radius of 1.67 $R_\odot$ and a luminosity of 25 $L_\odot$. What is the star's surface temperature?

\*23. Imagine a star whose diameter is 10 times larger than that of the Sun and whose surface temperature is 2900 K. Suppose that both the Sun and this star were located at the same distance from you. Which would be brighter? By what factor?

24. In Figure 5-12 you can see two distinct dark lines at the boundary between the orange and yellow parts of the Sun's spectrum (in the center of the third colored band from the top of the figure). The wavelengths of these dark lines are 588.99 and 589.59 nm. What do you conclude from this about the chemical composition of the Sun's atmosphere?

25. During the 1970s and 1980s, balloon-borne instruments detected 511 keV photons coming from the direction of the center of our Galaxy. (k means *kilo*, or thousand, so 1 keV = $10^3$ eV.) What is the wavelength of these photons? To what part of the electromagnetic spectrum do these photons belong?

26. Calculate the wavelength $H_\delta$, the fourth wavelength in the Balmer series. Draw a schematic diagram of the hydrogen atom and indicate the electron transition that gives rise to this spectral line.

\*27. An imaginary atom has just 3 energy levels: 0 eV, 1 eV, and 3 eV. Draw an energy-level diagram for this atom. Show all of the possible transitions between energy levels. For each transition, determine the photon energy and the photon wavelength. Which transitions involve the emission or absorption of visible light?

28. The wavelength of $H_\beta$ in the spectrum of the star Megrez in the Big Dipper (part of the constellation Ursa Major, the Great Bear) is 486.112 nm. Laboratory measurements demonstrate that the normal wavelength of this spectral line is 486.133 nm. Is the star coming toward us or moving away from us? At what speed?

29. The nebula NGC 2363 shown in Figure 5-16 is located within the galaxy NGC 2366 in the constellation Camelopardis (the Camel). This galaxy and the nebula within it are moving away from us at 252 km/s. At what wavelength does the red $H_\alpha$ line of hydrogen (which causes the color of the nebula) appear in the nebula's spectrum?

30. You are given a traffic ticket for going through a red light (wavelength 700 nm). You tell the judge that because you were approaching the light, the Doppler effect caused a blueshift that made the light appear green (wavelength 500 nm). How fast would you have had to be going for this to be true? Would the judge be justified in giving you a speeding ticket? Explain.

## Discussion Questions

31. The equation that relates the frequency, wavelength, and speed of a light wave, $\nu = c/\lambda$, can be rewritten as $c = \nu\lambda$. A friend who knows mathematics but not much astronomy or physics looks at this equation and says: "This equation tells me that the higher the frequency $\nu$, the greater the wave speed $c$. Since visible light has a higher frequency than radio waves, this means that visible light goes faster than radio waves." How would you respond to your friend?

32. Why do you suppose that ultraviolet light can cause skin cancer but ordinary visible light does not?

**33.** Compare chemical identification based on spectral line patterns with the identification of people by their fingerprints. In what ways is this a good or poor analogy?

**34.** Suppose you look up at the night sky and observe some of the brightest stars with your naked eye. Is there any way of telling which stars are hot and which are cool? Explain.

**35.** (a) If you could see ultraviolet radiation, what differences might there be in the appearance of the night sky? In the appearance of ordinary objects in the daytime? (b) What differences might there be in the appearance of the night sky and in the appearance of ordinary objects in the daytime if you could see infrared radiation?

**36.** The human eye is most sensitive over the same wavelength range at which the Sun emits the greatest intensity of radiation. Suppose creatures were to evolve on a planet orbiting a star somewhat hotter than the Sun. To what wavelengths would their "eyes" most likely be sensitive?

## Observing Projects

**37.** As soon as weather conditions permit, observe a rainbow. Confirm that the colors are in the same order as shown in Figure 5-3.

**38.** Turn on an electric stove or toaster oven and carefully observe the heating elements as they warm up. Relate your observations to Wien's law and the Stefan-Boltzmann law.

**39.** Obtain a glass prism (or a diffraction grating, which is probably more readily available and is discussed in the next chapter) and look through it at various light sources, such as an ordinary incandescent light, a neon sign, and a mercury vapor street lamp. **DO NOT LOOK AT THE SUN! Looking directly at the Sun can cause blindness.** Do you have any trouble seeing spectra? What do you have to do to see a spectrum? Describe the differences in the spectra of the various light sources you observed.

## Where to Learn More

*Books and magazine articles*

Gingerich, O. "Unlocking the Chemical Secrets of the Cosmos." *Sky & Telescope*, July 1981. This brief article offers some fascinating insights into the work of Kirchhoff and Bunsen.

Griffin, R. "The Radial-Velocity Revolution." *Sky & Telescope*, September 1989. This article describes how advances in our ability to determine radial velocities from spectra has expanded our knowledge of the universe.

Lynch, D. K., and Livingston, W. *Color and Light in Nature*. Cambridge University Press, 1995. This lucid, beautifully illustrated book describes many phenomena of light and color in the world around us, ranging from the straightforward to the exotic.

Minnaert, M. G. J. *Light and Color in the Outdoors*. Springer, 1993. A new edition of a classic book that explains the wide collection of optical effects, such as rainbows, the colors of shadows, haloes around the Sun and Moon, and more.

Schaaf, F. "Riches of the Rainbow." *Sky & Telescope*, August 1992. Nature's most dramatic expression of refraction is the rainbow. This beautifully illustrated article describes the different features of rainbows and explains how they are formed.

Snow, C. *The Physicists*. Macmillan, 1981. This book describes many of the people who contributed to our modern understanding of light and matter.

Trefil, J. S. *From Atoms to Quarks*. Anchor, 1994. Our understanding of the nature of light and matter was transformed dramatically during the twentieth century. This well-written book relates how this transformation took place.

Van Heel, A., and Velzel, C. *What Is Light?* McGraw-Hill, 1968. This clearly written book surveys our modern understanding of the phenomenon of light.

Weiss, R. J. A. *A Brief History of Light and Those that Lit the Way*. World Scientific, 1996. In a story that spans from the sixteenth to the twentieth century, this entertaining book tells the story of how scientists struggled to understand the nature of light.

**W** *World Wide Web*

A wealth of information about how the nature of light and matter came to be understood can be found at the World Wide Web Virtual Library on the History of Science, Technology, and Medicine (**http://www.asap. unimelb.edu.au/hstm/**). The discovery of the electron, the first of the basic constituents of matter to be identified, is the subject of a web site at the American Institute of Physics (**http://www.aip.org/history/electron/**). An on-line version of the periodic table of the elements (Box 5-4) has detailed information about each element (**http://www.shef.ac.uk/chemistry/web-elements/**).

One of the most beautiful manifestations of light is the rainbow, which is the subject of several informative on–line tutorials (examples include **http://www.unidata. ucar.edu/staff/blynds/rnbw.html** and **http://ww2010. atmos.uiuc.edu/(Gh)/guides/mtr/opt/home.rxml**). Another useful tutorial describes the photoelectric effect, one of the key phenomena that showed that light must be thought of as a stream of photons (**http://www. telusplanet.net/public/mroy/photoelectric_effect/ photoelectriceffect.html**).

As we have seen in this chapter and Chapter 4, Isaac Newton left his mark in many areas of science. You can learn more about this intellectual colossus at Newtonia, a web site hosted by the European Laboratory for Particle Physics (**http://wwwcn.cern.ch/~mcnab/n/**). Albert Einstein, perhaps the greatest physicist since Newton, is the subject of a highly educational web site based on a PBS television program (**http://www.pbs.org/wgbh/nova/einstein/**). Another unique site with much Einstein information, including photographs and audio clips, is hosted by the American Institute of Physics (**http://www.aip.org/ history/einstein/**). Some of the many ideas of Einstein's that have transformed modern physics are described at Physics 2000, a site maintained at the University of Colorado (**http://www.colorado.edu/physics/2000/einstein.html**). Biographies of many of the other scientists who have made their mark on our understanding of light and atoms can also be found on the Web (**http://www.astro.virginia.edu/ ~eww6n/bios/**). Many of these scientists won Nobel Prizes in either physics or chemistry (there is no Nobel Prize in astronomy). Information on these is available from the Nobel Foundation in Sweden (**http://www.nobel.se/**).

# Optics and Telescopes

**The Summit of Mauna Kea in Hawaii** This mountaintop (elevation 4205 m, or 13,796 ft) is the world's best observing site. The two largest domes house the twin 10-m Keck reflectors, the largest steerable optical telescopes in the world. Under the silvery dome is NASA's Infrared Telescope Facility. Behind the closed dome of Keck I is Japan's 8.3-m Subaru telescope, shown here under construction; at the far left is the British-Canadian-Dutch James Clerk Maxwell Telescope, which detects millimeter waves. More than half a dozen other telescopes are located elsewhere on the summit. (Richard J. Wainscoat)

*In this chapter you will find the answers to the following questions:*

6-1 Why is it important that telescopes be large?

6-2 Why do most modern telescopes use a large mirror rather than a large lens?

6-3 Why are observatories in such remote locations?

6-4 Do astronomers use ordinary photographic film to take pictures of the sky? Do they actually look through large telescopes?

6-5 How do astronomers use telescopes to measure the spectra of distant objects?

6-6 Why do astronomers need to have telescopes that detect radio waves and other nonvisible forms of light?

6-7 Why is it useful to put telescopes in orbit, like the Hubble Space Telescope?

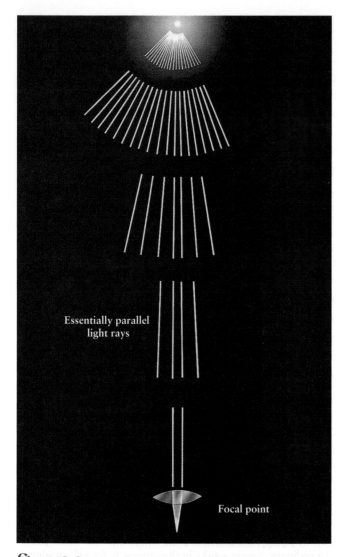

**ƒigure 6-3**

**Light Rays from Distant Objects Are Parallel** Light rays travel away in all directions from an ordinary light source. If a lens is located very far from the light source, only a few of the light rays will enter the lens, and these rays will be essentially parallel. This is why we drew parallel rays entering the lens in Figure 6-2b.

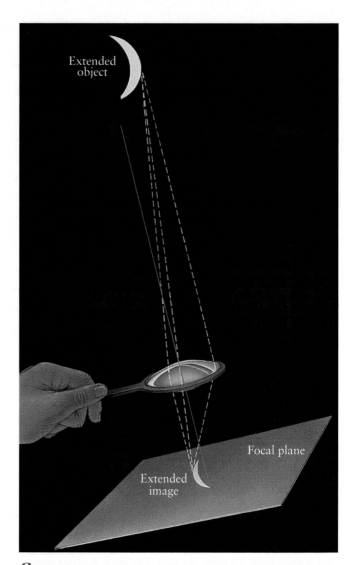

**ƒigure 6-4**

**A Lens Creates an Extended Image of an Extended Object** Light coming from each point on an extended object passes through a lens and produces an image of that point. All of these tiny images put together make an extended image of the entire object. The plane in which the image is formed is called the focal plane; for a very distant object, the distance from the lens to the focal plane is just the focal length.

or **refractor** (Figure 6-5). The large-diameter, long-focal-length lens at the front of the telescope, called the **objective lens,** forms the image; the smaller, shorter-focal-length lens at the rear of the telescope, called the **eyepiece lens,** magnifies the image for the observer. The refracting telescope was invented in the Netherlands in the early seventeenth century. Soon thereafter, Galileo used a small refracting telescope for his groundbreaking astronomical observations (see Section 4-5).

In addition to the focal length, the other important dimension of the objective lens of a refractor is the lens diameter. Compared to a small-diameter lens, a large-diameter lens captures more light, produces brighter images, and allows

astronomers to detect fainter objects. (For the same reason, the iris of your eye opens when you go into a darkened room to allow you to see dimly lit objects.) The **light-gathering power** of a telescope is directly proportional to the area of the objective, which in turn is proportional to the square of the lens diameter (Figure 6-6). Thus, if you double the diameter of the lens, the light-gathering power increases by a factor of $2^2 = 2 \times 2 = 4$. Box 6-1 describes how to compare the light-gathering power of different telescopes. Because light-gathering power is so important for seeing faint objects, the lens diameter is almost always given in describing a telescope. For example, the Lick telescope on Mount Hamilton in

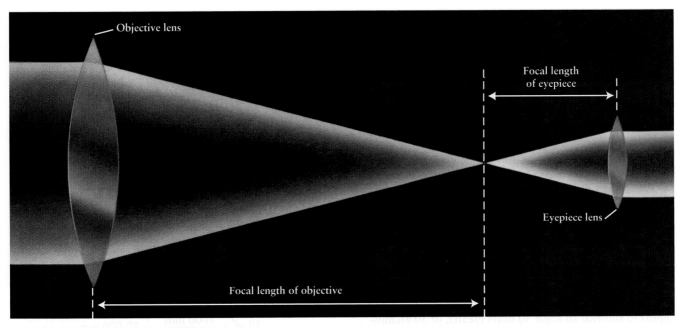

## figure 6-5

**A Refracting Telescope** A refracting telescope consists of a large-diameter objective lens with a long focal length and a small eyepiece lens of short focal length. The eyepiece lens magnifies the image formed at the focus of the objective lens.

California is a 90-cm refractor, which means that it is a refracting telescope whose objective is 90 cm in diameter. By comparison, Galileo's telescope of 1610 was a 3-cm refractor. The Lick telescope has an objective 30 times larger in diameter, and so has $30 \times 30 = 900$ times the light-gathering power of Galileo's instrument.

In addition to their light-gathering power, telescopes are useful because they magnify distant objects. As an example, the angular diameter of the Moon as viewed with the naked eye is about 0.5°. But when Galileo viewed the Moon through his telescope, its apparent angular diameter was 10°, large enough so that he could identify craters and mountain ranges. The **magnification,** or magnifying power, of a telescope is the ratio of an object's angular diameter seen through the telescope to its naked-eye angular diameter. Thus, the magnification of Galileo's telescope was 10°/0.5° = 20, usually written as 20×.

The magnification of a refracting telescope depends on the focal lengths of both of its lenses:

$$\text{Magnification} = \frac{\text{focal length of objective lens}}{\text{focal length of eyepiece lens}}$$

This formula shows that using a long-focal-length objective lens with a short-focal-length eyepiece gives a large magnification. Box 6-1 illustrates how this formula is used.

 Many people think that the primary purpose of a telescope is to give magnified images. While magnification is certainly important, it is *not* the most

## figure 6-6

**Light-Gathering Power** A large lens intercepts more starlight than a small one. Large objective lenses therefore produce brighter images than small objective lenses. The light-gathering power is proportional to the square of the lens diameter. The same principle applies to the curved mirrors used in reflecting telescopes, discussed in Section 6-2.

large lens that is entirely free of chromatic aberration. Fourth, because the lens can be supported only around its edges, it tends to sag and distort under its own weight. This has adverse effects on the image clarity. All these problems can be avoided by using a mirror instead of a lens to form an image.

## 6-2 A reflecting telescope uses a mirror to concentrate incoming light at a focus

Almost all modern telescopes form an image using the principle of **reflection.** To understand reflection, imagine drawing a dashed line perpendicular to the surface of a flat mirror at the point where a light ray strikes the mirror, as shown in Figure 6-9. The angle $i$ between the *incident* (arriving) light ray and the perpendicular is always equal to the angle $r$ between the *reflected* ray and the perpendicular. Knowing this, Isaac Newton realized that a concave mirror would cause parallel light rays to converge to a focus, as

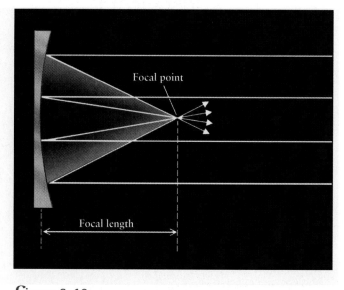

**ƒigure 6-10**

**Reflection by a Concave Mirror** A concave mirror causes parallel light rays to converge to a focus. The distance between the mirror and the focus is the focal length of the mirror.

shown in Figure 6-10. The distance between the reflecting surface and the focus is the focal length of the mirror. A telescope that uses a curved mirror to make an image of a distant object is called a **reflecting telescope,** or **reflector.** Using terminology similar to that used for refractors, the mirror that forms the image is called the **objective mirror** or **primary mirror.**

To make a reflector, an optician grinds and polishes a large slab of glass into the appropriate concave shape. The glass is then coated with silver, aluminum, or a similar highly reflective substance. Defects within the glass, which would have very negative consequences for the objective lens of a refracting telescope, have no effect on the optical quality of a reflecting telescope because light reflects off the surface of the glass rather than passing through it.

Another advantage of reflectors over refractors is that reflection is not affected by the wavelength of the incoming light. Because all wavelengths are reflected to the same focus, a mirror does not have chromatic aberration. (A small amount of chromatic aberration may arise if the image is viewed using an eyepiece lens.) Finally, the mirror can be fully supported by braces on its back, so that a large, heavy mirror can be mounted without much danger of breakage or surface distortion.

Even with the advantages of a reflecting telescope, we can improve on the arrangement shown in Figure 6-10. One problem with this arrangement is that the focal point is in front of the objective mirror. If you try to view the image formed at the focal point, your head will block part or all of the light from reaching the mirror. To get around this problem, Newton simply placed a small, flat mirror at a 45°

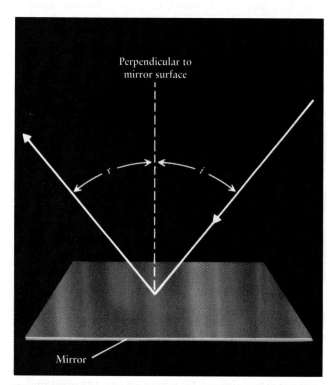

**ƒigure 6-9**

**Reflection by a Flat Mirror** The angle at which a beam of light approaches a mirror, called the angle of incidence ($i$), is always equal to the angle at which the beam is reflected from the mirror, called the angle of reflection ($r$). Reflection is accurately described by the equation $i = r$.

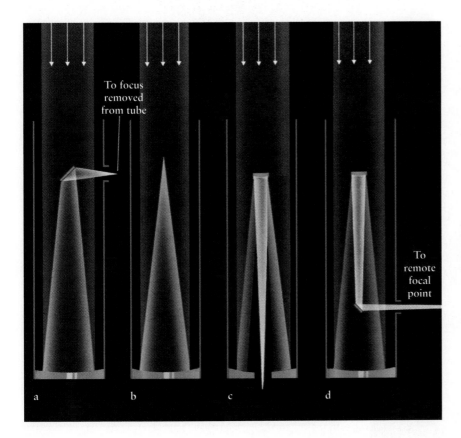

To focus
removed
from tube

To
remote
focal
point

a     b     c     d

figure 6-11

**Designs for Reflecting Telescopes** Four of the most popular optical designs for reflecting telescopes are shown here. **(a)** The Newtonian focus, devised by Isaac Newton, is used on some small reflecting telescopes for amateur astronomers. **(b)** Prime focus is used only on some large telescopes; an observer or instrument is placed directly at the focal point, within the barrel of the telescope. **(c)** The Cassegrain focus is used on reflecting telescopes of all sizes, from 20-cm (8-in.) reflectors used by amateur astronomers to the giant 10-m Keck telescopes on Mauna Kea (see the chapter opening figure). **(d)** The coudé focus is useful when large and heavy optical apparatus is to be used at the focal point. Light reflects off the objective mirror to a secondary mirror, then back down to an angled tertiary mirror. The light exits the side of the telescope along the same axis around which the telescope pivots up and down; hence, the heavy apparatus at the focal point does not have to move up and down with the telescope.

angle in front of the focal point, as sketched in Figure 6-11*a*. This secondary mirror deflects the light rays to one side, where Newton placed an eyepiece lens to magnify the image. A reflecting telescope having this optical design is appropriately called a **Newtonian reflector.** The magnifying power of a Newtonian reflector is calculated in the same way as for a refractor: The focal length of the objective mirror is divided by the focal length of the eyepiece (see Box 6-1).

Later generations of astronomers modified Newton's original design. The objective mirrors of some modern reflectors are so large that the astronomer could actually sit in an "observing cage" at the undeflected focal point, directly in front of the objective mirror. (In practice, riding in this cage on a winter's night is a remarkably cold and uncomfortable experience.) This arrangement is called a **prime focus** (Figure 6-11*b*). It usually provides the highest-quality image, because there is no need for a secondary mirror (which might have imperfections).

Another popular optical design, called a **Cassegrain focus,** also has a convenient, accessible focal point. A hole is drilled directly through the center of the primary mirror, and a convex secondary mirror placed in front of the original focal point reflects the light rays back through the hole (Figure 6-11*c*).

In a fourth design, a series of mirrors channels the light rays away from the telescope to a remote focal point. Heavy

optical equipment that could not be mounted directly on the telescope is then located at the resulting **coudé focus,** named after a French word meaning "bent like an elbow" (Figure 6-11*d*).

You might think that the secondary mirror in the Newtonian, Cassegrain, and coudé designs shown in Figure 6-11 would cause a black spot or hole in the center of the telescope image. But this does not happen. The reason is that light from every part of the object lands on every part of the primary, objective mirror. Hence, any portion of the mirror can itself produce an image of the distant object, as shown in Figure 6-12. The only effect of the secondary mirror is that it prevents part of the light from reaching the objective mirror and thus reduces somewhat the light-gathering power of the telescope.

A reflecting telescope must be designed to minimize a defect called **spherical aberration.** At issue is the precise shape of a mirror's concave surface. A spherical surface is easy to grind and polish, but different parts of a spherical mirror have slightly different focal lengths (Figure 6-13*a*). This results in a fuzzy image.

One common way to eliminate spherical aberration is to polish the mirror's surface to a parabolic shape, because a parabola reflects parallel light rays to a common focus (Figure 6-13*b*). Unfortunately, the astronomer then no longer

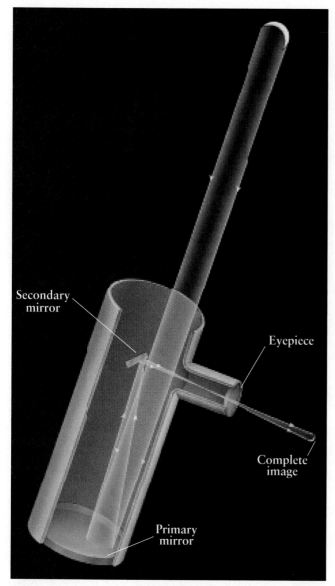

## figure 6-12

**The Secondary Mirror Does Not Cause a Hole in the Image**
Any part of the objective mirror of a reflecting telescope can form an image of a distant object. For example, this figure shows an image of the Moon being formed by a portion of the objective mirror of a Newtonian reflector. Thus, the secondary mirror does not cause a black spot or hole in the image. It merely makes the image a bit dimmer by reducing the total amount of light that reaches the objective.

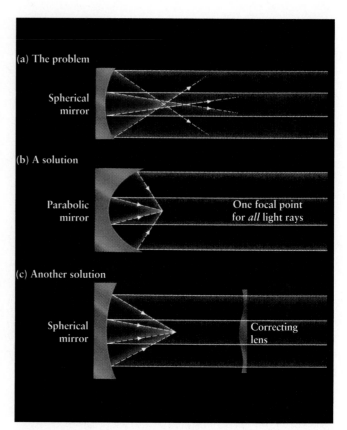

## figure 6-13

**Spherical Aberration** **(a)** Different parts of a spherically concave mirror reflect light to slightly different focal points. This effect, called spherical aberration, causes image blurring. This difficulty can be corrected by either **(b)** using a parabolic mirror or **(c)** using a correcting lens in front of the mirror.

has a wide-angle view. Furthermore, unlike spherical mirrors, parabolic mirrors suffer from a defect called **coma,** wherein star images far from the center of the field of view are elongated to look like tiny teardrops. A different approach is to use a spherical mirror, thus minimizing coma, and to place a thin correcting lens at the front of the tele-

scope to eliminate spherical aberration (Figure 6-13c). This approach is only used on relatively small reflecting telescopes, not on the large-diameter reflectors that have the greatest light-gathering power.

At the time of this writing, there are nine optical reflectors around the world with primary mirrors at least 4 meters (13.1 feet) in diameter (Box 6-2). Figure 6-14 shows one of the smallest of these, the 4.0-meter Blanco telescope at Cerro Tololo in Chile. The largest full-capability optical telescopes are the twin 10-meter Keck telescopes atop the dormant volcano Mauna Kea in Hawaii (see the photograph that opens this chapter). The mirror of the Keck I telescope is shown in Figure 6-15. Even larger is the 11-meter Hobby-Eberly Telescope, located in the Davis Mountains of western Texas; it cannot be pointed to all portions of the sky, but it cost only one-seventh as much as the $94 million price of either Keck telescope. A matching pair of telescopes, called Gemini, are being built atop Mauna Kea and Cerro Pachón, Chile. Both have mirrors 8 meters (26.2 feet) in diameter. These twins will allow astronomers to observe both the northern and southern parts of the celestial sphere with essentially the same state-of-the-art instrument.

In addition to the "giants" listed in Box 6-2, several other reflectors have objective mirrors between 3 and 4 meters in diameter, and dozens of smaller but still powerful telescopes have mirrors in the range from 1 to 3 meters. There are thousands of professional astronomers, each of whom has several ongoing research projects, and thus the demand for all of these telescopes is high. On any night of the year, nearly every research telescope in the world is being used to explore the universe.

## Figure 6-14

**The 4-m Telescope at Cerro Tololo** This telescope is located on a mountaintop near Santiago, Chile. The 4.0-meter-diameter objective mirror is located within the large white housing at the bottom of the telescope. The narrower black cylinder at the top of the telescope framework holds the secondary mirror for the Cassegrain focus (see Figure 6-11c). The triangular shapes projecting above the objective mirror and below the secondary mirror are dust covers; these fold over the mirrors to protect them when the telescope is not in use. The entire telescope is mounted in a large, horseshoe-shaped ring. Over the course of a night's observations, this rotates along with the apparent motion of the celestial sphere so that the astronomical object under study remains in the telescope's field of view. (Cerro Tololo Inter-American Observatory)

## Figure 6-15

**The W. M. Keck Telescope on Mauna Kea** This photograph is a view down the Keck I telescope, looking toward the 10-meter objective mirror. (Note the astronomers standing on platforms to either side of the telescope.) The objective itself is not a single mirror, but an arrangement of 36 hexagonal mirrors, each of which is 1.8 meters (6 feet) across. This system is much lighter than a single 10-meter mirror would be. Sensors and mechanical actuators on the back of each hexagonal mirror (not visible here) keep the arrangement in perfect alignment. The hole in the center of the mirror arrangement is for the Cassegrain focus (see Figure 6-11c); the 1.4-m secondary mirror is housed in the dark hexagonal structure above the center of the photograph. (Roger Ressmeyer-Starlight)

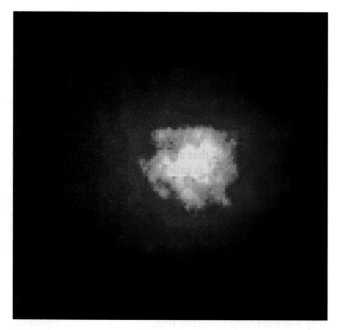
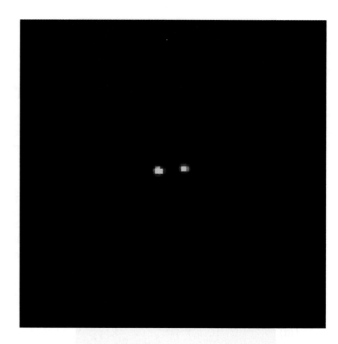

**figure 6-17**   R ◼ V U X G

**Using Adaptive Optics to "Unblur" Telescope Images**   Both of these images of the star
κ (Kappa) Pegasi (in the constellation Pegasus, the Winged Horse) were made with the same
3.5-m telescope. The left-hand image shows the blurring effects of the atmosphere; the star's
image is blurred across a seeing disk of more than 1 arcsec. The right-hand image was made with
an adaptive optics system turned on. With the greatly enhanced resolution, κ Pegasi is revealed
to be a binary star (two stars orbiting each other) with an angular separation between the stars
of only 0.3 arcsec. (R.Q. Fugate, USAF Research Laboratory, Starfire Optical Range)

**figure 6-18**

**The Very Large Telescope (VLT) Under Construction**   This
photograph shows the four domes of the European Southern
Observatory's Very Large Telescope (VLT) under construction at Cerro
Paranal in the Atacama desert of Chile. This site has very dark,
cloudless skies and provides excellent seeing. The four large structures
atop the mountain (elevation 2640 m, or 8660 ft) are the domes for
the four 8.2-meter telescopes. The first telescope was finished in
1998; all four should be in operation by 2001, and they should be able
to be used in concert by 2003. (European Southern Observatory)

Light from city street lamps and from buildings also degrades telescope images. This **light pollution** illuminates the
sky, making it more difficult to see the stars. You can appreciate the problem if you have ever looked at the night sky
from a major city. Only a few of the very brightest stars can
be seen, as against the thousands that can be seen with the
naked eye in the desert or the mountains. To avoid light pollution, observatories are built in remote locations far from
any city lights.

Unfortunately, the expansion of cities has brought light
pollution to observatories that in former times had none.
As an example, the growth of Tucson, Arizona, has had
deleterious effects on observations at the nearby Kitt Peak
National Observatory. Efforts have been made to have cities
adopt light fixtures that provide safe illumination for their
citizens but produce little light pollution. These efforts have
met with only mixed success.

One factor over which astronomers have absolutely no
control is the weather. Optical telescopes cannot see through
clouds, so it is important to build observatories where the
weather is usually good. One advantage of mountaintop observatories such as Mauna Kea is that most clouds form at
altitudes below the observatory, giving astronomers a better
chance of having clear skies.

In many ways the best location for a telescope is in orbit around Earth, where it is unaffected by weather, light pollution, or atmospheric turbulence. We will discuss orbiting telescopes in Section 6-7.

## 6-4 An electronic device is commonly used to record the image at a telescope's focus

Telescopes provide astronomers with detailed pictures of distant objects. The task of recording these pictures is called **imaging.**

Astronomical imaging really began in the nineteenth century with the invention of photography. It was soon realized that this new invention was a boon to astronomy. By taking long exposures with a camera mounted at the focus of a telescope, an astronomer can record features too faint to be seen by simply looking through the telescope. Such long exposures can reveal details in galaxies, star clusters, and nebulae that would not be visible to an astronomer looking through the telescope. Indeed, most large, modern telescopes do not have eyepieces at all.

Unfortunately, photographic film is not a very efficient light detector. Only about 1 out of every 50 photons striking photographic film triggers the chemical reaction needed to produce an image. Thus, roughly 98% of the light falling onto photographic film is wasted.

The most sensitive light detector currently available to astronomers is the **charge-coupled device (CCD).** A CCD is a thin piece of silicon (Figure 6-19), not unlike the silicon chips used in pocket calculators and programmable appliances. The silicon wafer of a CCD is divided into an array of small light-sensitive squares called picture elements or, more commonly, **pixels.** For example, some state-of-the-art CCDs have more than 4 million pixels arranged in 2048 rows by 2048 columns. When an image from a telescope is focused on the CCD, an electric charge builds up in each pixel in proportion to the number of photons falling on that pixel. When the exposure is finished, the amount of charge on each pixel is read into a computer, where the resulting image can be stored in digital form. Over certain wavelength ranges, more than 75% of the photons falling on a CCD can be recorded. The great sensitivity of CCDs also makes them very useful for **photometry,** which is the measurement of the brightnesses of stars and other astronomical objects. Because they are relatively inexpensive, CCDs are also used by many amateur astronomers.

In the modern world of CCD astronomy, astronomers need no longer spend the night in the unheated dome of a telescope. Instead, they operate the telescope electronically from a separate control room, where the electronic CCD images can be viewed on a computer monitor. The control room need not even be adjacent to the telescope. Although the Keck telescopes are at an altitude of 4200 m (13,800 feet), as shown in

**f**igure 6-19

**A Charge-Coupled Device (CCD)** This tiny silicon rectangle contains 163,840 light-sensitive electric circuits that store images. At the end of each exposure, additional circuits in the silicon chip control the transfer and readout of the data to a waiting computer. (Smithsonian Astrophysical Observatory)

the figure that opens this chapter, astronomers can now make observations from a facility elsewhere on the island of Hawaii that is much closer to sea level. This saves the laborious drive to the summit of Mauna Kea and eliminates the need for astronomers to acclimate to the high altitude.

Most of the images that you will see in this book were made with CCDs. Because of their extraordinary sensitivity and their ability to be used in conjunction with computers, CCDs have attained a role of central importance in astronomy.

## 6-5 Spectographs record the spectra of astronomical objects

We saw in Section 5-6 how the spectrum of an astronomical object provides a tremendous amount of information about that object. This is why the **spectrograph,** a device that records spectra on photographic plates or a CCD, is one of the

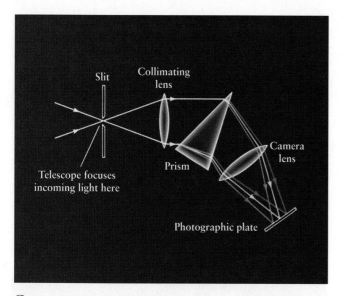

### figure 6-20

**A Prism Spectrograph** This optical device uses a prism to break up the light from a source into a spectrum. A lens, called the collimating lens, directs incoming light rays so that they enter the prism parallel to each other. A second lens, called a camera lens, then focuses the spectrum onto a photographic plate.

astronomer's most important tools. This optical device is mounted at the focus of a telescope.

An older design for a spectrograph uses a prism to form the spectrum of a star or galaxy, as shown in Figure 6-20. A slit, two lenses, and the prism are arranged to focus the spectrum onto a small photographic plate. Next, the exposed portion of the photographic plate is covered, and light from a hot gas (usually an element like iron or argon) is focused on the spectrograph slit. This result is a *comparison spectrum* above and below the spectrum of the star or galaxy, as shown in Figure 6-21. The wavelengths of the bright spectral lines of the comparison spectrum are already known from laboratory experiments and can therefore serve as reference markers.

There are drawbacks to this old-fashioned spectrograph. A prism does not disperse the colors of the rainbow evenly: Blue and violet portions of the spectrum are spread out more than the red portion. In addition, because the blue and violet wavelengths must pass through more glass than the red wavelengths (examine Figure 5-3), light is absorbed unevenly across the spectrum. Indeed, a glass prism is opaque to the near-ultraviolet part of the spectrum.

A better device for breaking starlight into the colors of the rainbow is a **diffraction grating**, or **grating** for short. This is a piece of glass on which thousands of closely spaced parallel lines have been cut. Some of the finest diffraction gratings have more than 10,000 lines per centimeter, which are usually cut by drawing a diamond back and forth across the glass. The spacing of the lines must be very regular, and the best results are obtained when the grooves are beveled. When light is shone on a diffraction grating, a spectrum is produced by the way in which light waves leaving different parts of the grating interfere with each other. (This same effect produces the rainbow of colors you see reflected from a compact disc or CD-ROM. Information is stored on the disc in a series of closely spaced pits, which act as a diffraction grating.) Figure 6-22 shows the design of a modern grating spectrograph.

In Figure 6-22 a CCD, instead of a photographic plate, is placed at the focus of the spectrograph. When the exposure is finished, electronic equipment measures the charge that has accumulated in each pixel. This data is used to graph light intensity against wavelength. Dark lines (absorption lines) appear as depressions or valleys on the graph, while bright lines (emission lines) in the spectrum appear as peaks. Figure 6-23 compares two ways of exhibiting a spectrum in which several absorption lines appear. Later in this book we shall see spectra presented in both of these ways.

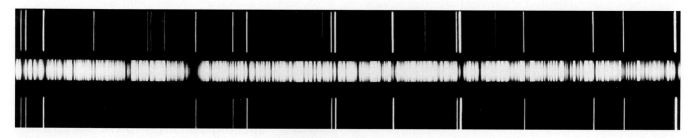

### figure 6-21

**A Spectrogram** The photographic record of a spectrum is called a spectrogram. This spectrogram shows the spectrum of a star. Numerous dark absorption lines can be seen against the brighter background of the spectrum. Above and below the star's spectrum are emission lines produced by an iron arc. These bright lines are the comparison spectrum. (Palomar Observatory)

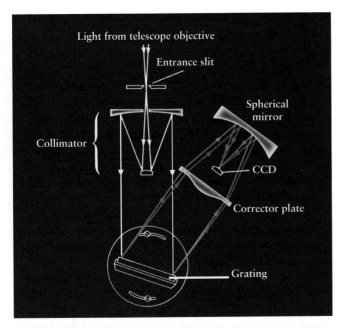

**figure 6-22**

**A Grating Spectrograph** This optical device uses a diffraction grating to break up the light from a source into a spectrum. The collimator ensures that light rays striking the grating are parallel. A corrector lens and mirror then focus the spectrum onto a CCD, which is far more sensitive than a photographic plate.

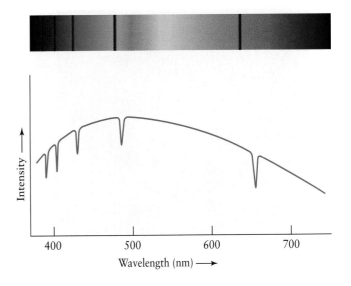

**figure 6-23**

**The Spectrum of Hydrogen** This drawing compares two ways of displaying a spectrum. When a photographic plate is placed at the focus of a spectrograph, a rainbow-colored spectrum is obtained. When a CCD is placed at the focus of a spectrograph, a graph of intensity versus wavelength is produced by a computer attached to the CCD. Note that absorption lines appear as dips in the curve of intensity versus wavelength.

---

| 6-6 | A radio telescope uses a large concave dish to reflect radio waves to a focus |
|---|---|

For thousands of years, all the information that astronomers gathered about the universe was based on ordinary visible light. In the twentieth century, however, astronomers first began to explore the nonvisible electromagnetic radiation coming from astronomical objects. The result has been an opening of astronomers' eyes to more of the cosmos than could ever be seen using optical telescopes alone. It is no exaggeration to say that today's astronomers learn as much about the universe using telescopes for nonvisible wavelengths as they do using visible light.

Radio waves were the first nonvisible part of the electromagnetic spectrum to be exploited for astronomy. This happened as a result of a research project seemingly unrelated to astronomy. In the early 1930s, Karl Jansky, a young electrical engineer at the Bell Telephone Laboratories, was trying to locate what was causing interference with the then-new transatlantic radio link. By 1932 he realized that one kind of radio noise is strongest when the constellation Sagittarius is high in the sky. The center of our Galaxy is located in the direction of Sagittarius, and Jansky concluded that he was detecting radio waves from an astronomical source.

At first only Grote Reber, a radio engineer living in Illinois, took up Jansky's research. In 1936 Reber built in his backyard the first **radio telescope**, a radio-wave detector dedicated to astronomy. He modeled his design after an ordinary reflecting telescope, with a parabolic "dish" (reflecting antenna) measuring 31 ft (10 m) in diameter. A radio receiver was placed at the focal point of the metal dish.

Reber spent the years from 1938 to 1944 mapping radio emissions from the sky at wavelengths of 1.9 m and 0.63 m. He found radio waves coming from the entire Milky Way, with the greatest emission from the center of the galaxy. These results, together with the development of improved radio technology during World War II, encouraged the growth of radio astronomy and the construction of new radio telescopes around the world. Radio observatories are as common today as major optical observatories.

Like Reber's prototype, a typical radio telescope has a large parabolic dish. A small antenna tuned to the desired frequency is located at the focus, and the incoming signal is relayed to amplifiers and recording instruments, typically located in a room at the base of the telescope's pier. Figure 6-24 shows a modern radio telescope.

Many radio telescope dishes, like the one in Figure 6-24, have visible gaps in them like a wire mesh. This does not

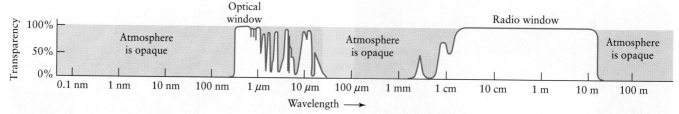

## figure 6-27

**The Transparency of the Earth's Atmosphere** This graph shows the percentage of radiation that can penetrate the Earth's atmosphere as a function of wavelength. The white regions, where the curve of transparency is high, show wavelengths at which the atmosphere is relatively transparent; for wavelengths in the gray regions, where the transparency curve is low, the atmosphere is opaque. Oxygen and nitrogen completely absorb all radiation with wavelengths shorter than 290 nm. Water vapor and carbon dioxide effectively block out all radiation from about 10 μm to 1 cm.

Figure 6-27 shows how the transparency of the Earth's atmosphere is different at different wavelengths. Two wavelength regions in which our atmosphere is most transparent are the **optical window,** through which we see visible light from space, and the **radio window,** through which we receive radio waves. The graph also shows several relatively transparent regions at infrared wavelengths between 1 and 10 m. Infrared radiation within these wavelength intervals does penetrate the Earth's atmosphere and can be detected with ground-based equipment. This wavelength range is called the *near-infrared*, because it lies just beyond the red end of the visible spectrum.

Water vapor is the main absorber of infrared radiation from space. Infrared observatories are therefore located at sites with low humidity. For example, the 14,000-ft summit of Mauna Kea in Hawaii is exceptionally dry. Making infrared observations is the primary function of several telescopes there. (The silvery dome of one of them, NASA's 3-meter Infrared Telescope Facility, can be seen in the photograph that opens this chapter.)

The best way of avoiding water vapor is to place a telescope in the Earth orbit. In 1983 NASA launched the Infrared Astronomical Satellite (IRAS) into a 900-km-high polar orbit (Figure 6-28). During its ten-month mission, this 60-cm reflector revealed the richness of the infrared sky over wavelengths from about 10 to 100 μm. For the first time, astronomers saw dust bands in our solar system and dust disks around nearby stars. This dust is too cold to emit much visible light, so it remained undetected by ordinary optical telescopes. IRAS also discovered distant galaxies that emit most of their radiation at infrared wavelengths.

The Infrared Space Observatory (ISO), a more advanced 60-cm reflector with better light detectors, was launched into orbit in 1995 by the European Space Agency (ESA). During its 2½-year mission it made a number of groundbreaking observations of very distant galaxies and of the thin, cold material between the stars of our own galaxy. Like IRAS, ISO

had to be cooled by liquid helium to temperatures just a few degrees above absolute zero. Had this not been done, the infrared blackbody radiation from the telescope itself would have outshone the infrared radiation from astronomical objects. The ISO mission came to an end when the last of the helium evaporated into space. Budget cuts have frustrated

## figure 6-28

**The Infrared Astronomical Satellite (IRAS)** This satellite contained a small reflecting telescope that gave astronomers their first in-depth look at the infrared sky. Launched in 1983, IRAS contributed to research topics running the gamut from asteroids to the large-scale structure of the universe. The follow-on Infrared Space Observatory (ISO), launched in 1995, used an objective mirror of the same diameter as IRAS but had much more sensitive and refined detectors for infrared light. (NASA)

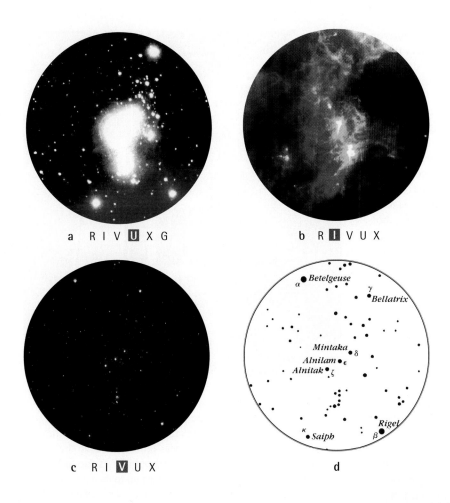

a R I V **U** X G

b R **I** V U X

c R I **V** U X

d

**figure 6-29**

**Orion Seen at Ultraviolet, Infrared, and Visible Wavelengths** **(a)** An ultraviolet view of the constellation of Orion was obtained during a brief rocket flight in 1975. **(b)** The 100-s exposure covers the wavelength range 125–200 μm. **(c)** The false-color view from IRAS displays infrared intensity according to color: red for strong 100-μm radiation, green for strong 60-μm radiation, and blue for strong 12-μm radiation. **(d)** For comparison, an ordinary optical photograph and a star chart are included. (Courtesy of G. R. Carruthers, NRL; NASA; R. C. Mitchell, Central Washington University)

NASA plans to develop and launch SIRTF (the Space Infrared Telescope Facility), an 0.85-m infrared telescope which would survey the infrared sky with unprecedented resolution. At present it is hoped that SIRTF will go into orbit in 2001 or 2002 for a mission lasting $2\frac{1}{2}$ years.

Earth's atmosphere is also transparent to the longest-wavelength ultraviolet light. This wavelength range, extending from about 400 nm down to 300 nm, is called the *near-ultraviolet,* because it lies just beyond the violet end of the visible spectrum. Astronomers can make ground-based observations in this wavelength range, but only if they do not use glass lenses in their telescopes. Glass is opaque to the near-ultraviolet, so lenses for ultraviolet astronomy must be made of quartz or some other nonabsorbing substance.

To see shorter-wavelength *far-ultraviolet* light, astronomers must again make their observations from space. During the early 1970s, Apollo and Skylab astronauts carried small ultraviolet telescopes above the Earth's atmosphere to give us our first views of the ultraviolet sky. Small rockets have also lifted ultraviolet cameras briefly above the Earth's atmosphere. A typical view is shown in Figure 6-29, along with infrared and visible views.

Some of the finest ultraviolet astronomy was accomplished by the International Ultraviolet Explorer (IUE), launched in 1978. This satellite (Figure 6-30) was built

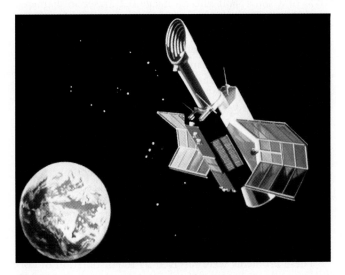

**figure 6-30**

**The International Ultraviolet Explorer (IUE)** Launched in 1978, this 671-kg satellite produced superb observations in the far-ultraviolet. The blue panels extending from either side of the midsection of the satellite are solar cell arrays that provide electrical power for the radio transmitters and other electronic equipment. (NASA)

around a Cassegrain telescope with a 45-cm (18-in.) mirror and a total focal length of 6.74 m (22 ft). Observations cover the range from 116 to 320 nm. IUE provided data until 1996; it could have lasted longer but was shut down for budgetary reasons.

To observe the sky at the very shortest ultraviolet wavelengths, NASA launched the Extreme Ultraviolet Explorer (EUVE) in 1992. This two-ton satellite carries four telescopes that focus high-energy ultraviolet photons by glancing them off highly polished surfaces of tapered cylinders. Three of the telescopes work together to survey the entire sky over a wavelength range of 7 to 76 nm. The fourth telescope is used to make observations of selected faint sources. Small, hot stars called white dwarfs are among the primary targets.

Although these satellites give excellent views of the heavens at selected wavelengths, astronomers dreamed for decades of having one large telescope that could be operated at any wavelength from the near-infrared through the visible range and out into the ultraviolet. This is the mission of the Hubble Space Telescope (HST), which was carried into orbit by the space shuttle in 1990 (Figure 6-31). HST has a 2.4-meter (7.9-ft) objective mirror and was designed to observe at wavelengths from 115 nm to 1 μm. Like most ground-based telescopes, HST uses a CCD to record images. These are then radioed back to Earth in digital form.

The dream of the astronomers who planned HST was a telescope whose angular resolution was limited only by diffraction. But soon after HST was placed in orbit, astronomers discovered that a manufacturing error had caused the telescope's 2.4-m primary mirror to suffer from spherical aberration. The mirror should have been able to concentrate 70% of a star's light into an image 0.1 arcsec in diameter. Instead, only 20% of a star's light is focused into the desired 0.1-arcsec spot; the remaining 80% is smeared out over an area about 1 arcsec in diameter. A star image therefore consists of a central spot of modest brightness surrounded by a hazy glow. Pictures taken using 100% of the incoming starlight gave star images about 1 arcsec in angular diameter, which was no better than the images achieved at major ground-based observatories.

On an interim basis, astronomers were able to cope partially with this problem by using only the 20% of incoming starlight that is properly focused and, with computer processing, discard the remaining poorly focused 80%. This procedure was practical only for brighter objects on which astronomers could afford to waste light. But many of the observing projects scheduled for the HST involved extremely dim galaxies and nebulae.

These problems were resolved by the installation of corrective optics by a second space shuttle mission in 1993. A set of small secondary mirrors were installed whose curvature exactly compensated for the error in curvature of the primary mirror. After these were installed, HST was finally able to make truly superb, sharp images of extremely faint objects (Figure 6-32). Many of the astronomical images that you will see in later chapters were made using HST.

## ƒigure 6-31

**The Hubble Space Telescope (HST)** The Hubble Space Telescope is a 2.4-m reflecting telescope designed to study the heavens at wavelengths from the near-infrared through the ultraviolet. Initially plagued by smeared-out images due to an improperly shaped mirror, HST finally lived up to its potential after corrective optics were installed in 1993. This photograph was taken from the space shuttle *Endeavour* during the 1993 repair mission. HST is being held by *Endeavour's* robotic arm (at upper left). The large solar panels on either side of the telescope provide electric power.

Because neither X rays nor gamma rays penetrate Earth's atmosphere, observations at these extremely short wavelengths must also be made from space. X rays are emitted by extremely hot gases (temperatures in the range of $10^6$ K), while gamma rays are produced by high-energy collisions between subatomic particles. Thus both of these kinds of radiation can be used to locate the most energetic places in the cosmos. To detect these high-energy photons, astronomers use electronic detectors similar to Geiger counters.

Astronomers got their first quick look at the X-ray sky from brief rocket flights during the late 1940s. By the early 1970s, small satellites had revealed hundreds of previously unknown X-ray sources, including several good black hole candidates. Since 1977 a number of increasingly sophisticated X-ray telescopes have been placed on board satellites. NASA's Einstein Observatory, one of three High-Energy Astrophysical Observatories (HEAO), was launched in 1978

**M100, a Spiral Galaxy in the Virgo Cluster**
**Hubble Space Telescope / Wide Field Planetary Camera**

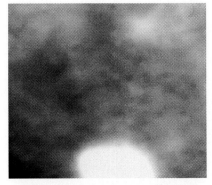

a Palomar 5m on a good night

b WFPC-1: Wide Field Camera

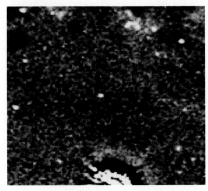

c WFPC-1: deconvolved

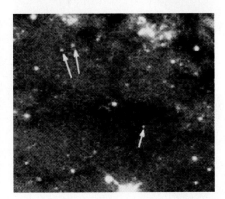

d WFPC-2: Wide Field Camera

**ƒigure 6-32** R I **V** U X G

**HST Images: Before and After Repair**

These four images show a portion of the galaxy M100 in the constellation Coma Berenices (Berenice's Hair), some 70 million light-years from Earth. **(a)** This image made with the Hale 5-meter telescope on Palomar Mountain shows little detail. Its resolution is limited by the seeing disk of about 1 arcsec. **(b)** This raw image from the original, uncorrected HST is substantially better than the ground-based image. **(c)** This is a computer-processed version of image (b). Some features are made sharper in this way. **(d)** This was one of the first images made with HST after the corrective optics was installed. The image is substantially sharper than before the repair. What is more, stars that were previously too dim to be seen (shown by the arrows) are now easily resolved. (Space Telescope Science Institute, NASA)

near the hundredth anniversary of Albert Einstein's birth. It observed shock waves from exploding stars and traced out the extremely hot gas that surrounds galaxies. Its successor is ROSAT, launched into a nearly circular Earth orbit in 1990. Its instruments include an X-ray telescope and a wide-field X-ray camera. ROSAT, which grew from a German venture into an international project, will continue to provide a wealth of information about the X-ray sky during the next few years.

ROSAT is an important evolutionary step on the way to building AXAF, the Advanced X-Ray Astrophysics Facility, which is scheduled to be launched in 1998. AXAF will carry out both imaging and spectroscopy with unprecedented sensitivity and precision.

One of several gamma-ray telescopes currently in orbit is the Compton Gamma Ray Observatory (Figure 6-33). Launched in 1991 and named in honor of Arthur Holly Compton, an American scientist who made important discoveries about gamma rays, this satellite carries four instruments that are performing a variety of observations. One particularly important task for this satellite has been the study of gamma-ray bursts, which are brief, unpredictable, and very intense flashes of gamma rays that occur unpredictably from

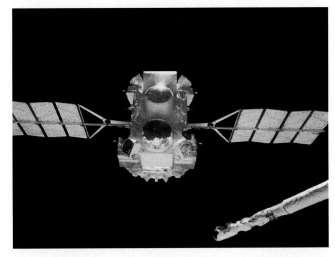

**ƒigure 6-33**

**The *Compton Gamma Ray Observatory*** This photograph of the Compton Observatory hovering above the cargo bay of the space shuttle *Atlantis* was taken as the spacecraft was being deployed in 1991. The best views of the high-energy gamma-ray sky come from this satellite, which carries four gamma-ray detectors. (NASA)

all parts of the sky. Data from the Compton GRO and other orbiting observatories strongly suggest that the sources of these gamma-ray bursts are billions of light-years away. To be able to see these bursts across such a distance, their sources must be almost among the most ener-getic objects in the universe.

The advantages and benefits of Earth-orbiting observatories cannot be overemphasized. We are no longer limited to the narrow ranges of whatever wavelengths manage to leak through our shimmering, hazy atmosphere (Figure 6-34). For the first time, we are really seeing the universe.

a  R I V U X G

b  R I V U X G

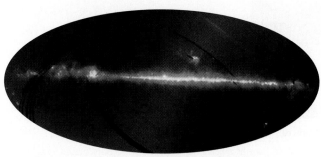

c  R I V U X G

d  R I V U X G

e  R I V U X G

## figure 6-34

**The Entire Sky at Five Wavelength Ranges** These five views show the entire sky at radio, infrared, visible, ultraviolet, and X-ray wavelengths. The Milky Way stretches horizontally across each picture.

**(a)** The radio view shows the sky at a wavelength of 73 cm (a frequency of 411 MHz), with the brightest regions in red and the dimmest in blue. The radio waves are emitted by charged particles spiraling in the Milky Way's magnetic field.

**(b)** The X-ray view from HEAO-1 covers 0.2 to 6 nm, corresponding to a range of photon energies from 6 to 0.2 keV. Some of this X-ray emission is from hot gas clouds, while some may be from compact, highly energetic regions around black holes. (Max Planck Institut für Radioastronomie; Jet Propulsion Laboratory; Griffith Observatory; Royal Observatory, Edinburgh; E. Boldt, NASA)

**(c)** The infrared view from IRAS shows several infrared wavelengths by color: 100 mm is yellow, 60 mm is red, and 12 mm is blue. The emission is dominated by dust particles in the plane of the Milky Way.

**(d)** In the visible view the constellation Orion is at the right, Sagittarius in the middle, and Cygnus toward the left. Many of the dark areas along the Milky Way are locations where the dust, shown in (b), is sufficiently thick to block visible light.

**(e)** The ultraviolet view covers 135 to 255 nm and shows numerous hot stars around Orion and Cygnus.

## Key Words

*Terms preceded by an asterisk (*) are discussed in the Boxes.*

adaptive optics, p. 143

angular resolution, p. 142

baseline, p. 148

Cassegrain focus, p. 139

charge-coupled device (CCD), p. 145

chromatic aberration, p. 136

coma, p. 140

coudé focus, p. 139

diffraction, p. 142

diffraction grating, p. 146

eyepiece lens, p. 134

focal length, p. 133

focal plane, p. 133

focal point, p. 133

focus (of a lens or mirror), p. 133

grating, p. 146

imaging, p. 145

interferometry, p. 148

light-gathering power, p. 134

light pollution, p. 144

magnification, p. 135

medium (*plural* media), p. 132

Newtonian reflector, p. 139

objective lens, p. 134

objective mirror (primary mirror), p. 138

optical telescope, p. 132

optical window (in Earth's atmosphere), p. 150

photometry, p. 145

pixel, p. 145

prime focus, p. 139

radio telescope, p. 147

radio window (in Earth's atmosphere), p. 150

reflecting telescope (reflector), p. 138

reflection, p. 138

refracting telescope (refractor), p. 133

refraction, p. 132

seeing disk, p. 143

spectrograph, p. 145

spherical aberration, p. 139

very-long-baseline interferometry (VLBI), p. 149

## Key Ideas

**Refracting Telescopes:** Refracting telescopes, or refractors, produce images by bending light rays as they pass through glass lenses.

• Chromatic aberration is an optical defect whereby light of different wavelengths is bent in different amounts by a lens.

• Glass impurities, chromatic aberration, opacity to certain wavelengths, and structural difficulties make it inadvisable to build extremely large refractors.

**Reflecting Telescopes:** Reflecting telescopes, or reflectors, produce images by reflecting light rays to a focus point from curved mirrors.

• Reflectors are not subject to most of the problems that limit the useful size of refractors.

**Angular Resolution:** A telescope's angular resolution, or ability to see fine details, is limited by two key factors.

• Diffraction is an intrinsic property of light waves. Its effects can be minimized by using a larger objective lens or mirror.

• The blurring effects of atmospheric turbulence can be minimized by placing the telescope atop a tall mountain with very smooth air. They can be dramatically reduced by the use of adaptive optics and can be eliminated entirely by placing the telescope in orbit.

**Charge-Coupled Devices:** Charge-coupled devices (CCDs) are often used at a telescope's focus to record faint images.

**Spectrographs:** A spectrograph uses a diffraction grating and lenses to form the spectrum of an astronomical object.

**Radio Telescopes:** Radio telescopes use large reflecting antennas or dishes to focus radio waves.

• Very large dishes provide reasonably sharp radio images. Higher resolution is achieved with interferometry techniques that link smaller dishes together.

**Transparency of the Earth's Atmosphere:** The Earth's atmosphere absorbs much of the radiation that arrives from space.

• The atmosphere is transparent chiefly in two wavelength ranges known as the optical window and the radio window. A few wavelengths in the near-infrared also reach the ground.

**Telescopes in Space:** For observations at wavelengths to which the Earth's atmosphere is opaque, astronomers depend on telescopes carried above the atmosphere by high-altitude airplanes, rockets, or satellites.

• Satellite-based observatories are giving us a wealth of new information about the universe and are permitting coordinated observation of the sky at all wavelengths.

## Review Questions

**1.** Describe refraction and reflection. Explain how these processes enable astronomers to build telescopes.

**2.** With the aid of a diagram, describe a refracting telescope.

**3.** What is chromatic aberration? For what kinds of telescopes does it occur? How can it be corrected?

**4.** With the aid of a diagram, describe a reflecting telescope. Describe four different ways in which an astronomer can access the focal plane.

**5.** Explain some of the advantages of reflecting telescopes over refracting telescopes.

**6.** What is spherical aberration? How can it be corrected?

**7.** Quite often advertisements appear for telescopes that extol their magnifying power. Is this a good criterion for evaluating telescopes? Explain your answer.

**8.** What kind of telescope would you use if you wanted to take a color photograph entirely free of chromatic aberration? Explain your answer.

**9.** No major observatory has a Newtonian reflector as its primary instrument, whereas Newtonian reflectors are extremely popular among amateur astronomers. Explain why this is so.

**10.** What is adaptive optics? Why is it useful? Would it be a good feature to include on a telescope to be placed in orbit?

**11.** Compare an optical reflecting telescope and a radio telescope. What do they have in common? How are they different?

**12.** Why can radio astronomers make observations at any time during the day, whereas optical astronomers are mostly limited to observing at night? (*Hint:* Does your radio work any better or worse in the daytime than at night?)

**13.** What are the optical window and the radio window? Why isn't there an X-ray window or an ultraviolet window?

**14.** Why must astronomers use satellites and Earth-orbiting observatories to study the heavens at X-ray and gamma-ray wavelengths?

## Advanced Questions

### Problem-solving tips and tools:

You may find it useful to review the small-angle formula discussed in Box 1-2. The area of a circle is proportional to the square of its diameter. Data on the planets are found in the appendixes at the end of this book. The relationship between frequency and wavelength is discussed in Section 5-2. Examples of how to calculate magnifying power and light-gathering power were given in Box 6-1.

**15.** Show by means of a diagram why the image formed by a simple refracting telescope is upside down.

**16.** Ordinary photographs made with a telephoto lens make distant objects appear close. How does the focal length of a telephoto lens compare to that of a normal lens? Explain your reasoning.

**17.** The observing cage in which an astronomer sits at the prime focus of the 5-m telescope on Palomar Mountain is about 1 m in diameter. Calculate what fraction of the incoming starlight is blocked by the cage.

**18.** Compare the light-gathering power of the Keck 10-m telescope with that of the Hubble Space Telescope (HST), which has a 2.4-m objective mirror. What advantages does HST have over Keck?

**19.** Suppose your Newtonian reflector has an objective mirror 25 cm (10 in.) in diameter with a focal length of 2 m. What magnification do you get with eyepieces whose focal lengths are (**a**) 9 mm, (**b**) 20 mm, and (**c**) 55 mm? What is the telescope's angular resolution?

**20.** Several telescope manufacturers build telescopes having a design they call a "Schmidt-Cassegrain," which use a correcting lens in an arrangement like that shown in Figure 6-13*b*. Consult advertisements in such magazines as *Sky & Telescope* and *Astronomy* to see their appearance and find out the cost of these telescopes. Why do you suppose they are very popular among amateur astronomers?

**21.** The four largest moons of Jupiter are roughly the same size as our Moon and are about 628 million ($6.28 \times 10^8$) kilometers from Earth at opposition. What is the size of the smallest surface features that the Hubble Space Telescope (resolution of 0.1 arcsec) can detect? How does this compare with the smallest features that can be seen on the Moon with the unaided human eye (resolution of 1 arcmin)?

**22.** Can the HST distinguish any features on Pluto? Justify your answer using calculations.

**23.** The Russian Space Agency plans to place a radio telescope into an even higher orbit than the Japanese HALCA telescope. Using this telescope in concert with the VLBA, baselines as long as 77,000 km may be obtainable. Astronomers want to use this combination to study radio emission at a frequency of 1665 MHz from distant objects called quasars. (**a**) What is the wavelength of this emission? (**b**) Taking the baseline to be the effective diameter of this radio-telescope array, what angular resolution can be achieved?

**24.** Consult such magazines as *Sky & Telescope* and *Science News,* or their web sites, to find out how some of the telescope projects mentioned in this chapter are going. Which of these telescopes are now in operation?

**25.** The Hubble Space Telescope is visited periodically by a space shuttle for refurbishing and upgrading of equipment. Consult such magazines as *Sky & Telescope* and *Astronomy,* or their web sites, to find out about the next space shuttle mission to HST. What changes will be made to HST on this mission? When is the launch date?

## Discussion Questions

**26.** Discuss the advantages and disadvantages of using a small telescope in Earth orbit versus a large telescope on a mountaintop.

**27.** If you were in charge of selecting a site for a new observatory, what factors would you consider important?

## Observing Projects

**28.** Obtain a telescope during the daytime along with several eyepieces of various focal lengths. If you can determine the telescope's focal length, calculate the magnifying powers of the eyepieces. Focus the telescope on some familiar object, such as a distant lamppost or tree. **DO NOT FOCUS ON THE SUN! Looking directly at the Sun can cause blindness.** Describe the image you see through the telescope. Is it upside down? How does the image move as you slowly and gently shift the telescope left and right or up and down? Examine the eyepieces, noting their focal lengths. By changing the eyepieces, examine the distant object under different magnifications. How does the field of view and the quality of the image change as you go from low power to high power?

**29.** On a clear night, view the Moon, a planet, and a star through a telescope using eyepieces of various focal lengths and known magnifying powers. (You may have to consult such magazines as *Sky & Telescope* and *Astronomy*, or their web sites, to determine the locations in the sky of the Moon and planets.) In what way does the image seem to degrade as you view with increasingly higher magnification? Do you see any chromatic aberration? If so, with which object and which eyepiece is it most noticeable?

**30.** Many towns and cities have amateur astronomy clubs. If you are so inclined, attend a "star party" hosted by your local club. People who bring their telescopes to such gatherings are delighted to show you their instruments and take you on a telescopic tour of the heavens. Such an experience can lead to a very enjoyable, lifelong hobby.

## Where to Learn More

*Magazine articles*

Berry, R. "Image Processing in Astronomy." *Sky & Telescope,* April 1994. Part of the usefulness of CCDs for astronomical imaging is that they deliver the image in digital form. This article describes how computers can process these images to extract as much information from them as possible.

Bowyer, S., Malina, R., and Haisch, B. "Observing a Partly Cloudy Universe." *Sky & Telescope,* December 1994. Written by the principal investigators for the Extreme Ultraviolet Explorer spacecraft, this article describes many of the results of EUVE. The title refers to hydrogen clouds in space, which are opaque to short-wavelength ultraviolet and pose a major challenge to observations with EUVE.

Dyer, A. "ROSAT's Penetrating X-Ray Vision." *Astronomy,* June 1991. Written shortly after ROSAT had completed a survey of the entire X-ray sky, this article includes some remarkable pictures from this orbiting X-ray observatory.

———. "Buying the Best Telescope." *Sky & Telescope,* December 1997. Shopping for a first telescope can be a frustrating and confusing experience. This informative article provides some useful pointers on choosing the best telescope.

Fugate, R. Q., and Wild, W. J. "Untwinkling the Stars." *Sky & Telescope,* May 1994 and June 1994. This two-part article gives an accessible introduction to ideas behind adaptive optics and the many problems involved in making it practical for astronomy. The two authors are pioneers in the field.

Hawley, S. A. "Hubble Revisited." *Sky & Telescope,* February 1997. Written by an astronomer-astronaut, this fascinating article explains how space shuttle astronauts service the Hubble Space Telescope.

Kellermann, K. I. "Radio Astronomy in the 21st Century." *Sky & Telescope,* February 1997. This article by a noted radio astronomer discusses plans for the future, with special emphasis on arrays of radio telescopes and on VLBI.

Kniffen, D. A. "The Gamma Ray Observatory." *Sky & Telescope,* May 1991. This article describes the Compton Gamma Ray Observatory and some features of the gamma-ray sky that it was designed to observe.

Ressmeyer, R. H. "Keck's Giant Eye." *Sky & Telescope,* December 1992. Beautiful photographs of the Keck Observatory illustrate this article.

W *World Wide Web*

Many observatories have sites on the World Wide Web that include pictures and descriptions of their telescopes, as well as of their latest astronomical discoveries. A master list of observatory web sites is available (**http://www.stsci.edu/astroweb/yp_telescope.html**), as is a list of every orbiting observatory ever launched (**http://www.seds.org/~spider/oaos/oaos.html**).

NASA's Jet Propulsion Laboratory maintains an educational web site about the basics of radio astronomy (**http://www.jpl.nasa.gov/radioastronomy/**). You can also read about the Next Generation Space Telescope, the planned successor to the Hubble Space Telescope (**http://ngst.gsfc.nasa.gov/**).

Glowing aurorae near the poles of Saturn, seen by the Hubble Space Telescope.
(J. Trauger, Jet Propulsion Laboratory; NASA)

# Planets and Moons

**f**orty years ago astronomers knew precious little about the planets. Even the best telescopes provided images of the planets that were frustratingly hazy and indistinct. Of asteroids, comets, and the satellites of the planets, we knew even less.

Today, our knowledge of the solar system has grown many thousandfold. Spacecraft have flown at close range past all the planets save Pluto, showing us detail that astronomers of an earlier generation could only dream about. We have landed spacecraft on the Moon, Venus, and Mars and dropped a probe into the immense atmosphere of Jupiter. This is truly the golden age of solar system exploration.

In the next 11 chapters we will explore the worlds that orbit the Sun. After an overview, we will start with Mercury—closest to the Sun of all the planets—and work our way outward to the cold, remote regions that border on interstellar space. We will learn, too, about the moons and other small but important bodies that populate our solar system.

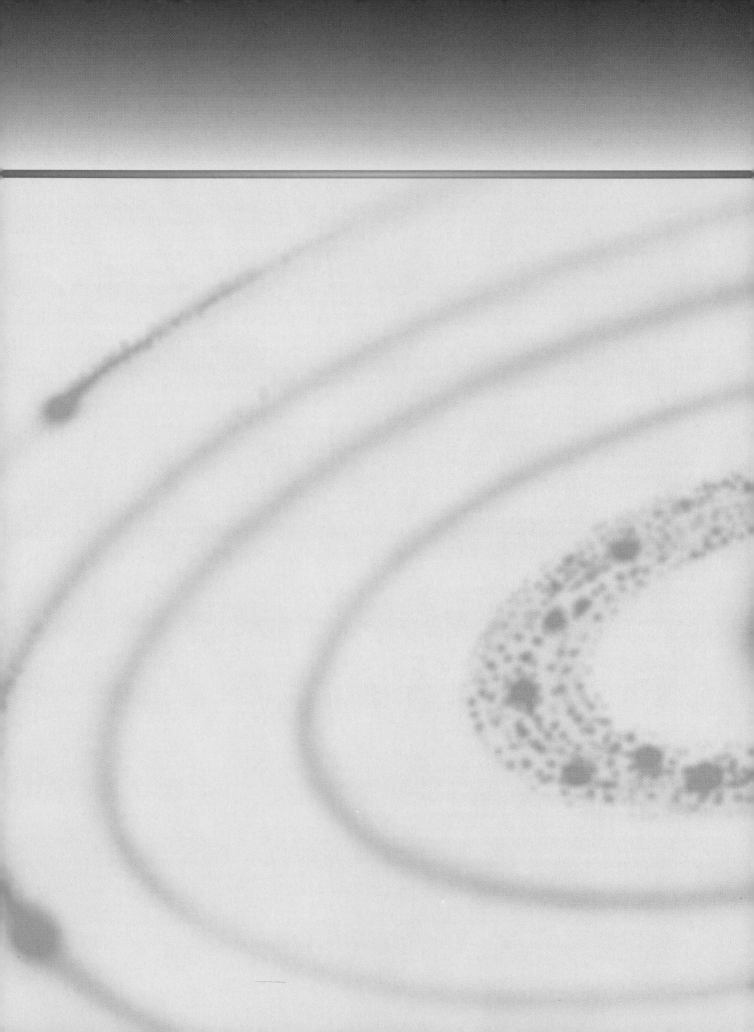

# Our Solar System

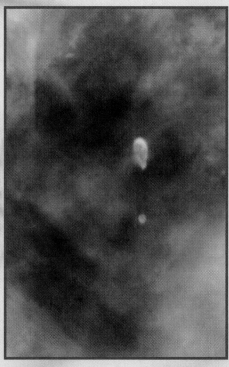

R I **V** U X G

**Stars and Planets Forming in the Orion Nebula**

The beautiful clouds of gas and dust in the left-hand image are a portion of the Orion Nebula, a giant star-forming region located some 1500 light-years from Earth. The closeup image on the right shows a region of the Orion Nebula only 0.14 light-years across. The light and dark blobs are stars in the process of formation, each surrounded by a disk of material within which planets may be forming. The planets of our own solar system are thought to have formed from such a disk surrounding the young Sun. (C. R. O'Dell and S. K. Wong, Rice University; NASA)

*In this chapter you will find the answers to the following questions:*

7-1 Are all the other planets similar to Earth, or are they very different?

7-2 Do other planets have moons like Earth's Moon?

7-3 How do astronomers know what the other planets are made of?

7-4 Are all the planets made of basically the same material?

7-5 What is the difference between an asteroid and a comet?

7-6 Why are some elements (like gold) quite rare, while others (like carbon) are more common?

7-7 How do astronomers think the solar system formed?

7-8 Did all of the planets form in the same way?

7-9 Are there planets orbiting other stars? How do astronomers search for such planets?

*S*o far we have learned about the concepts and tools that astronomers use to explore the universe. We are now ready to begin our own exploration. We start off close to home, with the solar system—the Sun and the objects that orbit the Sun. (One of those objects is our own Earth.) This chapter provides an overview of our local niche of the universe.

We will see that the planets divide naturally into two distinct classes. Small, rocky planets like Earth are found near the Sun, while huge, gaseous planets, like Jupiter, are located far from the Sun. Our solar system also contains seven large moons, as well as many smaller satellites, asteroids, and comets. They, too, have properties that depend on whether they come from the inner or outer solar system. Could the differences in objects reflect the different conditions under which they were formed? Can we find additional clues to the origin of the solar system from observing other stars as they form? The answer to both questions proves to be "yes."

As we will see, stars begin as huge clouds of interstellar gas and dust. Some $4.6 \times 10^9$ years ago, when one particular cloud contracted under the force of its own gravity, most of it condensed to form our Sun. But some surrounding gas and dust condensed into clumps, and these clumps in turn joined together to become planets. If this picture is correct, systems around other stars ought to have formed much the same way. Since 1995 astronomers have indeed discovered several planets orbiting other stars. These distant worlds may one day help us understand the origin of our own solar system.

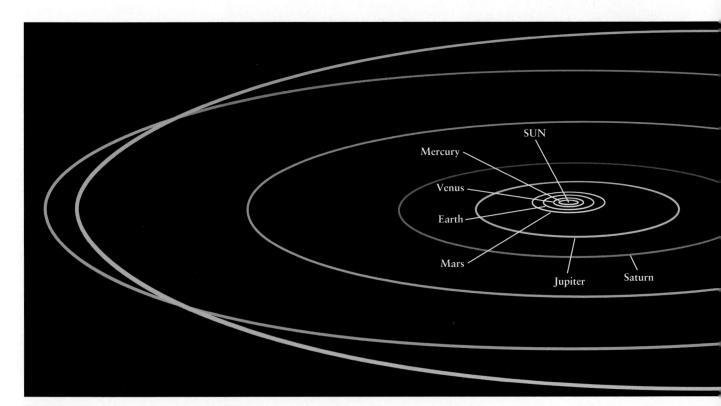

## figure 7-1

**The Solar System to Scale** This scale drawing shows a perspective view of the orbits of the planets around the Sun. The four inner planets are crowded in close to the Sun, while the five outer planets orbit the Sun at much greater distances. On the scale of this drawing, the planets themselves would be much smaller than the diameter of a human hair and too small to see.

## 7-1 The planets are classified as either terrestrial or Jovian by their physical attributes

Aside from the Sun, the largest objects in the solar system are the planets. (Pluto, the outermost planet, is something of an exception to this statement; it is actually smaller than our Moon, as well as certain of the moons of Jupiter and Saturn.) There are two useful ways in which we can compare the planets with each other. First, we can contrast their orbits around the Sun. Second, we can compare their individual physical properties, such as diameter, mass, average density, and chemical composition.

The planets fall naturally into two classes according to the sizes of their orbits. As Figure 7-1 shows, the orbits of the four inner planets (Mercury, Venus, Earth, and Mars) are crowded in close to the Sun. In contrast, the orbits of the next four planets (Jupiter, Saturn, Uranus, and Neptune) are widely spaced at great distances from the Sun. Table 7-1 lists the orbital characteristics of all of the planets.

**CAUTION!** Note that while Figure 7-1 shows the orbits of the planets, it does not show the planets themselves. The reason is simple: If Jupiter, the largest of the planets, were to be drawn to the same scale as the rest of this figure, it would be a dot just 0.0003 cm across—about $\frac{1}{200}$ of the width of a human hair and far too small to see without a microscope. The planets themselves are *very* small compared to the distances between the planets. Indeed, while an airliner traveling at 1000 km/h (620 mi/h) can fly around the Earth in less than two days, at this speed it would take 17 *years* to fly from the Earth to the Sun. The solar system is a very large and very empty place!

Most of the planets have orbits that are nearly circular. As discussed in Section 4-4, Kepler discovered in the sixteenth century that these orbits are actually ellipses. Astronomers denote the elongation of an ellipse by its *eccentricity* (see Figure 4-10*b*). The eccentricity of a circle is zero, and indeed most planets have orbital eccentricities that are very close to zero. The exceptions are Mercury and Pluto. In fact, Pluto's noncircular orbit sometimes takes it nearer the Sun than its neighbor, Neptune. (Happily, the orbits of Pluto and Neptune are such that the two planets will never collide.)

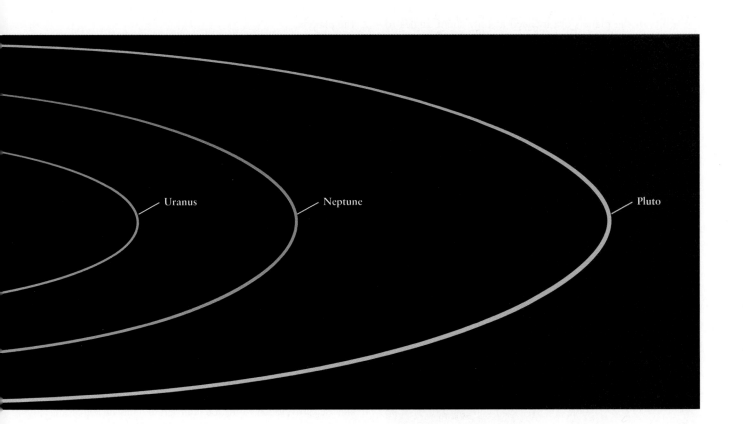

Uranus    Neptune    Pluto

| table 7-1 | Orbital Characteristics of the Planets |

| | Average distance from Sun | | Orbital period (yr) | Eccentricity | Inclination to the ecliptic (°) |
| --- | --- | --- | --- | --- | --- |
| | (AU) | (10⁶ km) | | | |
| Mercury | 0.387 | 57.9 | 0.241 | 0.206 | 7.00 |
| Venus | 0.723 | 108.2 | 0.615 | 0.007 | 3.39 |
| Earth | 1.000 | 149.6 | 1.000 | 0.017 | 0.00 |
| Mars | 1.524 | 227.9 | 1.881 | 0.093 | 1.85 |
| Jupiter | 5.203 | 778.3 | 11.86 | 0.048 | 1.30 |
| Saturn | 9.555 | 1429.0 | 29.42 | 0.056 | 2.49 |
| Uranus | 19.22 | 2875 | 83.75 | 0.046 | 0.77 |
| Neptune | 30.11 | 4504 | 163.7 | 0.009 | 1.77 |
| Pluto | 39.54 | 5916 | 248.0 | 0.249 | 17.14 |

If you could observe the solar system from a point several astronomical units (AU) above the Earth's north pole, you would see that all the planets orbit the Sun in the same counterclockwise direction. Furthermore, the orbits of the planets all lie in nearly the same plane. In other words, the orbits of the planets are inclined at only slight angles to the plane of the ecliptic, which is the plane of the Earth's orbit around the Sun. (Again, however, Pluto is an exception; the plane of its orbit is tilted at about 17° to the ecliptic.) Furthermore, the plane of the Sun's equator is very closely aligned with the orbital planes of the planets. As we will learn, these facts provide important clues about the origin of the planets.

When we compare the physical properties of the planets, we again find that they fall naturally into two classes—the four inner planets and the four outer ones—with Pluto as an exception. The four inner planets are called **terrestrial planets** because they resemble the Earth (in Latin, *terra*). They all have hard, rocky surfaces with mountains, craters, canyons, and volcanoes. You could stand on the surface of any of them, although you would need a protective spacesuit on Mercury, Venus, or Mars. The outer four planets are called **Jovian planets** because they resemble Jupiter. (Jove was another name for the Roman god Jupiter.) An attempt to land a spacecraft on the surface of any of the Jovian planets would be futile, because the materials of which these planets are made are mostly gaseous or liquid. The visible "surface" features of a Jovian planet are actually cloud formations in the planet's atmosphere. The montage of photographs in Figure 7-2 shows the distinctive appearances of the two classes of planets.

figure 7-2   R I **V** U X G

**The Planets (Not to Scale)** This montage of photographs taken by various spacecraft shows the different appearances of eight planets and the Earth's Moon. At the top (from left to right) are Mercury, Venus, Earth, and Mars, along with our Moon. At the bottom (from right to left) are Jupiter, Saturn, Uranus, and Neptune. The images are *not* reproduced to the same scale. If the planets were all shown to the same scale as the Earth, Jupiter (which is more than 11 times greater in diameter than the Earth) would cover most of this page. The planets are shown in proper scale in Figure 7-3. (NASA)

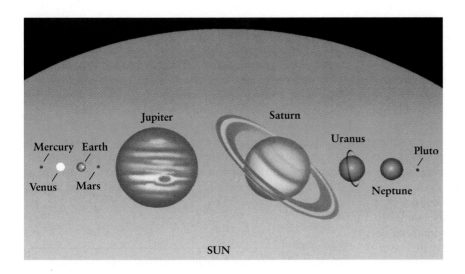

figure 7-3

**The Sun and the Planets to Scale** This drawing shows the nine planets in front of the disk of the Sun, with all ten bodies drawn to the same scale. The four terrestrial planets (Mercury, Venus, Earth, and Mars) have orbits nearest the Sun. They are small and dense, and made of rocky materials. The Jovian planets are the next four planets from the Sun (Jupiter, Saturn, Uranus, and Neptune). They are large and of low density and are composed primarily of hydrogen and helium. Pluto, an exceptional case, is a mixture of ice and rock.

The most apparent difference between the terrestrial and Jovian planets is in their diameters. The diameter of a planet can be computed from its angular diameter as seen from Earth and its distance from Earth. For example, on January 1, 1998, Venus was $4.44 \times 10^7$ km from Earth and had an angular diameter of 56.3 arcsec. Using the small-angle formula from Box 1-2, we can calculate the diameter of Venus to be 12,100 km (7520 mi). Similar calculations demonstrate that the Earth, with its diameter of about 12,756 km (7926 mi), is the largest of the four inner, terrestrial planets. In sharp contrast, the four outer, Jovian planets are much larger than the terrestrial planets. First place goes to Jupiter, whose equatorial diameter is about 142,984 km (88,846 mi). Pluto, always the exception, is even smaller than the inner planets, despite being the outermost planet. Its diameter is only about 2300 km (1400 mi). Figure 7-3 shows the Sun and the planets drawn to the same scale, and the diameters of the planets are given in Table 7-2.

The terrestrial and Jovian planets also have dramatically different masses. The mass of a planet is most easily determined if the planet has a satellite, because Newton's form of Kepler's third law (see Section 4-7 and Box 4-2 ) relates the planet's mass to a satellite's period and semimajor axis. If a planet does not have a satellite, as is the case for Mercury and Venus, astronomers can measure the mass of the planet by sending a spacecraft to pass near the planet. The planet's gravity, which is directly related to its mass, deflects the spacecraft's path. By measuring the amount of deflection

| table 7-2 | **Physical Characteristics of the Planets** | | | | |
|---|---|---|---|---|---|
| | **Equatorial diameter** | | **Mass** | | |
| | (km) | (Earth = 1) | (kg) | (Earth = 1) | **Average density (kg/m³)** |
| Mercury | 4,879 | 0.383 | $3.302 \times 10^{23}$ | 0.055 | 5430 |
| Venus | 12,104 | 0.949 | $4.869 \times 10^{24}$ | 0.815 | 5240 |
| Earth | 12,756 | 1.000 | $5.974 \times 10^{24}$ | 1.000 | 5515 |
| Mars | 6,794 | 0.533 | $6.419 \times 10^{23}$ | 0.107 | 3940 |
| Jupiter | 142,984 | 11.21 | $1.899 \times 10^{27}$ | 317.83 | 1330 |
| Saturn | 120,536 | 9.45 | $5.685 \times 10^{26}$ | 95.16 | 700 |
| Uranus | 51,118 | 4.01 | $8.662 \times 10^{25}$ | 14.50 | 1300 |
| Neptune | 49,528 | 3.88 | $1.028 \times 10^{26}$ | 17.20 | 1760 |
| Pluto | 2,302 | 0.180 | $1.500 \times 10^{22}$ | 0.0025 | 1100 |

and using Newton's laws, astronomers can determine the planet's mass. Using these techniques, astronomers have found that the four Jovian planets have masses that are tens or hundreds of times greater than the mass of any of the terrestrial planets. Again, first place goes to Jupiter, whose mass is 318 times greater than the Earth's.

Once a planet's diameter and mass are known, its **average density,** or mass divided by volume, can be determined. Average density, which is measured in kilograms per cubic meter (kg/m$^3$), is a physical property that depends on the composition of an object. For example, the average density of air near sea level on Earth is 1.2 kg/m$^3$, the average density of water is 1000 kg/m$^3$, and the average density of a piece of concrete is 2000 kg/m$^3$. Box 7-1 describes some applications of the idea of density to everyday phenomena on Earth. The four inner, terrestrial planets have very high average densities (see Table 7-2); the average density of the Earth, for example, is 5520 kg/m$^3$. By contrast, a typical rock found on the Earth's surface has a lower average density, about 3000 kg/m$^3$. Thus, the Earth must contain a large amount of material that is denser than rock. This information provides our first clue that terrestrial planets have dense iron cores.

In sharp contrast, the outer, Jovian planets have quite low densities. Saturn has an average density less than that of water. This information strongly suggests that the giant outer planets are composed primarily of such light elements as hydrogen and helium. All four Jovian planets probably have large cores of ice and rock buried beneath low-density atmospheres tens of thousands of kilometers thick.

Pluto is again an oddity. Although it is even smaller than the dense inner planets, its average density is closer to that of the giant outer planets. Pluto is probably composed of a mixture of rock and ice, because its average density of 2030 kg/m$^3$ is intermediate between the densities of ice and of rock.

| 7-2 | Seven large satellites are almost as big as the terrestrial planets |

All the planets except Mercury and Venus have satellites. More than 50 satellites are known (Jupiter, Saturn, and Uranus each have at least 15), and dozens of other small ones

---

| box 7-1 | The Heavens on the Earth |

## *Average Density*

Average density, which is the mass of an object divided by that object's density, is a useful quantity for describing the differences between planets in our solar system. This same quantity has many applications here on Earth.

A rock tossed into a lake sinks to the bottom, while an air bubble produced at the bottom of a lake (for example, by the air tanks of a scuba diver) rises to the top. These are examples of a general principle: An object sinks in a fluid if its average density is greater than that of the fluid but rises if its average density is less than that of the fluid. The average density of water is 1000 kg/m$^3$, which is why a typical rock (with an average density of about 3000 kg/m$^3$) sinks while an air bubble (average density of about 1.2 kg/m$^3$) rises.

At many summer barbecues, cans of soft drinks are kept cold by putting them in a container full of ice. When the ice melts, the cans of diet soda always rise to the top, while the cans of regular soda sink to the bottom. Why is this? The average density of a can of diet soda—which includes water, flavoring, artificial sweetener, and the trapped gas that makes the drink fizzy—is slightly less than the density of water, and so the can floats. A can of regular soda contains sugar instead of artificial sweetener, and the sugar is a bit heavier than the sweetener. The extra weight is just enough to make the average density of a can of regular soda slightly more than that of water, making the can sink. (You can test these statements by putting unopened cans of diet soda and regular soda in a sink or bathtub full of water.)

The concept of average density provides geologists with important clues about the early history of the Earth. The average density of surface rocks on Earth, about 3000 kg/m$^3$, is less than the Earth's average density of 5520 kg/m$^3$. The simplest explanation is that in the ancient past, the Earth was completely molten throughout its volume, so that low-density materials rose to the surface and high-density materials sank deep into the Earth's interior in a process called *chemical differentiation*. This series of events also suggests that the Earth's core must be made of relatively dense materials, such as iron and nickel. A tremendous amount of other geological evidence has convinced scientists that this picture is correct. We will see many other applications of average density as we explore the solar system.

| table 7-3 | The Seven Giant Satellites | | |
|---|---|---|---|
| **Satellite** | **Parent planet** | **Diameter (km)** | **Average density (kg/m³)** |
| Moon | Earth | 3476 | 3340 |
| Io | Jupiter | 3630 | 3570 |
| Europa | Jupiter | 3138 | 3020 |
| Ganymede | Jupiter | 5262 | 1940 |
| Callisto | Jupiter | 4800 | 1860 |
| Titan | Saturn | 5150 | 1880 |
| Triton | Neptune | 2700 | 2070 |

probably remain to be discovered. The known satellites fall into two distinct categories. Seven giant satellites are each roughly as big as the planet Mercury (Table 7-3). All the other satellites are much smaller, with diameters less than 2000 km.

Figure 7-4 shows these seven giant satellites along with Mercury, all to the same scale. Interplanetary spacecraft have made many surprising and fascinating discoveries about these giant satellites. We now know that Jupiter's satellite Io is the most geologically active world in the solar system, with numerous geyserlike volcanoes that continually belch forth sulfur-rich compounds. Europa, another of Jupiter's large satellites, shows evidence of having had a volcanic past, but with volcanoes that spewed ice rather than lava or sulfur. Saturn's largest satellite, Titan, has an atmosphere nearly twice as dense as the Earth's atmosphere. Like the terrestrial planets, all seven giant satellites have hard surfaces of either rock or ice. We discuss these remarkable satellites, each of them a world in its own right, in Chapters 9, 14, 15, and 16.

figure 7-4   R I **V** U X G

**The Smaller Terrestrial Worlds to Scale** This collection of photographs shows Mercury along with the seven largest satellites in our solar system: the Earth's Moon, the four largest satellites of Jupiter (Io, Europa, Ganymede, and Callisto), and the largest satellites of Saturn (Titan) and Neptune (Triton). The images are all reproduced to the same scale. Each of these satellites has its own unique characteristics, and all are comparable in size to the terrestrial planets. Only Titan possesses a substantial atmosphere. (NASA)

## 7-3 Spectroscopy reveals the chemical composition of the planets

To understand the nature of the planets and satellites, we need to know their chemical compositions. The most accurate determinations have come from direct chemical analyses of samples from the atmosphere and soil of planets. Unfortunately, we only have such direct information for four worlds: the Earth and the three worlds on which spacecraft have landed, Venus, the Moon, and Mars. In all other cases, astronomers must analyze sunlight reflected from the distant planets and their satellites. To do that, astronomers bring to bear one of their most powerful tools, **spectroscopy**, the systematic study of spectra and spectral lines. We discussed this branch of science in Sections 5-6 and 6-5

Spectroscopy is a sensitive probe of the composition of a planet's atmosphere. If a planet has an atmosphere, some of the sunlight reaching that planet will penetrate the atmosphere and reach the planet's surface. But some of the sunlight will only penetrate a distance into the atmosphere and then be reflected back into space. While passing through the planet's atmosphere, some of the wavelengths of sunlight will be absorbed. Hence, the spectrum of this reflected sunlight will have dark absorption lines. Astronomers look at the particular wavelengths absorbed and the amount of light absorbed at those wavelengths. Both of these depend on the kinds of chemicals present in the planet's atmosphere and the abundance of those chemicals.

For example, spectroscopy has been used to analyze the atmosphere of Saturn's largest satellite, Titan, depicted at the lower left of Figure 7-4. Figure 7-5a is a graph that shows the spectrum of visible sunlight reflected from Titan. The dips in this curve of intensity versus wavelength represent absorption lines. However, not all of these absorption lines are produced in the atmosphere of Titan (Figure 7-5b). Before reaching Titan, light from the Sun's glowing surface must pass through the Sun's hydrogen-rich atmosphere. This produces the hydrogen absorption line in Figure 7-5a at a wavelength of 656 nm. After being reflected from Titan, the light must pass through the Earth's atmosphere before reaching the telescope; this is where the oxygen absorption line in Figure 7-5a is produced. Only the two dips near 620 nm and 730 nm are caused by gases in Titan's atmosphere.

These two absorption lines are caused not by individual atoms in the atmosphere of Titan but by atoms combined to form **molecules**. For example, two hydrogen atoms can

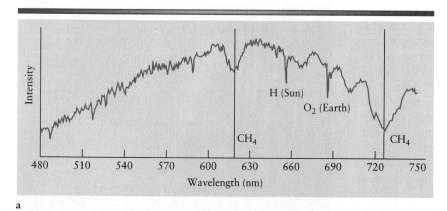

a

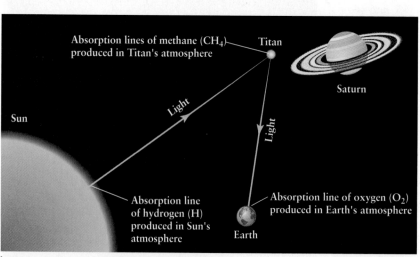

b

## figure 7-5

**The Spectrum of Saturn's Satellite Titan**
**(a)** This graph shows the spectrum of sunlight reflected from Titan, the largest satellite of Saturn and the only satellite in the solar system with an atmosphere. The dips in the curve denote absorption lines caused by atoms of hydrogen (H), molecules of oxygen ($O_2$), and molecules of methane ($CH_4$). Of these gases, only methane is actually present in Titan's atmosphere. **(b)** The spectrum in (a) was made by light that was emitted from the Sun's glowing surface, passed through the Sun's atmosphere, was reflected by Titan's atmosphere, and passed through the Earth's atmosphere before reaching a telescope. As the figure shows, the absorption lines of hydrogen and oxygen are produced in the atmospheres of the Sun and Earth respectively. Only the methane lines are caused by absorption in the atmosphere of Titan.

combine with an atom of oxygen to form a molecule of water. Water's chemical formula, $H_2O$, describes the atoms in a water molecule. In a similar way, two oxygen atoms can bond with a carbon atom to produce a molecule of carbon dioxide, whose formula is $CO_2$. Molecules, like atoms, also produce unique patterns of lines in the spectra of astronomical objects. The absorption lines in the spectrum of Titan shown in Figure 7-5a indicate the presence in Titan's atmosphere of molecules of methane, or $CH_4$ (a molecule made of one carbon atom and four hydrogen atoms). This shows that Titan is a curious place indeed, because on Earth methane is the primary ingredient in natural gas! The spectra of all of the planetary atmospheres in the solar system display absorption lines of molecules of various types.

In addition to visible-light measurements such as those shown in Figure 7-5a, it is very useful to study the *infrared* and *ultraviolet* spectra of planetary atmospheres. Many molecules have much stronger spectral lines in these nonvisible wavelength bands than in the visible. As an example, the ultraviolet spectrum of Titan shows that nitrogen molecules ($N_2$) are the dominant constituent of Titan's atmosphere. Furthermore, Titan's infrared spectrum includes spectral lines of a variety of molecules of carbon and hydrogen, indicating that Titan's atmosphere has a very complex chemistry. None of these molecules could have been detected using visible light alone.

Spectroscopy can also provide useful information about the solid surfaces of planets and satellites without atmospheres. Unlike a gas, a solid illuminated by sunlight does not produce sharp, definite spectral lines. Instead, only broad absorption features appear in the spectrum. By comparing such a spectrum with the spectra of samples of different substances on Earth, astronomers can infer the chemical composition of the surface of a planet or satellite. As an example, Figure 7-6 shows the infrared spectrum of light reflected from the surface of Jupiter's satellite Europa (Figure 7-4 includes a photograph of Europa). Because this spectrum is so close to that of ordinary ice, astronomers conclude that ice is the dominant constituent of Europa's surface. However, this technique does not give any information about what the material is like just below the surface. There is no substitute for sending a spacecraft to a planet and examining its surface directly.

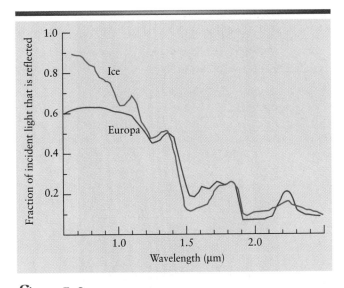

**figure 7-6**

**The Spectrum of Jupiter's Moon Europa** Infrared light from the Sun that is reflected from the surface of Europa, one of Jupiter's moons, has almost exactly the same spectrum as light reflected from ordinary ice. This shows that the surface of Europa is made predominantly of ice and not rock.

---

| 7–4 | Hydrogen and helium are abundant on the Jovian planets, whereas the terrestrial planets are composed mostly of heavy elements |

Spectroscopic observations from Earth and spacecraft demonstrate that the Jovian planets are composed primarily of the lightest gases, hydrogen and helium. In contrast, chemical analysis of soil samples from Venus, Earth, and Mars demonstrate that the terrestrial planets are made mostly of heavy

elements, such as iron, silicon, magnesium, sulfur, and nickel. Spacecraft images such as Figures 7-7 and 7-8 only hint at these striking differences in the chemical composition.

Temperature plays a major role in determining whether various substances exist as solids, liquids, or gases. This profoundly affects the characters of the planets. Hydrogen and helium are gaseous except at extremely low temperatures and extraordinarily high pressures. In contrast, rock-forming compounds such as iron and silicon are solids except at temperatures exceeding 1000 K. Between these two extremes are substances such as water ($H_2O$), carbon dioxide ($CO_2$), methane ($CH_4$), and ammonia ($NH_3$). At low temperatures (typically below 200 to 300 K), these common chemicals solidify into the solids called ices. At somewhat higher temperatures, they can exist as liquids or gases.

Virtually no hydrogen or helium is found in the atmospheres of the terrestrial planets. The atmospheres of Venus, Earth, and Mars are instead composed of heavier gases, such as carbon dioxide, oxygen, and nitrogen, all of which are made of atoms more massive than hydrogen or helium. To understand why the terrestrial planets do not have any hydrogen or helium in their atmospheres, we must look at the effects of their surface temperatures.

As you might expect, a planet's surface temperature is related to its distance from the Sun. (Review Box 5-1 with its discussion of temperature measurements.) The four inner planets are quite warm. For example, midday temperatures on Mercury may climb to 600 K (= 327°C = 621°F), and in the middle of summer on Mars, it is sometimes as warm as 300 K (= 27°C = 81°F). The outer planets, which receive

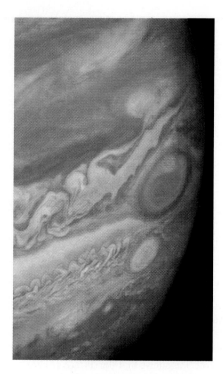

**figure 7-7** R I **V** U X G

**A Jovian Planet** This closeup view of Jupiter's turbulent cloudtops was taken by the *Voyager* spacecraft that flew past the planet in 1979. Because Jupiter is composed mostly of lightweight gases (hydrogen and helium), its average density is only 1330 kg/m³. Hydrogen and helium are colorless; the colors in the atmosphere are caused by trace amounts of other substances. (NASA)

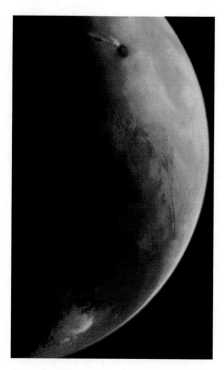

**figure 7-8** R I **V** U X G

**A Terrestrial Planet** This closeup view of Mars's rocky surface was taken by one of the *Viking* spacecraft that visited the planet in 1976. Mars is composed mostly of heavy elements (for example, iron, silicon, magnesium, sulfur), so its average density is 3950 kg/m³, which is greater than that for any of the Jovian planets. The red color is the true color of the Martian surface. The white area at the bottom of the image is one of the Martian polar caps; the dark circle near the top of the image is a giant, inactive volcano. (NASA)

much less solar radiation, are cooler. Typical temperatures range from about 150 K (= –123°C = –189°F) in Jupiter's cloudtops to 63 K (= –210°C = –346°F) on Neptune.

The temperature of a gas is directly related to the speeds with which atoms or molecules of a gas move: The higher the temperature of a gas, the greater the speed of its atoms or molecules. Furthermore, for a given temperature, lightweight atoms and molecules move more rapidly than heavy ones. On the four inner, terrestrial planets, where atmospheric temperatures are high, low-mass hydrogen and helium atoms move so swiftly that they can escape from the relatively weak gravity of these planets. Hence, the sparse atmospheres that surround the terrestrial planets are composed primarily of more massive, slower-moving molecules such as carbon dioxide ($CO_2$), nitrogen ($N_2$), oxygen ($O_2$), and water vapor ($H_2O$). On the four Jovian planets, low temperatures and relatively strong gravity prevent even lightweight hydrogen and helium gases from escaping into space. Box 7-2 discusses the thermal motion of atoms and the ability of a planet's gravity to retain gases.

## 7-5 Small chunks of rock and ice also orbit the Sun

In addition to the nine planets, many smaller objects orbit the Sun. Between the orbits of Mars and Jupiter are thousands of rocky objects called **asteroids.** There is no sharp dividing line between planets and asteroids, which is why asteroids are also called **minor planets.** The largest asteroid, Ceres, has a diameter of about 900 km. The next largest, Pallas and Vesta, are each about 500 km in diameter. Still smaller ones are increasingly numerous. There are thousands of kilometer-sized asteroids and millions of asteroids that are boulder-sized or smaller.

A closeup picture of an asteroid, taken by a spacecraft on its way to Jupiter, is shown in Figure 7-9. Because most asteroids orbit the Sun at distances of 2 to 3.5 AU, this region of the solar system between the orbits of Mars and Jupiter is called the **asteroid belt.**

| box 7-2 | Looking Deeper into Astronomy |

## Thermal Motion and the Retention of an Atmosphere

A moving object possesses energy as a result of its motion. The faster it moves, the more energy it has. Energy of this type is called **kinetic energy.** If an object of mass $m$ is moving with a speed $v$, its kinetic energy is given by

**Kinetic energy**

$$E_k = \frac{1}{2} mv^2$$

$E_k$ = kinetic energy of an object
$m$ = mass of object
$v$ = speed of object

This expression for kinetic energy is valid for all objects, both big and small, from atoms and molecules to planets and stars, as long as their speed is slow in relation to the speed of light. If the mass is expressed in kilograms and the speed in meters per second, the energy is expressed in joules (J).

**EXAMPLE:** An automobile of mass 1000 kg driving at a typical freeway speed of 30 m/s (= 108 km/h = 67 mi/h) has a kinetic energy of

$$E_k = \frac{1}{2} (1000 \times 30^2) = 450,000 \text{ J} = 4.5 \times 10^5 \text{ J}$$

Consider a gas, such as the atmosphere of a star or planet. The temperature of the gas is a direct measure of the average amount of kinetic energy in each atom or molecule of the gas. If the gas is cool, the atoms or molecules are moving relatively slowly; if the gas is hot, the atoms or molecules are moving at high speeds. If the gas temperature is sufficiently high, typically several thousand kelvins, molecules will be moving so fast that when they collide with each other, the energy of the collision can break the molecules apart into their constituent atoms. Hence, such high-temperature gases consist principally of individual atoms rather than molecules. Thus, the Sun's atmosphere, where the temperature is 5800 K, consists primarily of individual hydrogen atoms. By contrast, the hydrogen atoms in the Earth's atmosphere (temperature 290 K) are combined with oxygen atoms into molecules of water vapor ($H_2O$).

According to the kinetic theory of gases developed during the nineteenth century, in a gas of temperature $T$ (in kelvins), the average kinetic energy of an atom or molecule is

**Kinetic energy of an atom or molecule**

$$E_k = \frac{3}{2} kT$$

$E_k$ = average kinetic energy of a gas atom or molecule
$k$ = $1.38 \times 10^{-23}$ J/K
$T$ = temperature of gas in kelvins

The quantity $k$ is called the Boltzmann constant. Note that the higher the gas temperature, the greater the average kinetic energy of an atom or molecule of the gas. This average kinetic energy becomes zero at absolute zero, or $T = 0$, the temperature at which molecular motion is at a minimum.

At a given temperature, all atoms and molecules will have the same average kinetic energy. But the average *speed* of a given kind of atom or molecule depends on its mass. To see this, we note that the average kinetic energy of a gas atom or molecule can be written in two equivalent ways:

$$E_k = \frac{1}{2} mv^2 = \frac{3}{2} kT$$

where $v$ represents the average speed of an atom or molecule in a gas with temperature $T$. Rearranging this equation, we obtain

**Speed of a gas atom or molecule**

$$v = \sqrt{\frac{3kT}{m}}$$

$v$ = average speed of a gas atom or molecule in m/s
$k$ = $1.38 \times 10^{-23}$ J/K
$T$ = temperature of gas in kelvins
$m$ = mass of the atom or molecule in kilograms

For a given gas temperature, the greater the mass of a given type of gas atom or molecule, the slower its average speed. (The value of $v$ given by this equation is actually slightly higher than the average speed of the atoms or molecules in the gas, but it is close enough for our purposes here. If you are studying physics, you may know that $v$ is actually the root-mean-square speed.)

*(continued on following page)*

**box 7-2** *(continued)*

**EXAMPLE:** Suppose you want to know the average speed of the oxygen molecules that you breathe at a room temperature of 72°F (= 22°C = 295 K). From a reference book, you can find that the mass of an oxygen atom is $2.66 \times 10^{-26}$ kg. The mass of an oxygen molecule ($O_2$) is twice the mass of an oxygen atom, or $2(2.66 \times 10^{-26}$ kg) $= 5.3 \times 10^{-26}$ kg. Thus, the average speed is

$$v = \sqrt{\frac{3(1.38 \times 10^{-23})(295)}{5.3 \times 10^{-23}}} = 4.8 \times 10^2 \text{ m/s} = 0.48 \text{ km/s}$$

or almost exactly 1000 miles per hour. This example shows that the atoms and molecules in a gas are moving rapidly, even at moderate temperatures. The speeds may even be so great that the gas can overcome the attractive force of a planet's gravity and escape into space.

Astronomers find it useful to speak of the **escape speed** from a planet in trying to decide if an object can permanently leave the planet and escape into interplanetary space. The escape speed from a planet of mass $M$ and radius $R$ is given by

$$v_{escape} = \sqrt{\frac{2GM}{R}}$$

where $G$ is the universal constant of gravitation ($G = 6.67 \times 10^{-11}$ N m²/kg²).

The following table gives the escape speed for various objects in the solar system. For example, to get to another planet, a spacecraft must leave Earth with a speed greater than 11.2 km/s (25,100 mi/h).

A good rule of thumb is that a planet can retain a gas if the escape speed is at least six times greater than the average speed of the molecules in the gas. In such a case, very few molecules will be moving fast enough to escape from the planet's gravity.

**EXAMPLE:** Consider the Earth's atmosphere. We saw that the average speed of oxygen molecules is 0.48 km/s at room temperature. The escape speed from the Earth (11.2 km/s) is much more than six times the average speed of the oxygen molecules, so the Earth has no troublekeeping oxygen in its atmosphere.

A similar calculation for hydrogen molecules ($H_2$) gives a different result, however. At 295 K, the average speed of a hydrogen molecule is 1.9 km/s. Six times this speed is 11.4 km/s, which is slightly higher than the escape speed from the Earth. Thus, the Earth does not retain hydrogen in its atmosphere. Any hydrogen released into the air slowly leaks away into space.

| Planet | Escape speed (km/s) |
|--------|---------------------|
| Mercury | 4.3 |
| Venus | 10.4 |
| Earth | 11.2 |
| Moon | 2.4 |
| Mars | 5.0 |
| Jupiter | 59.5 |
| Saturn | 35.5 |
| Uranus | 21.3 |
| Neptune | 23.5 |
| Pluto | 1.3 |

 One common misconception about asteroids is that they are the remnants of an ancient planet that somehow broke apart or exploded, like the fictional planet Krypton in the comic book adventures of Superman. In fact, the asteroids were probably never part of any planet-sized body. The early solar system is believed to have been filled with asteroidlike objects, most of which coalesced to form the planets. For a variety of reasons, a relatively small number of these objects would have missed out on this process and would not have become part of a planet. These "leftover" objects make up our present-day population of asteroids.

Quite far from the Sun, beyond the orbit of Pluto, are chunks of very dirty ice called **comets**. Many comets have highly elongated orbits that occasionally bring them close to the Sun. When this happens, the Sun's radiation vaporizes some of the comet's **ices**, producing a long flowing tail (Figure 7-10).

Asteroids and comets are thought to be debris left over from the formation of the solar system. In the inner regions of the solar system, rocky fragments have been able to endure continuous exposure to the Sun's heat, but any ice originally present would have evaporated. Far from the Sun,

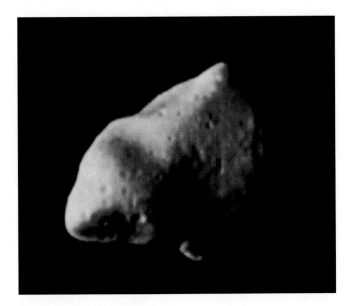

**figure 7-9**   R I **V** U X G

**An Asteroid** This image of the asteroid Gaspra was taken in 1991 by the *Galileo* spacecraft while on its way toward Jupiter. This asteroid measures 12 × 20 × 11 km. Like Gaspra, most asteroids have irregular shapes, unlike planets which are roughly spherical. The difference is that asteroids are much smaller than planets; hence, the gravitational attraction between different parts of an asteroid is too weak to compress it into a spherical shape. Several thousand asteroids orbit the Sun between the orbits of Mars and Jupiter. Of these, Gaspra was the first one ever viewed at close range by a spacecraft. (NASA)

**figure 7-10**   R I **V** U X G

**A Comet** The solid part of a comet is a chunk of dirty ice roughly 10 km in diameter. When a comet passes near the Sun, solar radiation vaporizes some of the comet's icy material, and the resulting gases form a tail millions of kilometers long. This photograph shows Comet Kohoutek as it appeared in January 1974. (NASA)

chunks of ice have survived for billions of years. Thus, debris in the solar system naturally divides into two families (asteroids and comets), which can be arranged according to distance from the Sun, just like the two categories of planets (terrestrial and Jovian). To better understand the chemical composition of the planets, their satellites, the asteroids, and the comets, we must look to the stars, for this is where most of the chemical elements are created.

## 7-6   The relative abundances of the elements are the result of cosmic processes

Some chemical elements are very common in our solar system, but others are very rare. Hydrogen is by far the most abundant substance in the solar system: It makes up nearly three-quarters of the combined mass of the Sun and planets. (We have seen that Jupiter, whose mass is greater than that of all the other planets combined, is composed mostly of hydrogen.) Helium is the second most abundant element.

Together, hydrogen and helium account for about 98% of the mass of all the material in the solar system. All of the other chemical elements combined make up the remaining 2%.

The dominance of hydrogen and helium is not merely a characteristic of our local part of the universe. Analysis of the spectra of stars and galaxies, out to the furthest distance attainable by the most powerful telescopes, shows the same pattern of chemical abundances. It is fair to say that the vast majority of the atoms in the universe are hydrogen and helium atoms. In the universe as a whole, the elements of which the Earth is made (mostly iron and nickel) and the elements that are the building blocks of all living organisms (carbon, oxygen, nitrogen, and phosphorus, among others) are relatively rare.

There is a good reason for this overwhelming abundance of hydrogen and helium. As we shall see in Chapters 28 and 29, a wealth of evidence leads most astronomers to think that the universe began roughly 15 billion years ago with a violent event called the Big Bang. Only the lightest elements—hydrogen, helium, and tiny amounts of lithium and beryllium—emerged from the enormously high temperatures following this cosmic event. All the heavier elements were manufactured by stars later, either by nuclear fusion reactions deep in their interiors or by the violent explosions that mark the end of massive stars. Were it not for these processes that take place only in stars, there would be no heavy elements in the universe today, no planets like Earth, and no living creatures.

Near the ends of their lives, some stars cast much of their matter back out into space. This process can be a comparatively gentle one, in which a star's outer layers are gradually expelled. Figure 7-11 shows the star HD 65750, which is losing material in this fashion. This ejected material appears as the cloudy region, or **nebulosity** (from *nubes,* Latin for "cloud"), that surrounds the star and is illuminated by it. Alternatively, a star may end its life with a spectacular detonation called a **supernova,** which blows the star apart. Either way, the interstellar gases in the galaxy become enriched with heavy elements dredged up from the dying star's interior, where they were created. New stars that form from this enriched material thus have an adequate supply of heavy elements from which to develop a system of planets, satellites, comets, and asteroids.

Stars create different heavy elements in different amounts. For example, the elements carbon, oxygen, silicon, and iron are readily produced in the interiors of massive stars, whereas gold is created only under special circumstances. Consequently, gold is rare in our solar system and in the universe as a whole, while carbon is comparatively abundant (although still much less abundant than hydrogen or helium).

A convenient way to express the relative abundances of the various elements is to say how many atoms of a particular element are found for every trillion (that is, $10^{12}$) hydrogen atoms. For example, for every $10^{12}$ hydrogen atoms in space, there are about 70 billion ($70 \times 10^9$, or $7 \times 10^{10}$) helium atoms. From spectral analysis of stars and chemical analysis of Earth rocks, Moon rocks, and meteorites, scien-

| Table 7-4 | Abundances of the Most Common Elements | | |
|---|---|---|---|
| Atomic number | Element | Symbol | Relative abundance |
| 1 | Hydrogen | H | $1 \times 10^{12}$ |
| 2 | Helium | He | $7 \times 10^{10}$ |
| 6 | Carbon | C | $4 \times 10^8$ |
| 7 | Nitrogen | N | $9 \times 10^7$ |
| 8 | Oxygen | O | $7 \times 10^8$ |
| 10 | Neon | Ne | $1 \times 10^8$ |
| 12 | Magnesium | Mg | $4 \times 10^7$ |
| 14 | Silicon | Si | $4 \times 10^7$ |
| 16 | Sulfur | S | $2 \times 10^7$ |
| 26 | Iron | Fe | $3 \times 10^7$ |

tists have determined the relative abundances of the elements in our part of the Galaxy today. Table 7-4 lists the ten most abundant elements, in order of their atomic number. The **atomic number** is equal to the number of protons in the atom's nucleus; it is also equal to the number of electrons orbiting the nucleus (see Box 5-4). In general, the greater the atomic number, the greater the mass of an atom.

In addition to these ten very common elements, five elements are moderately abundant: sodium, aluminum, argon,

**Figure 7-11** R I **V** U X G

**A Mass-Loss Star** The star HD 65750 in the southern constellation Carina (the Ship's Keel) is one of many stars that are shedding material rapidly. Some of the ejected material has condensed into tiny grains of dust, which reflect the star's light and make the cloud of ejected material (called IC 2220) visible in this image. In the same way, dust particles in Earth's atmosphere are made visible when illuminated by a shaft of sunlight. (Anglo-Australian Observatory)

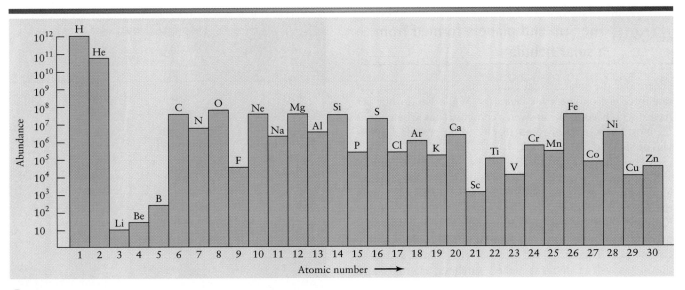

## figure 7-12

**Abundances of the Lighter Elements** This graph shows the abundances of the 30 lightest elements (hydrogen through zinc) compared to a value of $10^{12}$ for hydrogen. The elements are listed in order of increasing atomic number (equal to the number of protons in the atom's nucleus and the number of electrons orbiting the nucleus). Notice that the vertical scale is not a linear one; each division on the scale corresponds to a tenfold increase in abundance. All elements heavier than zinc (Zn) have abundances less than 1000 atoms per $10^{12}$ atoms of hydrogen.

calcium, and nickel. These elements have abundances in the range of $10^6$ to $10^7$ relative to the standard $10^{12}$ hydrogen atoms. Most of the other elements are much rarer. For example, for every $10^{12}$ hydrogen atoms in the solar system, there are only six atoms of gold. Figure 7-12 shows the relative abundances of the 30 lightest elements, arranged by atomic number.

The presence of heavy elements in our Sun and in the planets is powerful evidence that our solar system is made of matter created in stars that existed billions of years ago. The Sun is a fairly young star, only 5 billion years old. All the heavy elements in our solar system were created and cast off by ancient stars during the first 10 billion years of our Galaxy's existence. We are literally made of star dust (Figure 7-13).

## figure 7-13  R I **V** U X G

**A Dusty Region of Star Formation** Unlike Figure 7-11, which depicts an old star that is ejecting material into space, this image shows young stars that have only recently formed from a cloud of gas and dust. These stars in the constellation of Orion (the Hunter) are still surrounded by much of the material from which they formed. The bluish, wispy appearance of the nebulosity (called NGC 1973-1975-1977) is caused by starlight reflecting off interstellar dust grains, which are relatively abundant in the cloud. The grains are made of heavy elements produced by earlier generations of stars. (Anglo-Australian Observatory)

## 7-7 The Sun and planets formed from a solar nebula

Where did the solar system come from? This question has tantalized astronomers for centuries. While we do not yet have a wholly complete answer, a consensus has arisen about the most likely series of events that led to the present-day system of Sun and planets.

A key piece of evidence about the origin of the solar system is that all the planets orbit the Sun in the same direction and in nearly the same plane (Section 7-1). As long ago as the eighteenth century, the German philosopher Immanuel Kant and the French scientist Pierre-Simon de Laplace independently suggested that this state of affairs could not be a coincidence. They proposed that our entire solar system—the Sun as well as all of the planets, satellites, asteroids, and comets—formed from a vast, rotating cloud of gas and dust called the **solar nebula** (Figure 7-14a). In the modern version of their theory, the solar nebula is thought to have had a mass somewhat greater than that of our present-day Sun and to have been originally similar in character to the nebulosity shown in Figure 7-13.

Each part of the nebula exerted a gravitational attraction on the other parts, and these mutual gravitational pulls tended to make the nebula contract. As it contracted, the greatest concentration of matter occurred at the center of the nebula, forming a relatively dense region called the **protosun**. As its name suggests, the protosun is the part of the solar nebula that eventually developed into the Sun. The planets formed from the much lesser amount of material in the outer regions of the solar nebula. Indeed, the mass of all the planets together is only 0.1% of the mass of the Sun.

When you drop a ball, the gravitational attraction of the Earth makes the ball fall faster and faster as it falls; in the same way, material falling inward toward the protosun would have gained speed. As this fast-moving material ran into the protosun, the energy of the collision was converted into thermal energy, causing the temperature deep inside the solar nebula to climb. This process, in which the gravitational energy of a contracting gas cloud is converted into thermal energy, is called **Kelvin-Helmholtz contraction**, after the nineteenth-century physicists who first described it.

Temperatures within the newly created protosun soon climbed to several thousand kelvins. While the protosun's surface temperature stayed roughly constant, the temperature inside the protosun increased even more by means of further contraction. Eventually, after perhaps $10^8$ (100 million) years had passed since the solar nebula first began to contract, the central temperature of the protosun reached a few million kelvins (that is, a few times $10^6$ K), and nuclear reactions began in the protosun's interior. When this happened, the contraction stopped and a true star was born. As we will describe in Chapter 18, nuclear reactions in the interior of the present-day Sun are the source of all the energy that the Sun radiates into space.

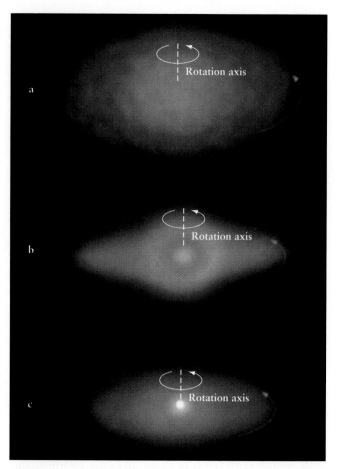

ƒigure 7-14

**The Birth of the Solar System** This sequence of drawings shows three stages in the formation of the solar system. In **(a)** a slowly rotating cloud of interstellar gas and dust begins to contract because of its own gravity. A central condensation forms in **(b)** as the cloud flattens and rotates faster around its rotation axis. In **(c)** the young Sun, which has begun to shine, is surrounded by a flattened disk of gas and dust. The planets will form out of the material in this disk.

If the solar nebula had not been rotating at all, everything would have fallen directly into the protosun, leaving nothing behind to form the planets. Instead, the solar nebula must have had an overall slight rotation, which caused its evolution to follow a different path. As the slowly rotating nebula collapsed inward, it would naturally have tended to rotate faster. Figure skaters use this same phenomenon; when a spinning figure skater pulls her arms and legs inward, close to her body, the rate at which she spins automatically increases. (If you're not a figure skater, you can demonstrate this yourself by sitting on an office chair. Sit with your arms outstretched and hold a weight, like a shoe or a full water bottle, in either hand. Now use your feet to start yourself rotating, lift your feet off the ground, and then pull your arms inward. Your rotation will speed up quite noticeably.) This relationship between the size of an object and its rotation speed is an example of a general principle called the **conservation of angular momentum.**

As the solar nebula began to rotate more rapidly, it also tended to flatten out (Figure 7-14*b*), just as a spinning ball of dough flattens out when it is spun rapidly by a pizza chef. From the perspective of a particle rotating along with the nebula, it felt as though there was a force pushing the particle away from the nebula's axis of rotation. (Likewise, passengers on a spinning carnival ride seem to feel a force pushing them outward and away from the ride's axis of rotation.) This apparent force was directed opposite to the inward pull of gravity, and so it tended to slow the contraction of material toward the nebula's rotation axis. But there was no such effect opposing contraction in a direction parallel to the rotation axis. The eventual result was the structure shown in Figure 7-14*c*, with a rotating, flattened disk surrounding a newly formed Sun. The planets formed later from this disk, which explains why their orbits all lie in essentially the same plane and why they all orbit the Sun in the same direction.

Because the Earth had not yet formed, there were no humans to observe these processes taking place during the formation of the solar system. But Earth astronomers have seen disks of material surrounding other stars that formed only recently. These are called **protoplanetary disks,** or **proplyds,** because it is thought that planets can eventually form from these disks around other stars. Thus, proplyds are planetary systems that are still "under construction." By studying these proplyds, astronomers are able to examine what our solar nebula may have been like some $5 \times 10^9$ years ago.

Figure 7-15 shows a number of such proplyds that have been found in the Orion Nebula, an active region of star formation, using the Hubble Space Telescope. (Other Orion proplyds can be seen in the figure that opens this chapter.) A star is visible at the center of each proplyd, which reinforces the idea that our Sun began to shine before the planets were fully formed. A study of 110 stars in the Orion Nebula detected proplyds around 56 of them, which suggests that systems of planets may form around a substantial fraction of stars. Later in this chapter we will see direct evidence for planets that have formed around stars other than the Sun.

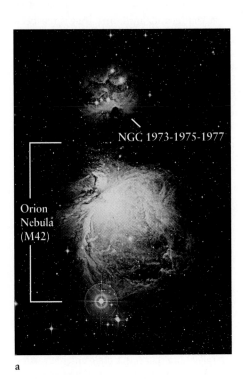

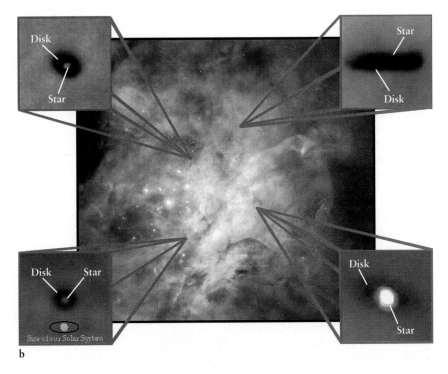

a                                                                 b

## ƒigure 7-15   R I **V** U X G

### Protoplanetary Disks (Proplyds) in the Orion Nebula

**(a)** The Orion Nebula (M42) is a star-forming region located some 1500 light-years from Earth. To the naked eye, the nebula is the middle star in Orion's "sword" (see Figure 2-2). The Orion Nebula is the large, reddish region in the center of this picture; the smaller, bluish nebula above it is the object seen close up in Figure 7-13. (Anglo-Australian Observatory) **(b)** At the center is a mosaic of Hubble Space Telescope images showing a closeup of the interior of the Orion Nebula. The four insets are extreme closeups of four proplyds, or protoplanetary disks, that lie within the nebula. The inset at the lower left also shows the size of our own solar system for comparison. The colors in the insets are false colors chosen to show the variations in brightness within the proplyd. In each proplyd there is a central bright spot, which is a young, recently formed star. (The proplyd at upper right is seen nearly edge-on.) The surrounding material in each proplyd may, with time, form itself into planets. (C. R. O'Dell and S. K. Wong, Rice University; NASA)

## 7-8 The planets formed by the accretion of planetesimals and the accumulation of gases in the solar nebula

We have seen how the solar nebula would have contracted to form a young Sun with a protoplanetary disk, or proplyd, rotating around it. But how did this disk coalesce into planets? How did the formation of the inner, rocky, terrestrial planets differ from that of the outer, largely gaseous, Jovian planets?

To understand how the planets, asteroids, and comets formed from the material of the solar nebula, we must look at how the substances that made up the nebula behaved. The density of the nebula was rather low, as was the pressure of the nebula's gas. Under such conditions of low pressure, a substance's **condensation temperature** determines whether it is a solid or a gas. If the temperature of a substance is above its condensation temperature, the substance is a gas; if the temperature is below the condensation temperature, the substance solidifies into tiny specks of dust or snowflakes. You can often see similar behavior on a cold morning. The air temperature can be above the condensation temperature of water, while the cold windows of parked cars may have temperatures below the condensation temperature. Thus, water molecules in the air remain as a gas (water vapor) but form solid ice particles on the car windows.

Substances such as water ($H_2O$), methane ($CH_4$), and ammonia ($NH_3$) have low condensation temperatures, ranging from 100 to 300 K. In the solid state, these substances are referred to as **ices**. Rock-forming substances have much higher condensation temperatures, typically in the range of 1300 to 1600 K. The gas cloud from which the solar system formed had an initial temperature near 50 K, so all of these substances could have existed in the solid form. Thus, the solar nebula would have been populated by an abundance of small ice particles and ice-covered dust grains (Figure 7-16). But hydrogen and helium, the most abundant elements in the solar nebula, have condensation temperatures so near absolute zero that these substances always existed as gases during the creation of the solar system. You can best visualize the initial state of the solar nebula as a thin gas of hydrogen and helium strewn with tiny dust particles.

As the central part of the solar nebula contracted to form the protosun, temperatures around the protosun climbed above 2000 K. Meanwhile, temperatures in the outermost regions of the solar nebula remained at less than 50 K. Figure 7-17 shows the probable temperature distribution throughout the solar nebula at this stage in the formation of our solar system. All the water, methane, and ammonia in the inner regions of the solar nebula were vaporized by the high temperatures. Only the rocky substances could have remained solid, which is why the four inner planets are today composed primarily of dense, rocky material. In contrast, ice particles and ice-coated dust grains were able to survive in the cooler, outer portions of the solar nebula. As a result, the Jovian planets possess abundant ice-forming substances, as evidenced by the spectral lines of methane and ammonia that have been identified in their spectra. Indeed, many of the satellites of the Jovian planets are at least partially composed of ices.

During the phase when the protosun was heated by Kelvin-Helmholtz contraction, it was actually quite a bit more luminous than the present-day Sun. As the protosun settled down to the more steady light output of today's Sun, temperatures within the solar nebula decreased below the values shown in Figure 7-17. As temperatures dropped, different substances that had been vaporized began to condense into microscopic solid grains. This condensation occurred at different times for different substances, depending on their condensation temperatures.

### Figure 7-16

**A Grain of Cosmic Dust** This greatly enlarged image shows a tiny dust grain from interplanetary space. This grain entered Earth's upper atmosphere and was collected by a high-flying aircraft. Dust grains of this sort are abundant in star-forming regions like that shown in Figure 7-13, and are believed to have been abundant in the nebula from which our solar system formed. Current theory says that these tiny grains were the building blocks of the planets. This grain is about 20 μm (0.02 mm) long. (Donald Brownlee, University of Washington)

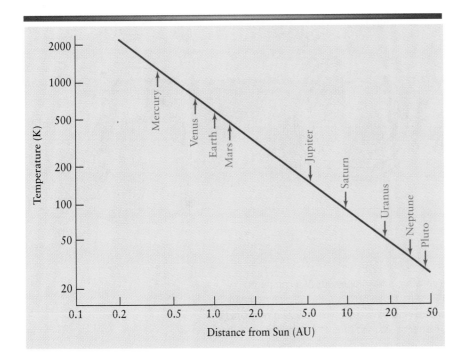

figure 7-17

**Temperature Distribution in the Solar Nebula** This graph shows how temperatures probably varied across the solar nebula as the planets were forming. For the inner planets, temperatures ranged from roughly 1200 K at Mercury to 500 K at the orbit of Mars. From the position of Jupiter outward, temperatures were everywhere less than 200 K.

In the inner solar system, where temperatures had been highest, some molecules containing aluminum condensed into grains at around 1600 K. Later, when the temperature had dropped to around 1400 K, iron and nickel condensed. Silicates, which are compounds containing silicon and oxygen (found in abundance in rocks on our present-day Earth), solidified later still when the temperature dropped to around 1300 K. Substances with even lower condensation temperatures formed grains still later. Thus, we might expect a variety of different grains, each with a different rocky composition, to have been floating around in the inner parts of the solar nebula. We can see evidence of this picture by looking at meteorites, which are small pieces of solar system material that escaped being incorporated into the planets and which orbited the Sun for billions of years before colliding with the Earth. Many meteorites are made up of different kinds of rocky materials, rather like bits of different colored tile that have been pieced together to make a mosaic.

Small, meteoritelike chunks would have formed in the inner part of the solar nebula from collisions between neighboring dust grains. Initially electric forces (chemical bonds) and gravitational forces held them together. Over a few million years, these accumulations of dust and pebbles coalesced into roughly a billion asteroidlike objects called **planetesimals,** with diameters of about 10 km. During the next stage, the gravitational attraction between the planetesimals caused them to collide and coalesce into still-larger objects called **protoplanets,** which were roughly the size and mass of our Moon. This accumulation of material to form larger and larger objects is called **accretion.** During the final stage, these Moon-sized protoplanets collided to form the

terrestrial planets. This final episode must have involved some truly spectacular, world-shattering collisions.

In recent years, astronomers have used computer simulations to learn more about how the inner planets formed from solid, rocky particles in the solar nebula. A computer is programmed to simulate a large number of particles circling a newborn sun along orbits dictated by Newtonian mechanics. As the simulation proceeds, the particles coalesce to form larger objects, which in turn collide to form planets. By performing a variety of simulations, each beginning with somewhat different numbers of planetesimals in different orbits, it is possible to see what kinds of planetary systems would have been created under different initial conditions. Such studies demonstrate that a wide range of initial conditions ultimately lead to basically the same result in the inner solar system: Accretion continues for roughly 100 million years and typically forms four or five terrestrial planets with orbits between 0.3 and 1.6 AU from the Sun.

Figure 7-18 shows one particular computer simulation. The calculations began with 100 planetesimals, each having a mass of $1.2 \times 10^{23}$ kg. This choice ensures that the total mass ($1.2 \times 10^{25}$ kg) equals the mass of the four terrestrial planets (Mercury through Mars) plus their satellites. The initial orbits of these planetesimals are inclined to each other by angles of less than 5° to simulate a thin layer of asteroidlike objects orbiting the protosun.

After an elapsed time simulating 30 million ($3 \times 10^7$) years, the 100 original planetesimals have coalesced into 22 protoplanets. After 79 million ($7.9 \times 10^7$) years, 11 larger protoplanets remain. Nearly another 100 million ($10^8$) years elapse before the total number of growing protoplanets is

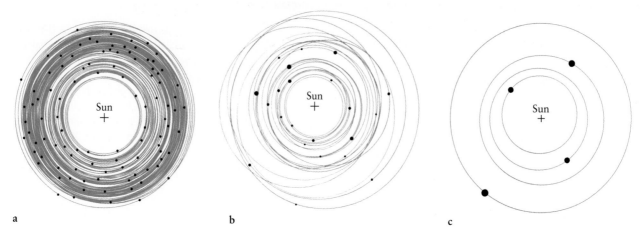

a          b          c

## ƒigure 7-18

**Accretion of the Terrestrial Planets** These three drawings show the results of a computer simulation of the formation of the inner planets. **(a)** The simulation begins with 100 planetesimals orbiting the Sun. **(b)** After 30 million years, these planetesimals have coalesced into 22 protoplanets. **(c)** This final view shows four planets orbiting the Sun after an elapsed time of 441 million years. In fact, in this simulation the inner planets were essentially formed after 150 million years. (Adapted from G. W. Wetherill)

reduced to six. Figure 7-18c shows four planets following nearly circular orbits after a total elapsed time of 441 million $(4.41 \times 10^8)$ years. In this particular simulation, the fourth planet from the Sun ends up being the most massive; in our own solar system, it is the third planet (Earth) that is the most massive of the terrestrial planets. Note that the four planets in the simulation end up in orbits that are nearly circular, just like the orbits of most of the planets in our solar system.

The material from which the inner protoplanets accreted was rich in substances with high condensation temperatures. Iron, silicon, magnesium, and sulfur were particularly abundant, followed closely by aluminum, calcium, and nickel. All of these elements combine with oxygen to form compounds called oxides, which also have high condensation temperatures and constituted an important fraction of the planetesimals. The violent impacts of the planetesimals on the growing protoplanets, as well as the decay of radioactive elements, melted much of this rocky material. The terrestrial planets therefore may have begun their existence as spheres of at least partially molten rock. During this period, gravity caused the denser, iron-rich minerals to sink to the centers of the planets, while the less dense silicon-rich minerals floated to their surfaces. This process is called **chemical differentiation** (see Box 7-1).

Like the inner planets, the outer planets may have begun to form by the accretion of planetesimals. The key difference is that ices as well as rocky grains were able to survive in the colder outer regions of the solar nebula. The elements of which ices are made are much more abundant than those that form rocky grains. Thus, much more solid material

would have been available to form planetesimals in the outer solar nebula than in the inner part. As a result, objects several times larger than any of the terrestrial planets formed in the outer solar nebula. Each such object, made up of a mixture of ices and rock, became the core of a Jovian planet and served as a "seed" around which the rest of the planet eventually grew. Thanks to the lower temperatures in the outer solar system, gas atoms (principally hydrogen and helium) were moving relatively slowly and so could more easily be captured by the strong gravity of these massive cores (see Box 7-2). Thus, according to calculations by Peter Bodenheimer of the University of California's Lick Observatory, James Pollack of the NASA Ames Research Center, and their colleagues, the core of a Jovian protoplanet began to capture an envelope of gas as it continued to grow by accretion. Both rock and gas slowly accumulated for about a million years, until the masses of the core and the envelope became equal. From that critical moment on, the envelope pulled in all the gas it could get, dramatically increasing the protoplanet's mass and size. This runaway growth of the protoplanet continued until all the available gas was used up. The result was a huge planet with an enormously thick, hydrogen-rich atmosphere surrounding a rocky core with five to ten times the mass of the Earth. This scenario occurred at four different distances from the Sun, thus creating the four Jovian planets. Figure 7-19 summarizes this story of the formation of the solar system.

While the planets were forming, the protosun was also evolving into a full-fledged star with nuclear reactions occurring in its core. The time required for this to occur was about the same as that required for the formation of the

a

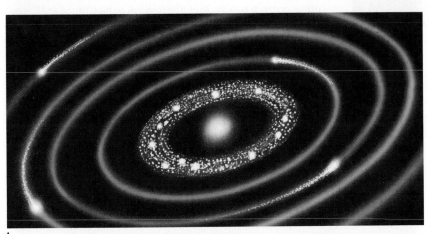

b

c

## figure 7-19

**Formation of the Planets** This series of sketches, spanning 100 million years, shows the major stages in the birth of the solar system. Terrestrial planets accrete from rocky material in the warm inner regions of the solar nebula. Meanwhile, the huge, gaseous Jovian planets form in the cold outer regions. **(a)** The solar nebula in its initial stages. **(b)** The early solar system after 50 million years. **(c)** Planetary formation is nearly complete after 100 million years.

inner, terrestrial planets, about $10^8$ years. Before the onset of nuclear reactions, however, instabilities can develop within a young star that cause its tenuous outermost layers to be vigorously expelled into space (Figure 7-20). This brief but intense burst of mass loss, observed in many young stars across the sky, is called a **T Tauri wind,** after the star in the constellation Taurus (the Bull) where it was first identified. The T Tauri wind that heralded the birth of the Sun may

have helped sweep the solar system clean of excess gases, thereby preventing further accretion onto the planets.

The Sun today continues to lose matter gradually in a mild fashion. This ongoing, gentle mass loss, called the **solar wind,** consists of high-speed protons and electrons leaking away from the Sun's outer layers. In later chapters, we will see how each planet carves out its own distinctive cavity in this solar wind.

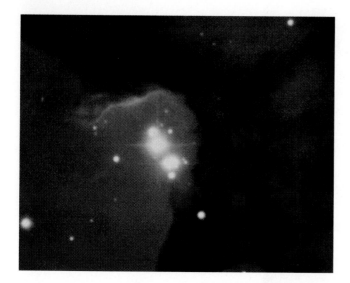

**figure 7-20** R I **V** U X G

**Newly Formed Stars** A full-fledged star is born when thermonuclear reactions ignite at its center. This ignition is often accompanied by an outpouring of particles and radiation from the star's surface that sweeps the surrounding space clean of dust and gas from which the star formed. This photograph shows dusty material being blown away from newly created stars at the center of the Trifid Nebula in the constellation Sagittarius (the Archer). (Anglo-Australian Observatory)

---

## 7-9 Astronomers have discovered planets orbiting other stars

If planets formed around our Sun, have they formed around other stars? That is, are there **extrasolar planets** orbiting stars other than the Sun? Our model for the formation of the planets would seem to suggest so. This model is based on the laws of physics and chemistry, which to the best of our understanding are the same everywhere in the universe. The discovery of a set of planets orbiting another star, with terrestrial planets in orbit close to the star and Jovian planets orbiting further away, would be a tremendous vindication of our theory of solar system formation. It would also be one of the most remarkable scientific discoveries of all time, for it would tell us that planetary systems like our own are not unique in the universe. Because at least one planet in our solar system (Earth) has the ability to support life, perhaps other planetary systems could also harbor living organisms.

Since 1995 evidence has accumulated for the existence of several planets orbiting stars other than the Sun. However, none of these is a terrestrial planet. A number of them appear to be more massive than Jupiter, and some are in eccentric, noncircular orbits quite unlike planetary orbits in our solar

system. There is even some controversy as to whether these objects orbiting other stars are really planets at all. To understand this controversy, we must look at the process that astronomers go through to search for extrasolar planets.

It is very difficult to make direct observations of planets orbiting other stars. The problem is that planets are small and dim compared to stars; at visible wavelengths, the Sun is $10^9$ times brighter than Jupiter and $10^{10}$ times brighter than the Earth. A hypothetical planet orbiting a distant star, even a planet ten times larger than Jupiter, would be lost in the star's glare as seen through even the largest telescope on Earth.

Instead, indirect methods must be used to search for extrasolar planets. One very powerful method is to search for stars that appear to "wobble." If a star has a planet, it is not quite correct to say that the planet orbits the star. Rather, both the planet and the star move in elliptical orbits around a point called the **center of mass.** Imagine the planet and the star as sitting at opposite ends of a very long seesaw; the center of mass is the point where you would have to place the fulcrum in order to make the seesaw balance. Because of the star's much greater mass, the center of mass is much closer to the star than to the planet. Thus, while the planet may move in an orbit that is hundreds of millions of kilometers across, the star will move in a much smaller orbit (Figure 7-21a). For example, the Sun and Jupiter both orbit their common center of mass with an orbital period of 5.2 years. (Jupiter has more mass than the other eight planets put together, so it's a reasonable approximation to consider the Sun's wobble as being due to Jupiter alone.) Jupiter's orbit has a semimajor axis of $7.78 \times 10^8$ km, while the Sun's orbit has a much smaller semimajor axis of 742,000 km. The Sun's radius is 696,000 km, so the Sun slowly wobbles around a point not far outside its surface. If astronomers elsewhere in the Galaxy could detect the Sun's wobbling motion, they could tell that there was a large planet (Jupiter) orbiting the Sun. They could even determine the planet's mass and the size of its orbit, even though the planet itself was unseen.

Detecting the wobble of other stars is not an easy task. One approach to the problem, called the **astrometric method,** involves making very precise measurements of a star's position in the sky relative to other stars. The goal is to find stars whose positions change in a cyclic way (Figure 7-21b). The measurements must be made with very high accuracy (0.001 arcsec or better) and, ideally, over a long enough time to span an entire orbital period of the star's motion.

A different approach to the problem is the **radial velocity method** (Figure 7-21c). This is based on the Doppler effect, which we described in Section 5-9. If a star wobbles because of the presence of a planet orbiting that star, the star will alternately move away from and toward the Earth. This will cause the dark absorption lines in the star's spectrum to change their wavelengths in a periodic fashion. When the star is moving away from us, the star's spectrum will undergo a redshift to longer wavelengths. When the star is approaching, there will be a blueshift of the spectrum to shorter wave-

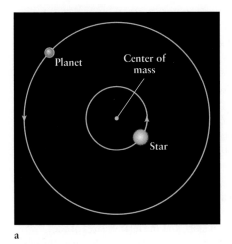 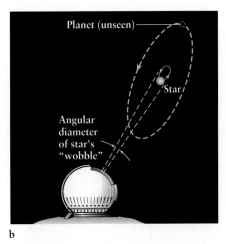 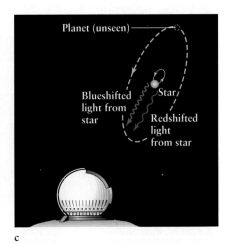

a        b        c

## ƒigure 7-21

**Detecting Planets Orbiting Other Stars** **(a)** Both a planet and its star orbit around their common center of mass, always staying on opposite sides of this point. Even if the planet cannot be seen, its presence can be inferred if the star's motion can be detected. **(b)** The astrometric method of detecting the unseen planet involves making direct measurements of the star's orbital motion. **(c)** In the radial velocity method, astronomers measure the Doppler shift of the star's spectrum as it orbits alternately toward and away from the Earth. The amount of Doppler shift can be used to determine the size of the star's orbit, which in turn allows a determination of the unseen planet's orbit.

lengths. These wavelength shifts are very small because the star's motion around its orbit is rather slow. As an example, the Sun moves around its small orbit at only 12.5 m/s (45 km/h, or 28 mi/h). If the Sun was moving directly toward an observer at this speed, the hydrogen absorption line at a wavelength of 656 nm in the Sun's spectrum would be shifted by only $2.6 \times 10^{-5}$ nm, or about 1 part in 25 million. Detecting these tiny shifts requires extraordinarily careful measurements and painstaking data analysis.

In 1995, Michel Mayor and Didier Queloz of the Geneva Observatory in Switzerland announced that they had used the radial velocity method to discover a planet orbiting the star 51 Pegasi, which is 50 light-years from Earth in the constellation Pegasus (the Winged Horse). Their results were soon confirmed by the team of Geoff Marcy of San Francisco State University and Paul Butler of the University of California, Berkeley, using observations made at the University of California's Lick Observatory. For the first time, solid evidence had been found for a planet orbiting a star like our own Sun. Marcy and Butler, along with other astronomers, have since discovered planets orbiting several other stars by means of the radial velocity method. Figure 7-22 summarizes these discoveries.

The extrasolar planets discovered by the radial velocity method all have masses comparable to or larger than that of Jupiter, and thus are presumably Jovian planets made primarily of hydrogen and helium. (This cannot be confirmed directly, because the spectra of the planets are too faint to be detected with present technology.) According to the picture

we presented in Section 7-8, such Jovian planets would be expected to orbit relatively far from their stars, where temperatures were low enough to allow the buildup of a massive envelope of hydrogen and helium gas. But as Figure 7-22 shows, many extrasolar planets are in fact found orbiting very *close* to their stars. For example, the planet orbiting 51 Pegasi has a mass at least 0.47 times as great as that of Jupiter but orbits only 0.051 AU from its star with an orbital period of just 4.2 days. In our own solar system, this orbit would lie well inside the orbit of Mercury, the innermost of the terrestrial planets.

Another surprising result is that about a third of the extrasolar planets found so far have very eccentric orbits. As an example, the planet around 16 Cygni B is in an orbit with an eccentricity of 0.57; its distance from the star varies between 0.7 AU and 2.7 AU. This is quite unlike planetary orbits in our own solar system, where no planet has an orbital eccentricity greater than 0.25.

Do these observations mean that our picture of how planets form is incorrect? If Jupiterlike extrasolar planets such as that orbiting 51 Pegasi formed close to their stars, the mechanism of their formation must have been very different from that which operated in our solar system. But another possibility is that extrasolar planets actually formed at large distances from their stars and have migrated inward since their formation. If there is enough material left in a protoplanetary disk after a planet forms, gravitational interactions between the disk material and the orbiting planet will cause the planet to lose energy and to spiral inward toward the

## PLANETS AROUND NORMAL STARS

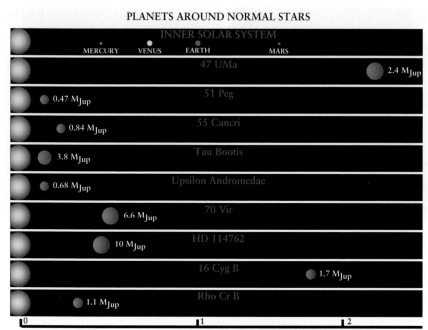

INNER SOLAR SYSTEM

MERCURY  VENUS  EARTH  MARS

47 UMa — 2.4 M$_{Jup}$

51 Peg — 0.47 M$_{Jup}$

55 Cancri — 0.84 M$_{Jup}$

Tau Bootis — 3.8 M$_{Jup}$

Upsilon Andromedae — 0.68 M$_{Jup}$

70 Vir — 6.6 M$_{Jup}$

HD 114762 — 10 M$_{Jup}$

16 Cyg B — 1.7 M$_{Jup}$

Rho Cr B — 1.1 M$_{Jup}$

0     1     2

**ORBITAL SEMIMAJOR AXIS (AU)**

**figure 7-22**

**Extrasolar Planets** This figure shows the arrangements of planets in orbit around other stars. For comparison, the arrangement of the inner planets in our own solar system is shown at the top of the figure. The scale at the bottom of the figure shows the average distance from each planet to its star (that is, the semimajor axis of each orbit). Note that several of these planets are in eccentric orbits, so the distance from the star to the planet can vary substantially over one orbit. The planets themselves are *not* shown to the same scale as their stars, and the relative sizes of the extrasolar planets are estimates only. The mass of each planet is given as a multiple of Jupiter's mass (M$_{Jup}$). The masses shown are lower limits; the actual masses of these extrasolar planets could be substantially higher. (Adapted from G. Marcy and R. P. Butler)

star around which it orbits. The planet's inward migration can eventually stop because of subtle gravitational effects from the disk or from the star. Gravitational interactions between the planet and the disk, or between planets in the same planetary system, could also have forced an extrasolar planet into a highly eccentric, noncircular orbit.

Yet a third possibility is that some of these extrasolar planets are not planets at all but objects that are more like stars in character. The problem is that the radial velocity method cannot give precise values for the masses of planets, only lower limits. (An exact determination of the mass would require knowing how the plane of the planet's orbit is inclined to our line of sight. Unfortunately, this angle is not known because the planets themselves are unseen.) Hence, the actual masses of some of the "planets" shown in Figure 7-22 may be much larger than the values shown in that figure. It is thought that planets more massive than roughly ten times the mass of Jupiter are actually **brown dwarfs,** objects like the Sun but that are insufficiently massive to ever have nuclear reactions begin in their cores (Figure 7-23). Brown

**figure 7-23**   R ▪ V U X G

**A Brown Dwarf** Brown dwarfs are objects like stars, but with masses too low for nuclear reactions to begin in their interiors. They are also substantially more massive than even the largest Jovian planets. Some of the extrasolar "planets" shown in Figure 7-22 may actually be brown dwarfs. This image, made with the Hubble Space Telescope, shows a brown dwarf (the small white dot) that orbits at a distance of 44 AU from the star Gliese 229 (the large bright object at the left). The brown dwarf is estimated to have a mass of 30 to 55 times the mass of Jupiter (between 0.030 and 0.055 of the Sun's mass) but about the same radius as Jupiter. Its surface temperature is about 1000 K. False colors are used to represent different levels of brightness (the image was actually made at infrared wavelengths). The spike running diagonally is an illusion caused by the telescope optics. Gliese 229 and its brown dwarf are 18 light-years from Earth in the constellation Lepus (the Hare). (S. Kulkarni, California Institute of Technology; D. Golimowski, Johns Hopkins University; NASA)

dwarfs are thought to form from gas, like the Sun. Therefore, these objects follow a different evolutionary path than do true planets, which form from dust grains that accrete into planetesimals, then protoplanets, and finally into planets.

Happily, some extrasolar planets appear to be of a sort that could indeed fit into our own solar system. The best example is the planet orbiting 47 Ursae Majoris, a star very similar to the Sun. This planet is in a nearly circular orbit with a radius of 2.1 AU, intermediate between the orbits of Mars and Jupiter, and has a mass of at least 2.4 times that of Jupiter. Some simulations of planet formation produce planets with orbital radii and masses in this range. Evidence for another solar system–like case has been presented by George Gatewood of the Allegheny Observatory. Applying the astrometric method to the star Lalande 21185, he finds that the wobble of that star indicates the presence of *two* planets, one with 0.9 Jupiter masses at a distance of 2.2 AU from the star and another with 1.1 Jupiter masses at 11 AU from the star.

(These masses are actual values, not lower limits, because the astrometric method makes it possible to determine the inclinations of the planets' orbits.)

The study of extrasolar planets is still in its infancy. Only a handful of extrasolar planets have yet been found, although more are being sought out. We do not yet know what extrasolar planets look like, nor do we know their chemical composition. No terrestrial planets have yet been discovered, because their masses are too small to be detected by either the radial velocity method or the astrometric method using current ground-based telescope technology. But advances in adaptive optics (Section 6-3) may eventually make it possible to record images of extrasolar planets the size of Jupiter, and orbiting telescopes are now being planned by both NASA and the European Space Agency to search for terrestrial planets around other stars. The next few decades may tell whether our own solar system is an exception or just one of a host of similar systems throughout the Galaxy.

## Key Words

*Terms preceded by an asterisk (\*) are discussed in the Boxes.*

accretion, p. 179

asteroid, p. 170

asteroid belt, p. 170

astrometric method (for detecting extrasolar planets), p. 182

atomic number, p. 174

average density, p. 166

brown dwarf, p. 184

center of mass, p. 182

chemical differentiation, p. 180

comet, p. 172

condensation temperature, p. 178

conservation of angular momentum, p. 177

\*escape speed, p. 172

extrasolar planet, p. 182

ices, p. 172

Jovian planet, p. 164

Kelvin-Helmholtz contraction, p. 176

\*kinetic energy, p. 171

minor planet, p. 170

molecule, p. 168

nebulosity, p. 174

planetesimal, p. 179

protoplanet, p. 179

protoplanetary disk (proplyd), p. 177

protosun, p. 176

radial velocity method (for detecting extrasolar planets), p. 182

solar nebula, p. 176

solar wind, p. 181

spectroscopy, p. 168

supernova, p. 174

T Tauri wind, p. 181

terrestrial planet, p. 164

## Key Ideas

**Properties of the Planets:** All of the planets orbit the Sun in the same direction and in almost the same plane. Most of the planets have nearly circular orbits.

• The four inner planets are called terrestrial planets. They are relatively small (with diameters of 5000 to 13,000 km), have high average densities (4000 to 5500 kg/m$^3$), and are composed primarily of rock.

• The four giant outer planets are called Jovian planets. They have large diameters (50,000 to 143,000 km) and low average densities (700 to 1700 kg/m$^3$) and are composed primarily of hydrogen and helium.

• Pluto appears to be a special case. It is smaller than any of the terrestrial planets and has an intermediate average density of 2030 kg/m$^3$, suggesting that it is composed of a mixture of ice and rock.

**Satellites and Small Bodies in the Solar System:** Besides the planets, the solar system includes satellites of the planets and asteroids and comets.

• There are seven large planetary satellites (one of which is the Moon) that are comparable in size to the planet Mercury. The remaining satellites of the solar system are much smaller.

• Asteroids are small, rocky objects, while comets are small objects made of dirty ice. Both are remnants left over from the formation of the planets.

**Spectroscopy and the Composition of the Planets:** Spectroscopy, the study of spectra, provides information about the chemical composition of objects in the solar system.

• By mass, 98% of the matter in the universe is composed of the two lightest elements, hydrogen and helium. These elements were probably formed shortly after the universe was created. The heavier elements were produced much later, in the centers of stars, and were cast into space when the stars died.

• The atoms of various elements can combine to form many different kinds of molecules.

• The basic planet-forming substances can be classified as gases, ices, or rock, depending on their condensation temperatures. The terrestrial planets are composed primarily of rock, whereas the Jovian planets are composed largely of gas.

**Formation of the Solar System:** The solar system formed from a disk-shaped cloud of hydrogen and helium that also contained ice and dust particles.

• The four inner planets formed through the accretion of dust particles into planetesimals, then into larger protoplanets.

• The four outer planets probably began as rocky protoplanetary cores, similar in character to the terrestrial planets. Gas then accreted onto these cores in a runaway fashion.

• The Sun formed by gravitational contraction of the center of the nebula. After about $10^8$ years, temperatures at the protosun's center became high enough to ignite thermonuclear reactions, thus forming a true star.

**Extrasolar Planets:** Astronomers have discovered planets orbiting other stars. The planets themselves are not visible; their presence is detected by the "wobble" of the stars around which they orbit.

• Most of the extrasolar planets discovered to date are quite massive and have orbits that are very different from planets in our solar system.

## Review Questions

1. What are the characteristics of a terrestrial planet?

2. What are the characteristics of a Jovian planet?

3. In what ways does Pluto not fit the usual classification of either terrestrial or Jovian planets?

4. In what ways are the largest satellites similar to the terrestrial planets? Which satellites are these?

5. What is an asteroid? What is a comet? In what ways are these minor members of the solar system like or unlike the planets?

6. Why do astronomers find it convenient to use the Kelvin temperature scale in their work rather than the Celsius or Fahrenheit scales?

7. What is meant by the average density of a planet? What does the average density of a planet tell us?

8. If hydrogen and helium account for 98% of the mass of all the material in the universe, why aren't the Earth and Moon composed primarily of these two gases?

9. What is meant by a substance's condensation temperature? What role did condensation temperatures play in the formation of the planets?

10. What is a planetesimal? How did planetesimals give rise to the planets?

11. What is meant by accretion?

12. What occurred during the formation of our solar system to ensure that the terrestrial planets formed close to the Sun and the Jovian planets far from the Sun?

13. Explain how our current understanding of the formation of the solar system can account for the following characteristics of the solar system: (**a**) All planetary orbits lie in nearly the same plane. (**b**) All planetary orbits are nearly circular. (**c**) The planets orbit the Sun in the same direction that the Sun itself rotates.

14. What techniques are used to detect planets orbiting other stars? Why are these techniques unable to detect planets like Earth?

15. Summarize the differences between the planets of our solar system and those found orbiting other stars.

## Advanced Questions

*Questions preceded by an asterisk (\*) involve topics discussed in the Boxes.*

| Problem-solving tips and tools: |

The volume of a sphere of radius $r$ is $\frac{4}{3}\pi r^3$. The average density of an object is its mass divided by its volume. To calculate escape speeds, you will need to review Box 7-2; to compute the mass of Mars or 70 Virginis, you may have to review Box 4-2 for the appropriate form of Kepler's third law. Be sure to use the same system of units (e.g., meters, seconds, kilograms) in all your calculations involving escape speeds, orbital speeds, and masses. Formulas relating various temperature scales are found in Box 5-1. The relationship between the angular diameter of an object and its actual diameter is given in Box 1-2.

**16.** Why do you suppose water ($H_2O$), methane ($CH_4$), and ammonia ($NH_3$) are comparatively abundant substances?

**\*17.** A spherical asteroid 2 km in diameter, and composed of rock with an average density of 2500 kg/m³, strikes the Earth with a speed of 25 km/s. (a) What is the kinetic energy of the asteroid at the moment of impact? (b) How does this energy compare with that released by a 20-kiloton nuclear weapon, like the device that destroyed Hiroshima, Japan, on August 6, 1945? (*Hint*: 1 kiloton of TNT releases $4.2 \times 10^{12}$ joules of energy.)

**\*18.** A hydrogen atom has a mass of $1.673 \times 10^{-27}$ kg, and the temperature of the Sun's surface is 5800 K. What is the average speed of hydrogen atoms at the Sun's surface?

**\*19.** The Sun's mass is $1.989 \times 10^{30}$ kg, and its radius is $6.96 \times 10^8$ m. What is the escape speed from the Sun's surface? Using your answer to Question 18, explain why the Sun has lost very little hydrogen over its entire 4.6 billion-year history.

**\*20.** Suppose a spacecraft landed on Jupiter's moon Europa, which has a diameter of 3138 km and a mass of $4.80 \times 10^{22}$ kg, and which moves around Jupiter in an orbit of radius 670,900 km. After collecting samples from the satellite's surface, the spacecraft prepares to return to Earth. (a) Calculate the escape speed from Europa. (b) Calculate the escape speed from Jupiter at the distance of Europa's orbit. (c) In order to begin its homeward journey, the spacecraft must leave Europa with a speed greater than either your answer to (a) or your answer to (b). Explain why this is so.

**21.** Figure 7-5 shows the spectrum of Saturn's largest satellite, Titan. Some of the spectral lines are produced in Earth's atmosphere, while others are produced in the Sun's atmosphere. Can you think of a way of distinguishing the actual spectral lines of Titan's atmosphere from the spectral lines of the atmospheres of the Sun and Earth?

**22.** Mars has two small satellites, Phobos and Deimos. Phobos circles Mars once every 0.31891 day at an average altitude of 5980 km above the planet's surface. The diameter of Mars is 6794 km. Using this information, calculate the mass and average density of Mars.

**23.** (a) What would the mass of the Earth be if it had retained hydrogen and helium in the same proportion to the heavier elements that exist elsewhere in the universe? Explain your reasoning. (b) How does your answer to (a) compare with the masses of the Jovian planets? What does your answer imply about how large the cores of the Jovian planets must be?

**24.** Suppose there was a planet with roughly the same mass as Earth but located 50 AU from the Sun. (a) What do you think this planet would be made of? Explain your reasoning. (b) On the basis of this speculation, assume a reasonable density for this planet and calculate its diameter. How many times bigger or smaller than Earth would it be?

**25.** Suppose you were to use the Hubble Space Telescope to monitor one of the protoplanetary disks shown in the figure that opens this chapter or in Figure 7-15b. Over the course of ten years, would you expect to see planets forming within the disk? Why or why not?

**\*26.** The planet discovered orbiting the star 70 Virginis, 59 light-years from Earth, moves in an orbit with semi-major axis 0.47 AU and eccentricity 0.40. The period of the orbit is 116.6 days. Find the mass of 70 Virginis. Compare your answer to the mass of the Sun.

**\*27.** Because of the presence of Jupiter, the Sun moves in a small orbit of radius 742,000 km with a period of 5.2 years. (a) Calculate the Sun's orbital speed in meters per second. (b) An astronomer on a hypothetical planet orbiting the star Vega, 26 light-years from the Sun, wants to use the astrometric method to search for planets orbiting the Sun. What would be the angular diameter of the Sun's orbit as seen by this alien astronomer? Would the Sun's motion be discernible if the alien astronomer could measure positions to an accuracy of 0.001 arcsec? (c) Repeat part (b), but now let the astronomer be located on a hypothetical planet in the Pleiades star cluster, 410 light-years from the Sun. Would the Sun's motion be discernible to this astronomer?

## Discussion Questions

**28.** Propose an explanation why the Jovian planets are orbited by terrestrial-like satellites.

**29.** Suppose that a planetary system is now forming around some protostar in the sky. In what ways might this planetary system turn out to be similar to or different from our own solar system? Explain your reasoning.

**30.** Suppose astronomers discovered a planetary system in which the planets orbit a star along randomly inclined orbits. How might a theory for the formation of that planetary system differ from that for our own?

**31.** By consulting the Space Telescope Science Institute web site (**http://www.stsci.edu/**), as well as recent issues of the magazines *Sky & Telescope* and *Astronomy*, find out what new observations have been made of protoplanetary disks. What insights have astronomers gained from these observations? Is there any evidence that planets have formed within these disks?

## Observing Projects

**32.** There are many young stars still embedded in the clouds of gas and dust from which they formed. In the winter evening sky, for example, is the famous Orion Nebula (Figure 7-15a). In the summer night sky, the Lagoon, Omega, and Trifid nebulae are found in the Milky Way. Examine some of these nebulae with a telescope. Describe their appearance. Can you guess which stars in your field of view are actually associated with the nebulosity? To help you find some of these nebulae, the table below gives their coordinates (right ascension and declination) for the year 2000.

| Nebula | Right ascension | Declination |
|--------|-----------------|-------------|
| Lagoon | 18$^h$ 03.8$^m$ | −24° 23′ |
| Omega | 18 20.8 | −16 11 |
| Trifid | 18 02.3 | −23 02 |
| Orion | 5 35.4 | −5 27 |

33. Consult such magazines as *Sky & Telescope* and *Astronomy* or their web sites (**http://www.skypub.com/** and **http://www.astronomy.com/**) to determine which planets are visible in your evening sky. Examine these planets through a telescope. Describe their appearance. From what you observe, is there any way of knowing whether you are looking at a planet's surface or cloud cover?

## Where to Learn More

*Books and magazine articles*

Angel, J. R. P., and Woolf, N. J. "Searching for Life on Other Planets." *Scientific American,* April 1996. This article describes plans to search for terrestrial planets orbiting other stars and, in particular, to locate planets that show evidence of life.

Beatty, J., and Chaikin, A., eds. *The New Solar System,* 3rd ed. Sky Publishing and Cambridge University Press, 1990. This excellent introduction to the solar system consists of 23 lavishly illustrated chapters, all written by noted scientists.

Black, D. C. "Other Suns, Other Planets?" *Sky & Telescope,* August 1996. A noted planetary astronomer casts a critical eye on recent discoveries of extrasolar planets.

Henry, T. J. "Brown Dwarfs Revealed—At Last!" *Sky & Telescope,* April 1996. This article describes how the existence of brown dwarfs was first theorized and how they were first detected.

Kross, J. F. "What's in a Name?" *Sky & Telescope,* May 1995. Every crater, mountain, and geological feature on the planets and moons of the solar system is eventually given a name. This amusing article describes how the names are chosen.

Marcy, G., and Butler, R. P. "The Diversity of Planetary Systems." *Sky & Telescope,* March 1998. Two leaders in the search for extrasolar planets describe how they made their discoveries and explain what these discoveries tell us about how planetary systems form.

Morrison, D. *Exploring Planetary Worlds.* Scientific American Library, 1993. This beautifully illustrated book by a noted planetary scientist summarizes our modern understanding of the solar system.

Morrison, D. and Owen, T. *The Planetary System,* 2nd ed. Addison-Wesley, 1996. This very clear book covers the entire solar system in great depth. The level of presentation is the same as the textbook you are now holding.

W *World Wide Web*

A complete multimedia tour of the solar system can be found at "The Nine Planets" (**http://www.seds.org/nineplanets/nineplanets/**), a web site maintained by the Students for the Exploration and Development of Space. Despite its name, "The Nine Planets" includes images and text about *all* the members of the solar system, including 187the Sun, satellites of the planets, comets, asteroids, and meteorites.

Two other excellent sites with similar intent are "Views of the Solar System" (**http://www.hawastsoc.org/solar/homepage.htm**) from the Hawaiian Astronomical Society and the Jet Propulsion Laboratory's "Welcome to the Planets" (**http://pds.jpl.nasa.gov/planets/**). It is well worth locating a pair of red and green 3-D glasses to experience the Lunar and Planetary Institute's 3-D tour of the solar system (**http://cass.jsc.nasa.gov/research/stereo_atlas/SS3D.HTM**).

From the main NASA web page (**http://www.nasa.gov/**), you can find information about all the spacecraft that are currently exploring our solar system. Well-written articles about the latest scientific developments about the planets can be found in an excellent online magazine called "Planetary Science Discoveries" (**http://www.soest.hawaii.edu/PSRdiscoveries/**).

Up-to-date information about extrasolar planets can be found at the web site for the San Francisco State University Planet Search Project (**http://cannon.sfsu.edu/~williams/planetsearch/planetsearch.html**).

# GEOFF MARCY

## Alien Planets

When Geoff Marcy sat down to write this essay, eight planets were known to orbit other stars. He leads the team that has found six. He is an astronomer at San Francisco State University and the University of California at Berkeley.

Dr. Marcy became interested in astronomy at age 14, when his parents bought him a small reflecting telescope. Since receiving his Ph.D. from the University of California, Santa Cruz, he has studied stars similar to our Sun. He helped show that magnetic regions on their surfaces cause dark "star spots" and stellar flares. He also showed that brown dwarfs—stars too little to burn hydrogen—rarely orbit other stars. His work may soon reveal whether a planetary system like our own is common or a quirk of the cosmos.

When we look up toward the twinkling stars, we see a future of human exploration that looms as challenging as the sixteenth-century voyages of Magellan and Neil Armstrong's first step on the Moon. We see the Milky Way, a galaxy filled with 100 billion stars, and more mysteries than all science-fiction novels combined.

Do stars commonly harbor planets, or are most stars isolated glowing globes in dark space? If planets do exist, are they mostly rock-strewn landscapes, giant balls of hydrogen gas like Jupiter, or warm earths covered with continents, rivers, and oceans? Will their liquid water provide the solvent for the biochemical origins of life? If so, evolution would surely lead to complex life forms and perhaps intelligent species.

We may yet be unique in the vast universe, however. Intelligence is not the only end- product of evolution. Then, too, humans often simplistically define "intelligence" to resemble our own. We think of the ability to build CD players and CAT scanners rather than international cooperation. We cleverly deem ourselves the necessary winners of the Darwinian sweepstakes here on the Earth, but we are the ones who make up the rules.

Recent advances may at last address the question of life elsewhere. I participated in one such discovery, planets orbiting stars like the Sun. The technique that I and my colleagues used to find these new worlds is simple. When a planet orbits its host star, it exerts a gravitational tug on the star. It pulls the star around in a small circle, like a leashed puppy yanked by its owner. The cosmic leash between a star and its planet is gravity.

A star's telltale reflex motion, a very tiny wobble, can be detected in its starlight, using the Doppler effect. Different wavelengths of the starlight become shorter as the star approaches us and longer as it recedes away. By measuring the Doppler shifts, we pick up the wobble, telling us the shape of the planet's orbit and its approximate mass. A more massive planet yanks its host star around more vigorously, and the Doppler shift reveals that vigorous motion. The graph on this page shows the wobble of one star, 47 Ursae Majoris. The points near the curve are our actual measurements.

Astronomers have examined the nearest 200 solar-type stars and found eight planets. Their masses are all about the same as Jupiter's (within a factor of, say, 3). At least 5% of the stars examined so far turn out to have giant planets. Sadly, current technology cannot detect small, rocky planets more like the Earth because of their low mass. Still, they probably accompany the larger worlds.

Already, the distant planets seem alien to our expectations. First, two of the eight new Jupiters follow a highly elongated, elliptical orbit around their star, rather than the nearly circular orbits of the planets in our solar system. Astronomers do not know why. Even more shocking, five of the planets lie extremely close to their host star, closer than Mercury to our Sun. Our own Jupiter resides some five times farther from the Sun than the Earth.

Several astronomers have guessed at the origin of these absurdly close Jupiters. Planets are formed, we know, in flattened clouds that swirl around a newborn star. The planets presumably condense out of the dust and gas in these disks. The theory explains why our nine planets all orbit in the same plane and in the same direction. Because of friction, however, the inner part of a newborn cloud must slowly drain inward, toward the star, like water going down the bathtub drain. When the disk dissipates away, like a wispy cloud on a hot summer day, a planet may be left high and dry, orbiting close to the star.

Are any planetary systems out there true kin of our solar system? New techniques for planet hunting should answer that question. Our latest search uses the world's largest telescope, the Keck 10-m telescope on the Big Island of Hawaii, located high atop a hopefully dormant volcano called Mauna Kea. From there, we are searching 400 nearby suns. Within the next decade, we hope to find most of the Jupiters and Saturns up to 50 light-years away.

We can only barely imagine what the next generation will uncover in our galactic neighborhood. But I feel confident that human destiny lies in exploring the Galaxy and in finding our roots, chemically and biologically, out among the stars.

# Our Living Earth

An Erupting Volcano Viewed from Orbit  This photograph from the space shuttle *Endeavour* shows several of the Earth's distinctive features. Our planet is geologically very active, as evidenced by the plume of ash from a volcano in Russia's Kamchatka Peninsula. The plume is being blown from west to east by our substantial atmosphere. The vast expanses of ocean are found nowhere else in the solar system.  Astronauts often report that our unique Earth is the most beautiful sight visible from space. (NASA)

*In this chapter you will find the answers to the following questions:*

8-1 How do ordinary rocks offer clues to the Earth's history?

8-2 Is the Earth completely solid inside? How can scientists tell?

8-3 How did the continents get their shapes?

8-4 What are conditions like high in the Earth's atmosphere?

8-5 Where does the Earth's magnetic field come from?

8-6 What are the greenhouse effect and the "ozone hole?" Why should they concern us?

*I*magine an alien spacecraft approaching the inner solar system. Its occupants pass Mars with its thin atmosphere and barren desert landscapes. Closer to the Sun, they see Venus, its shroud of corrosive clouds hiding a forbiddingly hot surface. Between these two relatively unpromising planets lies the Earth, where an ever-changing ballet of delicate white clouds plays against the darker browns and blues of its continents and oceans.

*They would encounter a geologically active world. The Earth's surface is a rocky crust divided into huge plates that rub against one another, creating mountain ranges, volcanoes, earthquakes, and oceanic trenches. Neither Venus nor Mars, the Earth's neighboring planets, exhibit any features indicative of widespread plate motions. The Earth's interior consists of an iron-rich core surrounded by a thick mantle of partly molten rock. Currents in the mantle propel the plate motions that shape the Earth's rigid crust. Electric currents within the iron-rich core generate a planetwide magnetic field, which extends far into space and shields us from the solar wind. The Earth's atmosphere is uniquely rich in nitrogen and in oxygen, a result of the activity of living organisms over billions of years. But today, by rapidly altering the chemical composition and thermal balance of the Earth's atmosphere, humans are threatening the survival of many species, perhaps even our own.*

*Because we walk on its surface, drink its water, and breathe its air, we know more about the Earth than any other object in the universe. (A summary of Earth data appears in Table 8-1.) In later chapters we use this knowledge as a point of reference for studying the other planets, especially our sister terrestrial planets, Venus and Mars (Figure 8-1). First, however, let us adopt the viewpoint of one of these aliens and explore this remarkable planet called Earth.*

## 8-1    Earth rocks contain clues about the history of our planet's surface

Were the aliens to land on Earth, they would find it radically different from either of its neighbors: The Earth is very wet. Nearly 71% of the Earth's surface is covered with water. An alien space probe to Earth might send back thousands of photographs such as Figure 8-2, a typical close-up view of the Earth's surface. Water is also locked into the chemical structures of many Earth rocks. In that sense, even the Sahara Desert is a veritable swamp compared to the arid, bone-dry surfaces of Venus and Mars.

Where land protrudes above the oceans, you find *rocks*, which are typical specimens of the Earth's outermost layer, or **crust.** Like all substances, rocks are composed of chemical elements (see the periodic table in Box 5-4), but individual chemical elements are rarely found in a pure state. The exceptions include nuggets of gold, silver, and copper, diamonds (crystals of carbon), and a few less valuable specimens, such as the sample of native sulfur shown in Figure 8-3. Somewhat more common are solids composed of a particular chemical combination of elements called a *compound*. Silica, for example, is a common substance composed entirely of the compound silicon dioxide, which is composed of silicon and oxygen atoms. Diamond and silica are both **minerals**—naturally occurring solids composed of a single element or compound. The most common mineral found in Earth's crust is feldspar, which contains aluminum, silicon, and oxygen. A **rock** is a solid part of the Earth's crust that is composed of one or more minerals. For example, granite contains both feldspar and quartz (a mineral containing silicon and oxygen).

In some minerals, the atoms are arranged in a jumbled, disordered way, and the mineral is rather shapeless. But the atoms in diamond, silica, and many other minerals are usually arranged in orderly rows. As a result, these minerals come in regular, symmetrical shapes called **crystals.** To identify

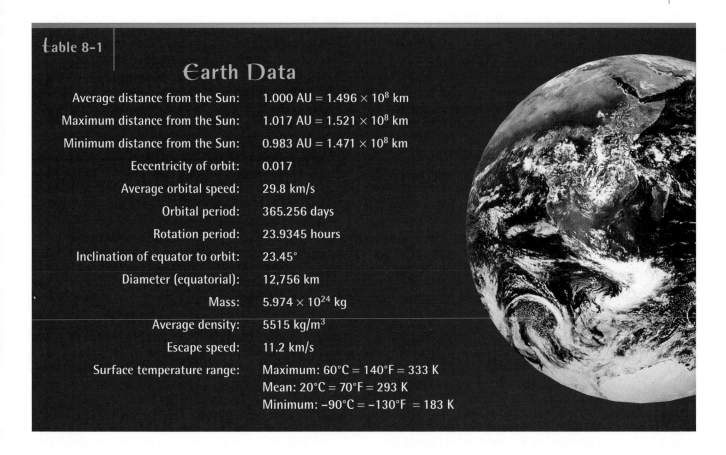

| table 8-1 | |
| --- | --- |
| **€arth Data** | |
| Average distance from the Sun: | $1.000$ AU $= 1.496 \times 10^8$ km |
| Maximum distance from the Sun: | $1.017$ AU $= 1.521 \times 10^8$ km |
| Minimum distance from the Sun: | $0.983$ AU $= 1.471 \times 10^8$ km |
| Eccentricity of orbit: | 0.017 |
| Average orbital speed: | 29.8 km/s |
| Orbital period: | 365.256 days |
| Rotation period: | 23.9345 hours |
| Inclination of equator to orbit: | 23.45° |
| Diameter (equatorial): | 12,756 km |
| Mass: | $5.974 \times 10^{24}$ kg |
| Average density: | 5515 kg/m$^3$ |
| Escape speed: | 11.2 km/s |
| Surface temperature range: | Maximum: 60°C = 140°F = 333 K<br>Mean: 20°C = 70°F = 293 K<br>Minimum: –90°C = –130°F = 183 K |

**figure 8-1**   R I **V** U X G

**The Three Largest Terrestrial Planets** The Earth and Venus have nearly the same mass, size, and surface gravity, but the surface of Venus is perpetually shrouded by a dense cloud cover containing poisonous, corrosive gases. Mars is only about half the size of Earth and has only one-tenth its mass. The thin, dry Martian atmosphere does not protect the planet from the Sun's ultraviolet radiation. Both the Venusian and the Martian environments are hostile to life forms that thrive on Earth. (NASA, ESA)

figure 8-2

**A Typical Close-up View of the Earth's Surface** Nearly three-quarters of the Earth's surface is covered with water. In contrast, there is no liquid water at all on Mercury, Venus, Mars, or the Moon.

figure 8-4

**Three Common Minerals** A mineral is a specific chemical combination of elements whose atoms are typically arranged in an orderly fashion to produce a crystal. A quartz crystal is at the left. Feldspar (center) is another silicate. Mica (right) is also a common rock-forming mineral.

figure 8-3

**Native Sulfur** Sulfur is one of the very few chemical elements found in a pure state. Carbon crystals, called diamonds, and gold nuggets are also pure elements found in nature. Most minerals are chemical combinations of various elements.

various minerals, geologists use a number of criteria, such as the shape, hardness, and color of the crystals. Three common rock-forming minerals are shown in Figure 8-4. Each has its own unique, easily identifiable characteristics. These particular minerals are found mixed together in the common kind of rock called granite.

Geologists cannot always classify rocks the same way they classify minerals, because rocks may contain a mixture of mineral characteristics. In a piece of granite, for example, the tiny quartz crystals have a different color and hardness from the shiny flecks of mica. Instead, geologists classify rocks according to how they were created.

The three major categories of rocks correspond to three ways in which rocks can form. **Igneous rocks** result when minerals cool from a molten state. The formation of igneous rock can be directly observed during a volcanic eruption. Molten rock is called **magma** when it is buried below the surface and **lava** when it flows out upon the surface. Two common igneous rocks, basalt and granite, are shown in Figure 8-5.

**Sedimentary rocks** are those produced by the action of wind, water, or ice. For example, as winds pile up layer after layer of sand, other minerals present amid the sand can gradually cement the sand grains together to produce sandstone. Minerals that precipitate out of the oceans can cover the ocean floor with layers of rock such as limestone. These two common sedimentary rocks are shown in Figure 8-6.

Sometimes igneous or sedimentary rocks become buried deep beneath the Earth's surface, where they are subjected to enormous pressure and high temperatures. These severe conditions change the rocks' structure, producing the third variety, called **metamorphic rock**. For example, when fine-grained igneous rock is subjected to pressure and heat, it becomes schist. When sedimentary limestone is metamorphosed, it becomes marble. These two common metamorphic rocks are shown in Figure 8-7.

By studying the kinds of rocks that are found at a particular location and seeing how they must have been created, geologists can deduce the history of that site. For example, layers of sedimentary rocks in the Grand Canyon indicate that this arid region was once covered by an ancient sea. By

## figure 8-5

**Igneous Rocks (Basalt and Granite)** Igneous rocks are created when molten minerals solidify. The sample on the left is basalt, which is a fine-grained mixture of feldspar along with other, iron-rich minerals that give the rock its dark color. A granite specimen, which contains feldspar as well as quartz, is on the right. The fine-grained basalt typically forms when molten minerals cool near the Earth's surface; the coarser-grained granite typically forms from slower cooling deeper within the Earth's crust.

## figure 8-7

**Metamorphic Rocks (Marble and Schist)** When igneous or sedimentary rocks are subjected to high temperatures and pressures deep in the Earth's crust, they are changed into metamorphic rock. Marble (left) is produced from sedimentary limestone (a mineral containing calcium, carbon, and oxygen); schist (right) is formed from fine-grained igneous rock.

contrast, the igneous rocks at Idaho's Craters of the Moon National Monument show that this region was once flooded by molten lava. Rocks tell us not only what the Earth's crust is made of but also how the crust has changed over the ages.

## figure 8-6

**Sedimentary Rocks (Limestone and Sandstone)** Sedimentary rocks are produced, often layer by layer, by the action of wind, water, or ice. The sample on the left is limestone, which is primarily calcite (a mineral containing calcium, carbon, and oxygen) with some impurities that give it a dark color. Sandstone, which is made primarily of grains of quartz and feldspar, is on the right. A sedimentary rock is typically formed when loose particles of soil or sand are buried a short distance below the Earth's surface and are cemented into rock by chemical changes.

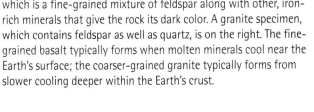

## 8-2 Studies of earthquakes reveal the Earth's layered interior structure

Although the specimens shown in Figures 8-5, 8-6, and 8-7 are typical samples of the Earth's crust, they cannot be representative of what makes up our planet's interior. The densities of typical crustal rocks are around 3000 kg/m$^3$, but the average density of the Earth as a whole (that is, its mass divided by its volume) is 5520 kg/m$^3$. The interior of the Earth must therefore be composed of a substance much denser than the crust. But what is this substance? How did it come to be that the Earth's interior is more dense than its crust? And is the Earth's interior solid like the crust or molten like lava?

Iron (chemical symbol Fe) is a good candidate for the substance that makes up most of the Earth's interior. This is so for two reasons: first, iron atoms are quite massive (a typical iron atom has 56 times the mass of a hydrogen atom), and second, iron is rather abundant. (Table 7-4 shows that iron is the tenth most abundant element in the universe.) Other elements have more massive atoms, but these other elements are quite rare. Hence, the solar nebula could not have had enough of these massive atoms to create the Earth's dense interior. Furthermore, iron is common in meteoroids that strike the Earth, which suggests that it was abundant in the planetesimals from which the Earth formed.

Geologists strongly suspect that the Earth was entirely molten soon after its formation, about 4.6 billion ($4.6 \times 10^9$) years ago. Energy released by the violent impacts of numerous

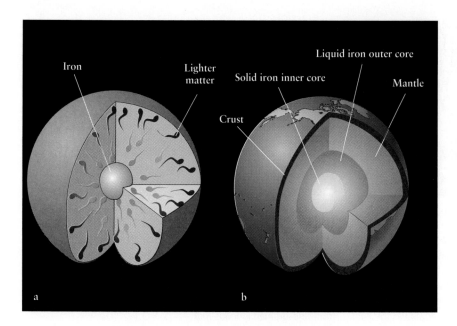

## figure 8-8

**Chemical Differentiation and the Earth's Internal Structure** **(a)** When the Earth first formed from planetesimals, it was probably molten throughout its volume. Within the molten interior of the planet, dense materials such as iron (shown in orange) sank toward the center while low-density materials (shown in blue) rose toward the surface. This process, in which the Earth's material becomes sorted by density, is called chemical differentiation. **(b)** The present-day Earth has cooled somewhat and is no longer completely molten inside. A dense, solid iron core lies at the center, surrounded by a less dense liquid core and an even less dense mantle. The crust, which includes the continents and ocean floors, is the least dense of all; it floats on top of the mantle, like the "skin" that forms on the surface of a cup of cocoa as it cools.

meteoroids and asteroids and by the decay of radioactive isotopes likely melted the solid material collected from the earlier planetesimals. Gravity caused abundant, dense iron to sink toward the Earth's center, forcing less dense material to the surface in the process called **chemical differentiation,** as shown in Figure 8-8*a*. (We discussed chemical differentiation in Section 7-8. In Box 7-1 we discussed how the density of a substance determines whether it sinks or rises in a liquid.) This process of chemical differentiation then produced a layered structure within the Earth (Figure 8-8*b*)—a central core composed of almost pure iron, surrounded by a mantle of dense, iron-rich minerals. This mantle, in turn, is surrounded by a thin crust of relatively light silicon-rich minerals. We live on the surface of this crust.

How have geologists determined that this layered structure is correct? The challenge is that the Earth's interior is as inaccessible as the most distant galaxies in space. The deepest wells go down only a few kilometers, barely penetrating the surface of our planet. Despite these difficulties, geologists have learned basic properties of the Earth's interior by studying earthquakes and the seismic waves that they produce.

Over the centuries, stresses build up in the Earth's crust. Occasionally, these stresses are relieved with a sudden motion called an **earthquake.** Most earthquakes occur deep within the Earth's crust. The point on the Earth's surface directly over an earthquake's location is called the **epicenter.**

Earthquakes produce three different kinds of **seismic waves,** which travel around or through the Earth in different ways and at different speeds. Geologists use sensitive instruments called **seismographs** to detect and record these vibratory motions. The first type of wave, which is analogous to water waves on the ocean, causes the rolling motion that

people feel around an epicenter. Seismic waves of this type travel only over the Earth's surface. The two other kinds of waves, called primary or **P waves** and secondary or **S waves,** travel through the body of the Earth. P waves are called *longitudinal* waves because their oscillations are parallel to the direction of wave motion, like a spring that is alternately pushed and pulled. In contrast, S waves are called *transverse* waves because their vibrations are perpendicular to the direction in which the waves move. S waves are analogous to waves produced by a person shaking a rope up and down (Figure 8-9).

What makes seismic waves useful for learning about the Earth's interior is that they do not travel in straight lines. Instead, the paths that they follow through the body of the Earth are bent because of the varying density and composition of the Earth's interior. We saw in Section 6-1 that light waves behave in a very similar way. Just as light waves bend, or refract, when they pass from air into glass or vice versa (Figure 6-2), seismic waves refract when they pass from one distinct region of the Earth's interior to another. By studying how the paths of these waves bend, geologists can map out the general interior structure of the Earth.

One key observation about seismic waves and how they bend has to do with the differences between S and P waves. When an earthquake occurs, seismographs relatively close to the epicenter record both S and P waves, but those on the opposite side of Earth record only P waves. The absence of S waves was first explained in 1906 by British geologist R. D. Oldham, who noted that transverse vibrations such as S waves cannot travel far through liquids. Oldham therefore concluded that our planet has a molten core. Furthermore, there is a region in which neither S waves nor P waves from

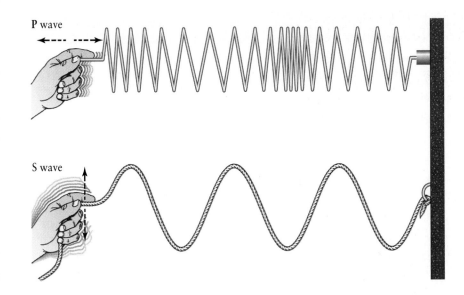

**figure 8-9**

**Seismic Waves** Earthquakes produce two kinds of waves that travel through the body of our planet. One kind, called P waves (for primary), are longitudinal waves. They are analogous to those produced by pushing a spring in and out. The other kind, S waves (for secondary), are transverse waves analogous to the waves produced by shaking a rope up and down.

an earthquake can be detected (Figure 8-10). This "shadow zone" results from the specific way in which P waves are refracted at the boundary between the solid mantle and the molten core. By measuring the size of the shadow zone, geologists have concluded that the core's radius is about 3500 km (2200 mi). For comparison, the overall radius of our planet is about 6400 km (4000 mi).

As the quality and sensitivity of seismographs improved, geologists discovered faint traces of P waves in an earthquake's shadow zone. In 1936, Danish seismologist Inge Lehmann explained that some of the P waves passing though the Earth are deflected into the shadow zone by a small, solid **inner core** at the center of our planet. The radius of this inner core is about 1300 km (800 mi).

The interior of our planet therefore has a curious structure—a liquid **outer core** sandwiched between a solid inner core and a solid mantle. This structure is summarized in Table 8-2. The explanation for this arrangement lies in how temperature and pressure inside the Earth affect the melting point of rock.

Both temperature and pressure increase with increasing depth below the Earth's surface. The temperature of the Earth's interior rises steadily from about 20°C on the surface to nearly 5000°C at our planet's center (Figure 8-11).

The Earth's outermost layer, the crust, is only about 30 km thick. It is composed of rocks for which the **melting point,** or temperature at which the rock changes from solid to liquid, is far higher than the temperatures actually found in the crust. Thus, the crust is solid.

The Earth's **mantle,** which extends to a depth of about 2900 km (1800 mi), is largely composed of minerals rich in iron and magnesium. On the Earth's surface, specimens of

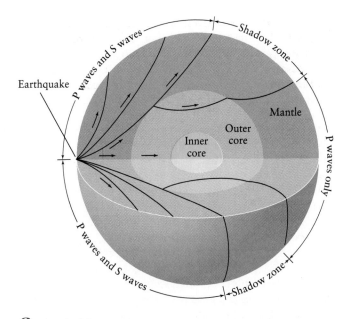

**figure 8-10**

**The Earth's Internal Structure and the Paths of Seismic Waves** Seismic waves follow curved paths because of differences in the density and composition of the material in the Earth's interior. Furthermore, only P waves can pass through the Earth's liquid outer core. From tracing the paths followed by seismic waves, geologists have deduced the dimensions of the Earth's mantle and core. For the most part, the paths curve only gradually, because the density and composition change only gradually. Sharp bends occur only where there is an abrupt change from one kind of material to another, such as at the boundary between the outer core and the mantle.

| Table 8-2 | The Earth's Internal Structure | | |
|---|---|---|---|
| Region | Depth below surface (km) | Distance from center (km) | Average density (kg/m$^3$) |
| Crust (solid) | 0–5 (under oceans) 0–35 (under continents) | | 3500 |
| Mantle (solid) | from bottom of crust to 2900 | 3500–6400 | 3500–5500 |
| Outer core (liquid) | 2900–5100 | 1300–3500 | 10,000–12,000 |
| Inner core (solid) | 5100–6400 | 0–1300 | 13,000 |

these minerals have melting points slightly over 1000°C. However, the melting point of a substance depends on the pressure to which it is subjected—the higher the pressure,

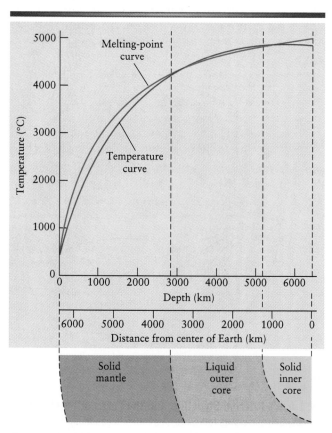

**figure 8-11**

**Temperature and Melting Point of Rock Inside the Earth**
The temperature (red curve) rises steadily from the Earth's surface to its center. The melting point of rock (blue curve) is also shown on this graph. Where the temperature is below the melting point, as in the mantle and inner core, the material is solid; where the temperature is above the melting point, as in the outer core, the material is liquid. (Adapted from F. Press and R. Siever)

the higher the melting point. As shown in Figure 8-11, the melting point of all the mantle's minerals is everywhere higher than the actual temperature in the mantle, so the mantle is primarily solid. The upper levels of the mantle are, however, able to flow slowly and are therefore referred to as being **plastic**.

It may seem strange to think of a plastic material as one that is able to flow. Normally we think of plastic objects as being hard and solid, like the plastic out of which audio and video cassettes are made. But to make these objects, the plastic material is heated so that it can flow into a mold, then cooled so that it solidifies. Strictly, the material is only "plastic" when it is able to flow.

At the boundary between the mantle and the outer core, there is an abrupt change in chemical composition, from iron-rich minerals to almost-pure iron with a small admixture of nickel. Because this iron-nickel material has a lower melting point than the iron-rich minerals, the melting-point curve in Figure 8-11 dips below the temperature curve as it crosses from the mantle to the outer core. The melting-point curve remains below the temperature curve down to a depth of about 5100 km (3200 mi). Hence, from depths of about 2900 to 5100 km, the core is liquid.

At depths greater than about 5100 km, the pressure is more than $10^{11}$ newtons per square meter. This is about $10^6$ times ordinary atmospheric pressure, or about $10^4$ tons per square inch. (The distinction between pressure and force, and the units used to measure pressure, are discussed in Box 8-1.) Because the melting point of the iron-nickel mixture under this pressure exceeds the actual temperature (see Figure 8-11), the Earth's inner core is solid.

Until the 1980s, geologists knew little more about the inner core than that it is solid and dense. Since then, evidence has accumulated that suggests the inner core is a single crystal of iron, just as a diamond in a wedding ring is a single crystal of carbon. If so, the inner core is the largest crystalline object anywhere in the solar system. Furthermore, there are strong indications that the inner core is rotating at a slightly faster rate than the rest of the Earth. Box 8-2 discusses these remarkable discoveries, which may transform our understanding of the Earth's interior.

## Box 8-1 | Looking Deeper into Astronomy

### *Atmospheric Pressure*

In everyday language, the words "pressure" and "force" mean pretty much the same thing. Think of two common phrases: "I'm under pressure to study for midterms," and "I'm forced to study for midterms." In science, however, these two words have different, but related, meanings.

*Force* is a push or pull. *Pressure* is force divided by the area over which the force acts. For a given force, the smaller the area, the greater the pressure. As an example, a sharp knife can cut through a raw potato easily, but a dull one probably cannot. In both cases you exert the same force on the knife handle, but with a sharp knife, the blade area that contacts the potato is much smaller, so the pressure on the potato is much greater.

Force is measured in newtons (or, in the English system, pounds). Pressure is measured in newtons per square meter ($N/m^2$) or, in the English system, pounds per square inch.

At sea level on Earth, the air weighs down with a pressure of 101,000 newtons per square meter (that is, $1.01 \times 10^5$ $N/m^2$), or 14.7 pounds per square inch. By definition, this pressure is called one atmosphere (1 atm). The various units of pressure are related to each other by

$$1 \text{ atm} = 14.7 \text{ lb/in.}^2 = 1.01 \times 10^5 \text{ N/m}^2$$

Because we are familiar with the pressure in our atmosphere, it is useful to also express the atmospheric pressure on other planets in (Earth) atmospheres. For example, the atmospheric pressure on Venus is 90 atm, while that on Mars is only 0.01 atm.

Scientists have defined another unit of pressure called a **bar**, which is precisely $10^5$ $N/m^2$. One bar is just slightly less than one atmosphere:

$$1 \text{ bar} = 0.987 \text{ atm}$$

As you go to higher altitudes in the Earth's atmosphere, the pressure decreases. In expressing low atmospheric pressures, it is often convenient to use the **millibar,** which is simply 0.001 bar. Thus, we say that the atmospheric pressure on Mars is about 10 millibars, approximately the same as the atmospheric pressure 30 km above the Earth's surface.

Just as pressure decreases with increasing altitude in the atmosphere, it increases with increasing depth below the surface of the Earth's oceans. You can feel this increased pressure on your skin and eardrums when you go swimming. The pressure increases by about 1 bar for every 10 meters below the surface of the water; at the average depth of the oceans, about 3 km, the pressure is 300 bars. Submersible vehicles designed to explore the ocean floor have to be made extraordinarily strong to withstand such pressures.

Pressures reach even larger values deep within the interior of the Earth. Although it cannot be measured directly, the pressure within the Earth can be inferred from the speed at which seismic waves travel through the Earth. The conclusion is that the pressure within the inner core is an astounding 3 *million* ($3 \times 10^6$) bars. Under normal atmospheric pressure, iron has a density of 7870 $kg/m^3$; in the inner core, the tremendous pressure compresses iron to almost twice the density, about 13,000 $kg/m^3$.

---

## 8-3 Plate tectonics produce earthquakes, mountain ranges, and volcanoes that shape the Earth's surface

One of the most important geological discoveries of the twentieth century is the realization that the Earth's crust is active and constantly changing. We have learned that the Earth's crust is divided into huge **plates** that constantly jostle each other, producing earthquakes, volcanoes, mountain ranges, and oceanic trenches.

Anyone who carefully examines a map of the Earth might come up with the idea of moving continents. South America, for example, would fit snugly against Africa were it not for the Atlantic Ocean. As Figure 8-12 shows, the fit between land masses on either side of the Atlantic Ocean is quite remarkable. This observation inspired the German meteorologist Alfred Wegener to advocate "continental drift"—the idea that the continents on either side of the Atlantic Ocean have simply drifted apart. After much research, in 1924 Wegener published the theory that there had originally been a single gigantic supercontinent, which he called Pangaea (meaning "all lands"), that began to break up and drift apart some 200 million years ago. Other geologists refined this theory,

## box 8-2 | Looking Deeper into Astronomy

### *The Surprising Inner Core of the Earth*

In the 1980s geologists studying seismic waves began to notice something quite surprising. When such waves pass through the Earth's inner core, their speed is 3 to 4% faster if they travel parallel to Earth's rotation axis (which runs from the north pole to the south pole) than if they travel parallel to the equator. Prior to this discovery, it was thought that the inner core must be *isotropic*—that is, the same in all directions—in which case the speed of seismic waves through the inner core would likewise be the same in all directions. The discovery that the inner core is *anisotropic* (has a preferred direction) came as a complete surprise.

The most promising explanation for this anisotropic character is that the inner core is made of a single crystal of iron. Iron crystals have a hexagonal structure, like the hexagonal cells of a beehive. An ordinary piece of iron is made of many tiny crystals oriented in different directions, so that the iron is isotropic. But if the tiny crystals are all aligned to form one large crystal, the piece of iron is anisotropic, with a crystal axis perpendicular to the plane of the hexagons. Seismic waves move faster along the axis of an iron crystal than perpendicular to it, so the crystal axis of the inner core is oriented roughly north to south.

The inner core probably acquired its crystal structure long ago, when it cooled and solidified. As it cooled, stresses caused by the rotation of the Earth would presumably have forced the crystal structure to align with the rotation axis. In fact, however, the inner core's crystal axis is tilted slightly away from the Earth's rotation axis (see part **a** of the accompanying figure). Careful measurements by Wei-jia Su and Adam Dziewonski of Harvard University show that the tilt may be as much as 10 degrees. To date there is no explanation for this tilt.

Another curious aspect of the inner core is its rotation. The surface of the Earth rotates from west to east once every 24 hours. To determine whether the inner core rotates as the same rate, Xiaodong Song and Paul Richards at the Lamont-Doherty Earth Observatory analyzed seismic data spanning almost 30 years. If the tilted inner core rotates at a different rate than the Earth's surface, then over time the crystal axis will change its orientation relative to the surface. If this happens, the direction in which seismic waves travel fastest through the core changes along with it (see part **b** of the figure). Song and Richards found that just such a change has taken place over the past few decades. It appears that the inner core spins faster than the Earth's surface by about a degree or two a year. In other words, more than 5000 km beneath your feet is an immense crystal the size of the Moon that is turning ever so slowly eastward.

The faster rotation of the inner core is thought to be caused by moving fluid in the outer core, which pushes the inner core along. Fluid motion in the outer core is also thought to be the cause of the Earth's magnetic field, as discussed in Box 8-3.

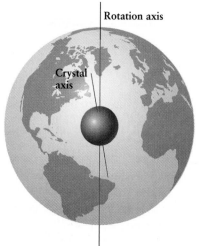

a

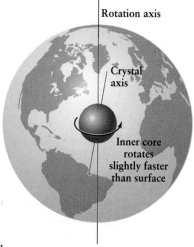

b

figure 8-12

**Comparing the Continents** Africa, Europe, Greenland, and North and South America fit together as though they had once been joined. The fit is especially convincing if the edges of the continental shelves (shown in blue) are used, rather than today's shoreline. This strongly suggests that these continents were in fact joined together at some point in the past. (Adapted from P. M. Hurley)

a  200 million years ago

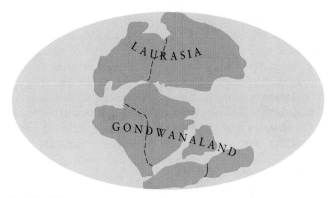

b  180 million years ago

c  Today

figure 8-13

**The Breakup of the Supercontinent Pangaea** **(a)** The shapes of the continents led Alfred Wegener to deduce that 200 million $(2 \times 10^8)$ years ago, the continents were merged into a single supercontinent, which he called Pangaea. **(b)** Pangaea first split into two smaller land masses, Laurasia and Gondwanaland. **(c)** Over the course of millions of years, the continents moved to their present-day locations. This picture of continental drift was considered to be a wild speculation in Wegener's day. Today, it has been confirmed by a variety of evidence. For example, nearly identical rock formations are found in locations that are thousands of miles apart today but would have been side by side on the ancient supercontinent Pangaea.

arguing that Pangaea must have first split into two smaller supercontinents, which they called Laurasia and Gondwanaland, separated by what they called the Tethys Sea. Gondwanaland later split into Africa and South America, with Laurasia dividing to become North America and Eurasia. According to this theory, the Mediterranean Sea is a surviving remnant of the ancient Tethys Sea (Figure 8-13).

Initially, most geologists ridiculed Wegener's ideas. Although it was generally accepted that the continents do "float" on the denser, somewhat plastic mantle beneath them, few geologists could accept the idea that entire continents could move around the Earth at speeds as great as several centimeters per year. The "continental drifters" could not explain what forces could be shoving the massive continents around.

Beginning in the mid-1950s, however, geologists found evidence that material is being forced upward to the crust from deep within the Earth. Bruce C. Heezen of Columbia

University and his colleagues began discovering long mountain ranges on the ocean floors, such as the Mid-Atlantic Ridge, which stretches all the way from Iceland to Antarctica (Figure 8-14). During the 1960s, Harry Hess of Princeton University and Robert Dietz of the University of Cali-fornia again carefully examined the floor of the Atlantic Ocean. They concluded that rock from the Earth's mantle is being melted and then forced upward along the Mid-Atlantic Ridge, which is in essence a long chain of underwater volcanoes. This helps to explain why the dominant type of rock found on the ocean floor is basalt, shown in Figure 8-5. Basalt is formed when molten rock cools rapidly, as would be expected for material from the mantle that was pushed upward into the cold ocean.

The upwelling of new material from the mantle to the crust is forcing the ocean floor to spread. In this **seafloor spreading**, the eastern floor of the Atlantic Ocean is moving eastward and the western floor is moving westward. By explaining what fills in the gap between continents as they move apart, seafloor spreading helps to fill out the theories of continental drift. Because of the seafloor spreading from the Mid-Atlantic Ridge, South America and Africa are moving apart at a speed of roughly 3 cm per year. Working backward, these two continents would have been next to each other some 200 million years ago—just as Wegener suggested (Figure 8-12).

In the early 1960s, geologists began to find additional evidence supporting the existence of large, moving plates. Thus was born the modern theory of crustal motion, which came to be known as **plate tectonics** (from the Greek *tekton*, meaning "builder"). Geologists today realize that earthquakes tend to occur at the boundaries of the Earth's crustal plates, where the plates are colliding, separating, or rubbing against each other. The boundaries of the plates therefore stand out clearly when the epicenters of earthquakes are plotted on a map (Figure 8-15). The vast majority of volcanoes also occur at plate boundaries. As an example, the eruption shown in the photograph that opens this chapter is at the boundary between the Pacific and Eurasian plates, near the upper left corner of Figure 8-15.

The flow of heat from the Earth's hot core outward to its cool crust is responsible for the movement of the plates. This heat transport is accomplished by **convection**, a process in which warm, low-density material rises upward and transfers heat to its surroundings. As a result, this material cools down and becomes denser. It then sinks downward to be heated again, and the process starts over. The overall pattern of motion is a circulating **convection current**.

**figure 8-14**

**The Mid-Atlantic Ridge** This artist's rendition shows the floor of the North Atlantic Ocean. The immense mountain range in the middle of the ocean floor is called the Mid-Atlantic Ridge. It is caused by lava seeping up from the Earth's interior along a rift that extends from Iceland to Antarctica. (Courtesy of M. Tharp and B. C. Heezen)

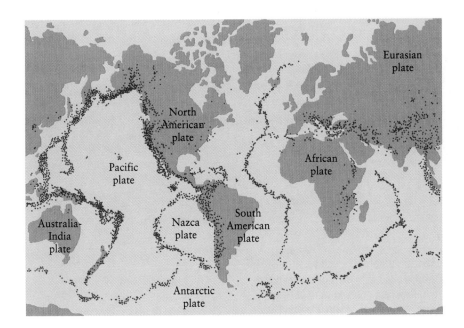

figure 8-15

**The Earth's Major Plates** The boundaries of the Earth's major plates are the scenes of violent seismic and geologic activity. Most earthquakes occur where plates separate, collide, or rub together. Plate boundaries are therefore easily identified simply by plotting earthquake epicenters (shown here as dots) on a map.

(You can see convection currents in action by boiling water on a stove, as shown in Figure 8-16.) The mantle is hot enough to permit an oozing, plastic flow throughout a soft region of the upper mantle called the **asthenosphere** (from the Greek *asthenia*, meaning "weak"). Atop the asthenosphere is a rigid layer, called the **lithosphere**, which is divided into plates that simply ride along the convection currents of the asthenosphere. The crust is simply the uppermost layer of the lithosphere.

Figure 8-17 shows how convection causes plate movement. Hot material (magma) seeps upward along **oceanic rifts**, where plates are separating. One such oceanic rift is the Mid-Atlantic Ridge, shown in Figure 8-14. Where plates collide, cool crustal material from one of the plates sinks back down into the mantle along a **subduction zone.** One such subduction zone is found along the west coast of South America, where the oceanic Nazca plate is being subducted into the mantle under the continental South American plate at a relatively speedy 10 centimeters per year. As the material from the subducted plate sinks, it pulls the rest of its plate along with it, thus helping to keep the plates in motion. New material is added to the crust from the mantle at the oceanic rifts, and is "recycled" back into the mantle at the subduction zones. In this way the total amount of crust remains essentially the same.

The boundaries between plates are the sites of some of the most impressive geological activity on our planet. Great mountain ranges, such as the Sierras and Cascades along the western coast of North America and the Andes along South America's west coast, are thrust up by ongoing collisions between continental plates and the plates of the ocean floor. Subduction zones, where old crust is pushed back down into the mantle, are typically the locations of deep oceanic trenches, such as the Peru-Chile Trench off the west coast of South America. Figures 8-18 and 8-19 show two

well-known geographic features that resulted from tectonic activity at plate boundaries.

In recent years, geologists have uncovered evidence that points to a whole succession of supercontinents that once broke apart and then reassembled. Pangaea is only the most recent supercontinent in this cycle, which repeats about every 500 million ($5 \times 10^8$) years. As a result, intense

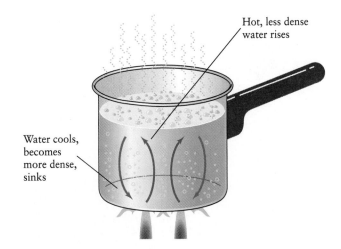

figure 8-16

**Convection in Boiling Water** When you heat a pot of water on the stove, the water at the bottom of the pot expands and its average density decreases. This heated water rises because of its decreased density, just as a balloon filled with helium (which is less dense than air) rises through the air. As it rises, the heated water transfers heat to its cooler surroundings, and in the process cools down and becomes denser. It then sinks back to the bottom of the pot to repeat the process. Convection also takes place in the plastic material of the Earth's upper mantle, but at a rate millions of times slower.

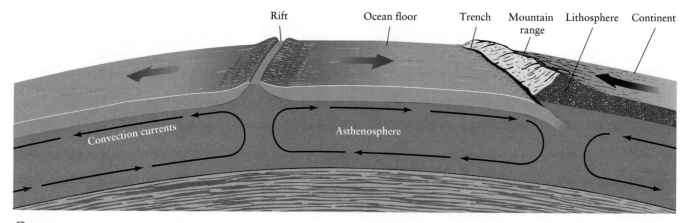

Rift · Ocean floor · Trench · Mountain range · Lithosphere · Continent

Convection currents · Asthenosphere

### figure 8-17

**The Mechanism of Plate Tectonics** Convection currents in the soft upper layer of the mantle (called the asthenosphere) are responsible for pushing around rigid, low-density crustal plates (called the lithosphere). New crust forms in oceanic rifts, where lava oozes upward between separating plates. Mountain ranges and deep oceanic trenches are formed where plates collide.

episodes of mountain building have occurred at roughly 500-million-year intervals. (In Chapter 9 we will see how scientists can determine the ages of rocks and thus put time scales on geological processes.)

Apparently, a supercontinent sows the seeds of its own destruction because it blocks the flow of heat from the Earth's interior. As soon as a supercontinent forms, temperatures beneath it rise, much as they do under a book lying on an electric blanket. As heat accumulates, the lithosphere

domes upward and cracks. Molten rock from the overheated asthenosphere wells up to fill the resulting fractures, which continue to widen as pieces of the fragmenting supercontinent move apart.

The convection currents associated with seafloor spreading (shown in Figure 8-17) carry heat outward from the Earth's interior to the crust. Here our planet differs radically from its neighbors. Venus and Mars do not have long mountain chains resembling the Earth's Mid-Atlantic Ridge and,

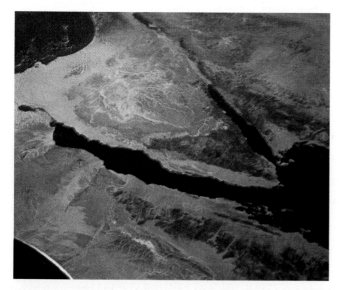

### figure 8-18

**The Separation of Two Plates** The plates that carry Egypt and Saudi Arabia are moving apart, leaving the trench that contains the Red Sea. This view, taken by astronauts in 1966, shows the northern Red Sea. Egypt is on the left, Saudi Arabia is on the right, and the Sinai Peninsula dominates the center of this view. (NASA)

### figure 8-19

**The Collision of Two Plates** The plates that carry India and China are colliding. As a result, the Himalayas have been thrust upward. In this photograph taken by astronauts in 1968, India is on the left and Tibet on the right; Mount Everest is one of the snow-covered peaks near the center. (NASA)

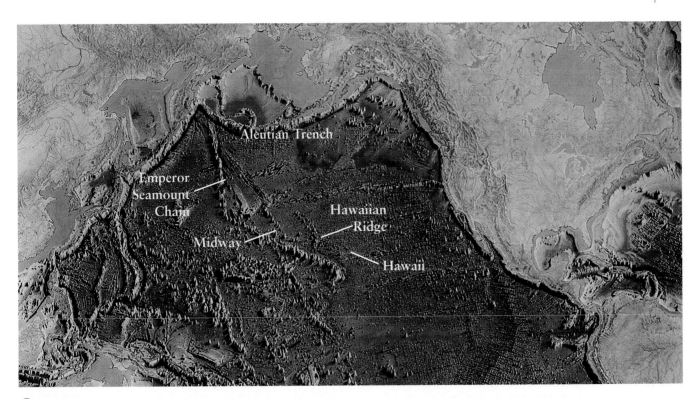

## ƒigure 8-20

**Hot-Spot Volcanism and the Hawaiian Islands**  For at least 70 million years, there has been a hot spot underneath the Pacific plate. The hot spot has remained essentially stationary, while the plate has moved to the northwest some 6000 km (4000 mi). Material upwelling from the hot spot has thus produced a long chain of volcanoes. The oldest volcanoes, at the northwest end of the chain, have been so eroded by wind and water that they no longer rise above the surface of the ocean. The young volcanoes on the island of Hawaii, at the southeast end of the chain are among the largest and most active on Earth. The hot spot is presently building a new volcano, called Loihi, to the south of the Big Island. When it grows tall enough to poke above the ocean surface, Loihi will become the youngest member of the Hawaiian chain of islands (and, no doubt, a tourist destination).

therefore, show no evidence of processes like seafloor spreading. This strongly suggests that there are no convection currents in the mantles of Venus or Mars, and hence the sort of plate tectonic activity that occurs on Earth does not happen on its sister planets.

It appears that on Venus and Mars heat is carried from the interior to the surface by a different mechanism, called **hot-spot volcanism.** Hot spots deep in the mantles of Venus and Mars squirt molten lava up through the crust. In the absence of any tectonic activity to move the crust around, millions of years of eruptions eventually build enormous volcanoes. Indeed, both Venus and Mars have huge volcanoes. Olympus Mons, the largest volcano on Mars, rises to an altitude of 24 km (15 mi) above the surrounding plains—nearly three times taller than Mount Everest. Similarly, the summit of Maxwell Montes on Venus attains an altitude of 11 km (7 mi).

Hot-spot volcanism also occurs on Earth. The Hawaiian Islands, which are in the middle of the Pacific plate, are a fine example. One hot spot in the Earth's mantle, beneath Hawaii, continuously pumps lava up through the crust. Unlike on Venus or Mars, however, the overlying crust moves; the Pacific plate is moving northwest at the rate of several centimeters per year. New volcanoes are thus being created over the hot spot, while older volcanoes move away from the magma source, become extinct, and eventually erode and disappear beneath the ocean. The Hawaiian Islands are the most recent additions to a long chain of extinct volcanoes that stretches all the way across the northern Pacific Ocean (Figure 8-20). This chain is a record of the movements of the Pacific plate over the past 70 million years.

The changes wrought by plate tectonics are very slow on the scale of a human lifetime, but they are very rapid in comparison to the age of the Earth. For example, the period over which the Hawaiian Islands formed is only about 0.1 percent of the Earth's age of $4.6 \times 10^9$ years. (To put this in perspective, 0.1 percent of your current age is about a week or two.) The lesson of plate tectonics is that the seemingly permanent face of the Earth is in fact dynamic and ever-changing.

## 8-4 Absorption of sunlight and the Earth's rotation govern the behavior of our nitrogen-oxygen atmosphere

An alien spacecraft exploring the inner solar system would find that the Earth differs from Venus and Mars in the type of geologic activity that occurs in the crust. The aliens would also find the Earth to be strikingly different from its neighbors in yet another way—the chemical composition of its atmosphere. The atmospheres of both Venus and Mars consist almost entirely of carbon dioxide, whereas carbon dioxide accounts for only 0.03% of the Earth's atmosphere. Our atmosphere is predominantly a 4-to-1 mixture of nitrogen and oxygen, two gases that are found in only small amounts on Venus and Mars (Table 8-3). Carbon dioxide is also abundant on Earth, but not in the atmosphere. Instead, it is dissolved in the oceans and chemically bound into carbonate rocks, such as limestone and marble.

As far as we know, large amounts of oxygen in a planetary atmosphere can arise only as a direct result of biological activity. Plants, for example, get energy from sunlight in a chemical process called photosynthesis that converts carbon dioxide into oxygen. From the study of geological records, we know that living organisms have been active on Earth for at least the past 3.8 billion ($3.8 \times 10^9$) years, long enough to account for the chemical composition of the Earth's atmosphere today.

Like the Earth's interior, its atmosphere has a layered structure. From bottom to top, these layers are called the *troposphere*, the *stratosphere*, the *mesosphere*, and the *thermosphere* (Figure 8-21). To understand these layers, we need to measure how two properties of the atmosphere, temperature and atmospheric pressure, vary with altitude.

Because of gravity, the air weighs down upon the Earth, producing **atmospheric pressure**. At sea level, the average atmospheric pressure is $1.01 \times 10^5$ N/m² or, in English units, 14.7 pounds per square inch. This pressure is defined to be 1 atmosphere, or 1 atm. (Box 8-1 describes in more detail the units used to measure pressure, as well as the distinction between *pressure* and *force*.) As you go up in the atmosphere, there is less air above you to weigh down on you. Hence, the pressure decreases smoothly with increasing altitude, falling by roughly half with every 5500 meters (18,000 feet).

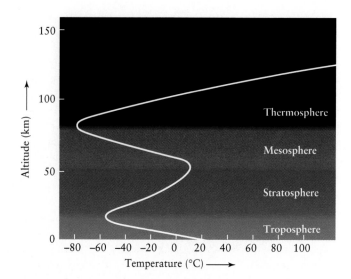

**figure 8-21**

**Temperature Profile of the Earth's Atmosphere** This graph shows how the temperature in the Earth's atmosphere varies with altitude. This variation is caused by differences in the way that sunlight interacts with atmospheric atoms and ions at different altitudes. In the troposphere and mesosphere, temperature decreases with increasing altitude; in the stratosphere and thermosphere, temperature actually increases with increasing altitude.

The variation of temperature within the atmosphere is more complex than that of pressure. Figure 8-21 shows that temperature decreases with altitude in some layers of the Earth's atmosphere. But in other layers, the temperature actually increases with altitude.

The lowest layer of the atmosphere, called the **troposphere**, extends from the surface to an average altitude of 12 km (roughly 7.5 mi, or 39,000 ft) above sea level. (The height of the troposphere is higher near the equator but lower at the poles.) This layer includes some 80% of the total mass of Earth's atmosphere. The temperature decreases with altitude within the troposphere, dropping to about –60°C at an altitude of 11 km. Thanks to this temperature variation, it is possible to have snow on top of Washington's Mount Rainier

| table 8-3 | **Chemical Compositions of Three Planetary Atmospheres** | | |
| --- | --- | --- | --- |
| | **Venus** | **Earth** | **Mars** |
| Nitrogen ($N_2$) | 4% | 77% | 3% |
| Oxygen ($O_2$) | Almost zero | 21% | Almost zero |
| Carbon dioxide ($CO_2$) | 96% | Almost zero | 95% |
| Other gases | Almost zero | 2% | 2% |

(elevation 4392 meters, or 14,410 ft) at the same time that the nearby city of Seattle (at sea level) is enjoying warm summer temperatures.

The temperature variation in the troposphere is caused by solar heating. Sunlight warms the Earth's surface, which in turn heats the low-lying, relatively dense air at the bottom of the troposphere. Air is a rather poor conductor of heat, so the warmth of the surface cannot effectively be transferred to higher altitudes. Thus, the lower part of the troposphere is warmer than the upper part.

Just as in a pot of water on the stove (Figure 8-16) or in the asthenosphere beneath the Earth's crust (Figure 8-17), the vertical temperature variation in the troposphere causes convection currents that move up and down. As moist air rises, it encounters lower temperatures and thus the water vapor can condense into droplets or ice crystals. In this way clouds are formed. If the moisture content is high enough, rain or snow falls from the clouds. Indeed, *all* of the Earth's weather is a result of convection in the troposphere, which in turn is driven by energy absorbed from sunlight by the Earth's surface.

Temperature variations within the troposphere, as well as the variation between the warm temperatures at the equator and the much cooler temperatures at the poles, are also responsible for the global circulation patterns of the atmosphere. If the Earth did not rotate, heated air would rise in the equatorial regions, flow toward the polar regions where it would cool and sink to lower altitudes, and then flow back toward the equator. Indeed, as we will see in Chapter 11, Venus's atmosphere exhibits this simple planetwide convection pattern because Venus rotates very slowly. However, the Earth's comparatively fast rotation breaks up the convection pattern into a series of *convection cells* in which air moves east and west as well as vertically and in a north-south direction (Figure 8-22). In Chapter 13 we will see a similar but even more complex structure in the atmosphere of the rapidly rotating planet Jupiter.

Above the troposphere is the region called the **stratosphere**, which extends from about 12 to 50 km (about 7.5 to 31 mi) above the Earth's surface. The key difference between the stratosphere and troposphere is the presence of ozone. An **ozone** molecule, $O_3$, is made of three oxygen atoms. (By contrast, molecules of ordinary oxygen, or $O_2$, are made of only two oxygen atoms.) Ozone molecules in the stratosphere are very efficient at absorbing solar ultraviolet rays. This causes enough heating of the stratosphere that temperature actually *increases* as you move upward in this layer, reaching a value of 0°C (= 32°F = 273 K) or higher at its top. Convection requires that the temperature decrease, not increase, with increasing altitude, so there are essentially no convection currents in the stratosphere. Jet airliners regularly cruise within the lower part of the stratosphere, in part because

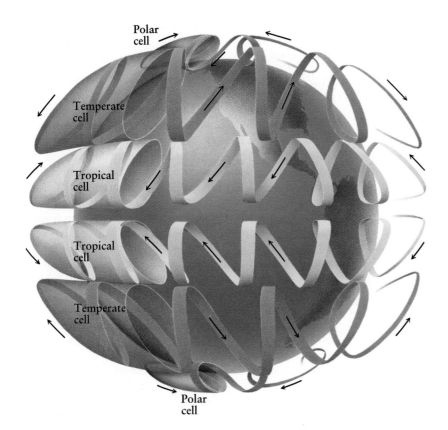

Polar cell

Temperate cell

Tropical cell

Tropical cell

Temperate cell

Polar cell

**ƒigure 8-22**

**Circulation Patterns in the Earth's Atmosphere** The dominant circulation in the Earth's atmosphere consists of three convection cells in the northern hemisphere and three cells in the southern hemisphere. The Earth's rotation deflects winds in the convection cells from a purely north-south flow. Thus, at ground level in the northern temperate region (for example, the continental United States), the prevailing winds—called "westerlies"—are from the southwest toward the northeast. Further south, within the northern tropical region (for example, Hawaii), the prevailing winds—called "trade winds"—are from northeast to southwest.

the lack of vertical atmospheric motion gives passengers a smoother ride.

Above the stratosphere lies the **mesosphere.** Very little ozone is found here, so solar ultraviolet radiation is not absorbed within the mesosphere, and atmospheric temperature again declines with increasing altitude. The temperature of the mesosphere reaches a minimum of about −75°C (= −103°F = 198 K) at an altitude of about 80 km (50 mi). This minimum marks the bottom of the **thermosphere,** in which temperature once again rises with increasing altitude. This is not, however, due to the presence of ozone molecules, because in this very low-density region oxygen and nitrogen are found in the form of individual atoms rather than molecules. Instead, the thermosphere is heated because these isolated atoms absorb very short-wavelength ultraviolet light from the Sun (which oxygen and nitrogen molecules cannot).

CAUTION! Figure 8-21 shows that temperatures in the thermosphere exceed those at the Earth's surface. Indeed, at altitudes near 300 km (190 mi) the temperature is near 1000°C (about 1800°F, or 1300 K). This is around the altitude at which the space shuttle and satellites orbit the Earth. Despite the high temperatures, however, a satellite in orbit does *not* risk being burned up as it moves through the thermosphere. The reason is that the thermosphere is very thin; the pressure at 300 km altitude is only $10^{-11}$ as much as at sea level. In a hot gas, the individual atoms or molecules of the gas are moving rapidly and have lots of *kinetic energy,* or energy of motion (see Box 7-2). When the atoms or molecules collide with an object, they transfer some of their kinetic energy to the object and make the object warmer. But the atoms in the thermosphere are so few and far between that only a very few of them hit an orbiting satellite; thus, they cause no noticeable warming. Instead, nearly all the heat that an orbiting satellite receives is from the sunlight that it absorbs.

While the thermosphere is very thin, it is nonetheless of importance to spacecraft. If a satellite is in orbit at a relatively low altitude of about 200 km, there is enough atmosphere to create a measurable amount of air resistance to the satellite's motion. Eventually, the cumulative effects of this tiny amount of air resistance cause satellites in relatively low orbits to reenter Earth's lower atmosphere. They burn up as they reenter, not because the atmosphere is hot but because of heating caused by air friction.

There is no definite upper limit to the thermosphere. At sufficiently high altitudes, it merges into the very thin gas between the planets, called the interplanetary medium.

The uppermost layers of the atmosphere play an important role in radio transmission. When an atom in the thermosphere absorbs an ultraviolet photon, it can lose an electron (see Box 5-5). An atom that carries an excess charge because it has gained or lost one or more electrons is called an **ion.** The ions in the thermosphere have a positive electric charge, because they have lost negatively charged electrons. The result is a layer of electrically charged particles, called the **ionosphere,** in which there are many negative electrons and positive ions. This layer extends from about 80 to 400 kilometers, and thus lies within the lower levels of the thermosphere. The electrons and ions in the ionosphere reflect radio waves over a wide range of frequencies. When you tune in a distant AM or short-wave radio station, you are receiving transmissions that bounced off the ionosphere and were reflected down to your radio's antenna. This reflection effect does not work for FM transmissions, which are at higher frequencies that pass easily through the ionosphere into space.

## 8-5 The Earth's magnetic field produces a magnetosphere that traps particles from the solar wind

As almost anyone with a compass knows, the Earth has a magnetic field. Magnetism arises whenever electrically charged particles are in motion. For example, a loop of wire carrying an electric current generates a magnetic field in the space around it. The magnetic field that surrounds an ordinary bar magnet (Figure 8-23a) is created by the motions of negatively charged electrons within the iron atoms of which the magnet is made. The Earth's magnetic field is much like that of a bar magnet, as shown in Figure 8-23b. Most geologists suspect that this magnetic field is caused by electric currents flowing in the liquid portions of the Earth's iron core. As our planet rotates, these currents produce a magnetic field that dominates space for tens of thousands of kilometers in all directions. Box 8-3 discusses recent attempts using supercomputers to simulate how the Earth's magnetic field is generated.

CAUTION! While the Earth's magnetic field is similar to that of a giant bar magnet, you should not take this picture too literally. The Earth is *not* simply a magnetized ball of iron. In an iron bar magnet, the electrons of different atoms orbit their nuclei in the same general direction, so that the magnetic fields generated by individual atoms add together to form a single, strong field. But at temperatures above 770°C (1418°F = 1043 K), the orientations of the electron orbits become randomized. The fields of individual atoms tend to cancel each other out, and the iron loses its magnetism. Figure 8-11 shows that almost all of the body of Earth is hotter than 770°C, so the iron in Earth's interior cannot be extensively magnetized. The correct picture is thought to be that the iron carries electric currents, and it is these currents that create the Earth's magnetic field.

The Earth's magnetic field interacts dramatically with charged particles from the Sun, called the **solar wind.** The solar wind is a flow of mostly protons and electrons that streams constantly outward from the Sun's upper atmosphere. Near the Earth, the particles in the solar wind move at speeds of roughly 400 km/s, or nearly a million miles per hour. Because this is considerably faster than sound waves

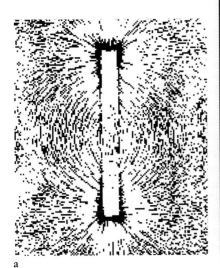

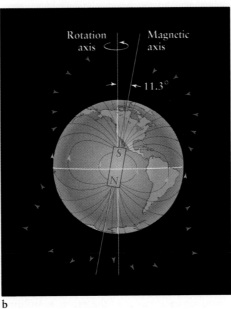

a  b

**figure 8-23**

**The Magnetic Fields of a Bar Magnet and of the Earth** **(a)** The magnetic field of a bar magnet is normally invisible. Here it is revealed by iron filings on a piece of paper around the magnet. Magnetic field lines seem to "stream" from one end of the magnet to the other. **(b)** The magnetic field of the Earth is created by the motion of electric currents in the liquid outer core. But the effect is much the same as if there were a giant bar magnet inside the Earth, tilted by a small angle (11.3°) from the Earth's rotation axis. Because of this tilt, the north and south magnetic poles (where the magnetic axis passes through the Earth's surface) are not at the same locations as the true, or geographic, north and south poles (where the rotation axis passes through the surface). A compass needle points toward the north magnetic pole, not the true north pole.

can travel in the very thin gas between the planets, the solar wind is said to be *supersonic*. (Because the gas between the planets is so thin, interplanetary sound waves carry too little energy to be heard by astronauts.)

When a planet, like the Earth, possesses a magnetic field, this field deflects the charged particles, thereby forming an elongated "cavity" in the solar wind. This cavity is called the

planet's **magnetosphere.** Figure 8-24 is a scale drawing of the Earth's magnetosphere, which was discovered in the late 1950s by the first satellites placed in orbit. When the supersonic particles in the solar wind first encounter the Earth's magnetic field, they abruptly slow to subsonic speeds. The boundary where this sudden decrease in velocity occurs is called a **shock wave.** Still closer to the Earth there is another

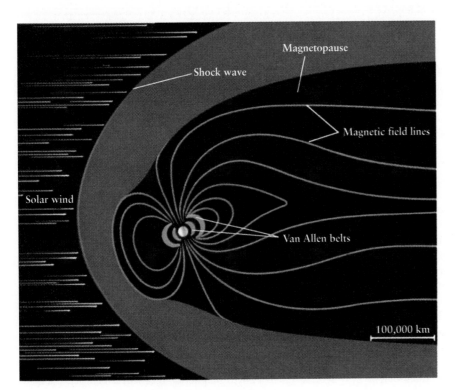

**figure 8-24**

**The Earth's Magnetosphere** The Earth's magnetic field carves out a cavity in the solar wind, shown here in cross-section. A shock wave marks the boundary where the supersonic solar wind is abruptly slowed to subsonic speeds. Most of the particles of the solar wind are deflected around the Earth in a turbulent region (colored purple in this drawing). Because of the strength of the Earth's magnetic field, our planet can trap charged particles in two huge, doughnut-shaped rings called the Van Allen belts. (This figure shows only a slice through the Van Allen belts.)

## box 8-3 | Looking Deeper into Astronomy

### *The Origin of the Earth's Magnetic Field*

Exactly how the Earth's magnetic field is generated has been a long-standing puzzle in science. Since the 1950s, there has been a general consensus that the field is caused by electric currents in the outer core, and that these currents are driven by convection of the liquid that makes up the outer core. Until recently, however, no one was able to show convincingly that this idea really could work.

Since 1995, Gary Glatzmaier at the Los Alamos National Laboratory and Paul Roberts of UCLA have been using a supercomputer to simulate the motions of the outer core and the magnetic field that it produces. They have simulated more than 300,000 years of these motions in several months of supercomputer time. Glatzmaier and Roberts have found that their simulated magnetic field is quite stable over time, just like the field of the Earth, and has the same kind of shape as the Earth's field. The accompanying figure shows the magnetic field lines at one moment during the simulation. The lines are colored gold where the field is directed outward from the Earth and blue where they are directed back into the Earth. This similarity to the Earth's actual field is compelling evidence that this simulation is on the right track.

One of the most striking results of Glatzmaier and Roberts' simulation was seen at a simulated elapsed time of 38,000 years. At that point, the north and south magnetic poles of the simulated Earth spontaneously changed places, and the direction of the magnetic field reversed itself. What makes this so impressive is that the real Earth's magnetic field also reverses direction. This can be deduced by measuring the magnetism of lava beds. When lava cools, it becomes magnetized in the same direction as the Earth's magnetic field. Measurements from around the world show that million-year-old lava beds are magnetized in the reverse direction from lava that has solidified recently; therefore, a million years ago a compass needle would have pointed south rather than north. By contrast, lava that is 2.5 million years old is magnetized in the same direction as recent lava. Observations like these show that the Earth's magnetic field reverses about every half-million years or so. The simulation of Glatzmaier and Roberts was the first to show this behavior.

The simulation also suggested that moving fluid in the outer core might drag on the solid inner core, thus speeding up its rotation. After learning of this prediction, Xiaodong Song and Paul Richards were motivated to search for and

---

boundary, called the **magnetopause,** where the outward magnetic pressure of the Earth's field is exactly counterbalanced by the impinging pressure of the solar wind. Most of the particles of the solar wind are deflected around the magnetopause, just as water is deflected to either side of the bow of a ship

Some charged particles of the solar wind manage to leak through the magnetopause. When they do, they are trapped by the Earth's magnetic field in two huge, doughnut-shaped rings around Earth called the **Van Allen belts.** These belts were discovered in 1958 during the flight of the United States' first successful Earth-orbiting satellite. They are named after the physicist James Van Allen, who insisted that the satellite carry a Geiger counter to detect charged particles. The inner Van Allen belt, which extends over altitudes of about 2000 to 5000 km, contains mostly protons. The outer Van Allen belt, about 6000 km thick, is centered at an altitude of about 16,000 km above the Earth's surface and contains mostly electrons. These charged particles all came from the solar wind.

The solar wind is an excellent electrical conductor. If a loop of electrically conducting wire moves through a mag-

netic field, an electric current begins to flow in the loop. This principle is used in electric generators, such as those at a hydroelectric power plant. At such a plant, the force of falling water is used to make a large coil turn through a magnetic field. This generates a substantial electric current, which is sent over transmission lines to homes and businesses. In the same way, as the solar wind sweeps past the Earth's magnetic field, it interacts with the entire magnetosphere to form a gigantic natural generator that produces enormous electrical currents of high-speed electrons and protons. These currents flow along magnetic field lines toward the north and south magnetic poles. When the electrons and protons enter the Earth's upper atmosphere, they collide with atoms and molecules of atmospheric gas, which excites the atoms and molecules to high energy levels. The atoms and molecules then release this energy as visible light, like the excited gas atoms in a neon light. The result is a beautiful, shimmering display called the **northern lights (aurora borealis)** or **southern lights (aurora australis),** depending on the hemisphere in which the phenomenon is observed. Figure 8-25 shows a glowing ring around Earth's north magnetic pole, where charged particles streaming down onto the atmosphere continually produce

eventually discover evidence that the Earth's inner core rotates faster than the surface (see Box 8-2).

Scientists are planning simulations that will be even more refined than those of Glatzmaier and Roberts. By running these simulations for a million years or more of simulated time, they hope to test as many aspects of the model as possible and to verify how the Earth's magnetic field is generated.

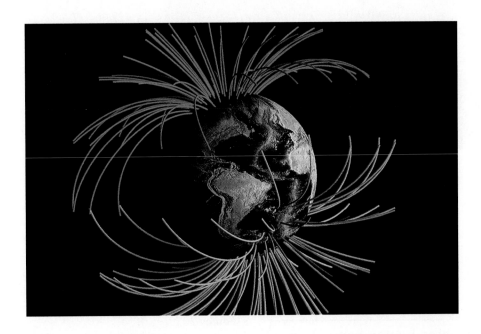

aurorae. As we will see in Chapters 13 and 15, the same effect is seen near the magnetic poles of Jupiter and Saturn.

Occasionally a violent event on the Sun's surface called a **solar flare** sends a burst of protons and electrons toward Earth. The resulting auroral display can be exceptionally bright (Figure 8-26) and can often be seen over a wide range of latitudes.

It is remarkable that the Earth's magnetosphere, including its vast belts of charged particles, was entirely unknown

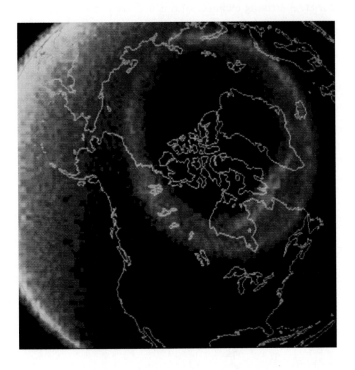

### ﬁgure 8-25 R I V **U** X G

**Aurorae Around the North Magnetic Pole** This false-color image was taken at ultraviolet wavelengths by the *Dynamics Explorer 1* spacecraft. Coastlines are overlaid in green. The glowing oval of auroral emission, about 4500 km in diameter, is predominantly from oxygen atoms and molecular nitrogen. These emissions are comparable in brightness to those from the sunlit hemisphere of Earth, shown at the far left of this image. (Courtesy of L. A. Frank and J. D. Craven, The University of Iowa)

**Figure 8-26**

**The Northern Lights (Aurora Borealis)**
A deluge of protons and electrons from a solar flare can produce aurorae that can be seen over a wide range of latitudes. Aurorae typically occur 100 to 400 km above the Earth's surface. (Courtesy of S.-I. Akasofu, Geophysical Institute, University of Alaska)

until a few decades ago. Such discoveries remind us of how little we truly understand and how much remains to be learned even about our own planet.

## 8-6 A burgeoning human population is profoundly altering the Earth's biosphere

One of the extraordinary characteristics of the Earth is that it is covered with life. Living organisms thrive in a wide range of environments, from ocean floors to mountaintops and from frigid polar caps to blistering deserts.

Living organisms subsist in a relatively thin layer enveloping the Earth called the **biosphere,** which includes the oceans, the lowest few kilometers of the troposphere, and the crust to a depth of almost 3 kilometers. A portrait of the Earth's biosphere, based on NASA satellite data, is shown in Figure 8-27. Ocean colors correspond to concentrations of microscopic, free-floating plants called phytoplankton. Land colors portray vegetation. The biosphere, which has taken billions of years to evolve to its present state, is a delicate, highly complex system in which plants and animals depend on each other for their mutual survival.

In recent years a skyrocketing human population has begun to alter the biosphere dramatically and irreversibly. Figure 8-28 shows the sharp rise in the human population that began in the late 1700s with the Industrial Revolution and the spread of modern ideas about hygiene. In the twentieth century, medical and technological advances ranging from antibiotics to high-yield grains (the so-called Green Revolu-

tion) fueled an unprecedented increase in the population. In 1960 there were 3 billion people on Earth, and in 1975, 4 billion. The 5-billion mark was passed in 1986, and, according to the United Nations Population Fund, there will be 6.2 billion people on Earth by the year 2000 and 10 billion by 2050.

Every human being has basic requirements: food, clothing, and housing. We all need fuel for cooking and heating. To meet these demands, we cut down forests, cultivate grasslands, and build sprawling cities, replete with roads and factories. A striking example of this activity is occurring in the Amazon rain forest. Tropical rain forests are vital to our planet's ecology because they absorb significant amounts of carbon dioxide and release oxygen through photosynthesis. Although rain forests occupy only 7% of the world's land areas, they contain at least 50% of all species of plants and animals on our planet. Nevertheless, to make way for farms and grazing land, people simply cut down the trees and set them on fire—a process called slash-and-burn. Figure 8-29 shows part of the Brazilian state of Rondônia, photographed from the space shuttle. So many trees have been felled that the forest now covers less than half the land. Similar deforestation, along with extensive lumbering operations, is occurring in Malaysia, Indonesia, and Papua New Guinea. The rain forests that once thrived in Central America, India, and the western coast of Africa are almost gone.

All this activity is adding carbon dioxide ($CO_2$) to the Earth's atmosphere faster than plants can extract it. Each year we put about 20 billion tons of $CO_2$ into the air we breathe. Figure 8-30 shows how the carbon dioxide content of the atmosphere increased between 1958 and 1995.

Carbon dioxide is transparent to visible light but not to infrared radiation. Consequently, sunlight has no trouble en-

## ƒigure 8-27

**The Earth's Biosphere** This computer-generated picture shows the distribution of plants over the Earth's surface. Ocean colors in the order of the rainbow correspond to phytoplankton concentrations, with red and orange for high productivity to blue and purple for low. Land colors designate vegetation: dark green for the rain forests, light green and gold for savannas and farmland, and yellow for the deserts. (NASA)

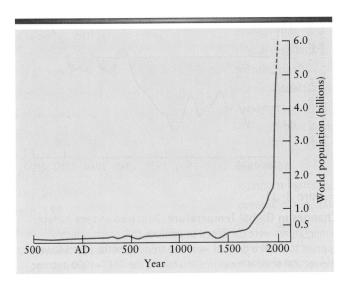

tering our atmosphere and warming the ground, which then radiates at infrared wavelengths. Carbon dioxide in the air absorbs this infrared radiation, thus warming the Earth's atmosphere. The trapping of energy from sunlight in the atmosphere is called the **greenhouse effect.**

A mild, naturally occurring greenhouse effect has warmed the Earth's surface for billions of years. A variety of gases,

## ƒigure 8-28

**The Human Population** Data and estimates from the U.S. Bureau of the Census, the Population Reference Bureau, and the United Nations Population Fund were combined to produce this graph showing the human population over the past 2500 years. Note how the population began to rise in the eighteenth century and the astonishing population increase in the twentieth century.

kinds of topography (for example, mountains, plains, seashore) dominate the geography of your hometown area? Do the topography and the frequency of earthquakes seem to be consistent with your hometown's proximity to a plate boundary?

## Where to Learn More

*Books and magazine articles*

Akasofu, S.-I. "The Dynamic Aurorae." *Scientific American,* May 1989. This detailed article by a renowned expert on aurorae describes how interactions of the Earth's magnetic field and the solar wind give rise to a vast generator that powers the aurorae.

Apt, J., ed. *Orbit: NASA Astronauts' Photographs of the Earth.* Random House, 1996. A beautiful book that contains many of the best photographs taken from orbit.

Bolt, B. A. *Earthquakes and Geological Discovery.* Scientific American Library, 1993. An excellent survey of how earthquakes occur, what they teach us about the Earth's interior, and how it may be possible to predict forthcoming earthquakes.

Clark, W. C. "Managing Planet Earth." *Scientific American,* September 1989. This leadoff article to a single-topic issue of *Scientific American* explores such questions as what kind of a planet do we want and what kind of a planet can we get.

Evans, D. L., Stofan, E. R., Jones, T. D., and Godwin, L. M. "Earth from Sky." *Scientific American,* December 1994. A special radar antenna carried on board the space shuttle has given scientists a new tool to observe Earth. This article summarizes the dramatic new insights obtained in this way.

Frederickson, J. K., and Onstott, T. C. "Microbes Deep Inside the Earth." *Scientific American,* October 1996. Until recently, it was thought that no single-cell organisms lived more than a few meters below the surface. This article describes how bacteria have been found living at depths of nearly 3 kilometers.

Karl, T. R., Nicholls, N., and Gregory, J. "The Coming Climate." *Scientific American,* May 1997. Three experts on climate describe how the Earth's weather patterns are expected to change over the next century as humans and their mechanical devices add more carbon dioxide to the atmosphere.

Murphy, J. B., and Nance, R. D. "Mountain Building and the Supercontinent Cycle." *Scientific American,* April 1992. This fascinating article describes the evidence for a 500-million-year cycle of the formation and breakup of supercontinents.

Press, F., and Siever, R. *Understanding Earth.* 2nd ed. W. H. Freeman, 1998. This comprehensive and clearly written text provides an introduction to all facets of geology.

Ruckelshaus, W. D. "Toward a Sustainable World." *Scientific American,* September 1989. This visionary article argues that moving people and nations toward a sustainable world will involve changes in society comparable in magnitude to the agricultural and industrial revolutions.

Taylor, S. R., and McLennan, S. M. "The Evolution of Continental Crust." *Scientific American,* January 1996. An up-to-date article on the processes by which continents are formed and evolve.

W *World Wide Web*

There is a vast amount of geological information available on the World Wide Web, with new items being added all the time. One good staring point for exploration is the Geology-Link site (**http://www.geologylink.com/**), which has a number of links to the latest developments and to sites specially designed for students. Space shuttle astronauts have taken a tremendous number of photographs of the Earth's surface from orbit. You can search an extensive collection of these at the San Diego Supercomputer Center (**http://earthrise.sdsc.edu/**).

An excellent introduction to plate tectonics, *This Dynamic Earth: The Story of Plate Tectonics* by W. J. Kious and R. I. Tilling, is available on the Web from the U. S. Geological Survey (**http://pubs.usgs.gov/publications/text/dynamic.html**). The text and graphics explain the history of plate tectonics (and the colorful characters behind the theory), as well as the latest research. USGS is also the place to find the most up-to-date earthquake information (**http://quake.usgs.gov/**).

An important mission of USGS is to maintain a network of volcano observatories at geologically active sites within the United States. From their volcano web page (**http://www.usgs.gov/themes/volcano.html**) you can get up-to-the-minute updates on volcanic activity. "The Electronic Volcano" web site at Dartmouth College (**http://www.dartmouth.edu/~volcano/**) is another excellent starting point for volcano information.

Current information about the growth of population in North America and the world can be found at the web sites for the U.S. Census Bureau (**http://www.census.gov/**) and Statistics Canada (**http://www.statcan.ca/**).

A web site at NASA's Goddard Space Flight Center (**http://pao.gsfc.nasa.gov/gsfc/earth/earth.htm**) is a good starting point for exploring the many ways that satellites monitor Earth's oceans, atmosphere, and biosphere. GSFC is also the headquarters for NASA's monitoring of ozone levels (**http://jwocky.gsfc.nasa.gov/index.html**).

Still more links to web sites about Earth can be found at the web sites "The Nine Planets" (**http://seds.lpl.arizona.edu/billa/tnp/**) and "Views of the Solar System" (**http://www.hawastsoc.org/solar/homepage.htm**).

# Our Barren Moon

**An Astronaut on the Moon**
The Moon is a desolate, lifeless world. This view of the lunar surface shows *Apollo 17* astronaut Harrison Schmitt near a large rock. Because the Moon has no atmosphere, lunar rocks have not been subjected to weathering and thus preserve information about the early history of our solar system. American astronauts brought back a total of 382 kg (842 lb) of moon rocks from six manned lunar landings between 1969 and 1972. (NASA)

*In this chapter you will find the answers to the following questions:*

9-1 Is the Moon completely covered with craters?

9-2 Has there been any exploration of the Moon since the Apollo program in the 1970s?

9-3 Does the Moon's interior have a similar structure to the interior of the Earth?

9-4 What keeps the same face of the Moon always pointed toward the the Earth?

9-5 How do Moon rocks compare to rocks found on the Earth?

9-6 How did the Moon form?

As we move beyond the Earth to explore other worlds, the first one we discover is our own Moon. The Moon is a familiar friend in the sky, so near to us that some of its surface features are readily visible to the naked eye. But the Moon is also a strange and alien place, with dramatic differences from our own Earth. A first glance shows that the Moon lacks an ocean and atmosphere (Figure 9-1). More careful examination shows that it also lacks both an atmosphere and a magnetic field. Unlike the Earth's surface, the surface of the Moon has remained essentially unaltered for billions of years. Long ago, debris left over from the formation of the planets rained down on the Moon, extensively cratering its rocky surface. Broad plains, called maria, bear silent witness to vast lava flows that flooded the largest impact basins, then ceased forever. Today, the Moon is devoid of the kind of geologic activity that we find on the Earth.

Over the past four decades, humans have landed on the lunar surface and placed spacecraft in orbit around the Moon. These voyages of discovery have given us many insights into the Moon's history. For one, we know from radioactive age-dating of lunar rocks brought back to the Earth that the Moon is 4.6 billion years old. Lunar rocks also reflect the Moon's chemical composition. By comparing the Moon's chemical composition to that of the Earth, and by studying the Moon's orbit, astronomers have even found evidence of how the Moon was formed. It appears that a huge object the size of Mars or Mercury may have struck the young Earth, ejecting a vast amount of molten rock into space. According to this plausible scenario, the ejected mass cooled and coalesced to form the Moon. As we continue our exploration of the solar system, we will find other evidence that catastrophic collisions have played an important role in shaping the solar system.

**Figure 9-1**   R I **V** U X G

**The Earth and Its Moon to Scale** While on its way to Jupiter, the *Galileo* spacecraft recorded this image of the Earth (at lower right) and the Moon (at upper left) on December 16, 1992. Both the Earth and the Moon were about 6.2 million kilometers (3.9 million miles) from the spacecraft, so this image shows them to scale. The Earth has blue oceans, an atmosphere streaked with white clouds, and continents continually being reshaped by plate tectonics. By contrast, the Moon has no oceans, no atmosphere, and no evidence of plate tectonics. (NASA)

## 9-1 The Moon's airless, dry surface is covered with plains and craters

Compared to the distances between planets, which are measured in tens or hundreds of millions of kilometers, the Moon is very close to the Earth: its average distance is just 384,400 km (238,900 mi). Despite its being so close, the Moon has an angular diameter of just ½°, which tells us that it is a rather small world. The diameter of the Moon is 3476 km (2160 mi), just 27% of the Earth's diameter. Table 9-1 lists some basic data about the Moon.

Even without binoculars or a telescope, you can easily see that dark gray and light gray areas cover vast expanses of the Moon's rocky surface. If you observe the Moon over many nights, you will see the Moon go through its phases as the dividing line between the illuminated and darkened portion of the Moon (the **terminator**) shifts. But you will always see the same pattern of dark and light areas, because the Moon keeps the same side turned toward the Earth; you never see the **far side** (or back side) of the Moon. As described in Section 3-1, this is the result of the Moon being in a state of **synchronous rotation**. It takes precisely the same amount of time for the Moon to rotate on its axis (one lunar "day") as it does to complete one orbit around the Earth. With patience, however, you can see slightly more

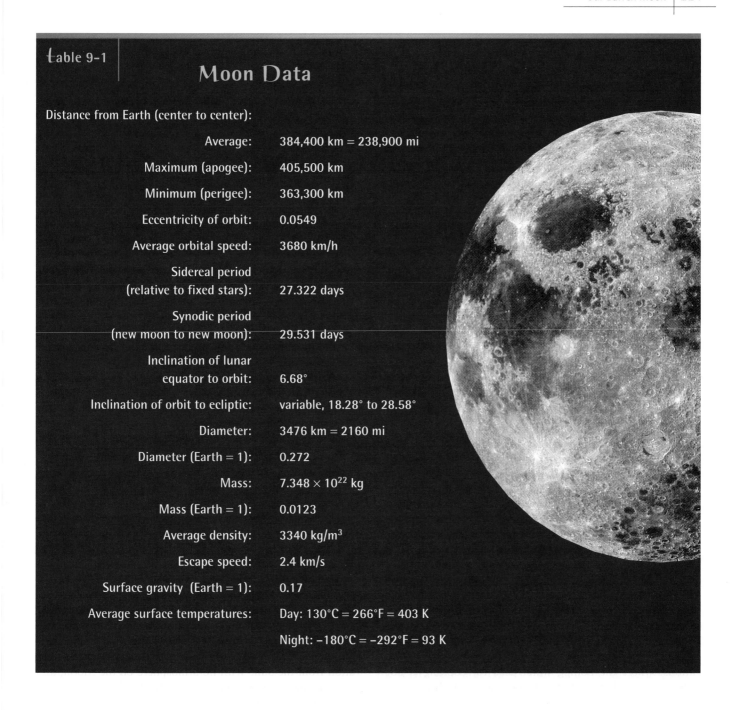

**table 9-1**

# Moon Data

| | |
|---|---|
| **Distance from Earth (center to center):** | |
| Average: | 384,400 km = 238,900 mi |
| Maximum (apogee): | 405,500 km |
| Minimum (perigee): | 363,300 km |
| Eccentricity of orbit: | 0.0549 |
| Average orbital speed: | 3680 km/h |
| Sidereal period (relative to fixed stars): | 27.322 days |
| Synodic period (new moon to new moon): | 29.531 days |
| Inclination of lunar equator to orbit: | 6.68° |
| Inclination of orbit to ecliptic: | variable, 18.28° to 28.58° |
| Diameter: | 3476 km = 2160 mi |
| Diameter (Earth = 1): | 0.272 |
| Mass: | $7.348 \times 10^{22}$ kg |
| Mass (Earth = 1): | 0.0123 |
| Average density: | 3340 kg/m³ |
| Escape speed: | 2.4 km/s |
| Surface gravity (Earth = 1): | 0.17 |
| Average surface temperatures: | Day: 130°C = 266°F = 403 K |
| | Night: –180°C = –292°F = 93 K |

than half the lunar surface, because the Moon appears to wobble slightly during an orbit. The Moon is not really wobbling; it just appears to do so because of its tilted, elliptical orbit around the Earth. This apparent periodic wobbling, called **libration,** permits us to view 59% of our satellite's surface. (We mentioned libration briefly in Box 3-2; Box 9-1 explores in more depth how the Moon's orbit causes this curious behavior.)

Clouds always cover some portion of the Earth's surface, but clouds are never seen on the Moon (Figure 9-1). This is because the Moon is too small a world to have an atmo-sphere. Its surface gravity is low, only about one-sixth as great as that of the Earth, and thus any gas molecules can easily escape into space (see Box 7-2). For the same reason, the Moon can have no liquid water on its surface. On the Earth, water molecules are kept in the liquid state by the pressure of the atmosphere pushing down on them. To see what happens if there is no atmosphere, as on the Moon, consider what happens on the Earth when the pressure of the atmosphere is reduced. In Denver, Colorado, at an elevation of 1650 m (5400 ft), the air pressure is only 83% as great as at sea level. Therefore, in Denver, less energy has to

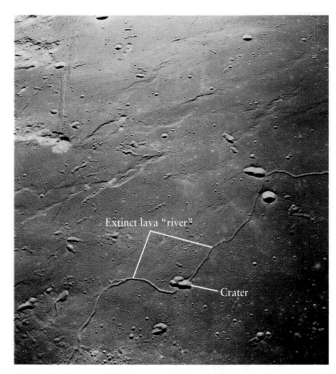

**figure 9-6** R I **V** U X G

**Details of Mare Tranquillitatis** This close-up view of the surface of Mare Tranquillitatis, a typical mare, reveals numerous tiny craters and cracks. Running from lower left to lower right on this image is what appears to be a dry riverbed. This was indeed once a river, but with flowing lava rather than flowing water. Some time after this river of lava cooled and solidified, a meteoroid landed on top of it and produced the crater at the lower middle of the image. This color photograph also shows that the Moon's surface is not a uniform gray color. (Figures 9-2, 9-3, and 9-5 are black-and-white photographs.) Mare Tranquillitatis was the site of the first manned landing on the Moon on July 20, 1969. This photograph was taken from lunar orbit in May of that year by astronauts on the *Apollo 10* spacecraft, during a dress rehearsal for the first landing. (NASA)

**figure 9-7** R I **V** U X G

**The Near and Far Sides of the Moon** This composite image, made by the *Galileo* spacecraft on December 9, 1990, while en route to the planet Jupiter, gives a perspective on the Moon that cannot be obtained from Earth. The right side of the image shows the Earth-facing side of the Moon with its large, dark, flat maria. On the left is the Moon's far side, which is almost entirely composed of light-colored lunar highlands. The isolated mare at the center of the image is Mare Orientale, which lies on the boundary between the near and far sides. The broad dark region at lower left is the South Pole–Aitken Basin, a scar some 2100 km (1300 mi) across that was left by an extraordinarily powerful impact. This impact is thought to have excavated all the way through the Moon's crust, exposing part of the mantle. (NASA)

---

| 9-2 | Manned exploration of the lunar surface was one of the greatest adventures in human history |
|---|---|

For thousands of years, storytellers have invented fabulous tales of voyages to the Moon. The seventeenth-century French satirist Cyrano de Bergerac wrote of a spacecraft whose motive power was provided by spheres of morning dew. (After all, reasoned Cyrano, dew rises with the morning sun, so a spacecraft would rise along with it.) In his imaginative tale *Somnium*, Cyrano's contemporary Johannes Kepler (the same Kepler who deduced the laws of planetary motion) used the astronomical knowledge of his time to envision what it would really be like to walk on the Moon's surface. In 1865 French author Jules Verne published *From the Earth to the Moon*, a story of a spacecraft sent to the Moon from a launching site in Florida.

Almost exactly a hundred years later, Florida was indeed the starting point of the first voyages made by humans to other worlds. Between 1969 and 1972, 12 astronauts walked on the Moon, and the fantasies of centuries became magnificent reality.

As in Verne's story, politics had much to do with the motivation for going to the Moon. But in fact there were excellent scientific reasons to do so. Because the Earth is a geologically active planet, typical terrestrial surface rocks are only a few hundred million years old, which is only a small fraction of the Earth's age. Thus, all traces of the Earth's origins have been erased. By contrast, rocks on the geologically inactive surface of the Moon have been essentially undisturbed for billions of years. Samples of lunar rocks have helped answer many questions about the Moon and have shed light on the origin of the Earth and the birth of the entire solar system.

Lunar missions began in 1959, when the Soviet Union sent three small spacecraft toward the Moon. American attempts to reach the Moon began in the early 1960s with Project Ranger. Equipped with six television cameras, the *Ranger* spacecraft transmitted close-up views of the lunar surface taken in the final few minutes before it crashed onto the Moon. Figure 9-8 shows a view from *Ranger 9* just before it hit the floor of the crater Alphonsus. A photograph of the same area taken through an Earth-based telescope demonstrates the dramatic improvement in resolution that was achieved by the spacecraft.

A complete high-resolution photographic survey of the lunar surface was the mission of the Lunar Orbiter program, under which five spacecraft were launched between August 1966 and August 1967. In all, the five *Lunar Orbiters* sent back a total of 1950 high-resolution pictures covering 99½% of the Moon. In some of these photographs, features as small as 1 meter across can be seen. Because these pictures revealed

the lunar surface in close detail, they were essential in helping NASA scientists choose landing sites for the astronauts.

Even after the Ranger missions, it was unclear whether the lunar surface was a safe place to land. It was thought that billions of years of impacts by meteoroids might have ground the Moon's surface to a fine dust, much like the pumice powder used by dental hygienists to polish teeth. Perhaps a spacecraft attempting to land would simply sink into a quicksand-like sea of dust! Thus, before a manned landing on the Moon could be attempted, it was necessary to try a soft landing of an unmanned spacecraft. This was the purpose of the Surveyor program. Between June 1966 and January 1968, five *Surveyors* successfully touched down on the Moon, sending back pictures and data directly from the lunar surface. These missions demonstrated convincingly that the Moon's surface was as solid as that of the Earth. Figure 9-9 shows an astronaut visiting *Surveyor 3,* two and a half years after it had landed on the Moon.

Six manned lunar landings were made between July 1969 and December 1972. The first two, *Apollo 11* and *Apollo 12,* set down in maria. The *Apollo 13* mission experienced inflight difficulties, which made it impossible for them to land on the Moon. The remaining four missions, *Apollo 14* through *Apollo 17,* were made in progressively more challenging terrain, culminating in rugged mountains east of

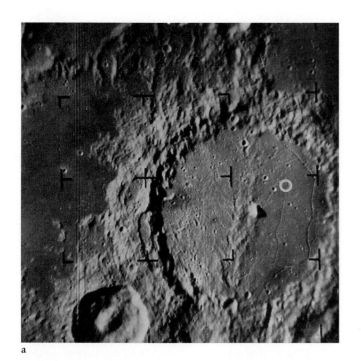

a

b

**figure 9-8**   R I **V** U X G

**The Moon from *Ranger 9* and from the Earth**  **(a)** Seconds before it crashed onto the Moon on March 24, 1965, *Ranger 9* sent back this picture of the crater Alphonsus. (The white circle indicates the impact target.) Note that the resolution is far better than that of the Earth-based telescopic view shown in **(b).** The diameter of Alphonsus is about 129 km (80 mi). (NASA)

## figure 9-9

**An Astronaut Visits *Surveyor 3*** *Surveyor 3* successfully soft-landed in Oceanus Procellarum (Ocean of Storms) in April 1967. Like the four other *Surveyors*, it sampled the lunar soil and sent back pictures to the Earth. *Surveyor 3* was visited two and a half years later by *Apollo 12* astronaut Alan Bean, who took pieces of the spacecraft back to the Earth for analysis. The retrieved pieces showed evidence of damage by micrometeoroid impacts, which provided scientists with information about the number of small objects that are still colliding with the Moon. Because there is no atmosphere to scatter sunlight, the daytime sky on the Moon appears as black as the nighttime sky on the Earth. (NASA)

Mare Serenitatis. These sites gave the astronauts the best opportunity to land safely and to explore a wide variety of geologically interesting features. Figure 9-10 shows one of the Apollo bases.

Between 1966 and 1976, a series of unmanned Soviet spacecraft also orbited the Moon and soft-landed on its surface. The first soft-landing, by *Luna 9*, came four months before the United States' *Surveyor 1* arrived. The first lunar satellite, *Luna 10*, orbited the Moon four months before *Orbiter 1*. In the 1970s, robotic spacecraft named *Luna 16*, *20*, and *24* brought samples of rock back to the Earth, and *Luna 17* and *21* landed vehicles that roamed the lunar surface. Since the successful mission of *Luna 24* in August 1976, no spacecraft has returned to the Moon's surface.

Unmanned spacecraft have continued the exploration of the Moon. The *Galileo* spacecraft was launched in 1989 on a mission to explore the planet Jupiter. On the way, it followed

a convoluted trajectory that took it past the Earth-Moon system in December 1990 and December 1992. During these two passes, *Galileo* recorded color images of the Moon with cameras far more advanced than those of the *Lunar Orbiters*. Figure 9-7 is an example of a *Galileo* image. In 1994 a small spacecraft named *Clementine* spent more than two months observing the Moon from lunar orbit. Although originally intended to test lightweight imaging systems for future defense satellites, *Clementine* proved to be a remarkable tool for lunar exploration. It carried an array of very high-resolution cameras sensitive to ultraviolet and infrared light as well as the visible spectrum. Different types of materials on the lunar surface reflect different wavelengths, so analysis of *Clementine* images made in different parts of the electromagnetic spectrum has allowed scientists to determine the composition of large areas of the Moon's surface (Figure 9-11).

Perhaps the most remarkable result of the *Clementine* mission was evidence for a patch of ice tens or hundreds of kilometers across at the Moon's south pole. The deep South Pole–Aitken Basin (Figure 9-7) has a region on its floor that remains perpetually shadowed from sunlight, and where temperatures are low enough to prevent ice from evaporating. Radio waves beamed from *Clementine* toward this region were detected by radio telescopes on the Earth and showed the characteristic signature of reflection from ice. The *Lunar Prospector* spacecraft, which went into orbit

## figure 9-10

**The *Apollo 15* Base** This photograph shows astronaut James Irwin and the *Apollo 15* landing site at the foot of the Apennine Mountains near the eastern edge of Mare Imbrium. The hill in the background is about 5 km behind the lunar module and rises about 3 km above the surrounding plain. At the right is a Lunar Rover, an electric-powered vehicle used by astronauts to explore a greater radius around the landing site. The last three Apollo missions carried Lunar Rovers. (NASA)

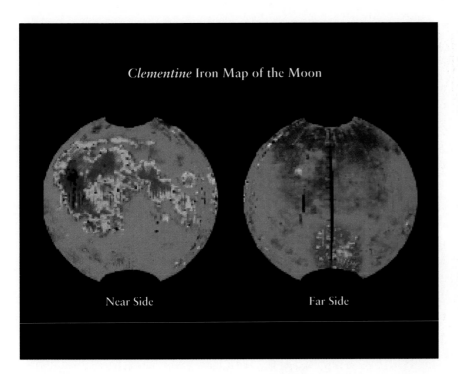

Clementine Iron Map of the Moon

Near Side

Far Side

**ƒigure 9-11** R **I** **V** **U** X G

**Clementine Maps the Moon's Surface Composition** Ultraviolet, infrared, and visible-light images from the Clementine spacecraft were used to make this map of the concentration of iron on the near side (left) and far side (right) of the lunar surface. The areas of highest iron concentration, shown in red, coincide closely with the maria (compare the view of the Moon's near side in Figure 9-2). The lowest concentration of iron is found in the blue areas, which correspond to the lunar highlands. (Areas shown in gray are those in which no data were collected.) The data confirm that the maria were formed by molten, iron-rich material that flowed from cracks in the lunar crust. Note that there are essentially no maria on the lunar far side. The green region of intermediate iron concentration on the far side is the South Pole–Aitken Basin, which can also be discerned in Figure 9-7. (The Clementine Project)

around the Moon in January 1998, confirmed this observation and discovered even more ice at the Moon's north pole. *Lunar Prospector* data suggest that there may be from 10 million to 300 million tons of ice at the lunar poles. The ice may have been deposited by comets (see Section 7-5) that crashed into the lunar surface.

## 9-3 The Moon has no global magnetic field but may have a small core beneath a thick mantle

The Moon landings gave scientists an unprecedented opportunity to explore an alien world. The Apollo landing modules were therefore packed with an assortment of apparatus and equipment that the astronauts deployed around the landing sites (Figure 9-12). For example, all the missions carried magnetometers, which confirmed that the present-day Moon has no magnetic field. Careful magnetic measurements of lunar rocks, however, indicated that the Moon did have a weak magnetic field when the rocks solidified billions of years ago, which implies that the Moon originally had a small, molten, iron-rich core. (We saw in Section 8-5 that the Earth's magnetic field is generated by fluid motions in its interior.) The core presumably solidified at least partially as the Moon cooled, so that the lunar magnetic field disappeared. If there is a solid core, it is probably less than 700 km (435 mi) in diameter.

Astronomers have used seismic waves to probe the Moon's interior, just as is done for the Earth (Section 8-2). The Apollo astronauts set up seismometers on the Moon's surface. Some of the seismic data suggested that the Moon's core is at least partially molten. It was also found that the Moon exhibits far less seismic activity than the Earth. Roughly 3000 **moonquakes** were detected per year, whereas a similar seismometer on the Earth would record hundreds of thousands of earthquakes per year. Furthermore, typical moonquakes are far weaker than typical earthquakes. A major moonquake, which typically measures 0.5 to 1.5 on the Richter scale, would go unnoticed by a person standing near the epicenter. For comparison, a major terrestrial earthquake is in the range of 6 to 8 on that scale.

Analysis of the feeble moonquakes reveals that most originate 600 to 800 km below the surface, deeper than most earthquakes. On the Earth, the deepest earthquakes mark the boundary between the solid lithosphere and the plastic asthenosphere. This is because the lithosphere is brittle enough to fracture and produce seismic waves, whereas the deeper rock of the asthenosphere oozes and flows rather than cracks. We may conclude that the Moon's lithosphere is about 800 km thick (Figure 9-13). By contrast, the Earth's lithosphere is no more than 70 km thick.

The lunar asthenosphere probably extends down to the Moon's relatively small iron-rich core, which is more than 1400 km below the lunar surface. The asthenosphere and the lower part of the lithosphere are presumably composed of relatively dense iron-rich rocks. The upper part of the lithosphere is a less dense crust, with an average thickness of

**figure 9-12**

**Seismic Equipment on the Moon** This view of the *Apollo 11* base shows astronaut Edwin Aldrin standing alongside a seismometer that detected moonquakes and transmitted data back to the Earth. By combining data from seismometers left on the Moon by Apollo astronauts, scientists have been able to deduce the structure of the Moon's interior. The *Apollo 11* lunar module and some additional equipment are in the background. (NASA)

about 60 km on the Earth-facing side and up to 100 km on the far side. For comparison, the thickness of the Earth's crust ranges from 5 km under the oceans to about 30 km under major mountain ranges on the continents.

On the whole, the interior of the Moon is much more rigid and less fluid than the Earth's interior. Plate tectonics requires that there be fluid motion just below a planet's surface, so it is not surprising that there is no evidence for plate tectonics on the lunar surface. But if there are no plate motions on the moon, how can there be moonquakes? The cause of these is actually the Earth's gravity, which not only keeps the Moon in orbit but also flexes the body of the Moon. (We describe how this happens in Section 9-4.)

Without tectonics, and without the erosion caused by an atmosphere or oceans, the only changes that are occurring today on the lunar surface are those due to meteoroid impacts. These impacts were also monitored by the seismometers left on the Moon by the Apollo astronauts. These devices, which remained in operation for several years, were sensitive enough to detect a hit by a grapefruit-sized meteoroid anywhere on the lunar surface. The data show that every year the Moon is struck by 80 to 150 meteoroids having masses between 100 g and 1000 kg (roughly ¼ lb to 1 ton).

Why is the Moon so much more solid in its interior than the Earth? One central reason is simply that the Moon is much smaller (Figure 9-1). A large turkey or roast taken from the oven will stay warm inside for hours, but a single meat-

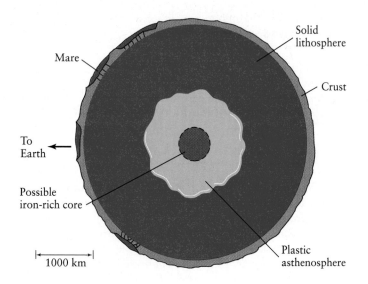

**figure 9-13**

**The Internal Structure of the Moon** Like the Earth, the Moon probably has a crust, a mantle, and a core. The lunar crust has an average thickness of about 60 km on the Earth-facing side but about 100 km on the far side. The crust and solid upper mantle form a lithosphere about 800 km thick. The plastic (nonrigid) asthenosphere extends all the way to the base of the mantle. If the Moon has an iron-rich core, it is probably less than 700 km in diameter.

ball will cool off much more rapidly. In a similar way, any planet will tend to cool down as it emits electromagnetic radiation into space, but a larger planet (which has less surface area relative to its volume) will cool down more slowly. Because the Moon is so much smaller than the Earth, it has lost much more of its internal heat.

As we explore other terrestrial worlds of the solar system we will find another useful rule: The smaller the world, the less internal heat it is likely to have retained, and, thus, the less geologic activity it will display on its surface. We will also find some remarkable exceptions to this rule, especially among the Moon-sized but geologically active satellites of Jupiter (Chapter 14).

## 9-4 Gravitational forces between the Earth and the Moon produce tides

The seismometers left at the Apollo landing sites revealed that moonquakes are more frequent at new moon and at full moon. Geologists concluded that moonquakes are influenced by tidal forces.

**Tidal forces** are differences in the gravitational pull at different points in an object. As an illustration, imagine that three billiard balls are lined up in space at some distance from a planet, as in Figure 9-14a. As Newton determined in the seventeenth century, the force of attraction between two objects is greater the closer the two objects are to each other (Section 4-7). Thus, the planet exerts more force on the 3-ball (in red) than on the 2-ball (in blue), and, in turn, the planet exerts more force on the 2-ball than on the 1-ball (in yellow). Now, imagine that the three balls are released and allowed to fall toward the planet. Figure 9-14b shows the situation a short time later. Because of the differences in gravitational pull, a short time later the 3-ball will have moved farther than the 2-ball, which will in turn have moved farther than the 1-ball. But now imagine that same motion from the perspective of the 2-ball. From this perspective, it appears as though the 3-ball is pulled toward the planet while the 1-ball is pushed away from the planet (Figure 9-14c). These apparent pushes and pulls are called tidal forces.

The Moon has a similar effect on the Earth as the planet in Figure 9-14 does on the three billiard balls. The yellow arrows in Figure 9-15a indicate the strength and direction of the gravitational force of the Moon at several locations on the Earth. The side of the Earth closest to the Moon feels a

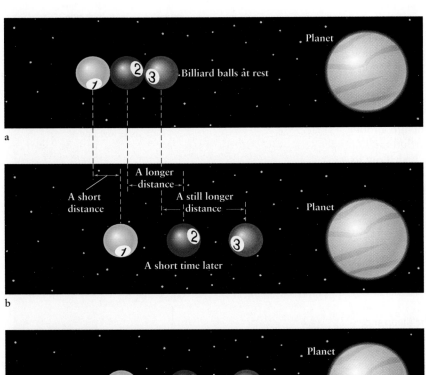

**Figure 9-14**

**The Origin of Tidal Forces** **(a)** To see where tidal forces come from, imagine three identical billiard balls placed some distance from a planet and released. **(b)** The closer a ball is to the planet, the more gravitational force the planet exerts on it. Thus, a short time after the balls are released, the yellow 1-ball has moved a short distance, the blue 2-ball has moved a longer distance, and the red 3-ball has moved a still longer distance. **(c)** From the perspective of the center ball (the 2-ball), a force seems to have pushed the 1-ball away from the planet, and a force seems to have pulled the 3-ball toward the planet. These forces are called tidal forces.

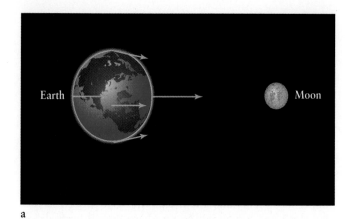

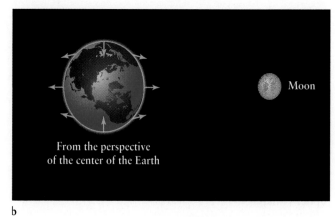

a

b

## figure 9-15

**Tidal Forces on the Earth** The Moon induces tidal forces on the Earth because the strength of the Moon's gravity varies over the Earth. **(a)** Red arrows indicate the strength and direction of the Moon's gravitational pull at selected points on the Earth. **(b)** Blue arrows indicate the strength and direction of the tidal forces acting on the Earth. At any location, the tidal force equals the Moon's gravitational pull at that point minus the gravitational pull of the Moon at the center of the Earth. These tidal forces tend to deform the Earth into a nonspherical shape.

greater gravitational pull than the Earth's center does, and the side of the Earth that faces away from the Moon feels less gravitational pull than does the Earth's center. Like the billiard balls in Figure 9-14, the result is that there are tidal forces acting on the Earth (Figure 9-15*b*). These forces try to elongate the Earth along a line connecting the centers of the Earth and the Moon and try to squeeze the Earth inward in the direction perpendicular to that axis. The net effect is that the Moon's gravity tries to deform the Earth into a football shape.

Because the body of the Earth is largely rigid, it cannot deform very much in response to the Moon's tidal forces. But the water in the oceans can and does deform into a football shape, as shown in Figure 9-16*a*. As the Earth rotates, a point on its surface goes from where the water is shallow to where the water is deep and back again. This is the origin of low and high ocean tides. (Note that in this simplified description we have assumed that the Earth is completely covered with water. The full story of the tides is much more

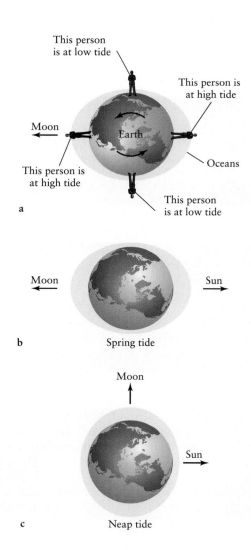

## figure 9-16

**Tides on the Earth** **(a)** The gravitational forces of the Moon and Sun deform the oceans. As the Earth rotates underneath the oceans, a given location experiences alternating low tides and high tides. **(b)** The greatest deformation (spring tides) occurs when the Sun, the Earth, and the Moon are aligned and the tidal effects of the Sun and Moon reinforce each other. **(c)** The least deformation (neap tides) occurs when the Sun, the Earth, and the Moon form a right angle. In this situation the tidal effects of the Sun and Moon partially cancel each other.

complex, because the shapes of the continents and the effects of winds must also be taken into account.)

Actually, the tides in the oceans result because *both* the Moon and the Sun act to distort the shape of the Earth. This tidal distortion is greatest when the Sun, Moon, and the Earth are aligned (at either new moon or full moon), so that the tidal effects of the Moon and Sun reinforce each other. This produces large shifts in water level that are called **spring tides** (Figure 9-16*b*). At first quarter and last quarter, when the Sun and Moon form a right angle with the Earth, the tidal effects of the Sun and Moon partially cancel each other. Hence, the tidal distortion of the oceans is the least pronounced, producing smaller tidal shifts called **neap tides** (Figure 9-16*c*).

**CAUTION!** Note that spring tides have nothing to do with the season of the year called spring. Instead, the name refers to the way that the ocean level "springs up" to a greater than normal height. Spring tides occur whenever there is a new moon or full moon, no matter what the season of the year.

We can now see how tidal forces might trigger moonquakes. Just as the gravitational forces of the Sun and Moon deform the Earth, the Sun and the Earth deform the Moon. The Moon is alternately squeezed and stretched as it orbits our planet. Because the Earth is more massive than the Moon, the Earth produces relatively large tidal deformations of the lunar surface. The average tidal deformation on the solid body of the Earth is about 30 cm (1 ft), but is about twice as great on the Moon. The strongest gravitational stresses occur when the Moon is nearest the Earth in its orbit—that is, at perigee (Section 3-4)—which is just the time when the Apollo seismometers reported the highest frequency of moonquakes.

The Earth's gravity also helps to maintain the Moon's synchronous rotation, so that the same side of the Moon always faces the Earth. Although the Moon is spherical, its mass is distributed slightly off center: One hemisphere contains a bit more mass than the other. (Figure 9-11 shows another way that the Moon's two hemispheres are different.) To see what effect this has, tape a coin to one end of a ruler. If you hold the ruler at its midpoint between your thumb and forefinger, the end with the coin—the more massive end—will swing downward toward the Earth. In the same way, the Earth's gravity keeps the Moon oriented with its more massive hemisphere pointed toward the Earth. This is the hemisphere that we see. As it maintains this orientation, the Moon spins once on its axis as it makes one orbit around the Earth. For this reason, synchronous rotation is also known by the more impressive name of **1-to-1 spin-orbit coupling**. As we will see in later chapters, most of the satellites in the solar system exhibit 1-to-1 spin-orbit coupling, and thus always keep the same side facing their planet.

Gravitational interactions between the Earth and the Moon also gradually enlarge the Moon's orbit and slow the Earth's rotation. The Earth rotates about its axis faster than the Moon revolves around the Earth. The Earth's rapid rota-

tion carries the tidal bulge of the oceans forward of the Moon's orbit. As viewed from the Moon, the tidal bulge on the Earth is always aimed slightly ahead of the Moon's own position (Figure 9-17). This bulge produces a small but constant gravitational force that tugs the Moon forward and lifts it into a more distant orbit around the Earth. In other words, the Moon is very slowly spiraling away from the Earth, at a rate of about 3.8 cm per year.

As the Moon moves away from the Earth, its sidereal period is becoming longer (recall the discussion of Kepler's third law in Section 4-4 and Section 4-7). Meanwhile, friction between the oceans and the Earth's surface is gradually slowing the Earth's rotation. The Earth's day is therefore becoming longer and longer, by approximately 0.002 second per century.

At some point in the distant future, the Earth will be rotating so slowly that a solar day will equal a lunar month. (Both will then equal 47 of our present solar days.) At that time, the Earth's tidal bulge will be aimed directly at the Moon, and the Moon will stop spiraling away from the Earth. The Earth will then keep the same side facing the Moon, just as the Moon presently keeps the same side facing the Earth. As we will see in Chapter 16, Pluto and its moon have already achieved this stable configuration.

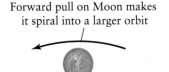

Forward pull on Moon makes
it spiral into a larger orbit

Moon

Friction between Earth and its
oceans makes Earth rotate more slowly

Earth

**ƒigure 9-17**

**The Moon's Tidal Recession** Because of the Earth's rapid rotation, the tidal bulge on the oceans is dragged ahead of a direct alignment with the Moon's position in the sky. The bulge on the leading side of the Earth exerts a small forward force on the Moon that causes it to spiral slowly away from the Earth. At the same time, friction between the body of the Earth and the oceans will cause the Earth's rotation rate to slow down.

<table>
<tr><td>

### 9-5 Lunar rocks were formed approximately 3 to 4½ billion years ago

</td></tr>
</table>

The Apollo astronauts brought back 2200 individual samples from six sites on the lunar surface, for a grand total of 382 kg (842 lb) of lunar material. In addition, the unmanned Soviet spacecraft *Luna 16, Luna 20,* and *Luna 24* returned 300 g ($\frac{2}{3}$ lb) from three other sites on the Moon. This collection of lunar material has provided important information about the early history of the solar system that could have been obtained in no other way. All of the lunar samples are igneous rocks; there are no metamorphic or sedimentary rocks. This suggests that the Moon's surface was once entirely molten. Indeed, Moon rocks are composed mostly of the same minerals that are found in terrestrial volcanic rocks.

Astronauts who visited the maria discovered that these dark regions of the Moon are covered with basaltic rock quite similar to the dark-colored rocks formed by lava from volcanoes on Hawaii and Iceland. The rock of these low-lying lunar plains is called **mare basalt** (Figure 9-18).

In contrast to the dark maria, the lunar highlands are composed of a light-colored rock called **anorthosite** (Figure 9-19). On the Earth, anorthositic rock is found only in such very old mountain ranges as the Adirondacks in the eastern United States. In comparison to the mare basalts, which have more of the heavier elements like iron, manganese, and titanium, anorthosite is rich in silicon, calcium, and aluminum. Anorthosite is therefore less dense than basalt. During the period when the Moon's surface was molten, the less dense anorthosite rose to the top, forming the terrae that make up the majority of the present-day lunar surface.

By carefully measuring the abundances of trace amounts of radioactive elements in lunar samples, geologists have determined that lunar rocks formed more than 3 billion ($3 \times 10^9$) years ago (Box 9-2). Of these ancient rocks, however, anorthosite is much older than the mare basalts. Typical anorthositic specimens from the highlands are between 4.0 and 4.3 billion years old, and one rock brought back by the *Apollo 17* astronauts is nearly 4.6 billion years old. All these extremely ancient specimens are probably samples of the Moon's original crust. In sharp contrast, all the mare basalts are between 3.1 and 3.8 billion years old.

Although lunar rocks bear a strong resemblance to terrestrial rocks, there are some important differences. Every terrestrial rock, even those from desert regions, has some water locked into its mineral structure. But lunar rocks are totally dry. With the exception of ice at the lunar poles (Section 9-2), there is no evidence that water has ever existed on the Moon. Because the Moon lacks both an atmosphere and liquid water, it is not surprising that the astronauts found no traces of life.

### figure 9-18

**Mare Basalt** This 1.531-kg (3.376-lb) specimen of mare basalt was brought back by the crew of *Apollo 15*. It is a vesicular basalt, so-called because of the holes, or vesicles, covering 30% of its surface. Gas must have dissolved under pressure in the lava from which this rock solidified. When the lava reached the airless lunar surface, bubbles formed as the pressure dropped. Some of the bubbles were frozen in place as the rock cooled. (NASA)

### figure 9-19

**Anorthosite** The light-colored lunar terrae (highlands) are composed of an ancient type of rock called anorthosite, which is believed to be the material of the original lunar crust. This sample, called the "Genesis rock" by the *Apollo 15* astronauts who picked it up at the base of the Apennine Mountains, has an age of approximately 4.1 billion years. (NASA)

Meteoritic bombardment is the only source of "weathering" to which rocks on the lunar surface are subjected. As a result of this bombardment, the entire lunar surface is covered with a layer of fine powder and rock fragments produced by billions years of relentless meteoritic bombardment. This layer, called the **regolith** (from the Greek for "blanket of stone"), ranges in depth from about 3 to 30 m.

In addition to churning up the regolith, meteoritic impacts can also release enough energy to melt rocks. Many lunar samples are coated with a thin layer of smooth, dark glass created when the surface of the rock suddenly melted and then rapidly solidified. Moreover, small black glass beads are common in the lunar regolith. Presumably, these glass spheres were formed from droplets of molten rock hurled skyward by the impact of a meteoroid. Finally, many lunar samples bear the scars of high-speed meteoritic dust. Dust grains traveling at thousands of kilometers per hour produce tiny craters on the exposed surfaces of moon rocks (Figure 9-20). Because of these tiny, glass-lined pits, lunar rocks often seem to sparkle when held in the sunshine.

By correlating the ages of Moon rocks with the density of craters at the sites where the rocks were collected, geologists

---

## box 9-2 | Looking Deeper into Astronomy

### *Radioactive Age-Dating*

The Apollo program gave geologists the opportunity to get their hands on extremely ancient rocks. By examining these lunar samples, scientists began to piece together a history of important events that happened shortly after our solar system was created. To get the story right, however, geologists had to determine the ages of the lunar rocks. Fortunately, most rocks contain trace amounts of radioactive elements such as uranium. The relative abundances of various radioactive isotopes and their decay products provided the key that geologists needed.

As we saw in Section 7-6, every atom of a particular element has the same number of protons in its nucleus. However, different isotopes of the same element have different numbers of neutrons in their nuclei. For example, the common isotopes of uranium are $^{235}U$ and $^{238}U$. Each isotope of uranium has 92 protons in its nucleus (correspondingly, uranium is element 92 on the periodic chart; see Box 5-4). However, a $^{235}U$ nucleus contains 143 neutrons, whereas a $^{238}U$ nucleus has 146 neutrons.

A radioactive nucleus is unstable because it contains too many protons or too many neutrons. It therefore ejects particles until it becomes stable. In doing so, a nucleus may change from one element to another. Physicists refer to this transmutation as "decay."

Some radioactive isotopes decay rapidly, while others decay slowly. Physicists find it convenient to talk about the decay rate in terms of an isotope's **half-life.** The half-life of an isotope is the time interval in which one-half of the nuclei decay. For example, the half-life of $^{235}U$ is 710 million ($710 \times 10^6$) years. This means that if you start out with 1 kg of $^{235}U$, after 710 million years you will have only ½ kg of $^{235}U$ left; the other ½ kg will have turned into other elements. Several isotopes useful for age-dating rocks are listed in the table.

To see how geologists date Moon rocks, consider the slow conversion of rubidium ($^{87}Rb$) into strontium ($^{87}Sr$). Over the years, the amount of $^{87}Rb$ in a rock decreases, while the amount of $^{87}Sr$ increases. Dating the rock is not simply a matter of measuring its ratio of rubidium to strontium, however, because the rock already had some strontium in it when it was formed. Geologists must therefore determine how much fresh strontium came from the decay of rubidium after the rock's formation.

To do this, geologists use as a reference another isotope of strontium whose concentration has remained constant. In this case, they use $^{86}Sr$, which is stable and is not created by radioactive decay; it is said to be "nonradiogenic." Dating a rock thus entails comparing the ratio of radiogenic and nonradiogenic strontium ($^{87}Sr/^{86}Sr$) in it to the ratio of radioactive rubidium to nonradiogenic strontium ($^{87}Rb/^{86}Sr$). Because the half-life for converting $^{87}Rb$ into $^{87}Sr$ is known, the rock's age can then be calculated from these ratios.

| Original radioactive isotope | Half-life (billions of years) | Final stable isotope |
|---|---|---|
| Potassium ($^{40}K$) | 1.3 | Argon ($^{40}Ar$) |
| Rubidium ($^{87}Rb$) | 47.0 | Strontium ($^{87}Sr$) |
| Uranium ($^{235}U$) | 0.7 | Lead ($^{207}Pb$) |
| Uranium ($^{238}U$) | 4.5 | Lead ($^{206}Pb$) |

have found that the rate of impacts on the Moon has changed over the ages. The ancient, heavily cratered lunar highlands are evidence of intense bombardment that dominated the Moon's early history. For nearly a billion years, rocky debris left over from the formation of the planets rained down upon the Moon's young surface. As Figure 9-21 shows, this barrage extended from 4.6 billion years ago, when the Moon's surface solidified, until about 3.8 billion years ago. At its peak, this bombardment from space would have produced a new crater of about 1 km radius somewhere on the Moon about once per century. (If this seems like a long time interval, remember that we are talking about a bombardment that lasted hundreds of millions of years and produced millions of craters over that time.)

The frequency of impacts gradually tapered off, however, as meteoroids and planetesimals were swept up by the newly formed planets. This crater-making era ended about the time when several large planetesimals gouged out the mare basins. A series of lava flows that occurred between 3.8 and 3.1 billion years ago flooded the mare basins with iron-rich magma that oozed up from the Moon's still-molten mantle. The relatively crater-free maria tell us that the impact rate over the past 3 billion years has been quite low.

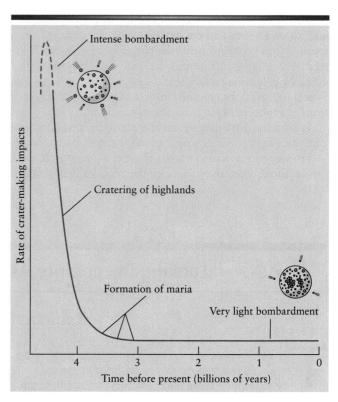

### Figure 9-21

**The Rate of Crater Formation on the Moon** This graph shows the rate at which craters have been formed on the Moon during different epochs in lunar history. The first 800 million ($8 \times 10^8$) years of the Moon's history were dominated by frequent crater-making impacts. (The earliest history of cratering is not well understood, because the most ancient craters have been covered up or erased by subsequent craters. To indicate this uncertainty, the earliest part of the graph at the far left is shown as a dashed curve.) Near the end of this intense bombardment, several large impacts gouged out the mare basins. For the past 3.5 billion years, the impact rate has been quite low, perhaps only one ten-thousandth as great as during the most intense bombardment.

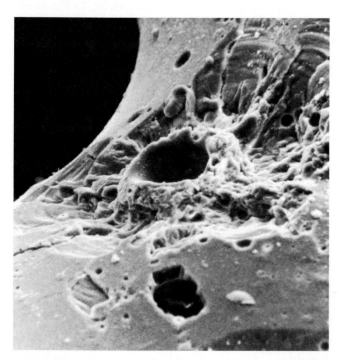

### Figure 9-20

**A Microscopic Crater** This photograph, made with a microscope, shows a tiny crater formed in a moon rock by the impact of a fast-moving grain of meteoritic dust. The upper surfaces of many moon rocks are covered with these glass-lined craters. Such "microcraters" are typically less than 1 mm in diameter. (NASA)

---

| 9-6 | The Moon probably formed from debris cast into space when a huge planetesimal struck the proto-Earth |
|---|---|

Prior to the Apollo program, there were three different theories about the origin of the Moon. First, the **fission theory** holds that the Moon was pulled out from a rapidly rotating proto-Earth. Second, the **capture theory** posits that the Moon was formed elsewhere in the solar system and then drawn into orbit about the Earth by gravitational forces.

Third, the **co-creation theory** proposes that the Earth and the Moon were formed at the same time but separately.

One fact used to support the fission theory was that the Moon's average density (3340 kg/m$^3$) is comparable to that of the Earth's outer layers, as would be expected if the Moon had been formed from material flung out by a proto-Earth's rapid rotation. However, most geologists counter this argument by pointing to the significant differences between lunar and terrestrial rocks—particularly the differences in water content and in the relative abundances of volatile and refractory elements.

**Volatile elements** (such as potassium and sodium) melt and boil at relatively low temperatures, whereas **refractory elements** (such as titanium, calcium, and aluminum) melt and boil at much higher temperatures. Compared to terrestrial rocks, the lunar rocks have slightly greater proportions of the refractory elements (and slightly lesser proportions of the volatile elements). This composition suggests that the Moon formed from material that was heated to a higher temperature than the rock from which the Earth formed. Some of the volatile elements boiled away, leaving the young moon relatively enriched in the refractory elements.

Proponents of the capture theory use these differences to argue that the Moon formed elsewhere and was later captured by the Earth. They also note that the plane of the Moon's orbit is close to the plane of the ecliptic (Box 9-1), as would be expected if the Moon had originally been in orbit about the Sun but was later captured by the Earth.

There are difficulties with the capture theory, however. If the Earth did capture the Moon intact, then a host of conditions must have been met entirely by chance. The Moon must have coasted to within 50,000 km of the Earth at exactly the right speed to leave a solar orbit, and must have adopted an Earth orbit without ever actually hitting our planet.

Because it is easier for a planet to capture swarms of tiny rocks, proponents of the co-creation theory argue that the Moon formed from just such rocky debris. Great numbers of rock fragments orbited the newborn Sun in the plane of the ecliptic along with the protoplanets. Heat from the protosun could have baked the volatile elements out of these smaller rock fragments before their capture into orbit about the proto-Earth. Then, just as planetesimals accreted to form the proto-Earth in orbit about the Sun, the fragments in orbit about the Earth accreted to form the Moon.

The present consensus among astronomers is that *none* of these three theories—fission, capture, and co-creation—is correct. Instead, the evidence points toward a more recently proposed theory called the **collisional ejection theory.** This theory suggests that the Earth was struck by an object perhaps as large as Mars and that this collision ejected debris from which the Moon formed. Clues provided by the lunar rocks along with what we know about the formation of the planets favor this most recent hypothesis. The collisional theory arose in the mid-1980s, when several teams of scientists began to appreciate what conditions were like during the early history of the solar system. As we have seen

(recall Figure 7-18), smaller objects collided and fused together to form the inner planets. Some of these collisions must have been quite spectacular, especially near the end of planet formation, when most of the mass of the inner solar system was contained in the protoplanets and a few dozen other large objects. According to the collisional ejection theory, one such object struck the proto-Earth 4.6 billion years ago. Figure 9-22 shows a supercomputer simulation of this cataclysm. In the simulation, energy released during the collision produces a huge plume of vaporized rock that squirts out from the point of impact. As this ejected material cools, it coalesces to form the Moon, as shown in Figure 9-22*f*.

The collisional ejection theory is in agreement with many of the known facts about the Moon. For example, rock vaporized by the impact would have been depleted of volatile elements and water, leaving the moon rocks we now know. If the collision took place after chemical differentiation had occurred on the Earth, when our planet's iron sunk to its center, then relatively little iron would have been ejected to become part of the Moon. This argument explains the Moon's low density and the small size of its iron core. It also explains certain measurements made by *Clementine* of the South Pole–Aitken Basin (Figure 9-7), where an ancient impact excavated the surface down to a depth of 12 km. The iron concentration in this basin is only 10%, far less than the 20–30% at a corresponding depth below the Earth's surface.

The surface of the newborn Moon was probably molten for many years, both from heat released during the impact of rock fragments falling into the young satellite and from the decay of short-lived radioactive isotopes. As the Moon gradually cooled, low-density lava floating on the Moon's surface began to solidify into the anorthositic crust that exists today. The barrage of rock fragments that ended about 3.8 billion years ago produced the craters that cover the lunar highlands.

Recorded among the final scars at the end of this crater-making era are the impacts of more than a dozen asteroid-size objects, each measuring at least 100 km across. As these huge rocks rained down on the young Moon, they blasted out the vast mare basins. Meanwhile, heat from the decay of long-lived radioactive elements like uranium and thorium began to melt the inside of the Moon. Then, from 3.8 to 3.1 billion years ago, great floods of molten rock gushed up from the lunar interior, filling the impact basins and creating the maria we see today.

Very little has happened on the Moon since those ancient times. A few relatively fresh craters have been formed, but the astronauts visited a world that has remained largely unchanged for more than 3 billion years. Our own planet has been through tremendous changes since 3 billion years ago, when the first organisms began populating the new oceans.

Many questions and mysteries still remain. The six American and three Soviet lunar landings have brought back samples from only nine locations, barely scratching the Moon's surface. We still know very little about the Moon's far side and poles. Is the Moon's interior molten? Does the Moon really have an iron core? How old are the youngest lunar

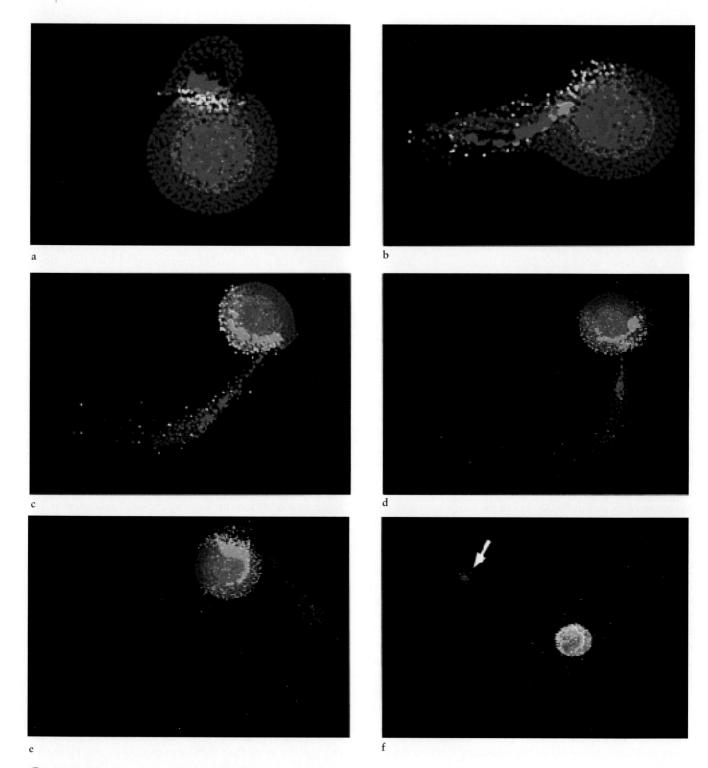

a

b

c

d

e

f

# ƒigure 9-22

**The Moon's Creation** This supercomputer simulation shows the creation of the Moon from material ejected by the impact of a large asteroid on the Earth. To follow the ejected material as it moves away from the Earth, successive views show increasingly larger volumes of space. Blue and green indicate iron from the cores of the Earth and the asteroid; red, orange, and yellow indicate rocky mantle material. In this simulation, the impact ejects both mantle and core material, but most of the iron falls back onto the Earth. The surviving ejected rocky matter coalesces to form the Moon (indicated by arrow) during its first orbit of the Earth. (Courtesy of W. Benz)

ƒigure 9-23

**The Moon—Site of Future Industry?** Volcanic material on the Moon's surface contains substantial amounts of trapped oxygen. This oxygen can be released by heating lunar material along with hydrogen gas. Oxygen could also be extracted from the deposits of ice at the lunar poles. These sources of oxygen could support a permanent lunar colony and long-term exploration of the Moon's surface. The Moon is also a useful launching pad for exploration of the solar system, since its weaker gravity and lack of an atmosphere make it easier to send payloads into space. The rocket fuel for this enterprise would use oxygen that was "mined" from the Moon. (NASA)

rocks? Did lava flows occur over western Oceanus Procellarum only 2 billion years ago, as crater densities there suggest? Is the Moon really geologically dead, or does it just look dead, simply because our examination of the lunar surface has been so cursory? Are there still any active volcanoes on the Moon? Are there mineral deposits on the lunar surface from which oxygen for life support and rocket fuel could be extracted, making the Moon a springboard for human exploration of the solar system (Figure 9-23)? Such questions can only be answered by returning to the Moon.

## Key Words

*Terms preceded by an asterisk (\*) are discussed in the Boxes.*

1-to-1 spin-orbit coupling, p. 223

anorthosite, p. 234

capture theory (of Moon's formation), p. 236

co-creation theory (of Moon's formation), p. 237

collisional ejection theory (of Moon's formation), p. 237

crater, p. 222

far side (of the Moon), p. 220

fission theory (of Moon's formation), p. 236

\*half-life, p. 235

impact crater, p. 223

libration, p. 221

lunar highlands, p. 225

mare (*plural* maria), p. 223

mare basalt, p. 234

moonquake, p. 229

neap tide, p. 223

refractory element, p. 237

regolith, p. 235

spring tide, p. 223

synchronous rotation, p. 220

terminator, p. 220

terrae, p. 224

tidal force, p. 231

volatile element, p. 237

## Key Ideas

**Appearance of the Moon:** The Earth-facing side of the Moon displays light-colored, heavily cratered highlands and dark-colored, smooth-surfaced maria. The Moon's far side has almost no maria. There is no evidence of plate tectonic activity on the Moon.

**Internal Structure of the Moon:** Much of our knowledge about the Moon came from lunar exploration in the 1960s and early 1970s. This has been supplemented by more recent observations from unmanned spacecraft.

• Analysis of seismic waves from moonquakes and meteoroid impacts indicates that the Moon has a crust thicker than that of the Earth (and thickest on the far side of the Moon), a mantle with a thickness equal to about 75% of the Moon's radius, and perhaps a small iron core.

• The Moon's lithosphere is far thicker than that of the Earth, and the Moon's asthenosphere probably extends from the base of the lithosphere to the core.

• The Moon has no magnetic field today, although it may have had a weak magnetic field billions of years ago.

**Tidal Forces:** Gravitational interactions between the Earth and the Moon produce tides in the oceans of the Earth and in the solid bodies of both worlds.

• Tidal interactions have locked the Moon into synchronous rotation with the Earth and have given rise to a small constant acceleration of the Moon in its orbit, thereby causing it to spiral slowly away from the Earth.

**Geologic History of the Moon:** The anorthositic crust exposed in the highlands was formed between 4.0 and 4.3 billion years ago, whereas the mare basalts solidified between 3.1 and 3.8 billion years ago.

• The Moon's surface has undergone very little change over the past 3 billion years.

• Meteoroid impacts have been the only significant "weathering" agent on the Moon; the Moon's regolith ("soil" layer) was formed by meteoritic action.

• All of the lunar rock samples are igneous rocks formed largely of minerals found in terrestrial rocks; the lunar rocks contain no water and also differ from terrestrial rocks in being relatively enriched in the refractory elements and depleted in the volatile elements.

**Origin of the Moon:** The collisional ejection theory of the Moon's origin holds that the proto-Earth was struck by a huge asteroid, and debris from this collision coalesced to form the Moon. This theory successfully explains most properties of the Moon.

• The Moon was molten in its early stages, and the anorthositic crust solidified from low-density magma that floated to the lunar surface; the mare basins were created later, by the impact of planetesimals, and filled with lava from the lunar interior.

## Review Questions

**1.** What kind of features can you see on the Moon with a small telescope?

**2.** Why do you suppose more lunar detail is visible through a telescope when the Moon is near first quarter phase or last quarter phase than when it is at full phase?

**3.** Why do you suppose temperature variations between day and night on the Moon are much more severe than on the Earth?

**4.** Why are Moon rocks so much older than Earth rocks, even though both worlds formed at nearly the same time?

**5.** On the basis of moon rocks brought back by the astronauts, explain why the maria are dark colored but the lunar highlands are light colored.

**6.** Why are there so few craters on the maria?

**7.** Briefly describe the main differences and similarities between Moon rocks and Earth rocks.

**8.** How do we know that the maria were formed *after* the lunar highlands?

**9.** Why do you suppose that no Apollo mission landed on the far side of the Moon?

**10.** What is a tidal force? How do tidal forces produce tides in the Earth's oceans?

**11.** What is the difference between spring tides and neap tides?

**12.** Explain why moonquakes occur more frequently when the Moon is at perigee than at other locations along its orbit.

**13.** In the distant future when a solar day will be the same length as a lunar month, how will the Moon appear to move across the sky over the course of a night on the Earth?

**14.** Rocks found on the Moon are between 3.1 and 4.6 billion years old. By contrast, the majority of the Earth's surface is made of oceanic crust that is less than 200 million years old, and the very oldest Earth rocks are about 3 billion years old. Why is there such a disparity in the ages of rocks on the two worlds?

**15.** Why do most scientists favor the collisional ejection theory for the Moon's formation?

**16.** Some people who supported the fission theory proposed that the Pacific Ocean basin is the scar left when the Moon pulled away from the Earth. Explain why this idea is probably wrong.

## Advanced Questions

### Problem-solving tips and tools

Recall that the average density of an object is its mass divided by its volume. The volume of a sphere is $\frac{4}{3}\pi r^3$, where $r$ is the sphere's radius. The surface area of a sphere of radius $r$ is $4\pi r^2$, while the surface area of a circle of radius $r$ is $\pi r^2$. Recall also that the acceleration of gravity on the Earth's surface is 9.8 m/s². You may find it useful to know that a 1-pound (1-lb) weight presses down on the Earth's surface with a force of 4.448 newtons. You might want to review Newton's universal law of gravitation in Section 4-7. Consult Table 9-1 and the appendices for any additional data.

**17.** Using the diameter and mass of the Moon given in Table 9-1, verify that the Moon's average density is 3340 kg/m³. Explain why this average density implies that the Moon's interior contains much less iron than the interior of the Earth.

**18.** How much would an 80-kg person weigh on the Moon? How much does that person weigh on the Earth?

**19.** During the period of most intense meteoritic bombardment, a new 1-km-radius crater formed somewhere on the Moon about once per century. During this same period, what was the probability that such a crater would be created within 1 km of a certain location on the Moon during a 100-year period? During a $10^6$-year period? (*Hint:* If you drop a coin onto a checkerboard, the probability that the coin will land on any particular one of the board's 64 squares is 1/64.)

**20.** Determine what the average distance between the Earth and the Moon will be when the length of the day and the lunar month will both be equal to 47 of our present days.

**21.** Can you think of any other weathering processes that might occur on the Moon besides those related to meteoritic impacts?

## Discussion Questions

**22.** Comment on the idea that without the presence of the Moon in our sky, astronomy would have developed far more slowly.

**23.** Compare the advantages and disadvantages of exploring the Moon with astronauts as opposed to using mobile, unmanned instrument packages.

**24.** Describe how you would empirically test the idea that human behavior is related to the phases of the Moon. What problems are inherent in such testing?

**25.** How would our theories of the Moon's history have been affected if astronauts had discovered sedimentary rock on the Moon?

**26.** Imagine that you are planning a lunar landing mission. What type of landing site would you select? Where might you land to search for evidence of recent volcanic activity?

**27.** The *Lunar 17* lunar module *Challenger* in December 1972. After lifting off from the lunar surface and returning to the command module *America*, in which they would return to the Earth, the *Apollo 17* astronauts sent the unoccupied *Challenger* to crash into the lunar surface. The seismic waves from this impact were detected by seismometers left by the crews of *Apollo 12, 14, 15, 16,* and *17.* Why was it useful to do this? Why was it not enough to detect seismic waves from naturally occurring moonquakes?

**28.** The *Lunar Prospector* spacecraft went into orbit around the Moon in January 1998. Check recent issues of *Sky & Telescope* and *Astronomy* magazines, as well as the *Lunar Prospector* web site (**http://lunar.arc.nasa.gov/**) for the latest news about this mission. What observations has the spacecraft made? What are the latest observations of ice at the lunar poles? What new insights have scientists gained into the nature and history of the Moon?

## Observing Projects

**Observing tips and tools:**

If you do not have access to a telescope, you can learn a lot by observing the Moon through binoculars. Note that the Moon will appear right-side-up through binoculars, but inverted through a telescope; if you're using a map of the Moon to aid your observations, you'll need to take this into account. Inexpensive maps of the Moon can be purchased from most good bookstores or educational supply stores. You can determine the phase of the Moon either by looking at a calendar (most of which tell you the dates of new moon, first quarter, full moon, and last quarter), by checking the weather page of your newspaper, by consulting the current issue of *Sky & Telescope* or *Astronomy* magazine, or by using the World Wide Web (**http://tycho.usno. navy.mil/vphase.html** or **http://www.fourmilab.ch/earthview/vplanet.html**). Many other observing tips can be found at the *Sky & Telescope* Touring the Moon with Binoculars web site (**http://www.skypub.com/whatsup/moontour.html**).

**29.** Use a telescope or binoculars to observe the Moon. Compare the texture of the lunar surface you see on the maria with that of the lunar highlands. How does the visibility of details vary with distance from the terminator (the boundary between day and night on the Moon)?

**30.** Observe the Moon through a telescope every few nights over a period of two weeks between new moon and full moon. Make sketches of various surface features, such as craters, mountain ranges, and maria. How does the appearance of these features change with the Moon's phase? Which features are most easily seen at a low angle of illumination? Which features show up best with the Sun nearly overhead?

**31.** If you live near the ocean, observe the tides to see how the times of high and low tides are correlated with the position of the Moon in the sky. You can also find tide information in your local newspaper.

## Where to Learn More

*Books and magazines*

Chaikin, A. *Man on the Moon: The Voyages of the Apollo Astronauts.* Penguin, 1994. An insightful history of how and why humans first went to the Moon.

Goldreich, P. "Tides and the Earth-Moon System." *Scientific American,* April 1972. This classic article explores the consequences of the gravitational interaction between the Earth and the Moon.

Moore, P. *The Moon.* Rand McNally, 1981. This comprehensive yet brief atlas covers the gamut, with many attractive illustrations and an excellent collection of maps and photographs.

Schmitt, H. "Exploring Taurus-Littrow: *Apollo 17.*" *National Geographic,* September 1973. This superb description of the final manned lunar landing near crater Littrow in the Taurus mountains was written by the astronaut-geologist who was there.

Spudis, P. D. *The Once and Future Moon.* Smithsonian Institution Press, 1996. An up-to-date history of man's study of the Moon, written by a lunar expert who was one of the principal scientists on the *Clementine* mission.

Weaver, K. "First Explorers on the Moon: The Incredible Story of *Apollo 11.*" *National Geographic,* December 1969. This beautifully illustrated article tells the story of the first manned lunar landing.

Wood, J. "The Moon." *Scientific American,* September 1975. This brief article gives an excellent summary of what we learned from the Moon rocks.

**W** *World Wide Web*

Two good starting places for exploring the Moon on the World Wide Web are "The Nine Planets" **http://www.seds.org.nineplanets/nineplanets/luna.html**) and "Views of the Solar System" (**http://www.hawastsoc.org/solar/eng/moon.htm**). Both of these have links to many images and to other related web pages.

NASA has a number of web pages devoted to all aspects of the Moon, including detailed descriptions of all manned and unmanned missions to the Moon (**http://www.hq.nasa.gov/office/pao/History/apollo.html** and **http://nssdc.gsfc.nasa.gov/planetary/planets/moonpage.html**), a tour of the Lunar Sample Laboratory Facility where moon rocks are stored and analyzed (**http://www.curator.jsc.nasa.gov/curator/lunar/tour/welcome.htm**), a description of how crater formation can be reproduced in experiments on the Earth (**http://sn-io.jsc.nasa.gov/outreach/craters/**), and an analysis of the political and economic factors that shaped the effort to land humans on the Moon (**http://www.hq.nasa.gov/office/pao/History/Apollomon/cover.html**).

The entire Apollo program is described in detail at a site maintained by the National Air and Space Museum (**http://www.nasm.edu/APOLLO/**).

Also available on the Web is information about the more recent unmanned explorations of the Moon by *Lunar Prospector* (**http://lunar.arc.nasa.gov/**) and *Clementine* (**http://nssdc.gsfc.nasa.gov/planetary/clementine.html**), as well as many images from *Clementine* (**http://www.nrl.navy.mil/clementine/clementine.html**).

# Sun-Scorched Mercury

RI **V** U X G

**Mercury and the Moon** Mercury (left), like our Moon (right), has a heavily cratered surface and virtually no atmosphere. Mercury's diameter is 4878 km; the Moon's is 3476 km. (Both worlds are shown here to the same scale.) For comparison, the distance from New York to Los Angeles is 3944 km (2451 mi). Daytime temperatures at the equator on Mercury reach about 430°C (700 K, or 800°F), hot enough to melt lead and zinc. One-half of Mercury's surface was photographed at close range by the *Mariner 10* spacecraft in the mid-1970s. (NASA)

*In this chapter you will find the answers to the following questions:*

10-1  What makes Mercury such a difficult planet to see?

10-2  What is so unique about Mercury's rotation?

10-3  How do the surface features on Mercury differ from those on the Moon?

10-4  Is Mercury's internal structure more like that of the Earth or the Moon?

The smallest planet to form in the warm inner regions of the solar nebula is also one of the least explored. Before the advent of space flight, information about Mercury was difficult to obtain for two simple reasons—Mercury's small size and its nearness to the Sun. In fact, Mercury is so close to the Sun that even many astronomers have never seen it. Earth-based visual observations reveal very few surface details, although radio and radar observations have provided information about the planet's temperature and rotation. It was not until 1974 that the unmanned Mariner 10 spacecraft gave us the first close-up images of Mercury. These images revealed that like the Moon, the planet is heavily cratered, still bearing the scars of countless impacts that occurred soon after the planets were formed.

Mercury is by no means a mere clone of the Moon, however. While Mercury has lava flows on its surface, they lack the characteristic dark color of the lunar maria. Mercury's chemical composition is different from that of the Moon or of the other terrestrial planets. The surface of Mercury has unusual wrinkles quite unlike anything found on the Moon or, for that matter, the Earth. And unlike the Moon, Mercury has a large iron core and a magnetic field. Mercury's interior is therefore more like that of the Earth than that of the Moon.

Despite its small size, Mercury is an important planet to astronomers. If we can understand how Mercury formed and acquired its unique personality, we may be able to understand how the other terrestrial planets—including our own Earth—came to be.

## 10-1 Earth-based optical observations of Mercury are difficult

Mercury is often one of the brightest objects in the sky. At its greatest brilliance, it appears brighter than any star, which is why Mercury has been known since ancient times. Its motions played a role in the religious beliefs of the ancient Mayans, Egyptians, Greeks, and Romans.

Like all the planets, Mercury shines by reflected sunlight. Mercury reflects about 12% of the sunlight that falls on its rocky surface. The fraction of incoming sunlight that a planet reflects is called its **albedo** (from the Latin for "whiteness"); Mercury's albedo is about 0.12, roughly comparable to that of weathered asphalt.

Although it may not be dim, Mercury is so close to the Sun that it is quite difficult to observe. Mercury orbits the Sun at an average distance of only 0.387 AU (57.9 million kilometers, or 36 million miles) along an orbit that is more eccentric than that of any other planet except Pluto. Figure 10-1 is a scale drawing of the orbits of Mercury and the Earth. Table 10-1 summarizes data about the planet.

Mercury can best be seen when it is as far from the Sun in the sky as it can be, at its greatest eastern or western elongation (Section 4-2). For a few days near the time of **greatest eastern elongation,** Mercury appears as an "evening star," hovering low over the western horizon for a short time after

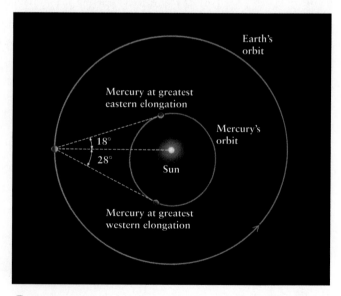

## figure 10-1

**Mercury's Orbit** Mercury moves around the Sun every 88 days in a rather eccentric orbit. The distance between Mercury and the Sun varies from 70 million kilometers (44 million miles) at aphelion to only 46 million kilometers (29 million miles) at perihelion. As seen from the Earth, the angle between Mercury and the Sun at greatest eastern or western elongation can be as large as 28° (when Mercury is near aphelion) or as small as 18° (near perihelion).

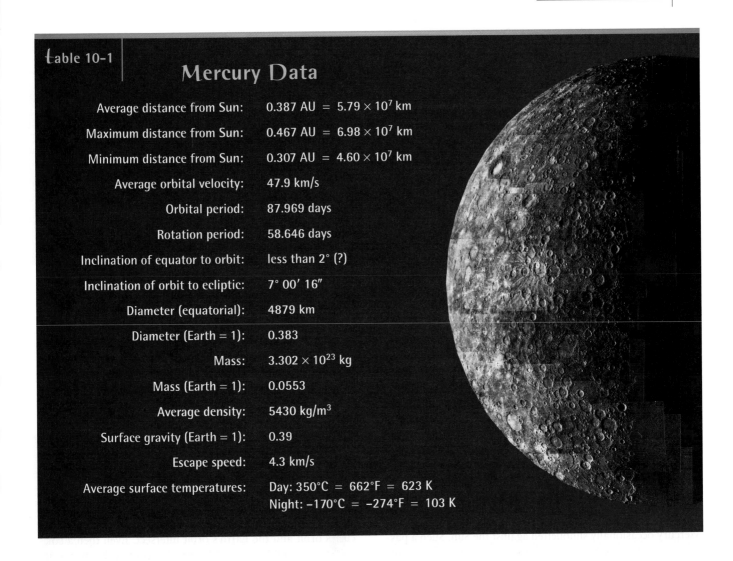

**table 10-1**

## Mercury Data

| | |
|---|---|
| Average distance from Sun: | 0.387 AU = $5.79 \times 10^7$ km |
| Maximum distance from Sun: | 0.467 AU = $6.98 \times 10^7$ km |
| Minimum distance from Sun: | 0.307 AU = $4.60 \times 10^7$ km |
| Average orbital velocity: | 47.9 km/s |
| Orbital period: | 87.969 days |
| Rotation period: | 58.646 days |
| Inclination of equator to orbit: | less than 2° (?) |
| Inclination of orbit to ecliptic: | 7° 00′ 16″ |
| Diameter (equatorial): | 4879 km |
| Diameter (Earth = 1): | 0.383 |
| Mass: | $3.302 \times 10^{23}$ kg |
| Mass (Earth = 1): | 0.0553 |
| Average density: | 5430 kg/m³ |
| Surface gravity (Earth = 1): | 0.39 |
| Escape speed: | 4.3 km/s |
| Average surface temperatures: | Day: 350°C = 662°F = 623 K<br>Night: −170°C = −274°F = 103 K |

sunset. Alternatively, near the time of **greatest western elongation**, Mercury can be glimpsed as a "morning star," heralding the rising Sun in the brightening eastern sky.

CAUTION! In the terms "greatest western elongation" and "greatest eastern elongation", remember that *western* or *eastern* refers to where in the sky Mercury is found relative to the Sun. When Mercury is at greatest western elongation, it is west of the Sun in the sky; at dawn, when the Sun is rising in the east, Mercury has already risen. When Mercury is at greatest eastern elongation, it is east of the Sun in the sky, and it will still be above the horizon when the Sun has set in the west.

Because its orbit is so close to the Sun, Mercury's maximum elongation is only 28°. The celestial sphere rotates at 15° per hour (360° divided by 24 hours), so Mercury never rises more than 2 hours before sunrise or sets more than 2 hours after sunset. Unfortunately, Mercury's elliptical orbit

and its inclination to the ecliptic often place Mercury much less than 28° from the horizon at sunset or sunrise. Some elongations are thus favorable for viewing Mercury and others are not, as shown in Figure 10-2. A total of six or seven greatest elongations (both eastern and western) occur each year (Table 10-2), but usually only two of these will be favorable for viewing the planet.

Naked-eye observations of Mercury are best made at dusk or dawn, but the best telescopic views are obtained at midday when the planet is high in the sky, far above the degrading atmospheric effects near the horizon. A yellow filter can eliminate much of the blue light from our sky. The photographs in Figure 10-3, which are among the finest Earth-based views of Mercury, were taken at midday.

Mercury travels around the Sun faster than any other object in the solar system. Its sidereal period, or time to complete one full orbit, is only 88 days. Its synodic period, which is the time to go through a complete cycle of configurations

While the Sun moves slowly across Mercury's sky, it does not always move from east to west like it does as seen from Earth. The reason is that Mercury's speed along its elliptical orbit varies substantially, in accordance with Kepler's second law (see Section 4-4, especially Figure 4-11). Its orbital velocity is greatest (59 km/s) at perihelion and least (39 km/s) at aphelion. As seen from Mercury's surface, the Sun rises in the east and sets in the west, just as it does on the Earth. When Mercury is near perihelion, however, the planet's rapid motion along its orbit outpaces its leisurely rotation about its axis (one rotation in 58.646 days). Hence, the usual east-west movement of the Sun across Mercury's sky is interrupted, and the Sun actually stops and moves backward (from west to east) for a few Earth days. If you were standing on Mercury watching a sunset occurring at perihelion, the Sun would not simply set. It would dip below the western horizon and then come back up, only to set a second time a day or two later.

---

## 10-3 | Photographs from *Mariner 10* reveal Mercury's heavily cratered, Moonlike surface

We acquired our first detailed knowledge about Mercury's surface in the spring of 1974, when the unmanned *Mariner 10* spacecraft coasted above the planet's surface. At its closest approach, on March 29, 1974, *Mariner 10* passed over Mercury's darkened, nighttime side, and so was not able to obtain images. The spacecraft's photography was therefore divided into two parts—incoming views, taken prior to this closest approach, and outgoing views, taken as *Mariner 10* receded from the planet. The best incoming and outgoing images have been synthesized to make the mosaic shown in Figure 10-8. Box 10-1 summarizes the *Mariner 10* mission.

As *Mariner 10* closed in on Mercury, scientists were struck by the Moonlike pictures appearing on their television monitors. It was obvious that Mercury, like the Moon, is a barren, heavily cratered world, with no evidence for plate tectonics. But craters are not the only features of Mercury's surface; there are also gently rolling plains, long, meandering cliffs, and an unusual sort of jumbled terrain.

Figure 10-9 shows a typical close-up view sent back from *Mariner 10*. The consensus among astronomers is that most of the craters on both Mercury and the Moon were produced during the first 700 million years after the planets formed. Debris remaining after planet formation would have rained down on these young worlds, gouging out most of the craters we see today. As discussed in Section 9-5, the strongest evidence for this chronology comes from analysis and dating of Moon rocks. *Mariner 10* did not land on the surface of Mercury, so we are not able to make the same kind of direct analysis of rocks from Mercury.

The differences between Mercury and the Moon are as striking as the similarities. As an example, Figure 10-10 shows the Moon's southern hemisphere. Notice how the lunar cra-

**f**igure 10-8   R I **V** U X G

**A *Mariner 10* Mosaic of Mercury** This mosaic of Mercury was composed from dozens of incoming and outgoing images recorded by *Mariner 10* as it approached and receded from the planet. The blank swatch running down from Mercury's north pole is a region that was not imaged by *Mariner 10*. The spacecraft also did not observe the hemisphere of Mercury opposite from that shown here. Although most of the images that make up this mosaic were recorded at distances of about 200,000 km (125,000 mi) from Mercury, their resolution is still vastly superior to that of the best Earth-based images (compare Figure 10-3). (Astrogeology Team, U.S. Geological Survey)

ters are densely packed, with one overlapping the next. In sharp contrast, Mercury's surface has extensive low-lying plains (examine the upper half of Figure 10-9). These large, smooth areas are about 2 km lower than the cratered terrain.

As described in Section 9-5, the lunar maria were produced by extensive lava flows between 3.1 and 3.8 billion years ago. Primordial lava flows probably also explain Mercurian plains. As large meteoroids punctured the planet's thin, newly formed crust, lava welled up from the molten interior to flood low-lying areas. Based on the number of craters that pit them, Mercury's plains appear to have been formed near the end of the era of heavy bombardment, a little more than 3.8 billion years ago. Mercury's plains are therefore older than most of the lunar maria. Analysis of *Mariner 10* images made at different wavelengths also shows that the material of which Mercury's plains are made is also lower in iron content than the lunar maria. This is presumably why the plains of Mercury do not have the dark color of the Moon's maria. (Contrast the photographs of Mercury and the Moon in the figure that opens this chapter).

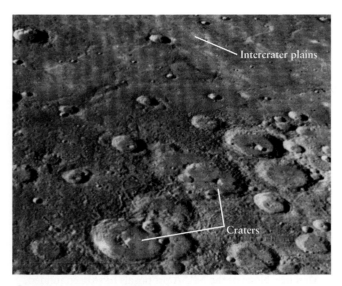

**figure 10-9**    R I **V** U X G

**Mercurian Craters and Plains** This view of Mercury's northern hemisphere was taken by *Mariner 10* at a range of 55,000 km (34,000 mi) from the planet's surface. Numerous craters and extensive intercrater plains appear in this photograph, which covers an area 480 km (300 mi) wide. (NASA)

**figure 10-10**    R I **V** U X G

**Lunar Craters** This Earth-based photograph shows a portion of the Moon's southern hemisphere during last quarter moon. Densely packed craters fill the view, which covers an area approximately 600 km (370 mi) wide. Note how these lunar craters are more densely packed than those on the surface of Mercury, shown in Figure 10-9. (Carnegie Observatories)

*Mariner 10* also revealed numerous long cliffs, called **scarps,** meandering across Mercury's surface (Figure 10-11). Some scarps rise as much as 3 km (2 mi) above the surrounding plains and are 20 to 500 km long. These cliffs probably formed as the planet cooled and contracted, causing its crust to wrinkle like the skin of a dried apple. Because the scarps seem to be younger than the lava flows that produced Mercury's plains, the episode of planetwide contraction began rather late in Mercurian history.

The most impressive feature discovered by *Mariner 10* was a huge impact basin called the Caloris Basin (from the Latin word for "hot"). The Sun is directly over the Caloris Basin during alternating perihelion passages, and thus it is the hottest place on the planet once every 176 days. *Mariner 10*'s cameras could only reveal about half of the Caloris Basin, because at the time that the spacecraft came by, it happened to lie on the terminator (the line dividing day from night, introduced in Section 9-1). Slightly more than half of the impact basin is hidden on the night side of the planet in Figure 10-12.

The Caloris Basin, which measures 1300 km (810 mi) in diameter, is filled and surrounded by smooth plains that resemble lunar maria. The Caloris Basin was probably gouged out by the impact of a large meteoroid that penetrated the planet's crust. Because relatively few craters pockmark the lava flows that filled the basin, the Caloris impact is relatively young. It must have occurred toward the end of the crater-making period that dominated the first 700 million years of our solar system's history.

The Caloris impact must have been a violent event that shook the planet. On the side of Mercury, opposite the Caloris

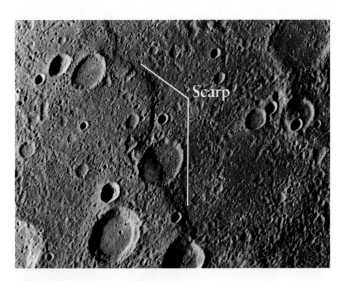

**figure 10-11**    R I **V** U X G

**A Scarp** A long, meandering cliff called Santa Maria Rupes runs from north to south across this *Mariner 10* image of a region near Mercury's equator. This cliff, called a scarp by geologists, is more than a kilometer high and extends for several hundred kilometers. Such scarps are thought to be "wrinkles" that formed when Mercury's crust cooled and shrank. The scarp in this photograph passes through an old, large crater at the center of the photograph; the crater was distorted vertically when the scarp formed. Another form of geologic activity can be seen at the upper right of the image, where a crater 70 km (45 mi) in diameter has apparently been flooded by lava that welled up from Mercury's interior. This photograph covers an area measuring approximately 200 km (120 mi) across. (NASA)

## Key Words

1-to-1 spin-orbit coupling, p. 247

3-to-2 spin-orbit coupling, p. 248

albedo, p. 244

dynamo effect, p. 264

greatest eastern elongation, p. 244

greatest western elongation, p. 245

magnetosphere, p. 254

scarp, p. 252

solar transit, p. 246

synchronous rotation, p. 247

## Key Ideas

**Motions of Mercury in the Earth's Sky:** At its greatest eastern and western elongations, Mercury is never more than 28° from the Sun, so it can be seen only briefly after sunset or before sunrise.

• Solar transits of Mercury occur about a dozen times per century.

**Mercury's Rotation:** Poor telescopic views of Mercury's surface led to the mistaken impression that the planet always keeps the same side toward the Sun, a configuration called synchronous rotation or 1-to-1 spin-orbit coupling.

• Radio and radar observations in the 1960s revealed that Mercury in fact has 3-to-2 spin-orbit coupling. For an observer on Mercury, the average time from sunset to sunrise (one-half of a solar day) is equal to the time for a complete orbit around the Sun (one Mercurian year).

**The Surface of Mercury:** The *Mariner 10* spacecraft made several passes near Mercury in the mid-1970s, providing pictures of its surface.

• The Mercurian surface is pocked with craters like those of the Moon, but there are extensive smooth plains between these craters.

• Long cliffs called scarps meander across the surface of Mercury. These scarps probably formed as the planet cooled, solidified, and shrank.

• The impact of a large object long ago formed the huge Caloris Basin and shoved up jumbled hills on the opposite side of the planet.

**Mercury's Interior and Magnetic Field:** Like the Earth, Mercury has an iron-rich core.

• The iron core of Mercury has a diameter equal to three-fourths of the planet's diameter, whereas the diameter of the Earth's core is only slightly more than one-half of the Earth's diameter.

• Mercury has a weak magnetic field, indicating that at least part of the iron core is liquid. The magnetic field produces a magnetosphere around Mercury that blocks the solar wind from reaching the surface of the planet.

## Review Questions

*Questions preceded by an asterisk (\*) involve topics discussed in Box 10-1.*

**1.** Why are naked-eye observations of Mercury best made at dusk or dawn, while telescopic observations are best made around noon?

**2.** Why can't you see any surface features on Mercury when it is closest to the Earth?

**3.** Explain why Mercury does not have a substantial atmosphere.

**4.** What is synchronous rotation? How is the rotation period of an object exhibiting synchronous rotation related to its orbital period?

**5.** How is Mercury's rotation rate related to its orbital period? What telescopic observations proved this?

**\*6.** Explain why *Mariner 10* was able to photograph only one side of Mercury even though the spacecraft returned to the planet three times.

**7.** Compare the surfaces of Mercury and our Moon. How are they similar? How are they different?

**8.** How can you tell an old crater from a new one?

**9.** What kind of tectonic features are found on Mercury? Why are they probably much older than tectonic features on the Earth?

**10.** Explain why the Sun is directly over the Caloris Basin on Mercury only every other time that the planet is at perihelion.

**11.** Why do astronomers think that Mercury has a very large iron core?

**12.** Briefly describe at least one possible history that would account for Mercury's large iron core.

**13.** With the aid of a drawing, describe Mercury's magnetosphere. Why do you suppose Mercury does not have Van Allen belts?

## Advanced Questions

### Problem-solving tips and tools:

You may need to refresh your memory about the small-angle formula, found in Box 1-2; about Kepler's third law, described in Section 4-4; and about Wien's law and the Doppler effect, discussed in Section 5-4 and Section 5-9, respectively. The circumference of a circle of radius $r$ is $2\pi r$. The criteria for a planet to be able to retain an atmosphere are discussed in Box 7-2.

**14.** Find the largest angular size that Mercury can have as seen from the Earth. In order for Mercury to have this apparent size, at what point in its orbit must it be?

**15.** Suppose you have a superb telescope that can resolve features as small as 1 arcsec across. What is the size (in kilometers) of the smallest surface features you should be able to see on Mercury? How does your answer compare with the size of the Caloris Basin? (*Hint:* Assume that you choose to observe Mercury when it is at greatest elongation, about 25° from the Sun. As Figure 10-1 shows, at this point in its orbit, the Earth-Mercury distance is about the same as the Earth-Sun distance.)

**16.** Explain why November solar transits of Mercury, which occur near the time of perihelion passage, are more common than May transits.

**17.** Find the value of $\lambda_{max}$ for blackbody radiation coming from the sunlit side of Mercury.

**18.** If the albedo of Mercury were increased, would the planet's surface temperature go up or down? Explain your answer.

**19.** (a) Mercury has a 58.65-day rotation period. What is the speed at which a point on the planet's equator moves due to this rotation? (*Hint:* Remember that speed is distance divided by time. What distance does a point on Mercury's equator travel as the planet makes one rotation?) (b) Use your answer to (a) to answer the following: As a result of rotation, what difference in wavelength is observed for a radio wave of wavelength 12.5 cm (such as is actually used in radar studies of Mercury) emitted from either the approaching or receding edge of the planet?

**20.** (a) Calculate the minimum molecular weight of a gas that could in theory be retained as an atmosphere by Mercury if the average daytime temperature were 620 K. b) Are there any abundant gases that meet this minimum criterion? Why doesn't Mercury have an atmosphere of these gases? (*Hint:* The mass of a molecule is equal to its molecular weight, $\mu$, multiplied by the mass of a hydrogen atom, $m_H = 1.67 \times 10^{-27}$ kg. The molecular weight of a molecule equals the sum of the atomic weights of its atoms, which can be looked up in a periodic table of the elements. Thus, for example, the molecular weight of $CO_2$ is $12.0 + 16.0 + 16.0 = 44.0$, and so the molecule's mass is 44.0 $m_H$.)

**21.** How much would an 80-kg person weigh on Mercury? How does that compare with that person's weight on the Moon? How much does that person weigh on the Earth?

**22.** When an impact crater is formed, material (called *ejecta*) is sprayed outward from the impact. (Look at the photograph of the Moon on the right-hand side of the figure that opens this chapter. At the lower right of this photograph, you can see the light-colored ejecta surrounding the crater Stevinus.) While ejecta are found surrounding the craters on Mercury, they do not extend as far from the craters as do ejecta on the Moon. Explain why, using the difference in surface gravity between the Moon and Mercury.

**23.** The orbital period of *Mariner 10* is twice that of Mercury. Use this fact to calculate the length of the semimajor axis of the spacecraft's orbit.

**24.** One theory for the origin of Mercury's magnetic field is that the planet has a solid iron-bearing mantle that is permanently magnetized like a giant bar magnet. Using the fact that iron demagnetizes at temperatures above 770°C, present an argument against this theory.

## Discussion Questions

**25.** If you were planning a return mission to Mercury, what features and observations would be of particular interest to you?

**26.** What evidence do we have that the surface features on Mercury were not formed during recent geological history?

## Observing Projects

### Observing tips and tools:

Remember that Mercury is visible in the morning sky when it is at or near greatest western elongation and in the evening sky when at or near greatest eastern elongation. You can consult such magazines as *Sky & Telescope* and *Astronomy*, or the web sites for these magazines (**http://www.skypub.com/** and **http://www.astronomy.com/**, respectively), for more detailed information about when and where to look for Mercury during a given month.

**27.** Refer to Table 10-2 to determine the dates of the next two or three greatest elongations of Mercury. Consult such magazines as *Sky & Telescope* and *Astronomy*, or the web sites for these magazines (see addresses above) to determine if any of these greatest elongations is going to be a favorable one. If so, make plans to be one of those rare individuals who has actually seen the innermost planet of the solar system. Set aside several evenings (or mornings) around the date of the favorable elongation to reduce the chances of being "clouded out." Select an observing site that has a clear, unobstructed view of the horizon where the Sun sets (or rises). If possible, make arrangements to have a telescope at your disposal. Search for the planet on the dates you have selected, and make a drawing of its appearance through your telescope.

 **28.** This observing project should be performed only under the direct supervision of an astronomer who knows how to point a telescope safely at Mercury. Make arrangements to view Mercury during broad daylight. This is best done by visiting an observatory where the coordinates (right ascension and declination) of Mercury's position can be used to point the telescope. **DO NOT LOOK AT THE SUN! Looking directly at the Sun can cause blindness.**

## Where to Learn More

*Books and magazines*

Davies, M., and others, eds. *Atlas of Mercury.* NASA SP-423, 1978. This oversized book from NASA includes many high-quality enlargements of the *Mariner 10* pictures of Mercury.

Hartmann, W. "The Significance of the Planet Mercury." *Sky & Telescope,* May 1976. This excellent article by a noted planetary scientist discusses the importance of what we have learned from the *Mariner 10* mission to Mercury.

Murray, B., and Burgess, E. *Flight to Mercury.* Columbia University Press, 1977. This exciting book, which includes an excellent selection of photographs, interweaves the history of the *Mariner 10* mission with newsworthy events that often overshadowed the mission.

Strom, R. *Mercury: The Elusive Planet.* Smithsonian Institution Press, 1987. This clear, well-written, and copiously illustrated book is a superb nontechnical introduction to the innermost planet.

————. "Mercury: The Forgotten Planet." *Sky & Telescope,* September 1990. This fascinating article explains some of the post-*Mariner 10* ideas about Mercury, including the theory that Mercury's large iron core resulted from a devastating impact 4.5 billion years ago.

**W** *World Wide Web*

Overview discussions of Mercury, as well as a number of images from *Mariner 10,* can be found at several web sites: "Views of the Solar System" (**http://www.hawastsoc.org/solar/eng/mercury.htm**), "The Nine Planets" (**http://www.seds.org/nineplanets/nineplanets/mercury.html**), and "Welcome to the Planets" (**http://pds.jpl.nasa.gov/ planets/welcome/mercury.htm**).

A detailed, illustrated history of the trials, tribulations, and ultimate rewards of the *Mariner 10* mission is available from NASA's Jet Propulsion Laboratory (**http://pdc.jpl.nasa.gov/Mariner10/Mariner10.html**). While there have been no new space missions to Mercury since *Mariner 10,* the wealth of data returned by that spacecraft is still being analyzed by scientists. Some recent results are discussed at the University of Hawaii's Planetary Science Research Discoveries web site (**http://www.soest.hawaii.edu/PSRdiscoveries/Jan97/MercuryUnveiled.html**).

# Cloud-Covered Venus

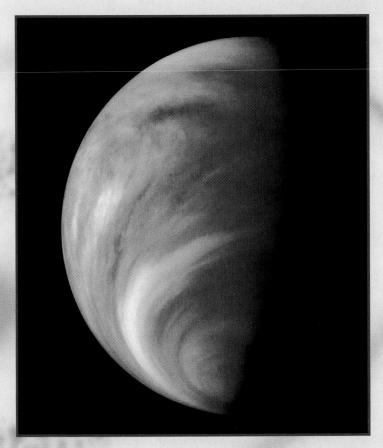

R I V U X G

Venus  Venus and the Earth have nearly the same size, mass, and surface gravity. However, Venus's dense carbon dioxide atmosphere efficiently traps energy from sunlight, resulting in a surface temperature on Venus of 750 K (= 480°C = 900°F), which is even hotter than Mercury. Unlike the Earth's clouds, which are made of water droplets, Venus's clouds are made of droplets of concentrated sulfuric acid. Active volcanoes may be responsible for maintaining these sulfur-rich clouds. This ultraviolet image was obtained by the *Pioneer Venus Orbiter* in 1979. (NASA)

*In this chapter you will find the answers to the following questions:*

11-1  What makes Venus such a brilliant "morning star" or "evening star?"

11-2  What is strange about the rotation of Venus?

11-3  Why is it unlikely that humans will ever visit the surface of Venus?

11-4  In what ways are the clouds in the Venusian atmosphere dramatically different from those on the Earth?

11-5  Why do astronomers suspect that there are active volcanoes on Venus?

11-6  Why is there almost no water on Venus today? Why do astronomers think that water was once very common on Venus?

11-7  Does Venus have the same kind of plate tectonic activity as the Earth?

At first glance, Venus looks like the Earth's twin. The two planets have almost the same mass, same diameter, same average density, and same surface gravity (see Figure 7-3 for a comparison of the Earth and Venus). Table 11-1 lists some basic data about Venus. But Venus is closer to the Sun than the Earth, so it is exposed to more intense sunlight. This has transformed this potentially Earthlike planet into a world that is utterly hostile to living organisms, with a crushingly thick atmosphere and an almost complete lack of water. Venus also has a perpetual cloud cover, which for many years kept astronomers from learning much about the planet. The great revolution in the study of Venus was the use of radio waves and radar, which can effortlessly penetrate the clouds. Radar observations not only demonstrated that Venus rotates backward but also gave us pictures of the planet's surface, which has two large "continents" and is mostly covered with gently rolling hills.

Beginning in the 1970s, spacecraft sent to Venus provided astronomers with their first high-resolution images of the Venusian surface. These Soviet and American missions also confirmed that Venus has an extremely hot, dense atmosphere of carbon dioxide with clouds of sulfuric acid droplets. Active volcanoes may be responsible for the high sulfur content of the Venusian clouds. The greenhouse effect is responsible for Venus's high temperature—the same greenhouse effect that, in a milder form, acts in our own atmosphere and that might someday cause our polar ice caps to melt. Hence, an understanding of our sister planet may give important insights into the history and future of our own world.

## 11-1 The surface of Venus is hidden beneath a thick, highly reflective cloud cover

Venus, familiar for centuries as the "morning star" or "evening star," is easy to identify. Its orbit makes it possible to see Venus without interference from the Sun's glare, and at times Venus is the brightest object in the night sky. These simple observations tell us quite a bit about Venus's orbit and atmosphere.

Venus can be seen fairly far from the Sun—at its greatest elongation, about 47° away (Figure 11-1)—because its orbit is almost twice as large as that of Mercury. At its greatest eastern elongation, when it is called an "evening star," Venus is seen high above the western horizon after sunset. At greatest western elongation, Venus is called the "morning star," because it rises nearly 3 hours before the Sun and is positioned high in the eastern sky at dawn. Table 11-2 lists the dates of greatest eastern and western elongation for Venus from 1998 to 2004.

At its greatest brilliance, Venus is 16 times brighter than the brightest star and is outshone only by the Sun and the Moon. There are three reasons why: Venus is close to the Sun, it is close to the Earth, and it reflects 76% of the sunlight that falls on the planet (its albedo is 0.76).

Through Earth-based telescopes, Venus appears almost completely featureless, with neither continents nor mountains. However, astronomers soon realized that they were not seeing the true surface of the planet. Rather, Venus must be covered by a thick, unbroken layer of clouds. A cloud layer would also explain why Venus reflects such a large fraction of the sunlight that falls on it. Direct evidence that Venus has a thick atmosphere came in the nineteenth century, when astronomers observed Venus near the time of inferior conjunction. This is when Venus lies most nearly between the Earth and the Sun so that we see the planet lit from behind. As Figure 11-2 shows, sunlight is scattered by Venus's atmosphere, producing a luminescent ring around the planet that would not otherwise be present.

One other aspect of Venus that can be observed in visible light is the phenomenon of a solar transit, in which Venus passes directly in front of the Sun at inferior conjunction. This does not happen at every inferior conjunction, because the plane of the orbit of Venus is inclined by 3.4° to the ecliptic (the plane of the Earth's orbit). Indeed, solar transits of Venus are exceedingly rare; the last two were in 1874 and 1882, and not one transit of Venus will occur during the

## table 11-1

# Venus Data

| | |
|---|---|
| Average distance from Sun: | 0.723 AU = $1.082 \times 10^8$ km |
| Maximum distance from Sun: | 0.728 AU = $1.089 \times 10^8$ km |
| Minimum distance from Sun: | 0.718 AU = $1.075 \times 10^8$ km |
| Eccentricity of orbit: | 0.0068 |
| Average orbital speed: | 35.0 km/s |
| Orbital period: | 224.70 days |
| Rotation period: | 243.01 days (retrograde) |
| Inclination of equator to orbit: | 177.3° |
| Inclination of orbit to ecliptic: | 3.39° |
| Diameter (equatorial): | 12,104 km |
| Diameter (Earth = 1): | 0.949 |
| Mass: | $4.869 \times 10^{24}$ kg |
| Mass (Earth = 1): | 0.815 |
| Average density: | 5240 kg/m³ |
| Surface gravity (Earth = 1): | 0.91 |
| Escape speed: | 10.4 km/s |
| Average surface temperature: | 480°C = 900°F = 750 K |

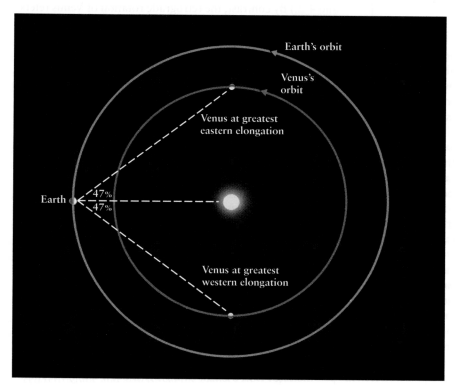

## figure 11-1

**Venus's Orbit** Venus travels around the Sun along a nearly circular orbit with a period of 224.7 days. The average distance between Venus and the Sun is 108 million kilometers (the average distance between the Earth and the Sun is 150 million kilometers). At its greatest eastern elongation, Venus appears as a prominent evening "star"; at its greatest western elongation, it is a prominent morning "star."

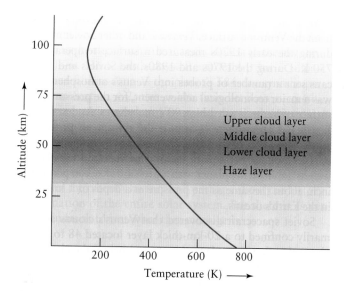

## Figure 11-5

**Temperature in the Venusian Atmosphere** This graph depicts how the temperature of Venus's atmosphere varies with altitude. As you descend into the atmosphere, the temperature increases smoothly from a minimum of about 170 K (= –100°C = –150°F) at an altitude of 100 km to a maximum of nearly 750 K (about 480°C, or 900°F) on the ground. During this descent you pass through Venus's cloud layers, which lie at altitudes between 48 and 68 km (30 to 42 mi) above the surface. By comparison, clouds in the Earth's atmosphere are seldom found at altitudes above 12 km (8 mi).

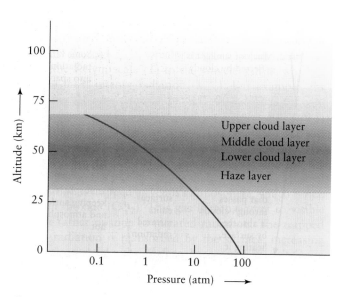

## Figure 11-6

**Pressure in the Venusian Atmosphere** The graph shows how the pressure of Venus's thick, low atmosphere depends on altitude. The Venusian atmosphere is so thick that the pressure at an altitude of 50 km (160,000 ft) is 1 atm, the pressure at the Earth's surface. Atmospheric pressure increases smoothly with decreasing altitude, reaching a crushing 90 atm (1300 pounds per square inch) at the surface.

substances known to chemists, capable of dissolving lead, tin, and most rocks. Imagine the difficulty in designing a spacecraft to descend through such a witches' brew!

The Venusian atmosphere differs from our own not only in its pressures, temperatures, and chemical composition, but also in the patterns of its global circulation. Astronomers have been able to study these patterns using spacecraft in orbit

around Venus. For example, in the early 1980s, the American *Pioneer Venus Orbiter* sent back numerous images of Venus taken at ultraviolet wavelengths, where its atmospheric markings stand out best (Figure 11-7). By following individual cloud markings, scientists determined that the upper levels of Venus's atmosphere rotate around the planet in just four days. This confirmed earlier observations made with ground-based

## Figure 11-7  R I V **U** X G

**Venus's Cloud Patterns** These four ultraviolet views of Venus were taken by the *Pioneer Venus Orbiter* in May 1980 at a distance of approximately 50,000 km. All the views show variations of the so-called V feature, produced by the rapid motion of the clouds around the planet. These motions are in the same retrograde direction as the rotation of the planet itself but at a much greater speed. (NASA)

# box 11-1 | Looking Deeper into Astronomy

## *Venus's Ionosphere*

The first successful mission to Venus was the flight of *Mariner 2* in 1962. During its three-month interplanetary voyage, this spacecraft verified the existence of the solar wind. The solar wind, which we introduced in Section 7-8, consists of charged particles escaping from the Sun at high speeds. Our first measurement of the density (about 1 particle/cm$^3$) and speed (roughly 400 km/s) of the solar wind came from *Mariner 2*. You can see a replica of *Mariner 2* on display at the National Air and Space Museum in Washington, D.C., or by visiting the museum on the World Wide Web (**http://www.nasm.edu/**).

As discussed in Sections 8-5 and 10-4, both the Earth and Mercury have magnetic fields, and these fields produce magnetospheres around the planets that protect their surfaces from the solar wind. By contrast, as *Mariner 2* passed Venus, its instruments failed to detect any magnetic field whatsoever. Subsequent missions confirmed this lack of a magnetic field. With no magnetic field, and hence no magnetosphere, the solar wind impinges directly on Venus's upper atmosphere.

As the fast-moving particles of the solar wind collide with atmospheric atoms, they strip many of the atoms of one or more electrons. An atom that carries an excess charge because it has gained or lost one or more electrons is called an **ion**. The ions in Venus's atmosphere have a positive electric charge, because they have lost negatively charged electrons. The electromagnetic interaction of these ions and the supersonic charged particles in the solar wind produces a shock wave where the supersonic flow of solar wind particles abruptly becomes subsonic (see illustration). Along a boundary called the **ionopause**, inside the shock wave, the pressure of the ions just counterbalances the pressure from the solar wind. The ionopause is analogous to the magnetopause that surrounds a planet with a magnetic field.

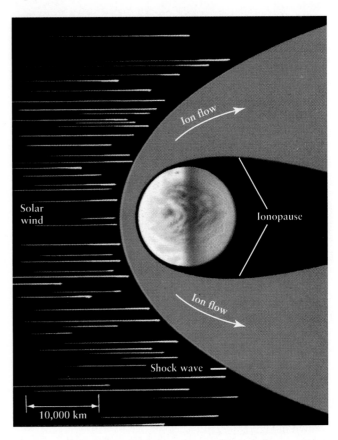

Solar wind

Ion flow

Ionopause

Ion flow

Shock wave

10,000 km

Because Venus has no magnetic field, the ions in Venus's ionosphere are not trapped but are driven away from Venus along with the solar wind. Thus, the layer of ionized atoms around Venus is continually both depleted and replenished by the effects of the Sun.

---

telescopes. The rapid motion of Venus's upper atmosphere is in sharp contrast to the slow rotation of the solid planet itself. Like the rotation of the planet, the motion of the atmosphere is in a retrograde direction (from east to west).

The American *Pioneer Venus Multiprobe* determined the dominant circulation patterns in the Venusian atmosphere after plunging into it in 1978. Warmed by the Sun, hot gases in the equatorial regions rise upward and travel in the upper cloud layer toward the cooler polar regions. At the polar latitudes, the cooled gases sink to the lower cloud layer, in which they are transported back toward the equator. Recall from Sections 8-3 and 8-4 that this process of heat transfer, whereby hot gases rise while cooler gases sink, is called **convection.**

The circulation of Venus's atmosphere is dominated by two huge **convection cells,** one in the northern hemisphere and another in the southern hemisphere, which circulate gases between the equatorial and polar regions of the planet (Figure 11-8). These convection cells, which are almost entirely contained within main cloud layers, are called "driving cells," because they propel similar circulation cells above and below the main cloud deck somewhat like a meshed set of gears.

As Venus's upper atmosphere rotates around the planet in only four days, strong prevailing winds with speeds of

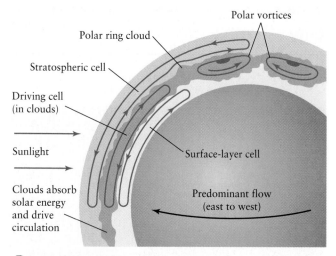

**ƒigure 11-8**

**The Circulation in Venus's Atmosphere** Strong convection cells within Venus's clouds drive the circulation patterns above and below the cloud layer. Circular wind patterns (vortices) in the polar regions force the cloud layer to bulge upward at polar latitudes, producing a polar ring cloud that is clearly visible in many *Pioneer Venus Orbiter* photographs. (Adapted from A. Seiff)

350 km/h (220 mi/h) blow from east to west at high altitude. These winds stretch out the driving cells and produce the characteristic V-shaped, chevron-like patterns that dominate the planet's appearance in Figure 11-7. (In a similar way, the Earth's rotation stretches out the convection cells in our atmosphere to produce the complicated circulation patterns shown in Figure 8-22.)

On the Earth, friction between the atmosphere and the ground causes wind speeds at the surface to be much less than at high altitude. The same is true on Venus: The greatest wind speed measured by spacecraft on Venus's surface is only about 5 km/h (3 mi/h).

## 11-5 Volcanic eruptions are probably responsible for Venus's clouds

Sulfur is found in the Earth's atmosphere in only trace amounts. Why, then, are sulfur and sulfur compounds found in abundance in the atmosphere of Venus? The answer appears to have been provided by long-term observations by *Pioneer Venus Orbiter*: Sulfurous gases are being injected into the Venusian atmosphere by processes that could only be volcanic. When the spacecraft arrived at Venus in 1978, its ultraviolet spectrometer recorded unexpectedly high levels of sulfur dioxide and sulfuric acid, which steadily declined over the next several years. A similar, anomalously high abundance of haze particles may have occurred in the late 1950s,

as indicated by telescopic observations made at that time from the Earth. In explanation, Larry W. Esposito of the University of Colorado proposed that in both the late 1950s and the late 1970s, energetic volcanic eruptions injected sulfur dioxide into the upper atmosphere. It therefore seems that Venus's sulfur-rich clouds may be regularly replenished by active volcanoes.

More strong evidence for volcanic activity on Venus is provided by the chemicals in the Venusian atmosphere. These include all of the major volcanic chemicals spewed by the Earth's volcanoes. Because many of these substances are highly reactive and very short-lived, they must be constantly replenished by new eruptions.

Antennas on *Pioneer Venus Orbiter* and the Jupiter-bound *Galileo* spacecraft have picked up further evidence of active volcanoes—radio bursts thought to be caused by strokes of lightning. Lightning discharges have often been seen in the plumes of erupting volcanoes on the Earth.

The first views of Venus's volcanoes came from Earth-based radar observations of the Venusian surface. Astronomers have often used the giant 305-m (1000-ft) dish at the Arecibo Observatory in Puerto Rico (see Figure 10-5) to map Venus, because microwaves easily penetrate the Venusian clouds and reflect off its surface. The Arecibo dish is used to transmit a powerful, brief burst of microwaves toward Venus. Different types of terrain reflect microwaves more or less efficiently, and by analyzing the radar echo, astronomers are able to construct a map of the Venusian surface. This method is successful only when Venus is near inferior conjunction. At other times, the Venus-Earth distance is too great and the radar echo too weak to produce good data.

Figure 11-9 is a radar view of Theia Mons, the largest volcano on Venus. It rises to an altitude of 6000 m (20,000 ft) and has gently sloping sides that extend over an area 1000 km (620 mi) in diameter. A volcano having this characteristic shape is called a **shield volcano,** because in profile it resembles an ancient Greek warrior's shield lying on the ground. Shield volcanoes also exist on both the Earth and Mars, the best-known example being the Hawaiian Islands. Shield volcanoes on Venus and Mars, as well as the Hawaiian Islands on the Earth, are thought to be the result of **hot-spot volcanism,** whereby a hot region beneath the planet's surface extrudes molten rock over a long period of time (see Section 8-3).

By far the best images of the Venusian volcanoes have come from the highly successful *Magellan* spacecraft that went into orbit around Venus in 1990. Between 1990 and 1992, *Magellan* used sophisticated radar techniques to map more than 98% of the planet with a resolution of 100 m. The spacecraft observed over 1600 major volcanoes and volcanic features on Venus. Figure 11-10 shows a typical *Magellan* image of a volcano.

From *Magellan*'s radar images, combined with radar height measurements, scientists have constructed a three-dimensional map of the Venusian surface. They have used this map and a supercomputer to create vivid perspective views, like the one in Figure 11-11. With such perspective images, scientists can learn much more about volcanism on Venus than was possible

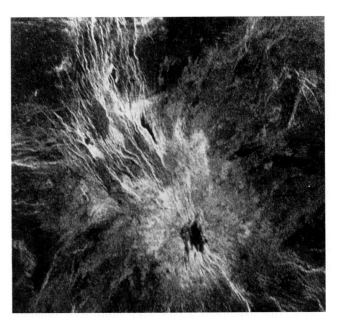

figure 11-9  R I V U X G

**An Earth-Based Radar Image of a Venusian Volcano** This image, produced by Earth-based astronomers using radar, shows details of the lava flows and rifts that surround a large volcano on Venus. The picture covers an area approximately 1700 by 1500 km (110 by 900 mi), roughly three times the size of the state of Texas. The volcano's summit is the darkish spot toward the bottom center, and lava has flowed from the summit toward the upper left of this image. (Courtesy of D. B. Campbell, Arecibo Observatory)

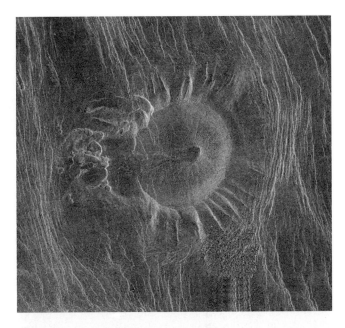

figure 11-10  R I V U X G

**A *Magellan* Radar Image of a Venusian Volcano** This *Magellan* radar image, made from an altitude of 720 km (450 mi) above the Venusian surface, shows the superior resolution of *Magellan* radar images of the Venusian surface as compared to Earth-based images, such as Figure 11-9. The picture covers an area of only 83 by 73 km (52 by 45 mi), and the smallest details visible are about 100 m across (the size of a football stadium). The flat summit of this volcano, about 35 km (22 mi) in diameter, is broken by a number of ridges and valleys, which give the volcano an insect-like appearance. (NASA)

figure 11-11  R I V U X G

**A *Magellan* Image of a Young Volcano on Venus** This volcano, called Maat Mons, is the second tallest on Venus. This computer-generated radar image has a vertical exaggeration of 23 to 1, equivalent to stretching your house into the shape of the Washington Monument. Maat Mons actually has gentle slopes, less steep than terrestrial shield volcanoes like those in Hawaii. Numerous cracks cross the gently rolling foothills at the base of the volcano. False color has been added to this radar image to suggest the actual color of sunlight that reaches the Venusian surface through the thick cloud layer. The dark and light areas are not indicative of the true colors but rather of whether those areas reflect radio waves weakly (dark areas) or strongly (light areas). The light-colored areas on the slopes of Maat Mons are lava flows that are estimated to be no more than 10 million years old. (NASA)

**figure 11-14**

**The Mount St. Helens Eruption of May 1980** A preponderance of data suggests that the terrestrial planets obtained their atmospheres from volcanic outgassing. Geologists study this outgassing process in eruptions of volcanoes on the Earth, such as that of Mount St. Helens in Washington State on May 18, 1980. (USGS)

ature, no amount of atmospheric pressure can keep liquid water from turning into vapor. Venus's oceans would have begun to evaporate, and the added water vapor in the atmosphere would have increased the greenhouse effect. This would have made the temperature even higher and caused the oceans to evaporate faster, producing more water vapor. That in turn would have further intensified the greenhouse effect and made the temperature climb higher still. This is an example of a **runaway greenhouse effect,** in which an increase in temperature causes a further increase in temperature, and so on.

**ANALOGY** A runaway greenhouse effect is like a house in which the thermostat has accidentally been connected backward. If the temperature in such a house gets above the value set on the thermostat, the heater comes on and makes the house even hotter.

The temperature in Venus's atmosphere would not have increased indefinitely. In time, solar ultraviolet radiation striking molecules of water vapor would have broken them into hydrogen and oxygen atoms. The light hydrogen atoms would have then escaped into space (see Box 7-2 for a discussion of why lightweight atoms can escape a planet's gravitational attraction more easily than heavy atoms). The remaining atoms of oxygen, which is one of the most chemically active elements, would have readily combined with other substances in Venus's atmosphere. Eventually, almost all of the water vapor would have been irretrievably lost from Venus's atmosphere, and the runaway greenhouse effect would have come to an end.

Although almost all water vapor has disappeared from Venus, its effects are still felt today. While the water vapor was still present and producing a runaway greenhouse effect, the high temperatures "baked out" any carbon dioxide that was trapped in carbonate rocks. This liberated carbon dioxide formed the thick atmosphere of present-day Venus. Because there is essentially no more carbon dioxide available to add to the atmosphere, the greenhouse effect on Venus is no longer a runaway one, and the infrared-absorbing properties of carbon dioxide have stabilized the surface temperature at its present value of 750 K. Thus, the state of today's Venusian atmosphere, in which only one in every 30,000 molecules is a water molecule, is a direct result of an earlier atmosphere that was predominantly water vapor.

## 11-7 The surface of Venus shows no evidence of plate tectonics

Venus is remarkably flat. More than 80% of the planet's surface is covered with volcanic plains that are the result of numerous lava flows. There are no mountain ranges of the sort found on the Earth, no subduction zones, and no evidence for continental drift. To make these discoveries about Venus, astronomers first had to create a comprehensive map of the planet.

*Magellan* mapped Venus using a radar altimeter that bounced microwaves off the ground directly below the spacecraft. By measuring the time delay of the radar echo, scientists could determine the heights and depths of Venus's hills and valleys. The results are shown in Figure 11-15, which uses a color code to denote elevation. Note that two large highlands, or "continents," rise well above the generally level surface of the planet.

The continent in the northern hemisphere (at the top of Figure 11-13) is Ishtar Terra, named after the Babylonian goddess of love. Ishtar Terra, approximately the same size as Australia, is dominated by a high plateau ringed by high mountains. The highest mountain is Maxwell Montes, whose summit rises to an altitude of 11 km (7 mi) above the average surface elevation. For comparison, Mount Everest on the Earth rises 9 km (6 mi) above sea level.

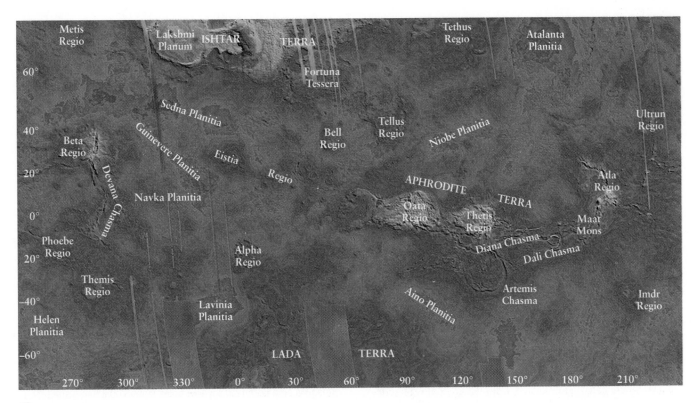

## figure 11-15

**A *Magellan* Topographic Map of Venus** Radar altimeter measurements by *Magellan* were used to produce this topographic map of Venus, which covers latitudes from 69° north to 69° south. The Venusian equator extends horizontally across the middle of the map. Color indicates elevation—red corresponds to the highest, blue to the lowest. Gray areas were not mapped by *Magellan*. The yellow and red area to the north (at the top of the map) is the elevated region Ishtar Terra. This region is dominated by Maxwell Montes, the planet's highest mountains, which are 11 km (36,000 ft) above the planet's average elevation. Southwest of Ishtar (to the left and down) are the highlands of Beta Regio and Phoebe Regio. The large, scorpion-shaped feature extending along the equator is Aphrodite Terra, a continent-like highland that contains several spectacular volcanoes. Maat Mons, the volcano shown in Figure 11-10, lies within the yellow area at the far right (that is, the eastern end) of Aphrodite Terra. The channel depicted in Figure 11-12 lies at the lower (southern) end of Thermis Regio, toward the lower left of this map. (Peter Ford and Gordon Pettengill, MIT)

The largest Venusian continent, Aphrodite Terra, is a vast belt of highlands just south of the equator (near the center of Figure 11-15). It is 16,000 km in length and 2000 km wide, giving it an area comparable to that of Africa. (Aphrodite was the Greek goddess whom the Romans called Venus.) The global radar image of Venus in Figure 11-16 shows that most of Aphrodite Terra is covered by vast networks of faults and fractures.

The presence of volcanism on Venus strongly suggests that Venus, like the Earth, has a molten interior. Unfortunately, we have no seismic data to back up this conclusion. None of the *Venera* spacecraft that landed on Venus carried seismometers, and, in any event, the instruments on these spacecraft only lasted a few hours before succumbing to Venus's harsh environment.

Before *Magellan*, geologists wondered if Venus's molten interior had given rise to plate tectonics, like those that have remolded the face of the Earth. If this were the case, then the same tectonic effects might also have shaped the surface of Venus. As described in Section 8-3, the Earth's hard outer shell, or lithosphere, is broken into about a dozen large plates that slowly shuffle across the globe. Long mountain ranges, like the Mid-Atlantic Ridge (Figure 8-14), are created where fresh magma wells up from the Earth's interior to push the plates apart.

Images beamed back from *Magellan* show *no* evidence of Earthlike tectonics on Venus. Specifically, the images failed to show a type of faulting that always occurs with seafloor spreading on the Earth. These faults, which are perpendicular to rifts on the ocean floor, give the Earth's ocean ridges

**figure 11-16** R I V U X G

**A Global View of Venus** In this radar image of Venus, Aphrodite Terra is the elongated, light-colored, wispy feature that wraps one-third of the way around the planet. Aphrodite is roughly parallel to Venus's equator, which extends horizontally across the middle of the picture. This image is actually a mosaic of many *Magellan* images, each of which shows only a small portion of the planet's surface. North is at the top. (NASA)

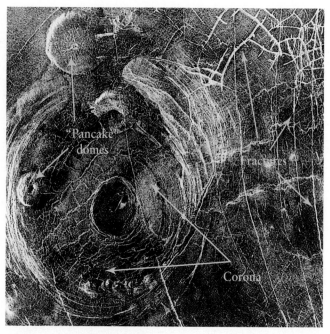

**figure 11-17** R I V U X G

**A Corona** The large circular structure near the center of this image is Aine Corona, a shallow dome some 200 km in diameter pushed upward by rising magma from Venus's mantle. The corona is surrounded by a number of smaller volcanic domes, called "pancake domes" for their shape, as well as by prominent and complex fracture patterns. Coronae (from the Latin for "crown") are unique to Venus; these geologic features have not been found on any other planet. Aine Corona lies in the vast plain to the south of Aphrodite Terra, shown in Figure 11-15. (NASA)

an appearance that distinctly resembles a staircase (examine Figure 8-14). Because Venus lacks features like these, we know that there can have been only limited horizontal displacement of its lithosphere. But there have been local, small-scale deformations and reshaping of the surface. As an example, roughly a fifth of Venus's surface is covered by folded and faulted ridges.

One possible explanation of these observations is that Venus's lithosphere is thinner and weaker than that of the Earth. A thin, weak crust would be unable to support plate tectonic activity, although it could withstand the localized deformations seen by *Magellan*. Alternatively, the lithosphere may actually be relatively thick, but Venus's molten interior may not be moving enough to push around huge blocks of the lithosphere. One class of Venusian surface feature, called coronae, supports the idea of a low level of activity in the interior of the planet. The **coronae** (Figure 11-17) are large oval or circular areas—up to 2000 km (1200 mi) across—where the ground has been pushed upward by the upward flow of magma from a hot spot in the mantle. This flow was relatively weak, however, so the magma never broke the sur-

face to form a large shield volcano. The coronae have no equivalent on the Earth, with its highly active interior.

*Magellan* has produced superb pictures of Venus's relatively few craters (Figure 11-18). Venus probably has only a thousand craters larger than a few kilometers in diameter, which is many more than have been found on the Earth but only a small fraction of the number on the Moon or Mercury. As discussed in Section 9-5, the number of craters is a clue to the age of a planet's surface. Craters formed at a rapid rate during the early history of the solar system, when considerable interplanetary debris still orbited the Sun. Since that time, impact rates have declined, as shown in Figure 9-21, and old craters have been obliterated as volcanism or tectonics renews a planet's surface (Figure 11-19). Consequently, the more craters a planet has, the older its surface. The average age of the Venusian surface seems to be roughly 400 million years. This is about twice the age of the Earth's surface but much younger than the surface of the Moon or Mercury, each of which is billions of years old.

One of the most remarkable observations about the surface of Venus is that the density of craters is the same over the

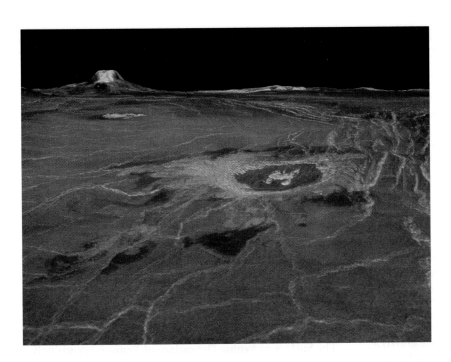

**figure 11-18**  R I V U X G

**A Venusian Crater** This three-dimensional view, constructed from *Magellan* data, shows one of Venus's large impact craters. This crater, named Cunitz for the eighteenth-century astronomer Maria Cunitz, is 48 km (30 mi) in diameter, about the same size as the city of St. Louis, Missouri, and its suburbs. Note the central peak rising out of the dark interior of the crater, a feature also seen in craters on the Moon (see Figure 9-4). The volcano near the horizon, known as Gula Mons, is 3000 m (10,000 ft) high. Note the cracks and folds in the Venusian surface toward the right side of the image. These surface features lie in the western part of Eistla Regio, just to the left of center in Figure 11-15. (NASA)

entire planet. If some regions were older than others, we would expect these older regions to be more heavily cratered because they have been exposed to bombardment for a longer time. Younger regions would be relatively free of craters. For example, the ancient highlands on the Moon are much more heavily cratered than the younger maria (see Section 9-5). Because these variations are not found on Venus, scientists conclude that the entire surface of the planet has essentially the *same* age. This is very different from the Earth, where geological formations of widely different ages can be found.

To explain this curious result, some scientists speculate that an intense period of eruptions about 400 million years ago may have resurfaced Venus with fresh lava. Another idea is that the entire surface of the planet melts periodically! This would occur if the crust is very thick compared to the Earth's crust. A thick Venusian crust would act as an insulating blanket enveloping the planet's interior, trapping heat until the mantle becomes hot enough to melt the crust. If this picture is correct, then every few hundred million years the entire face of Venus is erased at once, and there can be no very old rocks on the Venusian surface. By contrast, plate tectonics erases the Earth's surface in a piecemeal fashion, which is why some rocks on the Earth are billions of years old.

In a certain sense, Venus is the Earth's twin. Its surface tells us about our geologic past, and its atmosphere suggests that the Earth may have an awesome future. Five billion years from now, as our Sun swells to become a red giant star, the oceans will boil and shroud the Earth in a thick cloud cover. As the greenhouse effect drives temperatures toward 1000 K, vast amounts of carbon dioxide will be baked out of the Earth's rocks. By looking at Venus's atmosphere, perhaps we see the distant fate of our own world.

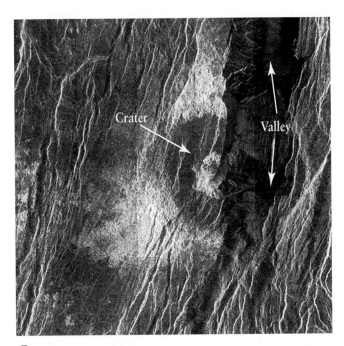

**figure 11-19**  R I V U X G

**A Partially Obliterated Crater** Over time, geologic activity on the Venusian surface erases impact craters. In this *Magellan* image, the right half of an old crater 37 km (23 mi) in diameter was destroyed when a fault in the crust formed a deep valley 20 km (12 mi) wide. The vertical streaks through the surviving left half of the crater are other fractures or faults that have occurred since a large asteroid formed this crater. This crater lies in Beta Regio, a region of Venus that appears near the left edge of Figure 11-15. (NASA)

29. Refer to Table 11-2 to see if Venus is near a greatest elongation. If so, view the planet through a telescope. Make a sketch of the planet's appearance. From your sketch, can you determine if Venus is closer to us or farther from us than the Sun?

30. Using a small telescope, observe Venus once a week for a month and make a sketch of the planet's appearance on each occasion. From your sketches, can you determine whether Venus is approaching us or moving away from us?

 31. This observing project should be performed only under the direct supervision of an astronomer who knows how to point a telescope safely at Venus. Make arrangements to view Venus during broad daylight. This is best done by visiting an observatory where the coordinates (right ascension and declination) of Venus's position can be used to point the telescope. **DO NOT LOOK AT THE SUN! Looking directly at the Sun can cause blindness.**

## Where to Learn More

*Books and magazines*

Bazilevskiy, A. T. "The Planet Next Door." *Sky & Telescope,* April 1989. A noted Soviet scientist describes the results and implications of the *Venera* missions to Venus.

Beatty, J. "A Soviet Space Odyssey." *Sky & Telescope,* October 1985. This article describes the Soviet *Vega* missions, which landed two spacecraft on Venus and floated balloon-borne instrument packages in its atmosphere.

Burnham, R. "What Makes Venus Go?" *Astronomy,* January 1993. This article contains recent ideas about Venus's surface features along with a nice collection of *Magellan* images.

Grinspoon, D. H. *Venus Revealed.* Addison-Wesley, 1997. In this entertaining book a Venus specialist describes the history of our understanding of the planet, with emphasis on the discoveries made by unmanned spacecraft.

Kargel, J. S. "The Rivers of Venus." *Sky & Telescope,* August 1997. On Venus, lava seems able to flow nearly as easily as water does on the Earth. This interesting article describes the riverlike features formed by flowing lava, including the long channel shown in Figure 11-12.

Luhmann, J. G., Pollack, J. B., and Colin, L. "The *Pioneer* Mission to Venus." *Scientific American,* April 1994. This article summarizes what was learned about the atmosphere of Venus from *Pioneer Venus,* a mission that successfully returned data to the Earth for 14 years.

Morrison, D. *Exploring Planetary Worlds.* Scientific American Library, 1993. Chapter 3 of this excellent popular book is a detailed comparison between Venus and the Earth, two worlds that the author calls "strange twins."

Powell, C. S. "Venus Revealed." *Scientific American,* January 1992. This article presents some of the first images to be returned by the *Magellan* spacecraft.

Prinn, R. C. "The Volcanoes and Clouds of Venus." *Scientific American,* March 1985. This excellent article discusses evidence for active volcanoes on Venus and describes how they affect the planet's atmosphere.

Stofan, E. R. "The New Face of Venus." *Sky & Telescope,* August 1993. This clear and copiously illustrated article describes many of the remarkable discoveries made by the *Magellan* spacecraft.

**W** *World Wide Web*

Our present knowledge of Venus is presented (along with links to a number of images) at the web sites "The Nine Planets" (**http://www.seds.org/nineplanets/nineplanets/venus.html**), "Welcome to the Planets" (**http://pds.jpl.nasa.gov/planets/welcome/venus.htm**), and "Views of the Solar System" (**http://www.hawastsoc.org/solar/eng/venus.htm**). The Face of Venus web site (**http://stoner.eps.mcgill.ca/bud/first.html**) allows you to learn about Venus's surface through hypertext documents and interactive databases of surface features. To learn about the *Magellan* mission and its scientific results, the best place to start is at NASA's Jet Propulsion Laboratory (**http://www.jpl.nasa.gov/Magellan/**). Extensive archives of *Magellan* images are available at the Venus Hypermap project (**http://www.ess.ucla.edu/hypermap/Vmap/top.html**), at the Massachusetts Institute of Technology (**http://delcano.mit.edu/http/midr-help.html**), and at the National Space Science Data Center (**http://nssdc.gsfc.nasa.gov/planetary/magellan.html**). *Magellan* showed the importance of volcanism on Venus, and you can learn more about volcanoes in general, including those on Venus, at the Volcano World site at the University of North Dakota (**http://volcano.und.nodak.edu/**).

# Red Planet Mars

R I **V** U X G

**Springtime on Mars** This Hubble Space Telescope image was made in March 1997, on the last day of spring in the Martian northern hemisphere. Like Earth, Mars has seasons that are caused by the tilt of the planet's rotation axis. But unlike Earth, surface temperatures on Mars can drop to –140°C (–220°F), and the polar ice caps (one of which is at the top of this image) are made of frozen carbon dioxide, known on Earth as "dry ice." It remains a mystery whether life exists, or has ever existed, under these harsh conditions. (David Crisp and the WFPC–2 Science Team, JPL/Caltech)

*In this chapter you will find the answers to the following questions:*

12-1  When is it possible to see Mars in the night sky?

12-2  Why was it once thought that there are canals on Mars?

12-3  How does Martian geology compare to the geology of the other terrestrial planets?

12-4  Is there water on Mars today? Was there ever water on Mars?

12-5  Why is the Martian atmosphere so thin?

12-6  What causes the seasonal color changes on Mars?

12-7  Have spacecraft that have landed on Mars found any evidence for life?

12-8  Do meteorites from Mars give conclusive proof that life once existed there?

12-9  What are the two Martian moons like?

12-10 What did scientists learn from *Mars Pathfinder* and its rover *Sojourner*?

ars Attacks! Invaders from Mars! Martians, Go Home! *Mars, the fourth planet from the Sun, has been the inspiration for many science-fiction films and novels about alien invasion. But why Mars? People have long speculated that life might exist there, because the red planet has many Earthlike characteristics. Around 1900 some astronomers claimed to have seen networks of linear features on the Martian surface, perhaps "canals" built by an advanced civilization. Seven decades later, spacecraft made many surprising discoveries about Mars—including an enormous volcano and a huge canyon—but found no canals and no signs of life. But life may have existed on Mars in the distant past. Spacecraft that landed on Mars, including the hugely successful Mars Pathfinder, found that water once flowed on this now-arid planet. And some scientists claim to have found fossil microorganisms within an unusual meteorite that came to the Earth from Mars. Is there now, or was there ever, life on Mars? How much liquid water once existed on Mars? How active were the planet's volcanoes? We may have the answers to these questions soon, because we are now in a golden age of Martian exploration, during which a series of spacecraft will observe Mars from orbit while others land on the planet's surface. Basic data about Mars are listed in Table 12-1.*

---

## 12-1    The best Earth-based observations of Mars are made during favorable oppositions

At the best times for observing, Mars is a brilliant red object in the nighttime sky that appears 3½ times brighter than Sirius, the brightest star in the sky. These optimal viewing conditions occur when Mars is simultaneously at opposition and near the perihelion of its elliptical orbit (Figure 12-1), a configuration called a **favorable opposition.** The Earth-Mars distance can then be as small as 56 million kilometers (35 million miles, or 0.37 AU). At a favorable opposition, Mars presents a disk nearly 26 arcsec in diameter, which is the same angular size as a moderate-sized (50-km) lunar crater viewed from the Earth.

Because the orbit of Mars is noticeably elongated, not all oppositions are so favorable. When Mars is at opposition but near aphelion, the Earth-Mars distance can be as large as 101 million kilometers (63 million miles, or 0.68 AU). And when Mars is not at opposition, the Earth-Mars distance can

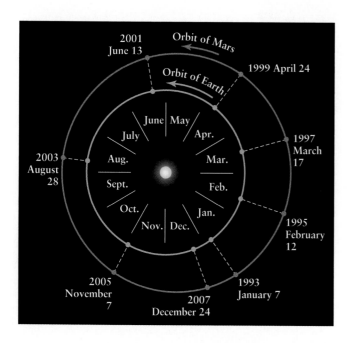

## figure 12-1

**The Orbits of the Earth and Mars** The Earth catches up with Mars at intervals of 2 years and 1 to 2 months. These close encounters, properly called oppositions, are the best times to observe the red planet. From one opposition to the next, the Earth completes a bit more than two complete orbits around the Sun, while Mars completes a bit more than one complete orbit. This figure shows the last four oppositions of the twentieth century and the first four of the twenty-first century. (The months shown inside the Earth's orbit refer to the time of year when the Earth is at each position around its orbit.) Because Mars's orbit is noticeably elongated, the Earth and Mars are closer at some oppositions than others. These are called favorable oppositions, because they afford the best views of Mars. The next favorable opposition will occur in 2003.

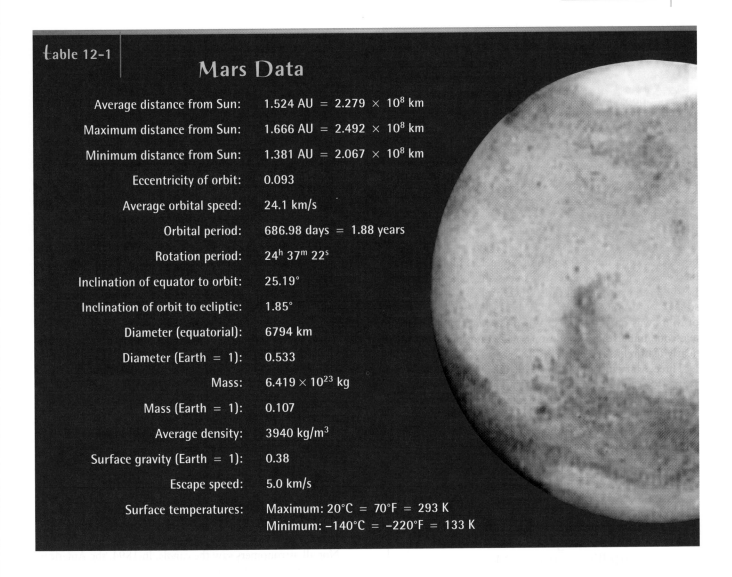

**table 12-1**

# Mars Data

| | |
|---|---|
| Average distance from Sun: | $1.524 \text{ AU} = 2.279 \times 10^8 \text{ km}$ |
| Maximum distance from Sun: | $1.666 \text{ AU} = 2.492 \times 10^8 \text{ km}$ |
| Minimum distance from Sun: | $1.381 \text{ AU} = 2.067 \times 10^8 \text{ km}$ |
| Eccentricity of orbit: | 0.093 |
| Average orbital speed: | 24.1 km/s |
| Orbital period: | 686.98 days = 1.88 years |
| Rotation period: | $24^h \ 37^m \ 22^s$ |
| Inclination of equator to orbit: | 25.19° |
| Inclination of orbit to ecliptic: | 1.85° |
| Diameter (equatorial): | 6794 km |
| Diameter (Earth = 1): | 0.533 |
| Mass: | $6.419 \times 10^{23} \text{ kg}$ |
| Mass (Earth = 1): | 0.107 |
| Average density: | 3940 kg/m³ |
| Surface gravity (Earth = 1): | 0.38 |
| Escape speed: | 5.0 km/s |
| Surface temperatures: | Maximum: 20°C = 70°F = 293 K |
| | Minimum: −140°C = −220°F = 133 K |

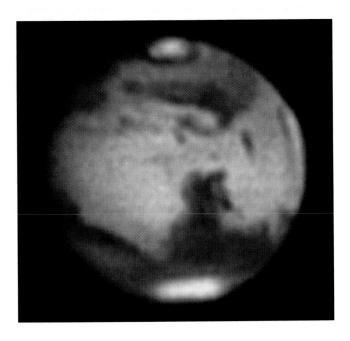

**figure 12-2**   R I **V** U X G

**Mars Viewed from the Earth**  During the opposition of 1997, when this Earth-based image was taken, the Earth-Mars distance was 100 million kilometers (62 million miles, or 0.669 AU). At that time, Mars presented a disk 14 arcsec in angular diameter. This high-quality image was made by an amateur astronomer using a 41-cm (16-in.) telescope and a CCD detector like those described in Section 6-4. The amount of detail is comparable to what you might see through a moderate-sized Earth-based telescope under excellent seeing conditions. (Even better observations will be possible during the favorable opposition of 2003, when Mars will appear about 80% larger through a telescope.) Compare this image to the Hubble Space Telescope image that opens this chapter, which was taken during the same month and shows the same face of Mars. (Courtesy of Donald Parker)

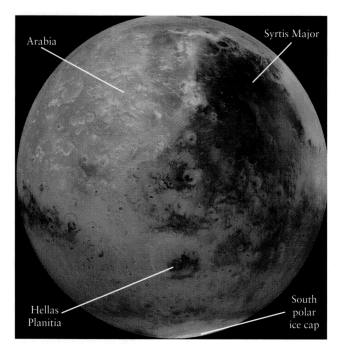

**figure 12-4**   R I **V** U X G

**Martian Craters** Numerous flat-bottomed craters are seen in this mosaic of about 100 images taken by the two *Viking Orbiter* spacecraft in 1980. North is at the top. The light-colored, heavily cratered area on the left side of this view is known as Arabia. The dark area at the upper right of the image, called Syrtis Major, is apparently composed of basalt of volcanic origin. Syrtis Major was first observed from the Earth by Huygens in 1659; it can be seen at the centers of Figure 12-2 and of the Hubble Space Telescope image that opens this chapter. The short, light-colored streaks across Syrtis Major are caused by the prevailing Martian winds. At the lower center of the image is the giant impact crater Hellas Planitia, some 6 km deep and 2000 km across. At the very bottom of the image, carbon dioxide snow covers the floors of craters near the Martian south pole. To make variations in surface coloration easier to see, the color contrast in this image was enhanced by a factor of two. (NASA, USGS)

Labels on figure: Arabia, Syrtis Major, Hellas Planitia, South polar ice cap

entire planet. Over the ages, deposits of dust from these storms have filled in the crater bottoms.

These storms, however, have not completely obliterated the craters because the Martian atmosphere is too thin to carry much dust. Measurements from spacecraft reveal an average pressure at the Martian surface of only 0.007 atmosphere. This is roughly the same as the air pressure at an altitude of 30 km (100,000 ft) above the Earth's surface. The material whipped up by the rarefied Martian winds must be extremely fine-grained powder, because 3 billion years of sporadic dust storms have not wiped out the craters.

Also conspicuously absent from spacecraft images of Mars is any evidence of vegetation. The dark surface markings are just different-colored terrain. Why, then, did many Earth-based observers report that these dark areas had the greenish hue of vegetation? The probable explanation is a quirk of human color vision. When a neutral gray area is viewed next to a bright red-orange area, the eye perceives the gray to have a blue-green color. Perhaps fantasies of Martian plant life also contributed to observations of greenish surface patterns on Mars.

The three *Mariner* spacecraft that flew past Mars in the 1960s observed just 10% of the planet's surface during their brief encounters with the planet. To get more complete observations required putting a spacecraft into orbit around Mars, which is just what was done in 1971 with the *Mariner 9* spacecraft. The images from *Mariner 9* showed important differences between the northern and southern hemispheres of the planet. For one thing, Mars is not cratered over its entire surface. Very few craters are found in the northern hemisphere, suggesting that it is geologically younger than the southern hemisphere. The northern hemisphere also has a lower average elevation than of the southern hemisphere. That is why the smooth, young, relatively crater-free terrain in

part of the Martian surface must therefore be extremely ancient, because it appears to have experienced few changes over the past 3 billion years. Thus, after the *Mariner* flyby missions, the perception of Mars changed from that of an Earthlike world to that of a somewhat larger version of the Moon.

There are key differences between craters on the Moon and Mars, however. The lunar maria were flooded with lava that solidified to a relatively flat, smooth surface (see Section 9-1 and especially Figure 9-5). By contrast, the flat, partially eroded bottoms of Martian craters probably resulted from dust storms. Winds in the Martian atmosphere stir up finely powdered dust (Figure 12-5) and blow it across the planet's surface. Astronomers can see these storms raging from the Earth, as the planet's surface markings disappear under a reddish-brown haze. Some dust storms actually obscure the

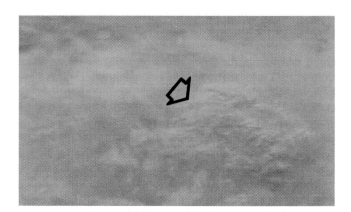

**figure 12-5**   R I **V** U X G

**A Martian Dust Storm** This *Viking Orbiter* image shows a 300-km-wide dust storm moving across the floor of the Argyre Planitia impact basin during early 1977. Such dust storms are thought to have filled in the floors of Martian craters, such as those in Figure 12-4. (NASA)

the northern hemisphere is called the **lowlands,** while the older, cratered terrain in the southern hemisphere is called the **highlands.** Thus, when we compare the terrestrial planets in terms of their geologic activity or inactivity, we see that Mars is an intermediate case. Only the northern hemisphere has been geologically active, and it is only there that ancient impact craters have been erased. Meanwhile, the southern hemisphere shows little sign of major geologic activity. Its surface is old and cratered, like that of the Moon or Mercury. The reason for these north-south differences is not known.

*Mariner 9* also gave astronomers detailed, high-resolution views of enormous volcanoes, deep chasms, and tremendous valleys. The largest Martian volcano, Olympus Mons, covers an area as big as the state of Missouri. It rises 24 km (15 mi) above the surrounding plains—nearly three times the height of Mount Everest (Figure 12-6). By comparison, the highest volcano on the Earth, Mauna Loa in the Hawaiian Islands, has a summit only 8 km (5 mi) above the ocean floor. None of the Martian volcanoes are active today.

The Martian volcanoes and the Hawaiian Islands, like Maxwell Montes and other volcanoes on Venus, are shield volcanoes formed by hot-spot volcanism (see Section 8-3 and Section 11-5). The Hawaiian Islands are part of a long chain of volcanoes formed as the Pacific tectonic plate moved over a hot spot, as illustrated in Figure 8-20. On Mars, by contrast, the huge size of Olympus Mons strongly suggests an absence of plate tectonics. Instead, a hot spot in Mars's mantle probably kept pumping lava upward through the same vent for millions of years, producing one giant volcano rather than a long chain of smaller ones.

Olympus Mons is one of several large volcanoes clustered together just north of the Martian equator. This volcanic region, some 2500 km (1500 mi) in diameter, is called the Tharsis rise. It is actually a dome-shaped bulge whose average

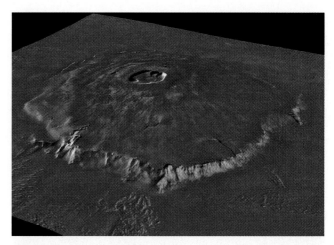

**Figure 12-6** R I **V** U X G

**Olympus Mons** This perspective view, created by combining a number of images from the two *Viking Orbiters,* shows the largest volcano on Mars as viewed from the northeast. Olympus Mons is nearly three times as tall as Mount Everest. Its base measures 600 km (370 mi) in diameter, and the scarps (cliffs) that surround the base are 6 km (4 mi) high in spots. The caldera, or volcanic crater, at the summit measures approximately 70 km across, large enough to contain the state of Rhode Island. (© 1998 Calvin J. Hamilton, Columbia, Maryland)

elevation is 5 to 6 kilometers higher than the rest of the planet (Figure 12-7). Apparently, a massive plume of material once welled upward through the Martian mantle, producing the Tharsis rise at the same time that it produced Olympus Mons and its sister volcanoes. Another, smaller bulge, with a

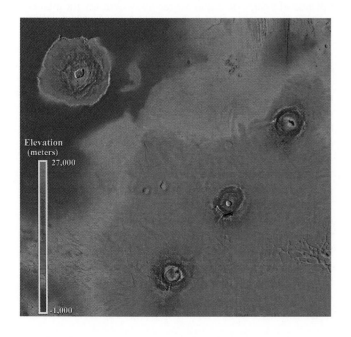

**Figure 12-7**

**Martian Volcanoes and the Tharsis Rise** *Viking Orbiter* images and elevation data have been combined to make this false-color image of the Tharsis region on Mars. Different colors represent different elevations relative to the average level of the Martian surface. North is at the top and west is to the left. The largest feature in this image is the Tharsis rise itself, whose average elevation of 5 to 6 km is shown by its green tint. Three large volcanoes—Arsia Mons at the lower left, Pavonis Mons in the middle, and Ascraeus Mons at the upper right—protrude upward from the rise. Olympus Mons is at the upper left, to the northwest of Tharsis. All of the features are presumed to have been formed by a tremendous upwelling of material in the Martian mantle underlying the Tharsis region. A similar but smaller bulge and volcano cluster is found in the Elysium region on the opposite side on Mars's northern hemisphere. This image shows an area about 2200 km (1400 mi) on a side. (© 1998 Calvin J. Hamilton, Columbia, Maryland)

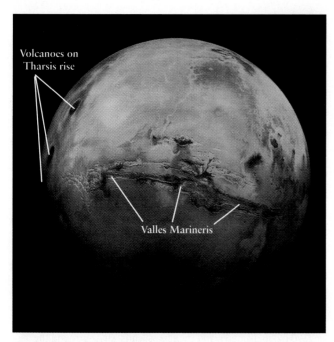

**Figure 12-8**    R I ☑ U X G

**Valles Marineris** This mosaic of *Viking Orbiter* images shows the side of Mars opposite that shown in Figure 12-4. North is at the top. The huge rift valley of Valles Marineris extends from west to east for more than 4000 km (2500 mi). At the far left is the Tharsis rise; note the three large volcanoes, which are also shown in Figure 12-7. At the center of Valles Marineris, the canyon is 600 km (400 mi) wide. The canyon floor has two major levels. The northern (upper) canyon floor is 8 km (5 mi) beneath the surrounding plateau, whereas the southern canyon floor is only 5 km (3 mi) below the plateau. (NASA, USGS)

less dramatic grouping of volcanoes, lies nearly on the opposite side of the planet.

Most of the volcanoes on Mars lie in the northern hemisphere, which has had a richer and more active geological history than the southern hemisphere. Between the two hemispheres and east of the Tharsis rise, *Mariner 9* discovered a vast canyon running roughly parallel to the Martian equator (Figure 12-8). In honor of the spacecraft that revealed so much of the Martian surface, this enormous chasm has been designated Valles Marineris.

Valles Marineris extends for more than 4000 km (2500 mi). It begins with heavily fractured terrain in the west and ends with ancient cratered terrain in the east. If this canyon were located on Earth, it would stretch from New York to Los Angeles. Many geologists suspect that Valles Marineris was caused by the same upwelling of material that formed the Tharsis rise. As the Martian surface bulged upward at Tharsis, there would have been tremendous stresses on the crust, which would have caused extensive fracturing. Thus, Valles Marineris may be a **rift valley**, a feature created when a planet's crust breaks apart

along a line. Rift valleys are found on Earth, where they include the Red Sea and the Rhine River valley in Europe. Other, smaller rifts in the Martian crust are found all around the Tharsis rise.

Valles Marineris is sometimes called the "Grand Canyon of Mars," but this term is actually a misnomer. In geology a *canyon* is formed by running water. The Earth's Grand Canyon is the result of 15 to 20 million years of erosion by the flowing waters of the Colorado River as well as by rain and wind. Valles Marineris, by contrast, is apparently the product of stresses in the Martian crust that caused stretching and cracking on a grand scale. The shape of the walls of Valles Marineris is quite different from what one would expect from erosion.

While the motion of material in the mantle has had a large effect on the Martian surface, plate tectonics has not. Mars lacks the global network of ridges and subduction zones that tectonic activity has produced on the Earth. The absence of plate tectonics on Mars is evidence that the red planet cooled more rapidly than the Earth when it was young. This is a direct consequence of Mars being only half the diameter of the Earth (see Figure 8-1). As we discussed in Section 9-3, small bodies tend to lose heat more rapidly than large bodies. (For this reason, newborn babies—the smallest of all humans—are kept bundled up to protect them against heat loss.) Plate tectonics on Mars presumably bogged down early in the planet's history as the rapidly cooling lithosphere became thick and sluggish.

## 12-4    Surface features indicate that water once flowed on Mars

Spacecraft orbiting Mars have not seen any evidence of rainfall nor any sign of liquid water anywhere on the planet's surface. But they have revealed many features that look like dried-up lakes and riverbeds (Figure 12-9). Intricate branched patterns and delicate channels meandering among flat-bottomed craters strongly suggest that water once flowed over the Martian surface. Such dried-up water channels are found throughout the old highland terrain of the southern hemisphere but in relatively few places on the younger lowlands of the northern hemisphere. Hence, large amounts of water flowed on the Martian surface in the distant past but not in the recent past.

In 1976 two spacecraft, called the *Viking Orbiters*, were placed in orbit around Mars as a follow-on to *Mariner 9*. Their in-depth observations revealed evidence of powerful flash floods that long ago raged across the surface of Mars (Figure 12-10). Based on the widths and depths of the Martian flash-flood channels, at their peak the flood discharges may have been greater than anything known on the Earth, with water speeds as high as 270 km/h (170 mi/h). In addition to flash floods, there is also evidence that water flowed for sustained periods on the Martian surface. Figure 12-11 shows a canyon that appears to have been formed gradually by water erosion. Part *b* of this image was made by the *Mars Global Surveyor* spacecraft, which went into orbit around the planet in 1997 armed with cameras far superior to those aboard the *Viking Orbiters*. The excellent resolution of these state-of-the-art cameras makes it possible to see fine details of Martian geology.

The discovery of Martian riverbeds (see Figure 12-11) and flash-flood channels was totally unexpected, because liquid water cannot exist today on Mars. Water is liquid over only a limited range of temperatures: If the temperature is too low, water becomes ice, and if the temperature is too high, it becomes water vapor. What determines this temperature range is the atmospheric pressure above a body of water. If the pressure is very low, molecules easily escape from the liquid's surface, causing the water to vaporize. Thus, at low pressures, water more easily becomes water vapor. The average surface temperature on Mars is only 218 K (−55°C, or −67°F) and the average pressure is only 0.007 atmosphere. (Recall from Section 8-4 that one atmosphere is the average sea-level pressure on the Earth.) With this combination of temperature and pressure, water can exist as a solid (ice) and as a gas (water vapor) but not as a liquid. You can see this

same situation inside a freezer, where water vapor swirls around over ice cubes.

In order to keep water on Mars in the liquid state, we would somehow have to increase both the temperature (to keep water from freezing) and the pressure (to keep the liquid water from evaporating). Therefore, during the past, when liquid water existed on Mars, the atmosphere must have been both thicker and warmer than it is today.

If water once flowed on Mars, where might that water be today? It is not in the Martian atmosphere, which is 95% carbon dioxide and contains only a trace amount of water vapor. Indeed, it never rains on Mars. If all the water vapor could somehow be squeezed out of the Martian atmosphere, it would not fill one of the five Great Lakes of North America.

Another possibility is that the Martian polar caps are reservoirs of frozen water. Because the winter temperature at the Martian poles is low enough (−140°C = −220°F) to freeze carbon dioxide as well as water, it was not clear how much of the polar caps consists of frozen carbon dioxide (dry ice) and how much is water ice.

Even before the *Viking Orbiters*, *Mariner 9* sent back solid evidence of water ice in the polar caps. In 1972 it observed spring and summer coming to the northern hemisphere of

**Figure 12-9** R I V U X G

**Ancient River Channels on Mars** This *Viking Orbiter* image shows a network of dried riverbeds extending across cratered terrain in Mars's southern highlands. The large craters at upper left and upper right are about 35 km (22 mi) across. The existence of such riverbeds suggests that the planet's atmosphere was thicker and its climate more Earthlike long ago, when liquid water flowed across its surface. Any water exposed to the sparse, present-day Martian atmosphere would rapidly boil away or freeze solid. (NASA)

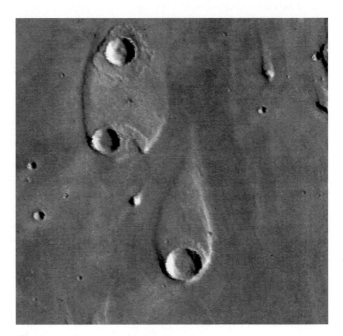

**Figure 12-10** R I V U X G

**Signs of Ancient Floods on Mars** These teardrop-shaped islands, each about 40 km (25 mi) long, rise above the floor of an ancient Martian flood plain called Ares Valles. A torrent of water (which flowed from the bottom of this image toward the top) carved away substantial amounts of surface material but was deflected around the walls of the large impact craters shown here. Similar flood-carved structures are found on the Earth in eastern Washington State. (NASA)

Mars. During the Martian spring, the north polar cap receded rapidly, strongly suggesting that a thin layer of carbon dioxide frost was evaporating in the sunlight. With the arrival of summer, however, the rate of recession slowed abruptly, suggesting that a thicker layer of water ice had been exposed. Scientists concluded that the **residual polar caps**—the portions that survive the Martian summers—contain a large quantity of frozen water (Figure 12-12). Calculating the volume of water ice in the residual caps is difficult, however, because we do not know the thickness of the layer of ice. Thus, the exact amount of frozen water stored in the Martian polar caps is still a matter of debate.

The residual ice caps cannot be the only site of Martian water, however. Close-up views such as Figure 12-10 show that flash-flood features usually emerge from craters or from collapsed, jumbled terrain. It therefore seems likely that frozen water is also stored in **permafrost** under the Martian surface, similar to the layer beneath the tundra in the far northern regions on the Earth. In the distant past, heat from a meteoroid impact or from volcanic activity occasionally caused a sudden melting of this subsurface ice. The ground then collapsed, and millions of tons of rock pushed the water to the surface.

The total amount of water on Mars is not known. But by examining flood channels at various locations around Mars, Michael Carr of the U.S. Geological Survey has estimated that there is enough water to cover the planet to a depth of 500 meters (1500 ft). By comparison, there is enough water on the Earth to cover our planet to a depth of 2700 meters (8900 ft). Thus, the total amount of water on Mars, while less than on the Earth, may indeed be quite substantial.

Even if we know where all of the water on Mars has gone, we must still ask why conditions on Mars were once so differ-

a

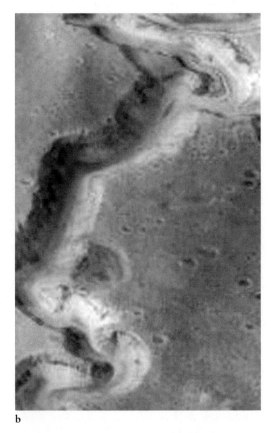

b

**ƒigure 12-11** R I **V** U X G

**Evidence for Sustained Water Flow on Mars** **(a)** This *Viking Orbiter 1* image shows a 2.5-km-wide (1.6-mi) canyon that running water may have carved out of the ancient, cratered plains of Mars. The area outlined in white is shown in more detail in **(b)**, which depicts an area of dimensions 9.8 by 18.5 km (6.1 by 11.5 mi). The terraces and small channels within the canyon are the same kinds of structures found within river beds on the Earth. In addition to erosion by running water, a localized collapse of the Martian surface (like a sinkhole on the Earth) may have helped shape this canyon. The image in (b) was made by the *Mars Global Surveyor* spacecraft in 1998. The smallest detail visible within this image is just 12 m (39 ft) across. Even sharper images will be possible when this spacecraft is moved into a low-altitude circular orbit (see Section 12-10). (NASA)

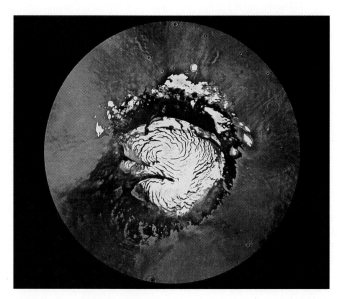

**ƒigure 12-12** R I **V** U X G

**The North Polar Ice Cap** This mosaic of *Viking Orbiter 1* images shows the northern polar cap of Mars. During the summer, the carbon dioxide frost that covers northern latitudes during the winter evaporates, exposing a residual polar cap made primarily of water ice. The reddish streaks consist of dust, and the dark bluish areas are sand dunes that form a huge "collar" encircling the north polar ice cap. The carbon dioxide frost never completely evaporates from the south pole, so it is not known for certain whether a similar layer of water ice exists there. (NASA, USGS)

ent—with a warmer and thicker atmosphere and with liquid water on the surface—than today. To find an answer, we must explore how the Martian atmosphere evolved.

## 12-5 The Earth and Mars began with similar atmospheres that evolved very differently

As nineteenth-century astronomers observed Mars through their telescopes, they were occasionally able to see cloud formations. This was the first evidence that Mars has an atmosphere (without an atmosphere, a planet can have no clouds). It was difficult to measure the atmosphere's spectrum, however, and for many decades no one knew for sure what the Martian atmosphere is made of. When the *Mariner 4* spacecraft flew past Mars in 1965, its measurements indicated that the Martian atmosphere is almost entirely carbon dioxide ($CO_2$). We now know that the atmosphere is 95% carbon dioxide and 3% nitrogen, with small amounts of argon, oxygen, carbon monoxide, and water vapor. As we have seen, the pressure at the planet's surface is very low, only 0.007 atmosphere.

As we have also seen, the Martian climate must have been much warmer and more Earthlike 4 billion years ago. What could have provided this extra warmth? The likeliest mechanism is the greenhouse effect (recall Section 8-6 and Section 11-3, in particular Figure 11-4).

Because the atmospheres of the Earth and Mars are transparent to sunlight, they are not heated directly by the Sun. Instead, they are heated from below by energy reradiated from the ground at longer, infrared wavelengths. Water vapor and carbon dioxide absorb these wavelengths, trapping the heat and warming the atmosphere. Without the greenhouse effect, the average temperature on Earth would be about 35°C lower than it is now, and our entire planet would be below freezing.

In the present-day Martian atmosphere, the greenhouse effect is much less efficient; it increases the temperature by only about 5°C. This is because the thin Martian atmosphere permits much of the infrared radiation to escape into space. If the greenhouse effect was once important on Mars, the Martian atmosphere must have held much more water vapor and carbon dioxide than today.

Extinct Martian volcanoes suggest how such an ancient atmosphere could have been created. The terrestrial planets obtained their atmospheres primarily through volcanic outgassing (see Figure 11-14). Water vapor, carbon dioxide, and nitrogen are among the most common gases in volcanic vapors. These three gases probably constituted the bulk of the atmospheres of both the Earth and Mars 4 billion years ago.

On Earth most of the water is in the oceans. Nitrogen, which is not very reactive, is still in the atmosphere. Carbon dioxide, however, dissolves in water; thus, rain washed carbon dioxide out of the air. The dissolved carbon dioxide reacted with rocks in rivers and streams, and the residue was ultimately deposited on the ocean floors, where it turned into carbonate rocks, such as limestone. Consequently, most of the Earth's carbon dioxide is tied up in the Earth's crust, and carbon dioxide makes up only a small fraction of the Earth's atmosphere. (About 1 in every 3,000 molecules in our atmosphere is a carbon dioxide molecule.) The amount of carbon dioxide in our atmosphere is sustained in part by plate tectonics. Tectonic activity causes carbonate rocks to be cycled through volcanoes, where they are heated and forced to liberate their trapped carbon dioxide. This liberated gas then rejoins the atmosphere. Without volcanoes, rain would wash all the carbon dioxide from the Earth's air in only a few thousand years. The truly unique feature of our atmosphere—the presence of oxygen—is the result of biological activity, in particular photosynthesis by plant life (Figure 12-13a).

On Mars, rainfall 4 billion years ago may have washed much of the planet's carbon dioxide from its atmosphere, perhaps creating carbonate rocks in which the carbon dioxide is today chemically bound. Thus, on both the Earth and Mars, carbonate rocks are significant reservoirs of the primordial carbon dioxide outgassed from ancient volcanoes. But plate tectonics did not occur on Mars, and its carbonate rocks were not recycled through volcanoes. The depletion of carbon dioxide from the Martian atmosphere into the rocks was therefore permanent. According to calculations by James B. Pollack of NASA's Ames Research Center, a 1-atmosphere carbon dioxide Martian atmosphere could have survived for only 10 to 100 million years. However, volcanic activity early in Mars's history may have recycled some carbon dioxide into the atmosphere, thus prolonging the greenhouse effect for perhaps half a billion years. Ultimately, however, rainfall succeeded in removing both water vapor and most of the carbon dioxide from the Martian atmosphere, thus largely destroying the greenhouse effect that had allowed liquid water to exist in the first place. With only a thin carbon dioxide

atmosphere remaining, surface temperatures on Mars would have begun to decline to their present frigid values.

As both water vapor and carbon dioxide became depleted, ultraviolet light from the Sun could penetrate the thinning Martian atmosphere to strip it of nitrogen. Nitrogen molecules normally do not have enough thermal energy to escape from Mars, but they can acquire that energy from ultraviolet photons, which break the molecules in two. Ultraviolet photons can also split carbon dioxide and water molecules, giving their atoms enough energy to escape (Figure 12-13b). Indeed, the Soviet *Mars 2* spacecraft that

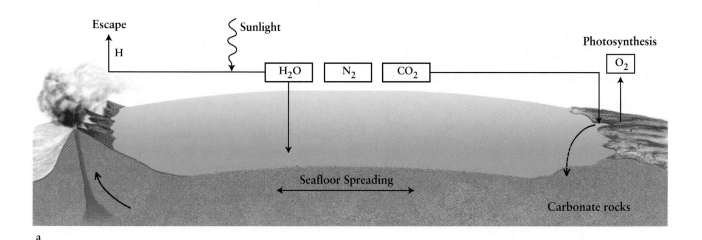

a

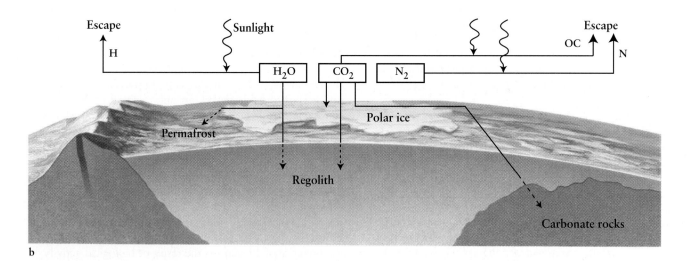

b

## ƒigure 12-13

**The Atmospheres of the Earth and Mars** The primitive atmospheres of the Earth and Mars were probably similar, but they evolved differently. **(a)** On the Earth, water condensed into oceans, and nitrogen remained in the air, while carbon dioxide was removed from the atmosphere by rainfall and then stored in carbonate rocks. **(b)** On Mars, rainfall destroyed the thick carbon dioxide atmosphere that kept the planet warm enough (through the greenhouse effect) for liquid water to exist. Today, water ice is found in the polar caps and the Martian permafrost. Some water, along with carbon dioxide, may also be chemically bound in the fine-grained Martian soil, or regolith. Ultraviolet light from the Sun breaks apart both carbon dioxide and nitrogen molecules into their constituent atoms, which then escape into space. (Adapted from R. M. Haberle)

went into orbit around Mars in 1971 found a stream of oxygen and hydrogen atoms (from the breakup of water molecules) escaping into space.

Oxygen atoms that did not escape into space could have combined with iron-bearing minerals in the surface, forming rust-like compounds. Such compounds have a characteristic red color and may be responsible for the overall red color of the planet.

These three effects—rainfall, loss of gases into space, and chemical reactions between gases and surface rocks—would have thinned the Martian atmosphere and cooled its surface. Perhaps in this way Mars came to have the sparse atmosphere we observe today.

## 12-6 The *Viking Landers* sent back close-up views of a barren Martian surface with seasonal variations

The discovery that water once existed on the surface of Mars rekindled speculation about Martian life. By the mid-1960s, although it was clear that Mars has neither civilizations nor fields of plants, some sort of microbial life forms still seemed possible. Searching for Martian microbes was one of the main objectives of the ambitious and highly successful *Viking* missions.

The two *Viking* spacecraft were launched during the summer of 1975 and arrived at Mars almost a year later. Each spacecraft consisted of two modules: an orbiter (introduced in Section 12-4) and a lander. After entering orbit about Mars, each lander separated from its orbiter and descended to the Martian surface with the aid of a heat shield, retrorockets, and a parachute. Each lander is roughly the size of a small automobile (Figure 12-14).

The *Viking Lander 1* landing site was a moderately cratered rocky plain about 20° north of the Martian equator, near the mouth of a large flash-flood channel. The upraised rim of a crater is visible on the horizon in Figure 12-15. Such craters are probably the source of the jagged rocks that litter the scene. Some light-colored bedrock is exposed near the horizon.

*Viking Lander 2* set down on nearly the opposite side of the planet from *Viking Lander 1* but about 1500 kilometers (930 miles) closer to Mars's north pole. As shown in Figure 12-16, the *Viking 2* site is remarkably crater-free, although many of the rocks in this view may have been ejected from a large crater about 200 kilometers east of the spacecraft.

Both *Viking Lander* spacecraft carried instruments to measure meteorological conditions on Mars. Because the Martian atmosphere is so thin, it provides little thermal insulation, and thus the *Viking Landers* measured large swings in air temperature, from 180 K (−93°C = −135°F) at dawn to 240 K (−33°C = −27°F) in the "heat" of the afternoon. More surprisingly, the *Viking Landers* also found substantial variations in atmospheric pressure.

**ſigure 12-14**

**A *Viking Lander*** From the base of its footpads to the top of its antenna, each *Viking Lander* is about 2 m (6 ft) tall and weighs 576 kg (1270 lb) without fuel. An extendable arm for retrieving soil samples is visible in the foreground. The two cylinders (one with an American flag) protruding from the top of the lander contain the two television cameras. The dish antenna was used to relay data to one of the *Viking* orbiters overhead, which in turn sent the information back to the Earth. This photograph was taken on the Earth, not Mars; it shows a full-size replica, or "test article," used on the Earth before and after the *Viking* missions to simulate the behavior of the spacecraft on the Martian surface. Such a test article is on display at the National Air and Space Museum in Washington, D.C. (NASA)

**ſigure 12-15**

**The *Viking Lander 1* Site** This view looks southward over the Chryse Planitia (the Golden Plains). The horizon is about 3 km (2 mi) from the spacecraft. Part of the upraised rim of a crater on the horizon is visible at the upper left. A few light-colored patches of exposed bedrock also appear in this picture. *Viking Lander 1* sent data back from the Martian surface from July 1976 until November 1982. It is now officially part of the National Air and Space Museum. (NASA)

### figure 12-16

**A Panorama of the *Viking Lander 2* Site** This mosaic of three images shows a panorama covering about 180° around the *Viking Lander 2*, which landed in a plains area called Utopia. Northwest is at the left and southeast on the right. The flat, featureless horizon is approximately 3 km (2 mi) away from the spacecraft. Both this landing site and that for *Viking Lander 1* (Figure 12-15) were chosen because they are relatively flat and, thus, more likely to provide the spacecraft with a smooth, survivable landing. *Viking Lander 2* landed on Mars in August 1976; it continued to transmit data to the Earth until April 1980. (NASA)

When they first landed in the northern hemisphere, both *Viking Landers* measured atmospheric pressures around 0.008 atmosphere. After only a few weeks on the Martian surface, however, atmospheric pressure at both landing sites was dropping steadily. Mars seemed to be rapidly losing its atmosphere, and some scientists joked that soon all the air would be gone! A straightforward explanation was, however, readily available—winter was coming to the southern hemisphere.

During the southern hemisphere's winter, the Martian south pole becomes so cold that large amounts of carbon dioxide condense out of the atmosphere there, covering the ground with flakes of dry ice. The formation of this "snow" removes gas from the Martian atmosphere, thereby lowering the atmospheric pressure across the planet. Several months later, when spring comes to the southern hemisphere, the dry-ice snow evaporates rapidly and the atmospheric pressure returns to its prewinter levels. The pressure drops again in the southern hemisphere's summer, because it is then winter in the northern hemisphere, and dry-ice snow condenses at northern latitudes (Figure 12-17). Therefore, both temperature and atmospheric pressure change significantly with the Martian seasons.

As on the Earth, the seasons on Mars are caused by the tilt of the planet's rotation axis. When the north pole is tilted toward the Sun, it is summer in the northern hemisphere and winter in the southern hemisphere; when the south pole is tilted toward the Sun, the seasons are reversed (see Figures 2-11 and 2-12). But because Mars has a rather elliptical orbit, the seasonal variations in the two hemispheres of Mars are somewhat different. Summer comes to the southern hemisphere of Mars at the same time that the

planet makes its closest approach to the Sun. In the northern hemisphere, by contrast, summer comes when Mars is farthest from the Sun. Hence, the temperature at the south pole during its summer should be higher than at the north pole when it is summer there. Paradoxically, the *Viking*

### figure 12-17

**Winter in Mars's Northern Hemisphere** This picture, taken during midwinter, shows a thin frost layer that lasted for about a hundred days at the *Viking Lander 2* site. Freezing carbon dioxide adheres to water-ice crystals and dust grains in the atmosphere, causing them to fall to the ground. The sky is therefore not as pink as it is in the summertime view of Figure 12-16. (NASA)

*Orbiters* measured just the opposite to be true! The north pole's temperature climbs to 200 K during the northern summer, while the midsummer temperature at the south pole never gets above 150 K.

Why is summer warmer in the northern hemisphere than in the southern hemisphere? This seemingly backward state of affairs is caused by dust in the atmosphere. When it is summer in the southern hemisphere, the Sun warms the soil surrounding the south polar ice cap, while the ice cap remains quite cold. This temperature contrast gives rise to strong winds that often produce swirling dust storms. A dusty atmosphere admits less sunlight than a clear one, so these storms help to keep the south polar ice cap cold even in midsummer. At the north pole, however, the solar heating during summer is weaker because the planet is far from the Sun. Hence, the temperature difference between the cold north polar ice cap and the surrounding sun-warmed soil is not as great, and the strong winds needed to produce dust storms do not appear. As a result, the north polar ice cap remains exposed to sunlight, the temperature there rises to levels never found at the south pole, and carbon dioxide frost is able to evaporate completely.

Seasonal winds can also explain why Mars changes in appearance over the course of a Martian year. As shown in Figure 12-15, some small sand drifts were seen at the *Viking Lander 1* site. Careful scrutiny of pictures taken over several months reveals that these drifts are pushed around slightly by the seasonal winds. New rocks are occasionally exposed and others covered up. Therefore, as seen from Earth, large areas of the planet vary from dark to light with the passage of the seasons. These variations mimic the color changes that would be expected from vegetation, which is why early observers thought they saw signs of plant life on the red planet (Section 12-2). Dust can also explain the distinctly pinkish-orange tint of the Martian sky

seen in *Viking Lander* images. This unusual coloration is probably caused by extremely fine-grained dust suspended in the thin Martian atmosphere.

## 12-7 The *Viking Landers* found abundant iron in the Martian regolith but failed to detect living organisms

In addition to instruments that analyzed the atmosphere, the *Viking Landers* studied the surface rocks and probed the soil for signs of life. The Martian soil is made of fine-grained, compacted sand that has been weathered by billions of years of wind. As on the Moon, this compacted, weathered material is called a regolith (see Section 9-5).

Each *Viking Lander* had a scoop at the end of a mechanical arm to dig in the Martian regolith and retrieve rock samples for analysis (Figure 12-18). Bits of the regolith were observed to cling to a magnet mounted on the scoop, indicating that the regolith contains iron. A device called an X-ray fluorescence spectrometer measured the chemical composition of the rocks at both lander sites and showed them to be rich in iron, silicon, and sulfur. The Martian regolith can best be described as an iron-rich clay. As mentioned in Section 12-5, the familiar reddish color of Mars may be caused by rust (iron oxides) in the regolith.

Despite the high iron content of its crust, Mars has a lower average density ($3950$ kg/m$^3$) than the other terrestrial planets (more than $5000$ kg/m$^3$ for Mercury, Venus, and the Earth). Mars must therefore have a lower total iron content than these other planets. Furthermore, Mars does not have a measurable magnetic field. Perhaps Mars's iron is more uniformly distributed throughout the body of the planet rather

**Figure 12-18**

**Digging in the Martian Soil** The mechanical arm with its small scoop protrudes from the right side of this view of the *Viking Lander 1* landing site. (Figure 12-14 shows the fully extended arm.) Several small trenches dug by the scoop in the Martian regolith appear near the left side of the picture. (NASA)

than being concentrated in a dense core. Because a seismometer on one of the *Viking Landers* failed, no seismic waves were measured and the structure of the Martian interior remains largely unknown.

Each *Viking Lander* carried a compact biological laboratory designed to perform three different tests for microorganisms in the Martian soil. The three biological experiments carried out by the *Viking Landers* were based on the general proposition that living things must alter their environment. They eat, they breathe, and they give off waste products. In each of the three experiments, a sample of the Martian regolith was placed in a closed container, with or without a certain nutrient substance. The container was then examined for any changes in its contents.

The *gas-exchange experiment* was designed to detect any processes that might be broadly considered as respiration. A small sample of regolith was placed in a sealed container along with a controlled amount of gas and nutrients. The gases in the container were then monitored to see if their chemical composition changed.

The *labeled-release experiment* was designed to detect processes resembling metabolism. A small sample of regolith was moistened with nutrients containing radioactive carbon atoms. Scientists assumed that any organisms in the regolith would eat the food and then emit gases containing the telltale radioactive carbon.

The *pyrolytic-release experiment* was designed to detect photosynthesis, the biological process by which plants on the Earth synthesize organic compounds from carbon dioxide, using sunlight as an energy source. In the *Viking* experiments, a regolith sample was placed in a container along with radioactive carbon dioxide and exposed to artificial sunlight. If plantlike photosynthesis occurred, some of the radioactive carbon from the gas would become incorporated into the microorganisms in the regolith.

The first data returned by the *Viking* biological experiments caused great excitement: In almost every case, rapid and extensive changes were detected inside the sealed containers. Further analysis of the data, however, led to the conclusion that these changes were due solely to nonbiological chemical processes. It appears that the Martian regolith is rich in chemicals that effervesce (fizz) when moistened. A large amount of oxygen is apparently tied up in the regolith in the form of unstable chemicals called peroxides and superoxides, which break down in the presence of water to release oxygen gas.

The unstable chemistry of the Martian regolith probably comes from ultraviolet radiation that beats down on the planet's surface. Ultraviolet photons easily break apart molecules of carbon dioxide ($CO_2$) and water vapor ($H_2O$) by knocking off oxygen atoms, which then become loosely attached to chemicals in the regolith. Ultraviolet photons also produce ozone ($O_3$) and hydrogen peroxide ($H_2O_2$), which also become incorporated in the regolith. In all these cases, the loosely attached oxygen atom makes the regolith extremely reactive.

Here on the Earth, hydrogen peroxide is commonly used as an antiseptic. When you pour this liquid on a wound, it fizzes and froths as the loosely attached oxygen atoms chemically combine with organic material, thereby destroying germs. The *Viking Landers* may have failed to detect any organic compounds on Mars because the superoxides and peroxides in the regolith make it literally antiseptic.

## 12-8 Meteorites from Mars have landed on Earth and have been scrutinized for life forms

As described in Section 9-5, spacecraft have visited the Moon and brought back several hundred kilograms of lunar rocks. The study of these rocks has revolutionized our understanding of the Moon and its evolution. As yet, however, no spacecraft has returned to Earth from Mars with any Martian rocks. But scientists have discovered the next best thing: A dozen meteorites that appear to have formed on Mars have been found at a variety of locations on the Earth.

These meteorites are called **SNC meteorites** after the names given to the first three examples found (Shegott, Nakhla, and Chassigny). They are igneous rocks similar in character to those found on the floors of the Earth's oceans. What identifies SNC meteorites as having come from Mars is the chemical composition of trace amounts of gas trapped within these meteorites. This composition is very different from that of the Earth's atmosphere, but is a nearly perfect match to the composition of the Martian atmosphere found by the *Viking Lander* spacecraft.

How could a rock have gotten from Mars to Earth? When a large piece of space debris collides with a planet's surface and forms an impact crater, most of the material thrown upward by the impact falls back onto the planet's surface. But some extraordinarily powerful impacts produce large craters—on Mars, roughly 100 km in diameter or larger. These tremendous impacts throw a small fraction of the ejected rocks upward with such speed that they escape the planet's gravitational attraction and fly off into space. There are numerous large craters on Mars, so a good number of Martian rocks have probably been blasted into space over the planet's history. These ejected rocks then go into elliptical orbits around the Sun. A few such rocks will have orbits that bring them close to the Earth, and these are the ones that scientists find as SNC meteorites.

Using radioactive age-dating (a technique described in Box 9-2), scientists find that most SNC meteorites are between 200 million and 1.3 billion years old, much younger than the 4.6-billion-year age of the solar system. These meteorites probably formed by volcanic action. But one SNC meteorite, denoted by the serial number ALH 84001 and found in Antarctica in 1984, was discovered in 1993 to be 4.5 billion years old (Figure 12-19). Thus, ALH 84001 is a truly ancient piece of Mars's original crust. Based on the relative amounts of different isotopes found within this meteorite, ALH 84001 is thought to have been fractured by an impact between 3.8 and 4.0 billion years ago, to have been ejected from Mars by

**figure 12-19**

**A Meteorite from Mars** This 1.9-kg meteorite, labeled ALH 84001, is thought to have been formed on Mars 4.5 billion years ago, dislodged from Mars by a massive impact about 16 million years ago, and landed on the Earth 13,000 years ago. The small cube with letters on it is 1 cm (0.4 in.) across. ALH 84001 was found in the Allan Hills ice field of Antarctica in 1984. It is preserved for study at the NASA Johnson Space Center Meteorite Processing Laboratory in Houston, Texas. (NASA, Johnson Space Center)

**figure 12-20**

**Fossil Martians?** This image, magnified some 100,000 times, shows tubular structures about 100 nanometers ($10^{-7}$ m) in length found within the Martian meteorite ALH 84001. One interpretation is that these are the fossils of microorganisms that lived on Mars billions of years ago. (*Science* and NASA)

another impact 16 million years ago, and to have landed in Antarctica around 11,000 B.C.

ALH 84001 is the only known specimen of a rock that was on Mars during the era when there was liquid water on the planet. Scientists have therefore investigated its chemical composition carefully, in the hope that this rock may contain clues to the amount of water that once flowed on the Martian surface. One such clue is the presence of rounded grains of minerals called carbonates, which can form only in the presence of water.

In 1996 David McKay and Everett Gibson of the NASA Johnson Space Center, along with several collaborators, reported the results of a two-year study of the carbonate grains in ALH 84001. They made three remarkable findings. First, in and around the carbonate grains were large numbers of elongated, tubelike structures resembling fossilized microorganisms (Figure 12-20). Second, the carbonate grains contain very pure crystals of iron sulfide and magnetite. These two compounds are rarely found together (especially in the presence of carbonates) but can be produced by certain types of bacteria. Third, the carbonates contain organic molecules—just the sort, in fact, that result from the decay of microorganisms. McKay and Gibson concluded that the structures seen in Figure 12-20 are fossilized remains of microorganisms. If so, these organisms lived and died on Mars billions of years ago, during the warm era when liquid water was abundant.

Are McKay and Gibson's conclusions correct? Their claims of ancient life on Mars are extraordinary, and they require extraordinary proof. With only one rock like ALH 84001 known to science, however, such proof is hard to come by, and many scientists are skeptical. They argue that the structures found in ALH 84001 could have been formed in other ways that do not require the existence of microorganisms. For now, the existence of microscopic life on Mars in the distant past remains an open question. What is without question, however, is that ALH 84001 has fired the imagination of scientists and the public and generated new excitement about the exploration of Mars.

## 12-9 The two Martian moons resemble asteroids

Two moons move around Mars in orbits close to the planet's surface. These satellites are so tiny that they were not discovered until the favorable opposition of 1877. While Schiaparelli was seeing canals, the American astronomer Asaph Hall spotted the two moons, which orbit almost directly above its equator. He named them Phobos ("fear") and Deimos ("panic"), after the mythical horses that drew the chariot of the Greek god of war.

Phobos is the inner and larger of the Martian moons. It circles Mars in only 7 hours and 39 minutes, moving over the Martian surface at an average distance of only 6000 kilometers (3700 miles). (Recall that our Moon orbits at an average distance of 376,280 km above the Earth's surface.) This orbital period is much shorter than the Martian day; hence Phobos rises in the *west* and gallops across the sky in only 5½ hours, as viewed by an observer near the Martian equator. During this time, Phobos appears several times brighter in the Martian sky than Venus does from the Earth.

Deimos, which is farther from Mars and somewhat smaller than Phobos, appears about as bright from Mars as Venus does from the Earth. Deimos's orbit, 20,000 kilometers (12,400 miles) above the Martian surface, is at almost the right distance for a synchronous orbit—an orbit in which a satellite seems to hover above a single location on the planet's equator. As seen from the Martian surface, Deimos rises in the east and takes about three full days to creep slowly from one horizon to the other.

The *Mariner 9* mission treated astronomers to their first close-up views of the Martian satellites. Numerous additional views provided by the *Viking Orbiters* revealed Phobos and Deimos to be jagged, heavily cratered, football-shaped rocks (Figure 12-21). Phobos was found to be approximately 28 by 23 by 20 km, with Deimos being slightly smaller, at roughly 16 by 12 by 10 km. It was further revealed that both Phobos and Deimos exhibit synchronous rotation. The gravity of Mars acting on these elongated moons ensures that they always keep the same sides facing the red planet, just as our Moon always keeps the same face toward the Earth.

The origin of the Martian moons is unknown, but they may be captured asteroids. Mars is quite near the asteroid belt, where thousands of large rocks orbit the Sun. Like Phobos and Deimos, many asteroids reflect very little sunlight (typically less than 10%) because of a high carbon content. Perhaps two of these asteroids wandered close enough to Mars to become permanently trapped by the planet's gravitational field. Alternatively, Phobos and Deimos may have formed in orbit around Mars out of debris left over from the formation of the planets. Some of this debris might have come from the asteroid belt, giving these two moons an asteroidlike character. Future measurements of the composition of the Martian moons—perhaps by landing spacecraft on their surfaces—may shed light on their origin.

## 12-10 A new campaign of Martian exploration is providing important insights about the planet

After the *Viking* spacecraft arrived at Mars in 1976, there were no successful missions to the red planet for twenty years. (Two Soviet and one NASA spacecraft were launched

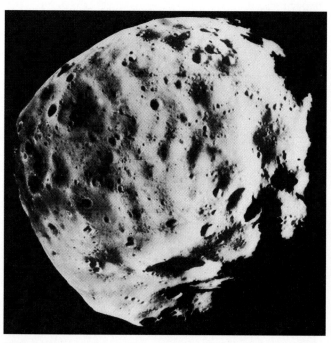

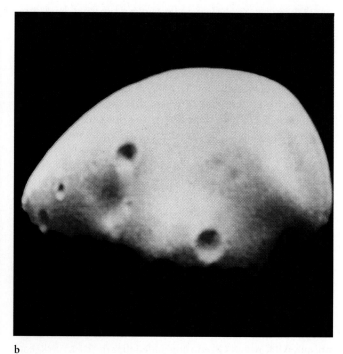

a                                                                   b

## ƒigure 12-21

**Phobos and Deimos** The *Viking Orbiters* provided these images of Mars's tiny moons.
**(a)** Phobos, the larger of Mars's two moons, is potato-shaped and measures approximately 28 by 23 by 20 km (about 17 by 14 by 12 mi). **(b)** Deimos seems to be less cratered than Phobos. It measures approximately 16 by 12 by 10 km (about 10 by 7 by 6 mi). (NASA)

in the late 1980s and early 1990s, but contact was lost with all three before they could begin their programs of observation.) In late 1996, however, NASA launched the first two spacecraft in a renewed, decade-long program of Martian exploration. These two spacecraft, *Mars Global Surveyor* and *Mars Pathfinder*, were emblematic of a major change in NASA philosophy.

Prior to the mid-1990s, most of NASA's planetary probes were large, heavy spacecraft that carried a dozen or more scientific instruments. Because of their size and complexity, these spacecraft could take a decade or more to design and build and required powerful and costly rockets to launch them into space. And when one of these planetary probes failed—as happened in 1993 to NASA's *Mars Observer* spacecraft, just three days before it was to enter Martian orbit—it was a catastrophic and very expensive loss.

Faced with declining budgets but spurred by growing scientific interest in the solar system, the motto of NASA's planetary exploration program has changed to "faster, cheaper, better." New, lighter spacecraft carry just a few highly capable scientific instruments. Thanks to their small size and relative simplicity, these miniaturized probes can be designed and built much more quickly and launched with less powerful and less expensive rockets.

*Mars Global Surveyor* (Figure 12-22) was the first of the new generation of spacecraft to be sent toward Mars. Designed to map Mars in greater detail than ever before, its mass is just 1060 kg, compared to 2450 kg for *Mars Observer*. The price tag to build and launch *Mars Global Surveyor* and operate it through its planned five-year mission is just $145 million—less than the cost of some recent Hollywood blockbusters.

*Mariner 9* and the *Viking Orbiters* produced detailed maps of only about 15% of the Martian surface. To provide more complete coverage, *Mars Global Surveyor* was placed in an orbit that passes over both poles of Mars. As Mars rotates beneath it, the spacecraft is able to view the entirety of the planet's surface. The state-of-the-art camera on *Mars Global Surveyor* can resolve features on Mars as small as 3 m (10 ft) across, a tremendous improvement over the 100 m (330 ft) resolution of the *Viking Orbiter* cameras (see Figure 12-11*b*). Beginning in March 1999, *Mars Global Surveyor* will use its camera to observe the entire surface of the planet over a full Martian year (1.88 Earth years). This will provide a detailed portrait of seasonal changes and daily weather patterns.

In addition to its visible-light camera, *Mars Global Surveyor* carries three other scientific instruments. First, a spectrometer is used to analyze infrared radiation coming from the Martian surface. Sunlight heats different materials by different amounts, which causes them to emit more or less infrared radiation. By studying variations in this radiation from different parts of Mars, scientists will learn about the chemical composition of the surface. Second, a laser altimeter can measure the heights of volcanoes, canyons, and other surface features with an accuracy of 2 meters. Third, a magnetometer is used to search for a Martian magnetic field. This instrument

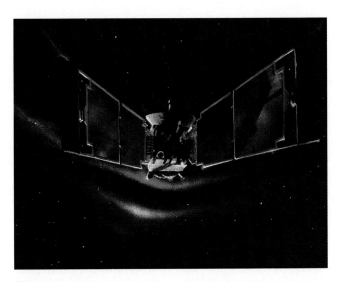

**figure 12-22**

***Mars Global Surveyor*** Placed in orbit around Mars in September 1997, *Mars Global Surveyor* will produce very precise three-dimensional maps of the planet's surface. It will also measure variations in the planet's surface composition and probe for remnants of an ancient Martian magnetic field. To save weight and minimize cost, *Mars Global Surveyor* carried only a small amount of fuel for its rocket engine. As a result, it was only able to place itself in a highly elliptical orbit around Mars, not the low-altitude circular orbit required to get the best close-up views of the planet. But each time the spacecraft's elliptical orbit brings it close to Mars, air resistance from the planet's very thin upper atmosphere slows *Mars Global Surveyor* down. (To aid in this, the blue solar panels act as air brakes.) Over time, this slowing makes the orbit become circular. (NASA, Painting by Michael J. Carroll)

has shown that while Mars as a whole does not have a magnetic field, it does have a few magnetized regions in its crust. These "magnetic anomalies" are thought to be relics of an earlier epoch when Mars had an active interior and a planetwide magnetic field.

Even more ambitious than *Mars Global Surveyor* is *Mars Pathfinder*, which on July 4, 1997, became the first spacecraft to land on Mars in two decades. Unlike the *Viking Landers*, *Mars Pathfinder* did not have a heavy and expensive rocket engine to lower it gently to the Martian surface. Instead, the spacecraft was surrounded by airbags like those used in automobiles. Had there been anyone on Mars to watch *Mars Pathfinder* land, they would have seen an oversized beach ball hit the surface at 50 km/h (30 mi/h) and make several bounces as it rolled a kilometer across the ground. Unharmed by its wild ride, the spacecraft then deflated its airbags and began to study the geology of Mars.

On board *Mars Pathfinder* was the small, wheeled robot called *Sojourner* (Figure 12-23). Following directions from

Earth, *Sojourner* spent three months exploring the rocks around its landing site. The rover carried a camera of its own, as well as an X-ray spectrometer (visible as a metal cylinder at the left-hand end of the rover in Figure 12-23a) that could measure the chemical composition of individual rocks.

*Mars Pathfinder* landed in an ancient flood plain called Ares Valles, a portion of which is shown in Figure 12-10. Flood waters can carry rocks of all kinds great distances from their original locations, so scientists expected that this landing site would show great geological diversity. They were not disappointed. Figure 12-24 is a view from the stationary portion of the spacecraft. Rocks of all sizes were found, from small pebbles to boulders like the one in Figure 12-23b. The rocks are covered with a layer of red dust, perhaps left when the flood waters retreated and left mud to dry in the Martian sun. Liquid water, however, may have been present for quite some time. A number of the pebbles are rather round, which could only have happened if they were exposed to running water for an extended period. To explain how water remained liquid on the Martian surface

long enough to have this effect, it may be necessary to imagine an ancient Mars that was even warmer and had an even thicker atmosphere than described in Section 12-4.

*Sojourner* measured the chemical compositions of many of the rocks visible in Figure 12-24. Most of the rocks are basalts, a type of igneous rock produced by volcanic action and also found on the Earth (see Figure 8-5). Because Mars has volcanoes, this was not a surprise. What was surprising, however, is that some of the rocks have an unusually high concentration of silicon. Silicon-rich rocks are less dense than basalts. They are formed when crustal material gets so hot that it becomes molten, allowing less dense materials to float to the top. Perhaps the interior of Mars was so hot in its early history that its heat melted the entire crust, thus allowing silicon-rich rocks to form. If so, Mars may have had a more dynamic past than heretofore suspected.

*Mars Global Surveyor* and *Mars Pathfinder* are just the beginning of an unprecedented effort to understand the red planet. In December 1998, *Mars Orbiter 98* will be launched toward Mars to map the planet with a color camera. A month

a

b

## ƒigure 12-23

**The *Sojourner* Rover (a)** The *Mars Pathfinder* spacecraft that landed on Mars on July 4, 1997, carried a miniature wheeled robot designed to navigate and explore the Martian surface on its own. This rover, shown here under construction on the Earth, is named *Sojourner* for Sojourner Truth, the nineteenth-century African-American woman who escaped from slavery to become a spokesperson for human rights. The rover is only 63 cm (25 in.) long and 48 cm (19 in.) wide, about the size of a child's little red wagon. The specially garbed technicians in this photo are working in a "clean room," which is kept clear of any small particles that might contaminate the rover and cause mechanical problems on Mars. (NASA, JPL) **(b)** This photograph shows *Sojourner* on the Martian surface. The six-wheeled design enabled *Sojourner* to roll over rough, rocky terrain, and the solar cell panel on top of the rover provided the power needed to cruise over the Martian surface at 1.4 km/h (0.9 mi/h). In this photograph, *Sojourner* is using its X-ray spectrometer to measure the chemical composition of a boulder nicknamed Yogi. (NASA)

**figure 12-24** R I **V** U X G

**The View from *Mars Pathfinder*** This view from the *Mars Pathfinder* landing site shows the rock field surrounding the spacecraft. In the foreground is one of the three solar panels used to power the spacecraft, along with the yellow ramp along which *Sojourner* rolled down onto the surface. The rover can be seen next to a rock nicknamed Moe. (Other whimsical names given to rocks by the *Mars Pathfinder* scientists include Jiminy Cricket, Lunchbox, and Pooh Bear.) Several of the rocks in this image have been tilted in the same direction, presumably by the flood waters that once raged over them. On the horizon is a hill about 1 kilometer (0.6 mi) away. The stationary portion of the spacecraft, from which this view was made, was christened the Carl Sagan Memorial Station, after the American astronomer who did so much to popularize astronomy and the study of the planets. (NASA)

later, *Mars Lander 98* will head for a landing site near the edge of the south polar ice cap. For the next several years, two spacecraft will be launched toward Mars every 25 months, which is when the Earth and Mars are in the best positions for sending a spacecraft from one planet to the other with the minimum thrust. Each pair of launches will include a lander and an orbiter. Japan, Russia, and the European Space Agency also plan to send spacecraft to Mars during this period. By 2005, a spacecraft may land on Mars, collect samples using an advanced rover, and return those samples to the Earth. This extensive campaign of exploration may answer at last many open questions about Mars, its formation, and its evolution.

## Key Words

favorable opposition, p. 280
highlands, p. 285
lowlands, p. 285

permafrost, p. 288
residual polar cap, p. 288
rift valley, p. 286

SNC meteorite, p. 294

## Key Ideas

**Earth-Based Observations of Mars:** Earth-based observers found that the Martian solar day is nearly the same length as that on the Earth, that Mars has polar caps that expand and shrink with the seasons, and that the Martian surface undergoes seasonal color changes.

• The best Earth-based views of Mars are obtained at favorable oppositions, when Mars is simultaneously at opposition and near perihelion.

• A few observers reported a network of linear features called canals. These observations, which proved to be illusions, led to many speculations about Martian life.

**Spacecraft Observation of Mars:** Spacecraft have returned close-up views showing that the Martian surface has numerous flat-bottomed craters, several huge volcanoes, a vast rift valley, and dried-up riverbeds—but no canals.

**Water on Mars:** Flash-flood features and dried riverbeds on the Martian surface indicate that water once flowed on Mars.

• No liquid water exists on Mars today, because it would quickly evaporate in the cold, thin atmosphere. But Mars's polar caps contain frozen water, and a layer of permafrost may exist beneath the Martian regolith.

Goldman, S. J. "A Sol in the Life on Pathfinder." *Sky & Telescope*, November 1997. This article relates the events of a typical day on Mars (one sol) during the *Mars Pathfinder* mission and gives an excellent sense of what it would be like to visit the Martian surface.

Kargel, J. S., and Strom, R. G. "Global Climatic Change on Mars." *Scientific American*, November 1996. This article discusses the changes that led to the present-day Martian climate, and presents evidence that there was once a large ocean in Mars's northern hemisphere.

McSween, H. Y. "Nor Any Drop to Drink." *Sky & Telescope*, December 1995. A leading meteorite scientist describes how the SNC meteorites provide information about the amount of liquid water than once flowed on the Martian surface.

Parker, S. "*Mars Global Surveyor:* You Ain't Seen Nothin' Yet." *Sky & Telescope*, January 1998. This article describes some of the exciting first results from the *Mars Global Surveyor* mission.

Petersen, C. C. "Welcome to Mars!" *Sky & Telescope*, October 1997. Illustrated with many beautiful images, this article describes many of the first results from the *Mars Pathfinder* spacecraft.

Pollack, J. "Atmospheres of the Terrestrial Planets." In Beatty, J. K., and Chaikin, A., eds., *The New Solar System*, 3rd ed. Cambridge University Press, 1990. This chapter compares the chemical composition, circulation, and evolution of the atmospheres of the Earth, Venus, and Mars.

**W** *World Wide Web*

Three excellent places to begin further study of Mars are the web sites "The Nine Planets" (**http://www.seds.org/nineplanets/nineplanets/mars.html**), "Welcome to the Planets" (**http://pds.jpl.nasa.gov/planets/welcome/mars.htm**), and "Views of the Solar System" (**http://www.hawastsoc.org/solar/ eng/mars.htm**).

An absolutely essential web site to visit is the Jet Propulsion Laboratory's main site for Martian exploration (**http://mars.jpl.nasa.gov/**). This site will lead you to all of the dramatic images from *Mars Pathfinder* and *Mars Global Surveyor*, including some in 3-D and some that use virtual reality. (The *Mars Pathfinder* site is *the* most visited science site on the World Wide Web. On the day that the spacecraft landed, this site logged 40 million visits!) This site is also the gateway to information about *Mars Surveyor 98, Mars Surveyor 2001*, and other missions planned for the future.

A web site maintained by the U.S. Geological Survey (**http://www-pdsimage.wr.usgs.gov/PDS/public/mapmaker/**) allows you to explore the entire catalog of *Viking Orbiter* images and to study any part of the Martian surface in close-up detail.

A web site related to Martian meteorites, including the SNC meteorite ALH 84001, is maintained by the Antarctic Search for Meteorites Program (**http://www.cwru.edu/affil/ansmet/**).

A useful NASA site details the search for life on Mars (**http://rsd.gsfc.nasa.gov/marslife/**).

# Jupiter: Lord of the Planets

R I **V** U X G

**Jupiter with Io and Europa**
Jupiter is the largest and most massive planet in the solar system. Jupiter's diameter is 11¼ times larger than that of the Earth, and its mass is 318 times greater. Jupiter, however, is composed primarily of hydrogen and helium, in abundances very similar to the Sun's chemical composition. This image, made by the *Voyager 1* spacecraft in 1979, shows the complex and colorful patterns of Jupiter's clouds. The Great Red Spot, a persistent storm in the Jovian atmosphere, about twice the size of the Earth, is on the left. This image also shows two of Jupiter's moons, Europa (right) and Io (in front of Jupiter). Both satellites are nearly the same size as our Moon. (NASA)

*In this chapter you will find the answers to the following questions:*

13-1 Can you stand on the surface of Jupiter, as you can on the Earth or the Moon?

13-2 What is going on in Jupiter's Great Red Spot?

13-3 What is the origin of Jupiter's multicolored clouds?

13-4 What happened to the comet that crashed into Jupiter?

13-5 How do astronomers know about Jupiter's interior?

13-6 What is the origin of Jupiter's strong magnetic field?

13-7 Where do you find the highest temperatures in the solar system?

*J*upiter is an active, multicolored world, more massive than all the other planets combined. Its rapid rotation stretches its weather systems completely around the planet. Some features in the Jovian cloud cover, like the Great Red Spot, endure year after year. Five spacecraft have given us revealing close-up views of Jupiter's colorful clouds and probed its enormous magnetosphere. Jupiter's hydrogen-rich interior, so highly compressed that the gas has become a metal, is capable of conducting electricity and supporting the planet's vast magnetic field.

Jupiter is quite different from Mercury, Venus, the Earth, and Mars in its size, mass, and chemical composition. More than any other factor, this is the result of temperature differences in the young solar nebula (recall Figure 7-17). In the warm inner regions of the nebula, the dust grains consisted primarily of rock-forming substances (metals, silicates, and oxides). The temperature was too high for volatile substances—those that easily vaporize, such as water, methane, and ammonia—to condense significantly. The four inner planets therefore formed almost entirely out of rock. Their gravity was too weak and their surface temperatures too high to retain hydrogen and helium, even though these lightweight gases made up most of the nebula.

At the orbit of Jupiter, however, sunlight is only 1/25 as bright as on Earth, and temperatures were much lower. Cold dust grains so far from the protosun had thick, frosty coatings of frozen water, methane, and ammonia. These volatile substances thus became important constituents of the outer planets.

As we saw in Section 7-8, Jupiter may have formed in two stages. First, ice-coated dust grains accreted to make a protoplanet several times more massive than Earth. The gravitational pull of this protoplanet then attracted and retained substantial quantities of hydrogen and helium. These gases would have gathered quite rapidly once the protoplanet had grown beyond about ten times the mass of the Earth. In this way Jupiter grew into the largest planet in our solar system.

Jupiter is also surrounded by an extensive retinue of moons. Worlds in their own right, they deserve a chapter of their own, and we will discuss them in Chapter 14.

## 13-1 Huge, massive Jupiter is composed largely of lightweight gases

The huge diameter and immense mass of Jupiter were known to astronomers long before they could send spacecraft there. Given the distance to the planet and its angular size, they used the small-angle formula (Box 1-2) to calculate that Jupiter's diameter is about 11 times bigger than that of the Earth. By observing the orbits of Jupiter's four large moons and applying Newton's form of Kepler's third law (see Section 4-7 and Box 4-2), astronomers also determined that Jupiter is 318 times more massive than the Earth. In fact, it has 2½ times the combined mass of all the other planets, satellites, asteroids, meteoroids, and comets in the solar system. A visitor from interstellar space might well describe our solar system as the Sun, Jupiter, and some debris! Table 13-1 lists some basic data about Jupiter.

As for any planet whose orbit lies outside the Earth's orbit, the best time to observe Jupiter is when it is at opposition. At such times, Jupiter can appear nearly three times brighter than Sirius, the brightest star in the sky. Among the planets, Jupiter is outshone only by Venus. Through a telescope, Jupiter presents a disk nearly 50 arcsec in diameter, approximately twice the angular diameter of Mars under the most favorable conditions. Because Jupiter takes almost a dozen years to orbit the Sun, this planet appears to meander slowly across the 12 constellations of the zodiac at the rate of approximately one constellation per year. Successive oppositions occur at intervals of about 13 months (Table 13-2).

As seen through an Earth-based telescope, Jupiter's atmosphere displays colorful bands that extend around the planet parallel to its equator (Figure 13-1). Figure 13-2, a more detailed close-up view from a spacecraft, shows the alternating dark and light bands parallel to Jupiter's equator in subtle tones of red, orange, brown, and yellow. The dark, reddish bands are called **belts,** and the light-colored bands are called

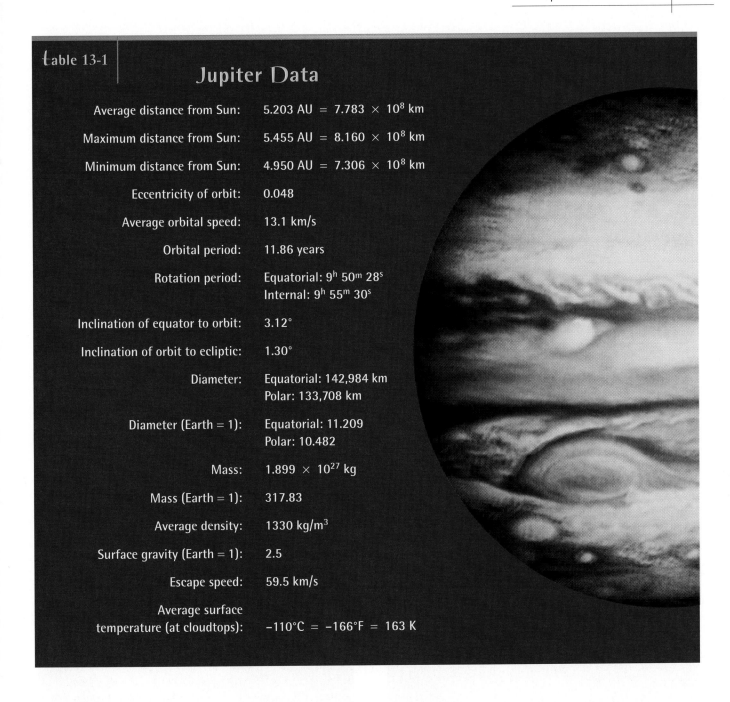

**table 13-1**

# Jupiter Data

| | |
|---|---|
| Average distance from Sun: | 5.203 AU = 7.783 × 10^8 km |
| Maximum distance from Sun: | 5.455 AU = 8.160 × 10^8 km |
| Minimum distance from Sun: | 4.950 AU = 7.306 × 10^8 km |
| Eccentricity of orbit: | 0.048 |
| Average orbital speed: | 13.1 km/s |
| Orbital period: | 11.86 years |
| Rotation period: | Equatorial: 9^h 50^m 28^s <br> Internal: 9^h 55^m 30^s |
| Inclination of equator to orbit: | 3.12° |
| Inclination of orbit to ecliptic: | 1.30° |
| Diameter: | Equatorial: 142,984 km <br> Polar: 133,708 km |
| Diameter (Earth = 1): | Equatorial: 11.209 <br> Polar: 10.482 |
| Mass: | 1.899 × 10^27 kg |
| Mass (Earth = 1): | 317.83 |
| Average density: | 1330 kg/m^3 |
| Surface gravity (Earth = 1): | 2.5 |
| Escape speed: | 59.5 km/s |
| Average surface temperature (at cloudtops): | −110°C = −166°F = 163 K |

zones. In addition to these conspicuous stripes, a huge, red-orange oval called the **Great Red Spot** is often visible in Jupiter's southern hemisphere. This remarkable feature was first seen by the English scientist Robert Hooke in 1664, but may be much older. It appears to be an extraordinarily long-lived storm in the planet's dynamic atmosphere. Many careful observers have also reported smaller spots and blemishes that last for only a few weeks or months in Jupiter's turbulent clouds.

Observations of Jupiter allow astronomers to determine how fast Jupiter rotates. At its equator, Jupiter completes a full rotation in only 9 hours, 50 minutes, and 28 seconds, making it not only the largest and most massive planet in the

solar system but also the one with the fastest rotation. However, Jupiter rotates in a strikingly different way from the Earth, the Moon, Mercury, Venus, or Mars. If Jupiter were a solid body like the terrestrial planets (or, for that matter, a billiard ball), all parts of Jupiter's surface would rotate through one complete circle in this same amount of time (Figure 13-3a). But by watching features in Jupiter's cloud cover, Giovanni Cassini discovered in 1690 that the polar regions of the planet rotate a little more slowly than do the equatorial regions. (You may recall this Italian astronomer from Section 12-2 as the gifted observer who first determined Mars's rate of rotation.) Near the poles, the rotation period of Jupiter's atmosphere is about 9 hours, 55 minutes, and 41 seconds.

| table 13-2 | Oppositions of Jupiter, 1998–2004 | | |
|---|---|---|---|
| | Earth-Jupiter distance | | Angular diameter |
| Date of opposition | (AU) | ($10^6$ km) | (arcsec) |
| 1998 September 16 | 3.96 | 593 | 49.7 |
| 1999 October 23 | 3.96 | 593 | 49.7 |
| 2000 November 28 | 4.05 | 606 | 48.6 |
| 2002 January 1 | 4.19 | 626 | 47.0 |
| 2003 February 2 | 4.33 | 647 | 45.5 |
| 2004 March 4 | 4.43 | 662 | 44.5 |

ANALOGY This behavior, called **differential rotation,** is like a pot of water being stirred on the stove; as you stir the water, different parts of the liquid take different amounts of time to make one "rotation" around the center of the pot (Figure 13-3b). Differential rotation shows that Jupiter cannot be solid throughout its volume but is at least partially fluid, like water in a pot.

If Jupiter has a partially fluid interior, it seems unlikely that it could be made of the same sort of rocky materials that constitute the terrestrial planets. An important clue to Jupiter's composition came from its average density, which is only 1330 kg/m³ (only one-fourth as much as the Earth's average density). To explain this low average density, Rupert Wildt of the University of Göttingen in Germany suggested in the

**figure 13-1**   R I **V** U X G

**Jupiter from the Earth**   Various dark-colored belts and light-colored zones are easily identified in this Earth-based view of the largest planet in our solar system. The Great Red Spot, which is about the same diameter as the Earth, was exceptionally prominent when this photograph was taken. (Courtesy of S. Larson)

**figure 13-2**   R I **V** U X G

**Jupiter from a Spacecraft**   This image was made by the *Voyager 1* spacecraft in 1979, at a distance of only 30 million kilometers from Jupiter. Features as small as 600 km across can be seen in the turbulent cloudtops. Complex cloud motions surround the Great Red Spot. (NASA)

1930s that Jupiter is composed mostly of hydrogen and helium atoms—the two lightest elements in the universe—held together by their mutual gravitational attraction to form a planet. Wildt was motivated in part by his observations of prominent absorption lines of methane and ammonia in Jupiter's spectrum. A molecule of methane ($CH_4$) contains four hydrogen atoms, and a molecule of ammonia ($NH_3$) contains three. The presence of these hydrogen-rich molecules was strong though indirect evidence of abundant hydrogen in Jupiter's atmosphere.

Direct evidence for hydrogen and helium in Jupiter's atmosphere, however, was slow in coming. The problem was that neither gas produces prominent spectral lines in the visible sunlight reflected from the planet. To show the presence of these elements conclusively, astronomers had to look for spectral lines in the ultraviolet part of the spectrum. These lines are very difficult to measure from the Earth, because almost no ultraviolet light penetrates our atmosphere (see Figure 6-27). The weak spectral lines of hydrogen molecules were first detected in Jupiter's spectrum in 1960. The presence of helium was finally confirmed in the 1970s, when spacecraft first flew past Jupiter and measured the hydrogen spectrum in detail. (Collisions between helium and hydrogen atoms cause small but measurable changes in the hydrogen spectrum, which is what the spacecraft instruments detected.)

Today, Jupiter's atmosphere is known to consist of approximately 80% hydrogen and 19% helium by mass, with very small amounts of methane, ammonia, water vapor, and other gases. As we will see in Section 13-5, there is good evidence that Jupiter has a large rocky core made of heavier elements. It is estimated that the breakdown by mass of the planet as a whole (atmosphere plus interior) is approximately 71% hydrogen, 24% helium, and 5% all heavier elements.

*CAUTION!* Because Jupiter is almost entirely hydrogen and helium, it would be impossible to land a spacecraft there. An astronaut foolish enough to try would notice the hydrogen and helium around the spacecraft becoming denser, the temperature rising, and the pressure increasing as the spacecraft descended. But the hydrogen and helium would never solidify into a surface on which the spacecraft could touch down. Our hypothetical astronaut would be well advised not to continue the descent in an attempt to land on Jupiter's rocky core. Long before reaching the core, the pressure of the hydrogen and helium would reach such unimaginably high levels that even a spacecraft made of the strongest possible materials would be crushed.

## 13-2 Pictures taken by spacecraft show many details in Jupiter's clouds

For many years astronomers realized that our understanding and appreciation of this extraordinary planet would be vastly improved if we could send spacecraft to examine Jupiter at close range. Such a program was carried out by four historic missions to Jupiter during the 1970s, and has been followed up by an even more capable spacecraft in the 1990s. They found striking evidence of stable, large-scale weather patterns in Jupiter's atmosphere, as well as evidence of dynamic changes on smaller scales.

The first four spacecraft to visit Jupiter each made a single flyby of the planet. *Pioneer 10* flew past Jupiter in December 1973. A nearly identical spacecraft, *Pioneer 11*, followed in December 1974. In 1979 another pair of spacecraft, *Voyager 1* and *Voyager 2*, sailed past Jupiter. These spacecraft sent back spectacular close-up color pictures of Jupiter's dynamic atmosphere. Most recently, in 1995, the *Galileo* spacecraft went into orbit around Jupiter and began a multiyear program of observations. By remaining in orbit, *Galileo* has been able to monitor changes in Jupiter's atmosphere on timescales of weeks or months.

Views of the entire planet show that while the general pattern of Jupiter's atmosphere stayed the same during the four years between the *Pioneer* and *Voyager* flybys, there were also some remarkable changes. These can be seen in Figure 13-4 in the area surrounding the Great Red Spot. During the *Pioneer* flybys, the Great Red Spot was embedded in a broad white zone that dominated the planet's southern hemisphere.

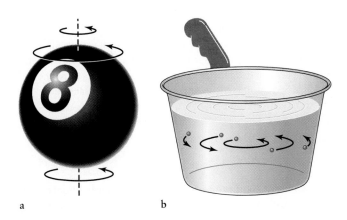

## Figure 13-3

**Solid Rotation versus Differential Rotation** (a) When a solid object like a billiard ball rotates, every part of the object takes exactly the same time to complete one rotation. (b) A rotating fluid behaves differently. To see this, put some grains of sand, bread crumbs, or other small particles in a pot of water, stir the water with a spoon to start it rotating, then take out the spoon. The particles near the center of the pot take less time to make a complete rotation than those away from the center. This same effect, called differential rotation, is seen in the atmosphere of Jupiter. The planet's equatorial regions take less time to complete one rotation than its polar regions.

**a** *Pioneer 10*, December 1973

**b** *Pioneer 11*, December 1974

**c** *Voyager 1*, March 1979

**d** *Voyager 2*, July 1979

**e** HST, February 1995

## figure 13-4   R I ⊠ U X G

**Five Views of Jupiter** These five images of Jupiter span more than 21 years and show major changes in the planet's upper atmosphere. Noticeable changes are apparent even between the two *Voyager* views **(c and d)**, which were separated by only four months. The Hubble Space Telescope image **(e)** was taken when this superb Earth-orbiting telescope was 960 million ($9.6 \times 10^8$) kilometers from Jupiter; it is comparable in quality to the *Voyager* images, which were made at a distance of only 30 million ($3 \times 10^7$) kilometers from Jupiter but with much smaller and less capable cameras. All five pictures show the Great Red Spot in Jupiter's southern hemisphere. The black dot on the *Pioneer 10* view **(a)** is the shadow of Jupiter's moon Io. (a-d, NASA; e, Reta Beebe and Amy Simon, New Mexico State University, and NASA)

By the time of the *Voyager* missions, a dark belt had broadened and encroached on the Great Red Spot from the north. In the same way that colliding weather systems in our atmosphere can produce strong winds and turbulent air, the interaction in Jupiter's atmosphere between the belt and the centuries-old Great Red Spot embroiled the entire region in greater turbulence.

Figure 13-4*e* is an image of Jupiter and the Great Red Spot in 1995, showing further changes that took place during the 16 years after the *Voyager* missions. This image was made not by a spacecraft approaching Jupiter, but by the Hubble Space Telescope (HST) (see Section 6-7). While HST is much further from Jupiter than the *Voyager* spacecraft were when they recorded the images in Figures 13-4*c* and 13-4*d*, its 2.4-m objective mirror has much better angular resolution than the relatively small cameras aboard *Voyager 1* and *Voyager 2*. Thus, the amount of detail in the HST image (Figure 13-4*e*) is comparable to the *Voyager* images shown in Figure 13-4*c* and Figure 13-4*d*. As Figure 13-5 shows, however, the amount of detail visible in *Galileo* images of Jupiter is far greater. In order to study Jupiter, there is no substitute for sending a spacecraft there!

Over the past three centuries, Earth-based observers have reported many long-term variations in the Great Red Spot's size and color. At its largest, it measured 40,000 by 14,000 km —so large that three Earths could fit side by side across it. At other times (as in 1976 and 1977), the spot almost faded from view. During the *Voyager* flybys of 1979, the Great Red Spot was comparable in size to the Earth (Figure 13-6).

**figure 13-5** R ▮ V U X G

***Galileo* Views the Great Red Spot** This June 1996 view from the *Galileo* spacecraft shows the Great Red Spot in unprecedented detail. It is actually a mosaic of 18 infrared images made through three different filters. Clouds at different levels in Jupiter's atmosphere reflect different wavelengths of infrared light, so using a variety of filters makes it possible to see the vertical structure of the Great Red Spot. Red or white areas are where there are high clouds, a green color indicates medium-level clouds, and blue or black denotes the lowest clouds. (These colors are not indicative of the true color of the Great Red Spot.) Most of the Great Red Spot is made of clouds at relatively high altitudes (shown in red and white). This is surrounded by a collar of very low-level clouds (shown in blue and black) at an altitude perhaps 50 km (30 miles, or 160,000 feet) below the high clouds at the center of the spot. This same kind of structure is seen in high-pressure areas in the Earth's atmosphere, although on a very much smaller scale. (NASA)

**figure 13-6** R I ▮ U X G

**Turbulence Around the Great Red Spot** Turbulence in the atmosphere surrounding the Great Red Spot is clearly seen in this view from *Voyager 2*. When this picture was taken in 1979, the Great Red Spot was about 20,000 km long and about 10,000 km wide. For comparison, the Earth's diameter is 12,756 km. The prominent white oval south of (below) the Great Red Spot has been observed since 1938. (NASA)

Careful examination of cloud motions in and around the Great Red Spot reveal that the spot rotates counterclockwise with a period of about six days. Furthermore, winds to the north of the spot blow to the west, and winds south of the spot move toward the east. The circulation around the Great Red Spot is thus like a wheel spinning between two oppositely moving surfaces (Figure 13-7). This surprisingly stable wind pattern has survived for at least three centuries. Weather patterns on the Earth tend to change character and eventually dissipate when they move between plains and mountains or between land and sea. Because there is no solid surface or ocean underneath Jupiter's clouds, however, no such changes can occur for the Great Red Spot—which may explain its persistence.

Other persistent features in Jupiter's atmosphere are the **white ovals**, like the one seen near the Great Red Spot in Figure 13-6. Several white ovals can also be seen in the images in Figure 13-4. As in the Great Red Spot, wind flow in white ovals is counterclockwise. White ovals are also apparently long-lived; Earth-based observers have reported seeing them in the same location since 1938.

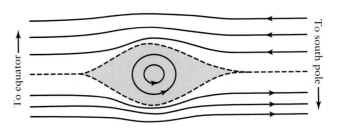

**figure 13-7** R I ▮ U X G

**Circulation Around the Great Red Spot** The winds to the north and south of the Great Red Spot blow in opposite directions. Winds within the Great Red Spot itself spin counterclockwise, completing a full revolution in about six days. Thus, the atmospheric motion within the spot resembles a wheel spinning between two oppositely moving surfaces. (Adapted from A. P. Ingersoll)

Most of the white ovals are observed in Jupiter's southern hemisphere, whereas **brown ovals** are more common in Jupiter's northern hemisphere (Figure 13-8). While brown ovals appear dark in a visible-light image like Figure 13-8, they appear *bright* in an image made with infrared light. For this reason, brown ovals are thought to be holes in Jupiter's cloud cover. They permit us to see into the depths of the Jovian atmosphere, where the temperature is higher and the atmosphere emits infrared light more strongly. White ovals, by contrast, have relatively low temperatures. They are probably areas with cold, high-altitude clouds that block our view of the lower levels of the atmosphere.

The *Galileo* and *Voyager* images might suggest a state of incomprehensible turmoil, but, surprisingly, there is also great regularity in the Jovian atmosphere. The computer-generated view in Figure 13-9 shows how Jupiter would look if you were located directly over either the planet's north pole or its south pole. Note the regular spacing of such cloud features as ripples, plumes, and light-colored wisps. The regular spacing of white ovals in the southern hemisphere is quite apparent. These regularities are probably the result of stable, large-scale weather patterns that encircle Jupiter.

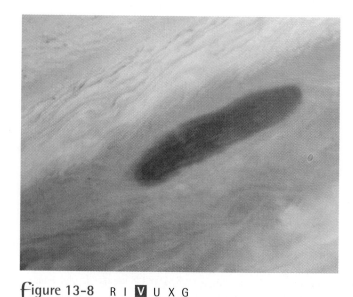

**f**igure 13-8   R I **V** U X G

**A Brown Oval** Oval-shaped openings in Jupiter's main cloud layer reveal warm, brownish gases below. The length of this oval is roughly equal to the Earth's diameter. *Voyager 1* was 4 million kilometers from Jupiter when this 1977 picture was taken. (NASA)

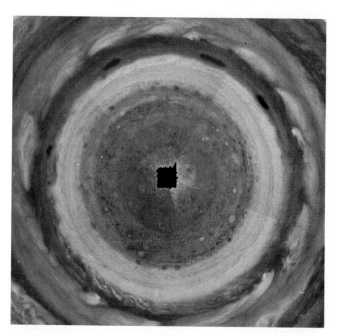

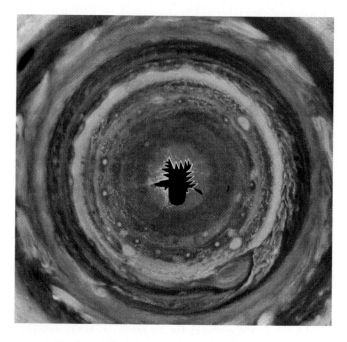

**f**igure 13-9   R I **V** U X G

**Jupiter's Northern and Southern Hemispheres** *Voyager* images were combined and computer processed to construct these views that look straight down onto Jupiter's north and south poles. **(a)** In this northern view, light-colored plumes are evenly spaced around the equatorial regions. Several brown ovals are visible. **(b)** In this southern view, the three biggest white ovals are separated by almost exactly 90° of longitude. In both the northern and southern views, the banded belt-zone structure is absent near the poles. The black spots at the poles themselves are areas not photographed by the spacecraft. (NASA)

## 13-3 The motions of Jupiter's atmospheric gases are affected by solar energy, the planet's internal heat, and differential rotation

Weather patterns on the Earth are the result of the motions of air masses. Winter storms in the midwest United States occur when cold, moist air moves southward from Canada, and tropical storms in the southwest United States are caused by the northeastward motion of hot, moist air from Hawaii. The source of energy that powers all such motions, as well the global circulation pattern in our atmosphere, is the energy of sunlight absorbed by the planet's surface and atmosphere (Section 8-4). Sunlight is also the energy source for the motions of the atmospheres of Venus (Section 11-4) and Mars (Section 12-6). In Jupiter's atmosphere, however, motions are powered both by solar energy and by the planet's own internal energy.

In the late 1960s, astronomers using Earth-based telescopes made the remarkable discovery that Jupiter emits more energy (in the form of infrared radiation) than it receives from sunlight—in fact, nearly twice as much. Many scientists suspect that this excess heat presently escaping from Jupiter is energy left over from the formation of the planet 4.5 billion years ago. As gases from the solar nebula fell into the protoplanet, vast amounts of gravitational energy were converted into thermal energy that heated the planet's interior. While Jupiter has cooled substantially since its formation, it still retains substantial thermal energy because of its size. It is this thermal energy that still radiates into space today.

In order for heat to flow upward through the Jovian atmosphere and radiate out into space, the temperature must be warmer deep inside the atmosphere than it is at the cloudtops. This follows from the everyday observation that heat flows from a hot place to a cold place, never the other way around. (If you put an ice cube on a hot frying pan, heat flows from the hot pan into the cold ice and melts it. If heat flowed the other way, the ice would get colder and the frying pan would get hotter!) Infrared measurements have confirmed that temperature rises with increasing depth in the Jovian cloud cover (Figure 13-10).

When a fluid is warm at the bottom and cool at the top—such as water being warmed in a pot or the fluid material in the Earth's mantle being heated by our planet's core—one effective way for energy to be distributed through the fluid is by the up-and-down motion called convection. (We first discussed convection in Section 8-3; see, in particular, Figure 8-16.) In a planet's atmosphere, convection causes warm gases from low levels to ascend and cool, while cooler gases at high altitude descend and heat up. On the Earth, the planet's rotation twists the rising and descending air into several convection cells that encircle the planet (see Figure 8-22). The same thing happens in Jupiter's atmosphere because of the planet's

rapid, differential rotation. The light-colored zones, where we see high-elevation clouds, are regions where gas is rising and undergoing a drop in temperature; the dark-colored belts, where we see deeper into Jupiter's cloud cover, are regions where gases are descending and being heated (Figure 13-11). This description is reinforced by comparing an infrared image of Jupiter with a visible-light image (Figure 13-12). The infrared image shows stronger emission from the belts, where we are looking at lower-lying, warmer levels of Jupiter's atmosphere.

In addition to the vertical motion of gases within the belts and zones, Jupiter's very rapid rotation gives rise to a global pattern of eastward and westward **zonal winds** with speeds that can exceed 500 km/h (300 mi/h). The atmospheric circulation pattern on the Earth has a similar pattern of eastward and westward flow, as shown in Figure 8-22, but with slower wind speeds. The faster, more energetic winds on Jupiter are presumably a result the planet's faster rotation, as well as the substantial flow of heat from the planet's interior.

As shown in Figure 13-11, Jupiter's zonal winds are generally strongest at the boundaries between belts and zones, and the wind reverses direction between the northern boundary of a belt or zone to the southern boundary. As shown in Figure 13-7, such reversals in wind direction are associated with circulating storms, such as the Great Red Spot.

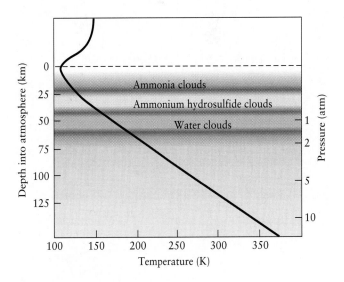

**figure 13-10**

**The Vertical Structure of Jupiter's Upper Atmosphere** This graph displays a temperature and pressure profile of Jupiter's upper atmosphere, as deduced from measurements at infrared and radio wavelengths. Three major cloud layers are shown, along with the colors that predominate at various depths. Data from the *Galileo* spacecraft indicates that these cloud layers are not found at all locations in the planet's atmosphere; there are some relatively clear, cloud-free areas.

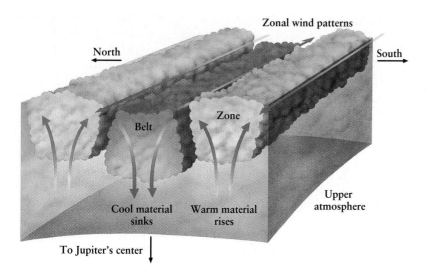

North                                    Zonal wind patterns                South

Belt        Zone

Cool material sinks        Warm material rises        Upper atmosphere

To Jupiter's center

## figure 13-11

### The Belt-Zone Structure of Jupiter's Clouds

The light-colored zones and dark-colored belts in Jupiter's atmosphere are regions of rising and descending gases, respectively. In the zones, gases warmed by heat from Jupiter's interior rise upward and cool, forming high-altitude clouds. In the belts, gases descend and undergo an increase in temperature; the cloud layers seen there are at lower altitudes than in the zones. Jupiter's rapid differential rotation shapes these regions of rising and descending gas into bands parallel to the planet's equator. The rotation also causes the wind velocities at the boundaries between belts and zones to be predominantly to the east or west.

From spectroscopic observations and calculations of atmospheric temperature and pressure, scientists conclude that Jupiter has three main cloud layers of differing chemical composition. The uppermost cloud layer is composed of crystals of frozen ammonia. About 25 km deeper in the atmosphere, ammonia ($NH_3$) and hydrogen sulfide ($H_2S$)—a compound of hydrogen and sulfur—combine to produce ammonium hydrosulfide ($NH_4SH$) crystals. Still farther down, the clouds are composed of crystals of frozen water.

The colors of Jupiter's clouds depend on the temperatures of the clouds and, therefore, on the depth of the clouds within the atmosphere. Brown clouds are the warmest and

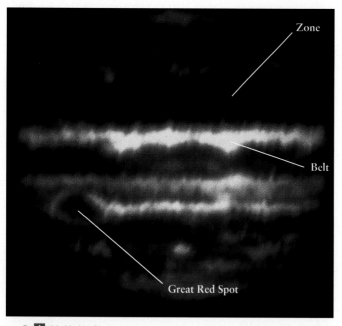

Zone

Belt

Great Red Spot

a  R **I** V U X G

Zone

Belt

Great Red Spot

b  R I **V** U X G

## figure 13-12

**Jupiter's Warm Belts and Cool Zones** Both of these images of Jupiter were made on January 10, 1979. **(a)** This infrared image was made with the Hale 5-meter telescope in California. Bright and dark areas in the image correspond to high and low temperatures, respectively. **(b)** This visible-light image is from the *Voyager 1* spacecraft. Comparison of the two images shows that the zones are regions where we see cold, light-colored, high-altitude clouds, while the belts are areas in which we see warm, dark-colored gases at lower levels within Jupiter's atmosphere. Note that the Great Red Spot appears dark in the infrared image, showing that it is a region of cool, high-altitude clouds. (Compare this image with Figure 13-5.) (NASA)

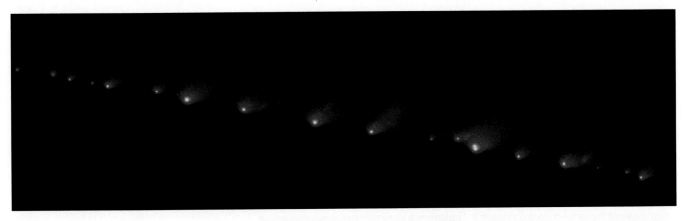

ƒigure 13-13   R I **V** U X G

**Comet Shoemaker-Levy 9**  This Hubble Space Telescope image shows 21 icy fragments of Comet Shoemaker-Levy 9. At the time that this image was made in May 1994, the fragments were spread over a distance of 1.1 million kilometers ($1.1 \times 10^6$ km = 710,000 miles), or approximately three times the distance from the Earth to the Moon. Two months later, these fragments impacted on Jupiter, producing a sequence of titanic fireballs. (H. A. Weaver, T. E. Smith [Space Telescope Science Institute], and NASA.)

are thus the deepest layers that we can see in the Jovian atmosphere. Whitish clouds form the next layer up, followed by red clouds in the highest layer. Jupiter's whitish zones are therefore somewhat higher than the brownish belts, while the red clouds in the Great Red Spot are among the highest found anywhere on the planet.

Despite all the data from the *Pioneer*, *Voyager*, and *Galileo* missions, we still do not know what gives Jupiter's clouds their colors. The crystals of ammonia, ammonium hydrosulfide, and frozen water in Jupiter's three main cloud layers are all white. Thus, other chemicals must cause the browns, reds, and oranges. Some scientists suspect that certain molecules closely related to ammonium hydrosulfide, which can form into long chains that have a yellow-brown color, play an important role in Jupiter's clouds. Others think that sulfur or phosphorus, which can assume many different colors depending on its temperature, might be involved. The Sun's ultraviolet radiation, which can induce chemical reactions, may also help give the Jovian clouds their beautiful colors.

## 13-4  Jupiter's atmosphere below the clouds has been explored by a comet and by a space probe

From observations of Jupiter's cloud layers, we have been able to learn a great deal about the atmosphere beneath the clouds. But there is no substitute for direct observation, and for many years scientists planned to send a spacecraft to explore deep into Jupiter's atmosphere. The first such probe was released by the *Galileo* spacecraft as it approached Jupiter in 1995, and it found a surprisingly dry atmosphere.

Even before this probe, however, nature provided an exploratory probe of its own—in the form of a comet that plunged into Jupiter.

In March 1993, professional astronomers Eugene and Carolyn Shoemaker and their amateur colleague David Levy discovered a remarkable comet that was actually a chain of more than 20 small pieces, strung out like pearls on a necklace (Figure 13-13). (We described comets briefly in Section 7-5.) It was officially named Comet Shoemaker-Levy 9, because it was the ninth comet that these three astronomers had discovered together. By tracing its orbit backward, it was deduced that this comet had originally been a single object until it passed close to Jupiter in July 1992. The planet's gravitational tidal forces then tore the comet into the fragments shown in Figure 13-13. (You may want to review the discussion of tidal forces in Section 9-4.) What is more, Jupiter's gravity had altered the trajectories of the comet fragments, so that they were doomed to collide with Jupiter.

As astronomers around the world watched, 23 fragments of Comet Shoemaker-Levy 9 crashed into Jupiter's atmosphere between July 16 and July 22, 1994. The fragments have been estimated to be no larger than 1 kilometer across, but were accelerated by Jupiter's gravitational pull to a speed on impact of 60 km/s (220,000 km/h = 130,000 mi/h). The largest fragment collided with Jupiter with an energy equivalent to 600,000,000 megatons of TNT, tens of thousands of times greater than the total destructive energy of all of the nuclear weapons on Earth. All of the impacts occurred on the side of Jupiter that faced away from Earth and, therefore, were not immediately visible to astronomers. Thanks to Jupiter's rapid rotation, however, each impact site came into view within a few minutes after the impact itself (Figure 13-14). As each fragment entered Jupiter's atmosphere, it generated a shock wave that vaporized the fragment

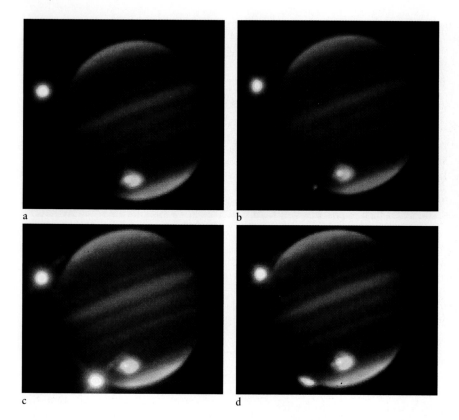

**figure 13-14** R ■ V U X G

**A Fragment of Comet Shoemaker-Levy 9 Impacts Jupiter** This sequence of infrared images of Jupiter shows the impact of fragment H (the eighth piece of Comet Shoemaker-Levy 9 to strike the planet) on July 18, 1994. **(a)** This image shows Jupiter just before the impact. Jupiter's moon Ganymede is the bright spot to the left of and above the planet, and the bright blob in Jupiter's southern hemisphere is the site where fragments D and G struck the planet several hours before. **(b)** The second image in the sequence was recorded just after the impact of fragment H on the far side of Jupiter. The fireball from this impact can be seen peeking around the lower left corner of the planet. **(c)** Seven minutes after the second image, the fireball is near its maximum brightness. **(d)** The fourth image, recorded 16 minutes after the third, shows the fireball flattening out and fading. These images were made using the 3.5-meter telescope at Spain's Calar Alto Observatory. (Tom Herbst, Max-Planck-Institut für Astronomie, Heidelberg, Germany)

itself and turned the fragment's debris, along with gases from Jupiter's atmosphere, into a huge, 10,000°C fireball. Each impact's fireball, which contained millions of tons of material, then erupted upward to as much as 3,000 km (1,900 mi) above Jupiter's clouds. As this material fell back downward, it left dark-colored blotches the size of the Earth in Jupiter's upper atmosphere. These "scars" from the different impacts persisted for several weeks after the impacts ended (Figure 13-15).

By examining the spectra of light emitted from the fireballs, astronomers hoped to learn more about the chemical composition of Jupiter's atmosphere. But analyzing the fireball spectra is complicated because of uncertainties about the composition of the comet as well as about how deep into Jupiter's atmosphere the comet fragments penetrated. As an example, spectral lines of sulfur were seen in the light from one of the larger impacts. This could mean that the comet fragment penetrated to the level of the ammonium hydrosulfide ($NH_4SH$) clouds shown in Figure 13-10, where the energy of the impact would have broken the $NH_4SH$ molecules apart and liberated sulfur atoms to be ejected upward with the fireball. Alternatively, the fragment might have penetrated to only a shallow depth above the $NH_4SH$ clouds, and the sulfur atoms whose spectrum was seen in the fireball might have been part of the fragment itself. Much more analysis of the Comet Shoemaker-Levy 9 impacts needs to be done before the full scientific potential of these dramatic events can be realized.

 The impact of the comet fragments on Jupiter led many people to worry about a comet or asteroid striking the Earth, with disastrous consequences. But, as we will see in Chapter 17, such impacts are extraordinarily rare; according to reliable estimates, an object the size of Comet Shoemaker-Levy 9 can be expected to collide with the Earth only about once in 300,000 years. Among the many worries of everyday life, comets crashing into the Earth are very far down the list!

Just one year after the comet impacted on Jupiter, a probe of human design began its own one-way trip into the giant planet's atmosphere. While still 81.52 million kilometers (50.66 million miles) from Jupiter, the *Galileo* spacecraft released the *Galileo Probe*, a cone-shaped body about the size of an office desk. While *Galileo* itself later fired its rockets to place itself into an orbit around Jupiter, the *Galileo Probe* continued on a course that led it on December 7, 1995, to a point in Jupiter's clouds just north of the planet's equator. Unlike the comet fragments that preceded it, the *Galileo Probe* did not burn up in a fireball. Instead, it used a heat shield to brake its descent, then floated down through the atmosphere on a parachute (Figure 13-16). This gave the probe an hour during which to observe its surroundings and radio its findings back to the main *Galileo* spacecraft, which in turn radioed them to the Earth. The mission ceased at a point some 200 kilometers (120 miles) below the upper cloud layer, where the tremendous atmospheric pressure (24 times that on Earth) and high

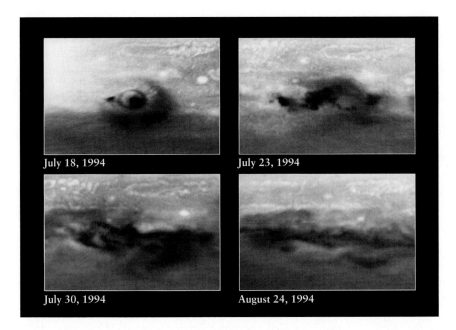

**figure 13-15**  R I V U X G

**Debris from an Impact on Jupiter** This sequence of Hubble Space Telescope images shows the evolution of one of the Comet Shoemaker-Levy impact sites. The first image, taken about 90 minutes after the impact of fragment G on July 18, 1994, shows a well-defined dark crescent. The other three images show how Jupiter's high-speed winds gradually erased this and other impact scars. (Heidi Hammel, MIT, and NASA.)

temperature (152°C = 305°F) finally overwhelmed the probe's electronics.

While the *Galileo Probe* did not have a camera, it did carry a variety of instruments that made several new discoveries. A radio emissions detector found evidence for lightning discharges that, while less frequent than on the Earth, are individually much stronger than lightning bolts on the Earth. Other measurements showed that Jupiter's winds, which are brisk in the atmosphere above the clouds, are even stronger beneath the clouds. In fact, the *Galileo Probe* mea-

sured a nearly constant wind speed of 650 km/h (400 mi/h) throughout its descent. This shows convincingly that the energy source for the winds is Jupiter's interior heat. If the winds were driven primarily by solar heating, as is the case on the Earth, the wind speed would have decreased with increasing depth.

A central purpose of the *Galileo Probe* mission was to test the three-layer cloud model depicted in Figure 13-10 by making direct measurements of Jupiter's clouds. But the probe saw only traces of the $NH_3$ and $NH_4SH$ cloud layers

**figure 13-16**

**The *Galileo Probe* Enters Jupiter's Atmosphere** Five months before arriving at Jupiter, the *Galileo* spacecraft released an atmospheric probe. Just before entering Jupiter's upper atmosphere on December 7, 1995, the *Galileo Probe* was moving at 171,000 km/h (106,000 mi/h); air friction then decelerated it with a force equal to 228 times the Earth's gravity, slowing the probe to only 40 km/h (25 mi/h) in just 3 minutes. The tremendous heat generated in the deceleration was absorbed by the probe's conical heat shield, shown glowing red-hot in this artist's impression. Once the heat shield had been released, the probe drifted down on a parachute. The probe made direct measurements and chemical analyses as it descended through the Jovian atmosphere. (NASA)

and no sign at all of the low-lying water clouds. One explanation of this surprising result is that the probe by sheer chance entered an unusually cloud-free part of Jupiter's atmosphere. Indeed, observations from the Earth showed strong infrared emission from the probe entry site, as would be expected where a break in the cloud cover allows a view of Jupiter's warm interior (see Figure 13-12).

The probe also made measurements of the chemical composition of Jupiter's atmosphere. The relative proportions of hydrogen and helium proved to be almost exactly the same as in the Sun, in line with the accepted picture of how Jupiter formed from the solar nebula (Section 7-8). A number of heavy elements, however, including carbon, nitrogen and sulfur, proved to be significantly more abundant on Jupiter than in the Sun. This was an expected result. Over the past several billion years, Jupiter's strong gravity should have pulled in many pieces of interplanetary debris. (Comet Shoemaker-Levy 9 is only the most recent.) This debris built up a small but appreciable fraction of heavy elements.

Water is a major constituent of small bodies in the outer system and thus should likewise have been added to Jupiter in noticeable quantities. Surprisingly, the amount of water actually detected by the *Galileo Probe* was less than half the expected amount. The explanation of this disparity remains a mystery. Has the water rained out of the atmosphere and coalesced at the planet's core? Or did the probe simply happen to enter the atmosphere at an unusually dry spot? As is so often the case when new scientific observations are made, the information from the *Galileo Probe* has posed many new questions at the same time that it provided answers to some old ones.

## 13-5 Jupiter's oblateness divulges the planet's rocky core

Neither Comet Shoemaker-Levy 9 nor the *Galileo Probe* went more than a few hundred kilometers into Jupiter's atmosphere. To learn about the planet's structure at greater depths, astronomers must use indirect clues. One such clue is the shape of Jupiter, which indicates the size of the rocky core buried deep beneath Jupiter's atmosphere.

Even a casual glance through a small telescope shows that Jupiter is not spherical but is slightly flattened or **oblate.** The diameter across Jupiter's equator (142,980 km) is 6.5% larger than its diameter from pole to pole (133,700 km). Thus, Jupiter is said to have an **oblateness** of 6.5%, or 0.065. By comparison, the oblateness of the Earth is just 0.34%, or 0.0034.

If Jupiter did not rotate, it would be a perfect sphere. A massive, nonrotating object naturally settles into a spherical shape so that every atom on its surface experiences the same gravitational pull aimed directly at the object's center. Because Jupiter is rotating, however, the body of the planet

tends to fly outward and away from the axis of rotation. You can demonstrate this effect for yourself. Stand with your arms hanging limp at your sides, then spin yourself around. Your arms will naturally tend to fly outward, away from the vertical axis of rotation of your body. Likewise, if you drive your car around a circular road, you feel thrown toward the outside of the circle; the car as a whole is rotating around the center of the circle, and you tend to move away from the rotation axis. (We discussed the physical principles behind this effect in Box 4-3.) For Jupiter, this effect means that the planet is deformed into its nonspherical, oblate shape.

The oblateness of a planet depends on how the planet's mass is distributed over its volume. The challenge to planetary scientists is to create a model of a planet's mass distribution that gives the right value for the oblateness. One such model for Jupiter's interior suggests that 4% of the planet's mass is concentrated in a dense, rocky core. The remaining 96% is predominantly hydrogen and helium. In this model, Jupiter's core is nearly 13 times more massive than the entire Earth. Much of this core was probably the original "seed" around which the proto-Jupiter accreted. However, additional amounts of rocky material were presumably added later by meteoritic material falling onto the planet.

Although Jupiter's rocky core is 13 times more massive than the Earth, it is probably not much bigger than our planet. Calculations suggest that the tremendous crushing weight of the remaining bulk of Jupiter—equal to the mass of 305 Earths—compresses Jupiter's core down to a sphere 20,000 km in diameter (by comparison, the Earth's diameter is 12,800 km). The pressure at Jupiter's center is estimated to be about 80 million atmospheres, meaning that the rocky material of Jupiter's core is squeezed to a density of about 20,000 $kg/m^3$. The corresponding temperature at the planet's center is probably about 25,000 K. In contrast, the temperature at Jupiter's cloudtops is only 165 K.

## 13-6 Metallic hydrogen in Jupiter's interior endows the planet with a strong magnetic field

The oblateness of Jupiter gives astronomers information about the size of Jupiter's rocky core. It does not, however, provide much insight into the hydrogen-rich bulk of Jupiter between its rocky core and its colorful cloudtops. Most of our knowledge of this part of the planet's interior comes from observations of Jupiter with radio telescopes. By using these telescopes to detect radio waves from Jupiter, astronomers have found evidence of electric currents flowing within the planet's hydrogen-rich interior.

Astronomers began discovering radio emissions coming from Jupiter in the 1950s. A small portion of these emissions is **thermal radiation,** with the sort of spectrum that would be expected from a warm object. (We discussed the electromagnetic radiation from heated objects in Sections 5-3 and 5-4.)

However, most of the radio energy emitted from Jupiter is **nonthermal radiation,** which has a very different sort of spectrum than radiation from a heated object—and is found in two distinct wavelength ranges.

At wavelengths of a few meters, Jupiter emits sporadic bursts of radio waves. This **decametric** (ten-meter) **radiation** is probably caused by electrical discharges associated with powerful electric currents in Jupiter's ionosphere. As we shall see in Chapter 14, these discharges result from electromagnetic interactions between Jupiter and its large satellite, Io. At shorter wavelengths of a few tenths of a meter, Jupiter emits a nearly constant stream of electromagnetic radiation. The properties of this **decimetric** (tenth-meter) **radiation** show that it is emitted by high-speed electrons moving through a strong magnetic field. The magnetic field deflects the electrons into spiral-shaped trajectories, and the spiraling electrons radiate energy like miniature radio antennas. Energy produced in this fashion is called **synchrotron radiation.** Astronomers now realize that synchrotron radiation is very important in a wide range of situations throughout the universe, from pulsars and quasars to the radio emissions of entire galaxies.

If some of Jupiter's radiation is indeed synchrotron radiation, then Jupiter must have a magnetic field. Measurements by the *Pioneer* and *Voyager* spacecraft verified this, and indicated that the field at Jupiter's equator is 14 times stronger than at the Earth's equator. As for the Earth, Jupiter's field is thought to be generated by motions of an electrically conducting fluid in the planet's interior (see Section 8-5). But wholly unlike Earth, the moving fluid within Jupiter is an exotic form of hydrogen that acts like a liquid metal.

A hydrogen atom consists of a single proton orbited by a single electron. Deep inside Jupiter, pressures are so great that the electrons are stripped from their protons. The result is a mixture of protons and electrons. The electrons, no longer bound to their protons, are free to wander around on their own. Their motion, directed by Jupiter's rotation and by convection within the planet, creates an electric current, just as the ordered movement of electrons in a copper wire constitutes an electric current. In other words, the highly compressed hydrogen deep inside Jupiter behaves like a metal; thus, it is called **liquid metallic hydrogen.**

Laboratory experiments show that hydrogen is transformed into a liquid metal when the pressure exceeds about 1.4 million atmospheres. This transition occurs at a depth of 10,000 to 20,000 km below Jupiter's cloudtops. Thus, the internal structure of Jupiter consists of three distinct regions (Figure 13-17): a rocky core, a layer of liquid metallic hydrogen 40,000 to 50,000 km thick, and a layer of molecular hydrogen 10,000 to 20,000 km thick. The colorful cloud patterns that we can see through telescopes are located in the outermost 100 km of the exterior layer of molecular hydrogen.

As Figure 13-17 shows, much of Jupiter's enormous bulk is electrically conductive liquid metal. Because of Jupiter's rapid rotation, electric currents in this metallic hydrogen generate a powerful magnetic field, in much the same way that liquid portions of the Earth's core produce the Earth's magnetic field.

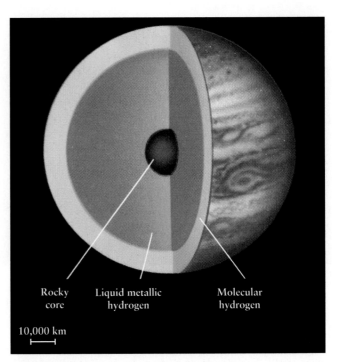

Rocky core    Liquid metallic hydrogen    Molecular hydrogen

10,000 km

## figure 13-17

**The Internal Structure of Jupiter** Jupiter probably has a rocky core approximately 20,000 km in diameter. This core is surrounded by a thick layer of liquid metallic hydrogen (shown in orange) and a relatively shallow outer layer composed primarily of ordinary hydrogen (shown in yellow). Recent experiments on hydrogen under very high pressures suggest that the boundary between these two layers may be rather indistinct, with a transition region in which molecular hydrogen behaves as a metal. Jupiter's magnetic field is caused by the motions of the metallic hydrogen; it is not yet clear whether the fluid motions primarily responsible for the field occur deep within Jupiter's interior or within the transition region.

## 13-7    Jupiter has an extensive magnetosphere

Jupiter's strong magnetic field surrounds the planet with a magnetosphere so large that it envelops the orbits of many of its moons. The *Pioneer* and *Voyager* spacecraft, which carried instruments to detect charged particles and magnetic fields, revealed the awesome dimensions of the Jovian magnetosphere. The shock wave that surrounds Jupiter's magnetosphere is nearly 30 million kilometers across. If you could see Jupiter's magnetosphere from the Earth, it would cover an area in the sky 16 times larger than the full Moon. Figure 13-18 shows the structure of the Jovian magnetosphere.

The *Pioneer* and *Voyager* spacecraft reported the outer boundary of Jupiter's magnetosphere at distances ranging from 3 to 7 million kilometers above the planet. Evidently,

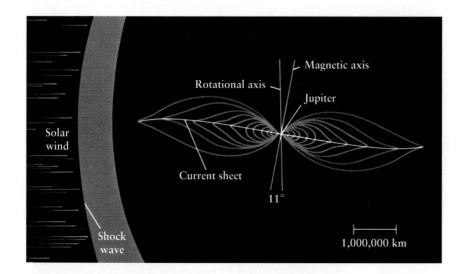

**figure 13-18**

**Jupiter's Magnetosphere** Jupiter's magnetosphere is enveloped by a shock wave, where the supersonic solar wind is abruptly slowed to subsonic speeds. Most of the particles of the solar wind are deflected around Jupiter in a turbulent region, colored purple in the scale drawing. Particles trapped inside Jupiter's magnetosphere are spewed out into a vast current sheet by the planet's rapid rotation. Jupiter's axis of rotation is inclined to its magnetic axis by about 11°.

Jupiter's magnetosphere is sensitive to fluctuations in the solar wind. It expands and contracts by a factor of two, rapidly and frequently. In contrast, dramatic changes in the size of the Earth's magnetosphere are extremely rare despite occasional "gusts" in the solar wind.

The inner regions of our magnetosphere are dominated by two huge Van Allen belts (recall Figure 8-24) that are filled with charged particles. Jupiter also entraps charged particles in a similar manner. In addition, however, Jupiter's rapid rotation spews the particles out into a huge electrically charged **current sheet** (Figure 13-18), which lies close to the plane of Jupiter's magnetic equator. Jupiter's magnetic axis is inclined 11° from the planet's axis of rotation, and the orientation of Jupiter's magnetic field is the reverse of the Earth's magnetic field—a compass would point toward the south on Jupiter.

Radio emissions from charged particles in the densest regions of Jupiter's magnetosphere are shown in Figure 13-19. These emissions vary slightly with a period of 9 hours, 55 minutes, and 30 seconds, as Jupiter's rotation changes the angle at which we view its magnetic field. Because the magnetic field is anchored deep within the planet, this variation reveals Jupiter's **internal rotation period**. This period, indicative of the rotation of the bulk of the planet's mass, is slightly slower than the atmospheric rotation at the equator (period 9 hours, 50 minutes, 28 seconds) but about the same as the atmospheric rotation at the poles (period 9 hours, 55 minutes, 41 seconds). The faster motion of the atmosphere at the equator relative to the interior is driven by Jupiter's internal heat.

The *Voyager* spacecraft discovered that the inner regions of Jupiter's magnetosphere contain a hot, gaslike mixture of charged particles called a **plasma**. A plasma is formed when a gas is heated to such extremely high temperatures that electrons are torn off the atoms of the gas. As a result, a plasma is a mixture of positively charged ions and negatively charged electrons. The plasma that envelops Jupiter consists primarily of electrons and protons, with some ions of helium, sulfur, and oxygen. These charged particles are caught up in Jupiter's rapidly rotating magnetic field and are accelerated to extremely high speeds.

The *Voyager* instruments revealed that the plasma surrounding Jupiter has an astoundingly high temperature of 300 million to 400 million K. The Jovian plasma is the hottest place in the solar system! Even the center of our Sun

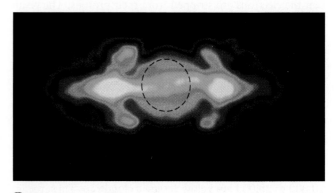

**figure 13-19**  R I V U X G

**A Radio View of Jupiter** The Very Large Array (VLA), shown in Figure 6-25, was used to produce this map of decimetric synchrotron emission from Jupiter at a wavelength of 21 cm. The image is about five Jupiter diameters wide and is elongated parallel to the planet's magnetic equator. Jupiter is at the center of the image, indicated by the dashed circle. Different colors are used to denote differences in the intensity of radiation. The strongest emission, shown in white, comes from electrons trapped in the densest regions of the current sheet that encircles Jupiter. (NRAO)

is only about 15 million K. However, the Jovian plasma is very thin. The average density in the hottest regions is about one charged particle per 100 cubic centimeters.

Because of its high temperature, Jupiter's plasma has a great deal of energy and exerts pressure that holds off the solar wind. The *Voyager* data suggest that the pressure balance between the solar wind and the hot plasma inside the Jovian magnetosphere is precarious. A gust in the solar wind can blow away some of the plasma, at which point the mag-

netosphere deflates rapidly to as little as one-half its original size. Charged-particle detectors carried by the *Voyager* spacecraft recorded several bursts of hot plasma that may have been associated with magnetospheric deflations. However, additional electrons and ions accelerated by Jupiter's rotating magnetic field soon replenish the plasma, and the magnetosphere expands again. As part of its ongoing mission, the orbiting *Galileo* spacecraft is exploring these and other aspects of Jupiter's dynamic magnetosphere.

## Key Words

belts, p. 304

brown ovals, p. 310

current sheet, p. 318

decametric radiation, p. 317

decimetric radiation, p. 317

differential rotation, p. 306

Great Red Spot, p. 305

internal rotation period, p. 318

liquid metallic hydrogen, p. 317

nonthermal radiation, p. 317

oblate, oblateness, p. 316

plasma, p. 318

synchrotron radiation, p. 317

thermal radiation, p. 316

white ovals, p. 309

zonal winds, p. 311

zones, p. 305

## Key Ideas

**Jupiter's Composition and Structure:** Jupiter, whose mass is equivalent to the mass of 318 Earths, is composed of 71% hydrogen, 24% helium, and 5% all other elements. This chemical composition has a higher percentage of heavy elements that of the Sun.

• Jupiter probably has a rocky core with a mass of about 13 Earth masses. This core is surrounded by a 40,000-km-thick layer of liquid metallic hydrogen and an outer layer of molecular hydrogen about 20,000 km thick. All of the visible features of Jupiter lie in the outermost 100 km of the molecular hydrogen layer.

• Jupiter has the shortest rotation period of any planet. The rotation is so rapid that the planet is flattened into a noticeably oblate shape. The rotation of Jupiter's interior is revealed by variations in Jupiter's radio emission.

• Jupiter emits about twice as much heat as it receives from the Sun. Presumably the planet is still cooling.

**Jupiter's Atmosphere:** The visible "surface" of Jupiter is actually the tops of its clouds. The planet's rapid rotation twists the clouds into dark belts (of warming descending gas) and light zones (where gas is rising and cooling) that run parallel to the equator. Strong zonal winds run along the belts and zones.

• The outer layers of the Jovian atmosphere show differential rotation, with the equatorial regions rotating

slightly faster than the polar regions. The polar rotation rate is nearly the same as the internal rotation rate.

• The colored ovals visible in the Jovian atmosphere represent gigantic storms. Some, such as the Great Red Spot, are quite stable and persist for many years.

• There are presumed to be three cloud layers in Jupiter's atmosphere. The reasons for the distinctive colors of these different layers are not yet known. The *Galileo Probe* spacecraft, so far the only one to directly explore Jupiter's atmosphere, detected only one of these cloud layers.

**Jupiter's Magnetic Field and Magnetosphere:** Jupiter has a strong magnetic field created by currents in the metallic hydrogen layer. Its huge magnetosphere contains a vast current sheet of electrically charged particles.

• Charged particles in the densest portions of Jupiter's magnetosphere emit synchrotron radiation at radio wavelengths.

• The Jovian magnetosphere encloses a plasma of charged particles that is hotter than the center of the Sun but has very low density. The magnetosphere exists in a delicate balance between pressures from the plasma and from the solar wind. When this balance is disturbed, the size of the magnetosphere fluctuates drastically.

## Review Questions

**1.** Which planet, Mars or Jupiter, passes closer to the Earth? On which planet is it easier to see details with an Earth-based telescope? Explain your answer.

**2.** Compare Jupiter's chemical composition with that of the Sun. What does this tell us about Jupiter's formation?

**3.** What is thought to be the source of Jupiter's excess internal heat?

**4.** What are the belts and zones in Jupiter's atmosphere? Is the Great Red Spot more like a belt or a zone?

**5.** Why is Jupiter oblate? What do astronomers learn from the size of Jupiter's oblateness?

**6.** If Jupiter does not have any observable solid surface and its atmosphere rotates differentially, how are astronomers able to determine the planet's internal rotation rate?

**7.** Did the Comet Shoemaker-Levy 9 impacts on Jupiter provide unambiguous information about the composition of Jupiter's atmosphere? Why or why not?

**8.** What data from the *Galileo Probe* were in agreement with astronomers' predictions? What data were surprising?

**9.** What is liquid metallic hydrogen? What is its significance for Jupiter?

**10.** What data and techniques have been used to determine the internal structure of Jupiter?

**11.** What is a plasma? Where are plasmas found in the vicinity of Jupiter?

**12.** Compare and contrast Jupiter's magnetosphere with the magnetosphere of a terrestrial planet like Earth. Why is the size of the Jovian magnetosphere variable, while that of the Earth's magnetosphere is not?

## Advanced Questions

### Problem-solving tips and tools:

Box 4-2 contains a very useful form of Kepler's third law. Newton's universal law of gravitation, discussed in Section 4-7, is the basic equation from which a planet's surface gravity can be calculated. Escape speed is discussed in Box 7-2, and the properties of thermal radiation (including the Stefan-Boltzmann law, which relates the temperature of a body to the amount of thermal radiation that it emits) are discussed in Sections 5-3 and 5-4.

**13.** The angular diameter of Jupiter at opposition varies little from one opposition to the next (see Table 13-2). By contrast, the angular diameter of Mars at opposition is quite variable (see Table 12-2). Why is there a difference between these two planets?

**14.** If Jupiter emitted just as much energy (as infrared radiation) as it receives from the Sun, the average temperature of the planet's cloudtops would be about 107 K. Given that Jupiter actually emits twice this much energy, find what the average temperature must actually be.

**15.** **(a)** Based on Figure 13-10, what is the temperature in Jupiter's atmosphere at the depth where the pressure is 1 atmosphere, the same as at the Earth's surface? Give your answer in the Kelvin, Celsius, and Fahrenheit scales. **(b)** At the depth where the temperature in Jupiter's atmosphere is the same as room temperature (20°C = 68°F), what is the atmospheric pressure? **(c)** At what depth in the Earth's oceans is the pressure the same as the value you found in (b)? (The pressure at the surface of the Earth's oceans is 1 atmosphere and increases by 1 atmosphere for every 10 meters that you descend below the surface.)

**16.** Using orbital data for a Jovian satellite of your choice (see Appendix 3), calculate the mass of Jupiter. How does your answer compare with the mass quoted in Table 13-1?

**17.** The *Galileo Probe* had a mass of 339 kg. On Earth, its weight (the gravitational force exerted on it by Earth) was 3,320 newtons, or 747 lb. What was the gravitational force that Jupiter exerted on the *Galileo Probe* when it entered Jupiter's clouds?

**18.** Consider a hypothetical future spacecraft that would float, suspended from a balloon, for extended periods in Jupiter's upper atmosphere. If it is desired to have this spacecraft return to the Earth after completing its mission, to what speed would the spacecraft's rocket motor have to accelerate it in order to escape Jupiter's gravitational pull? Compare to the escape speed from the Earth, equal to 11.2 km/s.

**19.** Estimate the wind velocities in the Great Red Spot, which rotates with a period of about six days.

**20.** What sort of experiment would you design in order to establish whether Jupiter has a rocky core?

## Discussion Questions

**21.** Describe some of the semipermanent features in Jupiter's atmosphere. Compare and contrast these long-lived features with some of the transient phenomena seen in Jupiter's clouds.

**22.** Suppose you were designing a mission to Jupiter involving an airplane-like vehicle that would spend many days (months?) flying through the Jovian clouds. What observations, measurements, and analyses should this aircraft be prepared to make? What dangers might the aircraft encounter, and what design problems would you have to overcome?

**23.** How is the orbiting *Galileo* spacecraft doing? What observations of Jupiter has it conducted? What discoveries has it made? Consult such magazines as *Sky & Telescope,*

*Astronomy,* and *Science News,* or the web site for the *Galileo* mission (**http://www.jpl.nasa.gov/galileo/**).

## Observing Projects

### Observing tips and tools:

Like Mars, Jupiter is most easily seen around opposition. The dates of Jupiter's oppositions are listed in Table 13-2. There is an opposition of Jupiter every 13 months, and the planet is visible in the night sky for several months before or after opposition. At other times, Jupiter may be visible in either the predawn morning sky or the early evening sky. Consult *Sky & Telescope* and *Astronomy* magazines, or their web sites (**http://www.skypub.com/** and **http://www.astronomy.com/**, respectively), for more detailed information about when and where to look for Jupiter during a given month. A relatively small telescope with an objective diameter of 15–20 cm (6–8 inches), used with a medium-power eyepiece to give a magnification of 25× or so, should enable you to see some of the dark belts.

**24.** If Jupiter is visible in the night sky, make arrangements to view the planet through a telescope. What magnifying power seems to give you the best view? Draw a picture of what you see. Can you see any belts and zones? How many? Can you see the Great Red Spot?

**25.** Make arrangements to view Jupiter's Great Red Spot through a telescope. During months when Jupiter is prominent in the night sky, the *Sky & Telescope* "What's Up in the Sky?" web page (**http://www.skypub.com/whatsup/whatsup.shtml**) lists the times when the center of the Great Red Spot passes across Jupiter's central meridian. The Great Red Spot is well placed for viewing for 50 minutes before and after this happens. You will need a refractor with an objective of at least 15 cm (6 in.) diameter or a reflector with an objective of at least 20 cm (8 in.) diameter. Using a pale blue or green filter can increase the color contrast and make the spot more visible. Other useful hints can be found in the article "Tracking Jupiter's Great Red Spot" by Alan MacRobert (*Sky & Telescope,* September 1997).

## Where to Learn More

*Books and magazine articles*

Beatty, J. K., and Levy, D. H. "Crashes to Ashes: A Comet's Demise." *Sky & Telescope,* October 1995. This article, coauthored by one of the discoverers of Comet Shoemaker-Levy 9, describes what was learned from the comet's collision with Jupiter as well as the controversies that still remain.

Burgess, E. *By Jupiter.* Columbia University Press, 1982. This nontechnical book reviews our understanding of Jupiter after the *Pioneer* and *Voyager* flybys.

Carroll, M. "Project Galileo: The Phoenix Rises." *Sky & Telescope,* April 1987. This article explains how the arrival of *Galileo* at Jupiter was delayed by several years because of the *Challenger* disaster.

Doody, D. F. "Jupiter at Last!" *Sky & Telescope,* December 1995. Written on the eve of *Galileo's* arrival at Jupiter, this article provides a good overview of the mission.

Dyer, A. "*Ulysses* Meets a Giant." *Astronomy,* July 1992. This one-page article describes the passage of the *Ulysses* spacecraft—whose primary mission is to study the Sun, not Jupiter—through the Jovian magnetosphere in February 1992, when Jupiter's gravity catapulted the tiny spacecraft into a polar orbit around the Sun.

Goldman, S. J. "*Galileo* in Retrospect." *Sky & Telescope,* December 1997. This article includes a remarkable collection of images from *Galileo's* flight to Jupiter and its first three years in orbit around the giant planet.

Ingersoll, A. "The Meteorology of Jupiter." *Scientific American,* March 1976. This article gives an exceptionally clear description of the dominant weather patterns on Jupiter.

Morrison, D., and Samz, J. *Voyage to Jupiter.* NASA SP-439, 1980. This superb book describes the *Voyager* missions to Jupiter and has an excellent selection of color photographs.

Rogers, J. H. *The Giant Planet Jupiter.* Cambridge University Press, 1995. This beautifully illustrated book gives clear and detailed descriptions of every aspect of Jupiter's atmosphere and internal structure.

Washburn, M. *Distant Encounters: The Exploration of Jupiter and Saturn.* Harcourt Brace Jovanovich, 1983. This entertaining book describes the *Pioneer* and *Voyager* missions to the outer solar system.

**W** *World Wide Web*

Start your exploration of Jupiter on the World Wide Web at the sites "Views of the Solar System" (**http://www.hawastsoc.org/solar/eng/jupiter.htm**), "The Nine Planets" (**http://www.seds.org/nineplanets/nineplanets/jupiter.html**), and "Welcome to the Planets" (**http://pds.jpl.nasa.gov/planets/welcome/jupiter.htm**).

An extensive collection of Jupiter images can be found at the National Space Science Data Center (**http://nssdc.gsfc.nasa.gov/photo_gallery/photogallery-jupiter.html**).

Each of the space missions that has explored Jupiter has its own web site: *Pioneer* (**http://pyroeis.arc.nasa.gov/pioneer/PNhome.html**); *Voyager* (**http://vraptor.**

jpl.nasa.gov/voyager/voyager.html); *Galileo* (http://www.jpl.nasa.gov/galileo/; and the *Galileo Probe* (http://ccf.arc.nasa.gov/galileo_probe/).

There are also two extensive web sites for Comet Shoemaker-Levy 9 (http://seds.lpl.arizona.edu/sl9/sl9.html and http://www.jpl.nasa.gov/sl9/).

Metallic hydrogen was predicted theoretically in 1935 but was only created in the laboratory in 1996. The story of the experiment that finally succeeded is available at the Lawrence Livermore National Laboratory web site (http://www.llnl.gov/str/Nellis.html).

# The Galilean Satellites of Jupiter

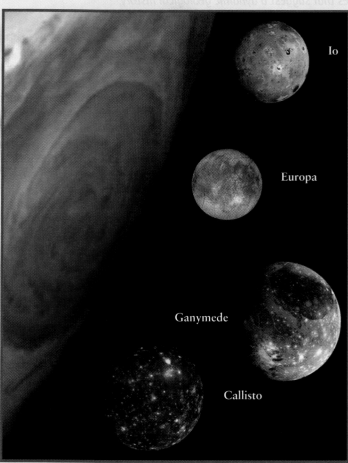

R I **V** U X G

**The Galilean Satellites with Jupiter** This montage shows part of the giant planet Jupiter with its four largest satellites. Io and Europa are both about the same size as our Moon, while Ganymede and Callisto are larger. Io has numerous active volcanoes, while Europa's icy surface may cover an ocean of liquid water. The pattern of colors on Ganymede suggests that it has had a geologically active history. Callisto, by contrast, is geologically dead. (NASA, JPL)

*In this chapter you will find the answers to the following questions:*

**14-1** What is special about the orbits of the Galilean satellites?

**14-2** Why do astronomers think that ice is an important part of some Galilean satellites?

**14-3** How did the Galilean satellites form?

**14-4** Why does Io have active volcanoes? How are they different from volcanoes on the Earth?

**14-5** How does Io act like an electric generator?

**14-6** Why do some scientists suspect that there might be an ocean below the surface of Europa?

**14-7** Do the icy satellites Ganymede and Callisto show any sign of geologic activity?

**14-8** If Jupiter has a ring like Saturn, why was it not discovered until only recently?

*Voyager 1* discovered a huge impact basin on Callisto's Jupiter-facing hemisphere, seen faintly in Figure 14-18. This basin, called Valhalla, was produced by the impact of an asteroid-sized object early in the satellite's history. Valhalla consists of a large number of concentric rings having diameters up to 3000 km.

The great age of Valhalla can be inferred both from the presence of impact craters there and from the absence of substantial vertical relief in the concentric rings. The Valhalla impact probably occurred around 4 billion years ago, when the satellite's young, relatively thin crust was still able to flow and thus reduce the height of the upraised rings in the ice.

In looking at the Galilean satellites, we see an orderly progression from a high-density, geologically active world (Io) nearest Jupiter to a low-density, geologically dead world (Callisto) farthest from Jupiter. The diagrams in Figure 14-19 summarize the probable interior structures of the Galilean satellites. These terrestrial worlds demonstrate the remarkable variety of which nature is capable.

## 14-8 The *Voyager* spacecraft discovered several small satellites and a ring around Jupiter

Besides the Galilean satellites, Jupiter has 12 other small satellites. (*Table 14-2* gives some details of the satellite orbits.) The four innermost of these satellites lie within the orbit of Io. The largest is Amalthea, which was imaged by both *Voyager 1* and Galileo. As Figure 14-20 shows, Amalthea has an irregular shape much like a potato (or a large asteroid). Its longest dimension is only 270 km—about ten times larger than the moons of Mars, but only about one-tenth the size of the Galilean satellites. Like the Galilean satellites, Amalthea orbits Jupiter with synchronous rotation, keeping its longest axis pointed toward the planet. The red color seen in Figure 14-20 is probably caused by sulfur that Jupiter's magnetosphere removed from Io's volcanic plumes (Section 14-4). This sulfur was later deposited onto Amalthea's surface.

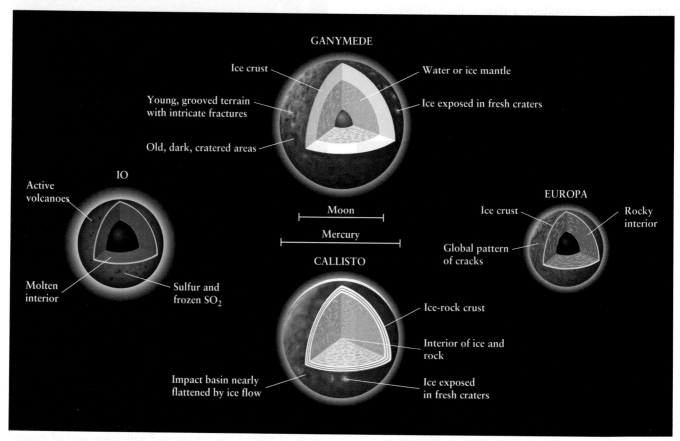

## Figure 14-19

**Interiors of the Galilean Satellites** These cross-sectional diagrams show the probable internal structure of the four Galilean satellites, based on their average densities and on information from the *Galileo* mission.

## Table 14-2  Jupiter's Family of Moons

| | Average radius of orbit | | Orbital period (days) | Year of discovery |
| | (km) | (Jupiter radii) | | |
|---|---|---|---|---|
| Metis | 127,960 | 1.7922 | 0.2948 | 1979 |
| Adrastea | 128,980 | 1.8065 | 0.2983 | 1979 |
| Amalthea | 181,300 | 2.539 | 0.4981 | 1892 |
| Thebe | 221,900 | 3.108 | 0.6745 | 1979 |
| Io | 421,600 | 5.905 | 1.769 | 1610 |
| Europa | 670,900 | 9.397 | 3.551 | 1610 |
| Ganymede | 1,070,000 | 14.99 | 7.155 | 1610 |
| Callisto | 1,883,000 | 26.37 | 16.689 | 1610 |
| Leda | 11,094,000 | 155.4 | 238.72 | 1974 |
| Himalia | 11,480,000 | 160.8 | 250.57 | 1904 |
| Lysithea | 11,720,000 | 164.2 | 259.22 | 1938 |
| Elara | 11,737,000 | 164.4 | 259.65 | 1905 |
| Ananke | 21,200,000 | 296.9 | 631$^R$ | 1951 |
| Carme | 22,600,000 | 316.5 | 692$^R$ | 1938 |
| Pasiphae | 23,500,000 | 329.1 | 735$^R$ | 1908 |
| Sinope | 23,700,000 | 331.9 | 758$^R$ | 1914 |

Note: The outermost four satellites move in a retrograde direction about Jupiter, indicated by a superscript R with the orbital period.

Jupiter's three other inner satellites are even smaller than Amalthea. They were first discovered in images from *Voyager 1* and *Voyager 2* but have since been observed from the Earth using infrared telescopes. Methane in Jupiter's atmosphere absorbs rather than reflects infrared light from the Sun, so Jupiter's glare is greatly reduced at infrared wavelengths. This makes it much easier to spot the satellites.

Figure 14-21, which is such an infrared image, shows two of the small inner satellites. It also shows that Jupiter is encircled by a very faint ring, which was seen for the first time in *Voyager* images. This ring is quite unlike the rings of Saturn, which can easily be seen with even a small telescope. The difference is that Saturn's rings are made up of a great many icy, reflective objects a centimeter or larger in size, as we will discuss in Chapter 15. By contrast, Jupiter's ring is composed of particles of rock that have an average size of only about 1 μm (= 0.001 mm = $10^{-6}$ m) and that reflect very little light, making this ring a challenge to observe. These ring particles may originate from meteorite impacts on the four small, inner satellites.

The ring, the four inner satellites, and the Galilean satellites all orbit Jupiter in the plane of the planet's equator. Furthermore, these are all **prograde orbits:** These objects orbit Jupiter in the same direction as Jupiter's rotation. This is

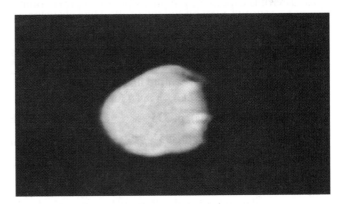

**f**igure 14-20  R I **V** U X G

**Amalthea** Tiny, reddish Amalthea is the largest of Jupiter's four innermost satellites. It is an irregularly shaped, asteroidlike object measuring 270 km along its longest dimension. This image taken by *Voyager 1* has a resolution of 8 km. Amalthea was discovered in 1892 by American astronomer E. E. Barnard at the University of California's Lick Observatory. It was the last satellite of any planet to be discovered without the aid of photography. The red color is thought to be caused by sulfur that originated on Io and came to Amalthea by way of Jupiter's magnetosphere. The two light-colored spots on the right side of Amalthea are craters. (NASA)

## Advanced Questions

**Problem-solving tips and tools:**

Kepler's third law tells us that the square of the period ($P^2$) divided by the cube of the size of the orbit ($a^3$) is a constant. The value of the constant depends on the mass of the body around which you are orbiting, as discussed in Box 4-2. The small-angle formula is discussed in Box 1-2. The best seeing conditions on the Earth give a limiting angular resolution of ¼ arcsec. Because the orbits of the Galilean satellites are almost perfect circles, the orbital speeds of these satellites can easily be calculated from the data listed in Table 14-1.

**20.** Using the orbital data in Table 14-1, demonstrate that the Galilean satellites obey Kepler's third law.

**21.** What is the size of the smallest feature you should be able to see on a Galilean satellite through a large telescope under conditions of excellent seeing when Jupiter is near opposition? How does this compare to the best *Galileo* images, which have resolutions around 25 meters?

**22.** Astronomers who observed Io with the Hubble Space Telescope in 1992 claimed that they can see features as small as 150 km across. When the observations were made, Io was 4.45 AU from the Earth. What was the angular resolution of the Hubble Space Telescope at this time? (Subsequent repair missions to the Hubble Space Telescope have further improved its resolution.)

**23.** Using the diameter of Io (3630 km) as a scale, estimate the height to which the plume of Prometheus rises above the surface of Io in Figure 14-5b. (You will need to make measurements on this figure using a ruler.)

**24.** Jupiter, its magnetic field, and the charged particles that are trapped in the magnetosphere all rotate together once every 10 hours. Io takes 1.77 days to complete one orbit. Using a diagram, explain why particles from Jupiter's magnetosphere hit Io primarily from behind (that is, on the side of Io that trails as it orbits the planet).

**25.** Assuming material is ejected from Io into Jupiter's magnetosphere at the rate of 1 ton per second (1000 kg/s), how long will it be before Io loses 10% of its mass? How does your answer compare with the age of the solar system?

**26.** How long does it take for Ganymede to enter or leave Jupiter's shadow? Assume that the shadow has a sharp edge.

**27.** Consult the *Galileo* web site (http://www.jpl.nasa.gov/galileo/) and magazines like *Science News* and *Sky & Telescope* to learn about the current status of the *Galileo* mission. Is all well with the spacecraft? What new discoveries has it made?

## Discussion Questions

**28.** Speculate on the possibility that Europa, Ganymede, or Callisto might harbor some sort of life. Explain your reasoning.

**29.** If you could replace our Moon with Io, and if Io could maintain its present amount of volcanic activity, what changes would this cause in our nighttime sky? Do you think that Io could in fact remain volcanically active in this case? Why or why not?

**30.** Suppose you were planning four missions that would land a spacecraft on each of the Galilean satellites. What kinds of questions would you want these missions to answer? What kinds of data would you want your spacecraft to send back? In view of the different environments on the four satellites, how would the designs of the four spacecraft differ? Be specific about the possible hazards and problems each spacecraft might encounter in landing on the four satellites. If only one of these missions could be funded, which one would you choose? Why?

## Observing Projects

**Observing tips and tools:**

An indispensable aid to observing the Galilean satellites is the "Celestial Calendar" section of *Sky & Telescope*. In months when Jupiter is visible in the night sky, a chart shows the positions of the satellites relative to Jupiter for that month. Note that this chart uses *universal time* (UT), formerly known as Greenwich Mean Time, and you must convert your local time to UT. Universal time is counted from 0 hours beginning at midnight in Greenwich, England. For example, to convert Pacific Standard Time (PST) to universal time, you must add eight hours. Thus, 8:00 P.M. PST on January 15 is the same as 4:00 UT on January 16. The apparent positions of the Galilean satellites can also be found very easily for any date and time using the *Starry Night*™ software on the CD-ROM that accompanies this textbook. You can also use the World Wide Web (http://ringside.arc.nasa.gov/www/tools/tools.html—note that you will need the capability to view Postscript files). For even more detailed information about satellite positions, consult the "Satellites of Jupiter" section in the *Astronomical Almanac* for the current year.

**31.** Observe Jupiter through a pair of binoculars. Can you see all four Galilean satellites? Make a drawing of what you observe. If you look again after an hour of two, can you see any changes?

**32.** Observe Jupiter through a small telescope on three or four consecutive nights. Make a drawing each night showing the positions of the Galilean satellites relative to Jupiter. Record the time and date of each observation. Consult the "Satellites of Jupiter" section in the *Astronomical Almanac* for the current year, the "Celestial Calendar" section in the current issue of *Sky & Telescope*, or the World Wide Web (**http://ringside.arc.nasa.gov/www/tools/tools.html**) to see if you can identify the satellites by name.

**33.** Make arrangements to observe an eclipse, a transit, or an occultation of one of the Galilean satellites. Consult a listing of such phenomena in the "Satellites of Jupiter" section in the *Astronomical Almanac* for the current year. Choose the phenomenon you would like to see and calculate its scheduled time by converting the universal time given in the *Astronomical Almanac* to your local time.

**34.** If you are fortunate enough to have access to a large telescope with a primary mirror 1 meter or more in diameter, make arrangements to view Jupiter through that telescope. Describe what you see. Under conditions of excellent seeing when Jupiter is near opposition, the Galilean satellites should look like tiny discs rather than starlike pinpoints of light.

## Where to Learn More

*Books and magazine articles*

Beatty, J. K. "Galileo: An Image Gallery." *Sky & Telescope*, November 1996 and March 1997. These articles include many *Galileo* images with explanatory text.

Beebe, R. *Jupiter: The Giant Planet*. Smithsonian Institution Press, 1997. This book, written for a popular audience by a leading astronomer who specializes in observations of Jupiter and its satellites, includes some of the early results from the *Galileo* mission.

Carroll, M. "Europa: Distant Ocean, Hidden Life?" *Sky & Telescope*, December 1997. Might there be an ocean beneath the icy surface of Europa? Could there be life in that ocean? These speculations are the subject of this stimulating article.

Griffin, R. "Barnard and His Observations of Io." *Sky & Telescope*, November 1982. This brief article discusses pre-*Voyager* visual observation of Io, including some attempts to map its surface.

Johnson, T., and Soderblom, L. "Io." *Scientific American*, December 1983. This article, which is devoted to volcanism on Io, includes an enlightening discussion of the properties of sulfur and sulfur dioxide.

Rogers, J. H. *The Giant Planet Jupiter*. Cambridge University Press, 1995. A richly illustrated, in-depth survey of everything that was known about Jupiter and its satellites prior to the *Galileo* mission.

Rothery, D. A. *Satellites of the Outer Planets: Worlds in Their Own Right*. Oxford, 1992. An excellent, introductory-level review of all of the satellites of the outer solar system, from the viewpoint of a geologist.

Soderblom, L. "The Galilean Moons of Jupiter." *Scientific American*, January 1980. An excellent selection of photographs accompany this fine article, which summarizes the discoveries made during the *Voyager* flybys.

W *World Wide Web*

By far the best place to find the latest information about the Galilean satellites is the *Galileo* mission web site (**http://www.jpl.nasa.gov/galileo/**). Two other good starting places are the Jupiter page of "Views of the Solar System" (**http://www.hawastsoc.org/solar/eng/jupiter.htm**) and "The Nine Planets" page for Io (**http://www.seds.org/nineplanets/nineplanets/io.html**), both of which have links to information about all of Jupiter's satellites.

Many spacecraft images of the Galilean satellites can be found at NASA's Planetary Photojournal web site (**http://photojournal.jpl.nasa.gov/**). Future images may come from *Europa Orbiter*, a proposed NASA mission that would use radar to explore under Europa's ice (**http://www.jpl.nasa.gov/pluto/europao.htm**).

Some of the mythological characters for whom the Galilean satellites are named appear in the Roman poet Ovid's classic *Metamorphoses*, which you can read on-line (**gopher://gopher.vt.edu:10010/02/128/1**).

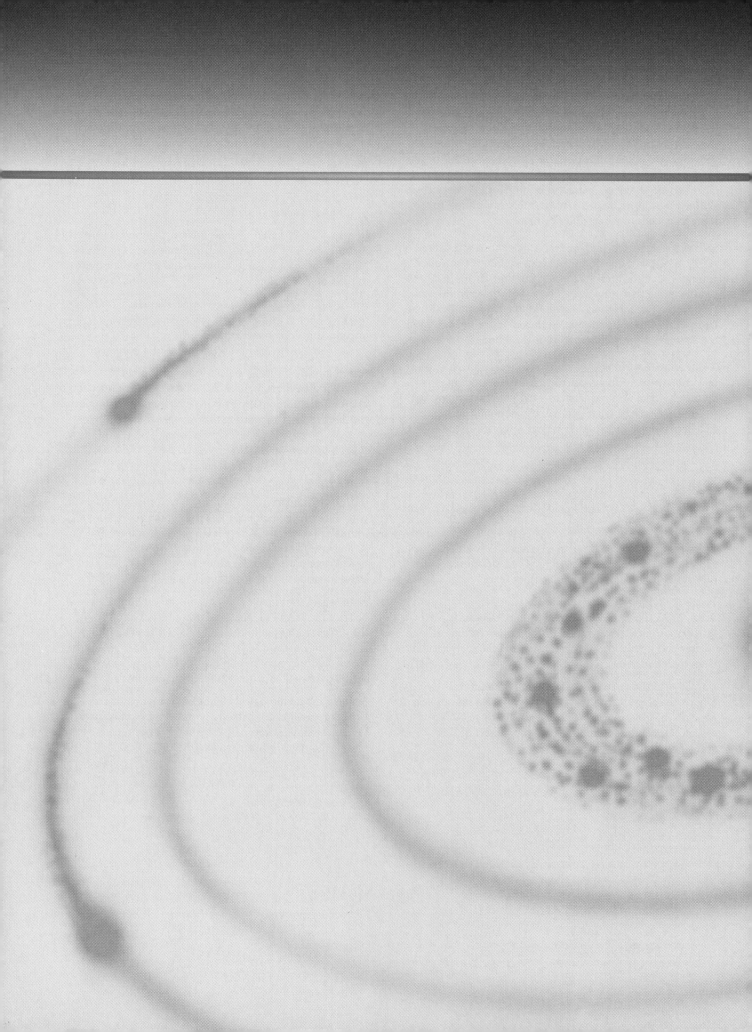

# The Spectacular Saturn-ian System

R I **V** U X G

**Saturn** These Hubble Space Telescope images were made in 1995, when Saturn's rings were almost edge-on to our line of sight. The upper image shows how thin the rings are. Saturn's largest moon, Titan, is to the left of the planet; the black dot on Saturn's surface is Titan's shadow. Four smaller moons appear on the right. The lower image, made three months later, shows the moons Tethys (upper left) and Dione (lower right). Dione casts a long shadow on the rings. (Erich Karkoschka, University of Arizona Lunar and Planetary Laboratory/NASA)

*In this chapter you will find the answers to the following questions:*

*S*aturn is a fascinating and uniquely beautiful world. Even through a small telescope, Saturn stands out because of the system of thin, flat rings that encircles it. Through a large telescope, details of the planet's clouds and rings become more evident (Figure 15-1). And when observed at close range, as was done by the Voyager spacecraft that flew past Saturn in the early 1980s, the ringed planet is revealed to be as complex and diverse as it is beautiful. Table 15-1 lists some basic data about Saturn.

The Voyager explorations of Saturn's hydrogen-rich atmosphere and cloud patterns suggest differences in how Saturn and neighboring Jupiter evolved. The Voyager spacecraft also glimpsed surface features on many of Saturn's satellites. But the most interesting of all the satellites is Titan, Saturn's largest moon, on which no surface features can be glimpsed at all. This is because Titan is enveloped by a thick, hazy, nitrogen-rich atmosphere where hydrocarbon rain may fall. Other, smaller satellites exert gravitational forces that profoundly affect the structure and appearance of the rings. The satellites and rings of Saturn, as well as the planet itself, will come under intense scrutiny when the Cassini spacecraft reaches Saturn in 2004.

## 15-1  Earth-based observations reveal three broad rings encircling Saturn

In 1610, Galileo became the first person to view Saturn through a telescope. He saw few details, but he did notice two puzzling lumps protruding from opposite edges of the planet's disk. Curiously, these lumps disappeared in 1612, only to reappear in 1613. Other observers saw similar appearances and disappearances over the next several decades. In 1655, Dutch astronomer Christiaan Huygens (who we first encountered in <u>Section 12-2</u>, regarding his observations of Mars) began to observe Saturn with a better telescope than any of his predecessors. On the basis of his observations, Huygens suggested that Saturn was surrounded by a thin, flattened ring. At times this ring was edge-on as viewed from the Earth, making it almost impossible to see. At other times Earth observers viewed Saturn from an angle either above or below the plane of the ring, and the ring was visible. (The lumps that Galileo saw were the parts of the ring to either side of Saturn, blurred by the poor resolution of his rather small telescope.) Over the next several years, astronomers confirmed this brilliant deduction as they watched the ring's appearance change in just the manner predicted by Huygens.

As the quality of telescopes improved, astronomers realized that Saturn's "ring" is actually a *system* of rings. In 1675, Italian astronomer Giovanni Cassini (who we also met in <u>Section 12-2</u>) discovered a dark division in the ring. The **Cassini division** is an apparent gap about 5000 kilometers wide that separates the outer **A ring** from the brighter **B ring** closer to the planet. In the mid-1800s, astronomers

**f**igure 15-1   R I **V** U X G

**Saturn from the Earth**  This image of Saturn is one of the best ever produced by an Earth-based observatory. Astronomer Stephen Larson combined 16 original color images, all taken on the same night with the 1.5-m telescope at the Catalina Observatory in Arizona, to make this photograph. Note the faint stripes in Saturn's cloudtops as well as the prominent gap in the rings (called the Cassini division). (NASA)

managed to identify the faint **C ring**, or crepe ring, that lies just inside the B ring. Figure 15-2 shows the diameters of these rings superimposed on a *Voyager 1* image of Saturn.

Saturn and its rings are best seen when the planet is at or near opposition (Table 15-2). Whereas a modest telescope

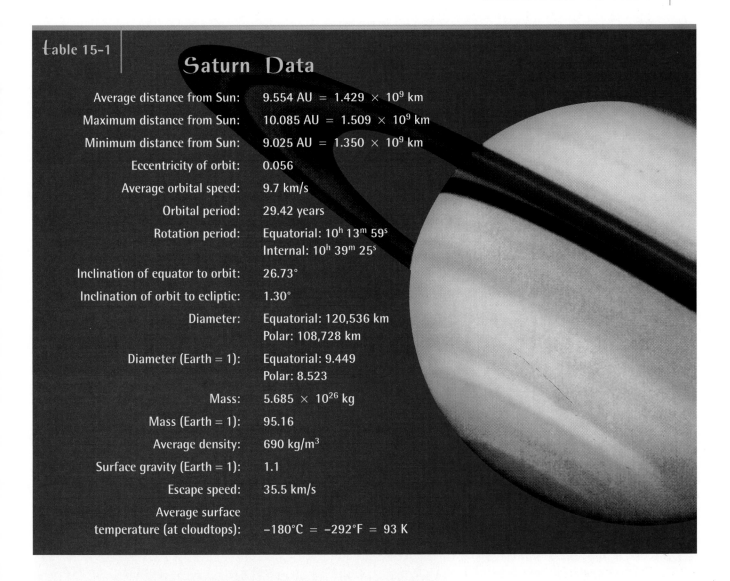

**table 15-1**

## Saturn Data

| | |
|---|---|
| Average distance from Sun: | 9.554 AU = 1.429 × 10⁹ km |
| Maximum distance from Sun: | 10.085 AU = 1.509 × 10⁹ km |
| Minimum distance from Sun: | 9.025 AU = 1.350 × 10⁹ km |
| Eccentricity of orbit: | 0.056 |
| Average orbital speed: | 9.7 km/s |
| Orbital period: | 29.42 years |
| Rotation period: | Equatorial: 10ʰ 13ᵐ 59ˢ<br>Internal: 10ʰ 39ᵐ 25ˢ |
| Inclination of equator to orbit: | 26.73° |
| Inclination of orbit to ecliptic: | 1.30° |
| Diameter: | Equatorial: 120,536 km<br>Polar: 108,728 km |
| Diameter (Earth = 1): | Equatorial: 9.449<br>Polar: 8.523 |
| Mass: | 5.685 × 10²⁶ kg |
| Mass (Earth = 1): | 95.16 |
| Average density: | 690 kg/m³ |
| Surface gravity (Earth = 1): | 1.1 |
| Escape speed: | 35.5 km/s |
| Average surface temperature (at cloudtops): | −180°C = −292°F = 93 K |

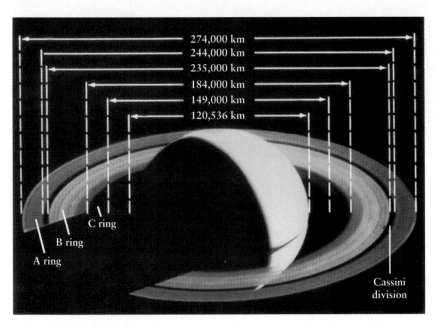

**figure 15-2** R I **V** U X G

**Saturn's System of Rings** Details of Saturn's rings are visible in this image sent back by *Voyager 1*. The equatorial diameter of Saturn as well as the diameters of the inner and outer edges of the rings are given in the overlay. Closest to the planet is the 17,500-km-wide C ring, so faint that it is almost invisible in this view. Next outward from Saturn is the broad, bright B ring, whose width is 25,500 km, or twice the Earth's diameter. The outermost ring that can be seen from the Earth is the A ring, which is 15,000 km wide. The 4500-km-wide Cassini division lies between the B ring and the A ring. This image shows that the rings are not solid, because Saturn is visible through the rings (look near the bottom of the image). Note the wide shadow that Saturn casts on the rings, and the narrow shadow that the rings cast on Saturn. (NASA)

gives a good view of the A and B rings, a large telescope and excellent observing conditions are needed to see the C ring.

Earth-based views of the Saturnian ring system change as Saturn slowly orbits the Sun. Huygens was the first to understand that this change occurs because the rings lie in the plane of Saturn's equator, and this plane is tilted 27° from the plane of Saturn's orbit. Saturn's rotation axis and the plane of its equator keep the same orientation in space as the planet goes around the orbit (just as does the Earth, as shown in Figure 2-11). Hence, over the course of a Saturnian year, the rings are viewed from various angles by an Earth-based observer (Figure 15-3). At certain times Saturn's north pole is tilted toward the Earth and the observer looks

"down" on the "top side" of the rings. Half a Saturnian year later, Saturn's south pole is tilted toward us and the "underside" of the rings is exposed to our Earth-based view.

When our line of sight to Saturn is in the plane of the rings, the rings are viewed edge-on and seem to disappear entirely (see the upper of the two images that open this chapter). This disappearance indicates that the rings are very thin. In fact, they are thought to be only a few tens of meters thick. In proportion to their diameter, Saturn's rings are thousands of times thinner than the sheets of paper used to print this book.

The last edge-on presentation of Saturn's rings was in 1995–1996, and the next will occur in 2008–2009. Until 2008, Earth observers will see the "underside" of the rings.

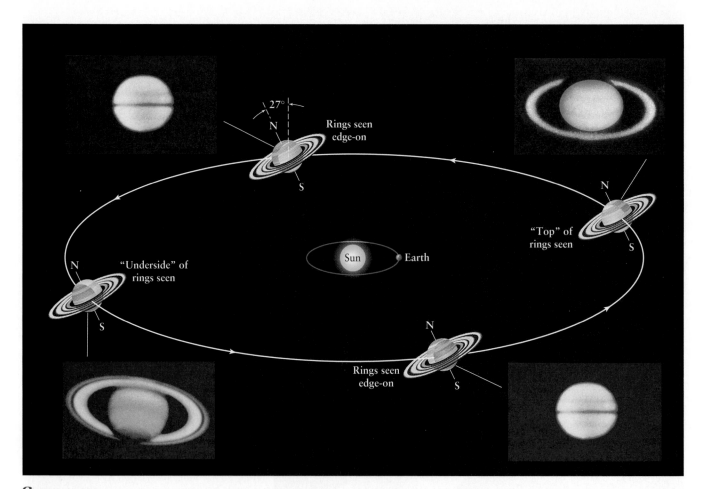

### figure 15-3

**The Changing Appearance of Saturn's Rings** Saturn's rings are aligned with its equator, which is tilted 27° from the plane of Saturn's orbit. Therefore, Earth-based observers see the rings at various angles as Saturn moves around its orbit. Note that the plane of Saturn's rings and equator keep the same orientation in space as the planet goes around its orbit, just as the Earth does as it orbits the Sun (compare Figure 2-11). The accompanying Earth-based photographs show Saturn at various points in its orbit. Note that the rings seem to disappear entirely when viewed edge-on, which occurs about every 15 years. (Lowell Observatory)

**table 15-2 | Oppositions of Saturn, 1998–2004**

| Date of opposition | Earth-Saturn distance | | Apparent diameter (arcsec) |
| | (AU) | ($10^6$ km) | |
| --- | --- | --- | --- |
| 1998 October 23 | 8.29 | 1241 | 20.1 |
| 1999 November 6 | 8.21 | 1228 | 20.3 |
| 2000 November 19 | 8.13 | 1217 | 20.5 |
| 2001 December 3 | 8.08 | 1209 | 20.6 |
| 2002 December 17 | 8.05 | 1204 | 20.7 |
| 2003 December 31 | 8.05 | 1204 | 20.7 |

## 15-2 Saturn's rings are composed of numerous icy fragments

Astronomers have long known that Saturn's rings could not possibly be solid, rigid, thin sheets of matter. In 1857, the Scottish physicist James Clerk Maxwell proved theoretically that tidal forces would tear such a broad, thin sheet apart. He concluded that Saturn's rings are made of "an indefinite number of unconnected particles." As we will see, these particles are bits of dirty ice that range in size from pebbles to boulders.

Observational confirmation that the rings are not rigid came in 1895, when James Keeler at the Allegheny Observatory in Pittsburgh photographed the spectrum of sunlight reflected from Saturn's rings. As the rings orbit Saturn, the spectral lines from the side approaching us are blueshifted by the Doppler effect (recall Section 5-9 and Figure 5-23). At the same time, spectral lines from the receding side of the rings are redshifted. Keeler noted that the size of the wavelength shift increased inward across the rings—the closer to the planet, the greater the shift. This variation in Doppler shift proved that the inner portions of Saturn's rings are moving around the planet more rapidly than the outer portions. Indeed, the orbital speeds across the rings are in complete agreement with Kepler's third law: The square of the orbit period about Saturn at any place in the rings is proportional to the cube of the distance from Saturn's center. (Kepler's three laws are discussed in Section 4-4, and the third law is examined in detail in Section 4-7 and Box 4-2.) This is exactly what would be expected if the rings consisted of numerous tiny "moonlets," or **ring particles,** each individually circling Saturn.

Saturn's rings are quite bright; they reflect 80% of the sunlight that falls on them. Astronomers therefore long suspected that the ring particles are made of ice and ice-coated rock. This hunch was confirmed in the 1970s, when American astronomers Gerard Kuiper and Carl Pilcher identified absorption features of frozen water in the rings' near-infrared spectrum. The *Voyager 1* and *Voyager 2* spacecraft made

even more detailed infrared measurements as they flew past Saturn in 1980 and 1981, respectively. These indicate that the temperature of the rings ranges from –180°C (–290°F) in the sunshine to less than –200°C (–300°F) in Saturn's shadow. Water ice is in no danger of melting or evaporating at these temperatures.

To determine the sizes of the particles that make up Saturn's rings, astronomers analyzed the radio signals received from the *Voyager* spacecraft as the spacecraft passed behind the rings. How effectively radio waves can travel through the rings depends on the relationship between the wavelength and the particle size. The results showed that most of the particles range in size from pebble-sized fragments about 1 cm in diameter to chunks about 5 m across, the size of large boulders. Most abundant are snowball-sized particles about 10 cm in diameter.

It seems reasonable to suppose that all this material is ancient debris that failed to accrete (fall together) into satellites. The total amount of material in the rings is quite small. If Saturn's entire ring system were compressed together to make a satellite, it would be no more than 100 km (60 mi) in diameter. But, in fact, the ring particles are so close to Saturn that they will never be able to form moons.

To see why, imagine a collection of small particles orbiting a planet. Gravitational attraction between neighboring particles tends to pull the particles together. However, because the various particles are at differing distances from the parent planet, they also experience different amounts of gravitational pull from the planet. This difference in gravitational pull is a **tidal force** that tends to keep the particles separated. (We discussed tidal forces in detail in Section 9-4. You may want to review that section, and in particular Figure 9-14.) At a certain distance from the planet's center, called the **Roche limit,** the disruptive tidal force of the planet exactly balances the attractive gravitational pull between neighboring particles. Inside the Roche limit, the tidal force overwhelms the gravitational pull between neighboring particles, and these particles cannot accrete to form a larger body. Instead, they tend to spread out into a ring around the planet. Indeed, most of Saturn's system of rings visible in Figure 15-2 lie within the planet's Roche limit. Likewise, the ring of Jupiter shown in Figure 14-21 lies within the Roche limit of Jupiter.

All large planetary satellites are found outside their planet's Roche limit. If any large satellite were to come inside its planet's Roche limit, the planet's tidal forces would cause the satellite to break up into fragments. As discussed in the next chapter, such a catastrophic tidal disruption may be the eventual fate of Neptune's large satellite, Triton, whose orbit is gradually decaying.

**CAUTION!** It may seem that it would be impossible for any object to hold together inside a planet's Roche limit. But the ring particles within Saturn's Roche limit survive and do not break apart. The reason is that the Roche limit applies only to objects held together by the gravitational attraction of each part of the object for the other parts. By contrast, the forces that hold a rock or a ball of ice together are chemical bonds between the object's atoms and molecules. Because these chemical forces are much stronger than the disruptive tidal force of a nearby planet, the rock or ball of ice does not break apart. In the same way, people walking around on the Earth's surface (which is inside the Earth's Roche limit) are in no danger of coming apart, because we are held together by comparatively strong chemical forces rather than gravity.

We will see in Chapter 16 that the rings of Uranus and Neptune are also made of many individual particles orbiting each planet within the Roche limit. Unlike the rings of Saturn, however, these rings are quite dim and difficult to see from the Earth.

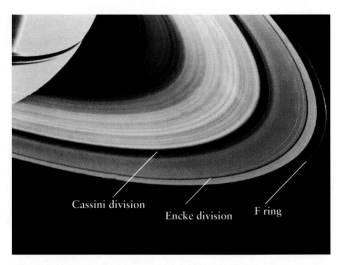

**Figure 15-4   R I V U X G**

**Details of Saturn's Rings** This view from *Voyager 1* shows that Saturn's rings are actually composed of thousands of closely spaced ringlets. The broad Cassini division is clearly visible, as is a narrow, 270-km-wide gap in the outer A ring called the Encke gap (after the German astronomer Johann Franz Encke, who reported seeing it in 1838). The very thin F ring is visible just beyond the outer edge of the A ring. The colors in this image have been enhanced to emphasize small differences between different portions of the rings. (NASA)

---

## 15-3   Saturn's rings consist of thousands of narrow, closely spaced ringlets

The photograph in Figure 15-1, which was made in 1974, indicates what astronomers understood about the structure of Saturn's rings in the mid-1970s. Each of the A, B, and C rings appeared to be rather uniform, with little or no evidence of any internal structure. Within a few years, however, close-up observations from spacecraft revealed the true complexity of the rings, as well as their chemical composition.

Three spacecraft—*Pioneer 11, Voyager 1,* and *Voyager 2*—flew past Saturn in 1979, 1980, and 1981, respectively, after each had first made a close flyby of Jupiter. (*Pioneer 10* also flew past Jupiter but did not visit Saturn. Instead, it was aimed so that Jupiter's gravity deflected its path out of the plane of the solar system and into interstellar space.) *Pioneer 11* had only a relatively limited capability to make images, but cameras on board the two *Voyager* spacecraft sent back a number of pictures showing the detailed structure of Saturn's rings. One such picture is shown in Figure 15-4. Scientists were amazed to find that the broad A, B, and C rings are not uniform at all, but instead consist of hundreds upon hundreds of closely spaced bands or **ringlets**. These ringlets are arrangements of ring particles that have evolved from the combined gravitational forces of neighboring particles, of Saturn's moons, and of the planet itself.

The *Voyager* cameras also sent back high-quality pictures of the narrow **F ring**, which was first detected by *Pioneer 11.* The F ring lies 4000 km beyond the outer edge of the A ring (examine Figure 15-4). Close-up views showed that the F ring actually consists of several intertwined strands, as shown in Figure 15-5. One *Voyager 1* image displayed a total of five strands, each about 10 km across. Planetary scientists suspect that gravitational forces from tiny moons within the F ring produce this curious and complex structure.

Spacecraft have allowed scientists not only to see the rings in greater detail, but also to view the rings from perspectives not possible from the Earth. Through Earth-based telescopes, we can see only the sunlit side of Saturn's rings. From this perspective, the B ring appears very bright, the A ring moderately bright, the C ring dim, and the Cassini division dark (see Figure 15-1 and Figure 15-2). The proportion of sunlight reflected back toward the Sun (the albedo) is directly related to the concentration and size of the particles in the ring. The B ring is bright because it has a high concentration of relatively large, icy, reflective particles, whereas the darker Cassini division has a lower concentration of such particles.

Figure 15-6 is an image made by *Voyager 1* from the shaded side of the rings. Because the spacecraft was looking back toward the Sun, this picture shows sunlight that has passed through the rings. The B ring looks darkest because little sunlight gets through its dense collection of particles.

If the Cassini division were completely empty, it would look black, because we would then be seeing through the division to the blackness of empty space, but the Cassini division looks *bright* in Figure 15-6. This shows that the Cassini division is not empty but contains a relatively few particles. (In the same way, dust particles in the air in your room become visible when a shaft of sunlight passes through them.)

The process by which light bounces off particles in its path is called **light scattering.** The proportion of light scattered in various directions depends upon both the size of the particles and the wavelength of the light. (See Box 15-1 for examples of light scattering on Earth.) Because visible light has a much smaller wavelength than radio waves, light scattering allows scientists to measure the sizes of particles too small to detect using the radio technique described in Section 15-2. As an example, by measuring the amount of light scattered from the rings at various wavelengths and from different angles as the *Voyager* spacecraft sped past Saturn, scientists showed that the F ring contains a substantial number of tiny particles about 1 μm (= $10^{-6}$ m) in diameter. This is about the same size as the particles found in smoke.

Subtle differences in color from one ring to the next also provide important clues about the composition of the particles in the different rings. These variations are clearly visible in Figure 15-7, in which the colors have been exaggerated by computer processing. Although the main chemical constituent

**figure 15-5**  R I **V** U X G

**Details of the F Ring** This *Voyager 1* image shows several strands, each measuring about 10 km across, that together comprise the F ring. The total width of the F ring is about 100 km. The narrowness of the F ring is the result of gravitational forces exerted by two small satellites, as we discuss in Section 15-4. (NASA)

**figure 15-6**  R I **V** U X G

**The View from the Far Side of the Rings** This false-color view of the side of Saturn's rings away from the Sun was taken by *Voyager 1.* Both the narrow F ring and the Encke gap in the A ring are clearly visible toward the right side of the picture. The white band is the Cassini division. Because this division does not appear black, as does the empty gap between the A and F rings, it follows that the Cassini division is not empty but contains a number of relatively small particles that scatter sunlight like dust motes. The purple band at far left is the outer part of the B ring. (NASA)

**figure 15-7**  R I **V U** X G

**An Enhanced-Color View of Ring Details** Computer processing has severely exaggerated the subtle color variations in this *Voyager 2* view of the sunlit side of Saturn's rings (actually a composite of visible-light and ultraviolet images). Note that the C ring and the Cassini division appear bluish. Also note the distinct color variations across both the A and the B rings. These color variations are indicative of slight differences in chemical composition between ring particles in different parts of the rings. (NASA)

## box 15-1 | The Heavens on the Earth

### *Light Scattering*

The scattering of light by particles is used by scientists to study Saturn's rings, but it also explains a number of phenomena that you can see here on the Earth.

The light that comes from the daytime sky is sunlight that has been scattered by the molecules that make up our atmosphere (see part **a** of the accompanying figure). Very small particles like molecules are quite effective at scattering blue, short-wavelength light but less effective at scattering long-wavelength red light. That's why the sky looks blue. Smoke particles are also quite small, which explains why the smoke from a cigarette or a fire has a bluish color.

Scattered blue light, however, does not necessarily imply small particles. Light scattered from Saturn's C ring and Cassini division appears blue, but the particles in these regions are actually relatively *large*, with sizes that range from centimeters to meters. In this case, the blue color is indicative of the actual color of the particles.

Distant mountains often appear blue thanks to light being scattered from the atmosphere between the mountains and your eyes. (This helps to explain the name of the Blue Ridge Mountains, which extend from Pennsylvania to Georgia.) Sunglasses often have a red or orange tint which blocks out blue light. This cuts down on the amount of scattered light from the sky reaching your eye and thus allows you to see distant objects more clearly.

Light scattering also explains why sunsets are red. The light from the Sun is a mixture of all colors, but as this light passes through our atmosphere, the blue component is scattered away from the straight-line path from the Sun to your eye. The red component undergoes relatively little scattering, so the Sun always looks a bit redder than it really is. When you look toward the setting sun, the sunlight that reaches your eye has had to pass through a relatively thick layer of atmosphere (part **b** of the accompanying figure). Hence, a large fraction of the blue light from the Sun has been scattered, and so the Sun appears quite red. The same effect also applies to sunrises, but sunrises seldom look as red as sunsets. The reason is that dust is lifted into the atmosphere during the day by the wind (which is typically stronger in the daytime than at night), and dust particles in the atmosphere help to scatter even more blue light.

If the particles that scatter light are sufficiently concentrated, there will be almost as much scattering of red light as of blue light, and the scattered light will appear white. This explains the white color of clouds, fog, and haze, in which the scattering particles are ice crystals or water droplets. Whole milk looks white because of light scattering from tiny fat globules; nonfat milk has only a very few of these globules and thus has a slight bluish cast.

Light scattering also has many applications to astronomy outside the solar system. In Chapter 20 we will see how light scattering explains the colors of certain clouds, or nebulae, in interstellar space.

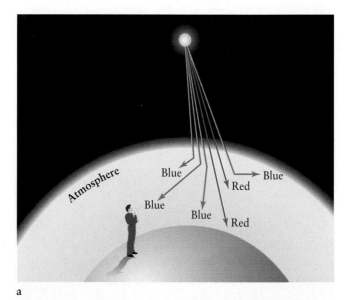

a

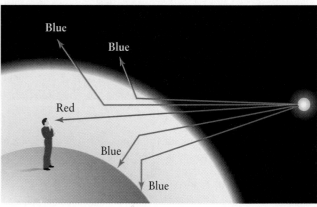

b

is frozen water, trace amounts of other chemicals—perhaps coating the surfaces of the ice particles—are probably responsible for the colors seen in the computer-enhanced view. These trace chemicals have not yet been identified.

The existence of color variations between different part of the rings suggest that the icy particles do not migrate substantially from one ringlet to another. Had such migration taken place, the color differences would have been smeared out over time. The color differences may also indicate that different sorts of material were added to the rings at different times. In this picture, the rings did not all form at the same time as Saturn but were added to over time. New ring material could have come from small satellites that shattered after being hit by a stray asteroid or comet, like the comet that ran into Jupiter in 1994 (Section 13-4).

In addition to revealing new details about the A, B, C, and F rings, the *Voyager* cameras also discovered three new ring systems: the D, E, and G rings. The drawing in Figure 15-8 shows the layout of all of Saturn's known rings along with the orbits of some of Saturn's 18 known satellites. The **D ring** is Saturn's innermost ring system. It consists of a series of extremely faint ringlets located between the inner edge of the C ring and the Saturnian cloudtops. The **E ring** and the **G ring** both lie far from the planet, well beyond the outer edge of the A ring. Both of these outer ring systems are extremely faint, fuzzy, and tenuous. Each lacks the ringlet structure so prominent in the main ring systems. The E ring lies along the orbit of Enceladus, one of Saturn's icy satellites. Some scientists suspect that water geysers on Enceladus are the source of ice particles in the E ring, much as Io's volcanoes produce a torus of material along its orbit around Jupiter (see Section 14-5).

## 15-4 Saturn's inner satellites affect the appearance and structure of its rings

Why do Saturn's rings have such a complex structure? The answer is that, like any object in the universe, the particles that make up the rings are affected by the force of gravity. Saturn's gravitational pull keeps the ring particles in orbit around the planet. But the satellites of Saturn also exert gravitational forces on the rings. These forces can shape the orbits of the ring particles and help give rise to the rings' structure.

Astronomers have long known that one of Saturn's moons, Mimas, has a gravitational effect on its ring system. Mimas is a moderate-sized satellite that orbits Saturn every 22.6 hours. According to Kepler's third law, particles

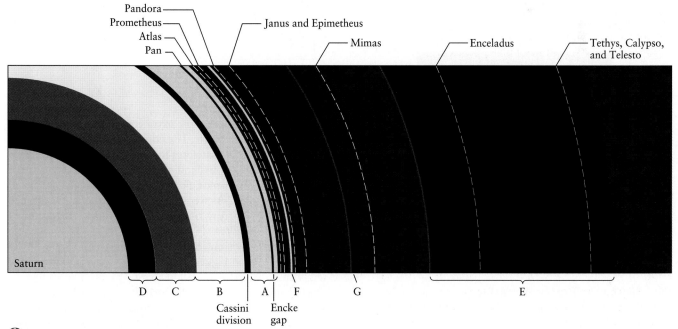

## ƒigure 15-8

**The Arrangement of Saturn's Rings** This scale drawing shows the location of Saturn's rings along with the orbits of the 11 inner satellites. Only the A, B, and C rings can readily be seen from the Earth. The remaining rings are very faint and were discovered during spacecraft flybys (although the D and E rings may have been glimpsed beforehand by Earth-based observers). Note that the satellites Janus and Epimetheus share the same orbit, as do the three satellites Tethys, Calypso, and Telesto.

in the Cassini division should orbit Saturn approximately every 11.3 hours. Consequently, on every second orbit, the particles in the Cassini division line up between Saturn and Mimas. Because of these repeated alignments, the combined gravitational forces of Saturn and Mimas cause small particles to deviate from their original orbits. This 2-to-1 resonance with Mimas tends to sweep ring particles out of the Cassini division, which helps explain why this portion of the ring is relatively empty (although not completely so, as shown in Figure 15-6). This effect is somewhat analogous to the rhythmic tidal effects that act among the Jovian satellites Io, Europa, and Ganymede (recall Sections 14-1 and 14-4).

Other gravitational effects are produced by two tiny satellites that follow orbits on either side of the narrow F ring (Figure 15-9). Peter Goldreich of the California Institute of Technology and Scott Tremaine of Princeton University pointed out that the gravitational forces of these two satellites keep the F ring particles in place. Pandora, the outer of the two satellites, moves around Saturn at a slightly slower speed than the particles in the ring. As the ring particles pass Pandora, they experience a tiny gravitational tug that tends to slow down the particles. As a result, these particles lose a little energy, causing them to fall into orbits a bit closer to Saturn.

At the same time, Prometheus, the inner satellite, orbits the planet somewhat faster than the F ring particles. The gravitational pull of Prometheus tends to speed the ring particles

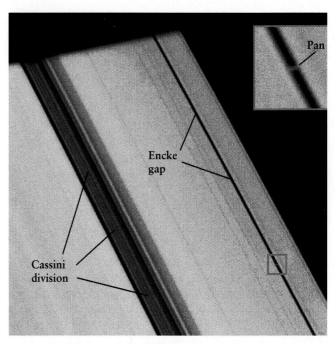

**figure 15-10** R I **V** U X G

**Pan and the Encke Gap** This *Voyager 2* image shows Saturn's tiny moon Pan orbiting within the Encke gap, a division in the A ring also visible in Figure 15-4. Pan exerts gravitational forces on the ring particles to either side that help to maintain the Encke gap's 270-km width. Pan's existence was predicted theoretically from the properties of the Encke gap; it was discovered in 1990 by Mark Showalter of NASA after an computer search of 30,000 *Voyager 1* and *Voyager 2* images made a decade earlier. (© 1998 by Calvin J. Hamilton, Columbia, Maryland)

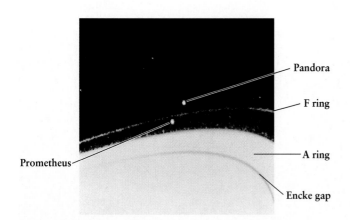

**figure 15-9** R I **V** U X G

**The F Ring and Its Two Shepherds** Two tiny satellites, each measuring only about 50 km across, move around Saturn on either side of its F ring. Prometheus orbits just inside the F ring, while Pandora orbits just outside this ring. Like the F ring itself, Prometheus and Pandora were discovered by the *Voyager* spacecraft. The gravitational effects of these two shepherd satellites focus and confine the particles in the F ring to a band about 100 km wide. *Voyager 2* captured this image while 10.5 million km (6.6 million mi) from Saturn. At the instant shown, Prometheus and Pandora were less than 1800 km (1100 mi) apart; they passed each other about 2 hours later. The satellites lap each other every 25 days. (NASA)

up, imparting to them some extra energy and nudging them into slightly higher orbits. The competition between the pulls exerted by these two satellites focuses the F ring particles into a well-defined, narrow band about 100 km wide. Because of their confining influence, Prometheus and Pandora are called **shepherd satellites.** As we will see in the next chapter, Uranus is encircled by narrow rings that are likewise kept confined by shepherd satellites.

A narrow division in Saturn's A ring, called the **Encke gap** and shown in Figure 15-4, actually has a satellite orbiting within it (Figure 15-10). This tiny moon, Pan, is just 20 km (12 mi) in diameter. In a fashion similar to the F ring's shepherds, Pan's gravitational forces cause ring particles outside its orbit to move farther out and cause those within its orbit to move farther inward toward Saturn. In this way, Pan helps to maintain a gap in the rings that is 270 km (170 mi) wide, many times wider than Pan itself. Gravitational forces from Pan also cause wavelike disturbances in the ring material on either side of the gap. Other small satellites presumably orbit within Saturn's rings and cause small gaps of their own, but their existence has yet to be confirmed.

## 15-5 Saturn's atmosphere extends to a greater depth and has higher wind speeds than Jupiter's atmosphere

Although Saturn's rings are unique in the solar system, the planet itself has many similarities to its larger cousin Jupiter. Both planets exhibit differential rotation, taking less time to complete one rotation at the equator than near the poles. The chemical compositions of their atmospheres are also similar. Earth-based spectroscopic observations along with data from spacecraft confirm that, like Jupiter, Saturn has a hydrogen-rich atmosphere with trace amounts of methane ($CH_4$), ammonia ($NH_3$), and water vapor ($H_2O$). These compounds are the simplest combinations of hydrogen with carbon, nitrogen, and oxygen. And like Jupiter, Saturn's atmosphere probably has three distinct cloud layers: an upper layer of frozen ammonia crystals, a middle layer of crystals of ammonium hydrosulfide ($NH_4SH$), and a lower layer of water ice crystals.

Although their atmospheres have similar structures and compositions, Saturn and Jupiter are by no means identical in appearance. Saturn's clouds lack the colorful contrast of Jupiter's clouds. Nevertheless, many Earth-based photographs and most spacecraft views show faint belts and zones (Figure 15-11).

**figure 15-12** R I **V** U X G

**A New Storm on Saturn** The white arrowhead-shaped feature near Saturn's equator is a giant storm that appeared in August 1994. The east-west dimension of this storm was about 13,000 km (8000 mi), or about the same as the diameter of Earth. The storm apparently formed when warm gases rose upward through Saturn's cold outer atmosphere. The lifted gases cooled as they rose, causing gaseous ammonia to solidify into crystals that appear as white clouds. On Earth, similar rapid lifting of air occurs in thunderstorms and can cause droplets of water to solidify into hailstones. The storm lasted several months; it had changed little when this Hubble Space Telescope image was made in December 1994. For reasons that are not understood, such large storms appear at intervals of about 57 years, or about two Saturnian years, and occur when it is summer in Saturn's northern hemisphere. (Reta Beebe, New Mexico State University; D. Gilmore and L. Bergeron, Space Telescope Science Institute; and NASA)

**figure 15-11** R I **V** U X G

**Belts and Zones on Saturn from *Voyager 1*** *Voyager 1* recorded this image of Saturn's cloudtops when the spacecraft was 1.8 million kilometers (1.1 million miles) from the planet. Belts and zones are clearly visible, but there is much less contrast between the belts and zones on Saturn than there is on Jupiter. (NASA)

Saturn has no long-lived storm systems analogous to Jupiter's Great Red Spot (described in Section 13-1). But on rare occasions, Earth-based observers have reported seeing storms in Saturn's clouds that last for several days or even months. Figure 15-12 shows a Hubble Space Telescope image of one such "white spot," which appeared near Saturn's equator in 1994. About 20 temporary white spots have been seen on Saturn over the past two centuries.

The different appearances of Saturn and Jupiter are related to the different masses of the two planets. Jupiter's strong surface gravity compresses its three cloud layers into a region just 75 km deep in the planet's upper atmosphere. Saturn's smaller mass and, thus, weaker surface gravity subjects its atmosphere to less compression, so the same three cloud layers are spread out over a range of nearly 300 km (Figure 15-13). The colors of Saturn's clouds are less dramatic than Jupiter's because deeper layers are partly obscured by the hazy atmosphere above them.

By following features in the Saturnian clouds, scientists have determined the wind speeds in the planet's upper atmosphere. As in Jupiter's atmosphere, there are counterflowing easterly and westerly currents in Saturn's upper atmosphere.

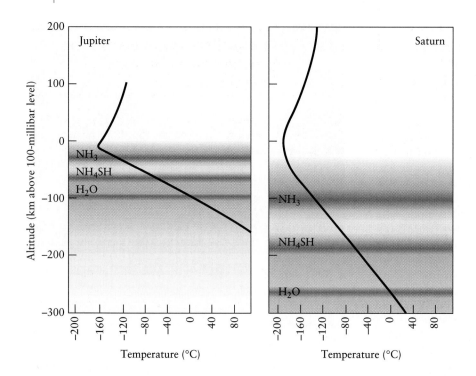

ƒigure 15-13

**The Upper Atmospheres of Jupiter and Saturn** The structures of the upper atmospheres of Jupiter and Saturn are displayed on these graphs of temperature versus depth. Note that Saturn's atmosphere is more spread out than that of Jupiter as a direct result of Saturn's weaker surface gravity. (Adapted from A. P. Ingersoll)

However, Saturn's equatorial jet is much broader—and much faster—than that of Jupiter. In fact, wind speeds near Saturn's equator approach 500 m/s (1800 km/h, or 1100 mi/h), which is approximately two-thirds the speed of sound in Saturn's atmosphere. The origin of these higher wind speeds on Saturn is not yet completely understood.

## 15-6 Saturn's internal structure is quite similar to that of Jupiter and includes a layer of liquid metallic hydrogen

Although Saturn's atmosphere has a composition and structure similar to Jupiter's, there are some significant differences between the two planets. The average density of Saturn, 690 kg/m³, is not only less than the density of water, it is the lowest of any object in the solar system, and only about half that of Jupiter. As for Jupiter, such a low density means that Saturn is composed mostly of hydrogen and helium. Saturn's smaller mass, less than a third that of Jupiter, means that there is less gravitational force tending to compress its hydrogen and helium. This explains Saturn's low density.

Saturn is also the most *oblate*—that is, the most flattened from pole to pole by its rotation—of all the planets (examine the figure that opens this chapter). Saturn's oblateness is about 0.10, which means the planet's equatorial diameter is about 10% larger than its polar diameter. As we explained in Section 13-5, the oblateness of a planet is directly related

to its speed of rotation and to how the planet's mass is distributed over its volume. In fact, Saturn rotates a little more *slowly* than Jupiter does, so the greater oblateness of Saturn cannot be caused by faster rotation. Instead, it must have a different mass distribution than Jupiter. Detailed calculations suggest that about 26% of Saturn's mass is contained in its core, whereas Jupiter's core contains only about 4% of the planet's entire mass.

Saturn's massive core and low average density means that its density must decrease rapidly outward from the core toward the surface. This suggests that Saturn generally resembles Jupiter in its internal structure: a rocky core surrounded by a mantle of liquid metallic hydrogen, which in turn is surrounded by a layer of liquid and gaseous molecular hydrogen (Figure 15-14).

Saturn's mantle of liquid metallic hydrogen, like Jupiter's, is thought to be the source of the planet's magnetic field. Saturn's magnetic field turns out to be somewhat weaker than that of Jupiter, perhaps because of Saturn's slightly slower rotation and much smaller volume of liquid metallic hydrogen.

Saturn's magnetosphere contains far fewer charged particles than Jupiter's. There are two reasons for this deficiency. First, Saturn lacks a moon like Io, which ejects a ton of material into Jupiter's magnetosphere each second from its surface and volcanic plumes. Second, and more important, many charged particles are absorbed by the icy particles that make up Saturn's rings. This depletes the concentration of particles in the inner magnetosphere. The charged particles that do exist in Saturn's magnetosphere are concentrated in radiation belts similar to the Van Allen belts in the Earth's magnetosphere.

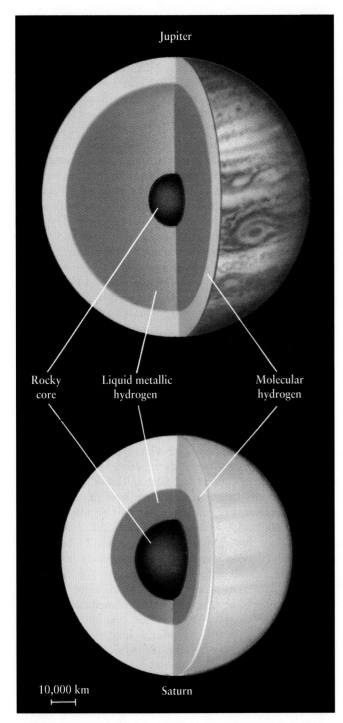

## figure 15-14

**The Internal Structures of Saturn and Jupiter** These diagrams of the interiors of Jupiter and Saturn are both drawn to the same scale. The interiors of both planets have three distinct layers. Each planet's rocky core is surrounded by a layer of liquid metallic hydrogen, which in turn is enveloped in a thick layer of liquid and gaseous molecular hydrogen. Note that Saturn's rocky core is larger than Jupiter's. Saturn also has a smaller volume of liquid metallic hydrogen, which helps explain why Saturn's magnetic field is weaker than that of Jupiter.

## 15-7 Like Jupiter, Saturn emits more radiation than it receives from the Sun

Both Jupiter and Saturn radiate more energy into space than they receive as sunlight. This means that both planets have internal sources of heat. As we saw in Section 13-3, a tremendous amount of thermal energy was trapped inside Jupiter as the planet accreted from the solar nebula 4.5 billion years ago. Since then, Jupiter has been slowly cooling off as this energy escapes in the form of infrared radiation.

Because Saturn is smaller than Jupiter, it should have begun with less internal heat trapped inside, and it should have radiated that heat away more rapidly. (We explained in Section 9-3 why a small world cools down faster than a large one.) Hence, we would expect Saturn to be radiating very little energy today. In fact, while Jupiter radiates about twice as much energy as it receives from sunlight, Saturn radiates almost *three* times as much energy as it receives. How is Saturn able to emit so much energy?

A second puzzle is Saturn's apparent deficiency of helium. Before the *Voyager* flybys in the 1980s, astronomers suspected both Jupiter and Saturn to have compositions similar to that of the original solar nebula, which is assumed to be the same as the Sun's composition. Each of these giant planets is both massive enough and cool enough to have retained all the gases that originally accreted from the solar nebula. The *Voyager* and *Galileo* missions have shown that Jupiter has an abundance of elements that is similar to the solar abundance (by mass, 80% hydrogen, 19% helium, and 1% all other elements). Surprisingly, however, the *Voyager* spacecraft reported that Saturn's atmosphere has less helium than expected. The chemical composition of Saturn's atmosphere is, by mass, 88% hydrogen, 11% helium, and 1% all other elements.

Prior to the *Voyager* missions, Edwin E. Salpeter of Cornell University and David J. Stevenson of the California Institute of Technology offered a brilliant hypothesis that turned out to explain both Saturn's excess heat output and its helium deficiency. According to their theory, Saturn did indeed cool more rapidly than Jupiter, triggering a process analogous to how rain develops here on the Earth. When the air is cool enough, humidity in the Earth's atmosphere condenses into raindrops that fall to the ground. Within Saturn, however, it is droplets of liquid helium that rain slowly downward through the planet's hydrogen-rich outer layers toward its core. As the helium droplets descend, their gravitational energy is converted into thermal energy (heat) that eventually escapes from Saturn's surface.

Helium is deficient in Saturn's upper atmosphere simply because it has fallen farther down into the planet. The precipitation of helium from Saturn's upper atmosphere is calculated to have begun 2 billion years ago. The energy released adequately accounts for the extra heat radiated by Saturn since that time.

## 15-8 Titan has a thick, opaque atmosphere rich in methane, nitrogen, and hydrocarbons

Unlike Jupiter, Saturn has only one large satellite that is comparable in size to our own Moon. This satellite, Titan, was discovered by Huygens in 1655. Titan, and the shadow that it casts on Saturn, can be seen in the upper of the two images that open this chapter. With a diameter of 5150 km, Titan is the second-largest satellite in the solar system. (Jupiter's satellite Ganymede, discussed in Section 14-7, is 5262 km in diameter.)

By the early 1900s, several scientists had begun to suspect that Titan might have an atmosphere, because it is cool enough and massive enough to retain heavy gases. (Box 7-2 explains the criteria for a planet or satellite to retain an atmosphere.) These suspicions were confirmed in 1944, when American astronomer Gerard P. Kuiper discovered the spectral lines of methane in sunlight reflected from Titan. (Titan's spectrum and its interpretation are discussed in Section 7-3.) Titan is now known to be the only satellite in the solar system with an appreciable atmosphere.

Because of its atmosphere, Titan was a primary target for the *Voyager* missions. To everyone's chagrin, however, *Voyager 1* spent hour after precious hour of its limited observation time sending back featureless images (Figure 15-15). Titan's atmosphere is so thick—about 200 km (120 mi) deep, or about ten times the depth of the Earth's troposphere—that it blocks any view of the surface. The haze surrounding Titan is so dense that little sunlight penetrates to the ground; noon on Titan is probably no brighter than a moonlit midnight on the Earth.

By examining how the radio signal from *Voyager 1* was affected by passing through Titan's atmosphere, scientists inferred that the atmospheric pressure at Titan's surface is 50% greater than the atmospheric pressure at sea level on the Earth. Titan's surface gravity is weaker than that of Earth, so to produce this pressure Titan must have considerably more gas in its atmosphere than Earth. Indeed, calculations show that about ten times more gas lies above each square meter of Titan's surface than lies above each square meter of Earth.

*Voyager* data suggest that roughly 90% of Titan's atmosphere is nitrogen. Most of this nitrogen probably came from ammonia ($NH_3$), a compound of nitrogen and hydrogen that is quite common in the outer solar system. Ammonia is easily broken into hydrogen and nitrogen atoms by the Sun's ultraviolet radiation. Titan's gravity is too weak to retain hydrogen atoms (see Box 7-2), so these atoms escape into space and leave the more massive nitrogen atoms behind.

The second most abundant gas on Titan is methane, which is the principal component of the "natural gas" used as a fuel on the Earth. The interaction of sunlight with methane induces chemical reactions that produce a variety of other carbon-hydrogen compounds, or **hydrocarbons**. For example, the *Voyager* infrared spectrometers detected small amounts of ethane ($C_2H_6$), acetylene ($C_2H_2$), ethylene ($C_2H_4$), and propane ($C_3H_8$) in Titan's atmosphere. Ethane is the most abundant of these by-products. It condenses into droplets as it is produced and falls to Titan's surface. Over Titan's entire history, enough ethane may have been produced to create rivers and lakes. By using an infrared camera, the Hubble Space Telescope has produced images of Titan's surface that reinforce this idea (Figure 15-16). Investigations of Titan's surface using radar, however, have so far not turned up evidence for liquid ethane.

Nitrogen in Titan's atmosphere can combine with airborne hydrocarbons to produce other compounds, such as hydrogen cyanide (HCN). Hydrogen cyanide, along with other molecules, can join together in long, repeating molecular chains to form substances called **polymers**. Droplets of some polymers remain suspended in the atmosphere, forming an **aerosol**. (Common aerosols on the Earth include fog, mist, and paint sprayed from a can.) The polymers in Titan's airborne aerosol are thought to be responsible for the reddish-brown color seen in Figure 15-15. The heavier polymer particles do not remain airborne but settle down onto Titan's solid surface, probably covering it with a thick layer of sticky, tarlike goo.

Some of the compounds of hydrogen, nitrogen, and carbon found in Titan's atmosphere and on its surface are the building blocks of the organic molecules on which life is based. There is little reason to suspect that life exists on Titan, however; its surface temperature of 95 K (–178°C = –288°F) is prohibitively cold. Nevertheless, a more detailed study of the chemistry of Titan may shed light on the origins of life on the Earth.

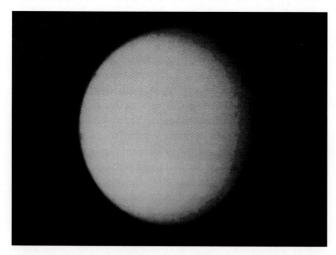

**ƒigure 15-15** R I **V** U X G

**Titan** This view of Titan was taken by *Voyager 2* from a distance of 4.5 million km (2.8 million mi). Hardly any features are visible in the thick, unbroken haze that surrounds this large satellite. The main haze layer is located 200 km (120 mi) above Titan's surface. The brown color is presumably caused by compounds of carbon, nitrogen, and hydrogen that condense into airborne droplets. (NASA)

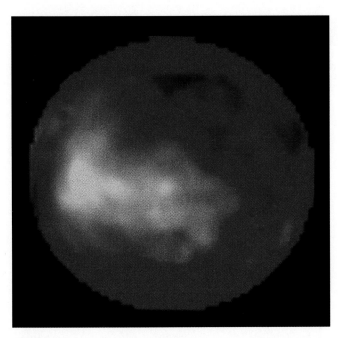

**figure 15-16**  R **I** V U X G

**Surface Features on Titan** This Hubble Space Telescope image of Titan was made using infrared wavelengths that can more easily penetrate Titan's thick atmosphere. False color is used to denote the amount of infrared radiation coming from different areas on the surface. The dark areas may be oceans of hydrocarbons, which effectively absorb infrared light. The prominent bright area, which is some 4000 km (2500 mi) long, may be a region of solid surface covered with reflective ices of water and ammonia. The smallest detail visible in this image is 580 km (360 mi) across. (Peter H. Smith, University of Arizona Lunar and Planetary Laboratory; NASA)

## 15-9  The icy surfaces of Saturn's six moderate-sized moons provide clues to their histories

Eighteen moons are known to orbit Saturn (Table 15-3). They can be placed in three groups according to size. First, planet-sized Titan is one of the terrestrial worlds of our solar system. With its diameter of 5150 km, Titan is intermediate in size between Mercury and Mars. Second, six moderate-sized satellites were all discovered before 1800. These six satellites have properties and diameters that lead us to group them in pairs: Mimas and Enceladus (400 to 500 km in diameter), Tethys and Dione (roughly 1000 km in diameter), and Rhea and Iapetus (about 1500 km in diameter). Third, several tiny moons may be either captured asteroids (as in the case of Phoebe, Saturn's outermost satellite) or jagged fragments of ice and rock produced by impacts and collisions. The shepherd satellites, for example, may be such fragments.

Saturn's six moderate-sized satellites share several characteristics. All six circle the planet in the plane of Saturn's equator and in the direction of the planet's rotation. These six satellites also exhibit synchronous rotation. All six have average densities between 1200 and 1400 kg/m$^3$, a value consistent with a composition largely of ice.

Mimas, the innermost of the six middle-sized moons, is heavily cratered. In fact, one of the craters (Figure 15-17) is so large that the impact that formed it must have come close to shattering Mimas into fragments.

In contrast to Mimas's heavily cratered surface, its neighbor and near twin, Enceladus, has extensive crater-free regions (Figure 15-18). These areas apparently have been resurfaced within the past 100 million years by some form of geological activity. Furthermore, Enceladus is the most highly reflective large object in our solar system, even more reflective than newly fallen snow. Its surface is composed of extremely pure ice, free of the rock and dust thought to explain the lower albedos of the other icy satellites. As we mentioned in Section 15-3, there is indirect evidence that some kind of eruptions on Enceladus supply particles to the E ring, but the *Voyager* spacecraft did not see any such eruptions.

What is the energy source for the geologic activity on Enceladus? One guess is that it might be tidal heating, which provides the internal heat for the inner Galilean satellites of Jupiter (see Section 14-4). The satellite Dione orbits Saturn in 65.7 hours, almost exactly twice Enceladus's orbital period of 32.9 hours, and the 1:2 ratio of orbital periods should set Enceladus up to be caught in a tidal tug-of-war between Saturn and Dione. However, tidal heating requires that the satellite be in an eccentric orbit around its planet, and the orbit of Enceladus is quite circular. Furthermore, there is also a 1:2 ratio between the orbital periods of Mimas and Saturn's

**figure 15-17**  R I **V** U X G

**Mimas** Mimas is the smallest and innermost of Saturn's six moderate-sized satellites. This view was taken by *Voyager 1* at a range of nearly 500,000 km (300,000 mi). The huge impact crater, named Herschel (after astronomer William Herschel), is 130 km (80 mi) in diameter. Mimas itself is only 400 km (250 mi) in diameter. (NASA)

## Table 15-3 | Saturn's Satellites

| | Distance from center of Saturn | | Orbital period (days) | Size (km) | Density (kg/m³) |
|---|---|---|---|---|---|
| | (km) | (Saturn radii) | | | |
| Pan | 133,570 | 2.22 | 0.573 | 20 | — |
| Atlas | 137,640 | 2.28 | 0.602 | 20 × 20 × 20 | — |
| Prometheus | 139,350 | 2.31 | 0.613 | 140 × 100 × 80 | — |
| Pandora | 141,700 | 2.35 | 0.629 | 110 × 90 × 70 | — |
| Epimetheus | 151,422 | 2.51 | 0.694 | 140 × 120 × 100 | — |
| Janus | 151,472 | 2.51 | 0.695 | 220 × 200 × 160 | — |
| Mimas | 185,520 | 3.08 | 0.942 | 392 | 1400 |
| Enceladus | 238,020 | 3.95 | 1.370 | 500 | 1200 |
| Tethys | 294,660 | 4.89 | 1.888 | 1060 | 1200 |
| Calypso | 294,660 | 4.89 | 1.888 | 34 × 28 × 26 | — |
| Telesto | 294,660 | 4.89 | 1.888 | 24 × 22 × 22 | — |
| Dione | 377,400 | 6.26 | 2.737 | 1120 | 1400 |
| Helene | 377,400 | 6.26 | 2.737 | 36 × 32 × 30 | — |
| Rhea | 527,040 | 8.74 | 4.518 | 1530 | 1300 |
| Titan | 1,221,850 | 20.25 | 15.945 | 5150 | 1880 |
| Hyperion | 1,481,000 | 24.55 | 21.277 | 410 × 260 × 220 | — |
| Iapetus | 3,561,300 | 59.02 | 79.331 | 1460 | 1200 |
| Phoebe | 12,952,000 | 214.7 | 550.48[R] | 220 | — |

*This table lists some facts about Saturn's 18 known satellites. (At least a dozen more have been reported, but their existence has not yet been confirmed.) All the larger satellites are spherical, and their diameters are listed in the "Size" column. Because the smaller satellites are not spheres, three dimensions— width, length, and height—are given for each of these satellites. The masses, and hence the densities, of the smaller satellites are not yet known.*

*Of the moons for which rotation rates are known, almost all rotate synchronously (that is, the rotation period is the same as the orbital period, so the moon always keeps the same face toward Saturn.) The exceptions are Phoebe and Hyperion.*

*The F ring shepherds are Prometheus and Pandora. Janus and Epimetheus are called co-orbital satellites, because they move in almost the same orbit. Tethys, Calypso, and Telesto are also co-orbital satellites, as are Dione and Helene. In the latter two cases, the tiny satellites occupy specific locations along the orbits of the larger moons, where a balance exists between the gravitational pulls of Saturn and the larger moon. We will discuss these locations, called the Lagrangian points, in Chapter 17.*

*Saturn's outermost satellite, Phoebe, moves in a retrograde direction about the planet, indicated by a superscript R with the orbital period.*

moon Tethys, but Mimas's cratered surface shows no sign that this moon was ever tidally heated. The origin of Enceladus's internal heat thus remains a mystery.

Tethys and Dione are the next-largest pair of Saturn's six moderate-sized satellites. Tethys, like Mimas, is heavily cratered and has one exceptionally huge impact crater. Dione is slightly larger than Tethys, and the surfaces of its leading hemisphere (that is, the hemisphere pointing toward its direction of orbital motion) and trailing hemisphere are quite different. Dione's leading hemisphere is heavily cratered, but its trailing hemisphere is covered by a network of strange, wispy markings (Figure 15-19). These wisps may be troughs and valleys in the icy surface, or they may be fresh deposits of frozen water formed by outgassings from Dione's interior.

Rhea and Iapetus are the largest and most distant of Saturn's moderate-sized moons. Rhea resembles Dione in that its trailing hemisphere is covered by a network of wispy mark-

ings and its leading hemisphere is heavily cratered (Figure 15-20). Whatever process produced the wisps on one side of Dione has apparently also been active on the corresponding side of Rhea.

Iapetus has been known to be unusual ever since its discovery, because its brightness varies greatly as it orbits Saturn. Pictures from the *Voyager* spacecraft show extreme differences between the leading and trailing hemispheres of Iapetus (Figure 15-21). The leading hemisphere is as black as asphalt, but the trailing hemisphere is highly reflective, like the surfaces of the other moderate-sized moons.

The dark material covering Iapetus's leading hemisphere may have come from Phoebe, the most distant of Saturn's known satellites. Phoebe moves in a retrograde direction along an orbit tilted well away from the plane of Saturn's equator. This motion would lead us to suspect that Phoebe is a captured asteroid. Its dark surface does indeed resemble

**figure 15-18** R I **V** U X G

**Enceladus** This high-resolution image of Enceladus was obtained by *Voyager 2* from a distance of 191,000 km (119,000 mi). Ice flows and cracks strongly suggest that Enceladus's surface has been subjected to recent geological activity. The youngest, crater-free ice flows are estimated to be less than 100,000 years old. (NASA)

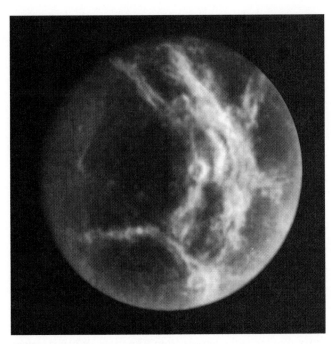

**figure 15-19** R I **V** U X G

**Dione** Bright, wispy streaks crisscross Dione's trailing hemisphere. (A similar network of light-colored wisps covers the trailing hemisphere of Rhea.) The nature and cause of these streaks are not known. This view of Dione was obtained by *Voyager 1* from a distance of 695,000 km. (NASA)

**figure 15-20** R I **V** U X G

**Rhea** The leading hemisphere of Rhea is heavily cratered, like the leading hemisphere of Dione. This view was taken by *Voyager 1* at a distance of 128,000 km (80,000 mi). Surface features as small as 3 km across are visible. (NASA)

**figure 15-21** R I **V** U X G

**Iapetus** *Voyager 2* recorded this image of Iapetus from a distance of 1.1 million kilometers (680,000 miles). The leading hemisphere of Iapetus is extremely dark. One theory suggests that this hemisphere may have been coated with material drifting inward from Saturn's outermost satellite, Phoebe. (NASA)

the surfaces of a particular class of carbon-rich asteroids. Bits and pieces of this dark charcoal-like material drifting toward Saturn may have been swept up onto the leading hemisphere of Iapetus.

## 15-10 A Saturn-orbiting spacecraft and a Titan lander will provide a wealth of new information

The data returned by the *Voyager* spacecraft revolutionized our view of Saturn's complex system of rings and satellites, but there is much still to be learned about the Saturnian system. The atmosphere and surface of Titan may even harbor important clues about the origin of life on Earth. The *Cassini* spacecraft is presently on its way to Saturn to provide us with this information (Figure 15-22).

Launched in October 1997, *Cassini* is a joint project involving NASA, the European Space Agency, the Italian Space Agency, and scientists from 17 nations. When it arrives at Saturn in 2004, *Cassini* will go into orbit around the giant planet. It will spend at least four years making close-up observations of Saturn's rings, its satellites, and the planet itself. Long-term monitoring of Saturn's atmosphere and rings will allow scientists to better understand their complicated dynamics, and high-resolution imaging of the satellites will allow us to explore their exotic geology. Titan will come in for special attention: *Cassini* will make more than 30 fly-bys of Titan and will use radar to map the satellite's surface to a resolution of 500 meters.

Prior to going into orbit around Saturn, *Cassini* will deploy a probe that will enter Titan's atmosphere. This probe, named *Huygens*, will spend 2½ hours descending through the atmosphere on a parachute and making measurements of its surroundings. As it descends, the probe will radio its findings to *Cassini* to be relayed on to Earth. If it survives the impact of a 25-km/h (15-mi/h) landing, *Huygens* will

**ƒigure 15-22**

***Cassini* and *Huygens*** A highly capable spacecraft named *Cassini* will go into orbit around Saturn in the year 2004. Approximately two stories tall and weighing (on Earth) more than 6 tons, *Cassini* will use its suite of instruments to examine Saturn, its rings, and satellites over a period of no less than four years. It will also deploy a probe called *Huygens* (seen at the bottom of this artist's conception), which will enter Titan's atmosphere and land on its surface. (NASA)

give an on-site report on the nature of Titan's surface—be it water ice, solid ammonia, liquid ethane, or hydrocarbon "goo." In this way this most mysterious of satellites may give up some of its secrets.

## Key Words

# Key Ideas

**Appearance of Saturn's Rings:** Saturn is circled by a system of thin, broad rings lying in the plane of the planet's equator. This system is tilted away from the plane of Saturn's orbit, which causes the rings to be seen at various angles by an Earth-based observer over the course of a Saturnian year.

**Structure of the Rings:** Three major, broad rings can be seen from the Earth. The faint C ring lies nearest Saturn. Just outside it is the much brighter B ring, then a dark gap called the Cassini division, and then the moderately bright A ring. Other, fainter rings were observed by the *Voyager* spacecraft.

• The principal rings of Saturn are composed of numerous particles of ice and ice-coated rock ranging in size from a few micrometers to about 10 m.

• Most of the rings exist inside the Roche limit of Saturn, where disruptive tidal forces are stronger than the gravitational forces attracting the ring particles to each other.

• Each of the major rings is composed of a great many narrow ringlets.

• The faint, narrow F ring, which is just outside the A ring, is kept narrow by the gravitational pull of shepherd satellites.

**Atmosphere and Internal Structure:** Saturn's internal structure and atmosphere are similar to those of Jupiter. However, Saturn's core makes up a larger fraction of its volume, and its liquid metallic hydrogen mantle is shallower than that of Jupiter.

• Saturn's atmosphere contains less helium than Jupiter's atmosphere. This lower abundance may be the result of precipitation of helium downward into the planet.

• If helium rain does fall, the resulting conversion of gravitational energy into thermal energy would account for Saturn's surprisingly strong heat output.

• Saturn has belts and zones like those of Jupiter. The cloud layers in Saturn's atmosphere are spread out over a greater range of altitude than those of Jupiter.

**Titan:** The largest Saturnian satellite, Titan, is a terrestrial world with a dense nitrogen atmosphere. A variety of hydrocarbons are produced there by the interaction of sunlight with methane. These compounds form an aerosol layer in Titan's atmosphere and possibly cover some of its surface with lakes of ethane.

**Other Satellites:** Six moderate-sized moons circle Saturn in regular orbits: Mimas, Enceladus, Tethys, Dione, Rhea, and Iapetus. They are probably composed largely of ice, but their surface features and histories vary significantly. Saturn also has more than a dozen much smaller satellites, some of which may be captured asteroids.

# Review Questions

1. Is there any way we can infer from naked-eye observations that Saturn is the most distant of the planets visible without a telescope? Explain. (*Hint:* Think about how Saturn's position on the celestial sphere must change over the course of weeks or months.)

2. Describe the structure of Saturn's rings. What evidence is there that ring particles do not migrate significantly between ringlets?

3. During the planning stages for the *Pioneer 11* mission, when relatively little was known about Saturn's rings, it was proposed to have the spacecraft fly through the Cassini division. Why would this have been a bad idea?

4. The space shuttle and other manned spacecraft orbit the Earth well within the Earth's Roche limit. Explain why these spacecraft are not torn apart by tidal forces.

5. Why is the term "shepherd satellite" appropriate for the objects so named? Explain how a shepherd satellite operates.

6. Compare the atmospheres of Jupiter and Saturn. Why does Saturn's atmosphere look "washed out" in comparison to that of Jupiter?

7. Compare the interiors of Jupiter and Saturn. What are the similarities? What are the differences?

8. It has been claimed that Saturn would float if one had a large enough bathtub. Using the mass and size of Saturn given in Table 15-1, confirm that the planet's average density is about 690 kg/m$^3$, and comment on this somewhat fanciful claim.

9. Explain why Saturn is more oblate than Jupiter, even though Saturn rotates more slowly.

10. Compare the internal sources of energy in Jupiter and Saturn that cause both planets to emit more energy than they receive from the Sun in the form of sunlight.

11. You can easily dissolve several tablespoons of sugar in a glass of hot water. But if the water is ice cold, the sugar is difficult to dissolve and sinks to the bottom of the glass. Relate this observation to the explanation of why there is relatively little helium in Saturn's atmosphere compared to the atmosphere of Jupiter.

12. Describe Titan's atmosphere. What effect has the Sun's ultraviolet radiation had on Titan's atmosphere?

13. Why do scientists suspect that there may be liquid ethane on Titan?

14. Which of Saturn's moderate-sized satellites show evidence of geologic activity? What might be the energy source for this activity?

**15.** Saturn's equator is tilted by 27° from the ecliptic, while Jupiter's equator is tilted by only 3°. Use this to explain why we see fewer transits, eclipses, and occultations of the Saturnian satellites than of the Galilean satellites.

## Advanced Questions

### Problem-solving tips and tools:

The small-angle formula is given in Box 1-2. The Doppler effect is described in Section 5-9, and factors affecting the angular resolution of a telescope are described in Section 6-3. Saturn's satellites, whose orbital parameters are given in Table 15-3, obey Newton's form of Kepler's third law, described in Section 4-7 and Box 4-2. A discussion of the retention of an atmosphere is found in Box 7-2, as is an explanation of escape speed.

**16.** As seen by Earth-based observers, the intervals between successive edge-on presentations of Saturn's rings alternate between 13 years, 9 months and 15 years, 9 months. Why do you think these two intervals are not equal?

**17.** It is well known that the Cassini division involves a 2-to-1 resonance with Mimas. Does the location of the Encke gap—133,500 km from Saturn's center—correspond to a resonance with one of the other satellites? If so, which one?

**18.** (a) Use Newton's form of Kepler's third law to calculate the orbital periods of particles at the outer edge of the A ring and at the inner edge of the B ring. (b) Saturn's rings orbit in the same direction as Saturn's rotation. If you were floating along with the cloudtops at Saturn's equator, would the outer edge of the A ring and the inner edge of the B ring appear to move in the same or opposite directions? Explain.

**19.** This *Voyager 2* close-up image of Saturn's rings shows a number of dark, straight features called *spokes*. As these features orbit around Saturn, they tend to retain

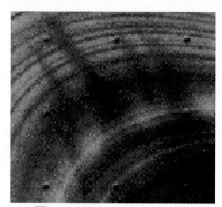

R I **V** U X G

their shape like the rigid spokes on a rotating bicycle wheel. (The black dots were added by the *Voyager* camera system to help scientists calibrate the electronic image.) The spokes rotate at the same rate as Saturn's magnetic field and are thought to be clouds of tiny, electrically charged particles that are kept in orbit by magnetic forces. Explain why the spokes could *not* maintain their shape if they were kept in orbit by gravitational forces alone.

**20.** Find the escape speed on Titan. Using the ideas presented in Box 7-2 and assuming an average atmospheric temperature comparable to that of Saturn, what is the limiting molecular weight of gases that could be retained by Titan's gravity?

**21.** Although Jupiter's satellite Ganymede is about the same size and mass as Titan, Ganymede does not have an appreciable atmosphere. Suggest a reason for this difference.

**22.** Many of the gases in the atmosphere of Titan, such as methane, ethane, and acetylene, are highly flammable. Why, then, doesn't Titan's atmosphere catch fire? (*Hint:* What gas in our atmosphere is needed to make wood, coal, or gasoline burn?)

**23.** Based on the information given in the caption for Figure 15-16, determine the angular resolution of the Hubble Space Telescope using infrared light. Should the angular resolution using visible light be better, worse, or the same? Explain your answer.

**24.** Consult the *Cassini* web site (**http://www.jpl.nasa. gov/cassini/**) or such magazines as *Sky & Telescope* and *Science News* to learn the current status of the *Cassini* mission. When will *Cassini* arrive at Saturn? What are the current plans for its tour of Saturn's satellites? What ideas are being considered for *Cassini*'s extended mission, to begin in 2008?

## Discussion Questions

**25.** Compare and contrast the internal structures and atmospheres of Jupiter and Saturn. Wherever possible, describe the reasons for these differences or similarities in terms of such physical parameters as their mass, chemical composition, and surface gravity.

**26.** Comment on the suggestion that Titan may harbor life forms.

**27.** Imagine that you are in charge of planning the *Cassini* orbital tour of the Saturnian system. In your opinion, what objects in the system should be examined, what data should be collected, and what kinds of questions should the mission attempt to answer?

# Observing Projects

### Observing tips and tools:

To determine the best time of night to view Saturn, consult the current issue of *Sky & Telescope* or *Astronomy* magazine or their web sites (**http://www.skypub.com/** and **http://www.astronomy.com/**, respectively). If your goal is to view Saturn's satellites, consult the section entitled "Satellites of Saturn" in the *Astronomical Almanac* for the current year. This includes a diagram showing the orbits of Mimas, Enceladus, Tethys, Dione, Rhea, Titan, and Hyperion. Plan your observing session by looking up the dates and times of the most recent greatest eastern elongations of the various satellites. You will have to convert from universal time (UT), which is the same as Greenwich mean time, to your local time zone. Then, using the tick marks along the orbits in the diagram, estimate the positions of the satellites relative to Saturn at the time you will be at the telescope. Another useful resource is the "Celestial Calendar" section of *Sky & Telescope*. During months when Saturn is visible in the night sky, this section of the magazine includes a chart of Saturn's satellites.

28. View Saturn through a small telescope. Make a sketch of what you see. Estimate the angle at which the rings are tilted to your line of sight. Can you see the Cassini division? Can you see any belts or zones in Saturn's clouds? Is there a faint, starlike object near Saturn that might be Titan? What observations could you perform to test whether the starlike object is a Saturnian satellite?

29. If you have access to a moderately large telescope, make arrangements to observe several of Saturn's satellites. At the telescope, you should have no trouble identifying Titan. Tethys, Dione, and Rhea are about one-sixth as bright as Titan and should be the next easiest satellites to find. Can you confidently identify any of the other satellites?

## Where to Learn More

*Books and magazine articles*

Beatty, J. K. "Report on the Voyager Encounters with Saturn." *Sky & Telescope*, January, October, and November 1981. This series of articles boasts a magnificent collection of the best images from the *Voyager* flybys of Saturn.

———. "Rings of Revelation." *Sky & Telescope*, August 1996. A wonderful collection of images graces this article about the edge-on presentation of Saturn's rings that occurred in 1995 and 1996.

Ingersoll, A. "Jupiter and Saturn." *Scientific American*, December 1981. Written shortly after the *Voyager* flybys, this enlightening article compares and contrasts the two largest planets.

Morrison, D. "The New Saturn System." *Mercury*, November/December 1981. This comprehensive article, written shortly after the *Voyager* flybys, gives an exciting description of what had just been learned.

———. *Voyages to Saturn*. NASA SP-451, 1982. This excellent book by a noted planetary scientist gives an insider's view of the Saturn flybys.

Owen, T. "Titan." *Scientific American*, February 1982. This article, which discusses many details learned from the *Voyager* missions, speculates that the chemistry of Titan's atmosphere may resemble that of the Earth before life arose.

———. "Titan." In Beatty, J. K., and others, eds., *The New Solar System*, 4th ed. Sky Publishing and Cambridge University Press, 1990. This brief, six-page chapter presents the latest ideas about Titan's atmosphere and surface.

Pollack, J., and Cuzzi, J. "Rings in the Solar System." *Scientific American*, November 1981. This excellent article on the structure and evolution of planetary rings focuses primarily on Saturn but also discusses the rings around Jupiter and Uranus.

Rothery, D. A. *Satellites of the Outer Planets: Worlds in their Own Right*. Oxford, 1992. This well-written book provides insight into the geology of Saturn's moderate-sized moons.

Soderblom, L., and Johnson, T. "The Moons of Saturn." *Scientific American*, January 1982. This article presents a coherent, comprehensive survey of Saturn's moons as revealed by the *Voyager* spacecraft.

W *World Wide Web*

A wealth of information about Saturn and its satellites can be found at the web sites "The Nine Planets" (**http://www.seds.org/nineplanets/nineplanets/saturn.html**), "Welcome to the Planets" (**http://pds.jpl.nasa.gov/planets/welcome/saturn.htm**), and "Views of the Solar System" (**http://www.hawastsoc.org/solar/eng/saturn.htm**).

Scientists at NASA's Ames Research Laboratory maintain a web site devoted to all aspects of planetary rings, including the rings of Saturn (**http://ringside.arc.nasa.gov/**). The edge-on presentation of Saturn's rings during 1995–1996 produced a number of exciting images, which can be viewed at web sites maintained by the Jet Propulsion Laboratory (**http://new products.jpl.nasa.gov/saturn/**) and by the Space Telescope Science Institute (**http://oposite.stsci.edu/pubinfo/SaturnRPC.html**).

To learn more about the *Cassini* mission, there are no better places than NASA's official *Cassini* web site (**http://www.jpl.nasa.gov/cassini/**) and the European Space Agency's *Huygens* web site (**http://www.estec.esa.nl/spdwww/huygens/html/index.html**). *Cassini* is a joint project of NASA and the European Space Agency (**http://www.esrin. esa.it/**).

Among the eighteen science instruments carried on board *Cassini* is a radar transmitter and receiver. This will be used to map the surface of Titan through that satellite's thick cloud layer, in much the same way that the *Magellan* spacecraft mapped the surface of Venus Section 11-5. A full description of this instrument, along with many simulated images and animations, can be found at the Cassini Radar web site (**http://cassini-radar.jpl.nasa.gov/**). Another web page describes the other instruments on board *Cassini* (**http://southport.jpl.nasa.gov/radar/bigpicture/si/science.html**).

*Cassini* carries a small amount of plutonium to serve as a power source. (Solar panels are of no use in the vicinity of Saturn, where sunlight is only 1% as intense as on Earth.) This has caused some consternation among antinuclear activists, who were concerned that this plutonium might accidentally be released into our atmosphere. An intelligent and balanced forum on this subject can be found

at the web site for the PBS television program *NewsHour* (**http://www.pbs.org/newshour/forum/october97/cassini.html**).

Both Saturn and Titan have been targets for the Hubble Space Telescope (HST), and the resulting images—including the one that opens Part II of this textbook, just before Chapter 7—can be found at the Space Telescope Science Institute web site (**http://oposite.stsci.edu/pubinfo/subject.html**). Additional HST images of Titan are available on the web page for Mark Lemmon, a planetary scientest at the University of Arizona's Lunar and Planetary Laboratory (**http://www.lpl.arizona.edu/~lemmon**).

As for Jupiter's satellites, the moons of Saturn are all named for characters from classical mythology. To learn more about these colorful characters, consult the on-line version of the classic book *Bulfinch's Mythology* (**http://www.webcom.com/shownet/medea/bulfinch/**).

# The Outer Worlds

Earth | Uranus | Neptune | Pluto

R I **V** U X G

Earth, Uranus, Neptune, and Pluto  These images of the Earth, Uranus, Neptune, and Pluto are to the same scale. Uranus and Neptune are quite similar in mass, size, and chemical composition. Both planets are surrounded by thin, dark rings. Almost everything we know about the outer two Jovian planets was gleaned from *Voyager 2*, which flew past Uranus in 1986 and Neptune in 1989. Pluto has never been visited by a spacecraft. The computer-generated image of Pluto shown here is based on data from the Hubble Space Telescope. (Alan Stein, Southwest Research Institute; Marc Buie, Lowell Observatory; NASA; ESA)

*In this chapter you will find the answers to the following questions:*

16-1  How did Uranus and Neptune come to be discovered?

16-2  What gives Uranus its distinctive greenish-blue color?

16-3  Why are the clouds on Neptune so much more visible than those on Uranus?

16-4  Are Uranus and Neptune merely smaller versions of Jupiter and Saturn?

16-5  What is so unusual about the magnetic fields of Uranus and Neptune?

16-6  Why are the rings of Uranus and Neptune so difficult to see?

16-7  Do the moons of Uranus show any signs of geologic activity?

16-8  Which of Neptune's moons has an atmosphere and evidence of geologic activity?

16-9  What plans are there to spend a spacecraft to Pluto?

16-10  Are there other planets beyond Pluto?

**B**eyond Saturn, in the cold, dark recesses of the solar system, orbit three planets that have long been shrouded in mystery. These planets—Uranus, Neptune, and Pluto—are so distant, so dimly lit by the Sun, and so slow in their motion against the stars that they were unknown to ancient astronomers and were only discovered after the invention of the telescope. Even then, little was known about Uranus and Neptune until Voyager 2 flew past these planets during the 1980s. Surrounded by a system of small moons and thin, dark rings, Uranus is tipped on its side so that its axis of rotation lies nearly in its orbital plane. This remarkable orientation suggests that Uranus may have been the victim of a staggering impact by a massive planetesimal. Neptune is, at first glance, a denser, more massive version of Uranus, but it is a far more active world. It has an internal energy source that Uranus seemingly lacks, as well as atmospheric bands and storm activity resembling those on Jupiter. Neptune also has dark rings, small, icy moons, and an intriguing large satellite, Triton, which is nearly devoid of impact craters and has geysers that squirt nitrogen-rich vapors. Triton's retrograde orbit suggests that this strange world may have been gravitationally captured by Neptune. A close cousin of Triton is Pluto, the smallest and most remote of the planets. Pluto still harbors many mysteries, because it has not yet been visited by a spacecraft. It may be just one of thousands of small, icy worlds that orbit the Sun at the outskirts of the solar system.

---

## 16-1    Uranus was discovered by chance, but Neptune's existence was predicted by applying Newtonian mechanics

Uranus was discovered by chance on March 13, 1781, by the then-little-known astronomer William Herschel. This German-born musician emigrated to England in 1757 and became fascinated by astronomy. Using a telescope that he built himself, Herschel was systematically surveying the sky when he noticed a faint, fuzzy object that he first thought to be a distant comet. By the end of 1781, however, he knew that the object had a planetlike orbit outside the orbit of Saturn, beyond where comets can normally be seen. Herschel had not only discovered the seventh planet from the Sun but in doing so had doubled the radius of the known solar system from 9.5 AU (the semimajor axis of Saturn's orbit) to 19.2 AU (the distance from the Sun to Uranus). Herschel originally named his discovery "Georgium Sidus" (Latin for "Georgian star") in honor of the reigning monarch, George III. The name Uranus—in Greek mythology, the personification of Heaven—came into currency only some decades later.

Although Herschel received the credit for discovering Uranus, he was by no means the first person to have seen it. At opposition, Uranus is just barely bright enough to be seen with the naked eye under good observing conditions, so it was probably seen by the ancients. Many other astronomers with telescopes had sighted this planet before Herschel; it is plotted on at least 20 star charts drawn between 1690 and 1781. But all of these other observers mistook Uranus for a dim star. Herschel was the first to track its motion relative to the stars and recognize it as a planet. This was no small task, because Uranus moves very slowly on the celestial sphere, just over 4° in the space of a year (compared to about 12° for Saturn and about 35° for Jupiter).

It was by carefully tracking Uranus's slow motions that astronomers were led to discover Neptune. By the beginning of the nineteenth century, it had become painfully clear to astronomers that they could not accurately predict the orbit of Uranus using Newtonian mechanics. By 1830 the discrepancy between the planet's predicted and observed positions had become so great (2 arcmin) that some scientists suspected that Newton's law of gravitation might not be accurate at great distances from the Sun.

In 1843, John Couch Adams, a 24-year-old recent graduate of Cambridge University in England, began to explore an earlier and sounder suggestion. Perhaps the gravitational pull of an as yet undiscovered planet was causing Uranus to deviate slightly from its predicted orbit. After spending two years attempting to calculate the orbit of this unknown object, Adams concluded that Uranus had indeed caught up with and had passed a more distant planet. Uranus had accelerated slightly as it approached the unknown planet, then decelerated slightly as it receded from the planet. Adams predicted that the unknown planet would be found at a certain position in the constellation of Aquarius (the Water Carrier).

In October 1845, Adams submitted his calculations to George Airy, the Astronomer Royal of Great Britain. Airy was unconvinced, and the matter was dropped. A few months later, however, the French astronomer Urbain Jean Joseph Le Verrier independently made the same calculations

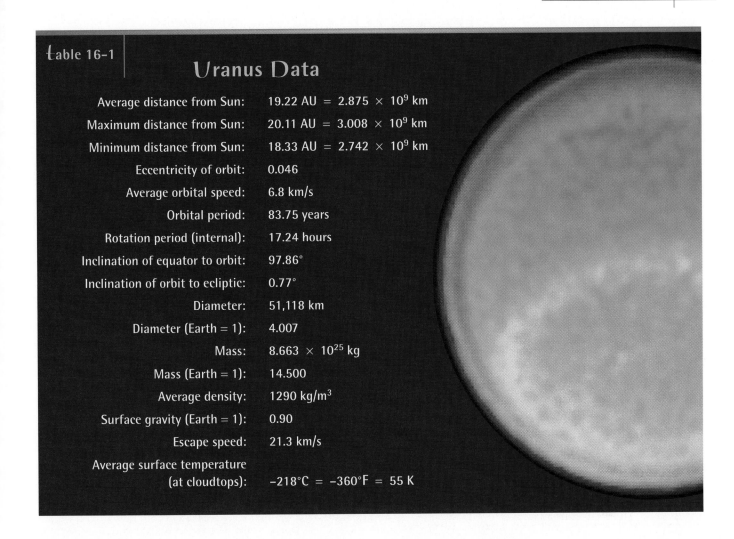

**Table 16-1**

## Uranus Data

| | |
|---|---|
| Average distance from Sun: | $19.22 \text{ AU} = 2.875 \times 10^9 \text{ km}$ |
| Maximum distance from Sun: | $20.11 \text{ AU} = 3.008 \times 10^9 \text{ km}$ |
| Minimum distance from Sun: | $18.33 \text{ AU} = 2.742 \times 10^9 \text{ km}$ |
| Eccentricity of orbit: | 0.046 |
| Average orbital speed: | 6.8 km/s |
| Orbital period: | 83.75 years |
| Rotation period (internal): | 17.24 hours |
| Inclination of equator to orbit: | 97.86° |
| Inclination of orbit to ecliptic: | 0.77° |
| Diameter: | 51,118 km |
| Diameter (Earth = 1): | 4.007 |
| Mass: | $8.663 \times 10^{25} \text{ kg}$ |
| Mass (Earth = 1): | 14.500 |
| Average density: | $1290 \text{ kg/m}^3$ |
| Surface gravity (Earth = 1): | 0.90 |
| Escape speed: | 21.3 km/s |
| Average surface temperature (at cloudtops): | $-218°C = -360°F = 55 \text{ K}$ |

and came to the same conclusion as Adams. Their predicted positions for the unknown planet differed by less than 1°.

Shocked into action, Airy promptly persuaded James Challis, the director of the observatory at Cambridge, to begin a thorough search. The search was hampered by Challis's lack of enthusiasm, as well as by his lack of an up-to-date star map for the part of the sky where Adams had predicted the new planet would be found. Many uncharted stars had to be observed, then followed over many nights to see whether any of them showed a planetlike motion with respect to other stars.

Meanwhile, Le Verrier wrote to Johann Gottfried Galle at the Berlin Observatory. Galle received the letter on September 23, 1846, and he sighted the planet that very night. Because he had the finest star charts available, Galle easily located an uncharted star with the expected brightness in the predicted location.

Le Verrier proposed that the planet be called Neptune. After years of debate between English and French astronomers, the credit for its discovery came to be divided equally between Adams and Le Verrier.

Interestingly, Galileo may have sighted Neptune more than two hundred years before anyone else, in January 1613. Galileo had been regularly observing the motions of the four large satellites of Jupiter. His drawings of Jupiter and its satellites show a "star" less than 1 arcmin from Neptune's location during those winter nights. At opposition, Neptune can be bright enough to be visible through small telescopes. Galileo even noted in his observation log that on one night this star seemed to have moved in relation to the other stars. But Galileo would have been hard pressed to identify Neptune as a planet, because its motion against the background stars is so slow (just over 2° per year).

Even today through a large telescope, both Uranus and Neptune are dim, uninspiring sights. Each planet appears as a hazy, featureless disk with a faint greenish-blue tinge. Although Uranus and Neptune are both about four times larger in diameter than Earth, they are so distant that their angular diameters as seen from the Earth are tiny—no more than 4 arcsec for Uranus and just over 2 arcsec for Neptune. To an Earth-based observer, Uranus is roughly the size of a golf ball seen at a distance of 1 kilometer.

From 1998 through 2004, Uranus and Neptune are together in the same part of the sky, moving slowly from Capricornus (the Sea Goat) into Aquarius. During these years, both planets are at opposition in either July or August, which are thus the best months to view these worlds with a telescope. Basic data about Uranus and Neptune are found in Tables 16-1 and 16-2.

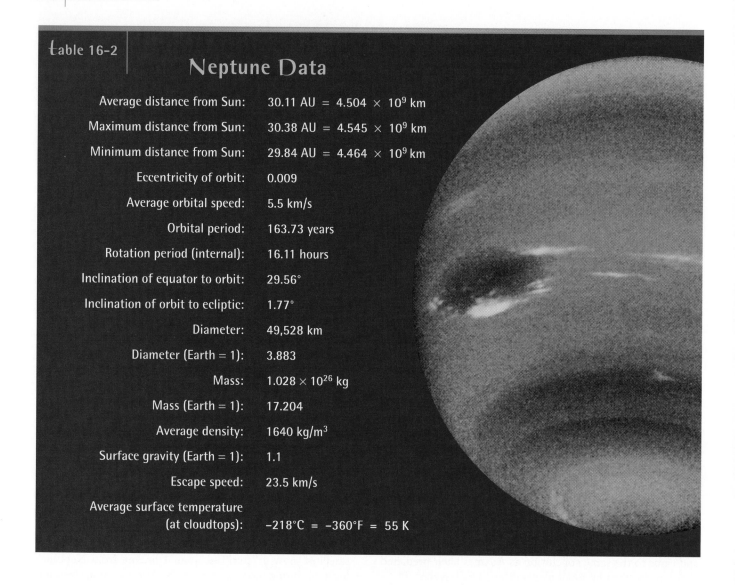

**table 16-2**

# Neptune Data

| | |
|---|---|
| Average distance from Sun: | $30.11 \text{ AU} = 4.504 \times 10^9 \text{ km}$ |
| Maximum distance from Sun: | $30.38 \text{ AU} = 4.545 \times 10^9 \text{ km}$ |
| Minimum distance from Sun: | $29.84 \text{ AU} = 4.464 \times 10^9 \text{ km}$ |
| Eccentricity of orbit: | 0.009 |
| Average orbital speed: | 5.5 km/s |
| Orbital period: | 163.73 years |
| Rotation period (internal): | 16.11 hours |
| Inclination of equator to orbit: | 29.56° |
| Inclination of orbit to ecliptic: | 1.77° |
| Diameter: | 49,528 km |
| Diameter (Earth = 1): | 3.883 |
| Mass: | $1.028 \times 10^{26} \text{ kg}$ |
| Mass (Earth = 1): | 17.204 |
| Average density: | 1640 kg/m³ |
| Surface gravity (Earth = 1): | 1.1 |
| Escape speed: | 23.5 km/s |
| Average surface temperature (at cloudtops): | −218°C = −360°F = 55 K |

## 16-2 Uranus is nearly featureless and has an unusually tilted axis of rotation

Scientists had hoped that *Voyager 2* would reveal cloud patterns in Uranus's atmosphere when it flew past the planet in January 1986. But even images recorded at close range showed Uranus to be remarkably featureless (Figure 16-1). Faint cloud markings only became visible in these images after severe computer enhancement.

*Voyager 2* data confirmed that the Uranian atmosphere is dominated by hydrogen (84% by mass) and helium (14%), similar to the atmospheres of Jupiter and Saturn. Uranus differs, however, in that 2% of its atmosphere is methane

($CH_4$), ten times the percentage found on Jupiter and Saturn. In fact, Uranus has a higher percentage of all heavy elements—including carbon atoms, which are found in molecules of methane—than Jupiter and Saturn. (We will discuss the reasons for this in Section 16-4.) Methane preferentially absorbs the longer wavelengths of visible light, so light that is reflected from Uranus's upper atmosphere is depleted of its reds and yellows. This gives the planet its distinct greenish-blue appearance. Like on Saturn's moon Titan (see Section 15-8), ultraviolet light from the Sun turns some of the methane gas into a hydrocarbon haze, making it difficult to see the lower levels of the atmosphere.

Ammonia ($NH_3$), which constitutes 0.03% of the atmospheres of Jupiter and Saturn, is completely lacking from the Uranian atmosphere. The reason is that Uranus is a very cold planet, with a temperature in its upper atmosphere of only 55 K (–218°C = –360°F). By comparison, the temperature of liquid nitrogen—used by physicians to remove moles from the skin, and the coldest temperature that you are ever likely to encounter personally—is a relatively balmy 77 K (–196°C = –321°F). Ammonia freezes at the very low temperatures found in Uranus's atmosphere, and so any ammonia has long since precipitated down to depths where we cannot detect its presence. For the same reason, Uranus's atmosphere is also lacking in water. Hence, the substances of which the clouds on Jupiter and Saturn are made—ammonia, ammonium hydrosulfide (NH$_4$SH), and water—are not available on Uranus. This helps to explain the bland, cloudless appearance of the planet shown in Figure 16-1. The few clouds found on Uranus are primarily made of methane, which only condenses into droplets if the pressure is sufficiently high. Therefore, Uranus's methane clouds appear fairly deep in the atmosphere, where they are difficult to see.

*Voyager 2* also confirmed a remarkable aspect of Uranus's rotation: Its rotation axis has a unusual tilt. Herschel found the first evidence of this tilt within a few years after his discovery of Uranus. He discovered two moons orbiting the planet in a plane that is almost perpendicular to the plane of Uranus's orbit around the Sun. Thirteen other satellites of Uranus have been discovered since then, and all of them orbit in this same tilted plane. (We will discuss these satellites in Section 16-5.) Because the large moons of Jupiter and Saturn were known to orbit in the same plane as their planet's equator and in the same direction as their planet's rotation, it was thought that the same must be true for the moons of Uranus. Thus, Uranus's axis of rotation lies very nearly in the plane of its orbit (Figure 16-2). Careful measurement shows that the axis of rotation is tilted by 98°, as compared to 23½° for Earth. Having a tilt angle greater than 90° means that Uranus exhibits retrograde rotation (see Figure 11-3). It is one of only three planets in the solar system to do so (Venus and Pluto are the other two). Modern astronomers suspect that Uranus acquired its large tilt angle billions of years ago, when another massive body collided with Uranus while the planet was still forming.

Thanks to this radical tilt, as Uranus moves along its 84-year orbit, its north and south poles alternately point toward or away from the Sun. This produces highly exaggerated seasonal changes on the planet. For example, near Uranus's south pole during the southern summer, the Sun remains above the horizon while northern latitudes are subjected to a continuous, frigid winter night. (This was the case when *Voyager 2* flew by in 1986.) Half a Uranian year later (42 of our years), the situation is reversed.

By following the motions of Uranus's faint clouds, scientists determined that the planet's winds flow primarily to the east (that is, in the same direction as the planet's rotation.) This is quite unlike the alternating easterly and westerly zonal winds found on Jupiter and Saturn (see Section 13-3 and Section 15-5). The Uranian winds move at speeds of 140 to 580 km/h (90 to 360 mi/h).

Although Uranus's equatorial region was receiving little sunlight at the time of the *Voyager 2* flyby, the atmospheric

**figure 16-1**   R I **V** U X G

**Uranus from *Voyager 2*** No distinctive cloud patterns are to be seen in any of the *Voyager 2* views of Uranus. The blue-green appearance of Uranus comes from methane in the planet's atmosphere, which absorbs red wavelengths from the sunlight. Severe computer enhancement of the Voyager pictures does reveal faint cloud features, from which the rotation period of Uranus's atmosphere was determined to be about 16½ hours. When this image was made, Uranus's south pole was aimed almost directly at the Sun. Thus, the face-on, fully illuminated view of Uranus's daytime hemisphere shown here looks nearly straight down onto the south pole. (NASA)

temperature there (about 55 K = –218°C = –359°F) was not too different from that at the sunlit pole. Heat must therefore be efficiently transported from the poles to the equator. This north-south heat transport, which is perpendicular to the wind flow, may also have mixed and homogenized the atmosphere to make Uranus nearly featureless.

The rotation period of Uranus's atmosphere is about 16 hours. Like Jupiter and Saturn, Uranus rotates differentially, so this period depends on the latitude. This can be measured by tracking the motions of clouds (Figure 16-3). To determine the rotation period for the underlying body of the planet, scientists looked to Uranus's magnetic field, which is presumably anchored in the planet's interior, or at least in the deeper and denser layers of its atmosphere. Data from *Voyager 2* indicate that Uranus's internal period of rotation is 17.24 hours.

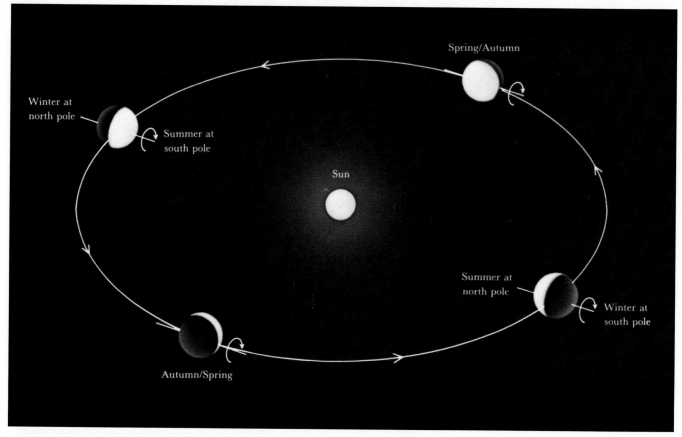

## figure 16-2

**Exaggerated Seasons on Uranus** Unlike most planets, whose rotation axes are roughly perpendicular to the plane of their orbits around the Sun, Uranus has its axis of rotation tilted so steeply that it lies just 8° from the planet's orbital plane. The seasonal changes on Uranus are thus severely exaggerated. For example, during midsummer at Uranus's south pole, the Sun appears nearly overhead for many Earth years, while the planet's northern regions are subjected to a long, continuous winter night.

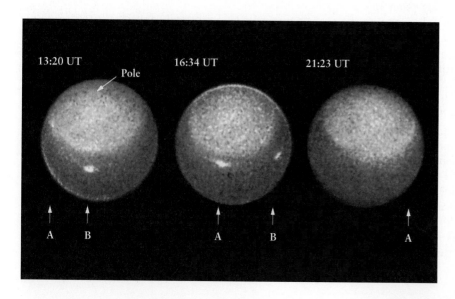

## figure 16-3   R I **V** U X G

**Cloud Motions on Uranus** These images from the Hubble Space Telescope show clouds in Uranus's southern hemisphere rotating along with the planet. The two clouds, labeled A and B, are about 4300 km (2500 miles) and 3100 km (1800 miles) in length, respectively. The first two images were recorded about 3 hours apart on August 14, 1994, and the second and third images, about 5 hours apart. (Kenneth Seidelmann, U. S. Naval Observatory, and NASA)

## 16-3 Neptune is a cold, bluish world with Jupiterlike atmospheric features

At first glance, Neptune appears to be the twin of Uranus, as can be seen from the image that opens this chapter and by comparing Tables 16-1 and 16-2. But these two planets are by no means identical. While Neptune and Uranus have almost the same diameter, Neptune is 18% more massive. Neptune's axis of rotation also has a more modest 30° tilt. When *Voyager 2* flew past Neptune in August 1989, it revealed that the planet has a more active and dynamic atmosphere than Uranus. This activity suggests a powerful source of energy within Neptune.

The *Voyager 2* data showed that Neptune has essentially the same atmospheric composition as Uranus: 84% hydrogen, 14% helium, 2% methane, and no ammonia or water vapor. As for Uranus, the presence of methane gives Neptune a characteristic bluish-green color. The temperature in the upper atmosphere is also the same as on Uranus, about 55 K. Unlike Uranus, however, Neptune has clearly visible cloud patterns in its atmosphere. At the time that *Voyager 2* flew past Neptune, the most prominent feature in the planet's atmosphere was a giant storm called the **Great Dark Spot** (Figure 16-4). The Great Dark Spot had a number of similarities to Jupiter's Great Red Spot (described in Sections 13-1 and 13-2): The storms on both planets were comparable in size to the Earth's diameter, both appeared at about the same latitude in the southern hemisphere, and the winds in both storms circulated in a counterclockwise direction. But Neptune's Great Dark Spot appears not to have been as long-lived as the Great Red Spot on Jupiter. When the Hubble Space Telescope first viewed Neptune in 1994, the Great Dark Spot had disappeared. Another dark storm appeared in 1995 in Neptune's northern hemisphere.

*Voyager 2* also saw a few conspicuous whitish clouds on Neptune. These clouds are thought to be produced when winds carry methane gas into the cool, upper atmosphere, where it condenses into visible crystals of methane ice. The high elevation of these clouds was confirmed in *Voyager 2* images that show these clouds casting shadows onto lower levels of the atmosphere (Figure 16-5).

Thanks to its greater distance from the Sun, Neptune receives less than half as much energy from the Sun as Uranus. But with less solar energy available to power atmospheric motions, why are there high-altitude clouds and huge, dark storms on Neptune but not on Uranus? At least part of the answer is probably that Neptune is still slowly contracting,

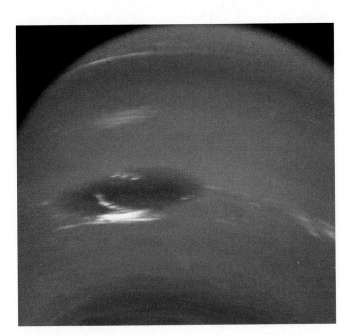

**figure 16-4**  R I **V** U X G

**Neptune from *Voyager 2*** This 1989 view from *Voyager 2* looks down on the southern hemisphere of Neptune. Near the center of the picture is the Great Dark Spot, which at the time measured about 12,000 by 8000 km—comparable in size to the Earth. Note the white, wispy clouds, thought to be composed of crystals of methane ice. Toward the lower left is a smaller, dark spot, located 54° south of Neptune's equator. The color contrast in this image has been exaggerated to emphasize differences between dark and light regions in the atmosphere. (NASA)

**figure 16-5**  R I **V** U X G

**Cirrus Clouds over Neptune** This picture shows vertical relief in Neptune's bright cloud streaks. *Voyager 2* recorded this image of clouds north of Neptune's equator near the terminator (the border between day and night on the planet). Note the shadows cast by the clouds onto the main cloud deck. The clouds are thought to be composed of crystals of methane ice. Hence, the whitish clouds in Neptune's atmosphere are similar to thin, wispy cirrus clouds in our own atmosphere, which are made of crystals of water ice. (NASA)

thus converting gravitational energy into thermal energy that heats the planet's core. The evidence for this is that Neptune, like Jupiter, emits more energy than it receives from the Sun. The combination of a warm interior and a cold outer atmosphere can cause convection in Neptune's atmosphere and the up-and-down motion of gases that produces clouds and storms. Uranus, by contrast, appears to have little or no internal source of thermal energy; measurements so far show that it radiates just as much energy into space as it receives from the Sun. Therefore, Uranus lacks the dynamic atmospheric activity found on Neptune and has a much blander appearance (compare Figure 16-1 and Figure 16-4).

Neptune also resembles Jupiter in having faint belts and zones parallel to the planet's equator. These can be seen in the *Voyager 2* image of Neptune that opens this chapter, as well as in the more recent Hubble Space Telescope image shown in Figure 16-6. As on Jupiter, the light-colored zones are regions where clouds have been lifted to relatively high altitudes, while the dark belts are where air is descending. As on Uranus, most of Neptune's clouds are probably made of droplets of liquid methane. Because these form fairly deep within the atmosphere, they are more difficult to see

than on Jupiter. Hence, Neptune's belts and zones are less pronounced than on Jupiter, although more so than on Uranus (thanks to the extra cloud-building energy from Neptune's interior).

Although Neptune displays more evidence of up-and-down motion in its atmosphere than Uranus, the global pattern of east and west winds is almost identical on the two planets. This is rather strange. The two planets are heated very differently by the Sun, thanks to their different distances from the Sun and the different tilts of their axes of rotation, so we might have expected the wind patterns on Uranus and Neptune to also be different. Perhaps the explanation of these wind patterns will involve understanding how heat is transported not only within the atmospheres of Uranus and Neptune but within their interiors as well.

## 16-4 Uranus and Neptune contain a higher proportion of heavy elements than Jupiter and Saturn

As we saw in Section 13-5 and Section 15-7, Jupiter and Saturn are composed primarily of hydrogen and helium, both in nearly the same abundance as the Sun. That composition, however, is incompatible with the relatively high average densities of Uranus and Neptune. Both of these outer planets are distinctly smaller and less massive than either Jupiter or Saturn (see Figure 7-3). If Uranus and Neptune also had solar abundances of the elements, the smaller masses of these planets would produce less gravitational compression, and Uranus and Neptune would both have lower average densities than Jupiter or Saturn. In fact, however, Uranus and Neptune have average densities (1290 kg/m$^3$ and 1640 kg/m$^3$, respectively) that are comparable to or greater than those of Jupiter (1330 kg/m$^3$) or Saturn (690 kg/m$^3$). Therefore, we can conclude that Uranus and Neptune contain greater proportions of the heavier elements than Jupiter and Saturn.

This picture is not what we might expect. According to our discussion in Section 7-8 of how the solar system formed, hydrogen and helium should be relatively more abundant as distance from the vaporizing heat of the Sun increases. But Uranus and Neptune contain a greater percentage of heavy elements, and, therefore, a smaller percentage of hydrogen and helium, than Jupiter and Saturn. Recall from Section 7-8 that the Jovian planets probably formed in a two-step process: first, planetesimals accreted to form each planet's core; second, hydrogen and helium was gravitationally attracted onto the core to form the planet's envelope. Apparently, the cores of Uranus and Neptune were only able to attract relatively little hydrogen or helium, leaving these planets deficient in these two elements.

One possible explanation is that Uranus and Neptune may have taken longer to form their cores than Jupiter and Saturn, because the outer regions of the solar nebula were thinner and planetesimals were more widely dispersed. By the time these cores had formed, strong winds from the young

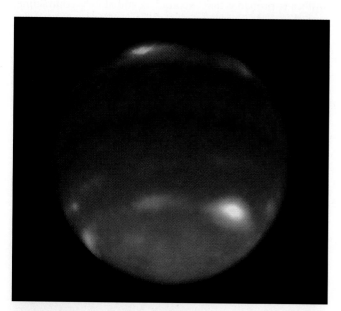

**figure 16-6** R I **V** U X G

**Neptune's Banded Structure** Several Hubble Space Telescope images made using different wavelengths were combined to create this enhanced-color view of Neptune. The planet's south pole is at the bottom of the image and tilted toward us. The dark blue and light blue areas are the belts and zones, respectively; the darkest belt, seen near the top of the image, lies just south of Neptune's equator. White areas denote high-altitude clouds, presumably of methane ice. The very highest clouds are shown in yellow-red, as seen at the very top of the image. The green belt near the south pole is a region where the atmosphere absorbs blue light, perhaps indicating some differences in chemical composition. (Lawrence Sromovsky, University of Wisconsin-Madison, and NASA)

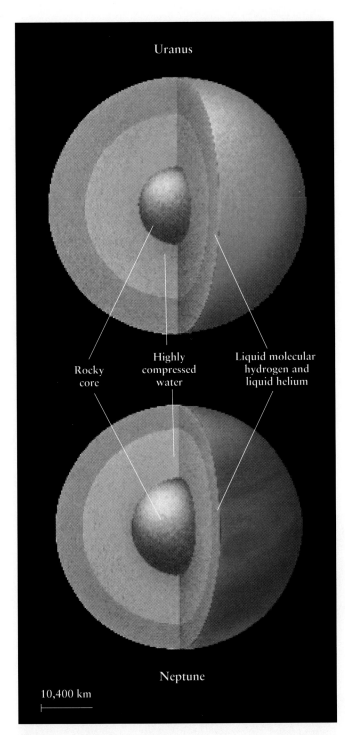

Uranus

Neptune

Rocky
core

Highly
compressed
water

Liquid molecular
hydrogen and
liquid helium

10,400 km

**ʄigure 16-7**

**The Interior Structures of Uranus and Neptune** This figure
shows one model for the interiors of Uranus and Neptune. In this
model, both planets have three regions in their interiors: a rocky
core, resembling a large terrestrial planet; a mantle of water, which
may be in a liquid or solid state, with ammonia dissolved in it; and
an outer layer of liquid molecular hydrogen and liquid helium. The
atmospheres (not shown) are thin gaseous shells on top of the outer
hydrogen-helium layer. The two planets have nearly the same
diameter, but because Neptune is more massive than Uranus, it
must have a somewhat larger core.

Sun would have already begun to dissipate the hydrogen and
helium gas in the solar nebula into interstellar space, leaving
relatively little to be incorporated into Uranus and Neptune.
One observation that reinforces this idea is that the masses of
Uranus and Neptune are comparable to the masses of the
*cores* of Jupiter and Saturn. All four Jovian planets may have
begun with cores of similar size, but Uranus and Neptune
failed to accumulate tremendous envelopes of hydrogen and
helium, as did Jupiter and Saturn.

Figure 16-7 shows one model for the internal structures of
Uranus and Neptune. In this model each planet has a rocky
core roughly the size of Earth, although more massive. Each
planet's core is surrounded by a mantle of liquid or solid
water. The mantle probably has dissolved in it the ammonia
that froze out of the planet's upper layers and descended into
the interior. (This means that the mantle is chemically similar
to household window cleaning fluid.) Around the mantle is a
layer of liquid molecular hydrogen and liquid helium, with a
small percentage of liquid methane. This layer is relatively
shallow compared to those on Jupiter and Saturn (see Figure
15-14), and the pressure is not high enough to convert the
liquid hydrogen into liquid metallic hydrogen.

## 16-5 The magnetic fields of both Uranus and Neptune are oriented at unusual angles

Astronomers were quite surprised by the data sent back
from *Voyager 2*'s magnetometer as the spacecraft sped past
Uranus and Neptune. These data, as well as radio emissions
from charged particles in their magnetospheres, showed that
the magnetic fields of both Uranus and Neptune are tilted at
steep angles to their axes of rotation. Uranus's **magnetic
axis,** the line connecting its north and south magnetic poles,
is inclined by 59° from its axis of rotation; Neptune's magnetic axis is tilted by 47°. By contrast, the magnetic axes
of Earth, Jupiter, and Saturn are all nearly aligned with their
rotation axes; the angle between their magnetic and rotation
axes is 12° or less (Figure 16-8). Scientists were also surprised
to find that the magnetic fields of Uranus and Neptune are
offset from the centers of the planets.

The drawing of the Earth's magnetic field at the far
left of Figure 16-8 may seem to be mislabeled, because it shows the *south* pole of a magnet at the
Earth's *north* pole. This is, in fact, correct, as you can understand by thinking about how magnets work. The north pole
of a magnet that is free to swivel, like the magnetized needle in
a compass, is called that because it tends to point north on
Earth. Likewise, a compass needle's south pole points toward
the south on Earth. Furthermore, opposite magnetic poles
attract. If you take two magnets and put them next to each
other, they try to align themselves so that one magnet's north
pole is next to the other magnet's south pole. Now, if you
think of a compass needle as one magnet and the entire Earth
as the other magnet, it makes sense that the compass needle's

north pole is being drawn toward a magnetic *south* pole—which happens to be located near the Earth's geographic north pole. The Earth's magnetic pole nearest its geographic north pole is called the "magnetic north pole." Note that the magnets drawn inside the four Jovian planets are all oriented opposite to the Earth. On any of the Jovian planets, the north pole of a compass needle would point south, not north!

Why are the magnetic axes and axes of rotation of Uranus and Neptune so badly misaligned? And why are the magnetic fields offset from the centers of the planets? One possibility is that their magnetic fields might be undergoing a reversal; geological data show that the Earth's magnetic field has switched north to south and back again many times in the past. Another possibility is that the misalignments resulted from catastrophic collisions with planet-sized bodies. As we noted in Section 16-2, the tilt of Uranus's rotation axis and its system of moons suggest that just such collisions occurred long ago. As we will see in Section 16-7, Neptune may have gravitationally captured its largest moon, Triton, but no one knows if that incident was responsible for offsetting Neptune's magnetic axis.

Because neither Uranus nor Neptune is massive enough to compress hydrogen to a metallic state, their magnetic fields cannot be generated in the same way as those of Jupiter and Saturn. One possibility is that under the high pressures found in the watery mantles of Uranus and Neptune, dissolved molecules such as ammonia lose one or more electrons and become electrically charged (that is, they become ionized; see Section 5-8). Water is a good conductor of electricity when it has such electrically charged molecules dissolved in it, so perhaps electric currents in this fluid give rise to the magnetic fields of the planets.

## 16-6 Uranus and Neptune each have a system of thin, dark rings

Like the planet itself, Uranus's rings were discovered by accident. On March 10, 1977, Uranus was scheduled to occult (move in front of) a faint star, as seen from the Indian Ocean. A team of astronomers headed by James L. Elliot of Cornell University observed the occultation from a NASA airplane equipped with a telescope. They hoped that by measuring how long the star was hidden, they could accurately measure Uranus's size. In addition, by carefully measuring how the starlight faded when Uranus passed in front of the star, they planned to deduce important properties of Uranus's upper atmosphere.

To everyone's surprise, the background star briefly blinked on and off several times just before the occultation and again immediately after (Figure 16-9). The astronomers concluded that Uranus must be surrounded by a series of narrow rings—nine in all. During the *Voyager 2* flyby, two more rings were discovered.

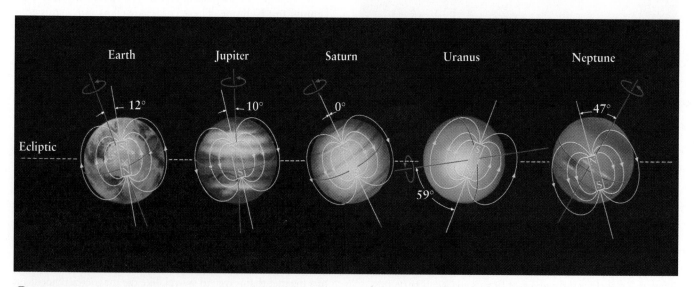

figure 16-8

**The Magnetic Fields of Five Planets** This drawing shows how the magnetic fields of Earth, Jupiter, Saturn, Uranus, and Neptune are tilted relative to their rotation axes. Note that the magnetic fields of all four Jovian planets are oriented opposite to that of Earth; on a Jovian planet, the north pole of a compass needle would point southward, not northward. Note also that the magnetic fields of Uranus and Neptune are offset from the center of the planets and steeply inclined to their rotation axes.

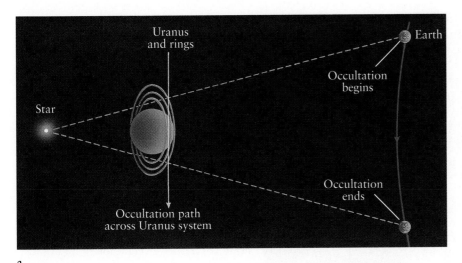

a

## figure 16-9

**How the Rings of Uranus Were Discovered** **(a)** As seen from the Earth, Uranus occasionally appears to move in front of a distant star. Such an event is called an occultation. **(b)** This graph suggests how the intensity of a star's light varies as Uranus passes in front of it. As expected, the star's light is completely blocked when it is directly behind the planet. But before and after the occultation by Uranus, the star's light dims and then brightens again. This is caused by light being blocked by Uranus's rings. It was in this way that the rings were discovered during an occultation in 1977.

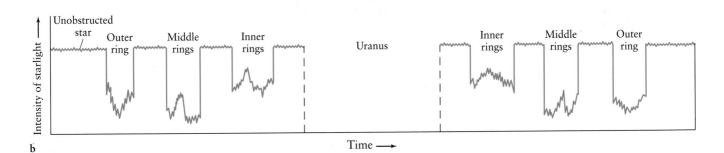

b

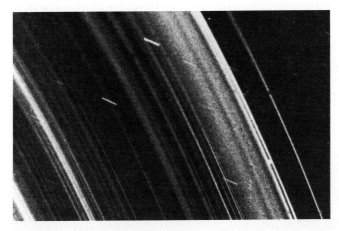

## figure 16-10   R I 🟥 U X G

**The Shaded Side of Uranus's Rings** This view, taken when *Voyager 2* was in Uranus's shadow, looks back toward the Sun. Numerous fine dust particles between the main rings gleam in the sunlight. This dust is probably fine-grained debris from collisions between larger particles in the main rings. Some planetary scientists suggest that the Uranian rings are being eroded by these collisions and will probably be gone in only a few hundred million years. The short streaks are star images, blurred because of the spacecraft's motion during the exposure. (NASA)

Unlike Saturn's rings, the rings of Uranus are dark and narrow, most less than 10 km wide. Typical particles in Saturn's rings are chunks of ice with the reflectivity and dimensions of snowballs, but typical particles in Uranus's rings are about 1 m in size and are no more reflective than lumps of coal. As a result, the Uranian rings are very dark; they reflect only a few percent of the sunlight that falls on them. It is not surprising that these narrow, dark rings escaped detection for so long. Figure 16-10 is a *Voyager 2* image of the rings from the side of the planet away from the Sun, where light scattering from the ring particles makes them more visible. (We discussed light scattering in Section 15-3 and Box 15-1.)

All of Uranus's rings are located less than 2 Uranian radii from the planet's center, well within the planet's Roche limit. Some sort of mechanism, possibly one involving shepherd satellites, efficiently confines particles to their narrow orbits. *Voyager 2* searched for shepherd satellites but found only two. The others may be so small and black that they have simply escaped detection. Or perhaps they aren't there, and our ideas need revision.

Like Uranus, Neptune is surrounded by a system of thin, dark rings that were first detected in stellar occultations. Figure 16-11 is a *Voyager 2* image of Neptune's rings. It is so cold at Uranus and Neptune that the ring particles can retain methane ice. Scientists speculate that eons of impacts by

**figure 16-11** R I **V** U X G

**Neptune's Rings** Two main rings are easily seen in this *Voyager 2* view, which blocks out an overexposed image of Neptune. This picture also reveals a faint, inner ring. A sheet of particles, whose outer edge is located between the two main rings, extends inward toward the planet. (NASA)\

electrons trapped in the magnetospheres of the planets have converted this methane ice into dark carbon compounds. This **radiation darkening** can account for the low reflectivity of the rings around both Uranus and Neptune.

| 16-7 | Uranus is orbited by satellites that bear the scars of many shattering collisions |
|------|-------------------------------------------------------------------------------------|

Prior to the *Voyager 2* flyby of Uranus, five moderate-sized satellites—Titania, Oberon, Ariel, Umbriel, and Miranda—were known to orbit the planet. All five are named after sprites and spirits in Shakespearean literature. Images from

*Voyager 2* (Figure 16-12) gave accurate information about the sizes of these five principal Uranian moons. They range in diameter from about 1600 km (1000 mi) for Titania and Oberon to less than 500 km (300 mi) for Miranda. All of these moons have average densities around 1500 kg/m³, which is consistent with a mixture of ice and rock. *Voyager 2* also discovered ten other small Uranian satellites, most of which have diameters of less than 100 km (60 mi). Several of these can be seen in the Hubble Space Telescope image shown in Figure 16-13. Information on all the moons of Uranus is summarized in Table 16-3.

The Uranian moons are all quite dark. This may be due to radiation darkening of methane ice on the surfaces of the moons, the same mechanism proposed to explain the darkness of Uranus's rings. Umbriel is one of the darkest moons in our solar system, but some of the craters on Oberon appear

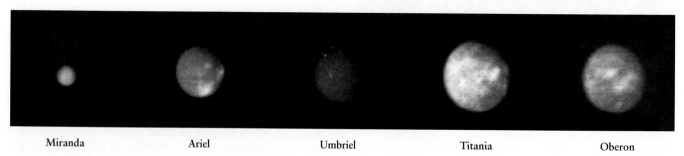

Miranda  Ariel  Umbriel  Titania  Oberon

**figure 16-12** R I **V** U X G

**The Principal Uranian Satellites** This "family portrait" of Uranus's largest moons—actually a montage of five *Voyager 2* images—shows the five satellites to the same scale and correctly displays their relative reflectivities. The surfaces of all five satellites vary only slightly in color, with the average color being nearly gray. Note how dark Umbriel is. (NASA)

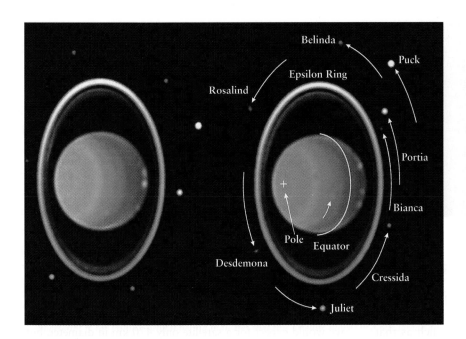

**figure 16-13** R **I** V U X G

**Uranus, Its Rings, and Eight of Its Satellites** This pair of false-color Hubble Space Telescope images, taken 90 minutes apart, shows eight of Uranus's small satellites orbiting around the planet. All of the satellites visible here were discovered by *Voyager 2* when it flew past Uranus in 1986. They all lie within 86,000 km of the planet's center, only about one-fifth of the distance from the Earth to our Moon. None is more than 150 km (90 mi) in diameter. Each image is actually a composite of three images taken at different infrared wavelengths; each wavelength is displayed in a different color. This technique also makes it easier to see the rings, which are much brighter in infrared light than in visible light. (Erich Karkoschka, University of Arizona, and NASA)

**table 16-3** | **Uranus's Satellites**

|  | Distance from center of Uranus (km) | Orbital period (d) | Diameter (km) | Density (kg/m$^3$) |
|---|---|---|---|---|
| Cordelia | 49,750 | 0.335 | 40 | — |
| Ophelia | 53,760 | 0.376 | 30 | — |
| Bianca | 59,160 | 0.435 | 40 | — |
| Cressida | 61,770 | 0.464 | 70 | — |
| Desdemona | 62,660 | 0.474 | 60 | — |
| Juliet | 64,360 | 0.493 | 80 | — |
| Portia | 66,100 | 0.513 | 110 | — |
| Rosalind | 69,930 | 0.558 | 60 | — |
| Belinda | 75,260 | 0.624 | 70 | — |
| Puck | 86,010 | 0.762 | 150 | — |
| Miranda | 129,780 | 1.414 | 470 | 1350 |
| Ariel | 191,240 | 2.520 | 1160 | 1660 |
| Umbriel | 265,790 | 4.144 | 1170 | 1510 |
| Titania | 435,840 | 8.706 | 1580 | 1680 |
| Oberon | 582,600 | 13.463 | 1520 | 1580 |

*Ten of Uranus's 15 known satellites were discovered during the 1986 Voyager 2 flyby. These ten tiny moons (listed first in the following table) orbit closer to the planet than the five moderate-sized satellites visible to the Earth-based observers. Of the five larger satellites, Titania and Oberon were discovered by William Herschel in 1789; Ariel and Umbriel were found by the English astronomer William Lassell in 1851; and Miranda was first detected in 1948 by the American astronomer Gerard P. Kuiper.*

*The diameters given for the ten smallest satellites are estimates. The masses of these satellites are unknown, so no densities are given.*

subsurface pockets of gas and creating fissures in the icy surface through which the gas can escape.

Triton's surface temperature is only 37 K (–236°C = –393°F), the lowest of any world yet visited by spacecraft. This temperature is low enough to solidify nitrogen, and indeed the spectrum of sunlight reflected from Triton's surface shows absorption lines due to nitrogen ice as well as methane ice. But Triton is also warm enough to allow some nitrogen to evaporate from the surface, like the steam that rises from ice cubes when you first take them out of the freezer. *Voyager 2* confirmed that Triton has a very thin nitrogen atmosphere with a surface pressure of only $1.6 \times 10^{-5}$ atmosphere, about the same as at an altitude of 100 km above the Earth's surface. Despite its thinness, Triton's atmosphere has noticeable effects. *Voyager 2* saw areas on Triton's surface where dark material has been blown downwind by a steady breeze. It also observed dark material from the geyserlike plumes being carried as far as 150 km by high-altitude winds.

Just as tidal forces presumably played a large role in Triton's history, they also determine its future. Triton's orbit around Neptune is decaying, and the culprit is tidal interaction between the satellite and the planet. Triton raises a tidal bulge on Neptune, just as our Moon distorts the Earth (recall Figure 9-17). In the case of the Earth-Moon system, the gravitational pull of the Earth's tidal bulge causes the Moon to spiral away from the Earth. But because Triton's orbit is retrograde, the tidal bulge on Neptune exerts a force on Triton that makes the satellite slow down rather than speed up. (In Figure 9-17, imagine that the moon is orbiting to the right rather than to the left.) This is causing Triton to spiral gradually in toward Neptune. In approximately 100 million years, Triton will be inside Neptune's Roche limit, and the satellite will eventually be torn to pieces by tidal forces. When this happens, the planet will develop a spectacular ring system—overshadowing by far the present system of narrow rings—as rock fragments gradually spread out along Triton's former orbit.

Prior to *Voyager 2*, only one other satellite was known to orbit Neptune. Nereid, which was first sighted in 1949, is in a prograde orbit. Hence, it orbits Neptune in the direction opposite to Triton. Nereid also has the most eccentric orbit of any satellite in the solar system; its distance from Neptune varies from 1.4 million to 9.7 million kilometers. One possible explanation is that when Triton was captured by Neptune's gravity, the interplay of gravitational forces exerted on Nereid by both Neptune and Triton moved Nereid from a relatively circular orbit (like those of Neptune's other, smaller moons) into its present elliptical one.

## 16-9 Pluto was discovered after a laborious search of the heavens

Speculations about a ninth planet date back to the late 1800s, when a few astronomers suggested that Neptune's orbit was being perturbed by an unknown object. Encouraged by the fame of Adams and Le Verrier, several people set out to become the discoverer of "Planet X." Two Boston gentlemen, William Pickering and Percival Lowell, were prominent in this effort. Modern calculations show that there are, in fact, no unaccounted perturbations of Neptune's orbit. It is thus not surprising that no planet was found at the positions predicted by Pickering, Lowell, and others. Yet the search continued.

Before he died in 1916, Lowell urged that a special widefield camera be constructed to help search for Planet X. After many delays, the camera was finished in 1929 and installed at the Lowell Observatory in Flagstaff, Arizona, where a young astronomer, Clyde W. Tombaugh, had joined the staff to carry on the project. On February 18, 1930, Tombaugh finally discovered the long-sought planet. It was disappointingly faint—a thousand times dimmer than the dimmest object visible with the naked eye and 250 times dimmer than Neptune at opposition—and presented no discernible disk. The planet was named for Pluto, the mythological god of the underworld, whose name has Percival

**f**igure 16-17   R I **V** U X G

**Pluto**  Pluto was discovered in 1930 by searching for a dim, starlike object that moves slowly in relation to the background stars. These two photographs were taken one day apart. Even when its apparent motion on the celestial sphere is fastest, Pluto moves only about 1 arcmin per day relative to the stars. (Lick Observatory)

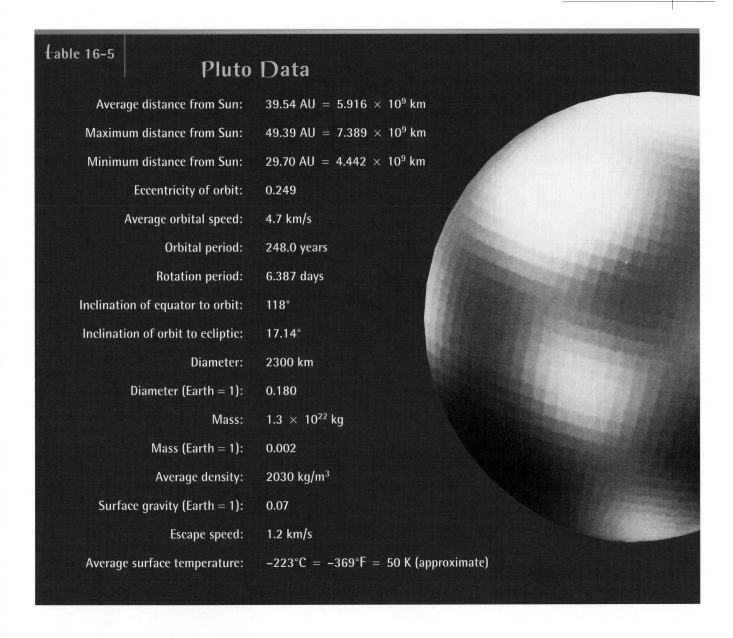

| table 16-5 | Pluto Data | |
| --- | --- | --- |
| Average distance from Sun: | 39.54 AU = $5.916 \times 10^9$ km | |
| Maximum distance from Sun: | 49.39 AU = $7.389 \times 10^9$ km | |
| Minimum distance from Sun: | 29.70 AU = $4.442 \times 10^9$ km | |
| Eccentricity of orbit: | 0.249 | |
| Average orbital speed: | 4.7 km/s | |
| Orbital period: | 248.0 years | |
| Rotation period: | 6.387 days | |
| Inclination of equator to orbit: | 118° | |
| Inclination of orbit to ecliptic: | 17.14° | |
| Diameter: | 2300 km | |
| Diameter (Earth = 1): | 0.180 | |
| Mass: | $1.3 \times 10^{22}$ kg | |
| Mass (Earth = 1): | 0.002 | |
| Average density: | 2030 kg/m$^3$ | |
| Surface gravity (Earth = 1): | 0.07 | |
| Escape speed: | 1.2 km/s | |
| Average surface temperature: | −223°C = −369°F = 50 K (approximate) | |

Lowell's initials as its first two letters. The discovery was publicly announced on March 13, 1930, the 149th anniversary of the discovery of Uranus. Two photographs showing one day's motion of Pluto appear in Figure 16-17.

Pluto's orbit about the Sun is more elliptical and more steeply inclined to the plane of the ecliptic than the orbit of any other planet. In fact, Pluto's orbit is so eccentric that this planet is sometimes closer to the Sun than Neptune. This was the case from 1979 until 1999; indeed, when Pluto was at perihelion in 1989, it was more than $10^8$ km closer to the Sun than was Neptune. Additional data about Pluto are listed in Table 16-5.

Because of Pluto's great distance, the planet subtends only a very small angle of 0.15 arcsec. Hence, it is extraordinarily difficult to resolve any surface features on the planet. But by observing Pluto with the Hubble Space Telescope over the course of a solar day on Pluto (6.3874 Earth days) and using computer image processing, Alan Stern of the Southwest Research Institute and Marc Buie of Lowell Observatory generated the maps of Pluto's surface shown in Figure 16-18. Bright polar ice caps can be seen, as can regions of different reflectivity near the planet's equator. Observations of Pluto's rotation confirm that the planet's rotation axis is tipped by more than 90°, so that Pluto has retrograde rotation like Uranus.

To determine the nature of Pluto's surface, higher resolution images are needed of the sort that can only be obtained with a spacecraft flyby. NASA is presently developing a project called the *Pluto-Kuiper Express* that would send a spacecraft past Pluto by the year 2013. This project has not yet been funded, however. If the *Pluto-Kuiper Express* is delayed too long, it will become increasingly difficult to reach Pluto, as the planet moves further away from the Sun on its elliptical orbit.

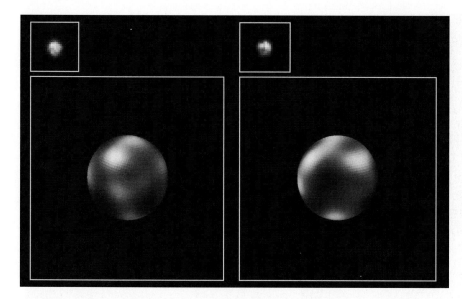

ƒigure 16-18   R I **V** U X G

**Surface Brightness Variations on Pluto**
The small upper images of Pluto show opposite hemispheres of the planet as viewed by the Hubble Space Telescope. The two large images show dark and bright areas on Pluto's opposite hemispheres as determined by computer processing of the Hubble images. The pattern of tiles, each about 160 km on a side, is an artifact of computer image processing. The bright regions at the top and bottom of each large image are thought to be polar ice caps, while the bright regions nearer Pluto's equator may be impact basins (like the lunar maria) or young craters where fresh, reflective, subsurface ice has been exposed. (Alan Stern, Southwest Research Institute; Marc Buie, Lowell Observatory; NASA; and ESA)

| 16-10 | Pluto and its moon Charon may be typical of a thousand icy objects that orbit the Sun at the outskirts of the solar system |
|---|---|

In 1978, while examining some photographs of Pluto, James W. Christy of the U.S. Naval Observatory noticed that the image of the planet on a photographic plate appeared slightly elongated. Pluto seemed to have a lump on one side (Figure 16-19). He promptly examined a number of other photographs of Pluto and found a series of images that showed the lump moving clockwise around Pluto with a period of about 6 days. He concluded that the lump was actually a satellite of Pluto. Christy proposed that the newly discovered moon be christened Charon (pronounced KAR-en), after the mythical boatman who ferried souls across the River Styx to Hades, the domain ruled by Pluto. (Christy also chose the name because of its similarity to Charlene, his wife's name.) The average distance between Charon and Pluto is a scant 19,640 km, less than 5% of the distance between the Earth and our Moon. The best pictures of Pluto and Charon have been obtained using the Hubble Space Telescope (Figure 16-20).

Observations show that Charon's orbital period of 6.3874 days is the same as the rotational period of Pluto *and* the rotational period of Charon. In other words, both Pluto and Charon rotate synchronously with their orbital motion, and so both Pluto and Charon always keep the same face toward each other. As seen from the Charon-facing side of Pluto, Charon neither rises nor sets but instead seems to hover in the sky, as if perpetually suspended above the horizon. Likewise, Pluto never rises or sets as seen from Charon.

Soon after the discovery of Charon, astronomers witnessed an alignment that occurs only once every 124 years.

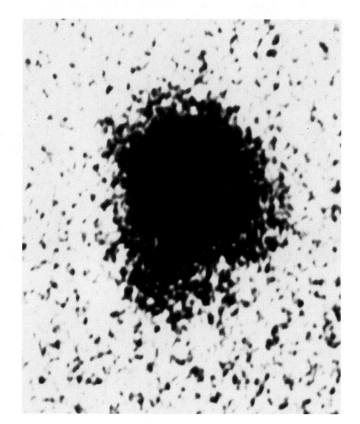

ƒigure 16-19   R I **V** U X G

**The Discovery of Pluto's Moon Charon** Pluto's moon Charon was discovered in 1978 when astronomer James Christy noted a "lump" protruding from the top of this greatly magnified image of the planet. This picture is actually a photographic negative, so black and white are reversed. The small, irregular blobs surrounding Pluto and Charon are part of the photographic emulsion. Fittingly, this image was made using a telescope located just a few kilometers from the Lowell Observatory, where Pluto itself was discovered. (U.S. Naval Observatory)

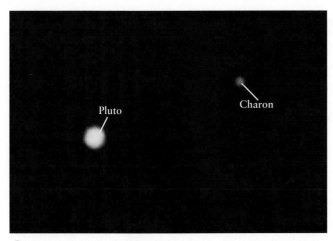

f igure 16-20   R I **V** U X G

**Pluto and Charon—A "Double Planet"** This image of Pluto and its moon Charon was taken by the Hubble Space Telescope. The Pluto-Charon system could be called a "double planet," because these two objects resemble each other in mass and size more closely than any other planet-satellite pair in the solar system. Pluto and Charon are separated by only 19,640 km (12,200 miles). When this image was made, the angle between Pluto and Charon as seen by the Hubble Space Telescope was only 0.9 arcsec. (R. Albrecht, ESA/ESO Space Telescope European Coordinating Facility, and NASA)

From 1985 through 1990, Charon's orbital plane appeared nearly edge-on as seen from the Earth, allowing astronomers to view mutual eclipses of Pluto and its moon. As the bodies passed in front of each other, their combined brightness diminished in ways that revealed their sizes and surface characteristics. Data from the eclipses gave Pluto's diameter as 2300 km and Charon's as 1190 km. For comparison, our Moon's diameter (3476 km) nearly equals the diameters of Pluto and Charon added together. The brightness variations during eclipses are consistent with the Hubble Space Telescope map of Pluto shown in Figure 16-18 and suggest that Charon has a bright southern polar cap.

The average densities of Pluto and Charon, at about 2000 kg/m$^3$, are essentially the same as that of Triton (2070 kg/m$^3$). All three worlds are therefore probably composed of a mixture of rock and ice. These average densities, notably higher than those of either Jupiter or Saturn, support the idea that some process favored the formation of rocklike materials in the outer solar system.

Pluto's spectrum shows absorption lines of methane, nitrogen, and carbon monoxide, which are ices that cover the planet's surface. Stellar occultation measurements have shown that Pluto, like Triton, has a very thin atmosphere. Exposed to a daytime temperature of around 50 K, nitrogen and carbon monoxide ices turn to gas more easily than frozen methane. For this reason, most of Pluto's tenuous atmosphere probably consists of these two gases. Charon's weaker gravity has allowed these three substances to escape, exposing a layer of water ice that covers its surface.

Pluto and Charon are remarkably like each other in mass, size, density, and, quite possibly, chemical composition. Throughout the rest of the solar system, planets are always many times larger and more massive than any of their satellites. The exceptional similarities between Pluto and Charon suggest to some astronomers that this binary system may have formed when Pluto collided with a similar body. Perhaps chunks of matter were stripped from the second body, leaving behind a mass, now called Charon, that was captured into orbit by Pluto's gravity. Alternatively, perhaps Charon was simply captured by Pluto during a close encounter between the two worlds.

For either of these scenarios to be feasible, there must have been many Plutolike objects in the outer regions of the solar system. Some astronomers estimate that there must have been at least a thousand Plutos in order for a collision or close encounter between two of them to have occurred at least once since the solar system formed 4.5 billion years ago.

An overview of the solar system reveals that small objects are vastly more common than large objects. In other words, the number of objects in the solar system increases dramatically with decreasing mass. This trend suggests that for a thousand Plutolike worlds, there also must have been 10 to 50 icy, Earth-sized worlds in the vicinity of Uranus and Neptune. A direct hit by an object with about the mass of the Earth could have knocked Uranus on its side. None of these objects remains near the orbits of Neptune or Uranus today, because gravitational deflections by these two giant planets long ago catapulted all the remaining Plutolike worlds far from the Sun.

In 1992 astronomers David Jewitt and Jane Luu discovered a small object 42 AU from the Sun. Named 1992 QB$_1$, its diameter is estimated to be only 240 km. Like Pluto, 1992 QB$_1$ has a reddish color, possibly because frozen methane has been degraded by eons of radiation exposure. As of this writing 60 such objects have been discovered orbiting beyond Neptune, and it has been estimated that there are of thousands of other tiny worlds yet to be discovered in the outer reaches of the solar system.

Observers may soon have the chance to test this theory by looking for these Plutolike worlds. Attempts to find distant Plutos with ordinary telescopes is frustrating because these tiny, remote bodies do not reflect enough sunlight to be seen. They would, however, be weak sources of infrared radiation. Even though these icy objects probably have surface temperatures of only about 60 K, they are nevertheless warmer than the blank background sky. Highly sensitive, wide-angle infrared telescopes may therefore be able to detect a population of Plutos at distances of about 100 AU from the Sun. Some excellent tools for this search will be the pair of 8-m telescopes Gemini North and Gemini South (one on Mauna Kea in Hawaii, the other on Cerro Pachón, Chile) and the 8-m Subaru (the Japan National Large Telescope on Mauna Kea). All three of these will become operational around 2000, and will be able to operate in the infrared as well as in the visible. If the search is successful, astronomers will have to account for a new class of objects in their theories about the formation of the solar system.

## Key Words

## Key Ideas

**Discovery of the Outer Planets:** Uranus was discovered by chance, while Neptune was discovered at a location predicted by applying Newtonian mechanics. Pluto was discovered after a long search. If another planet exists beyond Pluto, it is either very small or very far away.

**Atmospheres of Uranus and Neptune:** Both Uranus and Neptune have atmospheres composed primarily of hydrogen and helium, with about 2% methane.

• Absorption of red light by methane gives Uranus and Neptune their greenish-blue color.

• No white ammonia clouds are seen on Uranus or Neptune, presumably because the low temperatures have caused ammonia to precipitate into the interiors of the planets. All of the clouds are composed of methane.

• Much more cloud activity is seen on Neptune than on Uranus. This is because Uranus lacks a substantial internal heat source.

**Interiors and Magnetic Fields of Uranus and Neptune:** Both Uranus and Neptune may have a rocky core surrounded by a mantle of water and ammonia. Electric currents in these mantles may generate the magnetic fields of the planets.

• Uranus's magnetic axis is inclined by 59° from its axis of rotation, while Neptune's is inclined by 47°. The magnetic and rotational axes of all the other planets are more nearly parallel. The magnetic fields of Uranus and Neptune are also offset from the centers of the planets.

**Uranus's Unusual Rotation:** Uranus's axis of rotation lies nearly in the plane of its orbit, producing greatly exaggerated seasonal changes on the planet.

• The unusual orientation of Uranus's rotational and magnetic axes suggest that Uranus may have been knocked on its side by a collision with a planetlike object early in the history of our solar system.

**Ring Systems of Uranus and Neptune:** Uranus and Neptune are both surrounded by systems of thin, dark rings. The low reflectivity of the ring particles may be due to radiation-damaged methane ice.

**Satellites of Uranus and Neptune:** Uranus has five satellites similar to the moderate-sized moons of Saturn; ten more small Uranian satellites were discovered during the *Voyager 2* flyby. Neptune has eight satellites, one of which (Triton) is comparable in size to our Moon or the Galilean satellites of Jupiter.

• Triton has a young, icy surface indicative of tectonic activity. The energy for this activity may have been provided by tidal heating that occurred when Triton was captured by Neptune's gravity into a retrograde orbit.

• Triton has a tenuous nitrogen atmosphere.

**Pluto and Charon:** Pluto and its moon Charon move together in a highly elliptical orbit steeply inclined to the plane of the ecliptic. They are the only worlds in the solar system not yet visited by spacecraft.

• Pluto, Charon, and Triton may be typical of a thousand small, icy worlds that orbited far from the Sun shortly after the solar system was formed.

## Review Questions

1. Could astronomers in antiquity have seen Uranus? If so, why was it not recognized as a planet?

2. Why do you suppose that the discovery of Neptune is rated as one of the great triumphs of science, whereas the discoveries of Uranus and Pluto are not?

3. Why do you suppose the tilt of Uranus's rotation axis was deduced from the orbits of its satellites and not by observing the rotation of the planet itself?

4. Describe the seasons on Uranus. Why are the Uranian seasons different from those on any other planet?

5. Why are Uranus and Neptune distinctly blue-green in color, while Jupiter or Saturn are not?

6. Why are fewer white clouds seen on Uranus and Neptune than on Jupiter and Saturn?

7. How does the orientation of Uranus and Neptune's magnetic axes differ from those of other planets?

8. Briefly describe the evidence supporting the idea that Uranus was struck by a large planetlike object several billion years ago.

9. Compare the rings that surround Jupiter, Saturn, Uranus, and Neptune. Briefly discuss their similarities and differences.

10. As *Voyager 2* flew past Uranus, it only produced images of the southern hemispheres of the planet's satellites. Why do you suppose this was?

11. If you were floating in a balloon in Neptune's upper atmosphere, in what part of the sky would you see Triton rise? Explain your reasoning.

12. Briefly describe the evidence supporting the idea that Triton was captured by Neptune.

13. Why is it reasonable to suppose that Neptune will someday be surrounded by a broad system of rings, perhaps similar to those that surround Saturn?

14. How can astronomers distinguish a faint solar system object like Pluto from background stars within the same field of view?

15. Describe the circumstantial evidence supporting the idea that Pluto is one of thousands of similar icy worlds that once occupied the outer regions of the solar system.

## Advanced Questions

### Problem-solving tips and tools:

The small-angle formula is given in Box 1-2. Question 16 involves some geometry, and you will need to remember that the circumference of a circle is $\pi = 3.14159 \ldots$ times its diameter. Newton's formula for the gravitational force between two objects can be found in Section 4-7. Wien's law for blackbody radiation is described in Section 5-4. The transparency of the Earth's atmosphere to various wavelengths of light is discussed in Section 6-7.

16. Using the data given in Table 16-1, estimate the maximum duration of a stellar occultation by Uranus. (*Hint:* You will need to calculate how fast Uranus is moving in its orbit, then figure out how long it takes the planet to travel a distance equal to its own diameter. But you must also explain why these calculations are relevant to this question! To keep things simple, you can ignore the fact that the Earth moves during the occultation.)

17. At what planetary configuration is the gravitational force of Neptune on Uranus at a maximum? For this configuration, calculate the gravitational force exerted by the Sun on Uranus and by Neptune on Uranus. Then calculate the fraction by which the sunward gravitational pull on Uranus is reduced by Neptune at that configuration. Based on your calculations, do you expect that Neptune has a relatively large or relatively small effect on Uranus's orbit?

18. Suppose you were standing on Pluto. Describe the motions of Charon relative to the Sun, the stars, and your own horizon. Would you ever be able to see a total eclipse of the Sun? (*Hint:* You will need to calculate the angles subtended by Charon and by the Sun as seen by an observer on Pluto.) Under what circumstances would you *never* see Charon?

19. It is thought that Pluto's tenuous atmosphere may eventually disappear as it moves toward aphelion (which it will reach in 2113), then reappear as it again moves toward perihelion. Why should this be?

20. The brightness of sunlight is inversely proportional to the square of the distance from the Sun. For example, at a distance of 4 AU from the Sun, sunlight is only $(1/4)^2 = 1/16 = 0.0625$ as bright as at 1 AU. Compared to the brightness of sunlight on the Earth, what is its brightness (**a**) on Pluto at perihelion and (**b**) on Pluto at aphelion? (**c**) How much brighter is it on Pluto at perihelion compared to aphelion? (Even this brightness is quite low. Noon on Pluto is about as dim as it is on the Earth a half hour after sunset on a moonless night.)

21. Show that the ratio of the orbital periods of Neptune and Pluto is very close to 2:3. (This ratio is thought to result from gravitational interactions between Neptune and Pluto. These interactions keep the orbits of Neptune and Pluto from crossing.)

22. Suppose you wanted to search for planets beyond Pluto. Why might it be advantageous to do your observations at infrared rather than visible wavelengths? (Use Wien's law to calculate the wavelength range best suited for your search.) Could such observations be done at an Earth-based observatory? Explain.

23. If Earth-based telescopes can resolve angles down to 0.25 arcsec, how large could an object be at Pluto's average distance from the Sun and still not present a resolvable disk?

24. The observations of Pluto shown in Figure 16-18 were made using blue and ultraviolet light. What advantages does this have over observations made with red or infrared light?

25. Consult a list of small objects beyond Pluto on the World Wide Web (**http://cfa-www.harvard.edu/cfa/ps/lists/TNOs.html**) to learn about the current status of 1992 $QB_1$ and similar objects. What kinds of orbits do these objects have? How do these orbits compare with that of Pluto? What are the largest and smallest objects of this sort that have so far been found, and how large are they?

## Discussion Questions

26. Discuss the evidence presented by the outer planets which suggests that catastrophic impacts of planetlike objects occurred during the early history of our solar system.

27. Some scientists are discussing the possibility of placing spacecraft in orbit about Uranus and Neptune early in the twenty-first century. What kinds of data should be collected, and what questions would you like to see answered by these missions?

28. If Triton had been formed along with Neptune rather than having been captured, would you expect it to be in a prograde or retrograde orbit? Would you expect the satellite to show signs of tectonic activity? Explain your answers.

29. Would you expect the surfaces of Pluto and Charon to be heavily cratered? Explain why or why not.

**30.** The discovery image of Charon (Figure 16-19) was made by an astronomer at the United States Naval Observatory. Why do you suppose the U.S. Navy carries out work in astronomy? You may want to visit their web site to find out (**http://www.usno.navy.mil/**).

## Observing Projects

**Observing tips and tools:**

> During the period 1998–2004, Uranus and Neptune will both be opposition in late July and early August, and Pluto will be at opposition in May and June. You can find Uranus and Neptune with binoculars if you know where to look (a good star chart is essential), but Pluto is so dim that it can be a challenge to spot even with a 25-cm (10-inch) telescope. Each year, star charts that enable you to find these planets are printed in the issue of *Sky & Telescope* for the month in which each planet is first visible in the nighttime sky. You can also consult the web sites for *Sky & Telescope* and *Astronomy* magazines (**http://www.skypub.com/** and **http://www.astronomy. com/**, respectively).

**31.** Make arrangements to view Uranus through a telescope. The planet is best seen at opposition (that is, in July or August during the period 1998–2004). Use a star chart at the telescope to find the planet. Are you certain that you have found Uranus? Can you see a disk? What is its color?

**32.** If you have access to a large telescope, make arrangements to view Neptune. Like Uranus, during the years 1998–2004, Neptune is best seen in July and August and is most easily found using the star chart from *Sky & Telescope* referred to above. Can you see a disk? What is its color?

**33.** If you have access to a large telescope (at least 25 cm in diameter), make arrangements to view Pluto. Using the star chart from *Sky & Telescope* referred to above, view the part of the sky where Pluto is expected to be seen and make a careful sketch of all of the stars that you see. Repeat this process on a later night. Can you identify the "star" that has moved?

## Where to Learn More

*Magazine articles*

Bennett, J. "The Discovery of Uranus." *Sky & Telescope*, March 1981. This fascinating historical article describes Herschel's discovery of Uranus.

Binzel, R. "Pluto." *Scientific American*, June 1990. This superb article summarizes our current understanding of Pluto and Charon.

Cuzzi, J., and Esposito, L. "The Rings of Uranus." *Scientific American*, July 1987. The authors of this article argue that Uranus's rings are actually a fleeting phenomenon.

Elliot, J., and others. "Discovering the Rings of Uranus." *Sky & Telescope*, June 1977. The team of astronomers who discovered Uranus's rings describe their airborne observations.

Harrington, R., and Harrington, B. "The Discovery of Pluto's Moon." *Mercury*, January/February 1979. This article gives authoritative insights into the discovery of Pluto's moon.

Ingersoll, A. P. "Uranus." *Scientific American*, January 1987. This superb article summarizes the results of the *Voyager 2* flyby.

Johnson, T., and others. "The Moons of Uranus." *Scientific American*, April 1987. This beautifully illustrated article asserts that Uranus's moons had a geologically vigorous history.

Kinoshita, J. "Neptune." *Scientific American*, November 1989. Published a few months after the *Voyager 2* flyby, this article is illustrated with excellent pictures of Neptune and Triton.

Lumine, J. I. "Neptune at 150." *Sky & Telescope*, September 1996. This article summarizes current thinking about Neptune's internal structure, rings, and satellites and discusses Neptune's place in the history of the solar system.

Moore, P. "The Discovery of Neptune." *Mercury*, July/August, 1989. A prolific popularizer of astronomy gives a detailed history of how Neptune was found. A shorter article by this author on the same subject is "The Hunt for Neptune," *Sky & Telescope*, September 1996.

Tombaugh, C. "The Search for the Ninth Planet: Pluto." *Mercury*, January/February 1979. The discoverer of Pluto describes his historic observations.

Ⓦ *World Wide Web*

A tremendous number of images and up-to-date information on the outer planets and their satellites can be found at the web sites "The Nine Planets" (**http://www.seds.org/nineplanets/ nineplanets/**), "Welcome to the Planets" (**http://pds.jpl.nasa. gov/planets/welcome.htm**), and "Views of the Solar System" (**http://www.hawastsoc.org/solar/homepage.htm**).

Information on the ring systems of Uranus and Neptune can be found at NASA's Ames Research Laboratory (**http://ringside.arc.nasa.gov/**).

Marc Buie at Lowell Observatory, whose research led to the maps of Pluto shown in Figure 16-18, maintains an excellent web site concerning all aspects of Pluto and Charon (**http://www.lowell.edu/users/buie/pluto/ pluto.html**).

NASA's Jet Propulsion Laboratory has a web site for the *Pluto-Kuiper Express* project (**http://www.jpl.nasa.gov/ pluto/pkexprss.htm**), which includes Clyde Tombaugh's reminiscences about his discovery of Pluto (**http://www.jpl. nasa.gov/pluto/9thplant.htm**).

You can keep abreast of the latest discoveries of small bodies beyond Neptune at a site maintained by the Harvard-Smithsonian Center for Astrophysics (**http://cfa-www.harvard.edu/cfa/ps/lists/TNOs.html**).

# Vagabonds of the Solar System

R I **V** U X G

Comet Hale–Bopp  This image shows Comet Hale–Bopp on March 8, 1997, when it was 1.39 AU from Earth and 1.00 AU from the Sun. The heat of the Sun causes the comet's nucleus to evaporate, forming a white tail of dust particles and a bluish tail of ionized atoms and molecules. The ion tail was about 25 times the angular size of the full moon. The red object to the right is the North America Nebula, a star–forming region some 1500 light–years beyond the solar system. Alan Hale and Thomas Bopp discovered this comet independently on the same night in July 1995. (Courtesy of Tony and Daphne Hallas Astrophotos)

*In this chapter you will find the answers to the following questions:*

17-1  How and why were the asteroids first discovered?

17-2  Why didn't the asteroids coalesce to form a single planet?

17-3  What do asteroids look like?

17-4  How might an asteroid have caused the extinction of the dinosaurs?

17-5  What is the difference between meteoroids, meteors, and meteorites?

17-6  What do meteorites tell us about the way in which the solar system formed?

17-7  Why do comets have tails?

17-8  Where do comets come from?

17-9  What is the connection between comets and meteor showers?

On March 28, 1802, Heinrich Olbers discovered another faint, starlike object that moved against the background stars. He called it Pallas, after the Greek goddess of wisdom. Like Ceres, Pallas orbits the Sun every 4.6 years at an average distance of 2.77 AU, but its orbit is more steeply inclined from the plane of the ecliptic and is somewhat more eccentric. Pallas is even smaller and dimmer than Ceres, with an estimated diameter of only 522 km. Obviously, Pallas was also not the missing planet.

Did the missing planet even exist? Some astronomers speculated that perhaps there had once been such a full-size planet, but it somehow broke apart or exploded to produce a population of asteroids orbiting between Mars and Jupiter. The search was on to discover this population. Two more such asteroids were discovered in the next few years, Juno in 1804 and Vesta in 1807. Several hundred more were discovered beginning in the mid-1800s, by which time telescopic equipment and techniques had improved.

The next major breakthrough came in 1891, when German astronomer Max Wolf began using photographic techniques to search for asteroids. Before this, asteroids had to be painstakingly discovered by scrutinizing the skies for faint, uncharted, starlike objects whose positions move slightly from one night to the next. With photography, an astronomer simply aims a camera-equipped telescope at the stars and takes a long exposure. If an asteroid happens to be in the field of view, it leaves a distinctive, blurred trail on the photographic plate, because of its movement along its orbit during the long exposure (Figure 17-2). Using this technique, Wolf alone discovered 228 asteroids.

Today, dozens of new asteroids are discovered every month, mostly by amateur astronomers. After the discoverer reports a new find to the Minor Planet Center of the Smithsonian Astrophysical Observatory, the new asteroid is first given a provisional designation. For example, the designation 1980 JE means that this was the fifth ("E") asteroid discovered during the second half of May (the tenth half-month of the year, "J") 1980. If the asteroid is located again on at least four succeeding oppositions (a process than can take decades), the asteroid is assigned an official sequential number (Ceres is 1, Pallas is 2, and so forth), and the discoverer is given the privilege of suggesting a name for the asteroid. This suggestion must then be approved by the International Astronomical Union. For example, the asteroid 1980 JE was officially named 3834 Zappafrank (after the American musician Frank Zappa) in 1994.

Ceres—or, in modern nomenclature, 1 Ceres—is unquestionably the largest asteroid. With a diameter of about 900 km, 1 Ceres accounts for about 30% of the mass of all the asteroids combined. Only three asteroids—1 Ceres, 2 Pallas, and 4 Vesta—have diameters greater than 300 km. Thirty other asteroids have diameters between 200 and 300 km, and 200 more are bigger than 100 km across.

Astronomers estimate that approximately 100,000 asteroids are bright enough to appear on Earth-based photographs. The vast majority are smaller than 1 km across. Like Ceres, Pallas, Vesta, and Juno, most asteroids orbit the Sun at distances between 2 and 3.5 AU. This region of our solar system

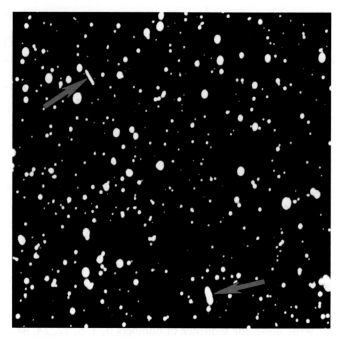

**f**igure 17-2   R I **V** U X G

**Two Asteroids** Telescopes used for photographing the sky have motor drives that allow them to follow the apparent motion of the stars across the sky. Because asteroids orbit the Sun, their positions change with respect to the stars. Thus, asteroids can be detected by their blurred trails on time exposure photographs of the stars. The images of two asteroids are seen in this picture. Astronomers sometimes find asteroids accidentally while photographing various portions of the sky for other purposes. (Yerkes Observatory)

between the orbits of Mars and Jupiter is called the **asteroid belt** (Figure 17-3). The asteroids whose orbits lie entirely within this region are called **belt asteroids**.

CAUTION! You may wonder how spacecraft such as *Voyager 1*, *Voyager 2*, and *Galileo* were able to make it to Jupiter, because they all had to pass through the asteroid belt. Wouldn't these spacecraft have been likely to run into an asteroid? Happily, the answer is no. While it is true that there are more than $10^5$ asteroids, they are spread over a belt with a total area (as seen from above the plane of the ecliptic) of about $10^{17}$ square kilometers—about $10^8$ times greater than the entire surface area of Earth. Hence, the average distance between asteroids in the ecliptic plane is about $10^6$ kilometers, about twice the distance between the Earth and Moon. Furthermore, many asteroids have orbits that are tilted out of the ecliptic. The *Galileo* spacecraft did pass within a few thousand kilometers of two asteroids, but these passes were intentional and required that the spacecraft be carefully aimed. Rather than being hazards to space navigation, as they are sometimes depicted in science fiction movies, asteroids are so small and so widely spaced that you would have to make a special effort to hit one!

What of the nineteenth-century notion of a missing planet? This idea has long since been discarded, because the

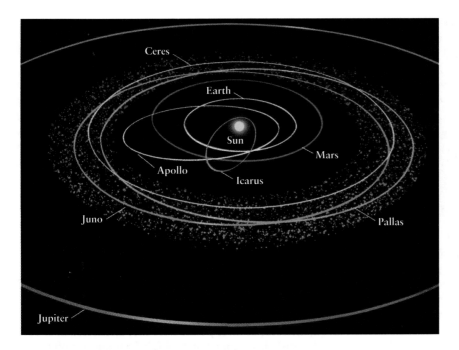

figure 17-3

**The Asteroid Belt** Most asteroids orbit the Sun in a belt about 1.5 AU wide between the orbits of Mars and Jupiter. The large asteroids 1 Ceres, 2 Pallas, and 3 Juno all have roughly circular orbits that lie within this belt. By contrast, some asteroids, such as 1862 Apollo and 1566 Icarus, have eccentric orbits that carry them inside the orbit of Earth.

combined matter of all the asteroids (including an estimate for those not yet officially known) would produce an object barely 1500 km in diameter. This is considerably smaller than anything that could be considered a missing planet. It seems more reasonable to suppose that asteroids are debris left over from the formation of the solar system.

## 17-2 | Jupiter's gravity helped shape the asteroid belt

Where did the asteroid belt come from? Why did asteroids, rather than a planet, form between Mars and Jupiter? Important insights into these questions are provided by supercomputer simulations of the formation of the terrestrial planets, like the one depicted in Figure 7-18. A typical simulation starts off with a billion ($10^9$) or so planetesimals, each with a mass of $10^{15}$ kg or more, so that their combined mass equals that of the terrestrial planets. The computer then follows these planetesimals as they collide and coalesce to form the planets. By selecting different speeds and positions for the planetesimals at the start of the simulation, a scientist can study various ways in which the inner solar system might have turned out.

If the effects of Jupiter's gravity are not included in a simulation, an Earth-sized planet usually forms in the asteroid belt, giving us five terrestrial planets instead of four. When Jupiter's gravity is added, however, this fifth terrestrial planet is less likely to survive. Jupiter's strong pull disrupts the orbits of planetesimals in the asteroid belt, ejecting most of them from the solar system altogether. As a result, the asteroid belt becomes depleted before a planet has a chance to form. The planetesimals that remained in the asteroid belt became the asteroids that we see today. Because the surviving asteroids make up only a fraction of the total mass that originally orbited the Sun within the asteroid belt, astronomers say that Jupiter "cleared" the asteroid belt as the solar system formed.

Jupiter's gravitational influence, however, probably cannot explain why asteroid orbits have a wide variety of semimajor axes, eccentricities, and inclinations to the ecliptic, as shown in Figure 17-3. Even with the influence of Jupiter, collisions between planetesimals in the solar nebula would have tended to leave the asteroids in roughly circular orbits close to the ecliptic plane, like the orbits of the planets. Therefore, something else must have "stirred up" the asteroids to put them into their current state. Recent analyses suggest that one or more Mars-sized objects—that is, terrestrial planets intermediate in size between the Earth and the Moon—did in fact form within the asteroid belt. Planetesimals that passed within close range of these Mars-sized bodies would have experienced strong gravitational forces, and these forces could have deflected the planetesimals into the eccentric or inclined orbits that many asteroids follow today. These types of deflections would also have helped clear out the asteroid belt by pushing planetesimals into orbits that crossed Jupiter's orbit. Once a planetesimal wandered close to Jupiter, it could be ejected from the solar system by the giant planet's gravitational force.

In this scheme, the Mars-sized objects disappeared when gravitational interactions from Jupiter either accelerated them out of the solar system entirely or caused them to slow down and fall into the Sun. A Mars-sized object colliding with the Earth is thought to have produced the Moon (see Section 9-6, especially Figure 9-22). Perhaps this object was an "escapee" from the asteroid belt.

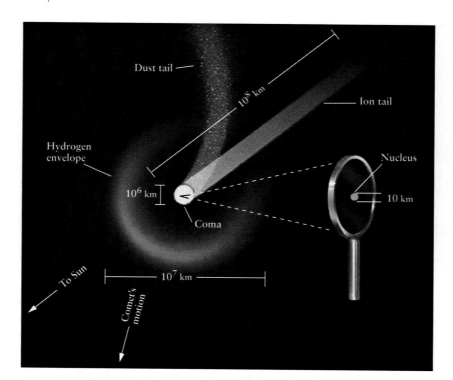

figure 17-20

**The Structure of a Comet** The solid part of a comet, the nucleus, is about 10 km in diameter. The coma typically measures 1 million kilometers in diameter, and the hydrogen envelope usually spans 10 million kilometers. A comet's tail can be as long as 1 AU.

Several jets of dust particles, about 15 km in length, protruded from bright areas on Halley's nucleus. These seem to be active only when exposed to the Sun. These bright areas, which probably cover about 10% of the surface of the nucleus, are presumably places where the dark layer covering the nucleus is particularly thin. When exposed to the Sun, these areas evaporate rapidly, producing the jets. When the nucleus's rotation brings the bright areas into darkness, away from the Sun, the jets shut off.

A part of a comet that is not visible to the human eye is the **hydrogen envelope**, a huge sphere of tenuous gas surrounding the nucleus. This hydrogen comes from water molecules ($H_2O$) that escape from the comet's evaporating ice and then break apart when they absorb ultraviolet photons from the Sun. The hydrogen atoms also absorb solar ultraviolet photons, which excite the atoms from the ground state into an excited state (as described in Section 5-8). When the atoms return to the ground state, they emit ultraviolet photons, and these can be detected using a camera above the Earth's atmosphere. (As we discussed in Section 6-7, the Earth's atmosphere is largely opaque to ultraviolet light.) Figure 17-22 shows two views of Comet Kohoutek, one as it appeared to the Earth-based observers and one as photographed by an ultraviolet camera aboard a rocket. This ultraviolet photograph showed astronomers the enormous extent of the hydrogen envelope, which can span 10 million kilometers.

As the diagram of a comet in Figure 17-20 suggests, comets' tails always point away from the Sun. This is true regardless of the direction of the comet's motion (Figure 17-23). The implication that something from the Sun was "blowing" the comet's gases radially outward led Ludwig Biermann to predict the existence of the solar wind a full

figure 17-21  R I **V** U X G

**The Nucleus of Comet Halley** In March 1986 five spacecraft passed near Comet Halley, which puts in an appearance every 76 years. This close-up picture, taken by a camera on board the European Space Agency spacecraft *Giotto*, shows the potato-shaped nucleus of the comet. The nucleus is darker than coal and measures 15 km in the longest dimension and about 8 km in the shortest. The Sun illuminates the nucleus from the left. Two bright jets of dust extend 15 km from the nucleus toward the Sun, suggesting that most of the vaporization takes place on the sunlit (left) side. (Max Planck Institut für Aeronomie)

R I **V** U X G

R I V **U** X G

**f**igure 17-22 **Comet Kohoutek and Its Hydrogen Envelope** These two photographs of Comet Kohoutek, which made an appearance in 1974, are reproduced to the same scale. The picture on the left shows the comet in visible light. The ultraviolet view on the right reveals a huge hydrogen cloud surrounding the comet's head. (Johns Hopkins University and Naval Research Laboratory)

decade before it was actually discovered in 1962 by instruments on the *Mariner 2* spacecraft.

In fact, as Figure 17-20 shows, the Sun usually produces two comet tails—an **ion tail** and a **dust tail.** (These two tails can be seen clearly in the photograph of Comet Hale-Bopp that opens this chapter.) Ionized atoms and molecules—that is, atoms and molecules missing one or more electrons—are swept directly away from the Sun by the solar wind to form the relatively straight ion tail. The distinct blue color of the ion tail is caused by emissions from methane ($CH_4$). The dust tail is formed when photons of light strike dust particles freed from the evaporating nucleus. Light exerts a pressure on any object that absorbs or reflects it. This pressure, called **radiation pressure,** is quite weak but is strong enough to make fine-grained dust particles in a comet's coma drift away from the comet, thus producing a dust tail. The particles in the dust tail respond to the Sun's gravity as well as to radiation pressure, and the combination of these forces gives the dust tail its curved shape.

The ion tail can exhibit a dramatic structure that changes from night to night (Figure 17-24). The dust tail, by contrast,

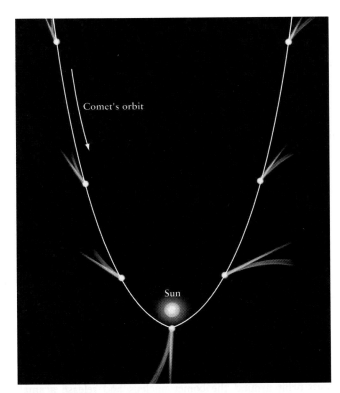

**f**igure 17-23

**The Orbit and Tail of a Comet** The solar wind and radiation pressure from sunlight blow a comet's dust particles and ionized atoms away from the Sun. Consequently, a comet's tail points generally away from the Sun. In particular, the tail does *not* always stream behind the nucleus. At the upper right of this figure, the comet is moving upward along its orbit and is literally chasing its own tail.

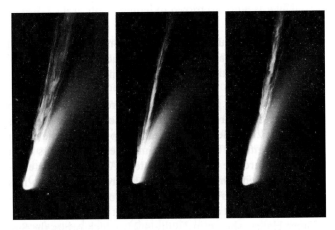

**f**igure 17-24 R I **V** U X G

**The Two Tails of Comet Mrkos** Comet Mrkos dominated the evening sky during August 1957. These three views, taken at two-day intervals, show dramatic changes in the comet's ion tail. In contrast, the slightly curved dust tail remained fuzzy and featureless. (Palomar Observatory)

## 17-8 | Comets originate either from a belt beyond Pluto or from a vast cloud in near–interstellar space

Why do comets have such elliptical orbits, while asteroids have relatively circular ones? The reason cannot have anything to do with comets being icy and asteroids being rocky, because the gravitational forces that shape an object's orbit do not depend on its chemical composition, only on its mass. Instead, the answer has to do with how the gravitational pulls of the planets helped to shape the orbits of asteroids and comets in different ways.

As we saw in Section 17-2, the asteroids formed in the space between Mars and Jupiter. The relatively small terrestrial planets, with their correspondingly small masses, exert only weak gravitational forces on asteroids. Hence, most asteroids, like the planets, remained in fairly circular orbits. (The exceptions are those few asteroids that wandered too close to Mars, whose gravitational force deflected these asteroids into becoming the near-Earth objects described in Section 17-3.) Jupiter's gravity acted primarily to bunch the asteroids away from the Kirkwood gaps.

Comets, by contrast, first formed in the outer solar system at the distances of Uranus and Neptune, where temperatures were low enough to permit ices to condense into chunks several kilometers across. Because Uranus and Neptune are much more massive than the terrestrial planets, their gravitational forces had a major effect, deflecting the comets into a wide variety of directions.

There are now two large reservoirs of comets, the Kuiper belt and the Oort cloud. The **Kuiper belt** lies in the plane of the ecliptic and extends from around the orbit of Pluto to about 500 AU from the Sun (Figure 17-27). The existence of this belt was first proposed by American astronomer Gerard Kuiper in 1951; the first Kuiper belt object to be discovered was seen in 1992. (This was the object 1992 QB$_1$, described in Section 16-10.) Dozens of these objects have now been observed (Figure 17-28). It is suspected that there could be as many as 35,000 objects in the Kuiper belt with diameters of 100 km or greater and many more smaller objects. Like other planetesimals, most Kuiper belt objects probably formed at the same distance from the Sun as we find them today.

Most of the **short-period comets**, which orbit the Sun in fewer than 200 years and return again and again to the inner solar system at predictable intervals (like Comet Halley), are thought to come from the Kuiper belt. Computer simulations

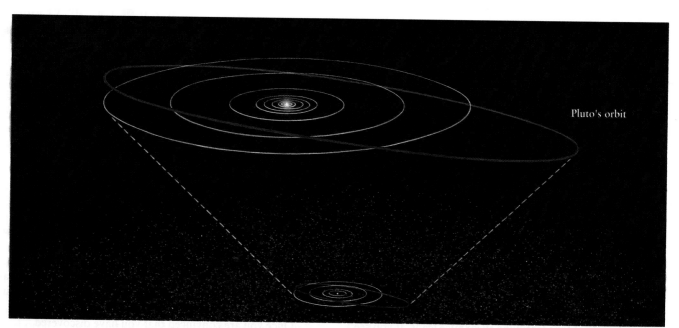

Pluto's orbit

## ƒigure 17-27

**The Kuiper Belt** The Kuiper belt of comets lies in the ecliptic, like the asteroid belt, and extends from around the orbit of Pluto to about 500 AU from the Sun. Gravitational perturbations from Neptune occasionally launch a Kuiper belt comet into a highly elliptical orbit that takes the comet close to the Sun, where it produces a visible tail. The Kuiper belt is thought to be the source of most of the comets that we see on the Earth. A few of these perturbed icy objects are thought to have been captured by the gravitational pulls of Jupiter and Saturn, becoming the small outer satellites of these planets. Some astronomers also regard Pluto and its moon Charon as Kuiper belt objects.

show that gravitational perturbations from the outer planets can deflect these comets from their roughly circular orbits within the Kuiper belt into more elongated orbits, such as that shown in Figure 17-26.

The majority of comets discovered each year are **long-period comets,** which take roughly 1 to 30 million years to complete one orbit of the Sun. These comets travel along extremely elongated orbits and consequently spend most of their time at distances of roughly $10^4$ to $10^5$ AU from the Sun, or about one-fifth of the way to the nearest star. Because these orbits extend far beyond the Kuiper belt, it is thought that these long-period comets come from a reservoir that extends from the Kuiper belt to some 50,000 AU from the Sun. This reservoir, first hypothesized by Dutch astronomer Jan Oort in 1950 and thus called the **Oort cloud,** has a striking difference from the Kuiper belt and from all other parts of the solar system: It does not lie in the ecliptic but is rather a spherical distribution centered on the Sun. This helps to explain why many long-period comets have orbits that are steeply inclined to the plane of the ecliptic.

Because astronomers discover long-period comets at the rate of about one per month, it is reasonable to suppose that there is an enormous population of comets in the Oort cloud. Estimates of the number of "dirty snowballs" in the Oort cloud range as high as 12 billion. Only such a large reservoir of comet nuclei would explain why we see so many long-period comets, even though each one takes several million years to travel once around its orbit. Because the Oort cloud is so distant, it has not yet been possible to detect objects in the Oort cloud directly.

The Oort cloud was probably created 4.5 billion years ago from numerous icy planetesimals that orbited the Sun in the vicinity of the newly formed Jovian planets. During near-collisions with the giant planets, many of these chunks of ice and dust were catapulted by gravity into highly elliptical orbits, in much the same way that *Pioneer* and *Voyager* spacecraft were flung far from the Sun during their flybys of the Jovian planets. Gravitational perturbations from nearby stars tilted the planes of the orbits in all directions, giving the Oort cloud its spherical shape.

The distinction between long-period and short-period comets can be blurred by the effects of gravitational perturbations. During a return trip toward the Sun, an encounter with a Jovian planet may force a long-period comet into a short-period orbit (Figure 17-29). There are several comets that have been perturbed in this way into orbits that always remain within the inner solar system. None of these has a prominent tail, however.

a  b

## figure 17-28   R I **V** U X G

**A Kuiper Belt Object** These 1993 images were the first to show 1993 SC, the sixth Kuiper belt object to be discovered. These two images of 1993 SC (shown between the white lines) were taken 4.6 hours apart, during which time the object moved from **(a)** to **(b)** against the background stars. The colors in this image are false: Blue represents dark areas, while green, yellow, red, and white denote successively brighter areas. (Alan Fitzsimmons, Queen's University of Belfast)

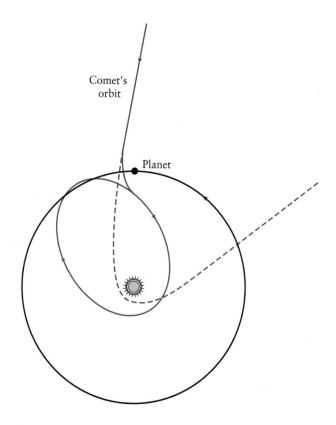

Comet's orbit

Planet

## figure 17-29

**Transforming a Long-Period Comet into a Short-Period Comet** The gravitational force of a planet can cause major changes in a comet's orbit. Initially on a highly elliptical orbit (dashed line), comets are sometimes deflected into less elliptical paths. In some cases these comets end up with orbits that keep them within the inner solar system.

## 17-9 Comets eventually break apart, and their fragments give rise to meteor showers

Because they evaporate away part of their mass each time they pass near the Sun, comets cannot last forever. A typical comet may lose about 0.5% to 1% of its ice each time it passes near the Sun. Hence, the ice completely vaporizes after about 100 or 200 perihelion passages, leaving only a swarm of dust and pebbles. On rare occasions, a comet's nucleus has been observed to fragment (Figure 17-30).

Comet Shoemaker-Levy 9, described in Section 13-4, broke apart for a different reason. As the comet swung by Jupiter in July 1992, the giant planet's tidal forces tore the nucleus into more than twenty fragments (Figure 13-13). Jupiter's gravity then deflected the trajectories of these fragments so that they plummeted into the planet's atmosphere (see Figures 13-14 and 13-15).

As a comet's nucleus evaporates, residual dust and rock fragments form a **meteoritic swarm,** a loose collection of debris that continues to circle the Sun along the comet's orbit (Figure 17-31). If the Earth's orbit happens to pass through this swarm, a **meteor shower** is seen as the dust particles strike the Earth's upper atmosphere.

Nearly a dozen meteor showers can be seen each year (see Box 17-3). Note that the **radiants** for these showers (that is, the places among the stars from which the meteors appear to come) are not confined to the constellations of the zodiac, which means that the swarms to do not all orbit in the plane of the ecliptic. Meteor showers can come from various parts of the sky, because the orbits of their parent comets were inclined at random angles to the plane of the ecliptic, as shown in Figure 17-31.

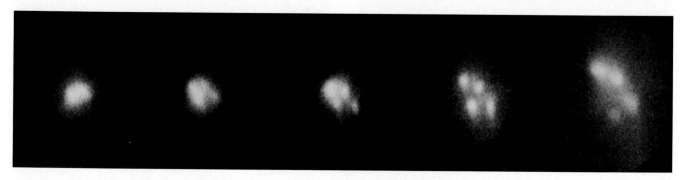

## figure 17-30  R I **V** U X G

**The Fragmentation of Comet West** Shortly after passing near the Sun in 1976, the nucleus of Comet West broke into four pieces. This series of five photographs clearly shows that disintegration. The remaining fragments of dust and rock continue to orbit the Sun along the same trajectory that the comet followed prior to its breakup. (New Mexico State University Observatory)

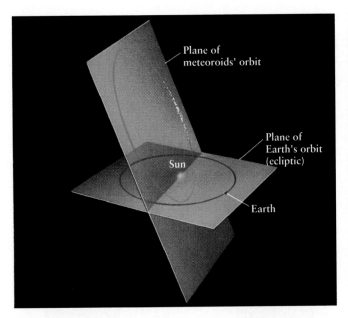

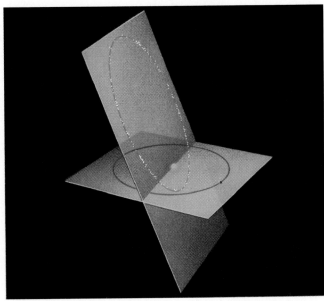

## figure 17-31

**Meteoritic Swarms** Rock fragments and dust from "burned out" comets continue to circle the Sun. **(a)** If the comet is only recently extinct, the particles will still be tightly concentrated in a compact swarm. The most spectacular meteor showers occur when the Earth happens to pass through such a swarm. **(b)** Over the ages, the particles gradually spread out along the old comet's elliptical orbit. This configuration produces the most predictable meteor showers, because the Earth must pass through the evenly distributed swarm on each trip around the Sun. If the plane of the original comet's orbit was steeply inclined to the plane of the Earth's orbit, as shown here, the meteor shower will strike the Earth from a direction well out of the ecliptic.

On June 30, 1908, a spectacular explosion occurred over the Tunguska region of Siberia that released about $10^{15}$ joules of energy, equivalent to a nuclear detonation of several hundred kilotons. The blast knocked a man off his porch some 60 km away and could be heard more than 1000 km away. Millions of tons of dust were injected into the atmosphere, causing a decrease in air transparency detected as far away as California.

Preoccupied with political upheaval and World War I, Russia did not send a scientific expedition to the site until 1927. At that time, Soviet researchers found that trees had been seared and felled radially outward in an area about 50 km in diameter (Figure 17-32). There was no clear evidence of a crater. In fact, the trees at "ground zero" were left standing upright, although they were completely stripped of branches and leaves. Because no significant meteorite samples

## figure 17-32

**Aftermath of the Tunguska Event** In 1908 a stony asteroid with a mass of about $10^8$ kilograms struck the Earth's atmosphere over the Tunguska region of Siberia. The asteroid's passage through the atmosphere made a blazing trail in the sky some 800 km long. The asteroid apparently exploded before reaching the surface, causing trees to be blown down over 2000 square kilometers around "ground zero." The force of the explosion was felt hundreds of kilometers away. (Courtesy of Sovfoto)

## box 17-3 | Looking Deeper into Astronomy

### *Meteor Showers*

At least ten notable meteor showers occur each year. They are listed with relevant information in the table below. The date of maximum is the best time to observe a particular shower, although good displays can often be seen a day or two before or after the maximum. The hourly rate is given for a single observer under optimum viewing conditions. The average speed refers to how fast the meteoritic material is moving as it strikes the atmosphere. The radiant is the location among the stars from which meteors seem to be coming; here we list only the constellation in which the radiant is located.

In order to see a fine meteor display, you need a view site away from city lights and a clear, moonless sky. The Moon's presence above the horizon can significantly detract from the number of faint meteors you will be able to see. In addition, the early morning hours (between approximately 2 A.M. and dawn) are the best times to make your observations. As sketched in the diagram, during the early morning hours you are on the side of the Earth that faces in the direction of our motion around the Sun. Hence, all meteor-producing particles in the Earth's path are swept into the atmosphere above you. On the other hand, during the evening hours you are on the

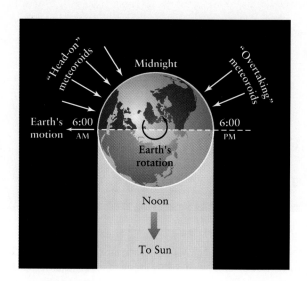

trailing side of the Earth, and only the fastest-moving meteoroids can manage to catch up with our planet to produce meteors.

| Shower | Date of maximum | Rate (meteors per hour) | Average speed (km/s) | Radiant constellation |
|---|---|---|---|---|
| Quadrantids | January 3 | 40 | 40 | Böotes |
| Lyrids | April 22 | 15 | 50 | Lyra |
| Aquarids | May 4 | 20 | 64 | Aquarius |
| Aquarids | July 30 | 20 | 40 | Aquarius |
| Perseids | August 12 | 50 | 60 | Perseus |
| Orionids | October 21 | 20 | 66 | Orion |
| Taurids | November 4 | 15 | 30 | Taurus |
| Leonids | November 16 | 15 | 70 | Leo |
| Geminids | December 13 | 50 | 35 | Gemini |
| Ursids | December 22 | 15 | 35 | Ursa Minor |

were found, for many years it was assumed that a small comet had struck Earth.

Recently, however, several teams of astronomers have argued that the Tunguska explosion was actually caused by a stony asteroid traveling at supersonic speed. They arrived at this conclusion after assessing the effects of various impactor sizes, speeds, and compositions. Even data from above-ground nuclear detonations of the 1940s and 1950s

were worked into the calculations. The Tunguska event is well-matched by a stony asteroid about 80 m (260 ft) in diameter entering the Earth's atmosphere at 22 km/s (80,000 km/h = 50,000 mph). A small comet would have broken up too high in the atmosphere to cause significant damage on the ground.

Large objects, such as that which formed the Chicxulub Crater and may have led to the demise of the dinosaurs

(Section 17-3), occasionally strike the Earth with the destructive force of as much as a million megatons of TNT. An asteroid 1 or 2 kilometers in diameter, striking the Earth at 30 km/s, would destroy an area the size of California. Such an impact would throw more than $10^{13}$ kg of microscopic dust particles into the atmosphere, which would decrease the amount of sunlight reaching the Earth's surface by enough to threaten the health of the world's agriculture. The loss of a year's crops could lead to the demise of a quarter of the world's population and would place our civilization—although perhaps not the survival of the species—in grave jeopardy.

The good news is that such large objects strike the Earth very infrequently. An asteroid 1 km in diameter might be expected to hit the Earth only once every 300,000 years. Statistically, the likelihood that an average North American would by killed by such an event is only 0.2% of the probability of dying in an automobile accident. While the threat of asteroid or comet impact exists, a catastrophic impact is very unlikely to occur in our lifetimes.

## Key Words

*Terms preceded by an asterisk (\*) are discussed in the Boxes.*

amino acids, p. 400

asteroid, p. 391

asteroid belt, p. 392

belt asteroid, p. 392

carbonaceous chondrite, p. 400

coma (of a comet), p. 401

comet, p. 401

differentiated asteroid, p. 399

dust tail, p. 403

fusion crust, p. 398

Hirayama families, p. 396

hydrogen envelope, p. 402

impact crater, p. 396

ion tail, p. 403

iron meteorite (or iron), p. 399

Kirkwood gaps, p. 394

Kuiper belt, p. 406

Lagrangian points, p. 394

long-period comet, p. 407

meteor, p. 397

meteorite, p. 397

meteoritic swarm, p. 408

meteoroid, p. 396

meteor shower, p. 408

minor planet, p. 391

near-Earth object (NEO), p. 395

nucleus (of a comet), p. 401

Oort cloud, p. 407

primitive asteroid, p. 400

radiant (of a meteor shower), p. 408

radiation pressure, p. 403

short-period comet, p. 406

stony iron meteorite, p. 398

stony meteorite (stone), p. 398

supernova, p. 400

tail (of a comet), p. 401

\*Titius-Bode law, p. 391

Trojan asteroid, p. 394

undifferentiated asteroid, p. 400

Widmanstätten patterns, p. 399

## Key Ideas

**Discovery of the Asteroids:** Astronomers first discovered the asteroids while searching for a "missing planet."

• Thousands of asteroids with diameters ranging from a few kilometers up to 1000 kilometers orbit within the asteroid belt between the orbits of Mars and Jupiter.

**Origin of the Asteroids:** The asteroids are the relics of planetesimals that failed to accrete into a full-sized planet, thanks to the effects of Jupiter and other Mars-sized objects.

• Even today, gravitational perturbations by Jupiter deplete certain orbits within the asteroid belt. The resulting gaps, called Kirkwood gaps, occur at simple fractions of Jupiter's orbital period.

• Jupiter's gravity also captures asteroids in two locations, called Lagrangian points, along Jupiter's orbit.

**Asteroid Collisions:** Asteroids undergo collisions with each other, causing them to break up into smaller fragments.

• Some asteroids, called near-Earth objects, move in elliptical orbits that cross the orbits of Mars and Earth. If such an asteroid strikes Earth, it forms an impact crater whose diameter depends on both the mass and the speed of the impinging object.

• An asteroid may have struck the Earth 65 million years ago, possibly causing the extinction of the dinosaurs and many other species.

**Meteroids, Meteors, and Meteorites:** Small rocks in space are called meteoroids. If a meteoroid entersthe Earth's atmosphere, it produces a fiery trail called a meteor. If part of the object survives the fall, the fragment that reaches the Earth's surface is called a meteorite.

• Meteorites are grouped into three major classes, according to composition: iron, stony iron, and stony meteorites. Irons and stony irons are fragments of the core of an asteroid that was large and hot enough to have undergone chemical differentiation, just like a terrestrial planet. Some stony meteorites come from the crust of such differentiated meteorites, while others are fragments of small asteroids that never underwent differentiation.

• Rare stony meteorites called carbonaceous chondrites may be relatively unmodified material from the solar nebula; these meteorites often contain organic material and may have played a role in the origin of life on Earth.

• Analysis of isotopes in certain meteorites suggests that a nearby supernova may have triggered the formation of the solar system 4.5 billion years ago.

**Comets:** A comet is a chunk of ice with imbedded rock fragments that generally moves in a highly elliptical orbit about the Sun.

• As a comet approaches the Sun, its icy nucleus develops a luminous coma, surrounded by a vast hydrogen envelope. An ion tail and a dust tail extend from the comet, pushed away from the Sun by the solar wind and radiation pressure, respectively.

• Fragments of "burned out" comets produce meteoritic swarms. A meteor shower is seen when the Earth passes through a meteoritic swarm.

**Origin of Comets:** Comets are thought to originate from two regions, the Kuiper belt and the Oort cloud.

• The Kuiper belt lies in the plane of the ecliptic at distances between 40 and 500 AU from the Sun. It is thought to contain many tens of thousands of comet nuclei. Many short-period comets probably come from the Kuiper belt, and a number of objects have been observed in the Kuiper belt.

• The Oort cloud contains billions of comet nuclei in a spherical distribution located between 10,000 and 100,000 AU from the Sun. Long-period comets are thought to originate in the Oort cloud. As yet no objects in the Oort cloud have been detected directly.

## Review Questions

1. How did asteroids come to be discovered? How did this discovery differ from what astronomers had expected to find?

2. Describe the asteroid belt. Does it lie completely within the plane of the ecliptic? What are its inner and outer radii?

3. What are Kirkwood gaps? What causes them?

4. Is there another place in our solar system where a phenomenon similar to Kirkwood gaps occurs? Explain.

5. What are the Trojan asteroids, and where are they located?

6. Why are there different kinds of asteroids with different chemical compositions?

7. Is there anywhere on the Earth where you might find large numbers of stony meteorites that are not significantly weathered? If so, where? If not, why not?

8. Suppose you found a rock you suspect might be a meteorite. Describe some of the things you could do to determine whether it was a meteorite or a "meteorwrong."

9. With the aid of a drawing, describe the structure of a comet.

10. Why is the phrase "dirty snowball" an appropriate characterization of a comet's nucleus?

11. Why do the ion tail and dust tail of a comet point in different directions?

12. Why do comets have prominent tails for only a short time during each orbit?

13. What is the Kuiper belt? How does it compare to the asteroid belt? What are the similarities? What are the differences?

14. What is the Oort cloud? How might it be related to planetesimals left over from the formation of the solar system?

15. Why do astronomers think that meteorites come from asteroids, while meteor showers are related to comets?

16. Why are asteroids, meteorites, and comets of special interest to astronomers who want to understand the early history of the solar system?

## Advanced Questions

*Questions preceded by an asterisk (*) involve topics discussed in the Boxes.*

**Problem-solving tips and tools:**

We discussed retrograde motion in Section 4-1 and described its causes in Section 4-2. You will need to use Kepler's third law, described in Section 4-4 and Box 4-2, in some of the problems below. The concept of escape speed is discussed in Box 7-2. A spherical object of radius $r$ intercepts an amount of sunlight proportional to its cross-sectional area, equal to $\pi r^2$. The volume of a sphere of radius $r$ is $\frac{4}{3}\pi r^3$.

17. When Olbers discovered Pallas in March 1802, the asteroid was moving from east to west relative to the stars. At what time of night was Pallas highest in the sky over Olbers's observatory? Explain your reasoning.

*18. What is the Titius-Bode law? Explain why it is not really a "law."

*19. Suppose that a planet were indeed 77.2 AU from the Sun, as predicted by the Titius-Bode law. How long would it take such an object to orbit the Sun? At what rate (in degrees per year) would this object appear to move in relation to the background stars? Would this motion be detectable with Earth-based telescopes?

20. Suppose that a double asteroid is observed in which one member is 16 times brighter than the other. Suppose that both members have the same albedo and that the larger of the two is 200 km in diameter. What is the diameter of the other member?

21. Consider the Kirkwood gap whose orbital period is two-fifths of Jupiter's period. Calculate the distance from the Sun to this gap. Does your answer agree with Figure 17-4?

22. Are there any examples in the solar system of objects being trapped at the $L_4$ and $L_5$ Lagrangian points other than in the Sun-Jupiter system?

*23. Assume that Ida's tiny moon Dactyl (Figure 17-8) has a density of 2500 kg/m$^3$. (a) Calculate the mass of Dactyl in kilograms. Express your answer as a fraction of the Earth's mass. (b) Calculate the escape speed from the surface of Dactyl. If you were an astronaut standing on Dactyl's surface, could you throw a baseball straight up so that it would never come down? Professional baseball pitchers can throw at speeds around 40 m/s (140 km/h, or 90 mi/h); your throwing speed is probably a bit less.

24. Use the percentages of stones, irons, and stony iron meteorites that fall to the Earth to estimate what fraction of their parent asteroids originally consisted of an iron core and a stony mantle. Compare to the Earth, which is about 35% iron. (*Note:* Assume that the percentages of stones and irons that fall to the Earth indicate the fractions of parent asteroids occupied by rock and iron, respectively.)

25. Find the orbital periods of comets whose perihelion distances are essentially zero (so-called *Sun-grazing comets*) and whose aphelion distances are (a) 100 AU, (b) 1000 AU, (c) 10,000 AU, and (d) 100,000 AU. Assuming that these comets can survive only a hundred perihelion passages, calculate their lifetimes. (*Hint:* Remember that the semimajor axis of an orbit is, as its name would suggest, one-half the length of the orbit's long axis.)

26. Comets are generally brighter a few weeks after passing perihelion than a few weeks before passing perihelion. Explain why might this be. (*Hint:* Water, including water ice, does an excellent job of retaining heat.)

## Discussion Questions

27. Suppose it were discovered that the near-Earth object 1994 XM$_1$ had been disturbed in such a way as to put it on a collision course with Earth. Describe what humanity could do within the framework of present technology to counter such a catastrophe.

28. Consult back issues of magazines such as *Scientific American* and *Sky & Telescope* to find out why some scientists disagree with the idea that a tremendous impact led to the demise of the dinosaurs. (They do not dispute that the impact took place, only what its consequences were.) What are their arguments? Based on what you learn, what is your opinion?

29. From the abundance of craters on the Moon and Mercury, we know that numerous asteroids and meteoroids struck the inner planets early in the history of our solar system. Is it reasonable to suppose that numerous comets also pelted the planets 3.5 to 4.5 billion years ago? Speculate about the effects of such a cometary bombardment, especially with regard to the evolution of the primordial atmospheres on the terrestrial planets.

30. Compare the consequences of a global thermonuclear war with that of an asteroid hitting Earth.

## Observing Projects

31. Make arrangements to view a meteor shower. Dates of major meteor showers are given in Box 17-3. Choose a shower that will occur near the time of a new moon. Informative details concerning upcoming meteor showers are often published a month ahead of time in such magazines as *Sky & Telescope* and *Astronomy* and are listed on the magazine's web sites (**http://www.skypub.com/** and **http://www.astronomy.com/**, respectively). Set your alarm clock for the early morning hours (1 to 3 A.M.). Get comfortable in a reclining chair or lie on your back so that you can view a large portion of the sky. Make note of how long you observe, how many meteors you see, and what location in the sky they seem to come from. How well does your observed hourly rate agree with published estimates, such as those in Box 17-3? Is the radiant of the meteor shower apparent from your observations?

32. Make arrangements to view a comet through a telescope. Because astronomers discover about a dozen comets each year, there is usually a comet visible somewhere in the sky. Unfortunately, they are often quite dim, and you will need to have access to a moderately large telescope. Consult recent issues of the *International Astronomical Union Circular,* published by the IAU's Central Bureau for Astronomical Telegrams, which contains the latest word on predicted positions and anticipated brightness of comets in the sky. You can also get up-to-date information from the Minor Planet Center web site (**http://cfa-www.harvard.edu/ cfa/ps/mpc.html**). Also, if there is an especially bright comet in the sky, useful information about it might be found in the latest issue of *Sky & Telescope*. Is a comet visible with a telescope at your disposal? If so, can you distinguish the comet from background stars? Can you see its coma? Can you see a tail?

33. Make arrangements to view an asteroid. At opposition, some of the largest asteroids are bright enough to be seen through a modest telescope. Check the "Minor Planets" section of the current issue of the *Astronomical Almanac* or the Minor Planet Center web site (**http://cfa-www.harvard. edu/cfa/ps/mpc.html**) to see if any bright asteroids are near opposition. If so, check the current issue as well as the most recent January issue of *Sky & Telescope* for a star chart showing the asteroid's path among the constellations. You will need such a chart to distinguish the asteroid from

# ALAN HALE

## Discovering a Comet

When Alan Hale shared in the discovery of Comet Hale-Bopp, he had been watching the sky for most of his life. He estimates that he has observed more than 200 comets since his childhood in New Mexico. His career in astronomy also began before graduate school. On leaving the U.S. Navy, he first worked at the Jet Propulsion Laboratory in California, where his projects included the *Voyager 2* encounter with Uranus in 1986. Dr. Hale is now director of the Southwest Institute for Space Research.

Hale's research looks at stars like our Sun and the likelihood of other solar systems, perhaps with planets similar to the Earth. He writes often for the general public, most recently *Everybody's Comet: A Layman's Guide to Comet Hale-Bopp* (High Lonesome Books, 1996), and in his weekly column for the *Alamogordo Daily News*.

O f the dozen or more comets discovered each year, most are dim and unremarkable. They pass by all but unnoticed, except by the handful of astronomers who study them. Perhaps once a year, however, a comet comes that is visible with the unaided eye, provided, of course, that we know when and where to look. And once every 10 to 20 years, on average, we are visited by one of the "Great Comets," bright enough to be seen by even a casual skywatcher.

I was privileged to discover one of our Great Comets and, as is often the case in science, it was a complete accident. It happened during the early morning hours of Sunday, July 23, 1995. I was taking a look at M70, a globular star cluster in the constellation Sagittarius, and I noticed a dim, diffuse object in the same field of view. An amateur astronomer in Arizona, Thomas Bopp, was also observing M70 that night, and he saw the same object. In accord with tradition, the new comet was named for its discoverers.

At its discovery, Comet Hale-Bopp was an unremarkable object about 100 times dimmer than the dimmest star visible to the naked eye, or of 11th magnitude.* It had little to distinguish it from all the other comets that are found all the time. Only after the first orbital calculations did astronomers realize how unusual it was. At the time, Hale-Bopp was halfway between the orbits of Jupiter and Saturn, at the enormous distance of 7.15 AU. For the comet to be visible at that distance in an ordinary backyard telescope, it must intrinsically be one of the brightest comets ever recorded. Even more dramatic, it would not reach its perihelion until April 1, 1997, at 0.91 AU from the Sun, just inside the Earth's own orbit. It therefore had

the potential to be one of the brightest comets to visit our skies in recent history.

The brightness of comets is notoriously difficult to predict, and it remained to be seen whether this potential would be realized. By May 1996, however, experienced observers were reporting naked-eye sightings, and the comet remained dimly visible to the unaided eye throughout the rest of the year. The next January it reappeared in the predawn sky, appearing much like a moderately bright, diffuse-looking star easily seen with the naked eye, and it brightened rapidly after that. When it was nearest to the Earth (1.3 AU), on March 23, it was rivaled by only a dozen of the brightest stars, and for a month thereafter it was visible in the northwestern sky in the early evening hours. At its peak in late March and early April, it reached about magnitude –1, outshining everything in the sky except one or two of the brightest stars and the brighter planets*. Its straight ion tail and broad, curving dust tail each spanned about 20° across the sky. Even skywatchers in major metropolitan areas thrilled to the celestial show above them, and from my home in the New Mexico mountains the sight was nothing less than glorious.

Modern telescopes and other instruments, on Earth and in space, should be able to follow the comet for several more years, perhaps for a decade into the twenty-first century. After that, Hale-Bopp will continue into the cold outer reaches of the solar system. Because it orbits the Sun, however, it will return someday. It apparently last visited the inner solar system some 4200 years ago, but this time a

*As discussed in more detail in Chapter 19, astronomers use the magnitude system to describe an object's apparent brightness.

close approach to Jupiter on its way in (0.77 AU) shortened its period to about 2380 years. We should see it again sometime around A.D. 4377.

In March 1996, Comet Hyakutake, another bright comet that recently visited our skies, passed only 0.1 AU from the Earth, close enough to bounce radar off it. A decade earlier, the *Giotto* spacecraft was sent to view Comet Halley. With Hale-Bopp, however, astronomers had to rely on indirect methods to study the nucleus, a "dirty snowball" shrouded within a dense cloud of gas and dust. Fortunately, Hale-Bopp's early discovery allowed astronomers around the world to plan extensive observations.

Photographs of Hale-Bopp, taken with the Hubble Space Telescope in late 1995, suggest a nucleus some 40 km in diameter. That compares to 15 km for Comet Halley and just 3 km for Hyakutake, and it seems consistent with the comet's high intrinsic brightness. Hale-Bopp's nucleus was found to rotate in 11.5 hours, although, like Halley's, it also exhibits a tumbling motion. Active jets of dusty material produce striking hoods, or ripples, in the inner coma that could be viewed with small telescopes.

A wide variety of organic molecules, including hydrogen sulfide, methyl alcohol, formaldehyde, and acetylene, were detected in the coma and nucleus of Comet Hale-Bopp. These and other organic compounds are thought to have played a major role in the development of life on the early Earth, which suggests that comets provided some of the materials that allowed life to begin. At the same time, observations have shown that the deuterium-to-hydrogen ratio in Hale-Bopp's water is about twice that inthe Earth's

seawater, challenging the rather widely held idea that the Earth has received a significant amount of its water from cometary impacts throughout its history. Clearly, much work remains to be done in order to determine the precise role that comets have played in the earth's geological and biological evolution.

Astronomers detected a third comet tail, composed of sodium atoms, and the Extreme Ultraviolet Explorer (EUVE) and BeppoSAX satellites both detected X rays from Hale-Bopp's coma. X-ray emissions were first spotted with Comet Hyakutake in 1996, and they have since been detected in a handful of other comets. There is still no complete explanation.

Comet Hale-Bopp provided a magnificent celestial spectacle and a scientific bonanza that will keep researchers busy for years to come. It also demonstrates the strong, continuing role of ignorance and superstition in modern society. Because its appearance roughly coincided with the end of the "second millennium," it fueled apocalyptic claims. Fears of the end of the world were heightened when some individuals claimed to detect a "companion," in reality, nothing more than a bright background star. In late March 1997, thirty-nine members of a religious cult in California committed suicide in order to "rendezvous" with this "companion."

These beliefs and actions, which harken back centuries, tell us that we still have a long way to go in our efforts to separate fact from fiction and reality from pseudoscience. Let us hope that when Comet Hale-Bopp comes our way again in the forty-fourth century, we will have learned the difference.

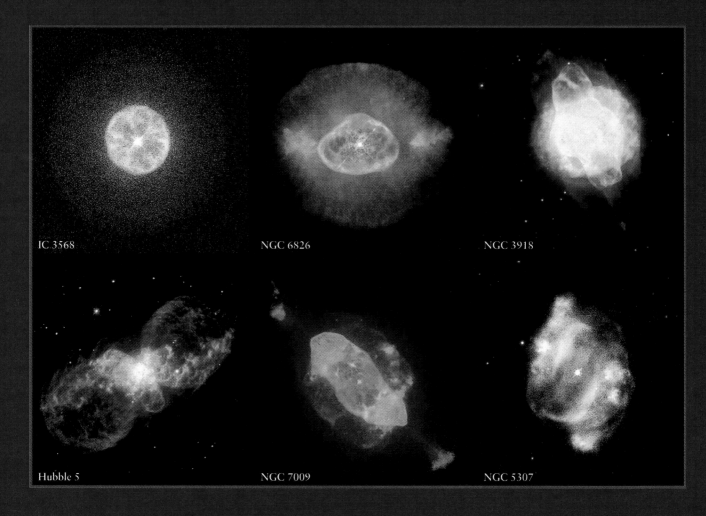

IC 3568

NGC 6826

NGC 3918

Hubble 5

NGC 7009

NGC 5307

Planetary nebulae, late stages in the lives of stars like the Sun.
(Space Telescope Science Institute and NASA)

# Stars and Stellar Evolution

To the great thinkers of ancient Greece, the stars in the heavens were bits of light embedded in a vast sphere with the Earth at its center. They thought the stars were composed of a mysterious "fifth element," quite unlike anything found on the Earth. To their thinking, the true nature of the stars would be forever hidden from our view.

Today we know that the stars are made of the very same chemical elements found on Earth. We know their sizes, their temperatures, and even something of their internal structures. And we have learned about how stars are born, mature, and die.

In the next seven chapters we will explore the nature of the stars. We begin with our own star, the Sun, before launching into the depths of interstellar space. We will see how astronomers weigh and measure the stars and how they have pieced together the story of stellar evolution. Most remarkable of all, we will learn how dying stars made it possible for humans like us to exist.

# Our Star, the Sun

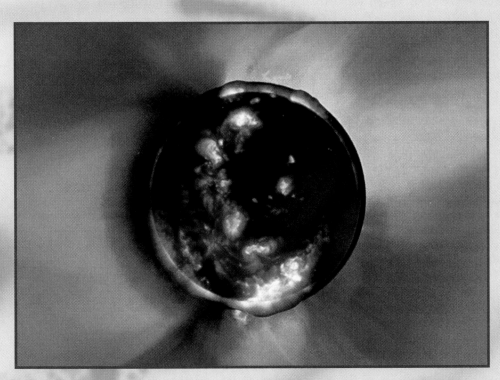

R I **V** U **X** G

The X-ray Sun and the Solar Corona This composite view of the Sun's outer atmosphere combines a white-light photograph taken during a solar eclipse with an X-ray image taken concurrently by a camera on board a rocket. The white-light photograph shows streamers extending far above the solar surface. The X-ray view shows hot gases near the solar surface in shades of yellow, orange, and red. By matching the two views, scientists can construct a three-dimensional picture of solar activity. (Courtesy of L. Golub and S. Koutchmy)

*In this chapter you will find the answers to the following questions:*

18-1  Does the Sun have a solid surface?

18-2  Since the Sun is so bright, how is it possible to see its dim outer atmosphere?

18-3  Where does the solar wind come from?

18-4  What are sunspots? Why are they dark?

18-5  What is the connection between sunspots and the Sun's magnetic field?

18-6  What is the source of the Sun's energy?

18-7  How do astronomers construct models of the Sun's internal structure?

18-8  How can astronomers measure the properties of the Sun's interior?

18-9  How can we be sure that thermonuclear reactions are happening in the Sun's core?

*The Sun is the most important member of our solar system, with almost a thousand times more mass than all of the planets, asteroids, comets, and meteoroids combined. But the Sun is also a star. In fact, it is a remarkably typical star, with a mass, size, surface temperature, and chemical composition that is approximately midway between the extremes exhibited by the myriad other stars in the heavens. Table 18-1 lists essential data on the Sun.*

*Within the solar atmosphere, ionized gases and magnetic fields interact on a colossal scale. These interactions create sunspots, relatively cool areas where the Sun's magnetic field is especially concentrated. Although the number of sunspots varies within an 11-year period, they are only part of a 22-year solar cycle that affects the entire solar atmosphere. The energy source that powers it all lies buried at the Sun's center, where thermonuclear reactions convert hydrogen into helium. These same thermonuclear reactions cause the Sun to shine.*

*Our existence depends on the Sun. Without the energy provided by sunlight, life on the Earth would not be possible. Furthermore, if the Sun were only slightly different in size or surface temperature, the polar ice caps could melt or another ice age could occur. Therefore, to understand why we are here, we must understand the Sun and its structure. In a quest for this understanding, astronomers have been able to use the laws of physics to construct a detailed model of the Sun, but we remain puzzled by many basic questions. Why do sunspots exist in the first place? Why are there so few solar neutrinos, particles that emanate from nuclear reactions in the Sun's core? These and other questions mark a major step in our study of the universe. They will be our bridge to understanding the nature of the stars.*

## 18-1 The photosphere is the lowest of three main layers in the Sun's atmosphere

Although astronomers often speak of the "solar surface," the Sun actually has no surface at all. The Sun is gaseous throughout its volume because of its high internal temperature. If you were somehow able to enter the Sun without vaporizing, you would encounter only denser and denser gases as you went to greater depths.

As shown in Figure 18-1, the Sun appears nonetheless to have a kind of surface. The reason is that essentially all of the Sun's visible light emanates from a single, thin layer of gas, appropriately called the Sun's **photosphere** ("sphere of light"). Just as you can see only a certain distance through the Earth's atmosphere before objects vanish in the haze, we can see only about 400 km into the photosphere. The Sun's radius is 696,000 km, so the photosphere appears to be a sharp, well-defined surface. Everything below the photosphere is called the **solar interior.**

The photosphere is actually the *lowest* of three layers that together constitute the **solar atmosphere.** Above it lie the *chromosphere* and the *corona*, which are transparent to visible light. We can see them only using special techniques, which we will discuss later in this chapter.

As in the lower regions of the Earth's atmosphere (Section 8-4), temperature decreases as you go up in the Sun's photosphere. We know this because the photosphere appears dark-

er around the edge, or **limb,** of the Sun than it does toward the center of the solar disk, an effect called **limb darkening** (examine Figure 18-1). This happens because when we look near the edge of the Sun, we do not see as deeply into the photosphere as we do when we look near the center of the disk (Figure 18-2). The limb looks dimmer because the high-altitude gas we observe there is not as hot and thus does not glow as brightly as the deeper, hotter gas seen near the disk center.

The spectrum of the Sun's photosphere confirms how its temperature varies with altitude. The photosphere shines like a nearly perfect blackbody with an average temperature of about 5800 K. (Blackbody spectra are described in Sections 5-3 and 5-4; in particular, see Figure 5-10.) However, superimposed on this spectrum are many dark absorption lines, shown in Figure 5-11. Hence, the net spectrum of the photosphere is an *absorption line spectrum,* which we introduced in Section 5-6. We see an absorption line spectrum whenever we view a hot, glowing object through a relatively cool gas. In this case, the hot object is the lower part of the photosphere; the cooler gas is in the upper part of the photosphere, where the temperature declines to about 4400 K. (You may find it hard to think of 4400 K as "cool," but keep in mind that the ratio of 4400 K to 5800 K, the temperature in the lower photosphere, is the same as the ratio of the temperature on a Siberian winter night to that of a typical day in Hawaii.) All the absorption lines in the Sun's spectrum are produced in this relatively cool layer, as atoms selectively absorb photons of various wavelengths streaming outward from the hotter layers below.

table 18-1

# Sun Data

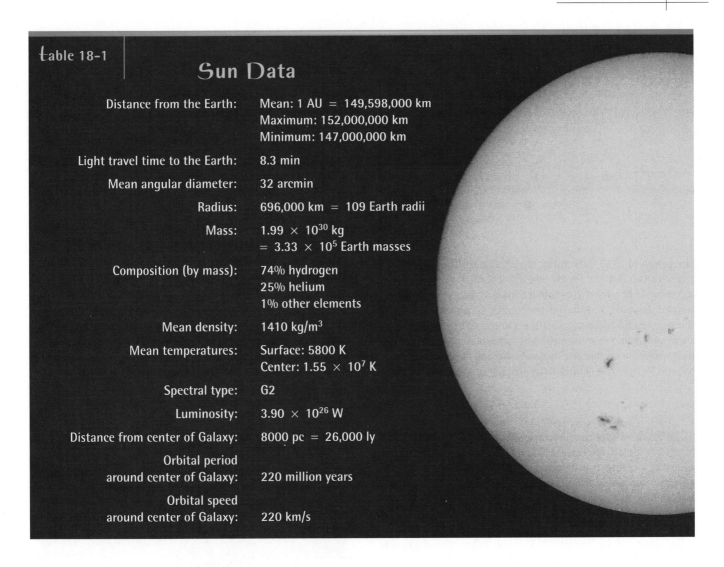

| | |
|---|---|
| Distance from the Earth: | Mean: 1 AU = 149,598,000 km |
| | Maximum: 152,000,000 km |
| | Minimum: 147,000,000 km |
| Light travel time to the Earth: | 8.3 min |
| Mean angular diameter: | 32 arcmin |
| Radius: | 696,000 km = 109 Earth radii |
| Mass: | $1.99 \times 10^{30}$ kg |
| | = $3.33 \times 10^5$ Earth masses |
| Composition (by mass): | 74% hydrogen |
| | 25% helium |
| | 1% other elements |
| Mean density: | 1410 kg/m³ |
| Mean temperatures: | Surface: 5800 K |
| | Center: $1.55 \times 10^7$ K |
| Spectral type: | G2 |
| Luminosity: | $3.90 \times 10^{26}$ W |
| Distance from center of Galaxy: | 8000 pc = 26,000 ly |
| Orbital period around center of Galaxy: | 220 million years |
| Orbital speed around center of Galaxy: | 220 km/s |

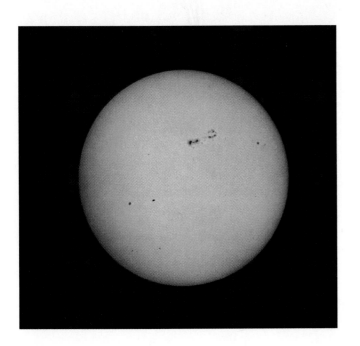

ƒigure 18-1   R I **V** U X G

**The Photosphere** The Sun is the only star whose surface details can be seen clearly using Earth-based telescopes. The layer in the Sun's atmosphere from which visible light is emitted is the photosphere. There is a not a sharp boundary dividing the photosphere from the solar interior that lies beneath it. Rather, the base of the photosphere is where the Sun's gases become sufficiently opaque to prevent light coming from lower levels to escape. Note that the photosphere appears darker near the limb, or edge, of the Sun; here we are seeing the upper photosphere, which is relatively cool and thus glows less brightly. The dark sunspots, which we discuss in Section 18-4, are also regions of relatively low temperature. When viewing the Sun, astronomers always take great care to avoid severe damage to their eyes by using extremely dark filters or by projecting the Sun's image onto a screen. The Sun is so bright that its rays, focused by the lens of an unprotected eye, can destroy the retina and cause permanent blindness. ***Never look directly at the Sun!*** (Celestron International)

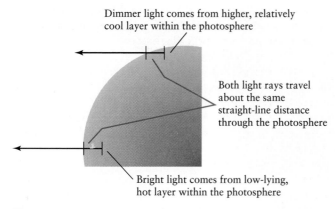

Dimmer light comes from higher, relatively cool layer within the photosphere

Both light rays travel about the same straight-line distance through the photosphere

Bright light comes from low-lying, hot layer within the photosphere

**figure 18-2**

**The Origin of Limb Darkening** The Sun's limb in Figure 18-1 appears darker than the center of the disk, a phenomenon called limb darkening. Light coming from both the limb and the disk center has traveled about the same straight-line distance through the photosphere to reach a telescope on Earth. Because of the Sun's curved shape, light from the limb comes from a greater height within the photosphere, where the temperature is lower and the gases glow less brightly. This gives the limb its darker appearance.

of granules, but descending downward along the boundaries between granules.

Time lapse photography reveals more of the photosphere's dynamic activity. Granules form, disappear, and reform in cycles lasting only a few minutes. At any one time, about 4 million granules cover the solar surface. A typical granule occupies about $10^6$ square kilometers, equal to the area of Texas and Oklahoma combined.

Although the photosphere is a very active place, there is actually relatively little material there. Careful examination of the spectrum shows that it has a density of only about $10^{-4}$ kg/m$^3$, roughly 0.01% the density of the Earth's atmosphere at sea level. It is made primarily of hydrogen and helium, the most abundant elements in the solar system (see Table 7-4).

Despite being such a thin gas, the photosphere is surprisingly opaque to visible light. If it were not so opaque, we could see into the Sun's interior to a depth of hundreds of thousands of kilometers, instead of the mere 400 km that we can see down into the photosphere. We also would see an emission spectrum, not a blackbody spectrum, if the photosphere's thin gas was transparent (see Section 5-6). What makes the photosphere so opaque is that its hydrogen atoms sometimes acquire an extra electron, becoming **negative**

Still more about the photosphere can be learned by examining it with a telescope—but *only* when using special dark filters to prevent eye damage. ***Looking directly at the sun without the correct filter, whether with the naked eye or with a telescope, can cause permanent blindness!*** Under good observing conditions, astronomers using such filter-equipped telescopes can often see a blotchy pattern in the photosphere, called **granulation** (Figure 18-3). Each light-colored **granule** measures about 1000 km across and is surrounded by a darkish boundary. The difference in brightness between the center and the edge of a granule corresponds to a temperature drop of about 300 K.

Granulation is caused by **convection,** the transport of heat by the movement of a fluid. Convection is easily observed on the Earth in a simmering pot of soup on a warm stove; heated water rises, cools off at the top of the pot, and then descends (see Section 8-3, especially Figure 8-16). Within the solar photosphere, the moving fluid is a gas rather than a liquid. Gas from lower levels rises upward in granules, cools off, spills over the edges of the granules, and then plunges back down into the Sun (Figure 18-4). This can only occur if the gas is heated from below, just as convection occurs in the pot of soup only when it is heated from below. Along with limb darkening and the Sun's absorption line spectrum, this shows that the upper part of the photosphere must be cooler than the lower part.

High-resolution spectroscopy reveals convection in the granules. Absorption lines are blueshifted in the spectrum of the central, bright regions of granules but are redshifted in the spectrum of the dark boundaries between granules. These Doppler shifts mean that gas is rising upward in the centers

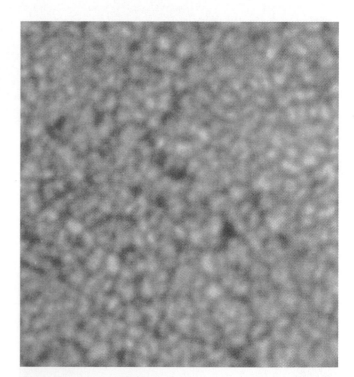

**figure 18-3**  R I **V** U X G

**Solar Granulation** High-resolution photographs of the Sun's surface, such as this one, reveal a blotchy pattern called granulation. The individual yellow blobs in this image are granules, each of which measures about 1000 km (600 mi) across. Granules are actually convection cells in the Sun's outer layers, as shown in Figure 18-4. (NOAO)

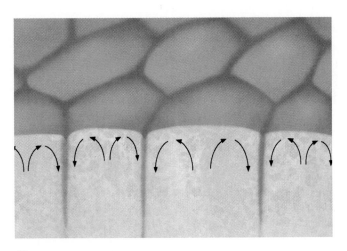

**figure 18-4**

**Convection in the Photosphere** Hot gas rising upward produces the bright granules shown in Figure 18-3. Cooler gas sinks downward along the boundaries between granules; this gas glows less brightly, giving the boundaries their dark appearance. The up-and-down convective motion shown here transports heat from the Sun's interior outward into the solar atmosphere.

hydrogen ions. The extra electron is only loosely attached and can be dislodged by absorbing a photon of any visible wavelength. Hence, negative hydrogen ions are very efficient light absorbers, and there are enough of these light-absorbing ions in the photosphere to make it quite opaque. Because it is so opaque, the photosphere's spectrum is very close to that of an ideal blackbody.

## 18-2 The chromosphere is located between the photosphere and the Sun's outermost atmosphere

Immediately above the photosphere is the tenuous **chromosphere** ("sphere of color"), which is the second of the three major levels in the Sun's atmosphere. The density of the chromosphere is only about $10^{-4}$ as great as the density of the photosphere, or about $10^{-8}$ as great as the density of our own atmosphere. No wonder it is normally invisible! During a total solar eclipse, however, the Moon blocks out the photosphere, and the chromosphere is visible as a pinkish strip around the edge of the dark Moon (Figure 18-5). This color gives the chromosphere its name.

Unlike the photosphere, which has an absorption line spectrum, the chromosphere has a spectrum dominated by *emission* lines. As described in Section 5-6 and Section 5-8, an emission line spectrum is produced by the atoms of a hot, thin gas. As their electrons fall from higher to lower energy levels, the atoms emit photons. This process gives the chromosphere

its reddish-pink color. This color has a definite wavelength of 656.3 nm, which is the wavelength of the photon emitted by a hydrogen atom when its single electron falls from the $n = 3$ level to the $n = 2$ level (recall Figure 5-21b). The emission line at this wavelength, also called the $H_\alpha$ Balmer line, is one of the strongest in the chromosphere's spectrum. The spectrum also contains the H and K emission lines of singly ionized calcium, as well as lines due to ionized helium and ionized metals.

The emission line spectrum of the chromosphere must originate from regions that are hotter than the underlying photosphere. (If they were cooler, the chromosphere would have an absorption line spectrum.) Analysis of these spectral lines shows that temperature *increases* with increasing height in the chromosphere. The temperature is about 4400 K at the top of the photosphere; 2000 km higher, at the top of the chromosphere, the temperature is nearly 25,000 K.

In order to observe the chromosphere at times other than during solar eclipses, astronomers exploit the difference between its spectrum and that of the photosphere. The photospheric spectrum is dominated by absorption lines at certain wavelengths, while the spectrum of the chromosphere has emission lines at these same wavelengths. In other words, the photosphere is dark at the wavelengths at which the chromosphere emits most strongly. By viewing the Sun through a special filter that is transparent to light only at the wavelength of $H_\alpha$, astronomers can screen out light from the photosphere and make the chromosphere visible.

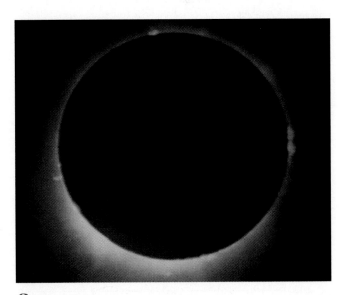

**figure 18-5**

**The Chromosphere Seen During a Solar Eclipse** This photograph, taken during a total solar eclipse, shows the chromosphere around the edge of the Moon. The chromosphere has an emission line spectrum, and thus emits light at only certain discrete wavelengths. Dominant among these is the $H_\alpha$ emission of hydrogen at a wavelength of 656.3 nm. This emission is at the red end of the visible spectrum, and gives the chromosphere its distinctive reddish-pink color.

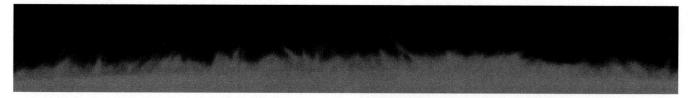

*f*igure 18-6   R I **V** U X G

**Spicules and the Chromosphere** This photograph of the Sun's limb was made using a filter that is transparent only to the wavelength of the hydrogen H$_\alpha$ spectral line. The photosphere has an absorption line at this wavelength, and so very little of its light went into making this image. Instead, most of the light recorded here comes from the chromosphere, which has an emission line at the wavelength of H$_\alpha$. The image shows numerous spicules, which are jets of cool gas that surge upward into the warmer regions of the Sun's outer atmosphere. (NOAO)

Figure 18-6 is a high-resolution image of the Sun's limb taken through an H$_\alpha$ filter, which reveals what is going on in the chromosphere. This image shows numerous vertical filaments, called **spicules**. These are actually jets of gas rising upward into the chromosphere. A typical spicule rises at the rate of 20 km/s, reaches a height of nearly 10,000 km, then collapses and fades away after 15 minutes or so. Approximately 300,000 spicules exist at any one time, covering about 1% of the Sun's surface. A schematic diagram of the solar atmosphere with a spicule is shown in Figure 18-7.

Spicules are generally located on boundaries between large, organized regions called **supergranules,** which are enormous convective cells in the photosphere (Figure 18-8). A typical supergranule is about 30,000 km in diameter and contains many hundreds of ordinary granules. Gases rise upward in the middle of a supergranule, move horizontally outward toward its edge, and descend back into the Sun. This large-scale convection also moves relatively slowly. The speed of supergranule motion is about 0.4 km/s, about one-tenth the speed of gases churning in a granule.

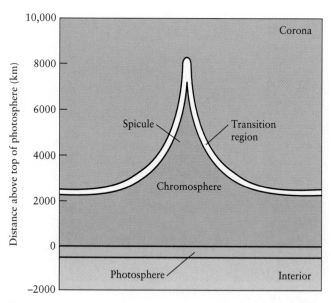

*f*igure 18-7

**The Solar Atmosphere** This schematic diagram shows the three layers of the solar atmosphere. The lowest, the photosphere, is about 400 km thick. The chromosphere above it extends to an altitude of about 2000 km, with spicules jutting up to nearly 10,000 km above the photosphere. The third, outermost layer is the corona (discussed in Section 18-3), which extends many millions of kilometers out into space. (Adapted from J. A. Eddy)

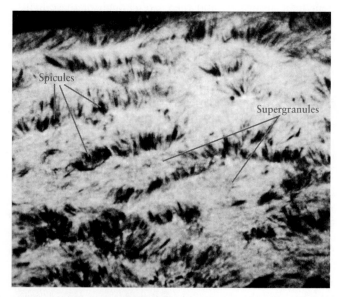

*f*igure 18-8   R I **V** U X G

**Spicules and Supergranulation** In this image made with an H$_\alpha$ filter, spicules appear as elongated, dark, vertical filaments. They appear dark here because they are seen against the bright background of the underlying atmosphere; in Figure 18-6, by contrast, spicules appear bright against the black background of space. Note that the spicules are not randomly distributed, but are located along the irregularly shaped boundaries between large organized cells called supergranules. (NOAO)

## 18-3 The corona is the outermost layer of the Sun's atmosphere

The outermost region of the Sun's atmosphere is called the **corona.** It extends from the top of the chromosphere out to a distance of several million kilometers, where it gradually becomes the *solar wind*. (The solar wind, as we described in Section 7-8, is a stream of high-speed protons and electrons that is constantly escaping from the Sun.)

The corona is only about one-millionth ($10^{-6}$) as bright as the photosphere—about as bright as the full moon. Hence, the corona can be viewed only when the light from the photosphere is blocked out, either during a total eclipse or when a specially designed telescope called a **coronagraph** is used. Figure 18-9 is an exceptionally detailed photograph of the Sun's corona taken during a solar eclipse. This photograph shows that the corona is not merely a spherical shell of gas surrounding the Sun; numerous streamers extend in different directions far above the solar surface. The shapes of these streamers vary on timescales of days or weeks. (Other views of the corona during solar eclipses are shown in Figure 3-11 and the photograph that opens Chapter 3.)

The spectrum of the corona is an emission line spectrum, characteristic of a hot, thin gas. When this spectrum was first measured in the nineteenth century, astronomers found a number of emission lines at wavelengths that had never been seen in the laboratory. Their explanation was that the corona contained elements that had not yet been detected on the Earth. However, laboratory experiments in the 1930s helped astronomers realize that these unusual emission lines were in fact caused by the same atoms found elsewhere in the universe—but in highly ionized states. For example, there is a prominent green line at 530.3 nm that is caused by highly ionized iron atoms, each of which has been stripped of 13 of its 26 electrons. Temperatures in excess of $2 \times 10^6$ K are required to strip that many electrons from atoms, so it follows that the corona must be at remarkably high temperatures. Figure 18-10 shows how temperature in both the chromosphere and corona varies with altitude.

 Despite its high temperatures, the corona is actually not very "hot," that is, it contains very little thermal energy. The reason is that the corona is nearly a vacuum. In the corona, there are only about $10^{11}$ atoms per cubic meter of gas. In comparison, there are about $10^{23}$ atoms per cubic meter in the Sun's photosphere and about $10^{25}$ atoms per cubic meter in the air that we breathe. Because of the corona's high temperature, the atoms there are moving at very high speeds, but because there are so few atoms in the corona, the total amount of energy in these moving atoms (a measure of how "hot" the gas is) is rather low.

**figure 18-9** R I **V** U X G

**The Solar Corona** This extraordinary photograph was taken during the total solar eclipse of July 11, 1991. Numerous streamers are visible, extending millions of kilometers above the solar surface. No photograph can capture the full grandeur of the corona during a total solar eclipse; it needs to be seen with one's own eyes. (Courtesy of R. Christen and M. Christen: Astro-Physics Inc.)

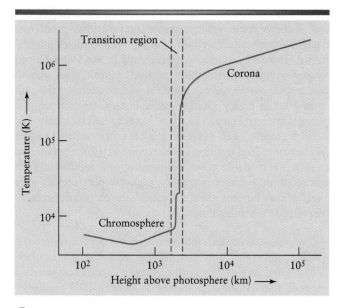

## figure 18-10

**Temperatures in the Sun's Upper Atmosphere** This graph shows how temperature varies with altitude in the Sun's chromosphere and corona. Note that both the vertical and horizontal scales are nonlinear, in order to allow a large range of values to be shown in the graph. The temperature rises abruptly by a hundredfold, from about $10^4$ K to about $10^6$ K, over a narrow transition region about 11,000 km above the Sun's photosphere. This transition region is the dividing line between the chromosphere and the corona. Astronomers are just beginning to understand how the corona is heated to such extraordinarily high temperatures. (Adapted from A. Gabriel)

**ANALOGY** If you flew a spaceship into the corona, you would have to worry about becoming overheated by the intense light coming from the photosphere, but you would notice hardly any heating from the high-temperature, yet ultrathin, gas of the corona. It is similar to what happens if you put your hand momentarily inside a conventional oven that is being used for baking. Both the walls of the oven and the air inside the oven are at the same high temperature, but the air contains very few atoms and thus carries little energy. The lion's share of the heat you feel is radiation from the oven walls.

The low density of the corona also explains why it is so dim compared to the photosphere. In general, the higher the temperature of a gas, the brighter it glows. But because there are so few atoms in the corona, the net amount of light that it emits is very feeble compared to the light from the much cooler, but also much denser, photosphere.

The temperatures in the corona and the chromosphere are not at all what we would expect. Just as you feel warm if you stand close to a campfire but cold if you move away, we would expect that the temperature in the corona and chromosphere would *decrease* with increasing altitude and, hence, increasing distance from the warmth of the Sun's photosphere. Why, then, does the temperature in these regions *increase* with increasing altitude? This has been one of the major unsolved mysteries in astronomy for the past half-century. In Section 18-5 we discuss recent observations that may provide an answer to this long-standing dilemma.

Because of the corona's high temperature, its ionized atoms move at high speeds. When ions collide, the energy of the impact boosts their electrons to upper energy levels. As the electrons fall back to lower levels, they emit radiation, much of which is in the ultraviolet and X-ray regions of the spectrum. Because the Earth's atmosphere is opaque to radiation in these wavelength ranges (see Section 6-7, especially Figure 6-27), the corona is best observed from spacecraft, and a number of space missions since the 1970s have been devoted to this purpose. Currently, the Sun is being continuously monitored at nonvisible wavelengths by the Japanese spacecraft *Yohkoh*, which detects X rays coming from the Sun, and by the spacecraft *SOHO* (short for *Solar and Heliospheric Observatory*), a joint project of the European Space Agency (ESA) and NASA. Several instruments on board *SOHO* are used to create ultraviolet images; two such images were used to make the composite view shown in Figure 18-11.

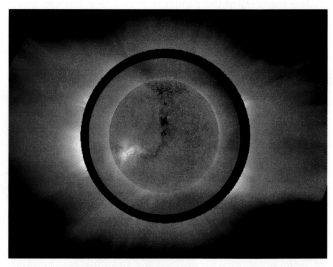

## figure 18-11   R I V **U** X G

**A Composite Ultraviolet Image of the Sun** This view of the Sun (shown here in false color) was put together from two ultraviolet images made by different instruments on board the *SOHO* spacecraft. The black ring separates the two separate images. The outer image, recorded using the light from highly ionized oxygen atoms, shows material from the corona streaming outward to become part of the solar wind. The inner image, made with ultraviolet light from highly ionized iron atoms, shows the origin of these streamers in the lower regions of the corona. The dark feature running from top to bottom across the disk of the Sun is a coronal hole, a region where the coronal gases are thinner and at a lower temperature than the average. Outward-flowing gases can most easily escape through such coronal holes. (Inner image courtesy of the *SOHO* Extreme-Ultraviolet Image Consortium; outer image courtesy of the *SOHO* UltraViolet Coronagraph Spectrometer Consortium, ESA, and NASA)

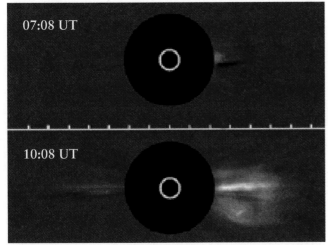

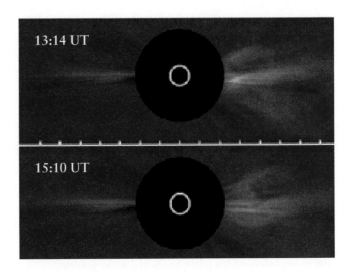

**figure 18-12**   R I **V** U X G

**A Coronal Mass Ejection Event** These four images were recorded over an eight-hour period on January 15, 1996, by the Large Angle and Spectrometric Coronagraph on board the *SOHO* spacecraft. To make it possible to see the corona's hot but thin gas in these visible-light images, the light from the photosphere has been blacked out by a disk. The yellow circle in each image shows the size of the Sun. The blob of hot, glowing gas ejected to the right expanded to the size of the Sun in the space of just a few hours. Note that hot gases appeared on the other side of the Sun at the same time. It is thought these eruptions are actually convulsions of the corona that extend all the way around the Sun's equator. (Courtesy of Guenter Brueckner, Naval Research Laboratory, the *SOHO* Large Angle and Spectrometric Coronagraph Consortium, ESA, and NASA)

Figure 18-11 shows that the corona is not uniform in temperature or density. The hottest, most dense regions appear bright, while the cooler, less dense regions are dark. Note the large dark area, called a **coronal hole** because it is almost devoid of luminous gas. Particles streaming away from the Sun can most easily flow outward through these particularly thin regions. Therefore, it is thought that coronal holes are the main corridors through which particles of the solar wind escape from the Sun.

The connection between coronal holes and the solar wind is perhaps most firmly established at the Sun's north and south poles, where there are apparently permanent coronal holes. In 1994 and 1995 the *Ulysses* spacecraft, another joint ESA/NASA mission, flew through the solar wind emanating from the poles and confirmed the connection with the coronal holes there. Curiously, this flow from the poles moves outward at a speed of 800 km/s (a remarkable $2.9 \times 10^6$ km/h, or $1.8 \times 10^6$ mi/h), twice as fast as the solar wind that flows from the Sun's equator. It is not known why there should be two components of the solar wind that travel at such different speeds.

In addition to the relatively steady flow of the solar wind, the corona is also the site of some truly stupendous eruptions. Figure 18-12 is a sequence of images showing a **coronal mass ejection,** an event in which more than $10^{12}$ kilograms of high-temperature coronal gas is suddenly blasted into space at speeds of hundreds of kilometers per second. When ionized gas ejected in this way reaches the Earth a few days later, it causes disturbances in our magnetosphere that can even interfere with electrical equipment on the Earth's surface. Such events occur every few months; smaller eruptions may occur almost daily.

X-ray images of the Sun, such as Figure 18-13, reveal the very hottest regions of the corona. The brightest areas in these images have temperatures that occasionally reach $4 \times 10^6$ K, twice the average temperature of the corona. Many of the hot spots seen in X-ray pictures are located over one of the Sun's most familiar features—sunspots.

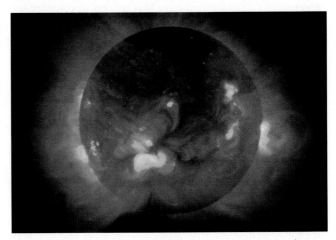

**figure 18-13**   R I V U **X** G

**The Sun in X Rays** This image, recorded by the *Yohkoh* satellite on May 8, 1992, shows X-ray emission from the corona in false color. Note the swirling, looping shape of the bright high-temperature regions. These regions, which change their appearance on time scales of days, often lie on top of sunspots in the photosphere. (Solar and Astrophysics Laboratory of the Lockheed-Martin Advanced Technology Center)

## 18-4 Sunspots are low-temperature regions in the photosphere

Spicules, granulation, supergranulation, and coronal mass ejections require specialized techniques or apparatus to observe, but there is one kind of activity in the solar atmosphere that can easily be seen with even a small telescope (*although only with an appropriate filter attached!*). **Sunspots** are irregularly shaped dark regions in the photosphere. Sometimes sunspots appear in isolation (Figure 18-14), but frequently they are found in clusters called **sunspot groups** (Figure 18-15; see also Figure 18-1). Although sunspots vary greatly in size, typical ones measure a few tens of thousands of kilometers across. Sunspots are not permanent features of the photosphere but last between a few hours and a few months.

Each sunspot has a dark central core, called the **umbra,** and a brighter border called the **penumbra.** We used these same terms in Section 3-3 to refer to different parts of the Earth's or the Moon's shadow. But a sunspot is not a shadow. Rather, it is a region where the temperature is less than that of the surrounding photosphere, which makes it appear dark. If the surrounding photosphere is blocked from view, a sunspot's umbra appears red and the penumbra appears orange. As we saw in Section 5-4, Wien's law relates the color of a blackbody (which depends on the wavelength at which it emits the most light) to the blackbody's temperature. Using this law, the colors of a sunspot indicate that the temperature of the umbra is typically 4300 K and that of the penumbra is typically 5000 K. While high by earthly standards, these temperatures are quite a bit less than the average photospheric temperature of 5800 K.

The Stefan-Boltzmann law (see Section 5-4) tells us that the energy flux from a blackbody is proportional to the fourth power of its temperature. This law lets us compare the amounts of light energy emitted by a square meter of a sunspot's umbra and by a square meter of undisturbed photosphere. The ratio is

$$\frac{\text{flux from umbra}}{\text{flux from photosphere}} = \left(\frac{4300\,\text{K}}{5800\,\text{K}}\right)^4 = 0.30$$

That is, the umbra emits only 30% as much light as an equally large patch of undisturbed photosphere, which is why sunspots appear so dark.

On rare occasions, a sunspot group is large enough to be seen without a telescope. Chinese astronomers recorded such sightings 2000 years ago, and a huge sunspot group visible to the naked eye (with an appropriate filter) was seen in 1989. Of course, a telescope gives a much better view, and so it was not until Galileo that anyone was able to examine sunspots in detail.

Galileo discovered that he could determine the Sun's rotation rate by tracking sunspots as they moved across the solar disk (Figure 18-16). He found that the Sun rotates once in about four weeks. A typical

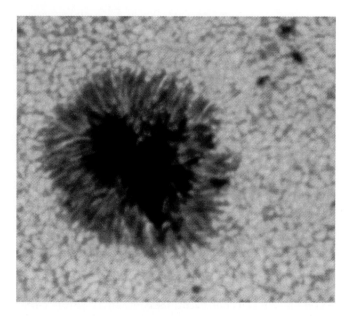

**ƒigure 18-14** R I **V** U X G

**A Sunspot** This high-resolution photograph shows a mature sunspot. A typical sunspot has about the same diameter as the Earth. The dark center of the spot is called the umbra. It is bordered by the penumbra, which is less dark and has a featherlike appearance. Granulation is visible in the surrounding, undisturbed photosphere. (NOAO)

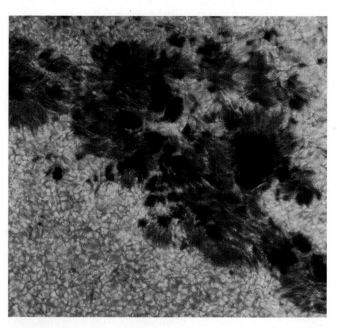

**ƒigure 18-15** R I **V** U X G

**A Sunspot Group** This image shows a typical sunspot group. Several sunspots are close enough to overlap, while others are nearby. As in Figure 18-14, granulation is visible in the photosphere around the sunspots.

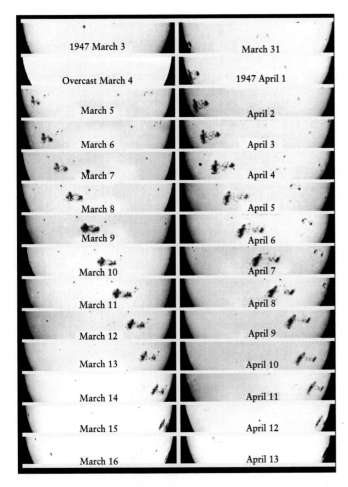

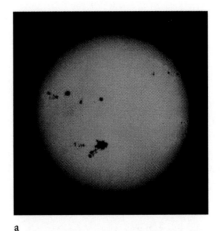

**figure 18-16**   R I **V** U X G

**Tracking the Sun's Rotation with Sunspots**  By observing the same group of sunspots from one day to the next, Galileo found that the Sun rotates once in about four weeks. The equatorial regions of the Sun actually rotate somewhat faster than the polar regions, a phenomenon called differential rotation. This series of photographs shows the same sunspot group over $1\frac{1}{2}$ solar rotations. (The Carnegie Observatories)

**figure 18-17**   R I **V** U X G

**The Sun at Sunspot Maximum and Sunspot Minimum**
(**a**) In this photograph taken in 1989, near sunspot maximum, the Sun shows a number of sunspots and large sunspot groups. The sunspot group visible near the bottom of the Sun's disk is about as large as the planet Jupiter. (**b**) Near sunspot minimum, as in this 1986 photograph, there are hardly any sunspots visible. (NOAO)

sunspot group lasts about two months, so a specific one can be followed for two solar rotations.

Further observations by the British astronomer Richard Carrington in 1859 demonstrated that the Sun does not rotate as a rigid body. Carrington found that the equatorial regions rotate more rapidly than the polar regions. A sunspot near the solar equator takes only 25 days to go once around the Sun, while at 30° north or south of the equator a sunspot takes 27½ days to complete a rotation. The rotation period at 75° north or south of the solar equator is about 33 days, while near the poles it may be as long as 35 days. This phenomenon is known as **differential rotation.**

Careful observations over many years revealed that the numbers of sunspots change periodically. In some years there are many sunspots, in other years almost none (Figure

18-17). This phenomenon, first reported by German astronomer Heinrich Schwabe in 1843 after many years of observing, is called the **sunspot cycle.** As shown in Figure 18-18, the average number of sunspots varies with a period of about 11 years. A period of exceptionally many sunspots is a **sunspot maximum,** as occurred in 1968, 1979, and 1989 and as is predicted to occur in 2000. Conversely, the Sun was almost devoid of sunspots at times of **sunspot minimum,** as occurred in 1965, 1976, 1986, and 1996.

The locations of the sunspots also vary over a sunspot cycle. At the beginning of a cycle, just after sunspot minimum, sunspots first start to appear at latitudes 30° north and south of the solar equator. Over the succeeding years, the sunspots occur closer and closer to the equator, until at the end of the cycle they are virtually all on the solar equator.

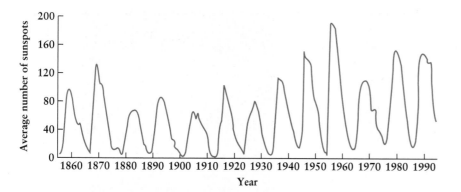

**ƒigure 18-18**

**The Sunspot Cycle** The number of sunspots on the Sun varies with a period of about 11 years. The most recent sunspot maximum occurred in 1990, and the most recent sunspot minimum occurred in 1996. The next sunspot maximum is predicted to occur in 2000.

## 18-5 Sunspots are produced by a 22-year cycle in the Sun's magnetic field

Why should the number of sunspots vary with an 11-year cycle? Why should their average latitude vary over the course of a cycle? And why should sunspots exist at all? The first step toward answering these questions came in 1908, when American astronomer George Ellery Hale discovered that sunspots are associated with intense magnetic fields on the Sun.

When Hale focused a spectroscope on sunlight coming from a sunspot, he found that many spectral lines appear to be split into several closely spaced spectral lines (Figure 18-19). This "splitting" of a single spectral line into two or more lines is called the **Zeeman effect** after Dutch physicist Pieter Zeeman, who first observed it in his laboratory in 1896. Zeeman showed that a spectral line splits when the atoms are subjected to an intense magnetic field. The more intense the magnetic field, the wider the separation of the split lines (see Box 18-1).

Hale's discovery showed that sunspots are places where a powerful, concentrated magnetic field projects through

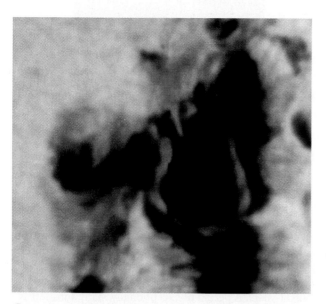

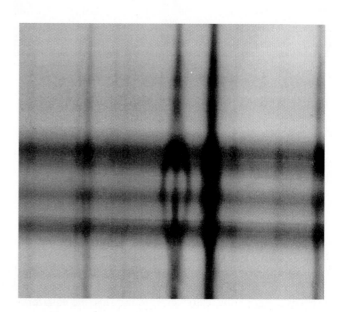

**ƒigure 18-19** R I **V** U X G

**How a Sunspot's Magnetic Field Affects Spectral Lines** **(a)** The magnetic field of a sunspot is determined from its spectrum, as measured using a spectrograph. A black line has been drawn across this image of a sunspot to indicate the location toward which the slit of the spectrograph was aimed. **(b)** This image shows a portion of the resulting spectrogram, including a dark absorption line caused by iron atoms in the photosphere. Within the sunspot, this line is split into three components. The separation in wavelength between these components can be used to calculate the strength of the magnetic field in the sunspot. Typical sunspot fields are more than 5000 times stronger than the Earth's magnetic field at its north and south poles. It is possible to generate even stronger fields in the laboratory, but this requires the use of powerful electromagnets. Such fields are used in industrial and medical applications. (NOAO)

the hot gases of the photosphere. Because of the high temperature of the Sun's atmosphere, many of the atoms there are ionized. The solar atmosphere is thus a gaseous mixture called a **plasma,** in which electrically charged ions and electrons can move freely. Like any moving, electrically charged objects, they can be deflected by magnetic fields. As an example, Figure 18-20 shows how a magnetic field in the laboratory bends a beam of fast-moving electrons into a curved trajectory. Similarly, the paths of moving ions and electrons in the photosphere are deflected by the Sun's magnetic field. In particular, magnetic forces act on the hot plasma that rises from the Sun's interior due to convection. Where the magnetic field is particularly strong, these forces push the hot plasma away. The result is a localized region where the gas is relatively cool and thus glows less brightly—in other words, a sunspot.

To get a more complete picture of the Sun's magnetic fields, astronomers take images of the Sun at two wavelengths, one just less than and one just greater than the wavelength of a magnetically split spectral line. From the difference between these two images, they can construct a picture called a **magnetogram,** which displays the magnetic fields in the solar atmosphere. The magnetogram in Figure 18-21 shows the entire solar surface, along with an ordinary photograph of the Sun taken at the same time. Dark blue indicates the areas of the photosphere with one magnetic polarity (north), and yellow indicates the area with the opposite (south) magnetic polarity. This figure shows that many sunspot groups are **bipolar,** meaning that they have roughly comparable areas covered by north and south magnetic polarities. This can be seen even more clearly in Figure 18-22, which is a close-up magnetogram of a single sunspot group. Thus, a sunspot group resembles a giant bar magnet, with a north magnetic pole at one end and a south magnetic pole at the other.

If different sunspot groups were unrelated to each other, their magnetic poles would be randomly oriented, like a bunch of compass needles all pointing in random directions. As George Ellery Hale discovered, however, there is a striking regularity in the magnetization of sunspot groups.

**figure 18-20**

**Magnetic Fields Deflect Moving, Electrically Charged Objects**
In this laboratory experiment, a beam of negatively charged electrons (shown by a blue arc) is aimed straight upward. However, the entire apparatus is inside a large magnet, and the magnetic field deflects the beam into a curved path. This is an important effect in the Sun's photosphere, where electrically charged electrons and ions are free to move and where there are strong magnetic fields. Magnetic deflection also has many technological applications. It is used to focus the electron beam in a television picture tube and to make electrons oscillate so that they produce microwaves in a microwave oven. (Courtesy Central Scientific Company)

In a sunspot group, the sunspots on the side of the group toward which the Sun is rotating are called the "preceding members" of the group. The spots that follow behind are referred to as the "following members." Hale compared the sunspot groups in the two solar hemispheres, north or south of the Sun's equator. He found that the preceding members in one solar hemisphere all have the same magnetic polarity, while the preceding members in the other hemisphere have the opposite polarity. Furthermore, in the

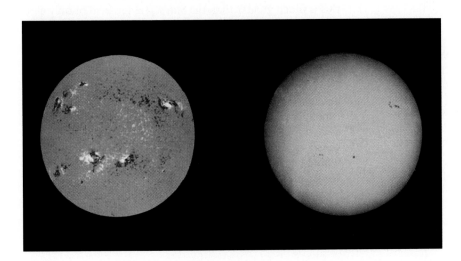

**figure 18-21** R I **V** U X G

**Mapping the Sun's Magnetic Field** This magnetogram of the Sun (left) was taken at the same time as the visible-light image (right). In the magnetogram, regions of north magnetic polarity and south magnetic polarity are shown in dark blue and in yellow, respectively. Regions with weak magnetic fields, which includes much of the Sun's disk, are shown in blue-green. Notice how regions with strong magnetic fields correlate with the locations of sunspots. Careful examination of the magnetogram also reveals that the magnetic polarity of the sunspots in the northern hemisphere is the opposite of those in the southern hemisphere. (NOAO)

## box 18-1 | Looking Deeper into Astronomy

### *The Zeeman Effect*

In 1896, the Dutch physicist Pieter Zeeman placed some hot gases between the poles of a powerful magnet and examined the spectra of the glowing gases through a spectroscope. He found that each spectral line was split into two, three, or more closely spaced lines, depending on the particular gases used. Zeeman shared the 1902 Nobel Prize in Physics for his discovery and interpretation of this curious effect, which turned out to be of tremendous importance to astronomy.

An electron orbiting around a nucleus is actually a miniature loop of electric current. When an atom is placed in a magnetic field, the interaction between the field and the electron current shifts the atom's energy. The amount of energy shift depends on how the electron's orbit is oriented relative to the field. According to the principles of quantum mechanics, only certain orientations are allowed. Hence, each one of these orientations corresponds to a slightly different energy level for the atom.

The diagram shows one example of how the Zeeman effect is produced. When the magnetic field is not present, an atom in a lower energy level can be excited to a higher level only by absorbing a photon of one specific energy and one specific wavelength. (The energy and wavelength of a photon are related, as we described in Section 5-5.) This produces a single absorption line in the spectrum. But suppose a magnetic field is turned on around the atom. Now the lower energy level is split into two slightly different energy levels, corresponding to two different orientations of a particular electron orbit. In our example, the higher energy level is also split, but into *four* slightly different energy levels. The combination makes *eight* different electron jumps possible. Actually, some of the jumps involve nearly the same energy, so we see only six distinct spectral lines can be resolved in the spectrum with the magnetic field turned on.

The stronger the magnetic field, the greater the difference in the energy levels. Thus, astronomers can deduce the strength of the magnetic field by measuring the separations of the lines. The strength of a magnetic field is often expressed in terms of a unit called the **gauss** (G), named after Karl Friedrich Gauss, a German mathematician who experimented with magnets in the early 1800s. (As we described in Section 17-1, Gauss also played an important role in the history of the first asteroid to be discovered.) For example, the strength of the Earth's magnetic field at its north and south poles is

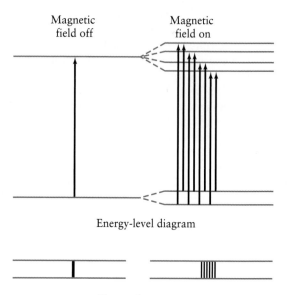

Magnetic field off     Magnetic field on

Energy-level diagram

Observed spectrum

about 0.7 gauss, and the field at the poles of a toy magnet may be 100 gauss. By comparison, the separation in wavelength between the spectral line components in Figure 18-19 corresponds to a magnetic field strength in excess of 4130 gauss. This tells us that the magnetic field of a sunspot is quite substantial by terrestrial standards.

The Zeeman effect is not just useful for studying the Sun. A number of stars also display Zeeman splitting in their spectra, indicating a magnetic field with an *average* strength across the star's surface of 100 to 1000 gauss. While less than the maximum field found in sunspots, this is much greater than the Sun's average field of only a few gauss. Because sunspots are associated with magnetic fields, this suggests that these *magnetic stars* may have much of their surfaces covered by "starspots." Indeed, the light output of magnetic stars can vary by a factor of 2 over a period of decades, apparently as giant spots form and dissipate on these stars' surfaces over the course of a "starspot cycle." (Unfortunately, these stars are too far away for us to see these spots directly.) As dramatic as sunspots are on our own Sun, these gigantic "starspots" would be truly impressive to any alien astronomers whose planet orbits such a magnetic star.

**figure 18-22**

**A Magnetogram of a Sunspot Group** This artificially colored picture displays the intensity and polarity of the magnetic field associated with a large sunspot group. The grayish background has a relatively weak magnetic field. Within the sunspot group, one side (shown in blue) has one magnetic polarity, and the other side (shown in yellow) has the opposite polarity. Thus, a sunspot group resembles a giant bar magnet. (NOAO)

hemisphere where the Sun has its north magnetic pole, the preceding members of all sunspot groups have north magnetic polarity. In the opposite hemisphere, where the Sun has its south magnetic pole, the preceding members all have south magnetic polarity.

Along with his colleague Seth B. Nicholson, Hale also discovered that the Sun's polarity pattern completely reverses itself every 11 years—the same as the time from one solar maximum to the next. The hemisphere that has preceding north magnetic poles during one 11-year sunspot cycle will have preceding south magnetic poles during the next 11-year cycle, and vice versa. The north and south magnetic poles of the Sun itself also reverse every 11 years. The Sun's magnetic pattern repeats itself only after two sunspot cycles, which is why astronomers speak of a **22-year solar cycle**.

In 1960, American astronomer Horace Babcock proposed a description that seems to account for many features of this 22-year solar cycle. Babcock's scenario, called a **magnetic-dynamo model**, makes use of two basic properties of the Sun's photosphere—differential rotation and convection. Differential rotation causes the magnetic field in the photosphere to become wrapped around the Sun (Figure 18-23). As a result, the magnetic field then becomes concentrated at latitudes on either side of the solar equator. Convection in the

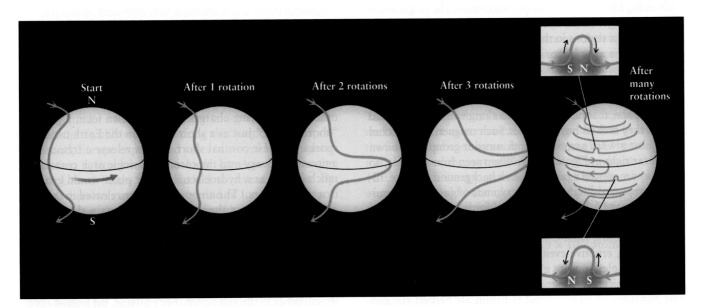

**figure 18-23**

**Babcock's Magnetic Dynamo** The sunspot cycle may in part result from the Sun's differential rotation. Magnetic field lines tend to move along with the electrically charged plasma in the Sun's outer layers. A field line that starts off running from the Sun's north magnetic pole (N) to its south magnetic pole (S), as in the drawing on the far left, soon becomes distorted because the Sun's equator rotates faster than the poles. After many rotations the magnetic field becomes wrapped around the Sun, like twine wrapped around a ball. Sunspot groups, shown as insets to the far right drawing, appear where the concentrated magnetic field breaks through the solar surface. Note that in each sunspot group, the preceding members—in this figure, the sunspots on the right in each group—have the same polarity as the Sun's magnetic pole in that hemisphere (in this drawing, N in the upper hemisphere and S in the lower hemisphere).

more energy per atom. Then the rate at which atoms would have to be consumed would be far less, and the lifetime of the Sun could be long enough to be consistent with the known age of the Earth. Albert Einstein discovered the key to such a process in 1905. According to his *special theory of relativity*, a quantity $m$ of mass can in principle be converted into an amount of energy $E$ according to a now-famous equation:

**Einstein's mass-energy equation**

$$E = mc^2$$

$m$ = quantity of mass, in kg

$c$ = speed of light = $3 \times 10^8$ m/s

$E$ = amount of energy into which the mass can be converted, in joules

Because $c$ is a large number, and thus $c^2$ is huge, a small amount of matter can release an awesome amount of energy.

Inspired by Einstein's mass-energy formula, astronomers began to wonder if the Sun's energy output might come from the conversion of matter into energy. In the 1920s, British astronomer Arthur Eddington showed that temperatures near the center of the Sun must be much greater than had previously been thought. Another British astronomer, Robert Atkinson, suggested that hydrogen nuclei near the Sun's center might fuse together to produce helium nuclei in a reaction that transforms a tiny amount of mass into a large amount of energy. This process of **nuclear fusion,** by which hydrogen is converted into helium at the Sun's center, is called **hydrogen burning,** even though nothing is actually burned in the conventional sense. (The chemical reactions involved in ordinary burning act to rearrange the outer electrons of atoms but have no effect on their nuclei.) Because it can occur only at high temperatures, hydrogen burning is called a **thermonuclear reaction,** or **thermonuclear fusion.**

Recall from Section 5-8 that the nucleus of a hydrogen atom (H) consists of a single proton. The nucleus of a helium atom (He) consists of two protons and two neutrons. In the nuclear process described by Atkinson, four hydrogen nuclei would combine to form one helium nucleus, with a concurrent release of energy:

$$4\,\text{H} \rightarrow \text{He} + \text{energy}$$

In several separate reactions, two of the four protons from hydrogen are changed into neutrons and eventually combine to produce a single helium nucleus. These reactions also release two positively charged electrons, called **positrons,** that carry off the electric charges relinquished by the protons that changed into neutrons. In addition, two curious particles called **neutrinos,** which have no electric charge and little or no mass, carry off some energy and momentum. All together, the release of these four supplementary particles ensures that the net amounts of electric charge, energy, and momentum present before the reaction are the same after the reaction.

When hydrogen is converted into helium, a small fraction of the mass is lost. The product (one helium nucleus) has slightly less mass than the ingredients (four hydrogen nuclei):

| | | |
|---|---|---|
| 4 hydrogen atoms | = | $6.693 \times 10^{-27}$ kg |
| −1 helium atom | = | $-6.645 \times 10^{-27}$ kg |
| Mass lost | = | $0.048 \times 10^{-27}$ kg |

Thus, a small fraction (0.7%) of the mass of the hydrogen going into the nuclear reaction does not show up in the mass of the helium. This lost mass is converted into energy, as predicted by Einstein's equation:

$$E = mc^2 = (0.048 \times 10^{-27} \text{ kg})(3 \times 10^8 \text{ m/s})^2$$
$$= 4.3 \times 10^{-12} \text{ joule}$$

This is only a tiny amount of energy, because it results from the creation of just a single helium atom. But it is about $10^7$ times larger than the amount of energy released in a typical chemical reaction, such as occurs in ordinary burning. Thus, this process is just what is needed to explain how the Sun could have been shining for billions of years.

To be specific, consider the conversion of 1 kg of hydrogen into helium. Although a kilogram of hydrogen goes into this reaction, only 0.993 kg of helium comes out. Using Einstein's equation, we find that the missing 0.007 kg of matter has been transformed into an amount of energy equal to

$$E = mc^2 = (0.007 \text{ kg})(3 \times 10^8 \text{ m/s})^2 = 6.3 \times 10^{14} \text{ joules}$$

For comparison, this equals the energy released by burning 20,000 metric tons ($2 \times 10^7$ kg) of coal.

To produce the Sun's luminosity of $3.9 \times 10^{26}$ watts ($3.9 \times 10^{26}$ joules per second), $6 \times 10^{11}$ kg of hydrogen must be converted into helium within the Sun each second. This rate is prodigious, but the Sun contains a vast amount of hydrogen. The Sun's total mass, usually designated $M_\odot$, is $2 \times 10^{30}$ kg (333,000 Earth masses). The Sun's core contains enough hydrogen to have been giving off energy at the present rate for as long as the solar system has existed, about 4.6 billion years, and to continue doing so for another 5 billion years.

The Sun's energy actually comes from a *series* of nuclear reactions, called the **proton-proton chain,** in which hydrogen nuclei combine to form helium. The proton-proton chain is also the energy source for many of the stars in the sky. In stars with central temperatures that are hotter than that of the Sun, however, hydrogen burning proceeds according to a different set of nuclear reactions, called the **CNO cycle,** in which carbon, nitrogen, and oxygen nuclei absorb protons to produce helium nuclei. Details of the proton-proton cycle and the CNO cycle are discussed in Box 18-2. In later chapters we will see that other thermonuclear reactions, such as helium burning, carbon burning, and oxygen burning, occur late in the lives of many stars.

## box 18-2 | Looking Deeper into Astronomy

## *The Proton-Proton Chain and the CNO Cycle*

There are two avenues by which hydrogen burning proceeds inside a star. Both yield the same result: Four hydrogen protons combine to form one helium nucleus, and a slight amount of mass is converted into energy. The temperature at the star's core determines which avenue is taken.

For stars with masses no greater than the Sun's mass, the central temperature does not exceed 16 million K, and hydrogen burning proceeds predominantly via the proton-proton chain. For stars more massive than the Sun, the central temperature is above 16 million K, and hydrogen burning occurs primarily through a series of reactions called the CNO cycle.

The proton-proton chain, originally proposed by the American physicist Charles Critchfield, has four branches. The primary branch, called PP I, produces 85% of the Sun's energy. It occurs in three steps (see the accompanying diagram). First, two of the most common kind of hydrogen nuclei—single protons, each denoted by $^1$H—combine to form an isotope of hydrogen called **deuterium**, which consists of one proton and one neutron bound together. This isotope is denoted $^2$H. (The superscript in nuclear designations such as $^1$H and $^2$H tells you the total number of protons and neutrons in the nucleus. We described this and many other facts about nuclei and their isotopes in Box 5-4.) In this step, one of the two protons turns into a neutron, releasing a positron ($e^+$) and a neutrino (denoted by the Greek letter $\nu$, or nu). A positron ($e^+$) is just like an ordinary electron ($e^-$), except that it has a positive rather than a negative electric charge. So, the first step in the proton-proton chain can be written as

$$^1H + {}^1H \rightarrow {}^2H + e^+ + \nu$$

The positron combines with an electron and annihilates into gamma-ray photons.

Most matter is virtually transparent to neutrinos. Consequently, neutrinos are very difficult (although not impossible) to detect, and their properties are a topic of ongoing research and debate. Neutrinos have no electric charge, but they may have a small mass. A neutrino is produced whenever a proton is converted into a neutron. Conversely, an antineutrino $\bar{\nu}$ is liberated when a neutron turns into a proton.

In the second step of the proton-proton chain, a third proton combines with the deuterium nucleus to produce a low-mass isotope of helium ($^3$He), whose nucleus contains two protons and one neutron. This reaction also releases

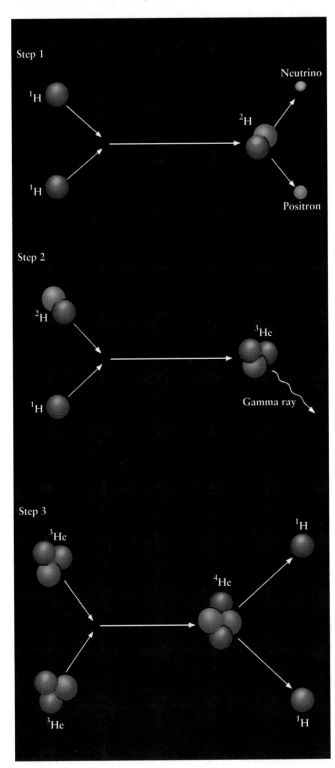

an equally large patch of umbra. Again, which patch is brighter?

**24.** Suppose that you want to determine the Sun's rotation rate by observing its sunspots. Is it necessary to take the Earth's orbital motion into account? Why or why not?

**25.** How much energy would be released if each of the following masses were converted *entirely* into its equivalent energy: (**a**) a carbon atom with a mass of $2 \times 10^{-26}$ kg, (**b**) 1 kilogram, and (**c**) a planet as massive as the Earth ($6 \times 10^{24}$ kg)?

**26.** Use the luminosity of the Sun (given in Table 18-1) and the answers to the previous question to calculate how long the Sun must shine in order to release an amount of energy equal to that produced by the complete mass-to-energy conversion of (**a**) a carbon atom, (**b**) 1 kilogram, and (**c**) the Earth.

**27.** The brightest-appearing star in the sky, Sirius, has a luminosity of 23.5 $L_\odot$, that is, it is 23.5 times as luminous as the Sun and burns hydrogen at a rate 23.5 times greater than the Sun. How many kilograms of hydrogen does Sirius convert into helium each second?

**28.** (Refer to the preceding question.) Sirius has 2.3 times the mass of the Sun. Do you expect that the lifetime of Sirius will be longer, shorter, or the same length as that of the Sun? Explain your reasoning.

**29.** Assuming that the current rate of hydrogen burning in the Sun remains constant, what fraction of the Sun's mass will be converted into helium over the next 5 billion years? How will this affect the overall chemical composition of the Sun?

**30.** What would happen if the Sun were not in a state of both hydrostatic and thermal equilibrium? Explain your reasoning.

**31.** In a typical solar oscillation, the Sun's surface moves up or down at a maximum speed of 0.1 m/s. An astronomer sets out to measure this speed by detecting the Doppler shift of an absorption line of iron with wavelength 557.6099 nm. What is the maximum wavelength shift that she will observe?

**32.** What is the status of Super Kamiokande, the Sudbury Neutrino Observatory, and Borexino? What are the most recent results from these experiments? What is the current thinking about the mystery of the missing solar neutrinos? (You should visit the web sites for these experiments, listed below under Where to Learn More.)

## Discussion Questions

**33.** Discuss the extent to which cultures around the world have worshipped the Sun as a deity throughout history. Why do you suppose there has been such widespread veneration?

**34.** Discuss some of the difficulties in correlating solar activity with changes in the terrestrial climate.

**35.** Describe some of the advantages and disadvantages of observing the Sun (**a**) from space and (**b**) from the South Pole. What kinds of phenomena and issues do solar astronomers want to explore from both the Earth-orbiting and the Antarctic observatories?

## Observing Projects

**Observing tips and tools:**

At the risk of repeating ourselves, we remind you to *never look directly at the sun, because this can easily cause permanent blindness.* You can view the Sun safely without a telescope just by using two pieces of white cardboard. First, use a pin to poke a small hole in one piece of cardboard; this will be your "lens," and the other piece of cardboard will be your "viewing screen." Hold the "lens" piece of cardboard so that it is face-on to the Sun and sunlight can pass through the hole. With your other hand, hold the "viewing screen" so that the sunlight from the "lens" falls on it. Adjust the distance between the two pieces of cardboard so that you see a sharp image of the Sun on the "viewing screen." So little light passes through the pinhole that this image is perfectly safe to view. It is actually possible to see sunspots with this low-tech apparatus.

For a better view, use a telescope with a solar filter that fits on the front of the telescope. A standard solar filter is a piece of glass coated with a thin layer of metal to give it a mirrorlike appearance. This coating reflects almost all of the sunlight that falls on it, so that only a tiny, safe amount of sunlight enters the telescope. An $H_\alpha$ filter, which looks like a red piece of glass, keeps the light at a safe level by admitting only a very narrow range of wavelengths. (Filters that fit on the back of the telescope are *not* recommended. The telescope focuses concentrated sunlight on such a filter, heating it and making it susceptible to cracking—and if the filter cracks when you are looking through it, your eye will be ruined instantly and permanently.)

To use a telescope with a solar filter, first aim the telescope away from the Sun, then put on the filter. Keep the lens cap on the telescope's secondary wide-angle "finder scope" (if it has one), because the heat of sunlight can fry the finder scope's optics. Next, aim the telescope toward the Sun, using the telescope's shadow to judge when you are pointed in the right direction. You can then safely look through the telescope's eyepiece. When you are done, make sure you point the telescope away from the Sun before removing the filter and storing the telescope.

Note that the amount of solar activity that you can see (sunspots, filaments, flares, prominences, and so on) will depend on where the Sun is in its 11-year sunspot cycle.

**36.** Use a telescope with a solar filter to observe the surface of the Sun. Do you see any sunspots? Sketch their appearance. Can you distinguish between the umbrae and penumbrae of the sunspots? Can you see limb darkening? Can you see any granulation?

**37.** If you have access to an $H_\alpha$ filter attached to a telescope

At what speed would a 10g marshmallow be traveling when it's dropped onto the surface of a neutron star from very far away? (Assume the neutron star has a mass 2x that of ☉ and a radius ≈ 10km) How much energy would be released in J when it hits the surface?

What is the Schwertzschild Radius of the ☉?

Plan at least-energy mission from Earth to Mars. What additional boost in speed (beyond the Earth's 30 km/s) would the probe have to be given? How long would it take to get the probe to Mars? How fast would it be traveling w/respect to Mars when it arrived there?

The Earth and Moon are stopped by an evil demon and then released. Describe what would happen next. How does the phase of the moon affect your answer?

helioseismology, detecting solar neutrinos, and understanding the corona.

———. "SOHO Reveals the Secrets of the Sun." *Scientific American,* March 1997. This lucid article describes the *SOHO* spacecraft; the twelve different scientific instruments that it carries to observe the Sun's interior, atmosphere, and solar wind; and some of the first scientific results from these observations.

Leibacher, J. W., and others. "Helioseismology." *Scientific American,* September 1985. This lucid article explains how astronomers can learn about the structure, composition, and dynamics of the Sun's interior from oscillations visible on its surface.

Marsden, R. G., and Smith, E. J. "*Ulysses:* Solar Sojourner." *Sky & Telescope,* March 1996. This article describes the *Ulysses* mission and its remarkable journey to examine the poles of the Sun.

Nesme-Ribes, E., Baliunas, S. L., and Sokoloff, D. "The Stellar Dynamo." *Scientific American,* August 1996. This article describes our current state of understanding of the Sun's magnetic field and its connection to the cycle of solar activity.

Schaefer, B. E. "Sunspots that Changed the World." *Sky & Telescope*, April 1997. As this article describes, sunspots and solar activity have had some surprising effects on human history.

Smith, E. J., and Marsden, R. G. "The *Ulysses* Mission." *Scientific American*, January 1998. Two scientists deeply involved with the *Ulysses* spacecraft—the first to view the polar regions of the Sun—describe the remarkable scientific discoveries made during the spacecraft's first few years in space.

**W** *World Wide Web*

As for the planets, two good places to start your exploration of the Sun on the World Wide Web are the sites "The Nine Planets" (**http://www.seds.org/nineplanets/nineplanets/sol.html**) and "Views of the Solar System" (**http://www.hawastsoc.org/ solar/eng/sun.htm**).

The Virtual Sun (**http://www.astro.uva.nl/michielb/sun/**) is a useful and informative multimedia tour. Images of the Sun and other solar information are made available on a daily basis at NASA's Solar Data Analysis Center (**http://umbra.nascom.nasa.gov/**). Some of the most striking images and movies are available at the web site for the *SOHO* spacecraft (**http://sohowww.nascom.nasa.gov/**); some of the most remarkable of these have come from the Large Angle and Spectrometric Coronagraph Experiment (LASCO) on board *SOHO* (**http://lasco-www.nrl.navy.mil/lasco.html**). Data from the *Ulysses* spacecraft can be found at either the mission's ESA web site (**http://helio.estec.esa.nl/ulysses/**) or its NASA site (**http://ulysses.jpl.nasa.gov/**), while *Yohkoh* images can be found at a site at Montana State University (**http://solar.physics.montana.edu/YPOP/**).

How the solar wind affects the Earth and its environment is the subject matter of the International Solar-Terrestrial Physics web site (**http://www-istp.gsfc.nasa.gov/**). The latest information about solar neutrino experiments can be found at the web sites for Super Kamiokande (**http://www.phys.washington.edu/~superk/**), the Sudbury Neutrino Observatory (**http://snodaq.phy.queensu.ca/SNO/sno.html**), and Borexino (**http://pupgg.princeton.edu/~borexino/welcome.html**).

There is also a web site for one of the most active groups in helioseismology, the Global Oscillation Network Group (**http://helios.tuc.noao.edu/gonghome.html**), as well as one for NASA's proposed mission to send a spacecraft into the solar corona (**http://www.jpl.nasa.gov/pluto/sprobe.htm**).

# JOHN N. BAHCALL

## Searching for Neutrinos Beyond the Textbooks

When John N. Bahcall went in search of solar neutrinos, he applied atomic and nuclear physics to the stars. In his many areas of expertise, in fact, Dr. Bahcall studies just that interplay—between the theories of physics and our understanding of the heavens. At the Institute for Advanced Study in Princeton, New Jersey, he has looked at emerging models of our Galaxy, dark matter, stellar evolution, and the spectra of quasars. He has also received the NASA Distinguished Public Service Medal for his work with the Hubble Space Telescope.

After graduate school at Harvard University, Dr. Bahcall joined the faculty of the California Institute of Technology, where he remained until 1971. A former president of the American Astronomical Society, he recently chaired the National Academy Design Survey Committee for astronomy and astrophysics. His research on solar neutrinos earned him the Heinemann Prize of the American Astronomical Society.

In attempting to understand how the Sun shines, physicists, chemists, and astronomers have been confronted with a mystery—the case of the missing neutrinos. These exotic particles travel at essentially the speed of light. They are so elusive that they can traverse a thousand light-years of lead before being stopped.

Physicists and astronomers believe that the Sun shines because of the conversion of hydrogen nuclei (protons) into helium nuclei (alpha particles), with the subsequent release of a substantial amount of nuclear energy. The same basic process, *nuclear fusion*, produces the explosion of a hydrogen bomb. We think that about 600 million tonnes (or $6 \times 10^{11}$ kg) of hydrogen are converted to helium every second in the Sun's central regions, providing the energy that we know as sunlight and making life on the Earth possible.

In the early 1960s, Ray Davis and I proposed to test the theory of how the Sun shines. Ray, a chemist at Brookhaven National Laboratory, had developed a neutrino detector that uses a cleaning fluid containing chlorine. Using standard theories of physics, I calculated the rate at which neutrinos are produced in the Sun. I could then predict the rate at which neutrinos should be captured in the largest detector Ray could build. If my calculations matched experiment, they would confirm that the Sun shines by nuclear fusion in its interior.

The actual experiment used 100,000 gallons of per-chloroethylene, about enough to fill an Olympic swimming pool. Ray Davis and his collaborators put their detector in a deep gold mine, to shield it from other particles that hit

the surface of the Earth. To everyone's surprise, Ray's chlorine detector captured many fewer neutrinos than I had predicted. The results were challenged and checked repeatedly over the following three decades, but always with the same result: many neutrinos appear to be missing!

The case of the missing neutrinos has grown stronger with time. Three other experiments, each with a different type of detector, searched for neutrinos from the Sun. They all found fewer than I predicted.

What could be wrong? Where have the neutrinos gone? There are three possibilities. Either the experiments are wrong, the standard model of how the Sun (and other stars) shine is wrong, or something happens to the neutrinos after they are produced. New experiments are now under way in Japan, in Italy, and in Canada to test these hypotheses.

Right now most people working in the field think that the last of the three explanations is most likely to be correct: Only physics beyond our textbooks can describe what has happened to solar neutrinos. Somehow, most physicists think, neutrinos created in the solar interior change into neutrinos that are more difficult to detect as they pass out of the Sun and travel to the Earth. They change their personalities, so to speak! If this is correct, it would be the first experimental demonstration of a process beyond the standard model of particle physics.

So far, evidence for the new physics is only circumstantial. We know only that the results differ markedly from predictions based on our understanding now of how the

Sun shines. Future experiments are designed to search for a "smoking gun"—unequivocal evidence of processes not in the physics textbooks. These new experiments will use the fact that neutrinos come in different types. Most easily detected are the so-called *electron-type* neutrinos; more difficult to detect are *muon-type* and *tau-type* neutrinos. Suppose some of these neutrinos from the Sun convert into neutrinos that are easier to detect as they pass through the Earth at night on their way to the detector. The change would make the Sun appear brighter at night than in the day. If that were seen, it would provide a dramatic demonstration that unconventional physics is occurring.*

If a smoking gun is found, it could offer a clue to new laws of particle physics, the laws governing the smallest scales of matter. Observations of solar neutrinos, together with laboratory experiments still to come, might reveal evidence that neutrinos have tiny masses. Conceivably, their masses could then account for much of the pervasive "dark matter" in the universe.

I do not know the correct explanation for the missing neutrinos. However, the particle-physics explanation has a mathematical beauty and simplicity that are very attractive. If the deity has not chosen this solution to the mystery, then he or she has missed an excellent opportunity to enrich the laws of the universe.

However, the real message of the experiments on solar neutrinos is even more remarkable. It is that working on the frontier of science, you may stumble across something that is beautiful and unexpected. We do not yet know if this has indeed happened in the study of solar neutrinos, but we do know that future experiments will solve the mystery for us. Their outcome might point the way to a better understanding of fundamental physics and the stars.

---

* *Note added in proof:* Just as this book was going to press, scientists at the Super Kamiokande neutrino observatory in Japan reported having observed an effect of just this type. Their results indicate that electron–type and muon–type neutrinos can indeed transform into each other as they pass through matter, an effect called *neutrino oscillation*. The most recent results from Super Kamiokande can be found on the World Wide Web (http://www.phys.washington.edu/~superk/ or http://www.phys.hawaii.edu/~superk/).

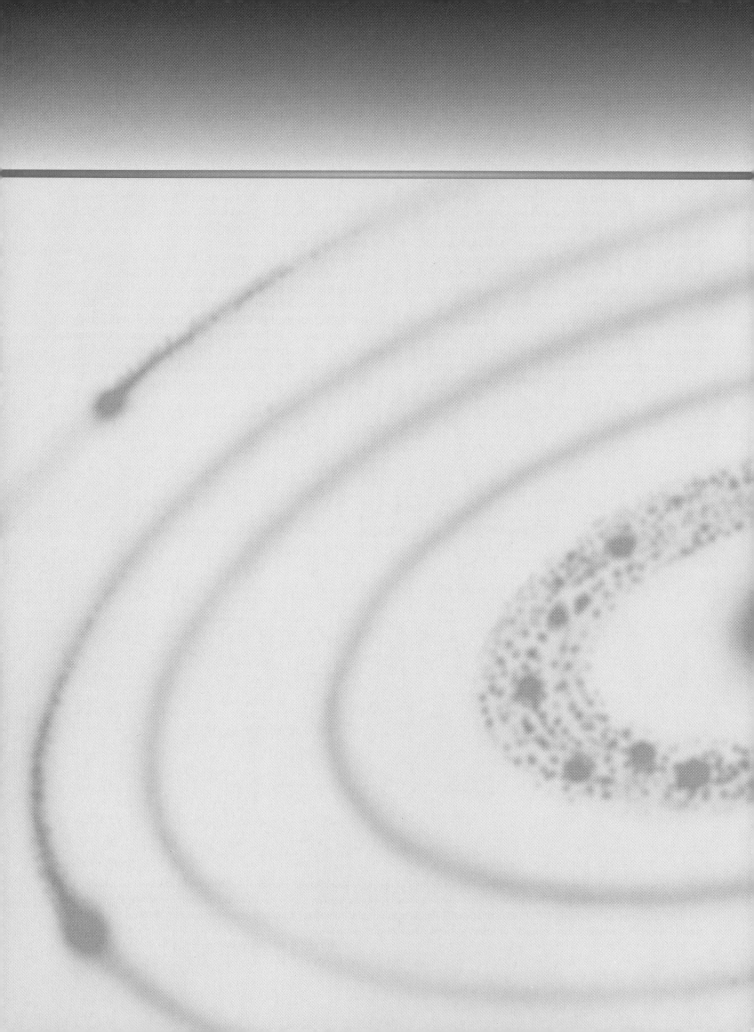

# The Nature of Stars

R I **V** U X G

**A Cluster of Stars** By analyzing its light, we can determine a star's surface temperature, chemical composition, and luminosity. This photograph shows color differences in the star cluster NGC 3293. The reddish stars visible in this image are of a type called red giants. They are comparatively cool, with surface temperatures around 3000 K. They are also quite luminous and have large diameters, typically 100 times larger than our Sun. The bluish and blue-white stars in this image have much higher surface temperatures (15,000 to 30,000 K) but are about the same size as the Sun. (Anglo-Australian Observatory)

*In this chapter you will find the answers to the following questions:*

19-1 How far away are the stars?

19-2 What evidence do astronomers have that the Sun is a typical star?

19-3 What is meant by a "first-magnitude" or "second-magnitude" star?

19-4 Why are some stars red and others blue?

19-5 What are the stars made of?

19-6 As stars go, is our Sun especially large or small?

19-7 What are giant stars, supergiant stars, and white dwarf stars?

19-8 How do we know the distances to remote stars?

19-9 Why are binary star systems important in astronomy?

19-10 How can a star's spectrum show whether it is actually a binary system?

19-11 What do astronomers learn from stars that eclipse each other?

To the unaided eye, the night sky is spangled with more than a thousand stars, each appearing as a bright pinpoint of light. With a pair of binoculars, you can see some 10,000 other, fainter stars; with a 15-cm (6-in.) telescope, the total rises to more than 2 million. Astronomers now know that there are in excess of 100 billion ($10^{11}$) stars in our galaxy alone—a few dozen stars for every man, woman, and child on the Earth. But what are these distant pinpoints? How can we know the nature of objects so remote that their light takes years, centuries, or millennia to reach us? Remarkably, astronomers have learned how to measure not only the distances to the stars and how stars move through space, but also the masses, radii, luminosities, temperatures, and chemical compositions of the stars. They have found that stars are huge, massive balls of hot gas much like our Sun, held together by their own gravity. Some stars are larger than our Sun and some smaller; some stars are brighter than the Sun and some dimmer. Some stars are hotter than the Sun, but others are cooler. Binary star systems, which consist of two stars orbiting each other, are quite common.

In this chapter we will learn about the measurements and calculations that astronomers make to determine these properties of stars. We will learn how astronomers can find the masses of stars by observing the motions of stars orbiting each other. We will also take a first look at the Hertzsprung-Russell diagram, an important tool that helps astronomers systematize the wealth of available information about the stars. In later chapters we will use this diagram to help us understand the ways in which stars are born, evolve, and eventually die.

## 19-1  Careful measurements of the parallaxes of stars reveal their distances

We now understand that the vast majority of stars are objects very much like the Sun. They are giant balls of hydrogen and helium that generate energy in their interiors by thermonuclear reactions and release this energy into space as electromagnetic radiation. This understanding followed from the discovery that the stars are tremendously far from us, at distances so great that their light takes years to reach us. Because the stars at night are clearly visible to the naked eye despite these huge distances, it must be that the **luminosity** of the stars—that is, how much energy they emit into space per second—is comparable to or greater than that of the Sun. Just as for the Sun, the only explanation for such tremendous luminosities is that thermonuclear reactions are occurring within the stars (see Section 18-6).

Clearly, then, it is important to know how far away the stars are. But how can we measure a star's distance? To answer this question, you might think that you need only look at how bright different stars appear in the nighttime sky. Just as the headlights of a car at night appear brighter the closer the car is, perhaps the bright star Betelgeuse in the constellation Orion is relatively close, while the dimmer and less conspicuous star Polaris (the North Star, in the constellation Ursa Minor) is farther away. But this line of reasoning is incorrect: Polaris is actually closer to us than Betelgeuse! How bright a star appears is *not* a good indicator of its distance. If you see a light ahead of you on a darkened road, it could be a motorcycle headlight a kilometer away or a person holding a flashlight just a few meters away. In the same way, a bright star might be extremely far away but have an unusually high luminosity, and a dim star might be relatively close but have a rather low luminosity. To determine the distances to the stars, astronomers must use other techniques.

The most straightforward way of measuring stellar distances uses an effect called **parallax.** This is the apparent displacement of an object because of a change in the observer's point of view (Figure 19-1). To see how parallax works, hold your arm out straight in front of you. Now look at the hand on your outstretched arm, first with your left eye closed, then with your right eye closed. When you close one eye and open the other, your hand appears to shift back and forth against the background of more distant objects. The closer the object you are viewing, the greater the parallax shift; to see this, repeat the experiment with your hand held closer to your face and notice the increase in the shift. In ordinary vision, your brain analyzes such parallax shifts constantly, as it compares the images that it receives from your left and right eyes. That is how your brain determines the distances to objects around you and provides you with depth perception.

## figure 19-1

**Parallax** Imagine looking at some nearby object (a tree) against a distant background (mountains). When you move from one location to another, the nearby object appears to shift with respect to the distant background scenery. This familiar phenomenon is called parallax.

To measure the distance to a star, astronomers measure the parallax shift of the star using two points of view that are as far apart as possible—at opposite sides of the Earth's orbit. The direction from the Earth to a nearby star changes as the Earth orbits the Sun, and the nearby star appears to move back and forth against the background of more distant stars (Figure 19-2). This motion is called **stellar parallax.** As shown in Figure 19-2, the parallax ($p$) of a star is equal to half the angle through which the star's apparent position shifts as the Earth moves from one side of its orbit to the other. The larger the parallax $p$, the smaller the distance $d$ to the star (compare Figure 19-2a with Figure 19-2b).

It is convenient to measure the distance $d$ in **parsecs;** a star with a parallax angle of 1 second of arc ($p = 1$ arcsec) would be at a distance of 1 parsec ($d = 1$ pc). (The word "parsec" is a contraction of the phrase "the distance at which a star has a *par*allax of one arc*sec*ond." Recall from Section 1-7 that 1 parsec equals 3.26 light-years, $3.09 \times 10^{13}$ km, or 206,265 AU; see Figure 1-13.) If the angle $p$ is measured in arcseconds, then the distance $d$ to the star in parsecs is given by the following equation:

**Relation between a star's distance and its parallax**

$$d = \frac{1}{p}$$

$d$ = distance to a star, in parsecs
$p$ = parallax angle of that star, in arcseconds

This simple relationship between parallax and distance in parsecs is one of the main reasons that astronomers usually measure cosmic distances in parsecs rather than light-years. For example, a star whose parallax is $p = 0.1$ arcsec is at a distance $d = 1/(0.1) = 10$ parsecs from the Earth. Barnard's star, named for American astronomer Edward E. Barnard, has a parallax of 0.545 arcsec. Hence, the distance to this star is:

$$d = \frac{1}{p} = \frac{1}{0.545} = 1.83 \text{ pc}$$

Because 1 parsec is 3.26 light-years, this can also be expressed as

$$d = 1.83 \text{ pc} \times \frac{3.26 \text{ ly}}{1 \text{ pc}} = 5.98 \text{ ly}$$

Eighteenth-century astronomers tried to measure the parallax angles of stars. They failed, because these angles are very small: All known stars have parallax angles less than one arcsecond, meaning that the closest star is more than 1 parsec away. It was not until 1838 that the German astronomer and mathematician Friedrich Wilhelm Bessel measured the parallax of the star 61 Cygni to be $\frac{1}{3}$ arcsec and thus determined that it is about 3 pc from the Earth. (Modern measurements of 61 Cygni give a parallax angle of 0.289 arcsec and a distance of 3.46 pc.) The star Proxima Centauri has the largest known parallax angle, 0.772 arcsec, and hence is the closest known star (other than the Sun); its distance is $1/(0.772) = 1.30$ pc. The parallax of Proxima Centauri is comparable to the angular diameter of a dime seen from a distance of about 3 kilometers (2 miles).

Appendix 4 at the back of this book lists all the stars within 4 pc of the Sun, as determined by parallax measurements. Most of these stars are far too dim to be seen with the naked eye, which is why their names are probably unfamiliar to you. By contrast, the majority of the familiar, bright stars in the nighttime sky (listed in Appendix 5) are so far away that their parallaxes cannot be measured from the Earth's surface. They appear bright not because they are close, but because they are far more luminous than the Sun. The brightest stars in the sky are *not* necessarily the nearest stars.

Because stellar parallax angles are so tiny, measuring stellar distances using the parallax method is one of the most challenging tasks for astronomers. Parallaxes smaller than about 0.01 arcsec are extremely difficult to measure from the Earth, in part because of the blurring effects of the atmosphere. Therefore, the parallax method used with ground-based telescopes can give fairly reliable distances only for stars nearer than about $1/0.01 = 100$ pc. But observations from an Earth-orbiting satellite are unhampered by our atmosphere, permitting astronomers to measure even smaller parallax angles and thereby determine the distances to more remote stars. In 1989 the European Space Agency (ESA) launched the satellite *Hipparcos*, an acronym for *High Precision Parallax Collecting Satellite* (and a commemoration of the ancient Greek astronomer Hipparchus, who created one of the first star charts and whose system for classifying stars by their brightness is still in use today). Over

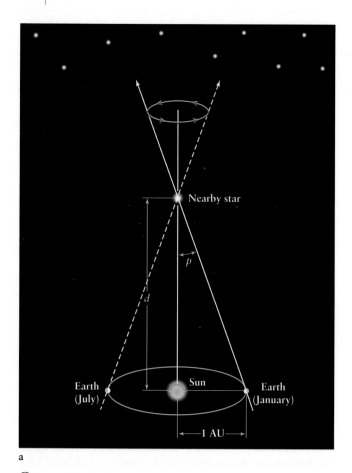

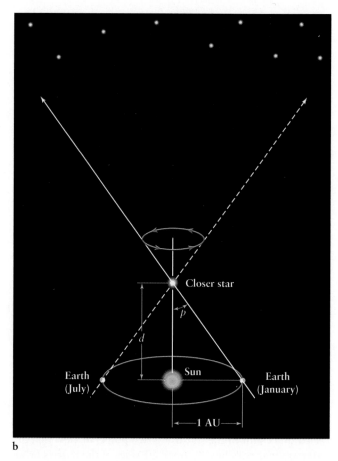

a

b

## figure 19-2

**Stellar Parallax** **(a)** As the Earth orbits the Sun, a nearby star appears to shift its position against the background of distant stars. The parallax ($p$) of the star is equal to the angular radius of the Earth's orbit as seen from the star. **(b)** The closer the star is to us, the greater the parallax angle $p$. The distance $d$ to the star (in parsecs) is found by taking the reciprocal of the parallax angle $p$ (in arcseconds): $d = 1/p$.

more than three years of observations, the telescope aboard *Hipparcos* was used to measure the parallaxes of 118,000 stars with an accuracy of 0.001 arcsecond. This has enabled astronomers to determine stellar distances out to several hundred parsecs, and with much greater precision than has been possible with ground-based observations. During the next century, astronomers will increasingly turn to space-based observations to determine stellar distances.

The measurement of the distances to the stars is tremendously important in modern astronomy. With such measurements, astronomers have charted the size and shape of the entire Milky Way Galaxy, as we will discuss in Chapter 25. Unfortunately, most of the stars in the Galaxy are so far away that their parallax angles are too small to be measured even with an orbiting telescope. Later in this chapter, we will discuss a technique that can be used to find the distances to these more remote stars. Still other techniques, to be discussed in Chapters 26 and 28, allow astronomers to determine not only the distances to other galaxies but also the overall size, age, and structure of the entire universe.

Because it can be used only on relatively close stars, stellar parallax might seem to be of limited usefulness. But in fact, parallax measurements are the cornerstone for all other methods of finding the distances to remote objects. The reason is that all these other methods require a precise and accurate knowledge of the distances to nearby stars, as determined by stellar parallax. This means that any inaccuracies in parallax measurements for nearby stars can translate into substantial errors in measurement for the whole universe. For this reason, astronomers are continually trying to perfect their parallax-measuring techniques.

Stellar parallax is an *apparent* motion of stars caused by the Earth's orbital motion around the Sun. But stars are not fixed objects and actually do move through space. As a result, stars change their positions on the celestial sphere (Figure 19-3), and they move either toward or away from the Sun. These motions are sufficiently slow, however, that changes in the positions of the stars are hardly noticeable over a human lifetime. Box 19-1 describes how astronomers study these motions, and what insights they gain from these studies.

# box 19-1 | Tools of the Astronomer's Trade

## *Stellar Motions*

Stars can move through space in any direction. The **space velocity** of a star describes how fast and in what direction it is moving. As the accompanying Figure shows, a star's space velocity $v$ can be broken into components parallel and perpendicular to our line of sight.

The component parallel to our line of sight—that is, across the plane of the sky—is called the star's **tangential velocity** ($v_t$). To determine it, astronomers must know the distance to a star ($d$) and its **proper motion** ($\mu$, the Greek letter mu), which is the number of arcseconds that the star appears to move per year on the celestial sphere. Proper motion does not repeat itself yearly, so it can be distinguished from the apparent back-and-forth motion due to parallax. In terms of a star's distance and proper motion, its tangential velocity (in km/s) is

$$v_t = 4.74\mu d$$

where $\mu$ is in arcseconds per year and $d$ is in parsecs. For example, Barnard's star has a proper motion of 10.358 arcseconds per year and a distance of 1.82 pc, so its tangential velocity is

$$v_t = 4.74(10.358)(1.82) = 89.4 \text{ km/s}$$

The component of a star's motion parallel to our line of sight is its **radial velocity** ($v_r$). It can be determined from measurements of the Doppler shifts of the star's spectral lines. As described in Section 5-9 and Box 5-6, if a star is approaching us, the wavelengths of all of its spectral lines are decreased (blueshifted); if the star is receding from us, the wavelengths are increased (redshifted). The radial velocity is related to the wavelength shift by the equation

$$\frac{\lambda - \lambda_0}{\lambda_0} = \frac{v_r}{c}$$

In this equation, $\lambda$ is the wavelength of light coming from the star, $\lambda_0$ is what the wavelength would be if the star were not moving, and $c$ is the speed of light. As an illustration, a particular spectral line of iron in the spectrum of Barnard's star has a wavelength ($\lambda$) of 516.438 nm. As measured in a laboratory on the Earth, the same spectral line has a wavelength ($\lambda_0$) of 516.629 nm. Thus, for Barnard's star, our equation becomes

$$\frac{516.438 - 516.629}{516.629} = -0.000370 = \frac{v_r}{c}$$

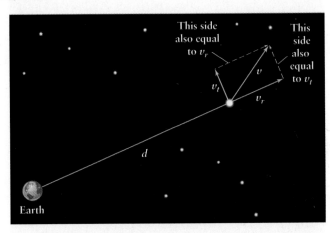

Solving this equation for the radial velocity $v_r$, we find

$$v_r = (-0.000370)c = (-0.000370)(3 \times 10^5 \text{ km/s}) = -111 \text{ km/s}$$

The negative sign means that Barnard's star is moving toward us. You can check this interpretation by noting that the wavelength $\lambda = 516.438$ nm received from Barnard's star is less than the laboratory wavelength $\lambda_0 = 516.629$ nm; hence, the light from the star is blueshifted, which indeed means that the star is approaching. If the star were receding, its radial velocity would be positive.

The illustration shows that the tangential velocity and radial velocity form two sides of a right triangle. The long side (hypotenuse) of this triangle is the space velocity ($v$). From the Pythagorean theorem, the space velocity is

$$v = \sqrt{v_t^2 + v_r^2}$$

For Barnard's star, the space velocity is

$$v = \sqrt{(-111 \text{ km/s})^2 + (89.4 \text{ km/s})^2} = 143 \text{ km/s}$$

Therefore, Barnard's star is moving though space at a speed of 143 km/s (515,000 km/h, or 320,000 mi/h) relative to the Sun.

Determining the space velocities of stars is essential for understanding the structure of the Galaxy. Studies show that the stars in our local neighborhood are moving in wide orbits around the center of the Galaxy, which lies some 8000 pc (26,000 light-years) away in the direction of the constellation Sagittarius (the Archer). While many of the orbits are roughly circular and lie in nearly the same plane, others are highly elliptical or steeply inclined to the galactic plane. We will see in Chapter 25 how the orbits of stars in our Galaxy are related to the Galaxy's spiral shape.

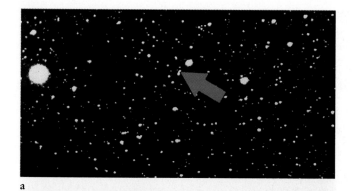

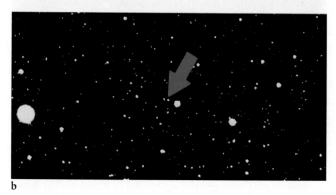

ƒigure 19-3   R I Ⓥ U X G

**The Motion of Barnard's Star** These two photographs of Barnard's star, which lies 1.82 parsecs away in the constellation Ophiuchus, were taken 22 years apart. Over this time interval, Barnard's star moved almost 230 arcseconds on the celestial sphere (about 4 arcminutes, or 0.06°). In 200 years this star travels the apparent angular diameter of the full moon. Most stars appear to move more slowly across the sky because they are farther from the Earth. Barnard's star emits only about $10^{-4}$ as much energy per second as the Sun, so despite its closeness, this star is too dim to be seen with the naked eye. (Yerkes Observatory)

---

| 19-2 | If a star's distance is known, its luminosity can be determined from its brightness |

Although all the stars you can see in the nighttime sky shine by thermonuclear fusion, just as the Sun does, they are by no means merely identical copies of the Sun. At the beginning of this chapter, we introduced one key property that differs from one star to the next—the *luminosity* (*L*), or amount of light energy emitted by the star each second. Luminosity is usually measured either in watts (1 watt is 1 joule per second) or as a multiple of the Sun's luminosity ($L_\odot$, equal to $3.90 \times 10^{26}$ W). Most stars are less luminous than the Sun, but some blaze forth with a million times the Sun's luminosity. Knowing a

star's luminosity is essential for determining the star's history, its present-day internal structure, and its future evolution.

To determine the luminosity of a star, we must first note that as light energy moves away from its source, it spreads out over increasingly larger regions of space. Imagine a sphere of radius *d* centered on the light source, as in Figure 19-4. The amount of energy passing through each square meter of the sphere each second is the total luminosity of the source (*L*) divided by the total surface area of the sphere (equal to $4\pi d^2$). The resulting quantity, called the **apparent brightness** of the light (*b*), is measured in watts per square meter:

**Inverse-square law relating apparent brightness and luminosity**

$$b = \frac{L}{4\pi d^2}$$

$b$ = apparent brightness of a star's light, in W/m$^2$
$L$ = star's luminosity, in W
$d$ = distance to star, in meters

This relationship is called the **inverse-square law,** because the apparent brightness of light that an observer can see or measure is inversely proportional to the square of the observer's distance (*d*) from the source. If you double your distance from a light source, its radiation is spread out over an area four times larger, so the apparent brightness you see is decreased by a factor of 4. Similarly, at triple the distance the apparent brightness decreases by a factor of 9, as demonstrated in Figure 19-4.

We can apply the inverse-square law to the Sun, which is $1.50 \times 10^{11}$ m from the Earth. Its apparent brightness ($b_\odot$) is

$$b_\odot = \frac{3.90 \times 10^{26} \text{ W}}{4\pi (1.50 \times 10^{11} \text{ m})^2} = 1370 \text{ W/m}^2$$

That is, a solar panel with an area of 1 square meter receives 1370 watts of power from the Sun.

The apparent brightness of a star can be measured using a telescope with an attached light-sensitive instrument, similar to the light meter in a camera that determines the proper exposure. Measuring a star's apparent brightness is called **photometry.**

To find the luminosity of a star, it is most convenient to express the inverse-square law in a somewhat different form. We first rearrange the inverse-square law:

$$L = 4\pi d^2 b$$

We then apply this equation to the Sun, that is, we write a similar equation relating the Sun's luminosity ($L_\odot$), the distance from the Earth to the Sun ($d_\odot$, equal to 1 AU), and the Sun's apparent brightness ($b_\odot$):

$$L_\odot = 4\pi d_\odot^2 b_\odot$$

If we take the ratio of these two equations, the unpleasant factor of $4\pi$ drops out and we are left with the following:

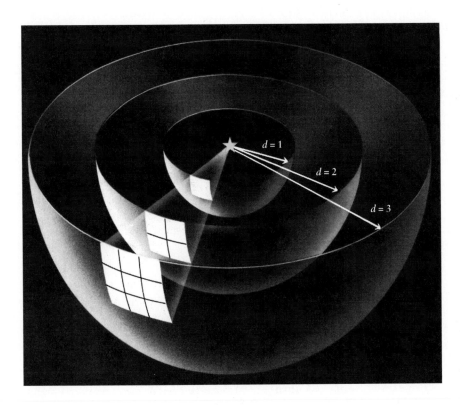

ƒigure 19-4

**The Inverse-Square Law** This drawing shows how the same amount of radiation from a light source must illuminate an ever-increasing area as distance from the light source increases. This area increases as the square of the distance from the source, so the apparent brightness decreases as the square of the distance; the brightness at $d = 2$ is $1/(2^2) = \frac{1}{4}$ of the brightness at $d = 1$, and the brightness at $d = 3$ is $1/(3^2) = \frac{1}{9}$ of the brightness at $d = 1$.

## Determining a star's luminosity from its apparent brightness

$$\frac{L}{L_\odot} = \left(\frac{d}{d_\odot}\right)^2 \frac{b}{b_\odot}$$

$L/L_\odot$ = ratio of the star's luminosity to the Sun's luminosity

$d/d_\odot$ = ratio of the star's distance to the Earth-Sun distance

$b/b_\odot$ = ratio of the star's apparent brightness to the Sun's apparent brightness

We need to know just two things to find a star's luminosity: the distance to a star as compared to the Earth-Sun distance (the ratio $d/d_\odot$), and how that star's apparent brightness compares to that of the Sun (the ratio $b/b_\odot$). Then, by using the above equation, we can determine how luminous that star is compared to the Sun (the ratio $L/L_\odot$). As an example, in Box 19-2 we use this equation to determine the luminosity of the nearby star ε (epsilon) Eridani, the fifth brightest star in the constellation Eridanus (named for a river in Greek mythology). Parallax measurements show that ε Eridani is 3.27 pc away, and photometry shows that the star appears only $6.73 \times 10^{-13}$ as bright as the Sun; using the above equation, we find that ε Eridani has only 0.31 times the luminosity of the Sun.

Calculations of this kind show that stars come in a wide variety of different luminosities, with values that range from about $10^6 L_\odot$ (a million times the Sun's luminosity) to only about $10^{-4} L_\odot$ (a mere ten-thousandth of the Sun's light output). This emphasizes that stars are a remarkably diverse lot. The most luminous star emits roughly $10^{10}$ times more energy each second than the least luminous! (To put this number in perspective, about $10^{10}$ human beings have lived on the Earth since the beginning of our species.) Our own Sun lies roughly in the middle of this luminosity range, suggesting that it is a rather ordinary, garden-variety star.

Studies of luminosity also show that most stars are actually less luminous than the Sun. Of the 34 stars within 4 pc of the Sun (Appendix 4), only three (α Centauri, Sirius, and Procyon) are more luminous than the Sun.

To better characterize a typical population of stars, astronomers count the stars out to a certain distance from the Sun and plot the number of stars that have different luminosities. The resulting graph is called the **luminosity function.** Figure 19-5 shows the luminosity function for stars in our part of the Milky Way Galaxy. Note how steeply the curve declines for the brightest stars toward the left side of the graph, indicating that they are much rarer than dimmer ones. For example, this graph shows that stars like the Sun are about 10,000 times more common than stars like Spica (with a luminosity of $2100 L_\odot$).

The exact shape of the curve in Figure 19-5 applies only to the vicinity of the Sun and similar regions in our Milky Way Galaxy. Other locations have somewhat different luminosity functions. In stellar populations in general, however, the overall tendency is for low-luminosity stars to be much more common than high-luminosity ones.

## box 19-2 | Tools of the Astronomer's Trade

### *Luminosity, Distance, and Apparent Brightness*

In Section 19-2 we presented an expression that is useful for relating a star's luminosity, distance, and apparent brightness to the corresponding quantities for the Sun:

$$\frac{L}{L_\odot} = \left(\frac{d}{d_\odot}\right)^2 \frac{b}{b_\odot}$$

A similar equation can be used to relate the luminosities, distances, and apparent brightnesses of any two stars, denoted 1 and 2:

$$\frac{L_1}{L_2} = \left(\frac{d}{d_2}\right)^2 \frac{b_1}{b_2}$$

Here are some examples of using these equations.

**EXAMPLE:** The star ε Eridani is at a distance $d = 3.27$ pc from the Earth. Because there are 206,625 AU in 1 parsec, this distance is $d = (3.27 \text{ pc})(206{,}625 \text{ AU/pc}) = 6.75 \times 10^5$ AU. The distance from the Earth to the Sun ($d_\odot$) is 1 AU, so the ratio of the two distances is $d/d_\odot = 6.75 \times 10^5$. As seen from the Earth, ε Eridani appears only $6.73 \times 10^{-13}$ as bright as the Sun, so the ratio of their apparent brightnesses ($b/b_\odot$) is $6.73 \times 10^{-13}$. From the first equation, the ratio of the luminosity of ε Eridani ($L$) to the Sun's luminosity ($L_\odot$) is

$$\frac{L}{L_\odot} = \left(\frac{d}{d_\odot}\right)^2 \frac{b}{b_\odot} = (6.75 \times 10^5)^2 \times (6.73 \times 10^{-13}) = 0.31$$

This means that ε Eridani has only 0.31 the luminosity of the Sun.

**EXAMPLE:** Suppose star 1 is at half the distance of star 2 (that is, $d_1/d_2 = 1/2$) and that star 1 appears twice as bright as star 2 (that is, $b_1/b_2 = 2$). We can use the second of the two equations above to see how the luminosities of these two stars compare:

$$\frac{L_1}{L_2} = \left(\frac{1}{2}\right)^2 \times 2 = 0.5$$

This says that star 1 has only half the luminosity of star 2. Despite this, star 1 appears brighter than star 2 because it is closer to us.

A variation of the above equation is used in the method of *spectroscopic parallax*, discussed in Section 19-8. It turns out that a star's luminosity can be determined simply by analyzing the star's spectrum. If the star's apparent brightness is also known, the star's distance can be calculated. The inverse-square law can be rewritten as an expression for the ratio of the star's distance from the Earth ($d$) to the Earth-Sun distance ($d_\odot$):

$$\frac{d}{d_\odot} = \sqrt{\frac{(L/L_\odot)}{(b/b_\odot)}}$$

Alternatively, this formula can be written as a relation between the properties of two stars, 1 and 2:

$$\frac{d_1}{d_2} = \sqrt{\frac{(L_1/L_2)}{(b_1/b_2)}}$$

**EXAMPLE:** The star Regulus has 140 times the luminosity of the Sun (so $L/L_\odot = 140$) but appears only $5.2 \times 10^{-12}$ as bright as the Sun (so $b/b_\odot = 5.2 \times 10^{-12}$). Its distance from the Earth, $d$, is given by the formula

$$\frac{d}{d_\odot} = \sqrt{\frac{(L/L_\odot)}{(b/b_\odot)}} = \sqrt{\frac{140}{5.2 \times 10^{-12}}} = \sqrt{2.7 \times 10^{13}} = 5.2 \times 10^6$$

Since $d_\odot = 1$ AU, Regulus is $5.2 \times 10^6$ AU from the Earth. Using the conversion 206,265 AU = 1 pc, the star's distance is $d = (5.2 \times 10^6 \text{ AU})(1 \text{ pc}/206{,}265 \text{ AU}) = 25$ pc.

**EXAMPLE:** The star δ (delta) Cephei, which lies 300 pc from the Earth, is thousands of times more luminous than the Sun. Thanks to this great luminosity, stars like δ Cephei can be seen in galaxies millions of parsecs away. As an example, the Hubble Space Telescope has detected stars like δ Cephei within the galaxy NGC 3351, which lies in the direction of the constellation Leo. These stars have only $9 \times 10^{-10}$ of the apparent brightness of δ Cephei, so if we let 1 denote the stars in NGC 3351 and 2 denote δ Cephei itself, we have $b_1/b_2 = 9 \times 10^{-10}$. Because δ Cephei has the same luminosity as its sister stars in NGC 3351, we also have $L_1/L_2 = 1$. Given these ratios, we can find the ratio $d_1/d_2$ (the distance from the Earth to NGC 3351 divided by the distance from the Earth to δ Cephei):

$$\frac{d_1}{d_2} = \sqrt{\frac{(L_1/L_2)}{(b_1/b_2)}} = \sqrt{\frac{1}{9 \times 10^{-10}}} = \sqrt{1.1 \times 10^9} = 33{,}000$$

Hence, NGC 3351 is 33,000 times further away than δ Cephei. The distance from the Earth to NGC 3351 is therefore $(33{,}000)(300 \text{ pc}) = 10^7$ pc, or 10 megaparsecs (10 Mpc). We will learn more about Cepheids in Chapter 21, and in Chapter 26 we will explore further how they are used to determine the distances to remote galaxies.

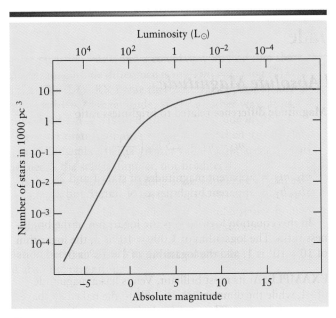

**figure 19-5**

**The Luminosity Function** This graph shows how many stars of a given luminosity lie within a representative 1000 cubic-parsec volume. Note that dim stars are far more numerous than bright stars. The steep decline of the curve toward the left (high-luminosity) side of the graph demonstrates that the most luminous stars are indeed quite rare. The scale at the bottom of the graph shows absolute magnitude, an alternative measure of a star's luminosity (described in Section 19-3). (Adapted from J. Bahcall and R. Soneira)

## 19-3 Astronomers often use the magnitude scale to denote brightness

Because astronomy is one of the most ancient of sciences, some of the tools used by modern astronomers are actually many centuries old. One such tool is a special scale, called the **magnitude scale,** which astronomers frequently use to denote the brightness of stars. This scale was invented in the second century B.C. by Greek astronomer Hipparchus, who called the brightest stars first-magnitude stars. Stars about half as bright as first-magnitude stars were called second-magnitude stars, and so forth, down to sixth-magnitude stars, the dimmest ones he could see. After telescopes came into use, astronomers extended Hipparchus's magnitude scale to include the dimmer stars now visible through their instruments.

These magnitudes are properly called **apparent magnitudes,** because they describe how bright an object *appears* to an Earth-based observer. Apparent magnitude is a measure of the light energy arriving at the Earth and is directly related to apparent brightness.

 The magnitude scale has a tendency to be confusing because it works "backward." Keep in mind that the *greater* the apparent magnitude, the *dimmer* the star. A star of apparent magnitude +3 (a third-magnitude star) is dimmer than a star of apparent magnitude +2 (a second-magnitude star).

In the nineteenth century, better techniques were developed for measuring the light energy arriving from a star. Astronomers then set out to define the magnitude scale more precisely. Their measurements showed that a first-magnitude star is about 100 times brighter than a sixth-magnitude star. In other words, it would take 100 stars of magnitude +6 to provide as much light energy as we receive from a single star of magnitude +1. To make computations easier, the magnitude scale was redefined so that a magnitude difference of 5 corresponds exactly to a factor of 100 in brightness. A magnitude difference of 1 corresponds to a factor of 2.512 in brightness, because

$$2.512 \times 2.512 \times 2.512 \times 2.512 \times 2.512 = (2.512)^5 = 100$$

Thus, it takes 2.512 third-magnitude stars to provide as much light as we receive from a single second-magnitude star.

Figure 19-6 illustrates the modern apparent magnitude scale. Note that the dimmest stars visible through a pair of

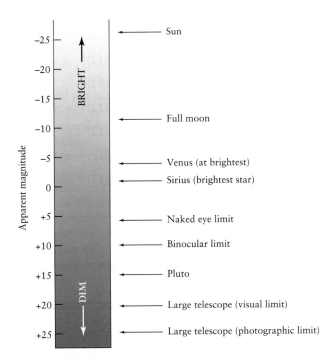

**figure 19-6**

**The Apparent Magnitude Scale** Astronomers denote the apparent brightness of objects in the sky by their apparent magnitudes. The stars visible to the naked eye have magnitudes in the approximate range from –1.5 to +6. Very long-exposure photography using the Hubble Space Telescope can reveal stars as faint as magnitude +27.

brightness rather than apparent magnitude. But if you go on to learn more about astronomy, you will undoubtedly encounter the ideas of apparent and absolute magnitude.

## 19-4 A star's color depends on its surface temperature

One of the first things you notice when comparing stars in the nighttime sky is their differences in apparent magnitude (that is, in apparent brightness). With a little more effort you may also notice that the stars also have different colors. For example, even with the naked eye you can easily note the difference between reddish Betelgeuse, the star in the "armpit" of the constellation Orion, and bluish Bellatrix at Orion's other "shoulder"; these colors are also shown in Figure 2-2. (Colors are most evident for the brightest stars, because your color vision doesn't work well at low light levels.)

CAUTION! It's true that the light from a star will appear red-shifted if the star is moving away from you and blueshifted if it's moving toward you. But even the fastest stars are not moving fast enough for these color shifts to be visible without sensitive instrumentation. The red color of Betelgeuse and the blue color of Rigel are not due to their motions; they are the actual colors of the stars.

As we discussed in Section 5-3, a star's color is directly related to its surface temperature. The intensity of light from a relatively cool star peaks at long wavelengths, making the star look red (Figure 19-7a). A hot star's intensity curve peaks instead at shorter wavelengths, so the star looks blue (Figure 19-7c). The maximum intensity of a star with an intermediate temperature, such as the Sun, occurs near the middle of the visible spectrum. This gives the star a yellowish color (Figure 19-7b).

To measure the surface temperatures of stars, astronomers must accurately measure star colors. This is done by using a light-sensitive device (such as a CCD) at the focus of a tele-

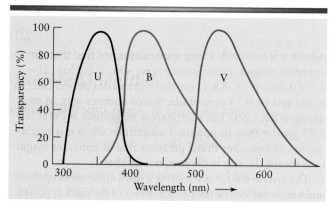

**figure 19-8**

**Light Transmission Through UBV Filters** This graph shows the wavelength ranges over which the standardized U, B, and V filters are transparent to light. The U filter is transparent to the near-ultraviolet. The B filter is transparent to violet, blue, and green light with wavelengths from about 380 to 550 nm. The V filter is transparent to green and yellow light with wavelengths from about 500 to 650 nm. By measuring the apparent brightness of a star with each of these filters and comparing the results, an astronomer can determine the star's surface temperature.

scope behind a standardized set of colored filters. The most commonly used filters are the **UBV filters,** and the technique is called **UBV photometry.** Each of the three UBV filters is transparent in one of three broad wavelength bands: the ultraviolet (U), the blue (B), and the central yellow (V, for *visual*) region of the visible spectrum (Figure 19-8). The transparency of the V filter mimics the sensitivity of the human eye.

To do UBV photometry, the astronomer aims a telescope at a star and measures the intensity of the starlight that passes through each of the filters individually. This procedure gives three apparent brightnesses for the star, designated by the symbols $b_U$, $b_B$, and $b_V$. The astronomer then compares the intensity of starlight in neighboring wavelength bands by taking the ratios of these brightnesses: $b_V/b_B$ and $b_B/b_U$. The

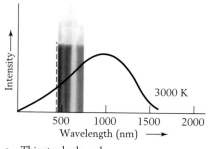

a  This star looks red

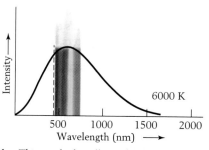

b  This star looks yellow-white

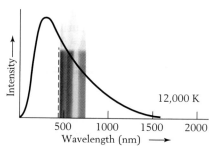

c  This star looks blue

**figure 19-7**

**Temperature and Color** This schematic diagram shows the relationship between the color of a star and its surface temperature. The intensity of light emitted by three hypothetical stars is plotted here against wavelength (compare Figure 5-10). The range of visible wavelengths is indicated. The star's apparent color is determined by whether the intensity curve has larger values at the short-wavelength or long-wavelength end of the visible spectrum.

values of these **color ratios** for several representative stars are given in Table 19-1.

The color ratios of a star are directly related to the star's surface temperature. If a star is very hot, its radiation is skewed toward the short-wavelength ultraviolet (as in Figure 19-7c). This makes the star dim through the V filter, brighter through the B filter, and brightest through the U filter. Hence, for a hot star $b_V$ will be less than $b_B$, which in turn will be less than $b_U$, so the ratios $b_V/b_B$ and $b_B/b_U$ will both be less than 1. One such star is Bellatrix (see Table 19-1), which has a surface temperature of 28,000 K. In contrast, if the star is cool, its radiation peaks at long wavelengths (as in Figure 19-7a). This makes the star brightest through the V filter, dimmer through the B filter, and dimmest through the U filter. So, for a cool star $b_V$ will be greater than $b_B$, which in turn will be greater than $b$, and the ratios $b_V/b_B$ and $b_B/b_U$ will both be greater than 1. The star Betelgeuse (surface temperature 2400 K) is an example. You can see these differences between hot and cool stars in <u>Figure 6-29</u>, which shows the constellation Orion (which includes both Bellatrix and Betelgeuse). <u>Figure 6-29a</u> shows that in ultraviolet light (actually at wavelengths a bit shorter than those transmitted by the U filter), Bellatrix is very bright but Betelgeuse is almost invisible. In <u>Figure 6-29c</u>, which approximates the transmission of a V filter, Bellatrix is relatively bright but is outshone by Betelgeuse.

The graph in Figure 19-9 gives the relationship between a star's $b_V/b_B$ color ratio and its temperature. If you know the value of the $b_V/b_B$ color ratio for a given star, you can use this graph to find the star's surface temperature. An example is the Sun, for which $b_V/b_B$ equals 1.77; this corresponds to a surface temperature of 5800 K.

CAUTION! A word of caution here is in order. As we will see in Chapter 20, dust and gas that pervade interstellar space cause distant stars to appear redder than they really are. (In the same way, particles in the Earth's atmosphere make the setting Sun look redder.) Astronomers must take this reddening into account whenever they attempt to determine a star's surface temperature from its color ratios. A

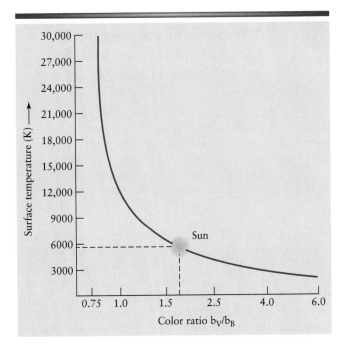

**Figure 19-9**

**Blackbody Temperature Versus Color Ratio** The $b_V/b_B$ color ratio is the ratio of a star's apparent brightness through a V filter to its apparent brightness through a B filter. If the star is hotter than about 10,000 K, it is a very bluish star with a value of $b_V/b_B$ less than 1. If a star is cooler than about 10,000 K, its value of $b_V/b_B$ is greater than 1. The Sun's $b_V/b_B$ color ratio is about 1.77, which corresponds to a temperature of 5800 K. After measuring a star's brightness with the B and V filters, an astronomer can estimate the star's surface temperature from a graph like this one.

star's spectrum provides a more precise measure of a star's surface temperature, as we will see next. Nevertheless, it is quicker and easier to observe a star's colors with a set of UBV filters than it is to take the star's spectrum with a spectrograph.

| table 19-1 | Colors of Selected Stars | | | |
|---|---|---|---|---|
| **Star** | **Surface temperature (K)** | $b_V/b_B$ | $b_B/b_U$ | **Apparent color** |
| Bellatrix (γ Orionis) | 28,000 | 0.82 | 0.45 | Blue |
| Regulus (α Leonis) | 22,000 | 0.90 | 0.72 | Blue-white |
| Sirius (α Canis Majoris) | 10,000 | 1.00 | 0.95 | Blue-white |
| Megrez (δ Ursae Majoris) | 8800 | 1.08 | 1.07 | White |
| Altair (α Aquilae) | 7400 | 1.22 | 1.08 | Yellow-white |
| Sun | 5800 | 1.77 | 1.10 | Yellow-white |
| Aldebaran (α Tauri) | 3700 | 4.13 | 5.75 | Orange |
| Betelgeuse (α Orionis) | 2400 | 5.50 | 6.67 | Red |

## 19-5 The spectra of stars reveal their chemical compositions as well as surface temperatures

Because stars are so distant, the only information that we can hope to obtain about their physical nature—their structure, chemical composition, temperature, and other properties—comes from analyzing the light that they emit. We have seen how the color of a star's light helps astronomers determine its surface temperature. To determine the other properties of a star, astronomers must carefully analyze the spectrum of its light. This field of investigation, called *stellar spectroscopy,* began in 1814 when German instrument maker Joseph Fraunhofer attached a spectroscope to a telescope and pointed it toward the stars. Fraunhofer had earlier observed that the Sun has an absorption-line spectrum—that is, a continuous spectrum with dark absorption lines, as described in Section 5-6. He found that stars have the same kind of spectra, which reinforces the idea that our own Sun is a rather typical star. But Fraunhofer also found that the pattern of absorption lines is different for different stars.

We see an absorption-line spectrum when a cool gas lies between us and a hot, glowing object (recall Figure 5-14). By itself, the hot, glowing object produces a continuous blackbody spectrum, or **continuum.** For the case of a star, the continuum is produced at low-lying levels of the star's atmosphere where the gases are hot and dense. The absorption lines are created when the continuum radiation flows outward through the cooler, less dense, upper layers of the star's atmosphere. Atoms in these upper layers absorb radiation at specific wavelengths, which depend on the specific kinds of atoms present—hydrogen, helium, or other elements—and on whether or not the atoms are ionized. This is the same way that absorption lines in the Sun's spectrum are produced (see Section 18-1).

Fraunhofer and the astronomers who continued his work made careful observations of the different absorption lines present in the spectra of different stars. Some stars have spectra in which the Balmer lines of hydrogen are prominent. But in the spectra of other stars, including the Sun, the Balmer lines are nearly absent and the dominant absorption lines are those of heavier elements such as calcium, iron, and sodium. Still other stellar spectra are dominated by broad absorption lines caused by molecules such as titanium oxide. To cope with this diversity, astronomers group similar-appearing stellar spectra into **spectral classes.** In a popular classification scheme that emerged in the late 1800s, a star was assigned a letter from A through P according to the strength or weakness of the hydrogen Balmer lines in the star's spectrum.

Nineteenth-century science could not explain why or how the spectral lines of a particular chemical are affected by the temperature and density of the gas. Nevertheless, a team of astronomers under the supervision of Edward C. Pickering at Harvard College Observatory forged ahead with a monumental project of examining the spectra of thousands of stars. Their goal was to develop a system of spectral classifi-

cation in which all spectral features (not just Balmer lines) change smoothly from one spectral class to the next.

Pickering's spectral classification project was financed by the estate of Henry Draper, a wealthy New York physician and amateur astronomer who in 1872 became the first person to photograph stellar absorption lines. Principal researchers on the project were Williamina Fleming, Antonia Maury, and Annie Jump Cannon. As a result of their efforts, many of the original A-through-P classes were dropped and others were consolidated. The remaining spectral classes were reordered in the sequence **OBAFGKM.** You can remember this sequence with the mnemonic: "Oh, *Be A Fine Girl* (or *Guy*), *Kiss Me!*"

Cannon found it useful to subdivide the original OBAF-GKM sequence into finer steps called **spectral types.** These additional steps are indicated by attaching an integer from 0 through 9 to the original letter. For example, the spectral class F includes spectral types F0, F1, F2, ... , F8, F9, which are followed by the spectral types G0, G1, G2, ... , G8, G9, and so on. Representative spectra of several spectral types are shown in Figure 19-10. From one spectral type to the next, the strengths of spectral lines change in a smooth fashion. For example, the Balmer absorption lines of hydrogen become increasingly prominent as you go from spectral type B0 to A0. From A0 onward through F and G types, the hydrogen lines weaken and almost fade from view. The Sun, whose spectrum is dominated by calcium and iron, is a G2 star.

The Harvard project culminated in the *Henry Draper Catalogue,* published between 1918 and 1924. It listed 225,300 stars, each of which Cannon had personally classified. Meanwhile, physicists had been making important discoveries about the structure of atoms. Ernest Rutherford had demonstrated that atoms have nuclei (recall Figure 5-17), and Niels Bohr made the remarkable hypothesis that electrons move along discrete orbits around the nuclei of atoms (recall Figure 5-20). These discoveries about atoms gave scientists the conceptual and mathematical tools required to interpret and understand stellar spectra.

In the 1920s, Harvard astronomer Cecilia Payne and Indian physicist Meghnad Saha succeeded in explaining how a star's surface temperature affects its spectrum. They demonstrated that the OBAFGKM spectral sequence is actually a sequence in temperature. The hottest stars are O stars. The absorption lines seen in the spectra of O stars can occur only if these stars have surface temperatures above 25,000 K. M stars are the coolest stars. The spectral features seen in the spectrum of M stars are consistent with stellar surface temperatures of about 3000 K.

To see why the appearance of a star's spectrum is profoundly affected by the star's surface temperature, consider the Balmer lines of hydrogen. Hydrogen is by far the most abundant element in the universe, accounting for about three-quarters of the mass of a typical star. Nonetheless, the Balmer lines do not necessarily show up in a star's spectrum. As discussed in Section 5-8, Balmer lines are produced when an electron in the $n = 2$ orbit of hydrogen is lifted into a higher orbit by absorbing a photon with the right amount of energy (see Figure 5-22). If the star is much hotter than 10,000 K, high-energy photons pouring out of the star's

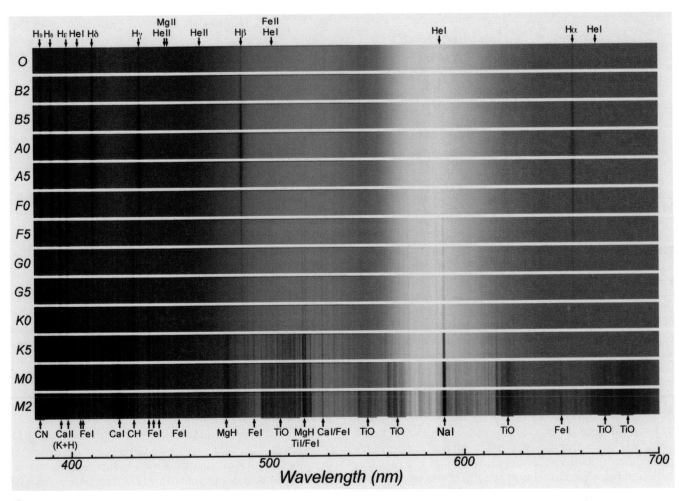

**Figure 19-10**

**Principal Types of Stellar Spectra** This figure shows the spectra for stars of several different spectral types, arranged so that surface temperature decreases from top to bottom. The Balmer lines of hydrogen ($H_\alpha$, $H_\beta$, $H_\gamma$, and $H_\delta$) are strongest in stars of spectral class A, which have surface temperatures of about 10,000 K. The spectra of G and K stars exhibit numerous lines caused by metals such as iron (Fe) and calcium (Ca), indicating surface temperatures in the range 4000 to 6000 K. The broad, dark bands in the spectra of M stars are caused by titanium oxide (TiO), which can exist only when the surface temperature is below about 3500 K. The roman numeral I after a chemical symbol means that the absorption line is caused by un-ionized atoms; a numeral II means that the absorption is caused by atoms that have each lost one electron. (R. Bell, University of Maryland, and M. Briley, University of Wisconsin at Oshkosh.)

interior will easily knock electrons out of the hydrogen atoms in the star's outer layers. This process ionizes the gas. When hydrogen's only electron is torn away, no hydrogen spectral lines can be produced. Hence, the Balmer lines will be relatively weak in the spectra of such hot stars, such as the hot O and B2 stars in Figure 19-10.

Conversely, if the star's atmosphere is much cooler than 10,000 K, almost all of the hydrogen atoms are in the lowest ($n = 1$) energy state. The majority of the photons passing through the star's atmosphere possess too little energy to boost electrons up from the $n = 1$ to the $n = 2$ orbit of the hydrogen atoms. Therefore, these unexcited atoms cannot absorb the photons characteristic of the Balmer lines, and

these lines are nearly absent from the spectrum of a cool star. (You can see this in the spectra of the cool M0 and M5 stars in Figure 19-10.) For the Balmer lines to be prominent in a star's spectrum, the star must be hot enough to excite the electrons out of the ground state but not so hot that all the hydrogen atoms become ionized. A stellar surface temperature of about 9000 K produces the strongest hydrogen lines; this is the case for the stars of spectral types A0 and A5 in Figure 19-10.

Every other element also has a characteristic temperature range in which it produces prominent absorption lines in the observable part of the spectrum. Figure 19-11 shows the relative strengths of absorption lines produced by different chemicals. For example, around 25,000 K the spectral lines

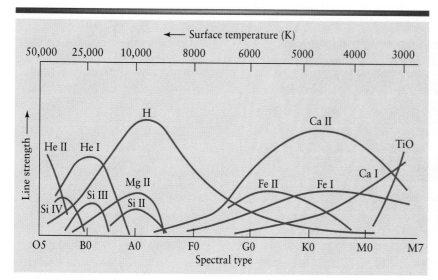

**The Strengths of Absorption Lines** Each variety of chemical in the outer layers of a star produces absorption lines in the star's spectrum. This graph shows how the relative strengths of different absorption lines depend on the star's surface temperature. For each chemical, the roman numeral indicates the degree of ionization; Si I is neutral, un-ionized silicon, Si II is silicon that has lost one electron, Si III is silicon that has lost two electrons, and so on. The peak in each curve occurs at the stellar surface temperature for which that chemical's absorption line is strongest. For example, the spectrum of the Sun has strong absorption lines of singly ionized calcium and iron (Ca II and Fe II), corresponding to a spectral type of G2 and a surface temperature of 5800 K.

of neutral (that is, un-ionized) helium are strong; at this temperature, photons have enough energy to excite helium atoms without tearing away the electrons altogether. In stars hotter than about 30,000 K, one of the two electrons in a helium atom is torn away. The remaining electron then produces a set of spectral lines that is recognizably different from the lines produced by neutral helium. When the spectral lines of singly ionized helium appear in a star's spectrum, we know that the star has a surface temperature greater than 30,000 K.

Astronomers use the term **metals** to refer to all elements other than hydrogen and helium. (This idiosyncratic use of the term "metal" is quite different from the definition used by chemists and other scientists. To a chemist, sodium and iron are metals but carbon and oxygen are not; to an astronomer, all of these substances are metals.) In this termi-

nology, metals dominate the spectra of stars cooler than 10,000 K. Ionized metals are prominent for surface temperatures between 6000 and 8000 K, while neutral metals are strongest between approximately 5500 and 4000 K. Below 4000 K, certain atoms combine to form molecules that can survive in a star's atmosphere. (At higher temperatures atoms move so fast that when they collide, they bounce off each other rather than "sticking together" to form molecules.) As these molecules vibrate and rotate, they produce bands of spectral lines that dominate the star's spectrum. Most noticeable are the lines of titanium oxide (TiO), which are strongest for surface temperatures of about 3000 K. Table 19-2 summarizes the relationship between the temperature and spectra of stars. Box 19-4 discusses how astronomers measure the strengths of spectral lines and use this information to accurately determine the surface temperature of a star.

| **t**able 19-2 | **The Spectral Sequence** | | | |
|---|---|---|---|---|
| Spectral class | Color | Temperature (K) | Spectral lines | Examples |
| O | Blue-violet | 28,000–50,000 | Ionized atoms, especially helium | Naos (ζ Puppis), Mintaka (δ Orionis) |
| B | Blue-white | 10,000–28,000 | Neutral helium, some hydrogen | Spica (α Virginis), Rigel (β Orionis) |
| A | White | 7500–10,000 | Strong hydrogen, some ionized metals | Sirius (α Canis Majoris), Vega (α Lyrae) |
| F | Yellow-white | 6000–7500 | Hydrogen and ionized metals such as calcium and iron | Canopus (α Carinae), Procyon (α Canis Minoris) |
| G | Yellow | 5000–6000 | Both neutral and ionized metals, especially ionized calcium | Sun, Capella (α Aurigae) |
| K | Orange | 3500–5000 | Neutral metals | Arcturus (α Boötis), Aldebaran (α Tauri) |
| M | Red-orange | 2500–3500 | Strong titanium oxide and some neutral calcium | Antares (α Scorpii), Betelgeuse (α Orionis) |

## box 19-4  Looking Deeper into Astronomy

### *Equivalent Widths and Line Strengths*

Most astronomers today use a light-sensitive CCD to record the spectra of stars because it is much more sensitive than photographic film. The large graph shows 14 stellar spectra obtained with such a device, which generates a plot of intensity versus wavelength. Absorption lines are seen as dips on the curve of the continuum, which has a shape quite like the blackbody curve at the temperature of the star's surface.

Astronomers designate an un-ionized atom with a roman numeral I; thus, H I is neutral hydrogen. A roman numeral II is used to identify an atom with one lectron missing; thus, He II is singly ionized helium (He⁺). This notation is used to label some of the prominent spectral lines on the 14 spectra, which cover spectral types from O5 to M5.

Using a plot of the intensity in a spectrum versus wavelength, an astronomer can study the shapes of individual spectral lines. The small graph shows a typical spectral line consisting of a "core" flanked by "wings." The detailed shape of a spectral line, called the **line profile,** contains important information about a star. For example, if a star is rotating, light from the approaching side is slightly blueshifted, while light from the receding side is comparably redshifted. As a result, the star's spectral lines are broadened in a characteristic fashion. By measuring the shape of the spectral lines, astronomers can deduce how fast the star is rotating.

The true shape of a spectral line reflects the properties of a star's atmosphere, but the observed line profile is also broadened somewhat by the astronomer's measuring instruments. Instrumental effects do not, however, significantly alter the *total* absorption of the line, which is a measure of the energy deleted from the continuum across the entire spectral line. Thus, the total absorption does not depend on the line's shape. Astronomers express the total absorption of a spectral line—that is, the *line strength*—in terms of its **equivalent width,** which is the width of a completely dark rectangular line with the same total absorption as the observed line. For example, the equivalent width of an iron line in the Sun's spectrum is about 0.01 nm. The darkest absorption lines in Figure 19-10 are those with the greatest equivalent widths; the faintest absorption lines have the smallest equivalent widths. Figure 19-11,

*(continued on the following page)*

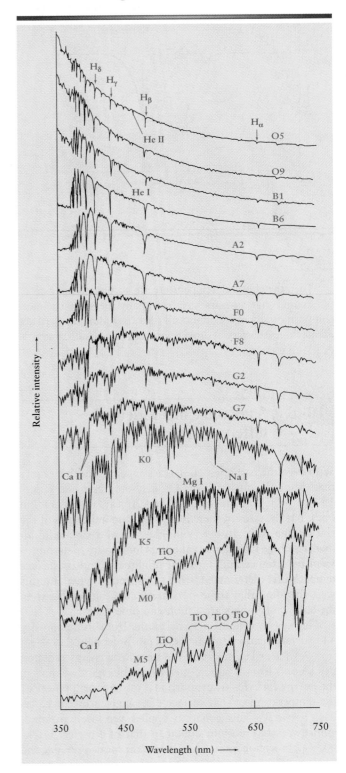

Relative intensity →

Wavelength (nm) ⟶

350    450    550    650    750

## box 19-4 *(continued from the previous page)*

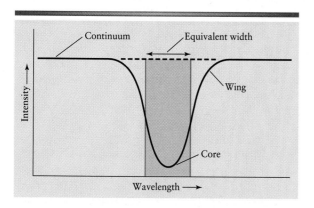

which shows how the line strengths of different chemicals depend on a star's surface temperature, is actually a graph of relative equivalent widths versus temperature.

The equivalent width of a spectral line depends on how many atoms in the star's atmosphere are in a state in which they can absorb the wavelength in question. For a given temperature, the more atoms there are, the stronger and broader the line is. Hence, by analyzing a star's spectrum, an astronomer can determine the star's chemical composition.

---

When the effects of temperature are accounted for, astronomers find that *all* stars have essentially the same chemical composition. By mass, almost all stars (including the Sun) are about three-quarters hydrogen, one-quarter helium, and 1% or less of metals.

## 19-6 | Stars come in a variety of different sizes

With even the best telescopes, stars appear as nothing more than bright points of light. On a photograph like Figure 19-3 or a CCD image, brighter stars appear larger than dim ones, but the sizes of these images give no indication of the star's actual size. To determine the size of a star, astronomers combine information about its luminosity (determined from its distance and apparent brightness) and its surface temperature (determined from its spectral type). In this way, they find that some stars are quite a bit smaller than the Sun, while others are a thousand times larger.

The key to finding a star's radius from its luminosity and surface temperature is the Stefan-Boltzmann law (Section 5-4). This law says that the amount of energy radiated per second from a square meter of a blackbody—that is, the energy flux (*F*)—is proportional to the fourth power of the temperature of that surface (*T*), as given by the equation $F = \sigma T^4$. This equation applies very well to stars, whose spectra are quite similar to that of a perfect blackbody. (Absorption lines, while important for determining the star's chemical composition and surface temperature, make

only relatively small modifications to a star's blackbody spectrum.) A star's luminosity, or amount of energy emitted per second from its entire surface, is equal to the energy flux *F* multiplied by the total number of square meters on the star's surface (that is, the star's surface area). We expect that most stars are nearly spherical, like the Sun, so we can use the formula for the surface area of a sphere. This is $4\pi R^2$, where *R* is the star's radius (the distance from its center to its surface). Multiplying together the formulas for energy flux and surface area, we can write the star's luminosity as:

**Relationship between a star's luminosity, radius, and surface temperature**

$$L = 4\pi R^2 \sigma T^4$$

*L* = star's luminosity, in watts
*R* = star's radius, in meters
$\sigma$ = Stefan-Boltzmann constant = $5.67 \times 10^{-8}$ W m$^{-2}$ K$^{-4}$
*T* = star's surface temperature, in kelvins

This equation says that a relatively cool star (low surface temperature *T*), for which the energy flux is quite low, can nonetheless be very luminous if it has a large enough radius *R*. Alternatively, a relatively hot star (large *T*) can have a very low luminosity if the star has only a little surface area (small *R*).

**ANALOGY** In a similar way, a roaring campfire can emit more light than a welder's torch. The campfire is at a lower temperature than the torch, but has a much larger surface area from which it emits light.

## box 19-5 | Tools of the Astronomer's Trade

### *Stellar Radii, Luminosities, and Surface Temperatures*

As we saw in Section 19-6, because stars emit light in almost exactly the same fashion as blackbodies, we can use the Stefan-Boltzmann law to relate the star's luminosity (*L*), surface temperature (*T*), and radius (*R*). The relevant equation is

$$L = 4\pi R^2 \sigma T^4$$

As written, this equation involves the Stefan-Boltzmann constant $\sigma$, equal to $5.67 \times 10^{-8}$ W m$^{-2}$ K$^{-4}$. In many calculations, it is more convenient to relate everything to the Sun, which is a typical star. Specifically, for the Sun we have $L_\odot = 4\pi R_\odot^2 \sigma T_\odot^4$, where $L_\odot$ is the Sun's luminosity, $R_\odot$ is the Sun's radius, and $T_\odot$ is the Sun's surface temperature (equal to 5800 K). Dividing the general equation for *L* by this specific equation for the Sun, we obtain

$$\frac{L}{L_\odot} = \left(\frac{R}{R_\odot}\right)^2 \left(\frac{T}{T_\odot}\right)^4$$

This is a more pleasant formula to use because the constant $\sigma$ cancels out, making the arithmetic much easier. We can also rearrange terms to arrive at a useful equation for the radius (*R*) of a star:

**Radius of a star related to its luminosity and surface temperature**

$$\frac{R}{R_\odot} = \left(\frac{T}{T_\odot}\right)^2 \sqrt{\frac{L}{L_\odot}}$$

$R/R_\odot$ = ratio of the star's radius to the Sun's radius
$T/T_\odot$ = ratio of the star's surface temperature to the Sun's surface temperature
$L/L_\odot$ = ratio of the star's luminosity to the Sun's luminosity

**EXAMPLE:** Consider the bright reddish star Betelgeuse in the constellation of Orion (see Figure 2-2 or Figure 6-29c). Betelgeuse has a luminosity $L = 10,000$ $L_\odot$ and a surface temperature $T = 3000$ K. Substituting these data into the previous equation, we get

$$\frac{R}{R_\odot} = \left(\frac{5800}{3000}\right)^2 \sqrt{10^4} = 370$$

In other words, Betelgeuse's radius is 370 times larger than that of the Sun. The radius of the Sun is $6.96 \times 10^5$ km, so the radius of Betelgeuse is $(370)(6.96 \times 10^5$ km$) = 2.6 \times 10^8$ km, or nearly 2 AU. If Betelgeuse were located at the center of our solar system, this star would extend beyond the orbit of Mars.

**EXAMPLE:** Sirius, the brightest star in the sky, is actually two stars orbiting each other (a *binary star*). The less luminous star, Sirius B, is too dim to see with the naked eye. Its luminosity is 0.0025 $L_\odot$ and its surface temperature is 10,000 K. Hence, the ratio of its radius to that of the Sun is

$$\frac{R}{R_\odot} = \left(\frac{5800}{10,000}\right)^2 \sqrt{0.0025} = 0.017$$

In other words, Sirius B is only 0.017 as large as the Sun: its radius is $(0.017)(6.96 \times 10^5$ km$) = 12,000$ km, or about twice the radius of the Earth.

The radii of some stars have been measured with other techniques, some of which are described elsewhere in this text. These other methods yield values consistent with those calculated by the method described in this box.

---

The above equation is important because it can be used to find a star's radius if the luminosity and surface temperature are known. The mathematical details are given in Box 19-5. In this way astronomers find that stars come in a wide range of sizes. The smallest stars are the same size as the Earth. Despite their tiny surface area (compared to stars like the Sun), many of these stars have such high surface temperatures that their luminosities are quite respectable. The largest stars are a thousand times larger in radius than the

Sun (and $10^5$ times larger than the smallest stars). If our own Sun were replaced by one of these supergiants, the Earth's orbit would lie completely inside the star!

Figure 19-12 summarizes how astronomers determine the distance from Earth, luminosity, surface temperature, chemical composition, and radius of a star close enough to us so that its parallax can be measured. Remarkably, all of these properties can be deduced from just a few measured quantities: the star's parallax angle, apparent brightness, and spectrum.

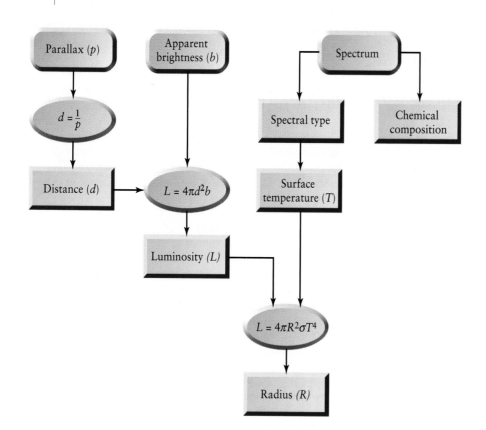

<figure>figure 19-12

**Finding Key Properties of Nearby Stars**
This flowchart shows how astronomers deduce the properties of relatively nearby stars (those close enough that their distance is known from parallax measurements). Given the star's distance, measuring the apparent brightness gives the star's luminosity from the inverse-square law (Section 19-2). The star's spectrum, in particular, the dark absorption lines, reveals the star's chemical composition and surface temperature (Section 19-5). And from luminosity and surface temperature, the Stefan-Boltzmann law allows the radius to be calculated (Section 19-6). To determine these same properties for more distant stars, different calculations have to be made (as we will describe in Section 19-8; see Figure 19-15).</figure>

---

## 19-7 Hertzsprung-Russell (H-R) diagrams reveal the different kinds of stars

Astronomers have collected a wealth of data about the stars, but merely having tables of numerical data is not enough. Astronomers, like all scientists, want to analyze their data to look for trends and underlying principles. One of the best ways to look for trends in any set of data, whether it comes from astronomy, finance, medicine, or meteorology, is to create a graph showing how one quantity depends on another. For example, investors consult graphs of stock market values versus dates, and weather forecasters make graphs of temperature versus altitude to determine whether thunderstorms will form. Astronomers have found that a particular graph of stellar properties shows that stars fall naturally into just a few categories. This graph, one of the most important in all astronomy, will in later chapters help us understand how stars form, evolve, and eventually die.

Which properties of stars should we include in a graph? Most stars have about the same chemical composition, but there are two properties of stars—their luminosities and surface temperatures—that differ substantially from one star to another. Stars also come in a wide range of radii, but

a star's radius is a secondary property that can be found from the luminosity and surface temperature (as described in Section 19-6 and Box 19-5). We also relegate the positions and space velocities of stars to secondary importance. (In a similar way, a physician is more interested in your weight and blood pressure than in where you live or how fast you drive.) We can then ask the following question: What do we learn when we graph the luminosities of stars versus their surface temperatures?

The first answer to this question was given in 1911 by the Danish astronomer Ejnar Hertzsprung. He pointed out that a regular pattern appears when the absolute magnitudes of stars (which measure their luminosities) are plotted against their colors (which measure their surface temperatures). Two years later, the American astronomer Henry Norris Russell independently discovered a similar regularity in a graph using spectral types (another measure of surface temperature) instead of colors. In recognition of their originators, graphs of this kind are today known as **Hertzsprung-Russell diagrams,** or **H-R diagrams.**

Figure 19-13 is a typical Hertzsprung-Russell diagram. Each dot represents a star whose spectral type and luminosity have been determined. The most luminous stars are near the top of the diagram, the least luminous stars near the bottom. Hot stars (O and B stars) are toward the left side of the graph and cool stars (M stars) are toward the right.

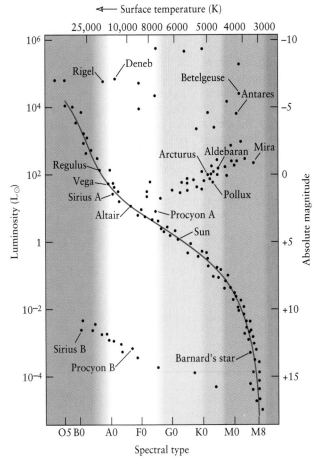

## figure 19-13

**A Hertzsprung-Russell (H-R) Diagram** On an H-R diagram, the luminosities of stars are plotted against their spectral types. Each dot on this graph represents a star whose luminosity and spectral type have been determined. Included on the diagram are some well-known stars. Note that the data points are grouped in specific regions on the graph. This pattern reveals the existence of different types of stars in the sky: main-sequence stars, giants, supergiants, and white dwarfs. The red curve indicates the location of the main sequence; giants and supergiants lie above this curve, and white dwarfs lie below it. The absolute magnitudes and surface temperatures are also shown at the right and top of the graph.

**CAUTION!** You are probably accustomed to graphs where the numbers on the horizontal axis increase as you move to the right. (For example, the business section of a newspaper includes a graph of stock market values versus dates, with later dates to the right of earlier ones.) But on an H-R diagram the temperature scale on the horizontal axis increases toward the *left*. This practice stems from the original diagrams of Hertzsprung and Russell, who placed

hot O stars on the left and cool M stars on the right. This arrangement is a tradition that no one has seriously tried to change.

The most striking feature of the H-R diagram is that the data points are not scattered randomly over the graph but are grouped in several distinct regions. The luminosities and surface temperatures of stars do *not* have random values; these two quantities are related! The band stretching diagonally across the H-R diagram includes the majority of the stars in the night sky. This band, called the **main sequence**, extends from the hot, bright, blue stars in the upper left corner of the diagram down to the cool, dim, red stars in the lower right corner. A star whose properties place it in this region of an H-R diagram is called a **main-sequence star**. The Sun (spectral type G2, luminosity 1 $L_\odot$, absolute magnitude +4.8) is such a star. Indeed, we will find that all main-sequence stars are like the Sun in that *hydrogen burning*—thermonuclear fusion that converts hydrogen into helium (Section 18-6)—is taking place in their cores.

The upper right side of the H-R diagram shows a second major grouping of data points. Stars represented by these points are both bright and cool. From the Stefan-Boltzmann law, we know that a cool object radiates much less light per unit of surface area than a hot object. In order for these stars to be as bright as they are, they must be huge, and so they are called **giants**. As explained in Box 19-5, these stars are around 10 to 100 times larger than the Sun. This is also shown by Figure 19-14, which is an H-R diagram to which dashed lines have been added to represent stellar radii. Most giant stars are around 100 to 1000 times more luminous than the Sun and have surface temperatures of about 3000 to 6000 K. Cooler members of this class of stars (those with surface temperatures from about 3000 to 4000 K) are often called **red giants** because they appear reddish. Aldebaran in the constellation of Taurus and Arcturus in Boötes are two red giants that can be easily seen with the naked eye.

A few rare stars are considerably bigger and brighter than typical red giants, with radii up to 1000 $R_\odot$. Appropriately enough, these superluminous stars are called **supergiants**. Betelgeuse in Orion and Antares in Scorpius are two supergiants you can find in the nighttime sky. Both giants and supergiants have thermonuclear reactions occurring in their interiors, but the character of those reactions and where in the star they occur can be quite different than for a main-sequence star like the Sun. We study these stars in more detail in Chapters 21 and 22.

Finally, there is a third distinct grouping of data points toward the lower left corner of the Hertzsprung-Russell diagram. Although these stars are hot, their luminosities are quite low; hence, they must be small. They are appropriately called **white dwarfs**. These stars, which can be seen only with the aid of a telescope, are approximately the same size as the Earth. As discussed in Chapter 22, no thermonuclear reactions take place within white dwarf stars. Rather, like embers left from a fire, they are the still-glowing remnants of what were once giant stars.

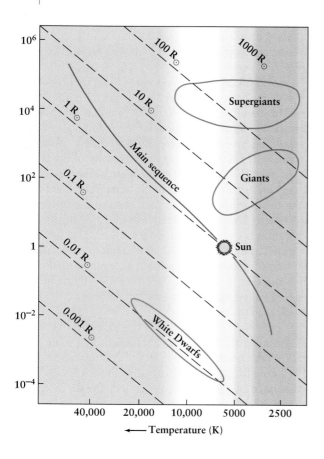

**The Sizes of Stars on an H-R Diagram** This H-R diagram has the same scale as Figure 19-13, but dashed lines have been added to show stars of different radii. The equation $L = 4\pi R^2 \sigma T^4$ relating a star's luminosity ($L$), radius ($R$), and surface temperature ($T$) was used to draw these lines. For a given stellar radius, as the surface temperature increases (corresponding to moving from right to left in the H-R diagram), the star glows more intensely and the luminosity increases (corresponding to moving upward in the diagram). That is why the dashed lines slope upward to the left. While individual stars are not plotted, we show the regions of the diagram in which main sequence, giant, supergiant, and white dwarf stars are found. Note that the Sun is intermediate in luminosity, surface temperature, and radius; it is very much a middle-of-the-road star.

The existence of fundamentally different types of stars is the first important lesson to come from the H-R diagram. These different kinds of stars represent various stages of a star's life. We will use the H-R diagram as an essential tool for understanding the life stories of stars.

## 19-8 Details of a star's spectrum reveal whether it is a giant, a white dwarf, or a main-sequence star

We have seen that a star's surface temperature largely determines which lines are prominent in its spectrum. Therefore, classifying stars by spectral type is essentially the same as categorizing them according to surface temperature. But as shown in the H-R diagram in Figure 19-14, stars of the same surface temperature can have very different luminosities. As an example, a star with surface temperature 5800 K could be either a white dwarf, a main-sequence star, a giant, or a supergiant, depending on the star's luminosity. By examining subtle features in a star's spectrum, however, astronomers can determine to which of these categories a star belongs. This gives astronomers a tool that can determine the distances to stars in galaxies outside the Milky Way, far beyond the maximum distance that can be measured from stellar parallax.

Figure 19-15 compares the spectra of two stars of the same spectral type but different luminosity: a B8 supergiant and a B8 main-sequence star. Note that the Balmer lines of hydrogen are narrow in the spectrum of the very luminous supergiant but quite broad in the spectrum of the less luminous main-sequence star. In general, for stars of spectral types B through F, the more luminous the star is, the narrower are its hydrogen lines.

Hydrogen lines are good indicators of luminosity because they are affected by the density and pressure of the gas in the star's atmosphere. The higher the density and pressure, the more frequently hydrogen atoms collide and interact with other atoms and ions in the atmosphere. These collisions shift the energy levels in the hydrogen atoms and thus broaden the hydrogen spectral lines.

The density and pressure in the atmosphere of a luminous giant star are quite low, because the star's mass is spread over a huge volume. Atoms and ions in the atmosphere are relatively far apart; hence, collisions between them are sufficiently infrequent that hydrogen atoms can produce narrow Balmer lines. A main-sequence star, however, is much more compact than a giant or supergiant. In the denser atmosphere of a main-sequence star, frequent interatomic collisions perturb the energy levels in the hydrogen atoms, thereby producing broader Balmer lines.

In the 1930s, W. W. Morgan and P. C. Keenan of the Yerkes Observatory of the University of Chicago developed a

$H_\gamma$ $H_\delta$

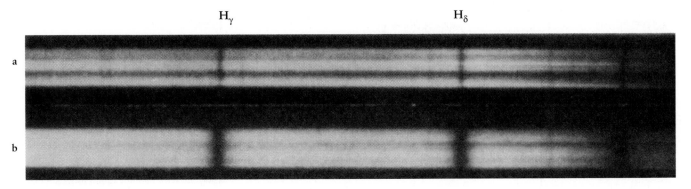

a

b

## figure 19-15

**How Luminosity Affects a Star's Spectrum** Shown here are the spectra of two stars of the same spectral type (B8) and surface temperature (13,400 K) but different luminosities: **(a)** The spectrum of the B8 supergiant Rigel (luminosity 58,000 $L_\odot$), one of the brightest stars in Orion, and **(b)** the spectrum of the B8 main-sequence star Algol (luminosity 100 $L_\odot$), the second brightest star in Perseus. Note the different widths of the hydrogen Balmer lines $H_\gamma$ and $H_\delta$ in the two spectra. (From *An Atlas of Stellar Spectra* by W. W. Morgan, P.C. Keenan, and E. Kellman)

system of **luminosity classes** based upon the subtle differences in spectral lines. When these luminosity classes are plotted on an H-R diagram (Figure 19-16), they provide a useful subdivision of the star types in the upper right half of the diagram. Luminosity classes Ia and Ib are composed of supergiants; luminosity class V includes all the main-sequence stars. The intermediate classes distinguish giant stars of various luminosities. As we will see in Chapters 21 and 22, different luminosity classes represent different stages in the evolution of a star. White dwarfs are not given a luminosity class of their own; as we mentioned in Section 19-7, they represent a final stage in stellar evolution in which no thermonuclear reactions take place.

Astronomers commonly use a shorthand description that combines a star's spectral type and its luminosity class. For example, the Sun is said to be a G2 V star. The spectral type indicates the star's surface temperature, and the luminosity class indicates its luminosity. Thus, an astronomer knows immediately that any G2 V star is a main-sequence star with a luminosity of about 1 $L_\odot$ and a surface temperature of about 5800 K. Similarly, a description of Aldebaran as a K5 III star tells an astronomer that it is a red giant with a luminosity of around 500 $L_\odot$ and a surface temperature of about 4000 K.

A star's spectral type and luminosity class, combined with the information on the H-R diagram, enable astronomers to estimate the star's distance from the Earth. As an example, consider the star Regulus (also called α Leonis, because it is the brightest star in the constellation Leo, the Lion). Its spectrum reveals Regulus to be a B7 V star (a hot, blue main-sequence star). By plotting a B7 V star on Figure 19-16, we can read off that its luminosity is 140 $L_\odot$. Given the star's luminosity and its apparent brightness—in the case of Regulus, $5.2 \times 10^{-12}$ the apparent brightness of the Sun—we can use the inverse-square law to determine its distance from the Earth. The mathematical details are worked out in Box 19-2. Because both the spectral type and luminosity class of a star are obtained by spectroscopy, this method for determining

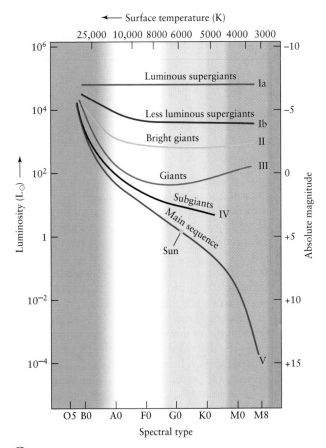

## figure 19-16

**Luminosity Classes** It is convenient to divide the H-R diagram into regions corresponding to luminosity classes. With this subdivision, finer distinctions can be made between giants and supergiants. Luminosity classes Ia and Ib include the supergiants, luminosity class V encompasses the main-sequence stars, and classes II, III, and IV denote giants of intermediate size. White dwarfs do not have their own luminosity class.

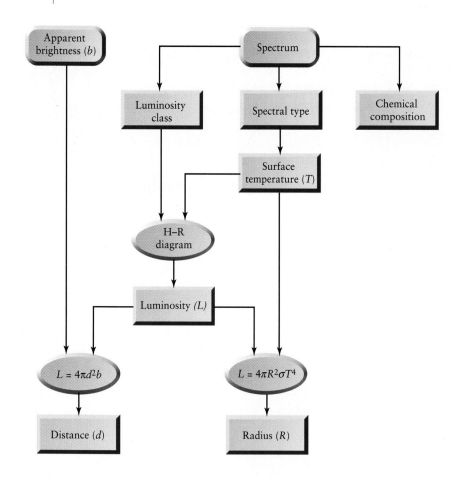

**The Method of Spectroscopic Parallax**
Astronomers can determine the distance to a star from its spectrum and apparent brightness. This technique, called the method of spectroscopic parallax, can be applied to stars at any distance. From the spectrum, it is possible to find the star's spectral type and luminosity class, from which the surface temperature and luminosity can be inferred using the H-R diagram. Given the star's luminosity and apparent brightness, the star's distance from the Earth can then be calculated using the inverse-square law. While this technique has some uncertainties, it has the advantage that it can be used for stars at very great distances. As for nearby stars (Figure 19-12), the star's chemical composition is determined from its spectrum, and the star's radius can be calculated from the luminosity and surface temperature by means of the Stefan-Boltzmann law.

distance is called **spectroscopic parallax.** (The name is a bit misleading, because no parallax angle is involved. A better name for this method, although not the one used by astronomers, would be "spectroscopic distance determination.") The method of spectroscopic parallax is shown schematically in Figure 19-17.

Spectroscopic parallax is an incredibly powerful technique. No matter how remote a star is, this technique allows astronomers to determine its distance, provided only that its spectrum and apparent brightness can be measured. Box 19-2 gives an example of how spectroscopic parallax has been used to find the distance to stars in other galaxies tens of millions of parsecs away. Because "real" stellar parallaxes can only be measured for stars within a few hundred parsecs (as described in Section 19-1), spectroscopic parallax expands the range of astronomical distance determinations by a factor of at least $10^5$. Unfortunately, spectroscopic parallax has its limitations; distances to individual stars determined using this method are only accurate to at best 10%. The reason is that the luminosity classes shown in Figure 19-16 are not thin lines on the H-R diagram but are moderately broad bands. This means that even if a star's spectral type and luminosity class are known, there is still some uncertainty in the luminosity that we read off an H-R diagram. Nonetheless, spectroscopic parallax is often the only means that an astronomer has to estimate the distance to remote stars.

What has been left out of this discussion is *why* different stars have different spectral types and luminosities. One key factor, as we shall see, turns out to be the mass of the star.

## 19-9 Binary star systems provide crucial information about stellar masses

We now know something about the sizes, temperatures, and luminosities of stars. To complete our picture of the physical properties of stars, we need to know their masses. In this section, we will see that some stars have just a tenth of the Sun's mass, while others can have 50 times the mass of the Sun. We will also discover an important relationship between the mass and luminosity of main-sequence stars. This relationship is crucial to understanding why some main-sequence stars are hot and luminous, while others are cool and dim. It will also help us understand what happens to a star as it ages and evolves.

Determining the masses of stars is not trivial, however. The problem is that there is no practical, direct way to measure the mass of an isolated star. Fortunately for astronomers, about half of the visible stars in the night sky are not isolated individuals. Instead, they are multiple-star systems, in which two or more stars orbit each other. By carefully observing the motions of these stars, astronomers can glean important information about their masses.

A pair of stars located at nearly the same position in the night sky is called a **double star.** William Herschel made the first organized search for such pairs. Between 1782 and 1821, he published three catalogs listing more than 800 double stars. Late in the nineteenth century, his son, John Herschel, discovered 10,000 more doubles. Some of these double stars are **optical double stars,** which are two stars that lie along nearly the same line of sight but are actually at very different distances from us. But many double stars are true **binary stars,** or **binaries**—pairs of stars that actually orbit each other.

When astronomers can actually see the two stars orbiting each other, a binary is called a **visual binary** (Figure 19-18). After many years of patient observation, astronomers can plot the orbits of the stars in a visual binary (Figure 19-19).

Because the two stars in a binary system orbit each other due to their mutual gravitational attraction, their orbital motions obey Kepler's third law as formulated by Isaac Newton (see Section 4-7 and Box 4-2). This law can be written as follows:

**Kepler's third law for binary star systems**

$$M_1 + M_2 = \frac{a^3}{P^2}$$

$M_1, M_2$ = masses of two stars in binary system, in solar masses
$a$ = semimajor axis of one star's orbit around the other, in AU
$P$ = orbital period, in years

Here, $a$ is the semimajor axis of the elliptical orbit that one star appears to describe around the other, plotted as in Figure 19-19. As this equation indicates, if we can measure this semimajor axis ($a$) and the orbital period ($P$), we can learn something about the masses of the two stars.

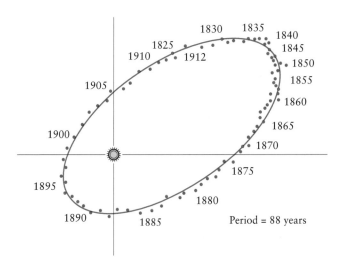

**figure 19-19**

**The Orbit of 70 Ophiuchi** After plotting observations of a visual binary over the years, astronomers can draw the orbit of one star with respect to the other. Once the orbit is known, Kepler's third law can be used to deduce information about the masses of the two stars. This illustration shows the orbit of a visual binary star in the constellation Ophiuchus. In plotting the orbit, either star may be regarded as the stationary one; the shape of the orbit will be the same in either case.

In principle, the orbital period of a visual binary is easy to determine. All you have to do is see how long it takes for the two stars to revolve once about each other. As Figure 19-19 suggests, however, the period may be so long that more than one lifetime is needed to complete the observations.

Determining the semimajor axis of an orbit is also a challenge. The *angular* separation between the stars can be determined by observation. To convert this angle into a physical distance between the stars, we need to know the distance between the binary and the Earth. This can be found from

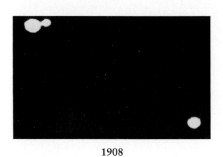
1908

1915

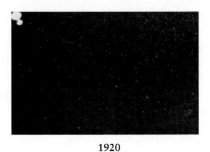
1920

**figure 19-18**   R I 〼 U X G

**The Binary Star Krüger 60** About half of the stars visible in the night sky are actually binary stars. This series of photographs shows the visual binary star Krüger 60 (in the constellation Cepheus) at the upper left of each photograph. The orbital motion of the two stars about each other is apparent. This binary system has a period of 44½ years. The maximum angular separation of the stars is about 2.5 arcsec. (Yerkes Observatory)

parallax measurements or by using spectroscopic parallax. The astronomer must also take into account how the orbit is tilted to our line of sight.

Once both $P$ and $a$ have been determined, Kepler's third law can be used to calculate $M_1 + M_2$, the sum of the masses of the two stars in the binary system. But this analysis tells us nothing about the *individual* masses of the two stars. To obtain these, more information about the motions of the two stars is needed.

Each of the two stars in a binary system actually moves in an elliptical orbit about the **center of mass** of the system. Imagine two children sitting on opposite ends of a seesaw (Figure 19-20a). For the seesaw to balance properly, they must position themselves so that their center of mass—an imaginary point that lies along a line connecting their two bodies—is at the fulcrum, or pivot point of the seesaw. If the two children have the same mass, the center of mass lies midway between them, and they should sit equal distances from the fulcrum. If their masses are different, the center of mass is nearer to the heavier child than to the lighter child.

Just as the seesaw naturally balances at its center of mass, the two stars that make up a binary system naturally orbit around their center of mass (Figure 19-20b). The center of mass always lies along the line connecting the two stars and is closer to the more massive of the two stars.

The center of mass of a visual binary is located by plotting the separate orbits of the two stars, as in Figure 19-20, using the background stars as reference points. The center of mass lies at the common focus of the two elliptical orbits. Comparing the relative sizes of the two orbits around the center of mass yields the ratio of the two stars' masses, $M_1/M_2$. Because the sum $M_1 + M_2$ is already known from Kepler's third law, the individual masses of the two stars can then be determined.

Years of careful, patient observations of binaries have slowly yielded the masses of many stars. As the data accumulated, an important trend began to emerge: For main-sequence stars, there is a direct correlation between mass and luminosity. The more massive a main-sequence star, the more luminous it is. This **mass-luminosity relation** can be conveniently displayed as a graph (Figure 19-21). Note that the range of stellar masses extends from about 0.1 of a solar mass to about 50 solar masses. The Sun's mass lies between these extremes.

The mass-luminosity relation shows that the main sequence on an H-R diagram is a progression in mass as well as in luminosity and surface temperature (Figure 19-22). The hot, bright, bluish stars in the upper left corner of an H-R diagram are the most massive main-sequence stars. Likewise, the dim, cool, reddish stars in the lower right corner of an H-R diagram are the least massive. Main-sequence stars of intermediate temperature and luminosity also have intermediate masses.

Figure 19-22 strongly suggests that main-sequence stars are objects like the Sun, with essentially the same chemical composition as the Sun but with different masses. In this picture, all main-sequence stars shine because of nuclear reactions at their cores that convert hydrogen to helium. The greater the total mass of the star, the greater will be the pressure and temperature at the core, the more rapidly nuclear reactions will take place in the core, and the greater will be the energy output—that is, the luminosity—of the star. In other words, the greater the mass of a main-sequence star, the greater its luminosity. This statement is just the mass-luminosity relation, which we can now recognize as a natural consequence of the nature of main-sequence stars.

There are no simple mass-luminosity relations for giant, supergiant, or white dwarf stars. Why these stars lie where

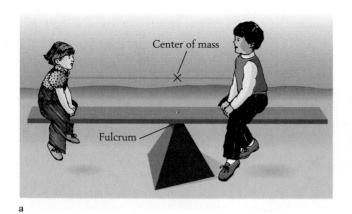

a

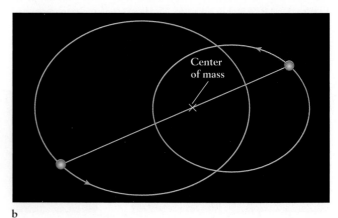

b

## Figure 19-20

**Center of Mass in a Binary Star System** **(a)** For a seesaw to balance, the center of mass of the two children must be at the fulcrum. The center of mass is closer to the more massive of the two children. **(b)** In the same way, the center of mass of a binary star system is always nearer the more massive of the two stars, in this case the reddish star. Both stars move in elliptical orbits about their common center of mass. Although their orbits cross each other, the two stars are always on opposite sides of the center of mass and thus never collide.

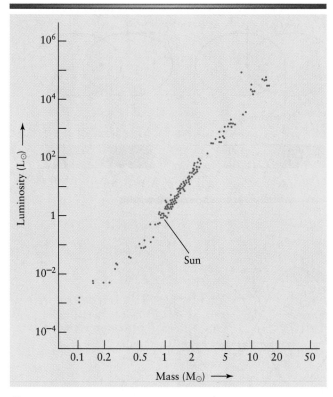

**Figure 19-21**

**The Mass-Luminosity Relation** For main-sequence stars, there is a direct correlation between mass and luminosity—the more massive a star, the more luminous it is. A main-sequence star of mass 10 M$_\odot$ (that is, 10 times the Sun's mass) has roughly 3000 times the Sun's luminosity; one with 0.1 the mass of the Sun has a luminosity of only about 0.005 L$_\odot$.

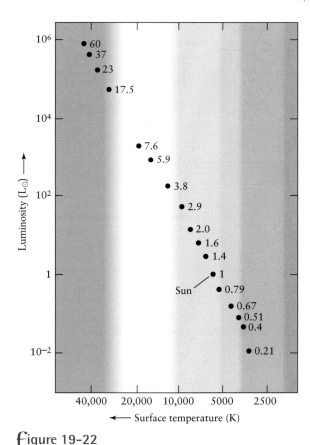

**Figure 19-22**

**The Main Sequence and Masses** On this H-R diagram, each dot represents a main-sequence star. The number next to each dot is the mass of that star in solar masses. As you move up the main sequence from the lower right to the upper left of the H-R diagram, the mass, luminosity, and surface temperature of main-sequence stars all increase.

they do on an H-R diagram will become apparent when we study the evolution of stars in Chapters 21 and 22. We will find that main-sequence stars evolve into giant and supergiant stars, and that some of these eventually end their lives as white dwarfs. As we will see, the mass-luminosity relation helps tell us how rapidly this evolution takes place for stars of different masses.

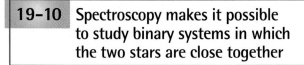

## 19-10 Spectroscopy makes it possible to study binary systems in which the two stars are close together

We have described how the masses of stars can be determined from observations of visual binaries, in which the two stars can be distinguished from each other. But if the two stars in a binary system are too close together, the images of the two stars can blend to produce the semblance of a single star. Happily, in many cases we can use spectroscopy to

decide whether a seemingly single star is in fact a binary system. Spectroscopic observations of binaries provide us with additional useful information about the masses of stars.

Some binaries are discovered when the spectrum of a star shows incongruous spectral lines. For example, the spectrum of what appears to be a single star may include both strong hydrogen lines (characteristic of a type A star) and strong absorption bands of titanium oxide (typical of a type M star). Because a single star cannot have the differing physical properties of these two spectral types, such a star must actually be a binary system that is too far away for us to resolve its individual stars. A binary system detected in this way is called a **spectrum binary.**

Other binary systems can be detected using the Doppler effect. If a star is moving toward the Earth, its spectral lines are displaced toward the short-wavelength (blue) end of the spectrum. Conversely, the spectral lines of a star moving away from us are shifted toward the long-wavelength (red) end of the spectrum. Figure 19-23 applies these ideas to a hypothetical binary star system with an orbital plane that is edge-on to our line of sight. As the two stars move around

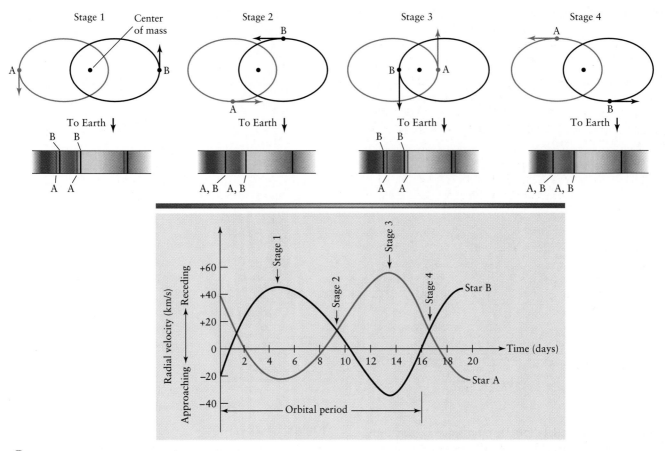

## figure 19-23

**Radial Velocity Curves** The lower graph displays the radial velocity curves of the binary system HD 171978. (The HD means that this is a star from the *Henry Draper Catalogue*, mentioned in Section 19-5.) The drawings at the top indicate the positions of the stars (labeled A and B) and the spectra of the binary at four selected moments (stages 1, 2, 3, and 4) during an orbital period. Note that at stages 1 and 3, the Doppler effect splits apart the absorption lines from stars A and B; compare Figure 19-24. The entire binary star system is moving away from us at 12 km/s, which is why the entire pattern of radial velocity curves is displaced upward from the zero-velocity line.

their orbits, they periodically approach and recede from us. Hence, the spectral lines of the two stars will be alternately blueshifted and redshifted. The two stars in this hypothetical system are so close together that they appear through a telescope as a single star with a single spectrum. Because one star shows a blueshift while the other is showing a redshift, the spectral lines of the binary system appear to periodically split apart and rejoin. Stars whose binary character is revealed by such shifting spectral lines are called **spectroscopic binaries.**

To analyze a spectroscopic binary, astronomers measure the wavelength shift of each star's spectral lines and use the Doppler shift formula (introduced in Section 5-9 and Box 5-6) to determine the **radial velocity** of each star—that is, how fast and in what direction it is moving along our line of sight. Figure 19-23 shows a graph of the radial velocity versus time, called a **radial velocity curve,** for the binary system HD 171978. The pattern of the curves repeats every 15 days, which is the orbital period of the binary. Also note that the entire wavy pattern is displaced upward from the zero-

velocity line by about 12 km/s, which is the overall motion of the binary system away from the Earth. Superimposed on this overall recessional motion are the periodic approaches and recessions of the two stars as they orbit about the center of mass.

Figure 19-24 shows two spectra of the spectroscopic binary κ (kappa) Arietis taken a few days apart. In Figure 19-24a, two sets of spectral lines are visible, offset slightly in opposite directions from the normal positions of these lines. This corresponds to stage 1 or stage 3 in Figure 19-23; one of the orbiting stars is moving toward the Earth and has its spectral lines blueshifted, and the other star is moving away from the Earth and has its lines redshifted. A few days later, the stars have progressed along their orbits so that neither star is moving toward or away from the Earth, corresponding to stage 2 or stage 4 in Figure 19-23. At this time there are no Doppler shifts, and the spectral lines of both stars are at the same positions. That is why only one set of spectral lines appears in Figure 19-24b.

spectral lines of stars split by Doppler effect

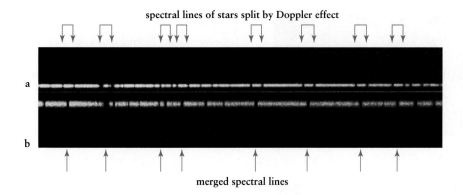

merged spectral lines

**figure 19-24**

**A Spectroscopic Binary** The spectrum of the double-line spectroscopic binary κ (kappa) Arietis has spectral lines that shift back and forth as the two stars revolve about each other. (**a**) The stars are moving parallel to the line of sight, as in stage 1 or stage 3 in Figure 19-23, producing two sets of spectral lines. (**b**) Both stars are moving perpendicular to our line of sight, as in stage 2 or stage 4 in Figure 19-23, and the spectral lines of the two stars merge. (Lick Observatory)

$^{CAUTION}$ It is important to emphasize that the Doppler effect applies only to motion along the line of sight. Motion perpendicular to the line of sight does not affect the observed wavelengths of spectral lines. Hence, the ideal orientation for a spectroscopic binary is to have the stars orbit in a plane that is edge-on to our line of sight. (By contrast, a *visual* binary is best observed if the orbital plane is face-on to our line of sight.) For the Doppler shifts to be noticeable, the orbital speeds of the two stars should be at least a few kilometers per second.

The binaries depicted in Figure 19-23 and Figure 19-24 are called **double-line spectroscopic binaries,** because the spectral lines of both stars in the binary system can be seen. In most spectroscopic binaries, however, one of the stars is so dim that its spectral lines cannot be detected. Such a double-star system, which shows only a single set of spectral lines, is called a **single-line spectroscopic binary.** The star is obviously a binary, however, because its spectral lines shift back and forth, thereby revealing the orbital motions of two stars about their center of mass.

As for visual binaries, spectroscopic binaries allow astronomers to learn about stellar masses. From a radial velocity curve, one can find the *ratio* of the masses of the two stars in a binary. The *sum* of the masses is related to the orbital speeds of the two stars by Kepler's laws and Newtonian mechanics. If both the ratio of the masses and their sum are known, the individual masses can be determined using algebra. However, determining the sum of the masses requires that we know how the binary orbits are tilted from our line of sight. This is because the Doppler shifts reveal only the radial velocities of the stars rather than their true orbital speeds. This tilt is often impossible to determine, because we cannot see the individual stars in the binary. Thus, the masses of stars in spectroscopic binaries tend to be uncertain.

There is one important case in which we can determine the orbital tilt of a spectroscopic binary. If the two stars are observed to eclipse each other periodically, then we must be viewing the orbit nearly edge-on. As we will see next, individual stellar masses—as well as other useful data—can be determined if a spectroscopic binary also happens to be such an *eclipsing* binary.

**19-11** | **Light curves of eclipsing binaries provide detailed information about the two stars**

Some binary systems are oriented so that the two stars periodically eclipse each other as seen from the Earth. These **eclipsing binaries** can be detected even when the two stars cannot be resolved visually as two distinct images in the telescope. The apparent brightness of the image of the binary dims briefly each time one star blocks the light from the other.

Using a sensitive detector at the focus of a telescope, an astronomer can measure the incoming light intensity quite accurately and create a **light curve,** such as those shown in Figure 19-25. The shape of the light curve for an eclipsing binary reveals at a glance whether the eclipse is partial or total (compare Figures 19-25*a* and 19-25*b*).

In fact, the light curve of an eclipsing binary can yield a surprising amount of information. For example, the ratio of the surface temperatures can be determined from how much their combined light is diminished when the stars eclipse each other. Also, the duration of mutual eclipse yields data on the relative sizes of the stars and their orbits.

If the eclipsing binary is also a double-line spectroscopic binary, an astronomer can calculate the mass and radius of each star from the light curves and the velocity curves. Unfortunately, very few binary stars are of this ideal type. Stellar radii determined in this way agree well with the values found using the Stefan-Boltzmann law, described in Section 19-6.

The shape of a light curve can reveal many additional details about a binary system. In some binaries, for example, the gravitational pull of one star distorts the other, much as the Moon distorts the Earth's oceans in producing tides (see Figure 9-16). Figure 19-25*c* shows how such tidal distortion affects the light curve. Another example is *hot-spot reflection*, produced when one of the stars in an eclipsing binary is so hot that its radiation creates a hot spot on its cooler companion star. Every time this hot spot is exposed to our view from the Earth, we receive a little extra light energy, which

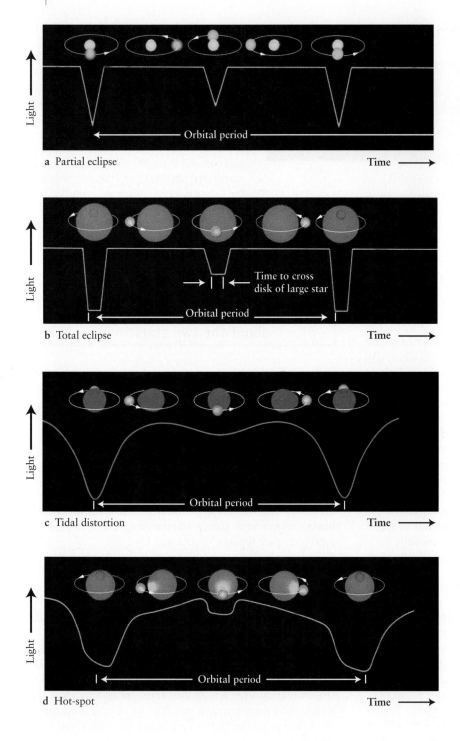

**Representative Light Curves of Eclipsing Binaries** The shape of the light curve of an eclipsing binary can reveal many details about the two stars that make up the binary. Illustrated here are **(a)** a partial eclipse, **(b)** a total eclipse, **(c)** tidal distortion, and **(d)** hot-spot reflection.

produces a characteristic bump on the binary's light curve (see Figure 19-25d).

Information about stellar atmospheres can also be derived from light curves. Suppose that one star of a binary is a luminous main-sequence star and the other is a bloated red giant. By observing exactly how the light from the bright main-sequence star is gradually cut off as it moves behind the edge of the red giant during the beginning of an eclipse, astrono-

mers can infer the pressure and density in the upper atmosphere of the red giant.

Binary systems are tremendously important because they enable astronomers to measure stellar masses as well as other key properties of stars. In the next several chapters, we will use this information to help us piece together the story of *stellar evolution*—how stars are born, evolve, and eventually die.

## Key Words

*Terms preceded by an asterisk (\*) are
discussed in the Boxes.*

absolute magnitude, p. 464

apparent brightness, p. 460

apparent magnitude, p. 463

binary star (binary), p. 479

center of mass, p. 480

color ratio, p. 467

continuum, p. 468

\*distance modulus, p. 465

double-line spectroscopic binary, p. 483

double star, p. 479

eclipsing binary, p. 483

\*equivalent width, p. 471

giant, p. 475

Hertzsprung-Russell diagram
   (H-R diagram), p. 474

inverse-square law, p. 460

light curve, p. 483

\*line profile, p. 471

luminosity, p. 456

luminosity class, p. 477

luminosity function, p. 461

magnitude scale, p. 463

main sequence, p. 475

main-sequence star, p. 475

mass-luminosity relation, p. 480

metals, p. 470

OBAFGKM, p. 468

optical double star, p. 479

parallax, p. 456

parsec, p. 457

photometry, p. 460

\*proper motion, p. 459

radial velocity, p. 482

radial velocity curve, p. 482

red giant, p. 475

single-line spectroscopic binary, p. 483

\*space velocity, p. 459

spectral classes, p. 468

spectral types, p. 468

spectroscopic binary, p. 482

spectroscopic parallax, p. 478

spectrum binary, p. 481

stellar parallax, p. 457

supergiant, p. 475

\*tangential velocity, p. 459

UBV filters, p. 466

UBV photometry, p. 466

visual binary, p. 479

white dwarf, p. 475

## Key Ideas

**Measuring Distances to Nearby Stars:** Distances to the nearer stars can be determined by parallax, the apparent shift of a star against the background stars observed as the Earth moves along its orbit.

• Parallax measurements made from orbit, above the blurring effects of the atmosphere, are much more accurate than those made with Earth-based telescopes.

• Stellar parallaxes can only be measured for stars within a few hundred parsecs.

**The Inverse-Square Law:** A star's luminosity (total light output), apparent brightness, and distance from the Earth are related by the inverse-square law. If any two of these quantities are known,. the third can be calculated.

**The Population of Stars:** Stars of relatively low luminosity are more common than more luminous stars. Our own Sun is a rather average star of intermediate luminosity.

**The Magnitude Scale:** The apparent magnitude scale is an alternative way to measure a star's apparent brightness.

• The absolute magnitude of a star is the apparent magnitude it would have if viewed from a distance of 10 parsecs. A version of the inverse-square law relates a star's absolute magnitude, apparent magnitude, and distance.

**Photometry and Color Ratios:** Photometry measures the apparent brightness of a star. The color ratios of a star are the ratios of brightness values obtained through different standard filters, such as the UBV filters. These ratios are a measure of the star's surface temperature.

**Spectral Types:** Stars are classified into spectral types (subdivisions of the spectral classes O, B, A, F, G, K, and M), based on the major patterns of spectral lines in their spectra. The spectral type of a star is directly related to its surface temperature: O stars are the hottest and M stars are the coolest.

**Hertzsprung-Russell Diagram:** The Hertzsprung-Russell (H-R) diagram is a graph plotting the absolute magnitudes of stars against their spectral types (or, equivalently, their luminosities against surface temperatures).

• The positions on the H-R diagram of most stars are along the main sequence, a band that extends from high luminosity and high surface temperature to low luminosity and low surface temperature.

• On the H-R diagram, giant and supergiant stars lie above the main sequence, while white dwarfs are below the main sequence.

• By carefully examining a star's spectral lines, astronomers can determine whether that star is a main-sequence star, giant, supergiant, or white dwarf. Using the H-R diagram and the inverse-square law, the star's luminosity and distance can be found without measuring its stellar parallax.

**Binary Stars:** Binary stars, in which two stars are held in orbit around each other by their mutual gravitational attraction, are surprisingly common. Those that can be resolved into two distinct star images by an Earth-based telescope are called visual binaries.

• Each of the two stars in a binary system moves in an elliptical orbit about the center of mass of the system.

• Binary stars are important because they allow astronomers to determine the masses of the two stars in a binary system. The masses can be computed from measurements of the orbital period and orbital dimensions of the system.

**Mass-Luminosity Relation for Main-Sequence Stars:** The mass-luminosity relation expresses a direct correlation between mass and luminosity for main-sequence stars. The greater the mass of a main-sequence star, the greater its luminosity.

**Spectroscopic Observations of Binary Stars:** Some binaries can be detected and analyzed, even though the system may be so distant or the two stars so close together that the two star images cannot be resolved.

• A spectrum binary appears to be a single star but has a spectrum with the absorption lines for two distinctly different spectral types.

• A spectroscopic binary has spectral lines that shift back and forth in wavelength. This is caused by the Doppler effect, as the orbits of the stars carry them first toward then away from the Earth.

• An eclipsing binary is a system whose orbits are viewed nearly edge-on from the Earth, so that one star periodically eclipses the other. Detailed information about the stars in an eclipsing binary can be obtained from a study of the binary's radial velocity curve and its light curve.

## Review Questions

1. Describe how the parallax method of finding a star's distance is similar to binocular (two-eye) vision in humans.

2. What is the inverse-square law? Use it to explain why a headlight on a car can appear brighter than a star, even though the headlight emits far less light energy per second.

3. Briefly describe how you would determine the luminosity of a nearby star. Of what value is knowing the luminosity of various stars?

4. Why is the magnitude scale called a "backward" scale? What is the difference between apparent magnitude and absolute magnitude?

5. Does the star Betelgeuse, whose apparent magnitude is 0.5, look brighter or dimmer than the star Pollux, whose apparent magnitude is 1.1?

6. Explain why the color ratios of a star are related to the star's surface temperature.

7. Which gives a more accurate measure of a star's surface temperature—its color ratios or its spectral lines? Explain.

8. What are the most prominent absorption lines you would expect to find in the spectrum of a star with a surface temperature of (a) 30,000 K, (b) 2800 K, and (c) 5800 K (like the Sun)? Briefly describe why these stars have such different spectra even though they have essentially the same chemical composition.

9. If a red star and a blue star both have the same radius and both are the same distance from the Earth, which one looks brighter in the night sky? Explain why.

10. Sketch a Hertzsprung-Russell diagram. Indicate the regions on your diagram occupied by (a) main-sequence stars, (b) red giants, (c) supergiants, (d) white dwarfs, and (e) the Sun.

11. What aspects of a star's spectrum would you concentrate on to determine the star's luminosity class? Explain what you would look for.

12. Suppose that you want to determine the temperature, diameter, and luminosity of an isolated star (not a member of a binary system). Which of these physical quantities require you to know the distance to the star? Explain.

13. What is the mass-luminosity relation? Does it apply to stars of all kinds?

14. Use Figure 19-21 to (a) estimate the mass of a main-sequence star that is 1000 times as luminous as the Sun, and (b) estimate the luminosity of a main-sequence star whose mass is one-fifth that of the Sun. Explain your answers.

15. Sketch the radial velocity curves of a binary consisting of two identical stars moving in circular orbits that are (a) perpendicular to and (b) parallel to our line of sight.

16. Sketch the light curve of an eclipsing binary consisting of two identical stars in highly elongated orbits oriented so that (a) their major axes are pointed toward the Earth and (b) their major axes are perpendicular to our line of sight.

## Advanced Questions

*Questions preceded by an asterisk (*) involve topics discussed in the Boxes.*

**Problem-solving tips and tools:**

Look carefully at the worked examples in Boxes 19-1, 19-2, 19-3 and 19-5 before attempting these exercises. Remember that a telescope's light-gathering power is proportional to the area of its objective or primary mirror. The volume of a sphere of radius $R$ is $4\pi R^3/3$. Make use of the H-R diagrams in this chapter to answer questions involving spectroscopic parallax. As shown in Box 19-3, some of the problems concerning magnitudes may require facility with logarithms.

17. Find the average distance from the Sun to Pluto in parsecs. Compared to Pluto, how many times farther away from the Sun is Proxima Centauri?

18. Suppose that a dim star were located 2 million AU from the Sun. Find (a) the distance to the star in parsecs and (b) the parallax angle of the star. Would this angle be measurable with present-day techniques?

19. Kapteyn's star, named after the Dutch astronomer who found it, has a parallax angle of 0.258 arcsec. How far away is the star?

*20. Kapteyn's star (see the previous question) has a proper motion of 8.65 arcsec per year and a radial velocity of +246 km/s. (a) What is the star's tangential velocity? (b) What is the star's actual speed relative to the Sun? (c) Is Kapteyn's star moving toward the Sun or away from the Sun? Explain.

*21. How far away is a star that has a proper motion of 0.08"/year and a tangential velocity of 50 km/s? For a star at this distance, what would its tangential velocity have to be in order for it to exhibit the same proper motion as Barnard's star (Box 19-1)?

*22. The space velocity of a certain star is 100 km/s and its radial velocity is 60 km/s. Find the star's tangential velocity.

*23. In the spectrum of a particular star, the Balmer line $H_\alpha$ has a wavelength of 656.39 nm. The laboratory value for the wavelength of $H_\alpha$ is 656.28 nm. (a) Find the star's radial velocity. (b) Is this star approaching us or moving away? Explain. (c) Find the wavelength at which you would expect to find $H_\beta$ in the spectrum of this star, given that the laboratory wavelength of $H_\beta$ is 486.13 nm. (d) Do your answers depend on the distance from the Sun to this star? Why or why not?

*24. Derive the equation given in Box 19-1 relating proper motion and tangential velocity. (Hint: For this problem, you need to understand the concept of a radian. You will also find it useful to know that 1 rad = 206,265 arcsec.)

25. How much dimmer does the Sun appear from Saturn than from the Earth? (Hint: The average distance between Saturn and the Sun is 9.54 AU.)

26. Stars A and B are both equally bright as seen from the Earth, but A is 5 pc away while B is 15 pc away. Which star has the greater luminosity? How many times greater is it?

27. Stars C and D both have the same luminosity, but C is 8 pc from the Earth while D is 32 pc from the Earth. Which star appears brighter as seen from the Earth? How many times brighter is it?

28. The solar constant, equal to 1370 W/m², is the amount of light energy from the Sun that falls on 1 square meter of the Earth's surface in 1 second (see Section 19-2). What would the distance between the Earth and the Sun have to be in order for the solar constant to be 1 watt per square meter (1 W/m²)?

29. Suppose two stars have the same apparent brightness, but one star is 10 times farther away than the other. What is the ratio of their luminosities? Which one is more luminous—the closer star or the farther star?

30. The star Procyon in Canis Minor (the Small Dog) is a prominent star in the winter sky, with an apparent brightness $1.3 \times 10^{-11}$ that of the Sun. It is also one of the nearest stars, being only 3.50 parsecs from the Earth. What is the luminosity of Procyon? Express your answer as a multiple of the Sun's luminosity.

31. The star Ross 248 is only 3.32 parsecs from the Earth, but can be seen only with a telescope because it is 300 times dimmer than the dimmest star visible to the unaided eye. How close to the Earth would Ross 248 have to be in order for it to be visible without a telescope? Give your answer in parsecs and in AU. Compare to the semimajor axis of Pluto's orbit around the Sun, given in Appendix 1.

*32. The star Krüger 60 B in the constellation Cepheus has an apparent magnitude of 11.3 and a parallax angle of 0.25 arcsecond. (a) Determine its absolute magnitude. (b) Give the approximate ratio of the luminosity of Krüger 60 B to the Sun's luminosity.

*33. Suppose you can just barely see a twelfth-magnitude star through an amateur's 6-inch telescope. What is the magnitude of the dimmest star you could see through a 60-inch telescope?

*34. A certain type of variable star is known to have an average absolute magnitude of 0.0. Such stars are observed in a particular star cluster to have an average apparent magnitude of +16.0. What is the distance to that star cluster?

35. The bright star Rigel in the constellation Orion has a temperature about 3 times that of the Sun, and its luminosity is 64,000 times that of the Sun. What is Rigel's radius compared to the radius of the Sun?

36. Suppose a star experiences an outburst in which its surface temperature doubles but its average density (its mass divided by its volume) decreases by a factor of 8. The mass of the star stays the same. By what factors do the star's radius and luminosity change?

37. The Sun experiences solar flares (see Section 18-5). The amount of energy radiated by even the strongest solar flare is not enough to have an appreciable effect on the Sun's luminosity. But when a flare of the same size occurs on a main-sequence star of spectral class M, the star's brightness can increase by as much as a factor of 2. Why should there be an appreciable increase in brightness for a main-sequence M star but not for the Sun?

38. The bright star Capella (α Aurigae) has a spectral type of G8 and a luminosity of 160 $L_\odot$. What are the star's approximate surface temperature (in kelvins) and diameter

(in kilometers)? How does Capella's diameter compare to that of the Sun?

**39.** Castor (α Geminorum) is an A1 V star with an apparent brightness of $4.4 \times 10^{-12}$ that of the Sun. Determine the distance from the Earth to Castor (in parsecs).

**40.** The visual binary 70 Ophiuchi (see Figure 19-19) has a period of 87.7 years. The parallax of 70 Ophiuchi is 0.2 arcsec, and the apparent length of the semimajor axis as seen through a telescope is 4.5 arcsec. (a) What is the distance to 70 Ophiuchi in parsecs? (b) What is the actual length of the semimajor axis in AU? (c) What is the sum of the masses of the two stars? Give your answer in solar masses.

**41.** An astronomer observing a binary star finds that one of the stars orbits the other once every 5 years at a distance of 10 AU. (a) Find the sum of the masses of the two stars. (b) If the mass ratio of the system $(M_1/M_2)$ is 4.0, find the individual masses of the stars. Give your answers in terms of the mass of the Sun.

## Discussion Questions

**42.** From its orbit around the Earth, the *Hipparcos* satellite could measure stellar parallax angles with acceptable accuracy—10% or better—only if the angles were larger than about 0.002 arcsec. Discuss the advantages or disadvantages of making parallax measurements from a space telescope in a large solar orbit, say at the distance of Jupiter from the Sun. Assuming that this space telescope can also measure parallax angles of 0.002 arcsec, what is the distance of the most remote stars that can be accurately determined? How much bigger a volume of space would be covered compared to the Earth-based observations? How many more stars would you expect to be contained in that volume?

**43.** Why do you suppose that stars of the same spectral type but of a different luminosity class exhibit slight differences in their spectra?

**44.** It is desirable to be able to measure the radial velocity of stars (using the Doppler effect) to an accuracy of 1 km/s or better. One complication is that radial velocities refer to the motion of the star relative to the Sun, while the observations are made using a telescope on the Earth. Is it important to take into account the motion of the Earth around the Sun? Is it important to take into account the Earth's rotational motion? To answer this question, you will have to calculate the Earth's orbital speed and the speed of a point on the Earth's equator (the part of the Earth's surface that moves at the greatest speed because of the planet's rotation). If one or both of these effects are of importance, how do you suppose astronomers compensate for them?

## Observing Projects

### Observing tips and tools:

> Even through a telescope, the colors of stars are sometimes subtle and difficult to see. To give your eye the best chance of seeing color, use the "averted vision" trick: When looking through the telescope eyepiece, direct your vision a little to one side of the star you are most interested in. This places the light from that star on a more sensitive part of your eye's retina.

**45.** The following table lists five well-known red stars. It includes their right ascension and declination (celestial coordinates described in Box 2-2), apparent magnitudes, and color ratios. As indicated by the apparent magnitude column, all these stars are somewhat variable.

| Star | Right ascension | Declination | Apparent magnitude | $b_V/b_B$ |
|---|---|---|---|---|
| Betelgeuse | $5^h$ $55.2^m$ | $+7° 24'$ | 0.4–1.3 | 5.5 |
| Y Canum Venaticorum | 12 45.1 | +45 26 | 5.5–6.0 | 10.4 |
| Antares | 16 29.4 | −26 26 | 0.9–1.8 | 5.4 |
| μ Cephei | 21 43.5 | +58 47 | 3.6–5.1 | 8.7 |
| TX Piscium | 23 46.4 | + 3 29 | 5.3–5.8 | 11.0 |

*Note: The right ascensions and declinations are given for epoch 2000.*

Observe at least two of these stars both by eye and through a small telescope. Is the reddish color of the stars readily apparent, especially in contrast to neighboring stars? (The Jesuit priest and astronomer Angelo Secchi named Y Canum Venaticorum "La Superba," and μ Cephei is often called William Herschel's "Garnet Star.")

**46.** The table of double stars which follows includes vivid examples of contrasting star colors. The apparent magnitudes, spectral types, and the angular separation between the stars of each binary are given along with star names and coordinates (epoch 2000). Observe at least four of these double stars through a telescope. Use the spectral types listed to estimate the difference in surface temperature of the stars in each pair you observe. Does the binary with the greatest difference in temperature seem to present the greatest color contrast? Based on what you see through the telescope and on what you know about the H-R diagram, explain why *all* the cool stars (spectral types K and M) listed are probably giants or supergiants.

**47.** Observe the eclipsing binary Algol (β Persei), using nearby stars to judge its brightness during the course of an eclipse. Algol has an orbital period of 2.87 days, and, with

| Star | Right ascension | Declination | Apparent magnitudes | Angular separation | Spectral types |
|---|---|---|---|---|---|
| 55 Piscium | 0$^h$ 39.9$^m$ | +21° 26′ | 5.4 and 8.7 | 6.5″ | K0 and F3 |
| γ Andromedae | 2 03.9 | +42 20 | 2.3 and 4.8 | 9.8″ | K3 and A0 |
| 32 Eridani | 3 54.3 | −2 57 | 4.8 and 6.1 | 6.8″ | G5 and A2 |
| ι Cancri | 8 46.7 | +28 46 | 4.2 and 6.6 | 30.5″ | G5 and A5 |
| γ Leonis | 10 20.0 | +19 51 | 2.2 and 3.5 | 4.4″ | K0 and G7 |
| 24 Coma Berenicis | 12 35.1 | +18 23 | 5.2 and 6.7 | 20.3″ | K0 and A3 |
| ν Boötis | 14 45.0 | +27 04 | 2.5 and 4.9 | 2.8″ | K0 and A0 |
| α Herculis | 17 14.6 | +14 23 | 3.5 and 5.4 | 4.7″ | M5 and G5 |
| 59 Serpentis | 18 27.2 | +0 12 | 5.3 and 7.6 | 3.8″ | G0 and A6 |
| β Cygni | 19 30.7 | +27 58 | 3.1 and 5.1 | 34.3″ | K3 and B8 |
| δ Cephei | 22 29.2 | +58 25 | 4* and 7.5 | 20.4″ | F5 and A0 |

*The brighter star in the δ Cephei binary system is a variable star of approximately the fourth magnitude.
Note: The right ascensions and declinations are given for epoch 2000.

the onset of primary eclipse, its apparent magnitude drops from 2.1 to 3.4. It remains this faint for about 2 hours. The entire eclipse, from start to finish, takes about 10 hours. Consult the "Celestial Calendar" section of the current issue of *Sky & Telescope* for the predicted dates and times of the minima of Algol. Note that the schedule is given in Universal Time (the same as Greenwich Mean Time), so you will have to convert the time to that of your own time zone. Algol is normally the second brightest star in the constellation of Perseus. Because of its position on the celestial sphere (R.A. = 3$^h$ 08.2$^m$, Decl. = +40° 57′), Algol is readily visible from northern latitudes during the fall and winter months.

## Where to Learn More

*Books and magazine articles*

Hearnshaw, J. B. "Origins of the Stellar Magnitude Scale." *Sky & Telescope*, November 1992. This delightful article describes the history of the archaic magnitude scale, with which astronomers still burden themselves.

Hirshfeld, A. "The Absolute Magnitudes of Stars." *Sky & Telescope*, September 1994. This article answers the question "What would the night sky look like if all the stars were the same distance from us?"

Kaler, J. *Stars*. Scientific American Library, 1992. This beautifully illustrated book gives a broad overview of our understanding of the stars.

———. *Stars and Their Spectra*. Cambridge University Press, 1997. Based on a series of articles that appeared in *Sky & Telescope*, this well-written and very accessible book explains the tremendous array of information that can be gleaned from the spectra of stars.

McAlister, H. A. "Twenty Years of Seeing Double." *Sky & Telescope*, November 1996. This article describes how a technique called speckle interferometry allows observations of very close binary stars.

Roth, J., and Sinnott, R. W. "Our Nearest Celestial Neighbors." *Sky & Telescope*, October 1996. Paradoxically, some of the nearest stars are also the hardest to detect because they are so dim. This article reviews recent searches of the Sun's neighborhood.

Steffey, P. C. "The Truth about Star Colors." *Sky & Telescope*, September 1992. You can see star colors using a telescope, but what you see can be deceiving. This article dispels misconceptions that have long surrounded the colors of stars.

Turon, C. "Measuring the Universe: From Hipparchus to *Hipparcos*." *Sky & Telescope*, July 1997. This article tells how the *Hipparcos* satellite was able to measure the distances to stars with unprecedented accuracy and how it narrowly escaped an orbital disaster.

W *World Wide Web*

The distribution of stars in space can best be visualized with 3-D images. A set of these showing the nearby stars can be found on the World Wide Web (**http://www.clockwk .com/stars/**).

The web site for the *Hipparcos* mission includes a wealth of information about the spacecraft and its scientific bounty and even includes 3-D images of various constellations (**http://astro.estec.esa.nl/SA-general/Projects/ Hipparcos/ hipparcos.html**).

Another web site describes how a group of astronomers headquartered at the University of Texas uses the Hubble Space Telescope to measure the positions and parallaxes of stars with great precision (**http://clyde.as.utexas.edu/**).

The color of a star or other astronomical objects can tell us much more than the object's surface temperature, as shown graphically by a web site at the University of Bonn (**http://aibn91.astro.uni-bonn.de/~skohle/colour.html**).

If your web browser is capable of using Java, a wonderful binary star simulation is available at Cornell University (**http:// instruct1.cit.cornell.edu/~tlh10/java/binary/binary.htm**).

# The Birth of Stars

R I **V** U X G

**Reflection and Emission Nebulae** Newly formed stars are found in and around interstellar clouds called nebulae. The two main bluish objects, NGC 6589 (top) and NGC 6590 (below), are reflection nebulae surrounding hot main–sequence stars of spectral types B5 and B6. These nebulae are made of dust that surrounds these stars and reflects their bluish light. The large reddish patch of ionized hydrogen gas is an emission nebula named IC 1283–4. Dust mixed with the gas dilutes the intense red emission with a soft blue haze. (Anglo–Australian Observatory)

*In this chapter you will find the answers to the following questions:*

20-1  Why do astronomers think that stars evolve?

20-2  What kind of matter exists in the spaces between the stars?

20-3  Where do new stars form?

20-4  What steps are involved in forming a star like the Sun?

20-5  When a star forms, why does it end up with only a fraction of the available matter?

20-6  What do star clusters tell us about the formation of stars?

20-7  Where in the Galaxy does star formation take place?

20-8  How can the death of one star trigger the birth of many other stars?

can be seen with a small telescope. As an example, Figure 20-2 shows emission nebulae that lie in a different part of the constellation Orion. Such nebulae are direct evidence of *gas* atoms in the interstellar medium.

Typical emission nebulae have temperatures around 10,000 K and masses that range from about 100 to about 10,000 solar masses. Because this mass is spread over a huge volume that is light-years across, the density is quite low by Earth standards. There are only a few thousand hydrogen atoms in each cubic centimeter. (By comparison, the air you are breathing contains more than $10^{19}$ atoms per $cm^3$!)

Emission nebulae are found near massive, hot, luminous main-sequence stars of spectral types O and B. Such stars have surface temperatures in excess of 10,000 K, and emit copious amounts of ultraviolet radiation. When atoms in the nearby interstellar gas absorb these energetic ultraviolet photons, the atoms become ionized. Indeed, emission nebulae are composed primarily of ionized hydrogen atoms, that is, free protons (hydrogen nuclei) and electrons. Astronomers use the notation H I for neutral, un-ionized hydrogen atoms and H II for ionized hydrogen atoms, which is why emission nebulae are also called **H II regions**. Two distinct H II regions can be seen in Figure 20-2.

H II regions emit visible light when some of the free protons and electrons manage to get back together to form hydrogen atoms, a process called **recombination**. When an atom forms by recombination, the electron is typically captured into a high-energy orbit. As the electron then cascades downward through the atom's energy levels toward the ground state, the atom emits visible-light photons. Particularly important is the transition from $n = 3$ to $n = 2$. It produces $H_\alpha$ photons with a wavelength of 656 nm, in the red portion of the visible spectrum (review the discussion in Section 5-8, and especially Figure 5-21). These emitted photons give H II regions their distinctive reddish color.

For each high-energy, ultraviolet photon absorbed by a hydrogen atom to ionize it, several photons of lower energy are emitted when a proton and electron recombine. As Box 20-1 describes, a similar effect takes place in an ordinary fluorescent light bulb. In this sense, H II regions are immense fluorescent light fixtures!

Further evidence of interstellar gas came in the early 1900s, as astronomers were studying the spectra of binary star systems. The two stars that make up the binary system orbit around their common center of mass (see Sections 19-9 and 19-10). As they do, their spectral lines shift back and forth as shown in Figure 19-23. But certain calcium and sodium lines were found to remain at fixed wavelengths. These **stationary absorption lines** are therefore not associated with the binary star. Instead, they must be caused by interstellar gas between us and the binary system.

In addition to the presence of gas atoms in H II regions, Figure 20-2 also shows two kinds of evidence for larger bits of matter, called **dust grains**, in the interstellar medium. These dust grains make their appearance in *dark nebulae* and *reflection nebulae*. A prominent **dark nebula**, called the Horsehead Nebula for its shape, protrudes in front of one of

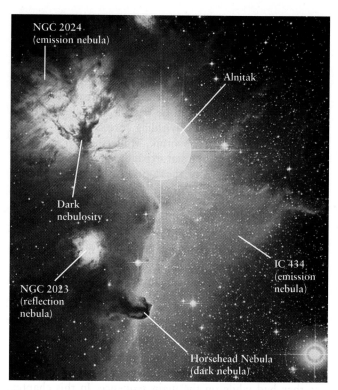

**f igure 20-2**   R I **V** U X G

**Emission, Reflection, and Dark Nebulae in Orion** A variety of different nebulae appear in the sky around Alnitak, also called ζ (zeta) Orionis, the easternmost star in the belt of Orion (see Figure 20-1a). To the left of Alnitak is a bright, red emission nebula called NGC 2024, while another, less dramatic emission nebula called IC 434 extends downward from the star. The glowing gases in these emission nebulae are excited by ultraviolet radiation from young, massive stars. Dust grains obscure part of NGC 2024, giving the appearance of black streaks, while the distinctively shaped dust cloud called the Horsehead Nebula blocks part of the light from IC 434. The Horsehead Nebula is part of a larger complex of dark nebulosity, seen at the lower left corner of the image. Just above and to the left of the Horsehead Nebula is the reflection nebula NGC 2023, whose dust grains scatter blue light more effectively than red. All of this nebulosity lies approximately 490 parsecs (1600 light-years) from Earth. The nebulae are actually nowhere near Alnitak; the star is much closer to us, only 41 parsecs (135 light-years) distant. The area shown in this photograph spans about 1.5° × 2° of the sky. (Royal Observatory, Edinburgh)

the H II regions. It is so opaque that it completely blocks our view of any stars that lie behind it. Another dark nebula, called Barnard 86, is shown in Figure 20-3. These nebulae are dark because they have a relatively dense concentration of microscopic dust grains, which scatter and *absorb* light much more efficiently than single atoms. A typical dark nebula has a temperature between 10 K to 100 K, which is low enough for hydrogen atoms to form molecules, and contains from $10^4$ to $10^9$ particles (atoms, molecules, and dust grains) per cubic centimeter.

The other evidence for dust in Figure 20-2 is the bluish haze surrounding the star immediately above and to the left of the Horsehead Nebula. A similar haze is shown in more detail in Figure 20-4. This haze, called a **reflection nebula,** is caused by fine grains of dust in a lower concentration than that found in dark nebulae. These dust grains scatter and *reflect* short-wavelength radiation more efficiently than long-wavelength radiation. When visible light from a star encounters interstellar dust, blue light is scattered by the dust grains; some of this blue light is deflected toward telescopes on the Earth, producing a reflection nebula. By contrast, the longer-wavelength red light from the star manages to get through the dust without much scattering or absorption. Hence, little red light is scattered toward the Earth observers, giving reflection nebulae their blue color. Box 15-1 explains how light scattering gives our own sky its familiar blue color.

In the 1930s, American astronomer Robert Trumpler discovered two other convincing pieces of evidence for the existence of interstellar matter—*interstellar extinction* and *interstellar reddening*. While studying the brightness and distances of certain star clusters, Trumpler noticed that remote clusters seem to be dimmer than would be expected from their distance alone. His observations demonstrated that the intensity of light from remote stars is reduced as the light passes through the sparse material in interstellar space. This process is called **interstellar extinction.** (In the same way, the headlights of oncoming cars appear dimmer when you are driving through smoke or fog.) The light from remote stars is also reddened as it passes through the interstellar medium, because the blue component of their starlight is scattered and absorbed by interstellar dust. This effect is called **interstellar reddening** (Figure 20-5). As we described in Box 15-1, the same effect causes sunsets to be red.

**ʃigure 20-3** R I V U X G

**A Dark Nebula** The black area near the top of this image (just below the bright yellow star) is a dark nebula called Barnard 86. When such nebulae were discovered in the late 1700s, they were thought to be "holes in the heavens" where very few stars are present. We now know that dark nebulae are opaque regions that block out light from the stars beyond them. The few stars that appear to be within the dark nebula are actually closer to us than the nebula itself. Despite their dark, dense appearance, nebulae such as Barnard 86 are actually very tenuous by earthly standards; the air that you breathe is some $10^{15}$ times denser. But a sufficient thickness of even such a tenuous cloud can prevent the passage of light, just as a sufficient depth of haze, smoke, or fog in our atmosphere can make it impossible to see distant mountains. Barnard 86 is located in the constellation Sagittarius and has an angular diameter of 4 arc-minutes, about one-seventh the angular diameter of the full moon. The cluster of bluish stars to the left of the nebula is called NGC 6520. (Anglo-Australian Observatory)

**ʃigure 20-4** R I V U X G

**Reflection Nebulae** Wispy, bluish reflection nebulae called NGC 6726-7 surround several stars in the southern constellation of Corona Australis (the Southern Crown). This nebulosity is produced by interstellar dust grains scattering starlight. These dust grains are no larger than a typical wavelength of visible light, about 500 nm across. Such small grains are very effective at scattering short-wavelength blue light from the stars but allow much more of the long-wavelength red starlight to pass straight through. Hence, the scattered starlight that we see coming from the reflection nebulae is quite blue in color. (Anglo-Australian Observatory)

## box 20-1 | The Heavens on the Earth

### *Fluorescent Lights*

The light that comes from the interstellar medium may seem to be, quite literally, otherworldly. But, in fact, the same principles that explain light emission from H II regions are also at the heart of common light phenomena that we see here on the Earth.

A fluorescent lamp produces light in a manner not too different from an emission nebula (H II region). In both cases, the physical effect is called **fluorescence:** High-energy ultraviolet photons are absorbed, and the absorbed energy is reradiated as lower-energy photons of visible light. Within the glass tube of a fluorescent lamp is a small amount of mercury, which is vaporized and excited when you turn on the lamp and makes an electric current pass through the tube. This excited mercury vapor emits light with an emission-line spectrum, including spectral lines in the ultraviolet. The white fluorescent coating that lines the inside of the glass tube absorbs these ultraviolet photons, exciting the coating's molecules to high energy levels. The molecules then cascade down through a number of lower levels before reaching the ground state. During this cascade, visible-light photons of many different wavelengths are emitted, giving an essentially continuous spectrum and a very white light. (By comparison, the hydrogen atoms in an H II region emit at only certain discrete wavelengths, because the spectrum of hydrogen is much simpler than that of the fluorescent coating's molecules. Another difference is that the molecules in the fluorescent tube never become ionized.)

Many common materials display fluorescence. Among them are teeth, fingernails, and certain minerals; when illuminated with ultraviolet light, these materials glow with a blue or green color. (Most natural history museums and science museums have an exhibit showing fluorescent minerals.) Laundry detergent also contains fluorescent material. After washing, your laundry absorbs ultraviolet light from the Sun and glows faintly, making it appear "whiter than white."

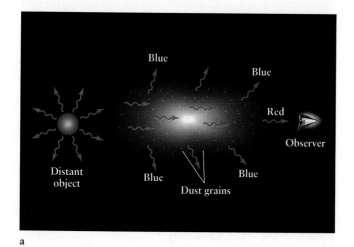

a

b   R  I  **V**  U  X  G

## figure 20-5

**Interstellar Reddening** (a) As a beam of light from a distant object passes through interstellar space, tiny dust grains scatter or absorb blue light but allow red light to pass through almost undisturbed. Thus, the light beam is deficient in blue light when it reaches us, and the object appears redder in color than it really is. This effect is called interstellar reddening. (b) The difference in color between these two emission nebulae (also called H II regions) is a result of interstellar reddening. The nebula on the left, NGC 3603, is about 7200 parsecs (23,000 light-years) from Earth, while the one on the right, NGC 3576, is only about 2400 parsecs (7800 light-years) distant. NGC 3603 has a ruddier color precisely because it is farther away. Its light has had to pass through more interstellar dust to reach us, so more interstellar reddening has occurred. The two nebulae lie about 1° apart in the sky in the southern constellation Carina (the Ship's Keel). (Anglo-Australian Observatory)

It is important to note the distinction between interstellar reddening and other, unrelated effects that can make a star, nebula, or galaxy appear red. Some objects are intrinsically red. Examples include a star with a low surface temperature (whose continuous spectrum contains more red light than any other visible wavelength, as shown in Figure 19-7) and an H II region (which has an emission-line spectrum dominated by certain wavelengths in the red part of the spectrum). Other objects can have their color shifted toward the red by the Doppler effect if they are moving away from us. However, these Doppler shifts are *far* too small to notice with the naked eye or even with a moderately large telescope. Interstellar reddening, by contrast, makes objects appear red not by shifting wavelengths but by filtering out short wavelengths. The effect is the same as if we looked at an object through red-colored glasses, which let red light pass but block out blue light.

Measurements of both extinction and reddening in various directions in space indicate that interstellar gas and dust are largely confined to the Milky Way—that faint, hazy band of stars you can see stretching across the sky on a dark, moonless night. Infrared views like Figure 20-6 clearly show clouds of gas and dust clumped along the Milky Way. As we will see in Chapter 25, the appearance of the Milky Way results from our inside view of the Galaxy, which is a disk-shaped collection of several hundred billion stars about 50,000 parsecs (160,000 light-years) in diameter. The Sun is located about 8000 parsecs (26,000 light-years) from the Galaxy's center. Astronomers know these dimensions because they can map the Galaxy from the locations of bright stars and nebulae and also by using radio telescopes. Radio observations indicate that bright stars, gas, and glowing nebulae are mostly located along arching spiral arms that wind outward from our Galaxy's center. If we could view our Galaxy from a great distance, it would look somewhat like the galaxy shown in Figure 20-7.

Interstellar gas and dust are the raw material from which stars are made. The disk of our Galaxy, where most of this matter is concentrated, is therefore the site of ongoing star formation. Our next goal is to examine the steps by which stars are formed.

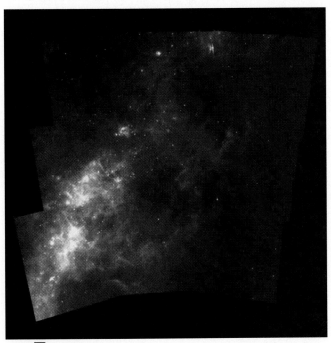

a  R **I** V U X G

b

## figure 20-6

**The Infrared Milky Way** **(a)** Because most stars are several light-years from their nearest neighbor, the dust that occupies the space between the stars is very cold. Thus, by Wien's law, this dust emits very long-wavelength, infrared light. This mosaic of infrared images, taken by the *Infrared Astronomical Satellite* (*IRAS*) in 1983, shows light emitted from cool interstellar dust at a wavelength of 60 μm (colored red in this picture). The blue dots are stars that shine brightly at 12 μm. This mosaic extends over the southern constellations of Puppis (the Ship's Stern) and Vela (the Ship's Sail) and covers a region of the sky approximately 35° × 35°. **(b)** This star chart shows the constellations covered by the infrared mosaic. The grayish areas indicate the visible Milky Way. Comparison of the infrared mosaic with the star chart shows that interstellar dust is concentrated within the plane of the Milky Way. Other observations show that the same is true for interstellar gas. (IPAC, NASA)

<figure 20-7   R I **V** U X G

**A Spiral Galaxy** Spiral galaxies, like our own Milky Way Galaxy, consist of stars, gas, and dust that are largely confined to a flattened, rotating disk. Spiral arms, delineated by their luminous stars and glowing gas, wind outward from the galaxy's center. The bright knots of gas strung along the spiral arms are H II regions. This galaxy, called NGC 2997, is oriented so that we see its disk nearly face-on. It lies some 11 million parsecs (35 million light-years) from Earth in the southern constellation of Antlia (the Air Pump) and spans 8.1 × 6.5 arcminutes. (Anglo-Australian Observatory)

## 20-3    Protostars form in cold, dark nebulae

How do stars form in the interstellar medium? In this section, we discuss how a contracting cloud gives birth to a clump called a *protostar*—a future main-sequence star.

In order for a star to condense, gravity—which tends to draw interstellar material together—must overwhelm the pressure pushing it apart. This means that stars will most easily form in regions where the interstellar material is relatively dense, so that atoms and dust grains are close together and gravitational attraction is enhanced.

To assist star formation, the pressure of the interstellar medium should be relatively low. This means that the interstellar medium should be as cold as possible, because the pressure of a gas goes down as the gas temperature decreases. (Conversely, increasing the temperature makes the pressure increase. That's why the air pressure inside automobile tires is higher after the auto has been driven a while and the tires have warmed up.)

The only parts of the interstellar medium with high enough density and low enough temperature for stars to form are the dark nebulae. Many of these, such as the Horsehead Nebula (Figure 20-2) and Barnard 86 (Figure 20-3), were discovered and catalogued around 1900 by Edward Barnard and are known as **Barnard objects.** (The Horsehead Nebula is also known as Barnard 33.) Other relatively small, nearly spherical dark nebulae are known as **Bok globules,** after the Dutch-American astronomer Bart Bok who first called attention to them during the 1940s (Figure 20-8). A Bok globule resembles the inner core of a Barnard object with the outer, less dense portions stripped away. The density of the gas and dust within a Barnard object or Bok globule is indeed quite high by cosmic standards, in the range from

100 to 10,000 particles per $cm^3$; by comparison, most of the interstellar medium contains only 0.1 to 20 particles per $cm^3$. Radio emissions from molecules within these clouds indicate that their internal temperatures are very low, only about 10 K.

A typical Barnard object contains a few thousand solar masses of gas and dust spread over a volume roughly 10 parsecs (30 light-years) across; a typical Bok globule is about one-tenth as large. The chemical composition of this material is the standard "cosmic abundance" of about 74% (by mass) hydrogen, 25% helium, and 1% heavier elements (review Table 7-4). Within these clouds, the densest portions can contract under their own mutual gravitational attraction and form clumps called **protostars.** Each protostar will eventually evolve into a main-sequence star. Because dark nebulae contain many solar masses of material, it is possible for a large number of protostars to form out of a single such nebula. Thus, we can think of dark nebulae as "stellar nurseries."

In the 1950s, astrophysicists such as Louis Henyey in the United States and C. Hayashi in Japan performed calculations that enabled them to describe the earliest stages of a protostar. At first, a protostar is merely a cool blob of gas several times larger than our solar system. The pressure inside the protostar is too low to support all this cool gas, and so the protostar contracts. As it does, gravitational energy is converted into thermal energy, making the gases heat up and start glowing (recall the discussion of Kelvin-Helmholtz contraction in Section 18-6). Energy from the interior of the protostar is transported outward by convection, warming its surface. After only a few thousand years of gravitational contraction, the surface temperature reaches 2000 to 3000 K. At this point the protostar is still quite large, so its glowing gases produce substantial luminosity. (The greater a star's radius and surface temperature, the greater its luminosity; you can review the details in Section 19-6 and Box 19-5.) For example, after only 1000 years of contraction, a protostar of 1 solar mass is 20 times larger in radius than the Sun and has

figure 20-8   R I **V** U X G

**Bok Globules** The black spots on this photograph, which shows a cluster of stars lying in front of a glowing, red H II region, are not rips or imperfections in the image; they are concentrations of dust and gas called Bok globules. These dusty, dark clouds are typically a parsec or less in size and contain anywhere from one to a thousand solar masses of material. Because of their relatively high density and low temperature, they are ideal sites for the formation of new stars. The foreground stars are part of a cluster called IC 2944, which lies in the constellation Centaurus (the Centaur). The image shows an area about $8 \times 10$ arcminutes, about one-third as wide as the apparent size of the full moon. (Anglo-Australian Observatory)

100 times the Sun's luminosity. Unlike the Sun, however, the luminosity of a young protostar is not the result of thermonuclear reactions, because the protostar's core is not yet hot enough for these reactions to begin. Instead, the radiated energy comes exclusively from the heating of the protostar as it contracts.

In order to determine the conditions inside a contracting protostar, astrophysicists use computers to solve equations similar to those for calculating the structure of the Sun (described in Section 18-7). The results tell how the protostar's luminosity and surface temperature change at various stages in its contraction. This information, when plotted on a Hertzsprung-Russell diagram, provides a protostar's **evolutionary track**. The track shows us how the protostar's appearance changes because of changes in its interior. These theoretical tracks agree quite well with actual observations of protostars.

CAUTION! When astronomers refer to a star "following an evolutionary track" or "moving on the H-R diagram," they mean that the star's luminosity, surface temperature, or both change. Hence, the point that represents the star moves on an H-R diagram. This is completely unrelated to how the star physically moves through space!

Figure 20-9 shows evolutionary tracks for protostars of seven different masses, ranging from 0.5 M$_\odot$ to 15 M$_\odot$. Because protostars are relatively cool when they begin to shine at visible wavelengths, all of these tracks begin near the right (low-temperature) side of the H-R diagram. The subsequent evolution is somewhat different, however, depending on the protostar's mass.

Observing the evolution of a protostar can be quite a challenge. Indeed, while Figure 20-9 shows that young protostars are much more luminous than the Sun, it is quite unlikely that you have ever seen a protostar shining in the night sky. The reason is that protostars form within clouds that contain

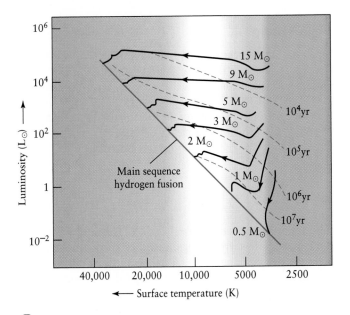

**figure 20-9**

**Pre–Main-Sequence Evolutionary Tracks** The evolutionary tracks of seven protostars of different masses are shown on this H-R diagram. The dashed lines indicate the stage reached after the indicated number of years of evolution. All of the tracks end on the main sequence, so these protostars all end up as main-sequence stars in whose cores hydrogen is converted into helium. Where on the main sequence a given track ends depends on the mass of the protostar and, therefore, the mass of the main-sequence star it becomes. The more massive the star, the greater its luminosity and surface temperature. This agrees with the mass-luminosity relation for main-sequence stars (Section 19-9). Protostars eject quite a bit of mass into space as they form (as described in Section 20-5), so the mass shown for each evolutionary track is that of the final main-sequence star. (Based on stellar model calculations by I. Iben)

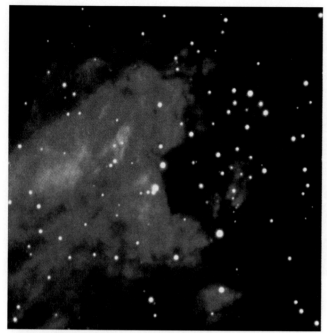

a  R I **V** U X G

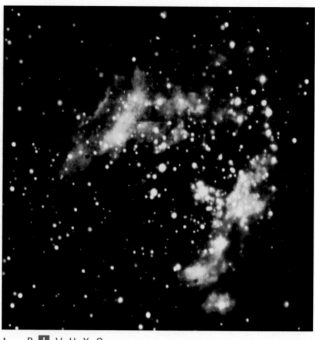

b  R **I** V U X G

## ƒigure 20-10

**Newborn Stars in the Omega Nebula** **(a)** This image at visible wavelengths shows an H II region called the Omega, Swan, or Horseshoe Nebula because of its characteristic shape. Also known as M17 and NGC 6618, it lies some 1700 parsecs (5500 light-years) from the Earth in the direction of the constellation Sagittarius (the Archer). Not visible in this picture is a large dark nebula that surrounds the glowing H II region. **(b)** This infrared view,

constructed from images taken at wavelengths of 1.2, 1.6, and 2.2 μm, reveals hundreds of stars that do not appear in the visible view. A comparison of the visible and infrared views demonstrates that many of the stars in this star-forming region are obscured by interstellar dust. The total mass of the nebula has been estimated at around 800 $M_\odot$, so there is plenty of material available to form even more stars. (NOAO)

substantial amounts of interstellar dust. The dust in a protostar's immediate surroundings, called its **cocoon nebula**, absorbs the vast amounts of visible light emitted by the protostar and makes the protostar very hard to detect using visible wavelengths.

Protostars can be seen, however, using infrared wavelengths. Because it absorbs so much energy from the protostar that it surrounds, a dusty cocoon nebula becomes heated to a few hundred kelvins. The warmed dust then reradiates its thermal energy at infrared wavelengths, to which the cocoon nebula is relatively transparent. So, by using infrared telescopes, astronomers can see protostars within the "stellar nursery" of a dark nebula.

Figure 20-10 shows views of such a stellar nursery taken at both visible and infrared wavelengths. The visible-light view (Figure 20-10*a*) shows an H II region, on the right (west) side of which is a dark, dusty nebula that almost completely blocks visible light. But the infrared image (Figure 20-10*b*) shows hundreds of previously unseen stars in this same area. These are protostars in the process of formation within the dark nebula, with properties that agree well with the theoretical ideas outlined in this section.

## 20-4 Protostars evolve into main-sequence stars

Evolutionary tracks show how a protostar evolves into a star as its gases contract. The details of the birth process depend on the star's mass. To follow these details, the basic principle to remember is that luminosity is proportional to the square of radius and to the fourth power of surface temperature.

For a protostar with the same mass as our Sun (1 $M_\odot$), the outer layers are cool and quite opaque (for the same reasons that the Sun's photosphere is opaque; see Section 18-1). This means that energy released from the shrinking inner layers in the form of radiation cannot reach the surface. Instead, energy flows outward by the slower and less effective method of convection. Hence, the surface temperature of the contracting protostar stays roughly constant, the luminosity decreases as the radius decreases, and the evolutionary track moves downward on the H-R diagram in Figure 20-9.

Although its surface temperature changes relatively little, the *internal* temperature of the shrinking protostar increases.

After a time, the interior becomes ionized, which makes it less opaque. Energy is then conveyed outward by radiation in the interior and by convection in the opaque outer layers, just as in the present-day Sun (see Section 18-7, especially Figure 18-28). This makes it easier for energy to escape from the protostar, so the luminosity increases. As a result, the evolutionary track bends upward (higher luminosity) and to the left (higher surface temperature, caused by the increased energy flow).

In time, the 1 $M_\odot$ protostar's interior temperature reaches a few million kelvins, hot enough for thermonuclear reactions to begin converting hydrogen into helium. As we saw in Section 18-6 and Box 18-2, these reactions release enormous amounts of energy. Eventually, these reactions provide enough heat and internal pressure to stop the star's gravitational contraction, and hydrostatic equilibrium is reached. The protostar's evolutionary track has now led it to the main sequence, and the protostar has become a full-fledged main-sequence star.

More massive protostars evolve a bit differently. If its mass is more than about 4 $M_\odot$, a protostar contracts and heats more rapidly, and hydrogen burning begins quite early. As a result, the luminosity quickly stabilizes at nearly its final value, while the surface temperature continues to increase as the star shrinks. Thus, the evolutionary tracks of massive protostars traverse the H-R diagram roughly horizontally (signifying approximately constant luminosity) in the direction from right to left (from low to high surface temperature). This can be seen most easily for the 9 $M_\odot$ and 15 $M_\odot$ evolutionary tracks in Figure 20-9.

Greater mass means greater pressure and temperature in the interior, which means that a massive star has an even larger temperature difference between its core and its outer layers than the Sun. This causes convection deep in the star's interior. By contrast, a massive star's outer layers are of such low density that energy flows through them more easily by

radiation than by convection. Therefore, main-sequence stars with masses more than about 4 $M_\odot$ have convective interiors but radiative outer layers (Figure 20-11a). By contrast, less massive main-sequence stars such as the Sun have radiative interiors and convective outer layers (Figure 20-11b).

The internal structure is also different for main-sequence stars of very low mass. As such a star forms from a protostar, the interior temperature is never high enough to ionize the interior. The interior remains too opaque for radiation to flow efficiently, so energy is transported by convection throughout the volume of the star (Figure 20-11c).

All of the protostar evolutionary tracks in Figure 20-9 end on the main sequence. The main sequence therefore represents stars in which thermonuclear reactions are converting hydrogen into helium. For most stars, this is a stable situation. For example, our Sun will remain on or very near the main sequence, quietly burning hydrogen at its core, for a total of some $10^{10}$ years. Each evolutionary track ends at a point along the main sequence that agrees with the mass-luminosity relation (recall Figure 19-21). The most massive stars are the most luminous, while the least massive stars are the least luminous.

The theory of how protostars evolve helps explain why the main sequence has both an upper mass limit and a lower mass limit. A protostar less massive than about 0.08 $M_\odot$ can never develop the necessary pressure and temperature to start hydrogen burning at its core. Instead, theory indicates that the failed star contracts to become a hydrogen-rich object called a **brown dwarf**. Such objects are intermediate in their properties between stars and Jovian planets (such as Jupiter, with a mass of 0.001 $M_\odot$). Because brown dwarfs are very dim and cool, they are a challenge to detect and escaped direct observation for many years. It was not until 1994 that a brown dwarf was confirmed to be orbiting the star Gliese 229 (see Section 7-9, especially Figure 7-23). Several other brown dwarfs have subsequently been identified.

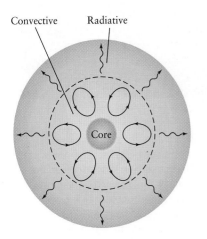

**a** Mass more than about 4 $M_\odot$

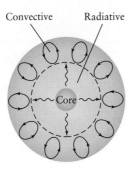

**b** Mass between about 4 $M_\odot$ and 0.8 $M_\odot$

**c** Mass less than 0.8 $M_\odot$

## Figure 20-11

**Main-Sequence Stars of Different Masses**
Stellar models show that when a protostar evolves into a main-sequence star, its internal structure depends on its mass. **(a)** In stars with masses in excess of about 4 $M_\odot$, energy is transported outward from the core by convection in the inner part of the star and by radiation in the outer part. **(b)** If the star's mass is the approximate range from 4 $M_\odot$ and 0.8 $M_\odot$, energy transport is by radiation in the interior and by convection in the outer regions. **(c)** Stars with masses below about 0.8 $M_\odot$ have convection taking place throughout their interiors. *Note:* the three stars shown here are *not* to scale. Compared to a 1-$M_\odot$ main-sequence star like that shown in (b), a 6-$M_\odot$ main-sequence star like that in (a) is more than 4 times larger in radius, and a 0.2-$M_\odot$ main-sequence star like that in (c) has only one-third the radius.

Protostars with masses greater than about 100 solar masses also do not become main-sequence stars. Such a protostar rapidly becomes very luminous, resulting in tremendous internal pressures. This pressure is so great that it overwhelms the effects of gravity, expelling the outer layers into space and disrupting the star. Main-sequence stars therefore have masses between about 0.08 and 100 $M_\odot$, although the high-mass stars are extremely rare.

**CAUTION!** Two words of caution are in order here. First, while the evolutionary tracks of protostars begin in the red giant region of the H-R diagram (the upper right), protostars are *not* red giants. As we will see in Chapter 21, red giant stars represent a stage in the evolution of stars that comes *after* being a main-sequence star. Second, it is worth remembering that stars live out most of their lives on the main sequence, after only a relatively brief period as protostars. A 15-$M_\odot$ protostar takes only 20,000 years to become a main-sequence star, and a 1-$M_\odot$ protostar takes about $2 \times 10^7$ years. By contrast, the Sun has been a main-sequence star for the past $4.6 \times 10^9$ years. By astronomical standards, pre–main-sequence stars are quite transitory.

## 20-5 During the birth process, stars both gain and lose mass

After the previous section, you may think that a main-sequence star forms simply by collapsing inward. In fact, much of the material of a cold, dark nebula is ejected into space and never incorporated into stars. As it is ejected, this material may help sweep away the dust surrounding a young star, making the star observable at visible wavelengths.

Mass ejection into space is a hallmark of **T Tauri stars,** which we first encountered in Section 7-8. These protostars with *emission* lines as well as absorption lines in their spectra and whose luminosity can change irregularly on time scales of a few days. T Tauri stars have masses less than about 3 $M_\odot$ and ages around $10^6$ years, so on an H-R diagram such as Figure 20-9 they appear above the right-hand end of the main sequence. The emission lines show that these protostars are surrounded by a thin, hot gas, and the Doppler shifts of these emission lines suggest that the stars eject gas at speeds around 80 km/s (300,000 km/h, or 180,000 mi/h).

T Tauri stars eject mass at an average rate of $10^{-8}$ to $10^{-7}$ solar masses per year. This may seem like a small amount, but by comparison the present-day Sun loses only about $10^{-14}$ $M_\odot$ per year. The T Tauri phase of a protostar may last $10^7$ years or so, during which time the protostar may eject roughly a solar mass of material. Thus, the mass of the final main-sequence star is quite a bit less than that of the cloud of gas and dust from which the star originated. (The stellar masses shown in Figure 20-9 are those of the final, main-sequence stars.)

Young stars that are more massive than about 3 $M_\odot$ do not vary in luminosity like T Tauri stars. They also lose mass,

however, because the pressure of radiation at their surfaces is so strong that it blows gas into space. Figure 20-12 is a dramatic photograph of a young, massive star experiencing mass loss driven by such stellar winds.

In the early 1980s, it was discovered that many young stars, including T Tauri stars, also lose mass by ejecting gas along two narrow, oppositely directed jets—a phenomenon called **bipolar outflow.** As this material is ejected into space at speeds of several hundred kilometers per second, it collides with the surrounding interstellar medium and produces knots of hot, ionized gas that glow with an emission-line spectrum. These glowing knots are called **Herbig-Haro objects** after the two astronomers, George Herbig in the United States and Guillermo Haro in Mexico, who discovered them independently. Figure 20-13 is a Hubble Space Telescope image of the Herbig-Haro objects HH 1 and HH 2, which are produced by the two jets from a single young star in the constellation Orion. Herbig-Haro objects like these change noticeably in position, size, shape, and brightness from year to year, indicating the dynamic character of bipolar outflows.

Observations suggest that most protostars at one time eject material in the form of jets. These bipolar outflows are very short-lived by astronomical standards, a mere $10^4$ to $10^5$ years, but they are so energetic that they typically eject into space more mass than ends up in the final protostar.

**f**igure 20-12   R I **V** U X G

**The Mass-Loss Star HD 148937** This unusual star in the constellation Norma (the Carpenter's Square) is the brightest member of a system of three stars in orbit about one another. This extremely hot, massive star is losing its outer layers continuously. Other vigorous outbursts in the past gave rise to two symmetric shells of material, called NGC 6164 and NGC 6165, on either side of this star. These shells absorb ultraviolet radiation from the star, causing them to glow with the characteristic red color of excited hydrogen gas. (Anglo-Australian Observatory)

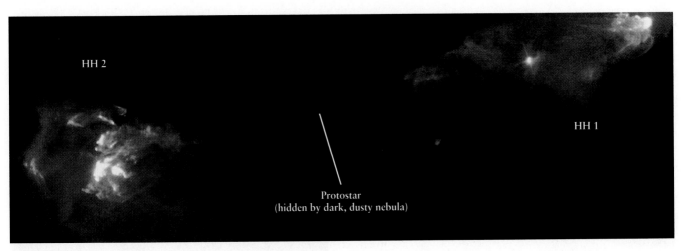

**figure 20-13** R I **V** U X G

**Bipolar Outflow and Herbig-Haro Objects** This Hubble Space Telescope image shows two bright knots of glowing, ionized gas, called Herbig-Haro objects. These are created when fast-moving material ejected from a protostar—which lies halfway between the two Herbig-Haro objects—slams into the surrounding interstellar medium, heating the gas to high temperature. The protostar itself cannot be seen in this visible-light image because it still lies within a dusty, opaque cocoon nebula. These two Herbig-Haro objects, HH 1 on the right and HH 2 on the left, are 0.34 parsec (1.1 light-year) apart and lie 470 parsecs (1500 light-years) from Earth in the constellation Orion. (J. Hester, Arizona State University; the WFPC-2 Investigation Definition Team; and NASA)

Protostars slowly add mass to themselves at the same time that they rapidly eject it into space. In fact, the two processes are related, because bipolar outflows are often found in protostars with **circumstellar accretion disks.** As a protostar's nebula contracts, it spins faster and flattens into a disk with the protostar itself at the center. The same flattening took place in the solar nebula, from which the Sun and planets formed (Section 7-7). Particles orbiting the protostar within this disk collide with each other, causing them to lose energy, spiral inward onto the protostar, and add to the protostar's mass. This process is called **accretion.** Figure 20-14 is an edge-on view of a circumstellar accretion disk, showing two oppositely directed jets emanating from a point at or near the center of the disk (where the protostar is located). For reasons that are not yet completely understood, infalling material from the accretion disk is ejected outward and away from the protostar along its rotation axis. Many astronomers suspect that the interaction between the protostar, the accretion disk, and the jets also helps to slow the protostar's rotation. If so, this would explain why main-sequence stars generally spin much more slowly than protostars of the same final mass.

In the 1990s astronomers using the Hubble Space Telescope discovered many examples of disks around newly formed stars in the Orion Nebula (Figure 20-1), one of the most prominent star-forming regions in the northern sky. The figure that opens Chapter 7 depicts the Orion Nebula as well as a number of **protoplanetary disks,** or **proplyds,** that surround young stars within the nebula. (See also the discussion

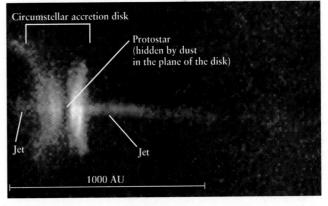

**figure 20-14** R I **V** U X G

**A Circumstellar Accretion Disk and Jets** This composite, false-color Hubble Space Telescope image shows material around a young star in the constellation Taurus. Red denotes emission from ionized gas, while green denotes the continuous spectrum of starlight scattered from dust particles. As the image shows, the star is surrounded by a circumstellar accretion disk, which we see nearly edge-on. It appears as though there are *two* disks around the star, but in fact there is only one—the midplane of this disk is so dusty and opaque that it appears dark. Two oppositely directed jets of ionized gas flow away from the star, perpendicular to the plane of the disk and along the rotation axis of the disk (and, presumably, the rotation axis of the central star). This star is associated with the Herbig-Haro object HH 30, and lies 140 parsecs (460 light-years) from Earth. (C. Burrows, STScI and ESA; the WFPC-2 Investigation Definition Team; and NASA)

of proplyds in Section 7-7, especially Figure 7-15.) As the name suggests, protoplanetary disks are thought to comprise the material from which planets form around stars. They are what remains of a circumstellar accretion disk after much of the material has either fallen onto the star or been ejected by bipolar outflows. Not all stars are thought to form protoplanetary disks; the exceptions probably include stars with masses in excess of about 3 $M_\odot$, as well as many stars in binary systems. But surveys of the Orion Nebula show that these disks are found with most young, low-mass stars. Thus, disk formation may be a natural stage in the birth of many stars.

## 20-6 Young star clusters give insight into star formation and evolution

Observations show that young stars tend to form in groups or **clusters**. This is because dark nebulae contain tens or hundreds of solar masses of gas and dust, enough to form many stars. Star clusters typically include stars with a range of different masses, all of which began to form out of the parent nebula at roughly the same time. Hence, star clusters are valuable laboratories for comparing the evolution of different stars. (In an analogous way, foot races are useful ways to compare the performance of sprinters because all the competitors start the race at the same time.) Star clusters are also objects of great natural beauty. One example is M16, shown in Figure 20-15; another is NGC 6520, depicted in Figure 20-3.

All the stars in a cluster may begin to form nearly simultaneously, but they do not all become main-sequence stars at the same time. As you can see from their evolutionary tracks (Figure 20-9), high-mass stars evolve more rapidly than low-mass stars. The more massive the protostar, the sooner it develops the central pressures and temperatures needed for steady hydrogen burning to begin. Let us compare what happens to high-mass and low-mass protostars in clusters.

Upon reaching the main sequence, *high-mass* protostars become hot, ultraluminous stars of spectral types O and B. As described in Section 20-2, these are the sorts of stars that emit copious amounts of ultraviolet radiation and ionize the surrounding interstellar medium to produce an H II region. Figure 20-15 shows an H II region called the Eagle Nebula, which surrounds the young star cluster M16. A few hundred thousand years ago, this region of space would have had a far less dramatic appearance. It was then a dark nebula, with protostars just beginning to form. Over the intervening millennia, mass ejection from these evolving protostars swept away the obscuring dust. The exposed young, hot stars heated the relatively thin remnants of the original dark nebula, creating the H II region that we see today.

When the most massive protostars to form out of a dark nebula have reached the main sequence, other *low-mass* protostars are still evolving nearby within their dusty cocoons. The evolution of these low-mass stars can be disturbed by their more massive neighbors. As an example, Figure 20-16

**figure 20-15** R I **V** U X G

**A Star Cluster with an H II Region** The swarm of bright stars at the middle of this image make up the brightest part of M16 (also called NGC 6611), a star cluster in the constellation Serpens (the Serpent). The cluster is thought to be no more than 800,000 years old and to contain several thousand stars with a total mass in excess of $10^4$ $M_\odot$. Several bright, hot O and B stars within the star cluster emit ultraviolet radiation that makes the surrounding gases glow, forming an H II region (emission nebula). The H II region, known as IC 4703 but also called the Eagle Nebula because of its shape, is laced with dark, dusty globules inside which star formation is still taking place. M16 and the Eagle Nebula are roughly 2200 parsecs (7000 light-years) from Earth. The bright central portion of the star cluster is some 6 parsecs (20 light-years) across; the H II region, which has an apparent size of $35 \times 28$ arcminutes, is about 20 parsecs (65 light-years) across. The region within the white box is shown in close-up in Figure 20-16. (Anglo-Australian Observatory)

is a close-up of part of the Eagle Nebula (the area shown by the box in Figure 20-15). Within these opaque pillars of cold gas and dust, protostars are still forming. At the same time, however, the pillars are being eroded by intense ultraviolet light from hot, massive stars that have already shed their cocoons. (These massive stars are off the upper edge of Figure 20-16.) As each pillar evaporates, the embryonic stars within have their surrounding material stripped away prematurely, limiting the total mass that these stars can accrete.

Individual stars in a young cluster tell us still more about stars in their infancy. Figure 20-17 shows the young star cluster NGC 2264 and its associated emission nebula. Astronomers have measured each star's apparent brightness and color ratio. Knowing the distance to the cluster, they have deduced the luminosities and surface temperatures of the stars (see Sections 19-2 and 19-4). Figure 20-17 shows all these stars on an H-R diagram. Note that the hottest stars, with surface temperatures around 20,000 K, are on the main sequence. These hot stars

figure 20-16   R I **V** U X G

**Exposing Young Stars in the Eagle Nebula**
This image from the Hubble Space Telescope shows a portion of the Eagle Nebula (Figure 20-15). Three dense, cold pillars of gas and dust are silhouetted against the glowing background of the emission nebula. Hot, luminous stars off the upper edge of the image emit copious amounts of ultraviolet radiation, which evaporates away the dark nebula of which the pillars are part. The pillars presumably lie in the shadow of the most dense parts of the nebula. This has retarded their evaporation, but not permanently; you can see tendrils of gas streaming away from the pillars. Evaporation gradually exposes young stars within each pillar. Each of the tiny "fingers" extending from the surface of the pillars has a young star and its cocoon nebula at its tip. Eventually, the cocoon nebula evaporates and the star itself is revealed. The pillar on the far left extends about one light-year from base to tip, and each "finger" is somewhat broader than our entire solar system. (Jeff Hester and Paul Scowen, Arizona State University; NASA)

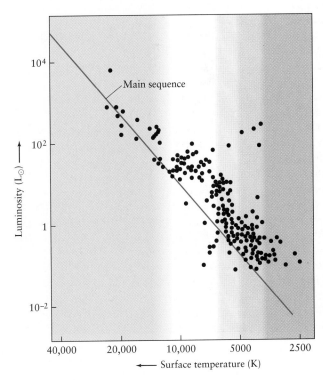

figure 20-17   R I **V** U X G

**A Young Star Cluster and Its H-R Diagram**  The photograph shows an H II region and the young star cluster NGC 2264 in the constellation of Monoceros (the Unicorn). It is located about 800 parsecs (2600 light-years) from Earth and contains numerous T Tauri stars. Each dot plotted on the H-R diagram represents a star in this cluster whose luminosity and surface temperature have been determined. Note that most of the cool, low-mass stars (on the right-hand side of the H-R diagram) have not yet arrived at the main sequence. This star cluster probably started forming only 2 million years ago. (Anglo-Australian Observatory)

are the rapidly evolving, massive ones whose radiation causes the surrounding gases to glow. Stars cooler than about 10,000 K, however, have not yet quite arrived at the main sequence. These less massive stars, which are in the final stages of pre–main-sequence contraction, are just now beginning to ignite thermonuclear reactions at their centers. To find their ages, we can compare Figure 20-17 with the theoretical calculations of protostar evolution in Figure 20-9. It turns out that this particular cluster is probably about 2 million years old.

Figure 20-18 shows another young star cluster called the Pleiades. The photograph shows that the Pleiades must be older than NGC 2264, the cluster in Figure 20-17. The gas that must once have formed an H II region around this cluster has dissipated into interstellar space, leaving only traces of dusty material that forms reflection nebulae around the cluster's stars. The H-R diagram for the Pleiades in Figure 20-18 bears out this idea. In contrast with the H-R diagram for NGC 2264, nearly all the stars in the Pleiades are on the main sequence. The cluster's age is about 50 million years, which is how long it takes for the least massive stars to finally begin hydrogen burning in their cores.

Note that the data points for the most massive stars in the Pleiades (at the upper left of the H-R diagram in Figure 20-18) lie above the main sequence. This is *not* because these stars have yet to arrive at the main sequence. Rather, these stars were the first members of the cluster to arrive at the main sequence some time ago and are now the first members to leave it. They have used up the hydrogen in their cores, so the steady process of core hydrogen burning that characterizes main-sequence stars cannot continue. In Chapter 21 we will see why massive stars spend a rather short time as main-sequence stars, and will study what happens to stars after the main-sequence phase of their lives.

A loose collection of stars such as NGC 2264 or the Pleiades is referred to as an **open cluster** or **galactic cluster**.

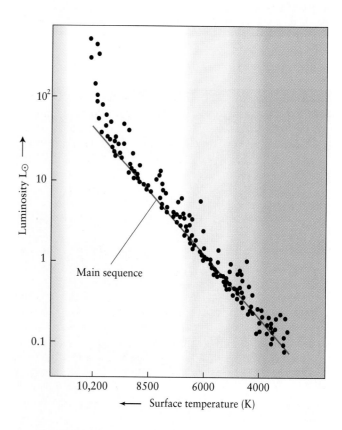

**figure 20-18**   R I **V** U X G

**The Pleiades and Its H-R Diagram**  The star cluster called the Pleiades is 117 parsecs (380 light-years) from Earth in the constellation Taurus. Spanning a full degree across the sky, this cluster is easily visible to the naked eye (and is commonly mistaken for the Little Dipper); it includes some 500 stars, of which a half dozen can be seen without a telescope. Each dot plotted on the H-R diagram represents a star in the Pleiades whose luminosity and surface temperature have been measured. (*Note:* The scales on this H-R diagram are different than those used in Figure 20-17.) Note that all of the cool low-mass stars have arrived at the main sequence, indicating that hydrogen burning has begun in their cores. This suggests that the Pleiades is about 50 million ($5 \times 10^7$) years old. (Anglo-Australian Observatory)

Such clusters possess barely enough mass to hold themselves together by gravitation. Occasionally, a star moving faster than average will escape, or "evaporate," from a cluster. Indeed, by the time the stars are a few billion years old, they may be so widely separated that a cluster no longer exists. If the group of stars is gravitationally unbound from the very beginning—that is, if the stars are moving away from one another so rapidly that gravitational forces cannot keep them together—then the apparent cluster that initially exists is called a **stellar association.** Because young stellar associations are typically dominated by luminous O and B main-sequence stars, they are also called **OB associations.**

## 20-7 Star birth can begin in giant molecular clouds

We have seen that star formation takes place within dark nebulae. But where within the Galaxy are these dark nebulae found? Does star formation take place everywhere within the Milky Way, or only in certain special locations? The answers to such questions can enhance our understanding of star formation and of the nature of our home Galaxy.

The challenge in locating dark nebulae is that they are dark—they do not emit visible light. Nearby dark nebulae can be seen silhouetted against background stars or H II regions (Figure 20-2), but sufficiently distant dark nebulae are impossible to see with visible light because of interstellar extinction. They can, however, be detected using longer-wavelength radiation that can pass unaffected through interstellar dust. In fact, dark nebulae actually emit radiation at millimeter wavelengths.

Such emission takes place because in the cold depths of interstellar space, atoms combine to form molecules. Molecules vibrate and rotate at specific frequencies that are predicted by the laws of quantum mechanics. As a molecule goes from one *vibrational state* or *rotational state* to another, it either emits or absorbs a photon. (In the same way, an atom emits or absorbs a photon as an electron jumps from one energy level to another.) Most molecules are strong emitters of radiation with wavelengths of around 1 to 10 mm. Consequently, observations with radio telescopes tuned to millimeter wavelengths make it possible to detect interstellar molecules of different types. Nearly 100 different kinds of molecules have so far been discovered in interstellar space, and the list is constantly growing.

Hydrogen is by far the most abundant element in the universe. Unfortunately, in cold nebulae much of it is in a molecular form ($H_2$) that is difficult to detect. The reason is that the hydrogen molecule is symmetric, with two atoms of equal mass joined together, and such molecules do not emit many photons at radio frequencies. In contrast, asymmetric molecules, such as carbon monoxide (CO), which consists of two atoms of unequal mass joined together, are easily detectable at radio frequencies. Carbon monoxide emits photons at wavelengths of 2.6 mm and shorter, which correspond to transitions between two rates of rotation of the molecule.

The ratio of carbon monoxide to hydrogen in interstellar space is reasonably constant. For every CO molecule, there are about 10,000 $H_2$ molecules. As a result, carbon monoxide is an excellent "tracer" for molecular hydrogen gas. Wherever astronomers detect strong emission from CO, they know molecular hydrogen gas must be abundant.

The first systematic surveys of our Galaxy looking for 2.6-mm CO radiation were undertaken in 1974 by Philip Solomon of the State University of New York at Stony Brook and Nicholas Scoville, now at the California Institute of Technology. In mapping the locations of CO emission, they discovered huge clouds, now called **giant molecular clouds,** that must contain enormous amounts of hydrogen. These clouds have masses in the range of $10^5$ to $2 \times 10^6$ solar masses and diameters that range from about 15 to 100 parsecs (50 to 300 light-years). Inside one of these clouds, the density is about 200 hydrogen molecules per cubic centimeter. This is several thousand times greater than the average density of matter in the disk of our Galaxy, but $10^{17}$ times less dense than the air we breathe. Astronomers now estimate that our Galaxy contains about 5000 of these enormous clouds.

Figure 20-19a is a map of the constellations Orion and Monoceros made with a radio telescope. The telescope was tuned to a wavelength of 2.6 mm in order to detect emissions from CO. Note the extensive areas of the sky covered by giant molecular clouds. This part of the sky is of particular interest because it includes several star-forming regions, shown in Figure 20-19b. By comparing the two parts of Figure 20-19, you can see that the areas where CO emission is strongest, and, thus, where giant molecular clouds are densest, are sites of star formation. Therefore, giant molecular clouds are associated with the formation of stars. Particularly dense regions within these clouds form dark nebulae, and within these stars are born.

By mapping the locations of giant molecular clouds, astronomers can find where in the Galaxy star formation occurs. American astronomer Thomas M. Dame and his colleagues at the Smithsonian Astrophysical Observatory have used CO emission to carry out this line of research, and the perspective drawing in Figure 20-20 shows the results of their work. They found that 17 molecular clouds clearly outline the spiral arm nearest the Sun, which is called the Sagittarius arm. These clouds lie roughly 100 parsecs (3000 light-years) apart, strung along the spiral arms like beads on a string. This arrangement is much like the spacing of H II regions in other galaxies, such as the galaxy NGC 2997 shown in Figure 20-7. Thus, the spiral arms are sites of ongoing star formation.

In Chapter 25 we will learn that spiral arms are locations where matter "piles up" temporarily as it orbits the center of the Galaxy. You can think of a spiral arm as analogous to a freeway traffic jam. Just as cars are squeezed close together in a traffic jam, a giant molecular cloud is compressed when it passes through a spiral arm. When this happens, vigorous star formation begins in the cloud's densest regions. As soon as

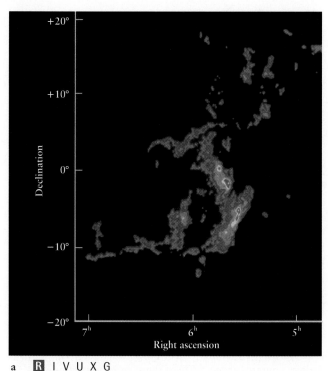

a  [R] I V U X G

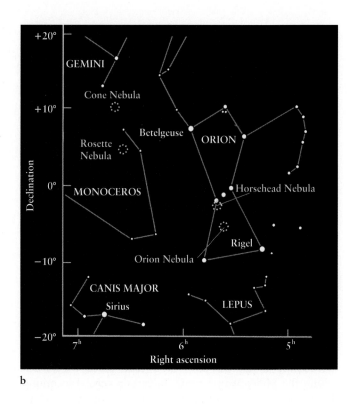

b

**ƒigure 20-19**

**A Map of Carbon Monoxide Features in Orion and Monoceros** **(a)** This color-coded map of a large section of the sky (about 40° × 40°) shows the extent of giant molecular clouds in the constellations Orion and Monoceros. The intensity of emission by carbon monoxide (CO) molecules is indicated by colors in the order of the rainbow, from violet for the weakest to red for the strongest. Black indicates no detectable emission. **(b)** This star chart covers the same area as (a). The locations of four prominent star-forming nebulae are indicated. Note that the Orion and Horsehead nebulae are located at sites of intense CO emission, indicating the presence of a particularly dense molecular cloud. (Courtesy of R. Maddalena, M. Morris, J. Moscowitz, and P. Thaddeus)

massive O and B stars form, they emit ultraviolet light that ionizes the surrounding hydrogen, and an H II region is born. An H II region is thus a small, bright "hot spot" in a giant molecular cloud. An example is the Orion Nebula, shown in Figure 20-1*b*. Four hot, luminous O and B stars at the heart of the nebula produce the ionizing radiation that makes the surrounding gases glow. The Orion Nebula is embedded in the edge of a giant molecular cloud whose mass is estimated at 500,000 M$_\odot$. The H II regions in Figure 20-2 are located at a different point on the edge of the same molecular cloud, some 25 parsecs (80 light-years) from the Orion Nebula.

Once star formation has begun and an H II region has formed, the massive O and B stars at the core of the H II region induce star formation in the rest of the giant molecular cloud. Vigorous stellar winds from the O and B stars carve out a cavity in the cloud, and the H II region, heated by the stars, expands into it. These winds travel faster than the speed of sound in the gas—that is, they are **supersonic.** Just as an airplane creates a shock wave (a sonic boom) when it flies faster than sound waves in our atmosphere, a shock wave forms where the expanding H II region pushes at supersonic speed into the rest of the giant molecular cloud. This shock wave compresses the hydrogen gas through which it passes, stimu-

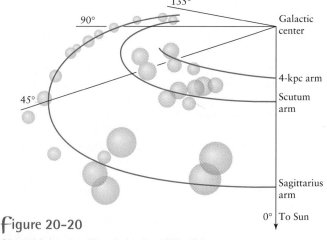

**ƒigure 20-20**

**Giant Molecular Clouds in the Milky Way** This perspective drawing shows the locations of giant molecular clouds in an inner part of our Galaxy as seen from a vantage point above the Sun. Note how the Sagittarius spiral arm is outlined by giant molecular clouds that lie along it like beads on a string. The locations of two other spiral arms, the 4-kiloparsec arm and the Scutum arm, are also indicated. The distance from the Sun to the galactic center is about 8000 parsecs (26,000 light-years). (Adapted from T. M. Dame and colleagues)

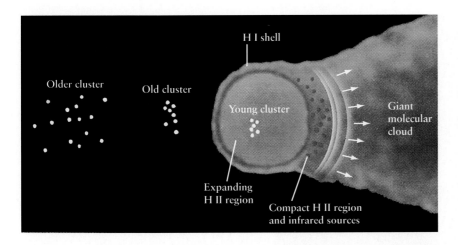

**figure 20-21**

**How O and B Stars Trigger Star Formation** Ultraviolet radiation from young O and B stars produces a shock wave that compresses gas farther into the giant molecular cloud, stimulating new star formation deeper into the cloud. This will result in more O and B stars, which will stimulate still more star formation, and so on. Meanwhile, older stars are left behind. (Adapted from C. Lada, L. Blitz, and B. Elmegreen)

lating more star birth. Newborn O and B stars further expand the H II region into the giant molecular cloud. Meanwhile, the older O and B stars, which were left behind, begin to disperse (Figure 20-21). In this way, an OB association "eats into" a giant molecular cloud, leaving stars in its wake.

Infrared observations reveal many features that resemble protostars in the swept-up layer immediately behind the shock wave from an OB association. For example, Figure 20-22 shows both optical and infrared views of the core of the Orion Nebula. Four O and B stars, called the Trapezium, and glowing gas and dust dominate the view at visible wavelengths. Infrared observations penetrate this obscuring material to reveal dozens of infrared objects. These are thought to be cocoons of warm dust enveloping newly formed stars.

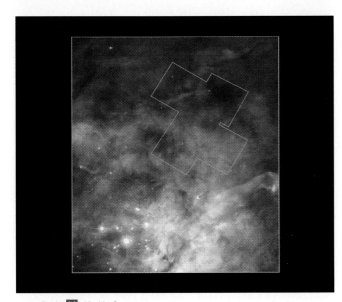

a  R I **V** U X G

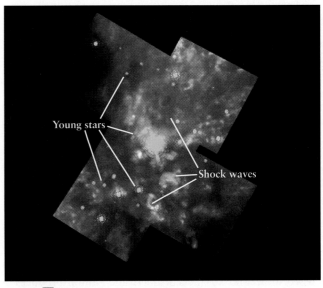

b  R **I** V U X G

**figure 20-22**

**The Core of the Orion Nebula** (a) This Hubble Space Telescope view at visible wavelengths shows the inner regions of the Orion Nebula (compare Figure 20-1b). At the lower left are the four massive stars, called the Trapezium, that cause the nebula to glow. These stars are separated from each other by only 0.1 light-year. (C. R. O'Dell and S. K. Wong, Rice University; NASA) (b) This infrared image is also from the Hubble Space Telescope. The region shown has a diagonal dimension of about 0.1 parsec (0.4 light-year) and is outlined in part (a). Infrared light can penetrate interstellar dust more easily than can visible light, revealing features that cannot be seen in part (a). Emission from young stars and glowing interstellar dust is colored yellow-orange, while emission from excited hydrogen molecules appears blue. Numerous infrared objects, many of which are probably new protostars, are seen in this view. Several curved arcs are thought to be shock waves caused by material flowing out of protostars faster than the speed of sound waves in the nebula. Shock waves from the Trapezium may have helped trigger the formation of the protostars in this view. (R. Thompson, M. Rieke, G. Schneider, and S. Stolovy, University of Arizona; E. Erickson, SETI Institute/Ames Research Center; D. Axon, STScI; NASA)

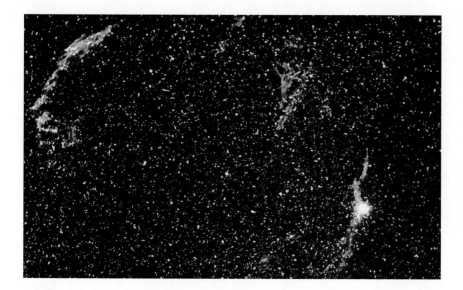

figure 20-23  R I **V** U X G

**A Supernova Remnant** This remarkable nebula, called the Cygnus Loop, is the remnant of a supernova that occurred about 15,000 years ago. This expanding spherical shell of gas, located about 800 parsecs (2,600 light-years) from Earth, now has a diameter of about 40 parsecs (120 light-years). The glowing material that comprises the Cygnus Loop is interstellar matter that has been excited by the passage of the supernova's shock wave, which is still spreading away from the location of the supernova even after thousands of years. (Courtesy of H. Vehrenberg)

| 20-8 | Supernovae compress the interstellar medium and can trigger star birth |
|------|---|

Spiral arms are not the only mechanism for triggering the birth of stars. Presumably, anything that compresses interstellar clouds will do the job. The most dramatic is a **supernova**, caused by the violent death of a massive star after it has left the main sequence. As we will see in Chapter 22, the core of the doomed star collapses suddenly, releasing vast quantities of particles and energy that blow the star apart. The star's outer layers are blasted into space at speeds of several thousand kilometers per second.

Astronomers have found many nebulae across the sky that are the shredded funeral shrouds of these dead stars. Such nebulae, like the Cygnus Loop shown in Figure 20-23, are known as **supernova remnants.** Many supernova remnants have a distinctly arched appearance, as would be expected for an expanding shell of gas. This wall of gas is typically moving away from the dead star faster than sound waves can travel through the interstellar medium. As discussed in Section 20-7, such supersonic motion produces a shock wave that abruptly compresses the medium through which it passes. Astronomers have studied the Cygnus Loop with the Hubble Space Telescope to examine in detail a supernova's shock wave plowing through the interstellar medium. Figure 20-24 shows the shock wave overrunning dense clumps of gas. This collision heats and compresses the gas, causing it to glow.

When the expanding shell of a supernova remnant slams into an interstellar cloud, it squeezes the cloud, stimulating star birth. This kind of star birth can be observed in the stellar association seen in Figure 20-25. This stellar nursery is located along a luminous arc of gas about 30 parsecs

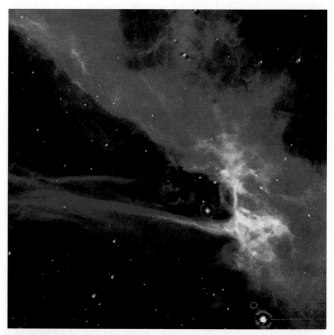

figure 20-24  R I **V** U X G

**A Supernova's Blast Wave** This image from the Hubble Space Telescope shows a small portion of the Cygnus Loop. The bluish ribbon stretching left to right across the picture may be a knot of gas ejected by the supernova some 15,000 years ago. This knot, traveling at about 1400 km/s (3 million miles per hour), is just now catching up with the shock wave, which has been slowed by plowing through the interstellar medium. A region of intense emission can be seen where this fast-moving knot has slammed into the red-colored arch of the Cygnus Loop itself. The different colors in this image are used to code emission from different gases. Blue denotes oxygen atoms, which emit strongly at temperatures of 30,000 K to 60,000 K; red labels sulfur atoms at a relatively cool 10,000 K; and green indicates hydrogen emission. (J. Hester, Arizona State University; NASA)

**figure 20-25** R I **V** U X G

**The Canis Major R1 Association** This luminous arc of gas, about 30 parsecs (100 light-years) long, is studded with numerous young stars. Both the luminous arc and the young stars can be traced to the same source, a supernova explosion. The shock wave from the supernova explosion is exciting the gas and making it glow; the same shock wave also compresses the interstellar medium through which it passes, triggering star formation. (Courtesy of H. Vehrenberg)

(100 light-years) in length that is presumably the remnant of an ancient supernova explosion. In fact, this arc is part of an almost complete ring of glowing gas with a diameter of about 60 parsecs (200 light-years). Spectroscopic observations of stars along this arc reveal substantial T Tauri activity. This activity results from newborn stars undergoing mass loss in their final stages of contraction before they become main-sequence stars.

As we saw in Section 17-6, there is strong evidence that the Sun was once a member of one of these loose associations created by a supernova. Recall that stellar associations do not remain intact for long. Individual stellar motions soon carry the stars in various directions away from their birthplaces. Nearly 5 billion years have passed since the birth of our star, so the Sun's brothers and sisters are now widely scattered across the Galaxy.

Many other processes can also trigger star formation. For example, a collision between two interstellar clouds can create new stars. When two such clouds collide, compression occurs at the interface and vigorous star formation follows. Similarly, stellar winds from a group of O and B stars may exert strong enough pressure on interstellar clouds to cause compression, followed by star formation. The Rosette Nebula, shown in Figure 20-26, is an example of this process.

Our understanding of star birth has improved dramatically in recent years, primarily through infrared- and millimeter-wavelength observations. Nevertheless, many puzzles and mysteries remain. For example, astronomers have generally assumed that there must be a lot of interstellar dust to shield a stellar nursery from the disruptive effects of ultraviolet light from nearby massive, hot stars. However, a neighboring galaxy, called the Large Magellanic Cloud (LMC), contains young OB associations with virtually no dust. Does the process of star birth differ slightly from one galaxy to another?

**figure 20-26** R I **V** U X G

**The Core of the Rosette Nebula** The Rosette Nebula (NGC 2237) is a large, circular emission nebula located near one end of a sprawling giant molecular cloud in the constellation of Monoceros. Radiation from young, hot stars has blown away the center of this nebula. Some of the remaining gas is clumped in dark globules that appear silhouetted against the glowing background gases. New star formation is taking place within these globules. The entire Rosette Nebula has an angular diameter of 1.3° and lies some 900 parsecs (3000 light-years) from Earth. (Anglo-Australian Observatory)

**16.** If you looked at a spectrum of a reflection nebula, would you see absorption lines, emission lines, or no lines? Explain your answer. As part of your explanation, describe how the spectrum demonstrates that the light was reflected from nearby stars.

**17.** In the direction of a particular star cluster, interstellar extinction allows only 15% of a star's light to pass through 1 kiloparsec (1000 parsecs) of the interstellar medium. If the star cluster is 3.0 kiloparsecs away, what percentage of its photons survive the trip to the Earth?

**18.** Find the density (in atoms per cubic centimeter) of a Bok globule having a radius of 1 light-year and a mass of 100 solar masses. How does your result compare with the density of a typical H II region, between 80 and 600 atoms per cm$^3$? (Assume that the globule is made purely of hydrogen atoms.)

**19.** The two false-color images below show a portion of the Trifid Nebula (see Question 15). The upper view is an infrared image from the Infrared Space Observatory, while the lower picture (shown to the same scale) was made with visible light. Explain why the dark streaks in the visible-light image appear bright in the infrared image.

**20.** At one stage during its birth, the protosun had a luminosity of 1000 $L_\odot$ and a surface temperature of about 1000 K. At this time, what was its radius? Express your answer in three ways: as a multiple of the Sun's present-day radius, in kilometers, and in astronomical units.

**21.** A newly formed protostar and a red giant are both located in the same region on the H-R diagram. Explain how you could distinguish between these two.

**22. (a)** Determine the radius of the circumstellar accretion disk in Figure 20-14. (You will need to measure this image with a ruler. Note the scale bar in this figure.) Give your answer in astronomical units and in kilometers. **(b)** Assume that the young star at the center of this disk has a mass of 1 $M_\odot$. What is the orbital period (in years) of a particle at the outer edge of the disk? **(c)** Using your ruler again, determine the length of the jet that extends to the right of the circumstellar disk in Figure 20-14. At a speed of 200 km/s, how long does it take gas to traverse the entire visible length of the jet?

**23.** The concentration or abundance of ethyl alcohol in a typical molecular cloud is about 1 molecule per $10^8$ cubic meters. What volume of such a cloud would contain enough alcohol to make a martini (about 10 grams of alcohol)? A molecule of ethyl alcohol has 46 times the mass of a hydrogen atom (that is, ethyl alcohol has a molecular weight of 46).

**24. (a)** Find the angular diameter of the Cygnus Loop, shown in Figure 20-23. **(b)** Assuming that the supernova shock wave that spawned the Cygnus Loop has always traveled outward at the same speed, determine that speed in meters per second and kilometers per hour. **(c)** How long would an Earth astronomer have to wait to see the bright regions of the Cygnus Loop move outward by an angle of 1 arcminute?

## Discussion Questions

**25.** What do you think would happen if our solar system were to pass through a giant molecular cloud? Do you think the Earth has ever passed through such clouds?

**26.** Speculate on why a shock wave from a supernova seems to produce relatively few high-mass O and B stars, compared to the lower-mass A, F, G, and K stars.

**27.** Many of the molecules found in giant molecular clouds are organic molecules (that is, molecules containing carbon). Speculate about the possibility of life forms and biological processes occurring in giant molecular clouds. In what ways might the conditions existing in giant molecular clouds favor or hinder biological evolution?

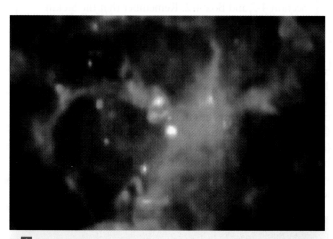

R **I** V U X G

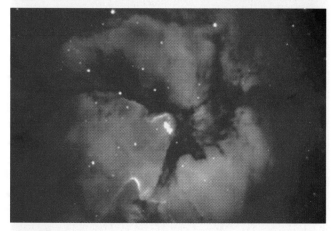

R I **V** U X G

(ESA/ISO, ISOCAM, and J. Cernicharo et al.; IAC, Observatorio del Teide, Tenerife)

# Observing Projects

**Observing tips and tools:**

After looking at the beautiful color photographs of nebulae in this chapter, you may find the view through a telescope a bit disappointing at first, but fear not. There is much that can be seen with even a small telescope. To get the best view of a dim nebula using a telescope, use the same "averted vision" trick described in Chapter 19. Direct your vision a little to one side of the object that you are looking at; this directs the light from that object onto a more sensitive part of the retina.

28. Use a telescope to observe at least two of the following H II regions. In each case, can you guess which stars are probably responsible for the ionizing radiation that causes the nebula to glow? Can you see any obscuration or silhouetted features that suggest the presence of interstellar dust? Draw a picture of what you see through the telescope and compare it with a photograph of the object. Take note of which portions of the nebula were not visible through your telescope.

| Nebula | Right Ascension | Declination |
|---|---|---|
| M42 (Orion) | 5$^h$ 35.4$^m$ | –5° 27′ |
| M43 | 5 35.6 | –5 16 |
| M20 (Trifid) | 18 02.6 | –23 02 |
| M8 (Lagoon) | 18 03.8 | –24 23 |
| M17 (Omega) | 18 20.8 | –16 11 |

*Note: The right ascensions and declinations are given for epoch 2000.*

29. On an exceptionally clear, moonless night, use a telescope to observe at least one of the following dark nebulae. These nebulae are very difficult to find, because they are recognizable only by the *absence* of stars in an otherwise starry part of the sky. Are you confident that you actually saw the dark nebula? Does the pattern of background stars suggest a particular shape to the nebula?

| Nebula | Right Ascension | Declination |
|---|---|---|
| Barnard 72 (The Snake) | 17$^h$ 23.5$^m$ | –23° 38′ |
| Barnard 86 | 18 02.7 | –27 50 |
| Barnard 133 | 19 06.1 | –6 50 |
| Barnard 142 and 143 | 19 40.7 | +10 57 |

*Note: The right ascensions and declinations are given for epoch 2000.*

30. A few fine objects cover such large regions of the sky that they are best seen with binoculars. If you have access to a high-quality pair of binoculars, observe the North American Nebula in Cygnus and the Pipe Nebula in Ophiuchus. Both nebulae are quite faint, so you should attempt to observe them only on an exceptionally dark, clear, moonless night. The North America Nebula is a cloud of glowing hydrogen gas located about 3° east of Deneb, the brightest star in Cygnus. While searching for the North America Nebula, you may glimpse another diffuse H II region, the Pelican Nebula, located about 2° southeast of Deneb. The Pipe Nebula is a 7°-long, meandering, dark nebula to the south and to the east of the star θ (theta) Ophiuchi, which is in a section of Ophiuchus that extends southward between the constellations of Scorpius and Sagittarius. Located about 12° east of the bright red star Antares, θ Ophiuchi is easily identified with the aid of star charts published during the summer months in such magazines as *Sky & Telescope* and *Astronomy*.

# Where to Learn More

*Books and magazine articles*

Burnham, R. *Burnham's Celestial Handbook: An Observer's Guide to the Universe Beyond the Solar System*. This three-volume compendium is an indispensable aid to anyone who wants to explore nebulae and star clusters with telescope or binoculars.

Iben, I., and Tutukov, A. V. "The Lives of Stars: From Birth to Death and Beyond." *Sky & Telescope*, December 1997.

———. "The Lives of Binary Stars: From Birth to Death and Beyond." *Sky & Telescope*, January 1998. Two renowned experts in stellar evolution summarize our understanding of how stars are born, evolve, and die.

Knapp, G. "The Stuff Between the Stars." *Sky & Telescope*, May 1995. The interstellar medium, and the role it plays in the evolution of our galaxy, is the topic of this lucid article.

Lada, C. "Deciphering the Mysteries of Stellar Origins." *Sky & Telescope*, May 1993. Beautiful illustrations and lucid text help to clarify our present understanding of how stars are formed from the interstellar medium..

Osterbrock, D. E., ed. *Stars and Galaxies: Citizens of the Universe*. W. H. Freeman, 1990. Several of the articles in this collection, selected from the pages of *Scientific American*, bear on the topics of the interstellar medium and star formation.

Parker, S. "The Eagle's Nest." *Sky & Telescope*, February 1996. This article goes into detail about the interpretation of the Hubble Space Telescope image of the Eagle Nebula shown in Figure 20-16.

———. "Diving into the Lagoon." *Sky & Telescope*, May 1997. Another of the "stellar nurseries" explored by the Hubble Space Telescope is the Lagoon Nebula, M8. This brief, illustrated article describes what has been found there.

Stahler, S. W. "The Early Life of Stars." *Scientific American*, July 1991. This article examines our current understanding of protostars.

Verschuur, G. L. "Interstellar Molecules." *Sky & Telescope*, April 1992. This well-written article examines some of the chemical and physical processes that occur in the interstellar medium.

————. *Interstellar Matters: Essays on Curiosity and Astronomical Discovery*. Springer Verlag, 1989. A classic book, written by an expert on the interstellar medium. It not only relates the history of how that medium was discovered and our modern understanding of its properties, but also gives insight into the nature of scientific discovery and curiosity.

### W *World Wide Web*

Nebulae in which stars are forming are some of the most beautiful objects in all of astronomy. Some of the best full-color images of these nebulae can be found at the web site of the Anglo-Australian Observatory (**http://www.aao.gov. au/images.html**) and at "The Web Nebulae," a site maintained by Bill Arnett of SEDS, the Students for the Exploration and Development of Space (**http://www.seds.org/billa/twn/**).

Two other sites with lots of images and textual information about nebulae, as well as links to other sites, are the SEDS Messier Database (**http://www.seds.org/messier/**) and the SEDS interactive NGC catalog (**http://www.seds.org/~spider/ngc/ngc.html**).

Another excellent site with lots of information about star formation is maintained at the University of Massachusetts (**http://www-astro.phast.umass.edu/images.shtml**).

The Hubble Space Telescope (HST) and the Infrared Space Telescope (ISO) have given astronomers improved insights into the formation of stars. Images and scientific results from these telescopes, arranged by subject matter, can be found at the Space Telescope Science Institute web site (**http://oposite.stsci.edu/pubinfo/Subject.html**) and at the ISO web site (**http://isowww.estec.esa.nl/science/**).

Another orbiting infrared telescope, the Space Infrared Telescope Facility (SIRTF), is scheduled to be launched in 2002 (**http://sirtf.jpl.nasa.gov/sirtf/**).

# Stellar Evolution: After the Main Sequence

R I **V** U X G

**A Mass–Loss Star** As stars age and become giant stars, they expand tremendously and shed matter into space. This star, HD 65750, is losing matter at a high rate and is surrounded by a reflection nebula (IC 2220) caused by starlight reflecting from dust grains. These dust grains may have condensed from material shed by the star. A typical red giant can lose $10^{-7}$ solar mass per year. Many red giants are surrounded by gas and dust they have ejected. (Anglo-Australian Observatory)

*In this chapter you will find the answers to the following questions:*

**21-1** Will our Sun always be as bright as it is now?

**21-2** Why are red giants larger than main-sequence stars?

**21-3** Do all stars evolve into red giants at the same rate?

**21-4** How do we know that many stars lived and died before our Sun was born?

**21-5** Why do some giant stars pulsate in and out?

**21-6** Why do stars in some binary systems evolve in unusual ways?

law (see Figure 19-7). The star is then appropriately called a **red giant** (Figure 21-3). Thus, we see that red-giant stars are stars that have finished their time as main-sequence stars and have evolved into a new stage of existence.

Red-giant stars undergo substantial **mass loss** because of their large diameters and correspondingly weak surface gravity. This makes it quite easy for gases to escape from the red giant into space. Mass loss can be detected spectroscopically, because gas escaping from a red giant toward a telescope on the Earth produces narrow absorption lines that are slightly blueshifted by the Doppler effect (review Figure 5-23). Typical observed blueshifts correspond to a speed of about 10 km/s, which is slightly greater than that needed to escape the star's gravitational attraction. A typical red giant loses roughly $10^{-7}$ $M_{\odot}$ of matter per year. For comparison, the Sun's present-day mass loss rate is only $10^{-14}$ $M_{\odot}$ per year. Hence, an evolving star loses a substantial amount of mass as it becomes a red giant. (An example is shown in the photograph that opens this chapter.)

Our own Sun will take about 5 billion more years to finish converting hydrogen into helium at its core. As the Sun's core contracts, its atmosphere will expand to envelop Mercury and reach most of the way to Venus. The red-giant Sun will eventually swell to a diameter of about 1 AU—roughly 100 times larger than its present size—and its surface temperature will decline to about 3500 K. Although the Sun's surface temperature will be much lower than it is at present, the Sun will be so huge that its luminosity will be much greater than it is today. As a full-fledged red giant, our star will shine with the brightness of 2000 present-day Suns. Some of the inner planets will be vaporized, and the thick atmospheres of the outer planets will boil away to reveal tiny, rocky cores. Thus, in its later years, the aging Sun will destroy the planets that have accompanied it since its birth.

## 21-2 Helium burning begins at the center of a red giant

When a star first becomes a red giant, its hydrogen-burning shell surrounds a small, compact core of almost pure helium. In a moderately low-mass red giant, which the Sun will be 5 billion years from now, the dense helium core is about twice the size of the Earth, and the star's bloated surface has a diameter of 1 AU.

Helium, the "ash" of hydrogen burning, is a potential nuclear fuel: The fusion of helium nuclei to make heavier nuclei releases energy. But this reaction does not take place within the core of our present-day Sun because the temperature there is too low. Each helium nucleus contains two protons, so it has twice the positive electric charge of a hydrogen nucleus, and there is a much stronger electric repulsion between two helium nuclei than between two hydrogen nuclei. Thus, for helium nuclei to get close enough to fuse together, they must be moving at very high speeds, which means that the temperature of the helium gas must be very high. (For more on the relationship between the temperature of a gas and the speed of the atoms that make up the gas, see Box 7-2.)

When a star first becomes a red-giant star, the temperature of the contracted helium core is still too low for helium

The Sun as a main-sequence star
(diameter $= 1.4 \times 10^6$ km $\approx \dfrac{1}{100}$ AU)

The Sun as a red giant
(diameter $\approx$ 1 AU)

## ƒigure 21-3

**The Sun Today and as a Red Giant** The Sun's energy today is produced in a hydrogen-burning core with a diameter of about 100,000 km. When the Sun becomes a red giant, in some 5 billion years, it will draw its energy from a hydrogen-burning shell surrounding a compact helium-rich core. The helium core will have a diameter of only about 30,000 km. The contraction of the core converts gravitational energy into thermal energy, heating the hydrogen-burning shell and making the reactions there proceed at a furious rate.

As a result, the Sun's luminosity as a red giant will be about 2000 times greater than today. The increased luminosity also increases the pressure in the Sun's outer layers, making them expand to approximately 100 times their present size.

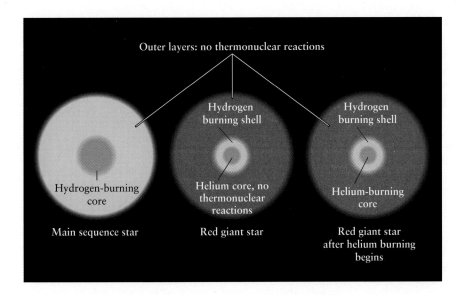

**Stages in Stellar Evolution** As a star evolves, thermonuclear reactions occur at different locations within its interior. In a main-sequence star, hydrogen is converted into helium in the core. When the core hydrogen is exhausted, reactions continue in a hydrogen-burning shell, and the star expands to become a red giant. When the temperature in the core becomes high enough because of contraction, core helium burning begins. (These three pictures are *not* drawn to scale. The star is much larger in its red-giant phase than in its main-sequence phase, then shrinks somewhat when core helium burning begins.)

to undergo thermonuclear reactions. But as the hydrogen-burning shell adds mass to the helium core, the core contracts even more, further increasing the star's central temperature. When the central temperature finally reaches 100 million ($10^8$) K, **helium burning** ignites at the star's center. As a result, the aging star again has a central energy source for the first time since leaving the main sequence (Figure 21-4).

Helium burning occurs in two steps. First, two helium nuclei combine to form an isotope of beryllium:

$$^4\text{He} + {}^4\text{He} \rightarrow {}^8\text{Be}$$

Because this particular beryllium isotope is very unstable, it breaks back down into two helium nuclei soon after it forms. However, in the star's dense core a third helium nucleus may strike the $^8$Be nucleus before it has a chance to fall apart. Such a collision creates a stable isotope of carbon:

$$^8\text{Be} + {}^4\text{He} \rightarrow {}^{12}\text{C} + \gamma$$

In this process, a gamma-ray photon ($\gamma$) is released.

During the pioneering days of nuclear physics, helium nuclei were called **alpha particles,** and the fusion of three helium nuclei to form a carbon nucleus (as we have just described) is still called the **triple alpha process.** Some of the carbon created in this process can fuse with an additional helium nucleus to produce a stable isotope of oxygen:

$$^{12}\text{C} + {}^4\text{He} \rightarrow {}^{16}\text{O} + \gamma$$

Thus, both carbon and oxygen make up the "ash" of helium burning. It is interesting to note that $^{12}$C and $^{16}$O are the most common isotopes of carbon and oxygen, respectively; the vast majority of the carbon atoms in your body are $^{12}$C, and almost all of the oxygen that you breathe is $^{16}$O. We will discuss the significance of this in the following section.

The second step in the triple alpha process and the process of oxygen formation both release energy. This energy source, properly called **core helium burning** because of its central location within the star, establishes thermal equilibrium and prevents any further gravitational contraction of the star's core. A mature red giant burns helium in its core for about 20% as long as the time it spent burning hydrogen as a main-sequence star. For example, in the distant future the Sun will consume helium in its core for about 2 billion years.

How helium burning begins at a red giant's center depends on the mass of the star. In high-mass stars (those with masses greater than 2 to 3 $M_\odot$), helium burning begins gradually as temperatures in the star's core approach $10^8$ K. In low-mass stars (those with masses less than 2 to 3 $M_\odot$), helium burning begins explosively and suddenly, in what is called the **helium flash.** These differences are summarized in Table 21-2.

The helium flash occurs because of unusual conditions that develop in the core of a low-mass star as it becomes a red giant. To appreciate these conditions we must first understand how an ordinary gas behaves. Then we can explore how the densely packed electrons at the star's center alter this behavior.

| Ŧable 21-2 | How Helium Burning Begins in Different Red Giants |
|---|---|
| **Mass of star** | **Onset of helium burning in core** |
| Less than 2–3 solar masses | Explosive (helium flash) |
| More than 2–3 solar masses | Gradual |

When a gas is compressed, it usually becomes denser and warmer. To describe this process, scientists use the convenient concept of an **ideal gas**, which has a simple relationship between pressure, temperature, and density. Specifically, the pressure exerted by an ideal gas is directly proportional to both the density and the temperature of the gas. Many real gases actually behave like an ideal gas over a wide range of temperatures and densities.

Under most circumstances, the gases inside a star act like an ideal gas. If the gas is compressed, it heats up, and if it expands, it cools down. This behavior serves as a safety valve, ensuring that the star does not explode. For example, if energy production overheats the star's core, the core expands, cooling the gases and slowing the rate of thermonuclear reactions. Conversely, if too little energy is being created to support the star's overlying layers, the core compresses, increasing the temperature and thus speeding up the thermonuclear reactions to increase the energy output.

In a low-mass red giant, the core must be extremely compressed in order to become hot enough for helium burning to begin. At these extreme pressures and temperatures, the atoms are completely ionized, and most of the matter in the star's core consists of nuclei and detached electrons. Eventually, the free electrons become so closely crowded that a limit to further compression is reached, as predicted by a remarkable law of quantum mechanics called the **Pauli exclusion principle**. Formulated in 1925 by the Austrian physicist Wolfgang Pauli, this principle states that two identical particles cannot simultaneously occupy the same quantum state. A quantum state is a particular set of circumstances concerning locations and speeds that are available to a particle. In the submicroscopic world of atoms and particles, the Pauli exclusion principle is analogous to saying that you can't have two things in the same place at the same time.

Just before the onset of helium burning, the electrons in the core of a low-mass star are so closely crowded together that any further compression would violate the Pauli exclusion principle. Because the electrons cannot be squeezed any closer together, they produce a powerful pressure that resists further core contraction.

This phenomenon, in which closely packed particles resist compression as a consequence of the Pauli exclusion principle, is called **degeneracy**. Astronomers say that the helium-rich core of a low-mass red giant is "degenerate" and is supported by **degenerate-electron pressure**. This degenerate pressure, unlike the pressure of an ideal gas, does not depend on temperature.

When the temperature in the core of a low-mass red giant reaches the high level required for the triple alpha reaction, energy begins to be released. The helium heats up, which causes the triple alpha process to proceed more rapidly. However, the pressure provided by the degenerate electrons is independent of the temperature, so the pressure does not change. Without the "safety valve" of increasing pressure, the star's core cannot expand and cool. The rising temperature causes the helium to burn at an ever-increasing rate, producing the helium flash.

Eventually the temperature becomes so high that the electrons in the core are no longer degenerate. The electrons then behave like an ideal gas and the star's core expands, terminating the helium flash. These events occur so rapidly that the helium flash is over in seconds.

CAUTION! The term "helium flash" might give you the impression that a star emits a sudden flash of light when the helium flash occurs. If this were true, a star undergoing the helium flash would be an incredible sight. During that brief time interval, the helium-burning core is about $10^{11}$ times more luminous than the present-day Sun, comparable to the total luminosity of all the stars in the Milky Way Galaxy! In fact, however, the helium flash has no immediately visible consequences—for two reasons. First, much of the energy released during the helium flash goes into heating the core and terminating the degenerate state of the electrons. Second, the energy that does escape the core is largely absorbed by the star's outer layers, which are quite opaque (just like the Sun's outer layers, as described in Section 18-1). Therefore, the explosive drama of the helium flash takes place where it cannot be seen directly.

Whether a helium flash occurs or not, the onset of core helium burning actually causes a *decrease* in the luminosity of the star. This is the opposite of what you might expect—after all, turning on a new energy source should make the luminosity greater, not smaller. What happens is that after the onset of core helium burning, a star's superheated core expands like an ideal gas. (If the star is of sufficiently low mass to have had a degenerate core, the increased temperature after the helium flash makes the core too hot to remain degenerate. Hence, these stars also have cores that behave like ideal gases.) Around the expanding core, temperatures fall, so the hydrogen-burning shell reduces its energy output and the star's luminosity decreases. This allows the star's outer layers to contract and heat up. Consequently, a post–helium-flash star is less luminous, hotter at the surface, and smaller than a red giant.

In its briefest form, the story of post–main-sequence evolution can be summarized as follows. Before the beginning of core helium burning, the star's core compresses and the outer layers expand, and just after core helium burning begins, the core expands and the outer layers compress. In Chapter 22 we will see that this behavior, in which the inner and outer regions of the star change in opposite ways, occurs again and again in the final stages of a star's evolution.

## 21-3 An H-R diagram and observations of star clusters reveal how red giants evolve

To see how stars evolve during and after their main-sequence lifetime, it is helpful to follow them on a Hertzsprung-Russell diagram. On such an H-R diagram

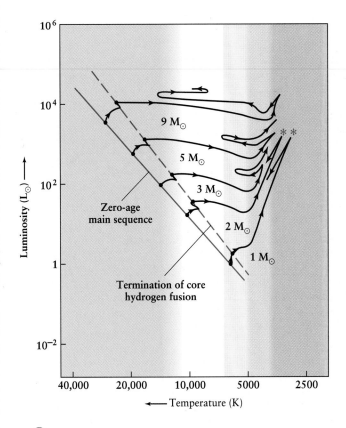

## figure 21-5

**Stellar Evolution on and off the Main Sequence** This H-R diagram shows the post–main-sequence evolutionary tracks of six stars of different mass. In the high-mass stars, core helium burning ignites where the evolutionary tracks make a sharp downward turn in the red-giant region of the diagram. The two lowest-mass stars shown here (1 $M_{\odot}$ and 2 $M_{\odot}$) are shown only up to the points, indicated by the red asterisks, where the helium flash occurs at their centers. (Adapted from I. Iben)

(Figure 21-5), zero-age main-sequence stars lie along a line called the **zero-age main sequence**, or **ZAMS**. These stars have just ended their protostar stage, are burning hydrogen steadily in their cores, and have attained hydrostatic equilibrium. With the passage of time, hydrogen in a main-sequence star's core is converted to helium, the luminosity slowly increases, the star slowly expands, and the star's position on the H-R diagram inches away from the ZAMS.

The dashed line in Figure 21-5 denotes stars whose cores have been exhausted of hydrogen and in which core hydrogen burning has ceased. From there, the points representing high-mass stars move rapidly from left to right across the H-R diagram. This means that, although the star's surface temperature is decreasing, its surface area is increasing at a rate that keeps its overall luminosity roughly constant. During this transition, the star's core contracts and its outer layers expand as energy flows outward from the star's hydrogen-burning shell.

Just before core helium burning begins, the evolutionary tracks of high-mass stars turn upward in the red-giant region of the H-R diagram (to the upper right of the main sequence). After core helium burning begins, however, the cores of these stars expand, the outer layers contract, and the evolutionary tracks back away from these temporary peak luminosities. The tracks then wander back and forth in the red-giant region while the stars readjust to their new energy sources.

The evolutionary tracks of two low-mass stars (1 $M_{\odot}$ and 2 $M_{\odot}$) also appear in Figure 21-5. The onset of core helium burning in these stars occurs with a helium flash, indicated by the red asterisks in the figure. As we described in the previous section, after the helium flash these stars shrink and become less luminous. The decrease in size is proportionately greater than the decrease in luminosity, and so the surface temperatures increase. Hence, after the helium flash, the evolutionary tracks move down and to the left.

We can summarize our understanding of stellar evolution from birth through the onset of helium burning by following the evolution of a hypothetical cluster of stars. The eight H-R diagrams in Figure 21-6 are from a computer simulation of the evolution of a hundred stars that differ only in initial mass. All the stars, which in this simulation all form at the same moment, start off as cool protostars on the right side of the H-R diagram (see Figure 21-6a). The protostars are spread out on the diagram according to their masses, and the greater the mass, the greater the protostar's initial luminosity. As discussed in Section 20-3, the source of a protostar's luminosity is its gravitational energy, which as the protostar contracts is converted to thermal energy and radiated into space.

The most massive protostars contract and heat up very rapidly, and after only 5000 years, they have already moved across the H-R diagram toward the main sequence (see Figure 21-6b). After 100,000 years, these stars have ignited hydrogen burning in their cores and settled down on the main sequence as O stars (see Figure 21-6c). After 3 million years, stars of moderate mass have also ignited core hydrogen burning and become main sequence stars of spectral classes B and A (see Figure 21-6d). Meanwhile, low-mass protostars continue to inch their way toward the main sequence as they leisurely contract and heat up.

After a simulated 30 million years (see Figure 21-6e), the most massive stars have depleted all the hydrogen at their cores and become red giants. These stars have "peeled off" the upper left end of the main sequence and are now found in the upper right corner of the H-R diagram. (This simulation follows stars only to the red-giant stage, after which they are simply deleted from the H-R diagram.) Intermediate-mass stars lie on the main sequence, while the lowest-mass stars are still in the protostar stage and lie above the main sequence.

After 66 million years (see Figure 21-6f), even the lowest-mass protostars have finally ignited core hydrogen burning and have settled down for a long sojourn as cool, dim M stars on the main sequence. These lowest-mass stars can continue to burn hydrogen in their cores for hundreds of billions of years.

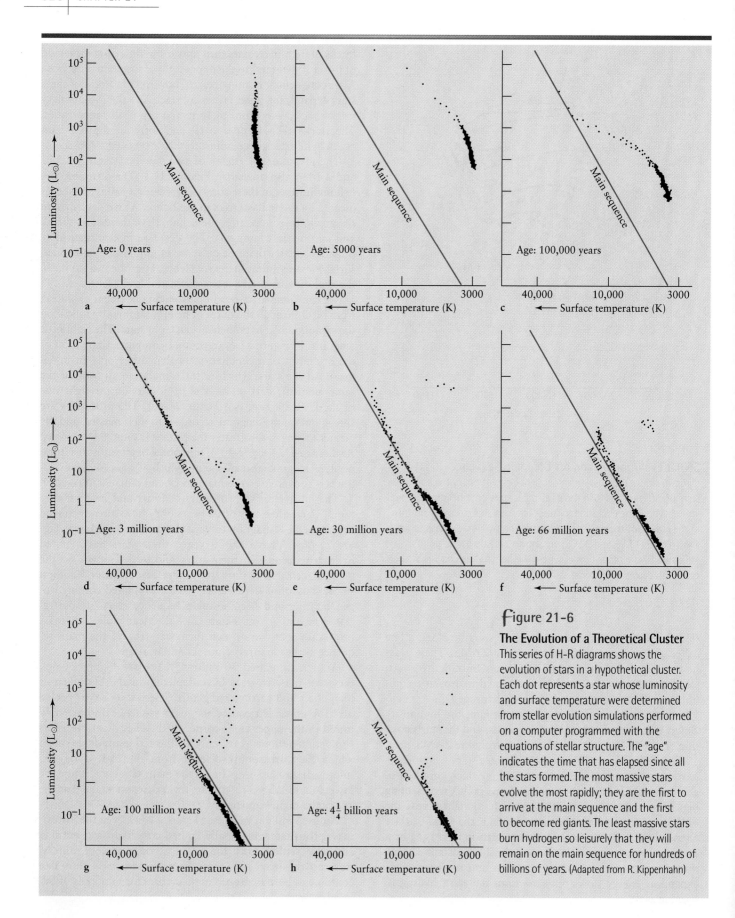

## figure 21-6

**The Evolution of a Theoretical Cluster**
This series of H-R diagrams shows the evolution of stars in a hypothetical cluster. Each dot represents a star whose luminosity and surface temperature were determined from stellar evolution simulations performed on a computer programmed with the equations of stellar structure. The "age" indicates the time that has elapsed since all the stars formed. The most massive stars evolve the most rapidly; they are the first to arrive at the main sequence and the first to become red giants. The least massive stars burn hydrogen so leisurely that they will remain on the main sequence for hundreds of billions of years. (Adapted from R. Kippenhahn)

In the final two H-R diagrams (Figures 21-6g and 21-6h), note how the main sequence gets shorter as stars exhaust their core supplies of hydrogen and evolve into red giants. The stars that leave the main sequence between Figure 21-6g and Figure 21-6h have masses between about 1 M$_\odot$ and about 3 M$_\odot$ and undergo the helium flash in their cores.

The hypothetical cluster whose evolution is displayed in Figure 21-6 can be compared with actual star clusters. The early stages of stellar evolution can be seen by observing open clusters, many of which are quite young (see Section 20-6, especially Figures 20-17 and 20-18). Examples of highly evolved post–main-sequence stars, by comparison, are commonly found in **globular clusters,** so called because of their spherical shape. A typical globular cluster, like that shown in Figure 21-7, contains up to 1 million stars in a volume less than 100 parsecs across.

Globular clusters must be old, because they contain no high-mass main-sequence stars. To show this, you would measure the apparent magnitude (a measure of apparent brightness, as discussed in Section 19-3) and color ratio of many stars in a globular cluster, then plot the data as shown in Figure 21-8. Such a **color-magnitude diagram** for a cluster is equivalent to an H-R diagram. The color ratio of a star tells you its surface temperature (as described in Section 19-

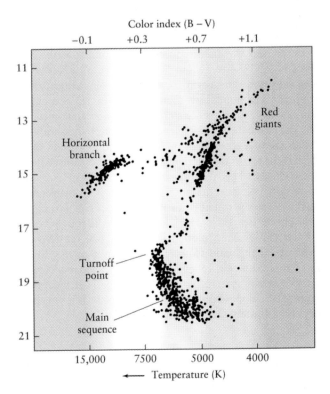

## Figure 21-8

**A Color-Magnitude Diagram of a Globular Cluster** Each dot in this diagram represents the apparent magnitude (measured through a V filter) and surface temperature (as measured by the color ratio $b_V/b_B$) of a star in the globular cluster M55 in Sagittarius. Because all the stars in M55 are at essentially the same distance from the Earth (about 6000 parsecs or 20,000 light-years), the apparent magnitude (a measure of apparent brightness) is a direct measure of luminosity. Note that the upper half of the main sequence is missing. The horizontal-branch stars are low-mass stars that recently experienced the helium flash and now exhibit core helium burning and shell hydrogen burning. (Adapted from D. Schade, D. VandenBerg, and F. Hartwick)

## Figure 21-7   R I V U X G

**A Globular Cluster** A globular cluster is a spherical cluster that typically contains a few hundred thousand stars. This cluster, called M13, is located in the constellation Hercules, approximately 7000 parsecs (23,000 light-years) from the Earth. The stars in M13 are all crowded within a diameter of only 45 parsecs (150 light-years), which means that the average density of stars in M13 is about 100 times greater than in the neighborhood of the Sun. M13 is bright enough to be seen with the naked eye on a dark, moonless night. (U. S. Naval Observatory)

4), and because all the stars in the cluster are at essentially the same distance from us, their relative brightnesses indicate their relative luminosities. What you would discover is that the upper half of the main sequence is missing. All the high-mass main-sequence stars evolved long ago into red giants, leaving behind only low-mass, slowly evolving stars that still have core hydrogen burning. (Compare Figure 21-8 with Figure 21-6h.)

The color-magnitude diagram of a globular cluster typically shows a horizontal grouping of stars in the left-of-center portion of the diagram (see Figure 21-8). These stars, called **horizontal-branch stars,** are post–helium-flash, low-mass stars with luminosities of about 50 L$_\odot$ and in which there is both core helium burning and shell hydrogen burning. In years to come, these stars will move back toward the red-giant region as their fuel is devoured.

An H-R diagram can be used to determine the age of a cluster. In the H-R diagram for a very young cluster, all the stars are on or near the main sequence. (An example is the open cluster NGC 2264, shown in Figure 20-17). As a cluster gets older, however, stars begin to leave the main sequence. The high-mass, high-luminosity stars are the first to become red giants as the main sequence starts to burn down like a candle. Over the years, the main sequence gets shorter and shorter.

The age of a cluster can be found from the **turnoff point,** which is the top of the surviving portion of the main sequence on the cluster's H-R diagram (see Figure 21-8). The stars at the turnoff point are just now exhausting the hydrogen in their cores, so their main-sequence lifetime is equal to the age of the cluster (see Table 21-1). For example, in the case of the globular cluster M55 plotted in Figure 21-8, 0.8-$M_\odot$ stars have just left the main sequence, indicating that the cluster's age is approximately 15 billion ($1.5 \times 10^{10}$) years. Data for several star clusters are plotted on a single H-R diagram in Figure 21-9, along with turnoff-point times from which the ages of the clusters can be estimated.

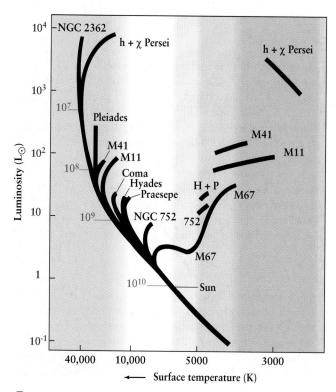

## figure 21-9

**An H-R Diagram for Open Star Clusters** The black bands indicate where data from various open star clusters fall on the H-R diagram. The ages of turnoff points (in years) are listed in red alongside the main sequence. The age of a cluster can be estimated from the location of the cluster's turnoff point, where the cluster's most massive stars are just now leaving the main sequence. (Adapted from A. Sandage)

## 21-4 Stellar evolution has produced two distinct populations of stars

Studies of star clusters reveal a curious difference between the youngest and oldest stars in our Galaxy. Stars in the youngest clusters (those with most of their main sequences still intact) are said to be **metal rich,** because their spectra contain many prominent spectral lines of heavy elements. (Recall from Section 19-5 that astronomers use the term "metal" to denote any element other than hydrogen and helium, which are the two lightest elements.) Such stars are also called **Population I stars.** The Sun is a relatively young, metal-rich, Population I star.

By contrast, the spectra of stars in the oldest clusters show only weak lines of heavy elements. These ancient stars are thus said to be **metal poor,** because they typically contain only about 3% of the abundance of heavy elements in the Sun. They are also called **Population II stars.** The stars in globular clusters are metal-poor, Population II stars. The difference in spectra between a metal-poor, Population II star and the Sun (a metal-rich, Population I star) is shown in Figure 21-10.

Why are there two distinct populations of stars? The explanation goes back to the Big Bang, the explosive origin of the universe that took place some 15 billion years ago. As we will discuss in Chapter 29, the early universe consisted almost exclusively of hydrogen and helium, with almost no heavy elements (metals). Thus, the first stars to form were likewise metal poor, and these are now the ancient stars of Population II. As these stars evolved, helium burning in their cores produced carbon and oxygen; as we will discuss in Chapter 23, further thermonuclear reactions in the most massive stars produce even heavier elements. As these stars age and die, they expel their metal-enriched gases into space, as is being done by the star in the photograph that opens this chapter. This expelled material joins the interstellar medium and is eventually incorporated into a later generation of stars that thus has a higher concentration of heavy elements. These "second-generation" stars are the metal-rich Population I stars, of which our Sun is an example.

 Be careful not to let the designations of the two stellar populations confuse you. Population *I* stars are members of a *second* stellar generation, while Population *II* stars belong to an older *first* generation.

The relatively high concentration of heavy elements in the Sun means that the solar nebula, from which both the Sun and planets formed (see Section 7-7), must likewise have been metal rich. The Earth is composed almost entirely of heavy elements, as are our bodies. Thus, our very existence is intimately linked to the Sun being a Population I star; a planet like the Earth probably could not have formed from the metal-poor gases that went into making Population II stars.

The concept of two stellar populations provides insight into our own origins. Recall from Section 21-2 that helium

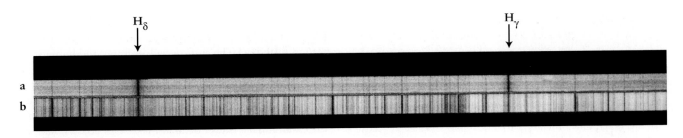

$H_\delta$ $H_\gamma$

a

b

## figure 21-10

**Spectra of a Metal-Poor Star and a Metal-Rich Star** These spectra compare **(a)** a metal-poor, Population II star and **(b)** a metal-rich, Population I star (the Sun) of the same surface temperature. Numerous spectral lines prominent in the solar spectrum are caused by elements heavier than hydrogen and helium. Note that the corresponding lines in the metal-poor star's spectrum are weak or absent. Both spectra cover a wavelength range that includes two strong hydrogen absorption lines, $H_\gamma$ (wavelength 434 nm) and $H_\delta$ (410 nm), described in <u>Section 5-8</u>. (Lick Observatory)

burning in red-giant stars produces the same isotopes of carbon and oxygen that are found most commonly on the Earth. The reason is that the Earth's carbon and oxygen atoms actually *were* produced by helium burning. These reactions occurred billions of years ago within an earlier generation of stars that died and gave up their atoms to the interstellar medium—the same atoms that later became part of our solar system, our planet, and our bodies. We are literally children of the stars.

## 21-5   Many mature stars pulsate

As we saw in <u>Section 18-8</u>, the surface of our Sun vibrates in and out, although by such a small amount that these vibrations were only noticed within the past few decades. For more than a century, however, astronomers have known of stars that undergo substantial changes in size, alternately swelling and shrinking. As these stars pulsate, they also vary dramatically in brightness. We now understand that these **pulsating variable stars** are actually evolved, post–main-sequence stars.

The first pulsating variable star was discovered in 1595 by the Dutch minister and amateur astronomer David Fabricius. He noticed that the star o (omicron) Ceti is sometimes bright enough to be easily seen with the naked eye but at other times fades to invisibility (Figure 21-11). By 1660, astronomers realized that these brightness variations repeated with a period of 332 days. Seventeenth-century astronomers were so enthralled by this variable that they renamed the star Mira ("wonderful").

Mira is an example of a class of pulsating stars called **long-period variables**. These stars are cool red giants that vary in brightness by a factor of 100 or more over a period of months or years. With surface temperatures of about 3500 K and

## figure 21-11   R I **V** U X G

**Mira—A Long-Period Variable Star** The star o (omicron) Ceti, also called Mira, is a variable star whose apparent brightness varies with a period of 332 days. At its dimmest, as in the upper photograph (taken in December 1961), Mira is less than 1% as bright as when it is at maximum, as in the lower image (taken in January 1965). These brightness variations occur because Mira pulsates and, thus, varies in both luminosity and surface temperature. (Lowell Observatory)

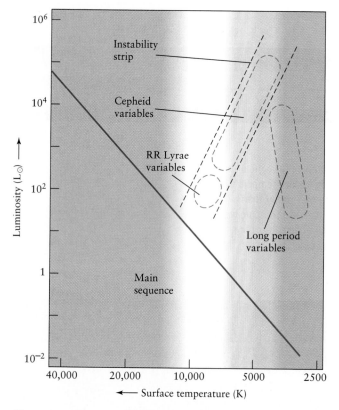

**Figure 21-12**

**Variable Stars on the H-R Diagram** This figure shows the locations on the H-R diagram where pulsating variable stars are found. Long-period variables are cool red giant stars that pulsate slowly, changing their brightness in a semiregular fashion over months or years. Cepheid variables and RR Lyrae variables are located in the instability strip, which occupies a region between the main sequence and the red-giant branch. A star passing through this region along its evolutionary track becomes unstable and pulsates.

average luminosities that range from about 10 to 10,000 L$_\odot$, they occupy the upper right side of the H-R diagram (Figure 21-12). Some, like Mira, are quite periodic, but others are irregular. Many are known to eject large amounts of gas and dust into space.

Astronomers do not fully understand why some cool red giants become long-period variables. Because the extended, tenuous atmospheres of these huge stars make it difficult to define quantities like the star's radius, it is difficult to calculate accurate stellar models.

Astronomers have a much better understanding of other pulsating stars, called **Cepheid variables,** or simply Cepheids. A Cepheid variable is recognized by the characteristic way in which its light output varies—rapid brightening followed by gradual dimming. They are named for δ (delta) Cephei, an example of this type of star discovered in 1784 by John Goodricke, a deaf, mute, 19-year-old English amateur astronomer. He found that at its most brilliant, δ Cephei is 2.3 times as bright as at its dimmest. The cycle of brightness variations repeats every 5.4 days. (Sadly, Goodricke paid for

his discoveries with his life; he caught pneumonia while making his nightly observations and died before his twenty-second birthday.) The surface temperatures and luminosities of the Cepheid variables place them in the upper middle of the H-R diagram (see Figure 21-12).

After core helium burning begins, mature stars move across the middle of the H-R diagram. Figure 21-5 shows the evolutionary tracks of high-mass stars crisscrossing the H-R diagram. Post–helium-flash, low-mass stars on the horizontal branch also cross the middle of the H-R diagram as they return to the red-giant region.

During these transitions across the H-R diagram, a star can become unstable and pulsate. In fact, there is a region on the H-R diagram between the upper main sequence and the red-giant branch called the **instability strip** (see Figure 21-12). When a star passes through this region as it evolves, the star pulsates and its brightness varies periodically. (The light curve in Figure 21-13a shows the brightness variations of δ Cephei.)

A Cepheid variable brightens and fades because of cyclic expansion and contraction of the star's outer envelope. This behavior has been deduced from spectroscopic observations. In 1894, Russian astronomer Aristarkh Belopol'skii noticed that spectral lines in the spectrum of δ Cephei shift back and forth with the same 5.4-day period as that of the magnitude variations. From the Doppler effect, we can translate these wavelength shifts into radial velocities and draw a velocity curve (Figure 21-13b). Negative velocities mean that the star's surface is expanding toward us; positive velocities mean that the star's surface is receding. Note that the light curve and velocity curve are mirror images of each other. The star is brighter than average while it is expanding and dimmer than average while contracting.

When a Cepheid variable pulsates, the star's surface oscillates up and down like a spring. During these cyclical expansions and contractions, the star's gases alternately heat up and cool down. The temperature changes in the star's surface that result are displayed in Figure 21-13c. The periodic changes in the star's diameter are shown in Figure 21-13d.

Just as a bouncing ball eventually comes to rest, a pulsating star would soon stop pulsating without some sort of mechanism to keep its oscillations going. In 1941, the British astronomer Arthur Eddington suggested that a Cepheid pulsates because the star is more opaque when compressed than when expanded. When the star is compressed, trapped heat increases the internal pressure, which pushes the star's surface outward. When the star expands, the heat escapes, the internal pressure drops, and the star's surface falls inward.

In the 1960s, the American astronomer John Cox followed up on Eddington's idea and proved that helium is what keeps Cepheids pulsating. Normally, when a star's helium is compressed, the temperature of the gas increases and the gas becomes more transparent. In certain layers near the star's surface, however, compression may ionize helium (remove one of its electrons) instead of raising its temperature. A gas of ionized helium is quite opaque, so these layers effectively trap heat, which makes the star expand, as Eddington suggested. This expansion cools the outer layers and makes the helium ions recombine with electrons, which makes the gas

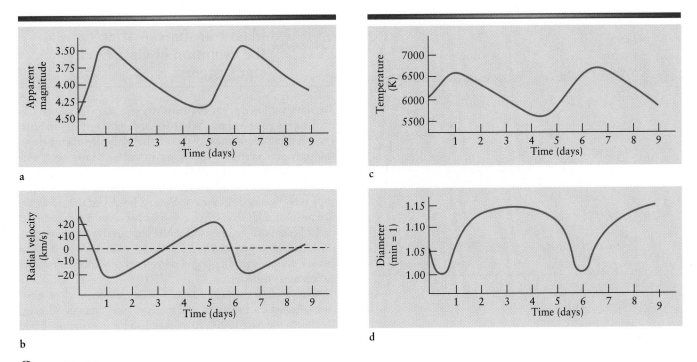

a

b

c

d

**ƒigure 21-13**

**The Details of δ Cephei** These four graphs display details about the pulsations of δ Cephei. **(a)** The star's light curve shows that the star brightens quickly but fades more slowly. **(b)** The velocity curve has positive values when the star is shrinking (and the surface is moving away from us) and negative values when it is expanding (and the surface is approaching us). **(c)** The star's surface temperature also varies periodically; the star is hottest around the time it is most luminous and coolest around the time it is dimmest. **(d)** This graph shows the periodic variations in the star's diameter. The star is at its brightest and hottest slightly after the time when the diameter is least and the star is at its smallest.

more transparent and releases the trapped energy. The star's surface then falls inward, recompressing the helium, and the whole cycle begins all over again.

Cepheid variables are very important, because they have two properties that allow astronomers to determine the distance to very remote objects. First, Cepheids can be seen even at distances of millions of parsecs. This is because they are very luminous, ranging from a few hundred times solar luminosity to more than $10^4$ $L_\odot$. Second, there is a direct relationship between a Cepheid's period and its average luminosity: The dimmest Cepheid variables pulsate rapidly, with periods of one to two days, while the most luminous Cepheids pulsate with much slower periods of about 100 days. This relationship, known as the **period-luminosity relation,** is plotted in Figure 21-14. By measuring the period of a distant Cepheid's brightness variations and using Figure 21-14, an astronomer can determine the star's luminosity. By also measuring the star's apparent brightness, the distance to the Cepheid can be found by using the inverse-square law (see Section 19-2). By applying the period-luminosity relation in this way to Cepheids in other galaxies, astronomers have been able to calculate the distances to those galaxies with great accuracy. (An example of such a calculation is given in Box 19-2.) Such measure-

ments play an important role in determining the overall size and structure of the universe, as we will see in Chapter 26.

The details of a Cepheid's pulsations depend on the amount of heavy elements in the star's outer layers, because even trace amounts of these elements can have a large effect on how opaque the stellar gases are. Hence, Cepheids are classified according to their metal content. If the star is metal rich, and thus a Population I star, it is called a **Type I Cepheid;** if it is a metal-poor, Population II star, it is called a **Type II Cepheid.** As shown in Figure 21-14, there are different period-luminosity relations for these two types of Cepheids. In order to know which period-luminosity relation to apply to a given Cepheid, an astronomer must determine the star's metal content from its spectrum (see Figure 21-10).

The evolutionary tracks of mature, high-mass stars pass back and forth through the upper end of the instability strip on the H-R diagram. These stars become Cepheids when the ionization layers occur at just the right depth to drive the pulsations. For stars on the high-temperature side of the instability strip, helium ionization is too close to the surface and involves only an insignificant fraction of the star's mass. For stars on the cool side of the instability strip, convection in the star's outer layers prevents the storage of the energy

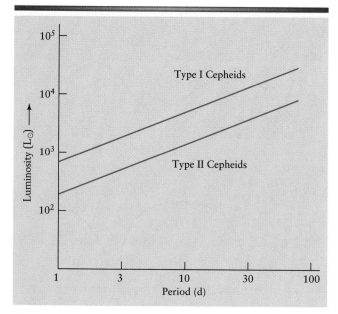

**figure 21-14**

**The Period-Luminosity Relation for Cepheids** As this graph shows, the period of a Cepheid variable is directly related to its average luminosity: the more luminous the Cepheid, the longer its period and the slower its pulsations. Note that there are actually two distinct period-luminosity relations—one for Type I Cepheids, which are metal-rich, Population I stars found in the plane of our Galaxy, and one for the less luminous Type II Cepheids, which are metal-poor, Population II stars found in globular clusters. (Adapted from H. C. Arp)

needed to drive the pulsations. Thus, Cepheids exist only in a narrow range of temperature on the H-R diagram.

Stars of lower mass do not become Cepheids. Instead, after undergoing the helium flash, they pass through the lower end of the instability strip as they move in the horizontal branch along their evolutionary tracks. Some of these stars become **RR Lyrae variables,** named after their prototype in the constellation Lyra. RR Lyrae variables all have periods shorter than one day and roughly the same average luminosity as the stars on the horizontal branch, about 100 $L_\odot$. In fact, as shown in Figure 21-12, the RR Lyrae region of the instability strip is actually a segment of the horizontal branch. RR Lyrae stars are all metal-poor, Population II stars. Many of them have been found in globular clusters, and they have been used to determine the distances to those clusters in the same way that Cepheids are used to find the distances to other galaxies. In Chapter 25 we will discuss how RR Lyrae stars helped astronomers determine the size of the Milky Way Galaxy.

In rare cases, stellar pulsations can be quite substantial. In some cases, the expansion velocity exceeds the star's escape speed and the star's outer layers are ejected completely. As we will describe in Chapter 22, dying stars eject a significant amount of mass, renewing and enriching the interstellar medium for future generations of stars.

## 21-6 Mass transfer can affect the evolution of close binary star systems

We have outlined what happens when a main-sequence star evolves into a red giant. What we have ignored so far, however, is that more than 50% of the stars are members of multiple-star systems, including binaries. Many such systems are made up of widely separated stars, in which case the individual stars follow the same course of evolution as if they were isolated. In a **close binary,** however, the two stars are near enough that when one star becomes a red giant, its outer layers can be gravitationally captured by the companion star. In other words, a bloated red giant in a close binary system can dump gas onto its companion, a process called **mass transfer.**

Our modern understanding of mass transfer in close binaries is based on the work of French mathematician Edouard Roche. In the mid-1800s, Roche studied the gravitational field produced by two stars in a double system. His work answered the question of what shapes stars in a binary system assume as a result of their rotation and mutual tidal interaction. This analysis assumes that the two stars in a close binary are in circular orbits about each other and keep the same side facing each other, just as our Moon keeps its same side facing the Earth. These conditions are normally produced by tidal interaction between the two stars in a close binary system.

Roche found that he could draw a series of curves, called *equipotential surfaces,* that describe the shapes of stars in a binary star system (Figure 21-15). In widely separated binaries, the stars are so far apart that tidal effects are small, and, therefore, the stars are nearly perfect spheres. In close bina-

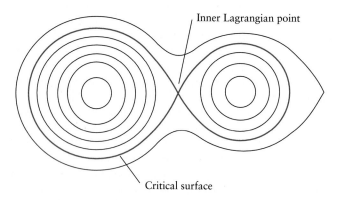

**figure 21-15**

**Equipotential Contours** This series of curves shows the shapes of stars in binary systems. As a star expands and fills its Roche lobe, it becomes somewhat egg-shaped because of the gravitational pull of its companion and the rotation of the system as a whole. If a star overflows its Roche lobe, gas flows across the inner Lagrangian point onto the companion star.

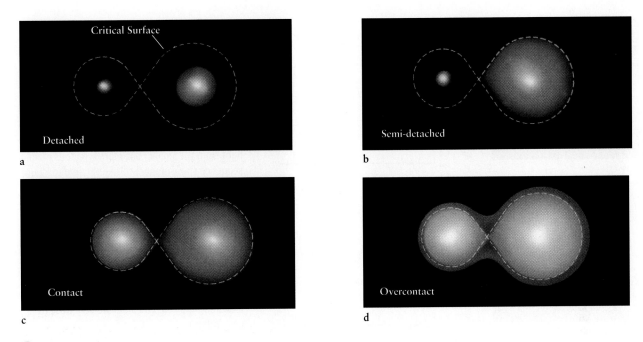

**figure 21-16**

**Detached, Semidetached, Contact, and Overcontact Binaries** This figure shows the various types of close binary star systems. **(a)** In a detached binary, neither star fills its Roche lobe. **(b)** If one star fills its Roche lobe, the binary is semidetached. Mass transfer is often observed in semidetached binaries. **(c)** In a contact binary, both stars fill their Roche lobes. **(d)** The two stars in an overcontact binary both overfill their Roche lobes. The two stars actually share the same outer atmosphere.

ries, where the separation between the stars is not much greater than their sizes, tidal effects are strong, causing the stars to be somewhat egg-shaped.

Roche discovered that one of the equipotential surfaces, a figure-eight curve that encloses both stars in a binary, marks the gravitational domain of each star. This figure-eight curve, colored red in Figure 21-15, is called the *critical surface*. Each half of the curve is known as a **Roche lobe**. The more massive star is always located inside the larger Roche lobe. If gas from a star leaks over its Roche lobe, it is no longer bound by gravity to that star and is free to fall onto the companion star—or to escape from the binary system.

The point where the two Roche lobes touch, called the **inner Lagrangian point**, is a kind of balance point between the two stars in a binary, and it is here that the effects of gravity and rotation cancel each other. When mass transfer occurs in a close binary, gases flow through the inner Lagrangian point from one star to the other.

In many binaries, the stars are so far apart that even during their red-giant stages the surfaces of the stars remain well inside their Roche lobes. As a result, little mass transfer can occur. Each star thus lives out its life as if it were single and isolated. A binary system in which each star is within its Roche lobe is referred to as a **detached binary** (Figure 21-16a).

However, if the two stars are relatively close together, when one star expands to become a red giant, it may fill or overflow its Roche lobe, in which case the system is called a **semidetached binary** (Figure 21-16b). If both stars happen

to fill their Roche lobes, the system is called a **contact binary,** because the two stars actually touch (Figure 21-16c). It is quite unlikely, however, that both stars exactly fill their Roche lobes at the same time. It is more likely that they overflow their Roche lobes, giving rise to a common envelope of gas. Such a system is called an **overcontact binary** (Figure 21-16d).

The binary star system Algol (from an Arabic term for "demon") provided the first clear evidence of mass transfer in close binaries. Also called β (beta) Persei, Algol can easily be seen with the naked eye in the constellation Perseus. Ancient astronomers knew that Algol varies periodically in brightness by more than a factor of 2. In 1782, John Goodricke (the discoverer of δ Cephei) first suggested that these brightness variations take place because Algol is an eclipsing binary, as we discussed in Section 19-11. Because the orbital plane of the two stars that make up the binary system is nearly edge-on to our line of sight, one star periodically eclipses the other. From Algol's light curve (Figure 21-17a) and spectrum, astronomers have confirmed Goodricke's brilliant hypothesis and determined that Algol is a semidetached binary. The detached star (on the right in Figure 21-17a) is a luminous blue main-sequence star, while its less massive companion is a red giant that fills its Roche lobe.

During the 1960s, astronomers labored to understand semidetached systems like Algol, which collectively came to be called Algol-type binaries. According to stellar evolution theory, the more massive a star, the more rapidly it should

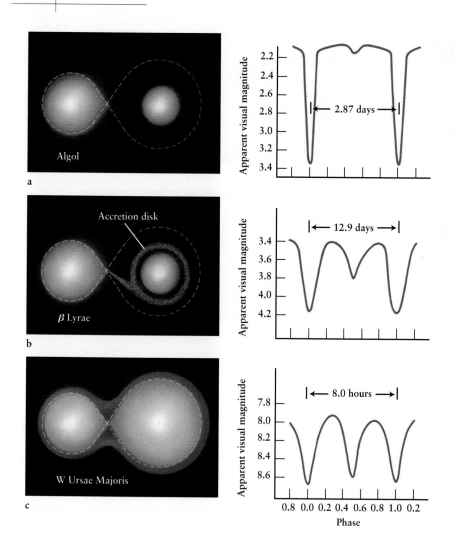

a

b

c

← 2.87 days →

← 12.9 days →

← 8.0 hours →

0.8 0.0 0.2 0.4 0.6 0.8 1.0 0.2
Phase

## figure 21-17

**Three Eclipsing Binaries** Sketches of three eclipsing binaries with their light curves are shown here. Compare these to the light curves shown in Figure 19-25. **(a)** Algol is a semi-detached binary. The deep eclipse occurs when the large red-giant star (on the left) blocks the light from the smaller but more luminous main-sequence star. Compare this light curve to Figure 19-25b. **(b)** β Lyrae is a semidetached binary in which mass transfer has produced an accretion disk around the secondary star. This disk is so thick and opaque that it renders the secondary star almost invisible. **(c)** W Ursae Majoris is an overcontact binary. Both stars overfill their Roche lobes, so that they share their outer atmospheres. The period of this binary is extremely short, indicating that the two stars are very close to each other..

evolve. But in Algol, the more massive star (on the right in Figure 21-17a) is still on the main sequence, whereas the less massive star (on the left in Figure 21-17a) has evolved to become a red giant. The apparent contradiction that the less massive star has evolved further was thus called the *Algol paradox*. The paradox was resolved when Zdenek Kopal at the University of Manchester and others proposed that the red giant in Algol-type binaries was originally the more massive star. As it left the main sequence to become a red giant, this star overflowed its Roche lobe, dumping gas onto the originally less massive companion. Because of the resulting mass transfer, that companion became the more massive star.

Mass transfer is also important in another class of semidetached binaries, called β (beta) Lyrae variables, after their prototype in the constellation of Lyra (the Harp). These variables provide evidence that matter transferred from one star can be captured into orbit about the other star to form a ring of material called an **accretion disk**. As with Algol, the less massive star in β Lyrae fills its Roche lobe, but its light curve (Figure 21-17b) and spectrum demonstrate that the more massive detached star radiates virtually no visible light at all. Furthermore, the spectrum of β Lyrae is unusual, in part because of gas flowing between the stars and around the system as a whole.

The mystery of β Lyrae was solved in 1963, when Su-Shu Huang of Northwestern University published his interpretation that the less-luminous star in β Lyrae is enveloped in a rotating disk of gas captured from its bloated companion. The disk is so thick that it completely shrouds the secondary star, making it very difficult to observe at visible wavelengths.

Observations made since the 1960s have largely confirmed Huang's model, although the accretion disk is now thought to be even thicker than Huang had supposed. As the primary star overflows its Roche lobe, gases stream across the inner Lagrangian point onto the disk at the rate of $10^{-5}$ M$_\odot$ per year. Other gas is constantly escaping altogether from the system.

What is the fate of an Algol or β Lyrae system? If the detached star is massive enough, it will evolve rapidly, expanding to also fill its Roche lobe. The result will be an overcontact binary in which the two stars share the gases of their outer layers. Such binaries are sometimes called W Ursae Majoris stars, after the prototype of this class (Figure 21-17c). Mass transfer can also continue when one of the stars in a close binary dies, so that all nuclear reactions in that star cease. In Chapters 22 and 23 we will see how mass transfer onto dead stars produces some of the most unusual objects in the sky.

## Key Words

accretion disk, p. 534

alpha particle, p. 523

Cepheid variable, p. 530

close binary, p. 532

color-magnitude diagram, p. 527

contact binary, p. 533

core helium burning, p. 523

core hydrogen burning, p. 518

degeneracy, p. 524

degenerate-electron pressure, p. 524

detached binary, p. 533

globular cluster, p. 527

helium burning, p. 523

helium flash, p. 523

horizontal-branch star, p. 527

ideal gas, p. 524

inner Lagrangian point, p. 533

instability strip, p. 530

long-period variable, p. 529

main-sequence lifetime, p. 518

mass loss, p. 522

mass transfer, p. 532

metal-poor star, p. 528

metal-rich star, p. 528

overcontact binary, p. 533

Pauli exclusion principle, p. 524

period-luminosity relation, p. 531

Population I and Population II stars, p. 528

pulsating variable star, p. 529

red giant, p. 522

Roche lobe, p. 533

RR Lyrae variable, p. 532

semidetached binary, p. 533

shell hydrogen burning, p. 520

triple alpha process, p. 523

turnoff point, p. 528

Type I and Type II Cepheids, p. 531

zero-age main sequence (ZAMS), p. 525

zero-age main-sequence star, p. 518

## Key Ideas

**The Main-Sequence Lifetime:** The duration of a star's main-sequence lifetime depends on the amount of hydrogen in the star's core and the rate at which the hydrogen is consumed.

• The more massive a star, the shorter is its main-sequence lifetime. The Sun has been a main-sequence star for about 4.6 billion years and should remain one for about another 5 billion years.

• During a star's main-sequence lifetime, the star expands somewhat and undergoes a modest increase in luminosity.

**Becoming a Red Giant:** Core hydrogen burning ceases when the hydrogen has been exhausted in the core of a main sequence star. This leaves a core of nearly pure helium surrounded by a shell through which hydrogen burning works its way outward in the star. The core shrinks and becomes hotter, while the star's outer layers expand and cool. The result is a red-giant star.

• As a star becomes a red giant, its evolutionary track moves rapidly from the main sequence to the red-giant region of the H-R diagram. The more massive the star, the more rapidly this evolution takes place.

**Helium Burning:** When the central temperature of a red giant reaches about 100 million K, the thermonuclear process of helium burning begins there. This process, also called the triple alpha process, converts helium to carbon and oxygen.

• In a more massive red giant, helium burning begins gradually; in a less massive red giant, it begins suddenly, in a process called the helium flash.

• After the helium flash, a low-mass star moves quickly from the red-giant region of the H-R diagram to the horizontal branch.

**Star Clusters and Stellar Populations:** The age of a star cluster can be estimated by plotting its stars on an H-R diagram.

• The cluster's age is equal to the age of the main-sequence stars at the turnoff point (the upper end of the remaining main sequence).

• As a cluster ages, the main sequence is "eaten away" from the upper left as stars of progressively smaller mass evolve into red giants.

• Relatively young Population I stars are metal rich; ancient Population II stars are metal poor. The metals (heavy elements) in Population I stars were manufactured by thermonuclear reactions in an earlier generation of Population II stars, then ejected into space and incorporated into a later stellar generation.

**Pulsating Variable Stars:** When a star's evolutionary track carries it through a region called the instability strip in the H-R diagram, the star becomes unstable and begins to pulsate.

• Cepheid variables are high-mass pulsating variables having a direct relationship between their periods of pulsation and their luminosities.

• RR Lyrae variables are low-mass, metal-poor pulsating variables with short periods.

• Long-period variable stars also pulsate but in a fashion that is less well understood.

**Close Binary Systems:** Mass transfer in a close binary system occurs when one star in a close binary overflows its Roche lobe. Gas flowing from one star to the other passes across the inner Lagrangian point. This mass transfer can affect the evolutionary history of the stars that make up the binary system.

## Review Questions

**1.** On what grounds are astronomers able to say that the Sun has about $5 \times 10^9$ years remaining in its main-sequence stage?

**2.** What will happen inside the Sun 5 billion years from now, when it begins to turn into a red giant?

**3.** What is the helium flash? Why does it happen in some stars but not in others?

**4.** How is a degenerate gas different from ordinary gases?

**5.** Explain why the majority of the stars in the sky are main-sequence stars.

**6.** What does it mean when an astronomer says that a star "moves" from one place to another on an H-R diagram?

**7.** On an H-R diagram, main-sequence stars do not lie along a single narrow line but are spread out over a band (see Figure 19-13). On the basis of how stars evolve during their main-sequence lifetime, explain why this should be so.

**8.** Using the same horizontal and vertical scales as in Figure 21-5, make points on an H-R diagram for each of the stars listed in Table 21-1. Label each point with the star's mass and its main-sequence lifetime. Which of these stars will remain on the main sequence after $10^9$ years? After $10^{11}$ years?

**9.** Explain how and why the turnoff point on the H-R diagram of a cluster is related to the cluster's age.

**10.** There is a good deal of evidence that our universe is no more than 15 billion years old (see Chapter 28). Explain why no main-sequence stars of spectral class M have yet evolved into red-giant stars.

**11. (a)** The main-sequence stars Sirius (spectral type A1), Vega (A0), Spica (B1), Fomalhaut (A3), and Regulus (B7) are among the 20 brightest stars in the sky. Explain how you can tell that all of these stars are younger than the Sun. **(b)** The third-brightest star in the sky, although it can only be seen south of 29° north latitude, is α (alpha) Centauri A. It is a main-sequence star of spectral type G2, the same as the Sun. Can you tell whether α Centauri A is younger than the Sun, the same age, or older? Explain your reasoning.

**12.** How do astronomers know that globular clusters are made of old stars?

**13.** What is the difference between Population I and Population II stars? In what sense can the stars of one population be regarded as the "children" of the other population?

**14.** Why do astronomers attribute the observed Doppler shifts of a Cepheid variable to pulsation, rather than to some other cause, such as orbital motion?

**15.** Why do Cepheid stars pulsate? Why are these stars important to astronomers who study galaxies beyond the Milky Way?

**16.** What is a Roche lobe? What is the inner Lagrangian point? Why are Roche lobes and the critical surface important in close binary star systems?

**17.** What is the difference between a detached binary, a semidetached binary, a contact binary, and an overcontact binary?

**18.** What is the Algol paradox? How was it resolved?

## Advanced Questions

*Questions preceded by an asterisk (*) involve topics discussed in the Boxes in this chapter or in Chapter 19.*

> **Problem-solving tips and tools:**
>
> Recall that $6 \times 10^{11}$ kg of hydrogen is converted into helium each second at the Sun's center, as explained in Section 18-6. Recall that you must use absolute (Kelvin) temperatures when using the Stefan-Boltzmann law. The orbits of stars in binary systems are described by Newton's form of Kepler's third law, as reviewed in Section 19-9. You may find it helpful to review the discussion of apparent magnitude, absolute magnitude, and luminosity found in Box 19-3.

**19.** The Sun has increased in radius by 6% over the past 4.6 billion years. Its present-day radius is 696,000 km. What was its radius 4.6 billion years ago? (*Hint:* The answer is *not* 654,000 km.)

**20.** As shown in Figure 21-2, the radius of the Sun has changed over the past several billion years. As discussed in Section 9-4, the Earth-Moon distance has changed as well. A few billion years ago, were annular eclipses of the Sun (Figure 3-14) more or less common than they are today? Explain.

**21.** What mass of hydrogen will the Sun convert into helium during its entire main-sequence lifetime of $10^{10}$ years? What fraction does this represent of the total mass of hydrogen that was originally in the Sun? Assume that the Sun's luminosity remains nearly constant during the entire $10^{10}$ years.

**22.** When the Sun becomes a red giant, its luminosity will be 2000 times greater than it is today. Assuming that this luminosity is caused *only* by the burning of the Sun's remaining hydrogen, calculate how long our star will be a red giant. (In fact, only a fraction of the remaining hydrogen will be consumed, so your answer will be an *over*estimate.)

**23. (a)** When the Sun has swollen to its full red-giant size, what will be its angular diameter as seen from the Earth? Assume that the size of the Earth's orbit remains unchanged.

**(b)** As a red giant, the Sun's luminosity will be about 2000 times greater than it is now, so the amount of solar energy falling on the Earth will increase by a factor of 2000. Hence, to maintain thermal equilibrium, each square meter of the Earth's surface will have to radiate 2000 times as much energy into space as it does now. Use the Stefan-Boltzmann law to determine what the Earth's surface temperature will be under these conditions.

**\*24.** What are the main-sequence lifetimes of **(a)** an 8-$M_\odot$ star and **(b)** a 0.6-$M_\odot$ star? Compare these lifetimes with that of the Sun.

**\*25.** The earliest fossil records indicate that life appeared on the Earth about a billion years after the solar system was formed. What is the most mass that a star could have in order that its lifetime on the main sequence be long enough to permit life to form on one or more of its planets? Assume that the evolutionary processes would be similar to those that have occurred on the Earth.

**26.** What observations would you make of a star to determine whether its primary source of energy is hydrogen or helium burning?

**27.** The two stars that make up the overcontact binary W Ursae Majoris (Figure 21-17*c*) have estimated masses of 0.99 $M_\odot$ and 0.62 $M_\odot$. **(a)** Find the average separation between the two stars. Give your answer in kilometers. **(b)** The radii of the two stars are estimated to be 1.14 $R_\odot$ and 0.83 $R_\odot$. Show that these values and your result in part (a) are consistent with the statement that this is an overcontact binary.

**\*28.** The star X Arietis is an RR Lyrae variable. Its apparent magnitude varies between 8.97 and 9.95 with a period of 0.65 day. Interstellar extinction dims the star by half a magnitude. Approximately how far away is the star?

**29.** Polaris (the North Star, also known as α Ursae Minoris) is a Type II Cepheid variable. Its apparent brightness varies with a period of 3.97 days, and its average apparent brightness is $3.1 \times 10^{-12}$ that of the Sun. Approximately how far away is Polaris?

**30.** Consult recent issues of *Sky & Telescope* and *Astronomy* to find out when Mira will next reach maximum brightness. Look up the star's location on a star chart. Why is it unlikely that you will be able to observe Mira at maximum brightness?

## Discussion Questions

**31.** The half-life of the $^8$Be nucleus is $2.6 \times 10^{-16}$ second, which is the average time that elapses before this unstable nucleus decays into two alpha particles. How would the universe be different if instead the $^8$Be half-life were zero? How would the universe be different if the $^8$Be nucleus were stable?

**32.** Suppose that an oxygen nucleus were to be fused with a helium nucleus. What element would be formed? Look up the relative abundance of this element in, for example, the *Hand-*

*book of Chemistry and Physics* or on the World Wide Web at the Brookhaven National Laboratory site (**http://necs01. dne.bnl.gov/CoN/index.html**). Comment on whether such a process is likely.

**33.** If the universe has a finite age (as a broad array of evidence indicates to be the case), what observational consequences does this have for H-R diagrams of star clusters? Could we use these consequences to establish constraints on the possible age of the universe? Explain.

## Observing Projects

> **Observing tips and tools:**
>
> An excellent resource for learning how to observe variable stars is the web site of the American Association of Variable Star Observers (**http://www.aavso.org/ educational/hoa/learning.html**). A wealth of data about specific variable stars can be found in the three volumes of *Burnham's Celestial Handbook: An Observer's Guide to the Universe Beyond the Solar System* (Dover, 1978).

**34.** Observe several of the red giants and supergiants listed below with the naked eye and through a telescope. (Note that γ Andromedae, α Tauri, and ε Pegasi are all multiple-star systems. The spectral type and luminosity class refer to the brightest star in these systems.) These stars are most easily found with the aid of star charts published every month in such magazines as *Sky & Telescope* and *Astronomy*. Is the reddish color of these stars apparent when they are compared with neighboring stars?

| Star | Spectral type | R.A. | Decl. |
|---|---|---|---|
| Almach (γ Andromedae) | K3 II | 2$^h$ 03.9$^m$ | +42° 20′ |
| Aldebaran (α Tauri) | K5 III | 4 35.9 | +16 31 |
| Betelgeuse (α Orionis) | M2 I | 5 55.2 | +07 24 |
| Arcturus (α Boötis) | K2 III | 14 15.7 | +19 11 |
| Antares (α Scorpii) | M1 I | 16 29.5 | –26 26 |
| Eltanin (γ Draconis) | K5 III | 17 56.7 | +51 29 |
| Enif (ε Pegasi) | K2 I | 21 44.2 | +09 52 |

*Note: The right ascensions and declinations are given for epoch 2000.*

**35.** Several of the open clusters listed in Figure 21-9 can be seen quite well with a good pair of binoculars. Observe as many of these clusters as you can, using both a telescope and binoculars. (Some are actually so large that they will not fit in the field of view of many telescopes.) You can learn more about all of these clusters on the Web at "The Messier Catalog" (**http://www.seds.org/messier/**). Note that in the star clusters listed below the M prefix refers to the Messier Catalog, NGC refers to the New General Catalogue, and Mel refers to the Melotte Catalog.

In making your own observations, note the overall distribution of stars in each cluster. Which clusters are seen better through binoculars than through a telescope? Which clusters can you see with the naked eye?

| Star cluster | Constellation | R.A. | Decl. |
|---|---|---|---|
| h Persei (NGC 869) | Perseus | 2$^h$ 19.0$^m$ | +57° 09′ |
| χ Persei (NGC 884) | Perseus | 2 22.4 | +57 07 |
| Pleiades (M45) | Taurus | 3 47.0 | +24 07 |
| Hyades (Mel 25) | Taurus | 4 27 | +16 00 |
| Praesep (M44) | Cancer | 8 40.1 | +19 59 |
| Coma (Mel 111) | Coma Berenices | 12 25 | +26 00 |
| Wild Duck (M11) | Scutum | 18 51.1 | −06 16 |

*Note: The right ascensions and declinations are given for epoch 2000.*

**36.** There are many beautiful globular clusters scattered around the sky that can be easily seen with a small telescope. Several of the brightest and nearest globulars are listed below. A wealth of information about all of these clusters, including many great images, can be found on the World Wide Web (**http://www.seds.org/messier/**).

Observe as many of these globular clusters as you can. Can you distinguish individual stars toward the center of each cluster? Do you notice any differences in the overall distribution of stars between clusters?

| Star cluster | Constellation | R.A. | Decl. |
|---|---|---|---|
| M3 (NGC 5272) | Canes Venatici | 13$^h$ 42.2$^m$ | +28° 23′ |
| M5 (NGC 5904) | Serpens | 15 18.6 | +2 05 |
| M4 (NGC 6121) | Scorpius | 16 23.6 | −26 32 |
| M13 (NGC 6205) | Hercules | 16 41.7 | +36 28 |
| M12 (NGC 6218) | Ophiuchus | 16 47.2 | −1 57 |
| M28 (NGC 6626) | Sagittarius | 18 24.5 | −24 52 |
| M22 (NGC 6656) | Sagittarius | 18 36.4 | −23 54 |
| M55 (NGC 6809) | Sagittarius | 19 40.0 | −30 58 |
| M15 (NGC 7078) | Pegasus | 21 30.0 | +12 10 |

*Note: The right ascensions and declinations are given for epoch 2000.*

## Where to Learn More

*Books and magazine articles*

Croswell, C. "The First Cepheid." *Sky & Telescope*, October 1997. The first Cepheid variable to be discovered was not δ Cephei, but η (eta) Aquilae. This brief article describes how the discovery was made as well as the remarkable individuals involved in the discovery.

Hack, M. "Epsilon Aurigae." *Scientific American,* October 1984. This article, which focuses on a particular eclipsing binary, demonstrates how astronomers use observations and an understanding of stellar evolution to deduce details of elaborate stellar systems.

Iben, I., and Tutukov, A. V. "The Lives of Stars: From Birth to Death and Beyond." *Sky & Telescope*, December 1997 and January 1998. These lucidly written articles give a mod-

ern perspective on the evolution of stars after they leave the main sequence.

Isles, J. "Watch an Evolving Binary Star." *Sky & Telescope,* June 1993. An experienced variable- star observer provides tips on how to observe β Lyrae and its brightness variations using only the naked eye.

Kafatos, M., and Michalitsianos, A. "Symbiotic Stars." *Scientific American*, July 1984. This article discusses fascinating phenomena associated with binary stars that consist of a red giant and a hot, compact companion.

Kaler, J. *Stars.* Scientific American Library, 1992. Chapter 5 of this clear, well-illustrated book has an excellent discussion of stellar evolution.

Kippenhahn, R. *100 Billion Suns: The Birth, Life, and Death of Stars.* Basic Books, 1983. This classic book by a noted German astrophysicist describes the life cycles of stars.

Meylan, G., and Brandl, B. "30 Doradus: Birth of a Star Cluster." *Sky & Telescope*, March 1998. All the globular clusters in our Galaxy are very old. This article describes how a *new* globular cluster may be forming in the Large Magellanic Cloud, a small galaxy some 50 kiloparsecs (160,000 light-years) away.

Percy, J. "Observing Variable Stars for Fun and Profit." *Mercury,* May/June 1979. This entertaining article contains many important tips for a person interested in observing eclipsing binaries, Cepheids, and other variable stars.

———. "Pulsating Stars." *Scientific American,* June 1975. Although this article was written more than 20 years ago, it is still the best popular description of the physics of stellar pulsations.

Tomkin, J., and Lambert, D. L. "The Strange Case of Beta Lyrae." *Sky & Telescope,* October, 1987. This fascinating article explores many of the mysteries associated with β Lyrae.

Wyckoff, S. "Red Giants: The Inside Scoop." *Mercury,* January/February 1979. This superb article, which includes many interesting details about red giants, emphasizes the effects of varying abundances of carbon and oxygen.

**W** *World Wide Web*

Two web sites with animations of stellar evolution are "Life Cycles of the Stars" (**http://www.eia.brad.ac.uk/ btl/m2.html**) and "The Electronic Universe" (**http://zebu. uoregon.edu/textbook/se.html**).

A tremendous number of images of star clusters, along with much useful data, can be found at a web site dedicated to the Messier Catalog of deep-sky objects (**http://www.seds. org/messier/**).

The web site for the American Association of Variable Star Observers (**http://www.aavso.org/**) explains much about variable stars and describes how amateur astronomers play an essential role in the study of these members of the stellar population.

The Hubble Space Telescope has been used to study stars in all stages of their evolution. The Space Telescope Science Institute web site has a list of these images by subject (**http://oposite.stsci.edu/pubinfo/Subject.html**).

# Stellar Evolution: The Deaths of Stars

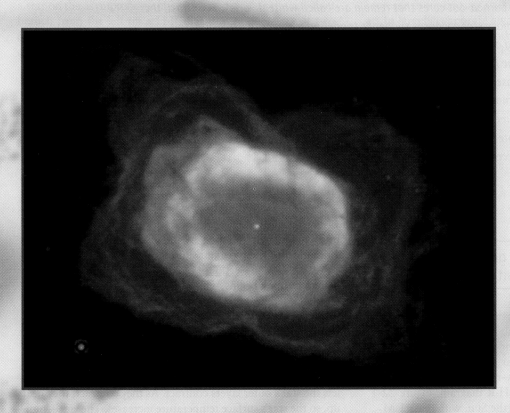

R **I** V U X G

**A Planetary Nebula** As a low–mass star ages, it ejects much of its mass into space to form a planetary nebula. Ultra-violet light from the remaining hot stellar core causes the ejected gases to glow. In this infrared image of the planetary nebula NGC 7027, made with the Hubble Space Telescope, the core appears as a bright spot at the nebula's center. NGC 7027 is about 900 pc (3000 light–years) from Earth in the constellation Cygnus. The nebula is roughly 14,000 AU across. (William B. Latter, SIRTF Science Center/Caltech; NASA)

*In this chapter you will find the answers to the following questions:*

**22-1** What kinds of nuclear reactions occur within a star like the Sun as it ages?

**22-2** Where did the carbon atoms in our bodies come from?

**22-3** What is a planetary nebula, and what does it have to do with planets?

**22-4** What is a white dwarf star?

**22-5** Why do high-mass stars go through more evolutionary stages than low-mass stars?

**22-6** What happens within a high-mass star to turn it into a supernova?

**22-7** Why was SN 1987A an unusual supernova?

**22-8** What was learned by detecting neutrinos from SN 1987A?

**22-9** How do white dwarf stars give rise to certain types of supernovae?

**22-10** What vestiges are left after a supernova explosion?

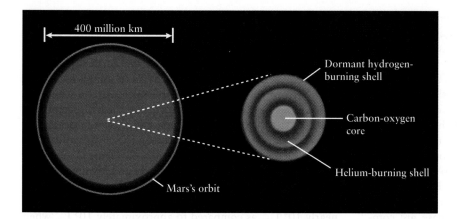

**figure 22-2**

**The Structure of an Old, Low-Mass AGB Star** Near the end of its life, a low-mass star such as the Sun becomes a red asymptotic giant branch (AGB) star whose diameter is almost as large as the diameter of the orbit of Mars. The star's inert carbon-oxygen core, active helium-burning shell, and dormant hydrogen-burning shell are all contained within a volume roughly the size of the Earth. Thermonuclear reactions within the helium-burning shell take place at a furious rate, giving the star a luminosity thousands of times that of the present-day Sun.

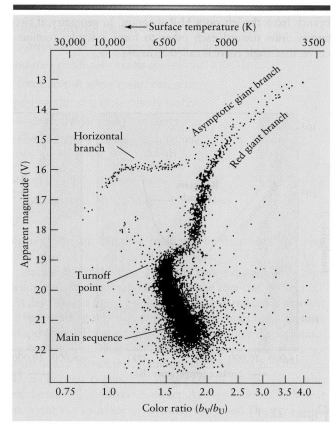

**figure 22-3** R I **V** U X G

**The Globular Cluster M3 and Its Color-Magnitude Diagram** The globular cluster M3 in the constellation Canis Venatici (the Hunting Dogs) is about 10,000 parsecs (32,000 light-years) from Earth. With an angular diameter of 16 arcminutes, the actual diameter of M3 is about 50 parsecs (150 light-years). Data for nearly 11,000 stars in M3 are plotted on the color-magnitude diagram. The main-sequence turnoff point indicates that M3 is about 10 billion ($10^{10}$) years old. The red giant branch stars are low-mass stars that have recently left the main sequence. They are all undergoing shell hydrogen burning and are all destined to experience the helium flash. The stars on the horizontal branch have already undergone the helium flash and are burning helium in their cores. Asymptotic giant branch (AGB) stars are low-mass stars undergoing helium shell burning; they are all post–horizontal-branch stars going through a second red giant phase. (NOAO; color-magnitude diagram adapted from R. Buonanno and colleagues.)

## 22-2 Dredge-ups bring the products of nuclear fusion to a giant star's surface

As we saw in Section 18-7, energy is transported outward from a star's core by one of two processes—radiative diffusion or convection. The first is the passage of energy in the form of electromagnetic radiation, and it dominates only when a star's gases are relatively transparent. The second involves up-and-down movement of the star's gases. Convection plays a very important role in giant stars, because it helps supply the cosmos with the elements essential to life.

In the Sun, convection dominates only the outer layers, from around 0.71 solar radius (measured from the center of the Sun) up to the photosphere (recall Figure 18-28). During the final stages of a star's life, however, the convective zone occasionally becomes so broad that it extends down to the star's core. At these times, convection can "dredge up" the heavy elements produced in and around the core by thermonuclear fusion, transporting them all the way to the star's surface.

The *first* dredge-up takes place after core hydrogen burning stops, when the star becomes a red giant for the first time. Convection dips so deeply into the star that material processed by the CNO cycle of hydrogen burning (see Section 18-6 and Box 18-2) is carried up to the star's surface, changing the relative abundances of carbon, nitrogen, and oxygen. A *second* dredge-up occurs after core helium burning ceases, further altering the abundances of carbon, nitrogen, and oxygen. Still later, during the AGB stage, a *third* dredge-up can occur if the star has a mass greater than about 2 $M_\odot$. This third dredge-up transports large amounts of freshly synthesized carbon to the star's surface, and the star's spectrum thus exhibits prominent absorption bands of carbon-rich molecules like $C_2$, CH, and CN. For this reason, an AGB star that has undergone a third dredge-up is called a **carbon star.**

All AGB stars have very strong stellar winds that cause them to lose mass at very high rates, up to $10^{-4}$ $M_\odot$ per year (a thousand times greater than that of a red giant, and $10^{10}$ times greater than the rate at which our present-day Sun loses mass). The surface temperature of AGB stars is relatively low, around 3000 K, so any ejected carbon-rich molecules can condense to form tiny grains of soot. Indeed, carbon stars are commonly found to be obscured in sooty cocoons of ejected matter (Figure 22-4).

Carbon stars are important because they enrich the interstellar medium with carbon and some nitrogen and oxygen. The triple alpha process that occurs in helium burning is the *only* way that carbon can be made, and carbon stars are the primary avenue by which this element is dispersed into interstellar space. Indeed, most of the carbon in your body was produced long ago inside a star by the triple alpha process, ejected into space, and later incorporated into the solar nebula from which our Earth—and all of the life on it—eventually formed.

**Figure 22-4** R I **V** U X G

**Death Throes of a Red Giant Star** This Hubble Space Telescope image shows the Egg Nebula, which lies some 900 parsecs (3000 light-years) from the Earth in the constellation Cygnus. At the center of the Egg Nebula is a highly evolved low-mass star that has ejected some of its mass into space in the form of a stellar wind with a speed of about 20 km/s. The ejected material forms the roughly spherical nebula, which is made visible because dust particles reflect light coming from the central star. The star itself is hidden within a dense, dusty cocoon that appears as a dark band at the center of the nebula. The star is also ejecting material at about 100 km/s in oppositely directed jets, traced by the "searchlight beams" that point to the upper right and lower left on this image. These may be similar to the jets seen in forming stars (see Section 20-5, and especially Figure 20-13), but their origin is not yet understood. (R. Sahai and J. Trauger, Jet Propulsion Laboratory; the WFPC-2 Science Team; and NASA.)

## 22-3 Low-mass stars die by gently ejecting their outer layers, creating planetary nebulae

The AGB stage in the evolution of a star of low mass (less than about 4 $M_\odot$) is a dramatic turning point. Before this stage, a star loses mass only gradually through steady stellar winds. But as it evolves during its AGB stage, a star divests itself completely of its outer layers. The aging star undergoes a series of bursts in luminosity, and in each burst it ejects a shell of material into space. Eventually all that remains of a

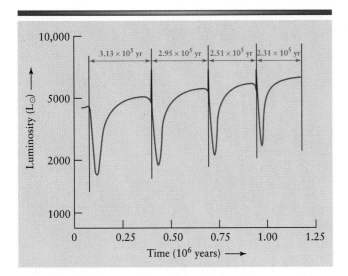

### figure 22-5

**Thermal Pulses** This graph shows how the luminosity of an 0.7-$M_\odot$ AGB star varies with time. The spikes in luminosity, called thermal pulses, are caused by brief periods of runaway helium burning. These sudden increases in luminosity eventually cause the star to eject its outer layers completely. (Adapted from I. Iben)

hot, burned-out core acts on the specks of dust, propelling them further outward, and the star sheds its outer layers altogether. In this way an aging 1-$M_\odot$ star loses as much as 40% of its mass. More massive stars eject even greater fractions of their original mass.

As a dying star ejects its outer layers, the star's hot core becomes exposed. With a surface temperature of about 100,000 K, this exposed core emits ultraviolet radiation intense enough to ionize and excite the expanding shell of ejected gases. These gases therefore glow and emit visible light, producing the so-called planetary nebula. (As described in Box 20-1, the process in which a substance absorbs ultraviolet light and emits visible light is called *fluorescence*. It also occurs in fluorescent light fixtures, as well as in the H II regions described in Section 20-2.)

**CAUTION!** Despite their name, planetary nebulae have nothing to do with planets. This misleading term was introduced in the nineteenth century because these glowing objects looked like distant Jovian planets when viewed through the small telescopes then available. The difference between planets and planetary nebulae became obvious with the advent of spectroscopy: Planets have absorption line

low-mass star is a fiercely hot, exposed core, surrounded by glowing shells of ejected gas. This late stage in the life of a star is called a **planetary nebula.**

To understand how this process works, consider the internal structure of an AGB star as shown in Figure 22-2. As the helium in the helium-burning shell is used up, the pressure that holds up the dormant hydrogen-burning shell decreases. Hence, the dormant hydrogen shell contracts and heats up, and hydrogen burning begins anew. This revitalized hydrogen burning creates helium, which rains downward onto the temporarily dormant helium-burning shell. As the helium shell gains mass, it shrinks and heats up. When the temperature of the helium shell reaches a certain critical value, it reignites in a **helium shell flash** that is similar to (but less intense than) the helium flash that occurred earlier in the evolution of a low-mass star (Section 21-2). The released energy pushes the hydrogen-burning shell outward, making it cool off, so that hydrogen burning ceases and this shell again becomes dormant. The process then starts over again.

When a helium shell flash occurs, the luminosity of an AGB star increases substantially in a relatively short-lived burst called a **thermal pulse.** Figure 22-5 shows a theoretical calculation of how the luminosity of a 0.7-$M_\odot$ AGB star varies because of helium shell flashes. The calculations predict that thermal pulses occur at ever-shorter intervals of about 300,000 years.

During these thermal pulses, the dying star's outer layers can separate completely from its carbon-oxygen core. As the ejected material expands into space, dust grains condense out of the cooling gases. Radiation pressure from the star's

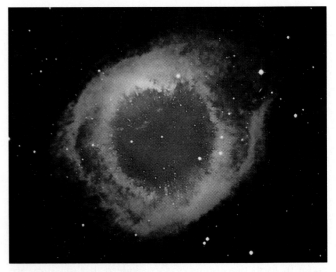

### figure 22-6   R I **V** U X G

**The Planetary Nebula NGC 7293** This beautiful object, often called the Helix Nebula, covers an area of the sky equal to half the full moon. The star that ejected these gases is at the center of the glowing shell. The bluish-green color comes from ionized oxygen, the pink and red from ionized nitrogen and hydrogen. The small radial blobs in the red shell (each of which is about 150 AU across) give the object its alternate name, the Sunflower Nebula. The ejected shell is actually spherical; it appears as a ring because we see a substantial thickness of the shell when we look near the shell's rim. The closest of all known planetary nebulae, NGC 7293 lies 140 parsecs (450 light-years) from Earth in the constellation Aquarius. (Anglo-Australian Observatory)

## Table 22-1 | Some Planetary Nebulae

| Planetary nebula | Distance (light-years) | Angular size (arcmin) | Constellation | Right ascension | Declination |
|---|---|---|---|---|---|
| Dumbbell (M27, NGC 6853) | 490–3500 | 8.0 × 5.7 | Vulpecula | 19$^h$ 59.6$^m$ | +22° 43′ |
| Ring (M57, NGC 6720) | 1300–4100 | 1.4 × 1.0 | Lyra | 18 53.6 | +33 02 |
| Little Dumbbell (M76, NGC 650) | 1700–15,000 | 2.7 × 1.8 | Perseus | 01 42.4 | +51 34 |
| Owl (M97, NGC 3587) | 1300–12,000 | 3.4 × 3.3 | Ursa Major | 11 14.8 | +55 01 |
| Saturn (NGC 7009) | 1600–3900 | 0.4 × 1.6 | Aquarius | 21 04.2 | –11 22 |
| Helix (NGC 7293) | 450 | 41 × 41 | Aquarius | 22 29.6 | –20 48 |
| Eskimo (NGC 2392) | 1400–10,000 | 0.5 | Gemini | 7 29.2 | +20 55 |
| Blinking Planetary (NGC 6826) | 3300 (?) | 2.2 × 0.5 | Cygnus | 19 44.8 | +50 31 |

*Note: The right ascensions and declinations are given for epoch 2000. Note also that many of the distances are quite uncertain.*

spectra, as discussed in Section 7-3, but the excited gases of planetary nebulae have emission line spectra.

Planetary nebulae are quite common. Astronomers estimate that there are 20,000 to 50,000 planetary nebulae in our Galaxy alone. Several well-known examples are listed in Table 22-1. Spectroscopic observations of these nebulae show bright emission lines of ionized hydrogen, oxygen, and nitrogen. From the Doppler shifts of these lines, astronomers have concluded that the expanding shell of gas moves outward from a dying star at speeds from 10 to 30 km/s. A typical planetary nebula has a diameter of roughly 1 light-year, indicating that it must have begun expanding about 10,000 years ago.

Many planetary nebulae, such as those in the figure that opens this chapter and in Figure 22-6, are relatively spherical in shape. This is a result of the symmetrical way in which the gases were ejected. In other cases, if the rate of expansion is not the same in all directions, the resulting nebula takes on an hourglass or dumbbell appearance, as shown in Figure 22-7 and in the image on the cover of this book.

By astronomical standards, a planetary nebula is very short-lived. After about 50,000 years, the nebula has spread out so far from the cooling central star that its nebulosity simply fades from view. The gases then mix with the surrounding interstellar medium. Astronomers estimate that all the planetary nebulae in the Galaxy return a total of about 5 M$_\odot$ to the interstellar medium each year. This amount is about 15% of all the matter expelled by all the various sorts of stars in the Galaxy each year. Because of this significant contribution, planetary nebulae play an important role in the chemical evolution of the Galaxy as a whole.

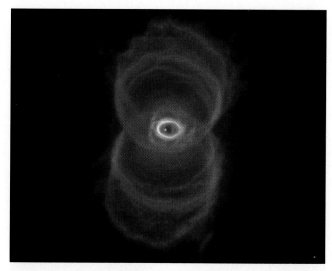

**figure 22-7** R I **V** U X G

**The Planetary Nebula MyCn18** This planetary nebula, called the Etched Hourglass Nebula because of its shape, lies some 2500 parsecs (8000 light-years) from Earth in the southern constellation Musca (the Fly). The hourglass shape of the nebula may result from ejected material trying to spread through a dense cloud around the star's equator. This false-color view is a combination of three Hubble Space Telescope images: red shows emission from ionized nitrogen, green shows hydrogen, and blue shows doubly ionized oxygen. The central star is the white dot slightly off the center of the greenish "eye." (R. Sahai and J. Trauger, Jet Propulsion Laboratory; the WFPC-2 Science Team; and NASA.)

## 22-4 The burned-out core of a low-mass star cools and contracts until it becomes a white dwarf

Stars less massive than about 4 $M_\odot$ never develop the necessary central pressures or temperatures to ignite thermonuclear reactions that use carbon or oxygen as fuel. Instead, as we have seen, the process of mass ejection just strips away the star's outer layers and leaves behind the hot carbon-oxygen core. With no thermonuclear reactions taking place, the core simply cools down like a dying ember. These burnt-out relics of a star's former glory are called **white dwarfs.** White dwarfs are quite small by stellar standards—approximately the same size as the Earth.

You might think that without thermonuclear reactions to provide it with internal heat and pressure, a white dwarf should keep on shrinking under the influence of its own gravity as it cools. Actually, however, a cooling white dwarf maintains its size, because the burnt-out stellar core is very dense. In fact, most of the electrons in the core are degenerate, and, thus, degenerate-electron pressure supports the star against further collapse. This pressure does not depend on temperature, so it continues to hold up the star even as the temperature drops.

Many white dwarfs are found in the solar neighborhood, but all are too faint to be seen with the naked eye. One of the first white dwarfs to be discovered is a companion to the bright star Sirius. The binary nature of Sirius was first deduced in 1844 by the German astronomer Friedrich Bessel, who noticed that this star was moving back and forth slightly, as if it was being orbited by an unseen object. This companion, designated Sirius B (Figure 22-8), was first glimpsed in

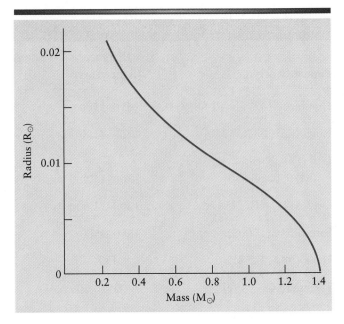

**figure 22-9**

**The Mass-Radius Relationship for White Dwarfs** The more massive a white dwarf is, the smaller it is. This unusual relationship between mass and radius is a result of the degenerate-electron pressure that supports the star. The maximum mass of a white dwarf, called the Chandrasekhar limit, is 1.4 $M_\odot$. Incidentally, 0.01 $R_\odot$ is 6960 km, just slightly larger than the Earth's radius of 6378 km.

1862 by the American astronomer Alvan Clark. Recent satellite observations at ultraviolet wavelengths, where hot white dwarfs emit most of their light, show that the surface temperature of Sirius B is about 30,000 K.

Observations of white dwarfs in binary systems like Sirius allow astronomers to determine the mass, radius, and density of these stars (see Sections 19-9, 19-10, and 19-11). Such observations show that the density of the degenerate matter in a white dwarf is typically $10^9$ kg/m³ (a million times denser than water). A teaspoonful of white dwarf matter brought to the Earth would weigh nearly 5.5 tons, as much as an elephant! As we learned in Section 21-2, degenerate matter has a very different relationship between its pressure, density, and temperature than that of ordinary gases. Consequently, white dwarf stars have an unusual **mass-radius relation:** The more massive a white dwarf star is, the *smaller* it is.

Figure 22-9 displays the mass-radius relation for white dwarfs. Note that the more degenerate matter you pile onto a white dwarf, the smaller it becomes. However, there is a limit to how much pressure degenerate electrons can produce. As a result, there is an upper limit to the mass that a white dwarf can have. This maximum mass is called the **Chandrasekhar limit,** after the Indian-American scientist Subrahmanyan Chandrasekhar, who pioneered theoretical studies of white dwarfs. The Chandrasekhar limit is equal to 1.4 $M_\odot$, meaning that all white dwarfs must have masses less than 1.4 $M_\odot$.

**figure 22-8** R I **V** U X G

**Sirius and Its White Dwarf Companion** Sirius, the brightest-appearing star in the sky, is actually a binary star: The secondary star, called Sirius B, is a white dwarf. Sirius B is seen in this photograph at the five o'clock position, almost obscured by the glare of the primary star. The spikes and rays around Sirius are the result of optical effects within the telescope. (Courtesy of R. B. Minton)

The material inside a white dwarf consists mostly of ionized carbon and oxygen atoms floating in a sea of degenerate electrons. As the dead star cools, the particles in this material slow down, and electric forces between the ions begin to prevail over the random thermal motions. The ions no longer move freely. Instead, according to calculations by Hugh Van Horn and Malcolm Savedoff at the University of Rochester, they arrange themselves in orderly rows, like an immense crystal lattice. From this time on, you could say that the star is "solid." The degenerate electrons move around freely in this crystal material, just as electrons move freely through an electrically conducting metal like copper or silver. Because a diamond is crystallized carbon, a cool carbon-oxygen white dwarf resembles an immense spherical diamond!

Figure 22-10 shows the evolutionary tracks followed by three burned-out stellar cores as they become white dwarfs. The initial red giants had masses of 0.8, 1.5, and 3.0 $M_\odot$. Mass ejection strips these dying stars of up to 60% of their

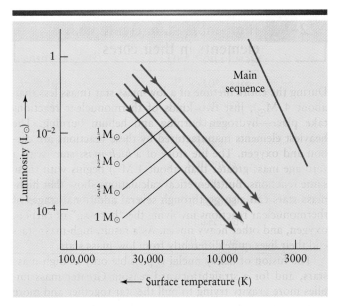

**figure 22-11**

**White Dwarf "Cooling Curves"** The evolutionary tracks of four white dwarfs of different masses are shown here. As these dead stars radiate their internal energy into space, they become dimmer and cooler, moving toward the lower right on the H-R diagram. The white dwarf's radius remains constant as it cools, however. To see this, compare these "cooling curves" with the lines of constant radius shown in Figure 19-14. The more massive a white dwarf is, the smaller and hence fainter it is.

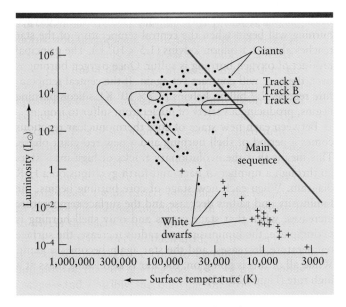

| Evolutionary track | Mass ($M_\odot$) | | |
|---|---|---|---|
| | Giant star | Ejected nebula | White dwarf |
| A | 3.0 | 1.8 | 1.2 |
| B | 1.5 | 0.7 | 0.8 |
| C | 0.8 | 0.2 | 0.6 |

**figure 22-10**

**Evolution from Giants to White Dwarfs** The evolutionary tracks of three low-mass giant stars are shown as they eject planetary nebulae. The table gives the extent of mass loss in each case. The dots on the graph represent the central stars of planetary nebulae whose surface temperatures and luminosities have been determined. The crosses are white dwarfs for which similar data exist. (Adapted from B. Paczynski)

matter. During their final spasms, the appearance of these stars changes quite rapidly. Because of the changes in luminosity and surface temperature, the points representing these stars on a H-R diagram race along their evolutionary tracks, sometimes executing loops corresponding to thermal pulses. Finally, as the ejected nebulae fade and the stellar cores cool, the evolutionary tracks of these dying stars take a sharp turn downward toward the white dwarf region of the H-R diagram.

Although a white dwarf maintains the same size as it cools, its luminosity and surface temperature both decrease with time. Consequently, the evolutionary tracks of aging white dwarfs point toward the lower right corner of the H-R diagram. You can see this in Figure 22-10; it is shown in more detail in Figure 22-11. The energy that the white dwarf radiates into space comes only from the star's internal heat, which is a relic from the white dwarf's past existence as a stellar core. Over billions of years, white dwarfs grow dimmer and dimmer as their surface temperatures drop toward absolute zero.

After ejecting much of its mass into space, our own Sun will eventually evolve into a white dwarf star about the size of the Earth and with perhaps one-tenth of its present luminosity. It will become even dimmer as it cools. After 5 billion years as a white dwarf, the Sun will be no more than one ten-thousandth of its present brilliance. With the passage of eons, our Sun will simply fade into obscurity.

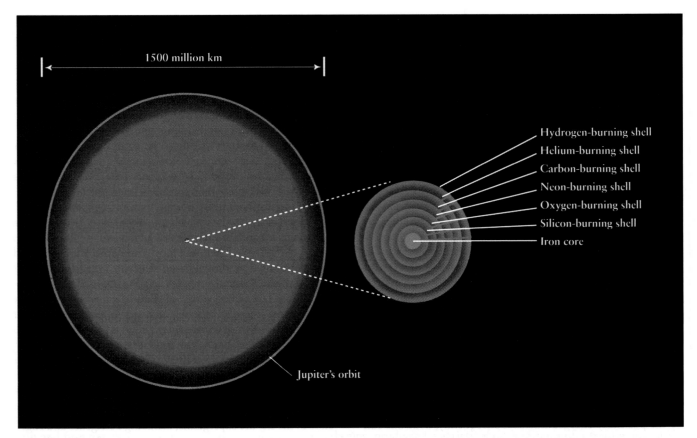

**figure 22-13**

**The Structure of an Old High-Mass Star** Near the end of its life, a high-mass star becomes a supergiant whose overall size can be as large as the orbit of Jupiter around the Sun. The star's energy comes from six concentric burning shells, all contained within a volume roughly the same size as the Earth. Thermonuclear reactions cannot occur within the iron core, because fusion reactions that involve iron absorb energy rather than releasing energy.

## 22-6 High-mass stars die violently by blowing themselves apart in supernova explosions

Our present understanding is that all stars of about 8 $M_\odot$ or less divest most of their mass in the form of planetary nebulae. The burned-out core that remains settles down to become a white dwarf star. But the truly massive stars—stellar heavyweights that begin their lives with more than 8 solar masses of material—do not pass through a planetary nebula phase. Instead, they die in spectacular explosions called *supernovae* (the plural of *supernova*) that can be seen hundreds of millions of light-years away.

To understand what happens in a supernova explosion, we must look deep inside a massive star at the end of its life. Of course, we cannot do this in actuality, because the interi-

ors of stars are opaque. But astronomers have developed theoretical models based on what we know about the behavior of gases and atomic nuclei. The story that follows, while largely theoretical, describes fairly well our observations of supernovae. And as we will see in Section 22-8, a special kind of "telescope" has allowed us to glimpse the interior of at least one famous supernova.

The core of an aging, massive star gets progressively hotter as it contracts to ignite successive stages of thermonuclear burning. As discussed in Sections 5-3, 5-4, and 5-5, temperature governs the properties of radiation emitted by matter. Specifically, Wien's law and Planck's law together tell us that as the temperature of a blackbody increases, so does the energy of the photons it emits. When the temperature in the core of a massive star reaches a few hundred million kelvins, the photons are energetic enough to initiate a host of nuclear reactions that create neutrinos. These neutrinos, which carry off energy, escape from the star's core, just as solar neutrinos flow freely out of the Sun (see Section 18-9 and the essay "Searching for Neutrinos Beyond the Textbooks" at the end of Chapter 18).

To compensate for the energy drained by the neutrinos, the star must provide energy either by burning more thermonuclear fuel, by contracting, or both. But when the star's core is converted into iron (as shown in Figure 22-13), no more energy-producing thermonuclear reactions are possible, and the only source of energy is contraction and rapid heating. Once a star with an original mass of about 8 $M_\odot$ or more develops an iron core—a process that takes only about $10^7$ years or less from the time that the star first arrived on the main sequence—the core contracts very rapidly, so that the core temperature skyrockets to $5 \times 10^9$ K within a tenth of a second. The gamma-ray photons emitted by the intensely hot core have so much energy that when they collide with iron nuclei, they begin to break the nuclei down into alpha particles ($^4$He nuclei). This process is called **photodisintegration.** As shown in Table 22-2, it takes a high-mass star millions of years and several stages of thermonuclear reactions to build up an iron core; within a short period of time, photodisintegration undoes much of those millions of years of effort.

Within another tenth of a second, the core becomes so dense that the negatively charged electrons within the core are forced to combine with the positively charged protons to produce electrically neutral neutrons. This process also releases a flood of neutrinos (denoted by the Greek letter ν, or nu):

$$e^- + p \rightarrow n + \nu$$

These neutrinos carry a substantial amount of energy away from the core, causing the core to cool down and, therefore, to condense even further.

CAUTION! Somewhat confusingly, astronomers also use the Greek letter ν to denote the frequency of a light wave (see Section 5-2). Fortunately, the two meanings of the symbol—one to refer to a kind of particle, as in the above expression, and the other to refer to the properties of a wave—are sufficiently different that you should never see them in the same mathematical expression or the same discussion.

At about 0.25 second after its rapid contraction begins, the core is less than 20 km in diameter and its density is in excess of $4 \times 10^{17}$ kg/m$^3$. This is **nuclear density,** the density with which neutrons and protons are packed together inside nuclei. (If the Earth were compressed to this density, it would be only 300 meters, or 1000 feet, in diameter.) Matter at nuclear density or higher is extraordinarily difficult to compress. Thus, when the density of the neutron-rich core begins to exceed nuclear density, the core suddenly becomes very stiff and rigid. The core's contraction comes to a sudden halt, and the innermost part of the core actually bounces back and expands somewhat. When it does so, it sends a powerful wave of pressure, like an unimaginably intense sound wave, outward into the outer core.

During this critical stage, the cooling of the core has caused the pressure to decrease profoundly in the regions surrounding the core, and the material from these regions is plunging inward at speeds up to 15% of the speed of light. When this inward-moving material crashes down onto the rigid core, it encounters the outward-moving pressure wave. In just a fraction of a second, the material that fell onto the core begins to move back out toward the star's surface, propelled in part by the flood of neutrinos trying to escape from the star's core. This wave accelerates as it encounters less and less resistance and soon reaches a speed greater than the speed of sound waves in the star's outer layers. When this happens, the outward-moving wave becomes a shock wave, like the sonic boom produced by a supersonic airplane. After a few hours, this shock wave reaches the star's surface, by which time the star's outer layers have begun to lift away from the core. The energy released in this titanic event is an incomprehensibly large $10^{46}$ joules—a hundred times more energy than the Sun has emitted over the past 4.6 billion years. When the star's outer layers thin out sufficiently, a portion of this energy escapes in a torrent of light. The star has become a **supernova** (plural **supernovae**).

Supercomputer simulations provide many insights into the violent, complex, and rapidly changing conditions deep inside a star as it is torn apart by a supernova explosion. For example, Figure 22-14 shows the first 20 milliseconds after the stiffening of the inner core. Note the turbulent swirls and eddies that grow behind the shock wave that moves outward from the star's core. Recent research suggests that this turbulence is an important feature of a supernova explosion. Without the turbulence, the shock wave might not be able to plow through the bulk of the doomed star, which continues to fall inward. Turbulence, which involves plumes of matter moving at speeds up to half the speed of light, can revitalize the explosion and ensure that the star blows up.

Figure 22-14 depicts the death of a 25-$M_\odot$ star. (The final stages in the evolution of other massive stars probably follow similar scenarios, although details may differ.) Detailed computer calculations suggest that a 25-$M_\odot$ star ejects about 96% of its material to the interstellar medium for use in producing future generations of stars. (Less massive stars eject proportionally less of their mass into space when they become supernovae.) Before this material is ejected into space, it is compressed so much by the passage of the shock wave through the star's outer layers that a new wave of thermonuclear reactions set in. These reactions produce many more chemical elements, including *all* the elements heavier than iron. These reactions require a tremendous input of energy, and thus cannot take place during the star's pre-supernova lifetime. The energy-rich environment of a supernova shock wave is the *only* place where such heavy elements as zinc, silver, tin, gold, mercury, lead, and uranium can be produced. Remarkably, all of these elements are found on the Earth. This means that our solar system, our Earth, and our bodies include material that long ago was part of a star that lived, evolved, and died as a supernova.

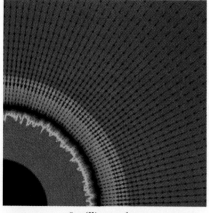

5 milliseconds

10 milliseconds

15 milliseconds

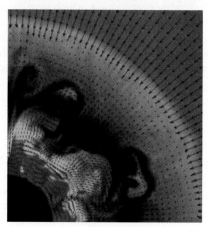

20 milliseconds

### figure 22-14

**The Core of a Supernova** These four images show the core of a massive star during the first 20 milliseconds of a supernova explosion. Color indicates temperatures that range from $2 \times 10^{10}$ K (red) to $6 \times 10^8$ K (blue). Matter that bounces off the stiffened core of the star (colored black) produces a shock wave. Instabilities generate turbulent eddies that grow as the shock wave moves outward. This simulation required 6 hours of computation on a supercomputer. (Courtesy of Adam Burrows, University of Arizona, and Bruce Fryxell, Goddard Space Flight Center)

---

## 22-7 In 1987 a nearby supernova gave us a close-up look at the death of a massive star

Our discussion of supernovae in the previous section was based on theoretical models and supercomputer calculations. But supernovae are not just theory, and several dozen of them are seen each year. For several months after they explode, supernovae have luminosities as great as $10^9$ $L_\odot$, rivaling the light output of an entire galaxy. This makes it possible to see supernovae in other galaxies far beyond our own Milky Way Galaxy. These observations provide a crucial test of the theoretical ideas in Section 22-6.

On February 23, 1987, a supernova was discovered in the Large Magellanic Cloud (LMC), which is a companion galaxy to our Milky Way. The supernova, designated SN 1987A because it was the first discovered that year, occurred near an enormous H II region in the LMC called the Tarantula Nebula because of its spiderlike appearance (Figure 22-15). The supernova was so bright that observers in the southern hemisphere could see it with the naked eye.

A bright supernova is a rare event. In 1885 a supernova in the Andromeda Galaxy (M31), which lies some 770,000 par-

secs ($2.5 \times 10^6$ light-years) from Earth, was just barely visible without a telescope. Before that, the last supernova bright enough to be seen with the naked eye (and the last one to have been observed in the Milky Way Galaxy) was in 1604. Although outside our galaxy, SN 1987A occurred relatively close to us—a mere 51,000 parsecs (165,000 light-years) away—in a part of the heavens that is obscured only slightly by dust within our own galaxy. Furthermore, SN 1987A appeared at a fortuitous time. The light from the supernova reached the Earth in the middle of the southern hemisphere summer, with its dry, clear skies. Not long after the explosion of SN 1987A, new orbiting telescopes were placed into service that enabled astronomers to study the supernova's subsequent evolution with unprecedented resolution and in wavelength ranges not accessible from the Earth's surface (see Section 6-7). As a result, SN 1987A has given astronomers a tremendous amount of data about the violent death of a massive star.

Much has been learned by observing how the brightness of SN 1987A changes with time. The light from a supernova does not all come in a single brief flash; as the outer layers expand into space, they continue to glow. For the first 20 days after the detonation of SN 1987A, the primary energy source for this radiation was the tremendous heat deposited in the star's outer layers by the passage of the shock wave. As the expanding gases cooled, the light energy began to be provided

**figure 22-15** R I **V** U X G

**Supernova 1987A** In 1987 a supernova was discovered in a nearby galaxy called the Large Magellanic Cloud (LMC), which lies some 51,000 parsecs (165,000 light-years) from Earth. This photograph shows a portion of the LMC that includes the supernova and a huge H II region called the Tarantula Nebula (also known as 30 Doradus or NGC 2070). The Tarantula Nebula itself is the largest and most luminous H II region known, and at its present distance, spans 40 × 25 arcminutes, slightly larger than the full moon. If it were placed where the Orion Nebula is (see the figure that opens Chapter 7), a mere 460 parsecs (1500 light-years) from the Earth, it would cover 30% of the sky and be 3 times brighter than the planet Venus. (European Southern Observatory)

by a different source—the decay of radioactive isotopes of cobalt, nickel, and titanium produced in the supernova explosion. Astronomers have been able to pinpoint the specific isotopes involved because different radioactive nuclei emit gamma rays of different wavelengths when they decay. These emissions have been detected by orbiting telescopes, such as the Compton Gamma Ray Observatory (Figure 6-33). Thanks to these radioactive decays, the brightness of the supernova actually increased for the first 85 days after the detonation, then settled into a slow decline as the radioactive isotopes were used up (Figure 22-16). The supernova remained visible to the naked eye for several months after the detonation.

Ideally, SN 1987A would have confirmed the theories of astronomers about typical supernovae. But SN 1987A was *not* typical. Its luminosity peaked at roughly $10^8$ $L_\odot$, only a tenth of the maximum luminosity observed for other, more distant supernovae. Because it was low, the peak luminosity of SN 1987A seemed to disagree with theoretical calculations like the one described in Table 22-2. Fortunately, the doomed star had been observed prior to becoming a supernova, and these observations helped explain why SN 1987A was an exceptional case. The star that exploded and became SN 1987A, called the **progenitor star,** was identified as a B3 I supergiant (see Figure 22-17). When this star was on the main sequence its mass was about 20 $M_\odot$, although by the

time it exploded—some $10^7$ years after it first formed—it probably had shed a few solar masses.

SN 1987A attained a lower luminosity than expected because the progenitor star was a *blue* supergiant when it exploded, rather than a *red* supergiant. The evolutionary track for an aging 20-$M_\odot$ star wanders back and forth across the top of the H-R diagram, and so the star alternates between being a hot (blue) supergiant and a cool (red) supergiant. The star's size changes significantly as the surface temperature changes. A blue supergiant is 10 times larger in diameter than the Sun, but a red supergiant of the same luminosity has a diameter 1000 times that of the Sun. Because the progenitor star was relatively small when it exploded, the star's outer layers were relatively close to the core and thus held more strongly by the core's gravitational attraction. When the detonation occurred, a relatively large fraction of the shock wave's energy had to be used against this gravitational attraction to push the outer layers into space. Hence, the amount of shock wave energy available to be converted into light was smaller than for most supernovae. The result was that SN 1987A reached only a tenth of the brightness of an exploding red supergiant.

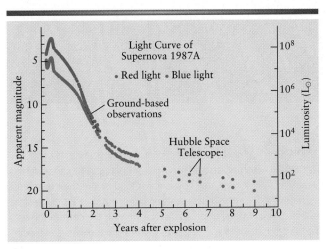

**figure 22-16**

**The Light Curve of SN 1987A** Using ground-based telescopes and the Hubble Space Telescope, astronomers have been able to keep track of changes in the brightness of SN 1987A. This graph of the supernova's brightness as a function of time is called a light curve, the same terminology used for eclipsing binary stars (see Figure 19-25) and for pulsating variable stars (see Figure 21-13a). The sudden rise in luminosity following detonation stalled at only a tenth of the anticipated maximum. The gradual decline that began after day 85 is nevertheless typical of supernovae. The supernova's energy source at these late times is the decay of radioactive nuclei in the expanding debris from the star; as these nuclei are used up, fewer decays occur each second and the luminosity decreases. At present, most of the energy radiated from SN 1987A comes from the slow decay of an isotope of titanium, $^{44}$Ti. This particular isotope will continue to be the main energy source for SN 1987A for decades to come. (Peter Challis and Robert Kirshner, Harvard-Smithsonian Center for Astrophysics)

1994, soon after the optics of the Hubble Space Telescope were repaired (as described in Section 6-7), SN 1987A was observed again and a set of *three* glowing rings was revealed (Figure 22-18*a*). These rings are relics of a hydrogen-rich outer atmosphere that was ejected by gentle stellar winds from the doomed star when it was a red supergiant, about 20,000 years ago. The diffuse gas was subsequently compressed into a narrow, hourglass-shaped shell by high-speed stellar winds that flowed from the star when it became a blue supergiant (Figure 22-18*b*). The gaseous rings seen in Figure 22-18*a* were ionized by the initial flash of ultraviolet radiation from the supernova; as electrons recombine with the ions, the rings emit visible-light photons. (We described this process of *recombination* in Section 20-2.)

At this writing, the shock wave from the supernova is just beginning to reach the inner ring of the hourglass-shaped shell (the "waist" of the hourglass shown in Figure 22-18*b*). Over the next several years, this collision will cause the shell of gas to brighten by a factor of 100 in visible wavelengths—not enough, unfortunately, to make the supernova again visible to the naked eye—and emit copious radiation at X-ray, ultraviolet, and visible wavelengths. By studying the radiation from this event, astronomers hope to learn more about the shock wave, which will provide more information about the supernova explosion. They will also learn about the matter in the inner ring, which will give us insight into the stellar winds that blew thousands of years ago. Because it provides a uniquely convenient laboratory for studying the dynamic evolution of a supernova, SN 1987A will be carefully monitored by astronomers for many years to come.

a

b

## ƒigure 22-17  R I **V** U X G

**SN 1987A—Before and After** The photograph on the top shows a small section of the Large Magellanic Cloud before the explosion of SN 1987A, with the supernova's progenitor star—a B3 blue supergiant—identified by an arrow. The view on the bottom shows the same region of the sky a few days after the supernova exploded into brightness. (Anglo-Australian Observatory)

Calculations suggest that the interior of the doomed star contained 6 $M_\odot$ of helium and 2 $M_\odot$ of metals (elements heavier than helium). At the time of detonation, the star had a 1.5-$M_\odot$ iron core at a temperature of $10^{10}$ K. Since the detonation, the gaseous debris of the supernova has cooled to below 300 K as it has expanded into space. (Box 21-1 gives other examples of how a gas cools as it expands.)

Three and a half years after SN 1987A exploded, astronomers used the newly launched Hubble Space Telescope to obtain a picture of the supernova. To their surprise, the image showed a ring of glowing gas around the exploded star. In

## 22-8  Neutrinos emanate from supernovae like SN 1987A

In addition to electromagnetic radiation, supernovae also emit a brief but intense burst of neutrinos from their collapsing cores. In fact, theoretical studies of supernovae such as SN 1987A show that *most* of the energy released by the exploding star is in the form of neutrinos. If it were possible to detect the flood of neutrinos from an exploding star, astronomers would have direct evidence of the nuclear processes that occur within the star during its final seconds before becoming a supernova.

Unfortunately, detecting neutrinos is a difficult, tricky business. The problem is that under most conditions matter is transparent to neutrinos (see Section 18-9). Consequently, when a massive star explodes as a supernova, the neutrinos that it emits pass easily through any gas, dust, or nebulae they happen to encounter. When supernova neutrinos encounter the Earth, almost all of them pass completely through the planet as if it were not there. The challenge to scientists is to detect the tiny fraction of neutrinos that *do* interact with the matter through which they pass.

a

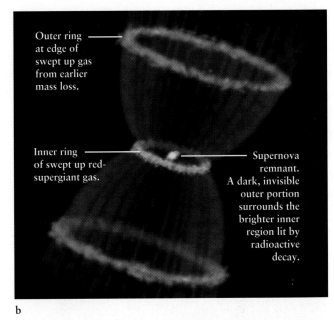

Outer ring at edge of swept up gas from earlier mass loss.

Inner ring of swept up red-supergiant gas.

Supernova remnant. A dark, invisible outer portion surrounds the brighter inner region lit by radioactive decay.

b

## Figure 22-18   R I **V** U X G

**SN 1987A and Its "Three-Ring Circus"**  **(a)** This true-color view of SN 1987A, assembled from four different Hubble Space Telescope images, clearly shows three bright rings around the supernova (the bright dot at the center of the rings). The inner ring has an angular size of just 1.7 arcseconds, corresponding to an actual diameter of about 1.3 light-years (0.4 parsec); only the Hubble Space Telescope has the resolution to detect structures of such small angular size. The two stars lying near the outer rings are also in the Large Magellanic Cloud but are not part of the supernova. **(b)** This drawing shows the supernova and its environment from a perspective not available from the Earth.

When the progenitor star of SN 1987A was still a red supergiant, a slow-moving wind from the star filled the surrounding space with a thin gas. When the star contracted into a blue supergiant, it produced a faster-moving stellar wind. The interaction between the fast and slow winds somehow caused gases to pile up along an hourglass-shaped shell surrounding the star. The burst of ultraviolet light from the supernova detonation ionized the gas in the rings, causing them to glow. The supernova itself, at the center of the hourglass, glows because of energy released from radioactive decay. (Robert Kirshner and Peter Challis, Harvard-Smithsonian Center for Astrophysics; STScl)

During the 1980s two "neutrino telescopes" uniquely suited for detecting supernova neutrinos went into operation—the Kamiokande detector in Kamioka, Japan (a joint project of the University of Tokyo and the University of Pennsylvania), and the IMB detector (an acronym for the University of California, Irvine, the University of Michigan, and Brookhaven National Laboratory, which collaborated in the development and operation of the detector). Both detectors consisted of large tanks containing thousands of tons of water. A brief flash of light is produced on the rare occasions when a neutrino collides with one of the water molecules. The ultrapure water used in these detectors was very transparent to light, which allowed any light flash to be recorded by photomultiplier tubes lining the walls of the tank (Figure 22-19). Because other types of subatomic particles besides neutrinos could also produce similar light flashes, the detectors were placed deep underground so that hundreds of meters of earth would screen out almost all particles except neutrinos.

A single neutrino from a supernova carries a relatively large amount of energy, typically 20 MeV or more. (We introduced the unit of energy called the electron volt, or eV, in Section 5-5. One MeV is equal to $10^6$ electron volts. Only the most energetic nuclear reactions produce particles with energies of more than 1 MeV.) If such a high-energy neutrino hits a proton in the water-filled tank of a neutrino telescope, the collision produces a positron. The positron then recoils at a speed greater than the speed of light in water, which is $2.3 \times 10^8$ m/s. (Such motion does not violate the ultimate speed limit in the universe, $3 \times 10^8$ m/s, which is the speed of light in a *vacuum*.) Just as an airplane that flies faster than sound produces a shock wave that we call a sonic boom, a positron that moves through a substance such as water faster than the speed of light in that substance produces a shock wave of light. This type of radiation is called **Cerenkov radiation** after the Russian physicist Pavel A. Cerenkov, who first observed it in 1934. It is this radiation that is detected by the photomultiplier tubes that line the

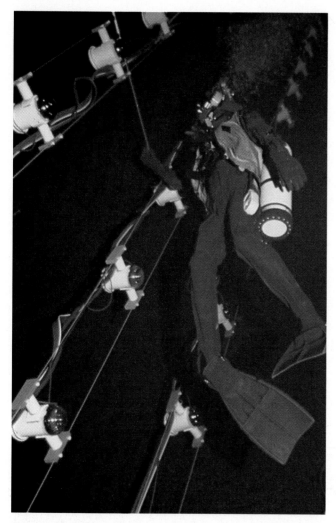

**figure 22-19**

**Inside a Neutrino Detector** A diver is shown servicing one of the light-sensitive photomultiplier tubes lining the walls of the Irvine-Michigan-Brookhaven (IMB) neutrino detector. Eight neutrinos from SN 1987A were detected by this apparatus when they struck protons in the water, producing brief flashes of light that were picked up by the photomultipliers. (Courtesy of K. S. Luttrell)

detector walls. By measuring the properties of the Cerenkov radiation from a recoiling positron, scientists can determine the positron's energy and, therefore, the energy of the neutrino that created the positron. This allows them to tell the difference between the high-energy neutrinos from supernovae and neutrinos from the Sun, which typically have energies of 1 MeV or less.

By happy coincidence, both the Kamiokande and IMB detectors were operational on February 23, 1987, when SN 1987A exploded in the Large Magellanic Cloud. Soon after the explosion, the physicists working with these detectors

excitedly reported that they had detected Cerenkov flashes from a 12-second burst of neutrinos that reached the Earth 3 hours before astronomers saw the light from the exploding star. The total number of neutrinos seen was quite small. The Kamiokande detector saw flashes from 12 neutrinos at about the same time that 8 were detected by the IMB detector. But when the physicists factored in the sensitivity of their detectors to neutrinos, they calculated that their detectors had actually been exposed to a torrent of more than $10^{16}$ neutrinos.

Given the flux of neutrinos measured by the detectors and the 165,000–light-year distance from SN 1987A to the Earth, physicists used the inverse-square law to determine the total number of neutrinos that had been emitted from the supernova. (This law applies to neutrinos just as it does to electromagnetic radiation; see Section 19-2.) They found that over a 10-second period, SN 1987A emitted $10^{58}$ neutrinos with a total energy of $10^{46}$ joules. This is more than 100 times as much energy as the Sun has emitted in its entire history and more than 100 times the amount of energy that the supernova emitted in the form of electromagnetic radiation. So, as impressive as SN 1987A was in its light output, its energy output in the form of neutrinos was more impressive by far. Indeed, for a few seconds the supernova's neutrino luminosity—that is, the *rate* at which it emitted energy in the form of neutrinos—was 10 times greater than the total luminosity in electromagnetic radiation of all of the stars in the observable universe! Such comparisons give a hint of the incomprehensible violence with which a supernova explodes.

Why did the neutrinos from SN 1987A arrive 3 hours *before* the first light was seen? As discussed in Section 22-6, neutrinos are produced when thermonuclear reactions cease in the core of a massive star and the core collapses. These neutrinos encounter little delay as they pass through the volume of the star. The massive increase in the star's output of light, by contrast, only occurs when the shock wave reaches the star's outermost layers (which are thin enough to allow light to pass through them, just like the Sun's photosphere described in Section 18-1). It took 3 hours for this shock wave to travel outward from the star's core to its surface, by which time the neutrino burst was already billions of kilometers beyond the dying star. For the next 165,000 years, the neutrinos that would eventually produce light flashes in Kamiokande and IMB remained in front of the photons emitted from the star's surface, and so the neutrinos were detected before the supernova's light. Thus, the neutrino data from SN 1987A has given astronomers a look at processes taking place within a supernova and has provided direct confirmation of theoretical ideas about how supernova explosions take place.

The data from SN 1987A also provided a clue about a long-standing puzzle: What is the mass of the neutrino? As we discussed in Section 18-9, there is some suspicion that neutrinos have a small mass (unlike the photon, which has no mass at all). If neutrinos were massless, they would travel at the speed of light, and all the supernova's neutrinos would arrive at the Earth at essentially the same moment. The more massive the neutrino is, however, the slower the neutrinos can travel and the greater the spread in their arrival times. The interval over which the bursts were observed suggests that the mass of a neutrino is no more than 16 eV. For comparison, the mass of an electron is 511,000 eV. Astronomers use a good deal of physics in their study of the universe; in the case of the neutrino observations of SN 1987A, astronomy was able to give back some key information to physicists.

Both the Kamiokande and IMB detectors have been replaced by a new generation of neutrino telescopes, as described in Section 18-9. While these new detectors are intended primarily to observe neutrinos from the Sun, they are also fully capable of measuring neutrino bursts from nearby supernovae. Astronomers have identified a number of supergiant stars in our Galaxy that are likely to explode into supernovae, among them the bright red supergiant Betelgeuse in the constellation Orion (see Figure 2-2 and Figure 6-29).

Unfortunately, astronomers do not yet know how to predict precisely when such stars will explode into supernovae, and it may be many thousands of years before Betelgeuse explodes. Then again, it could happen tomorrow. If it does, the neutrino telescopes will be ready to record the collapse of its massive core.

## 22-9 White dwarfs in close binary systems can also become supernovae

Astronomers discover dozens of supernovae in distant galaxies every year (Figure 22-20), but not all of these are the result of massive stars dying violently. A totally different, but no less luminous, type of supernova occurs when a white dwarf star in a binary system blows itself completely apart. The first clue that two entirely distinct chains of events could produce supernovae was a rather subtle one. Some supernovae have prominent hydrogen emission lines in their spectra but others do not. These different types of supernova spectra are shown in Figure 22-21.

Supernovae with hydrogen emission lines, called **Type II supernovae**, are the sort described in Section 22-6. They are caused by the death of highly evolved massive stars that still have ample hydrogen in their atmospheres when they explode. When the star explodes, the hydrogen atoms are excited and glow prominently, producing hydrogen emission lines (Figure 22-21*d*). SN 1987A, described in Sections 22-7 and 22-8, was an example of a Type II supernova.

a

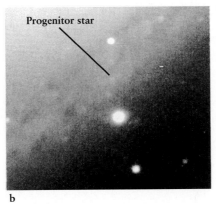

Progenitor star

b

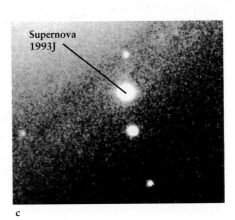

Supernova 1993J

c

### ƒigure 22-20 R I **V** U X G

**A Supernova in a Distant Galaxy** On the night of March 28, 1993, Francisco Garcia Diaz, a Spanish amateur astronomer, discovered a supernova in the galaxy M81 in Ursa Major. This galaxy lies some 3.6 million parsecs (12 million light-years) from Earth. The supernova was the tenth seen in 1993 and hence was named SN 1993J. **(a)** This photograph shows the appearance of M81 before the supernova explosion. The galaxy has an angular size of 10 × 21 arcminutes, roughly half that of the full moon. The supernova occurred within the region outlined by the yellow box; this region is shown in close-up in parts (b) and (c). **(b)** This close-up image made before the supernova explosion shows the progenitor star of SN 1993J, a K0 I red supergiant. **(c)** This image shows the same part of the sky as (b), but after the explosion of SN 1993J. This supernova, like SN 1987A, resulted from the explosion of a massive star after thermonuclear reactions ceased in the star's core. (Image a, Palomar Observatory; b and c, D. Jones and E. Telles, Isaac Newton Telescope)

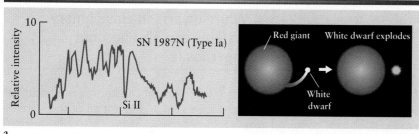

a

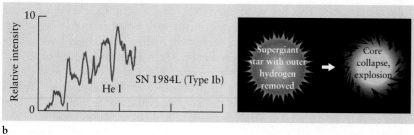

b

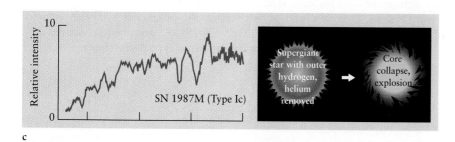

c

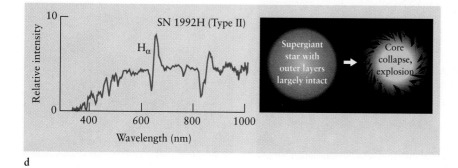

d

## ƒigure 22-21

**Supernova Types** Different types of supernovae have noticeably different spectra when they first erupt into brilliance. **(a)** The spectra of Type Ia supernovae have no hydrogen or helium lines. They do, however, have a prominent absorption line of ionized silicon (Si II). These supernovae are thought to be produced by the catastrophic explosion of a white dwarf star in a close binary system. This is caused by runaway carbon burning within the dwarf, triggered by mass transfer from the companion star. The silicon line comes from one of the by-products of carbon burning. **(b)** Type Ib supernovae lack hydrogen in their spectra, but have a strong absorption line of unionized helium (He I). These supernovae may result from core collapse in a massive star that lost the hydrogen from its outer layers. **(c)** Type Ic supernovae lack both hydrogen and helium lines in their spectra. These are also thought to be massive stars that undergo core collapse but lost both hydrogen and helium from their outer layers prior to the supernova explosion. **(d)** Type II supernovae, like their cousins Type Ib and Type Ic, result from core collapse within a massive star. In this case, however, enough of the hydrogen from the outer layers remained with the star before exploding to produce strong hydrogen lines (such as $H_\alpha$) in the supernova's spectrum. (Spectra courtesy Alexei V. Filippenko, University of California, Berkeley).

Hydrogen lines are missing in the spectrum of a **Type I supernova,** which tells us that little or no hydrogen is left in the debris from the explosion. Type I supernovae are further divided into three important subclasses. **Type Ia** supernovae have spectra that include a strong absorption line of ionized silicon. **Type Ib** and **Type Ic** supernovae both lack the ionized silicon line. The difference between them is that the spectra of Type Ib supernovae have a strong helium absorption line, while those of Type Ic supernovae do not.

Astronomers suspect that Type Ib and Ic supernovae are caused by dying massive stars, just like Type II supernovae. The difference is that the progenitor stars of Type Ib and Ic supernovae have been stripped of their outer layers before they explode. A star can lose its outer layers to a strong stellar wind (recall the photograph that opens Chapter 21) or, if it is part of a close binary system, by transferring mass to its companion star (see Figure 21-17b). If enough mass remains for the star's core to collapse, the star dies as a Type Ib supernova. Because the outer layers of hydrogen are absent, the supernova's spectrum exhibits no hydrogen lines but many helium lines (Figure 22-21b). Type Ic supernovae have apparently undergone even more mass loss prior to their explosion; their spectra show that they have lost much of their helium as well as their hydrogen (Figure 22-21c).

Evidence that Type II, Type Ib, and Type Ic supernovae all begin as massive stars is that all three types are found only near sites of recent star formation. The life span of a massive star from its formation to its explosive death as a supernova is only about $10^7$ years, less time than it took our own Sun to condense from a protostar to a main-sequence star (see Figure 20-9) and a mere blink of an eye on the time scale of stellar evolution. Because massive stars are so very short-lived, it makes sense that these stars should meet their demise very close to where they were formed.

Type Ia supernovae, by contrast, are found even in galaxies where there is no ongoing star formation. Hence, they are probably *not* the death throes of massive supergiant stars. Instead, Type Ia supernovae may result from the thermonuclear explosion of a white dwarf star. This may seem like an odd notion, because we learned in Section 22-4 that white dwarf stars have no thermonuclear reactions going on in their interiors. But these reactions *can* occur if a carbon-oxygen–rich white dwarf is in a close, semidetached binary system with a red giant star (see Figure 21-16b).

As the red giant evolves and its outer layers expand, it overflows its Roche lobe and dumps gas from its outer layers onto the white dwarf (Figure 22-21a). When the total mass of the white dwarf approaches the Chandrasekhar limit, the increased pressure applied to the white dwarf's interior causes carbon burning to begin there. Hence, the interior temperature of the white dwarf increases. If the white dwarf were made of ordinary matter, the temperature increase would cause a further increase in pressure, the white dwarf would expand and cool, and the carbon-burning reactions would abate. But because the white dwarf is composed of degenerate matter, this "safety valve" between temperature and pressure does not operate. Instead, the increased temperature just makes the reactions proceed at an ever-increasing rate, in a catastrophic runaway process reminiscent of the helium flash in low-mass stars. Soon the temperature gets so high that the electrons are no longer degenerate and the white dwarf blows apart, dispersing all of its mass into space.

Before exploding, the white dwarf contained primarily carbon and oxygen and almost no hydrogen or helium, which explains the absence of hydrogen and helium lines in the spectrum of the resulting supernova. Silicon is a by-product of the carbon-burning reaction and gives rise to the silicon absorption line characteristic of Type Ia supernovae.

CAUTION! It is worth emphasizing that different types of supernovae have fundamentally different energy sources. Type II, Ib, and Ic supernovae are powered by *gravitational* energy. The star's iron-rich core collapses inward under the influence of its own gravity, releasing a flood of energy in the form of neutrinos; the outer layers of the star also fall inward under gravity, providing the energy to power the nuclear reactions that generate the supernova's electromagnetic radiation. Type Ia supernovae, by contrast, are powered by *nuclear* energy released in the explosive thermonuclear burning of a white dwarf star. While Type Ia supernovae typically emit more energy in the form of electromagnetic radiation than supernovae of other types, they do not emit copious numbers of neutrinos because there is no core collapse. If we include the energy emitted in the form of neutrinos, the most luminous supernovae by far are those of Type II.

In addition to the differences in their spectra, different types of supernovae can be distinguished by their light curves (Figure 22-22). All supernovae begin with a sudden rise in brightness that occurs in less than a day. After reaching peak luminosity, Type Ia, Ib, and Ic supernovae settle into a steady, gradual decline in luminosity. By contrast, the Type II light curve has a steplike appearance caused by alternating periods of steep and gradual declines in brightness. For all supernova types, the energy source during this period of decline is the decay of radioactive isotopes produced during the supernova explosion. Because a different set of thermonuclear reactions occur for each type of supernova, each type produces a unique set of isotopes that decay at different rates. This helps explains the distinctive light curves for different supernova types.

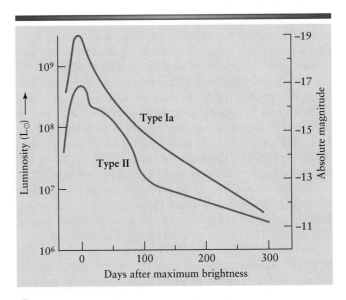

**figure 22-22**

**Supernova Light Curves** A Type Ia supernova reaches maximum brightness in about a day, followed by a gradual decline in brightness. Such supernovae are caused by an exploding white dwarf in a close binary system. A Type II supernova reaches a maximum brightness only about one-fourth that of a Type Ia supernova and usually has alternating intervals of steep and gradual decline. Type II supernovae are caused by the death of massive stars. For all supernovae, the energy source during the period of declining brightness is the decay of radioactive isotopes created in the supernova explosion.

For the same reason, a somewhat different mix of elements is ejected into the interstellar medium by different types of supernovae. As an example, Type Ia supernovae are primarily responsible for the elements near iron in the periodic table, because they generate these elements in more copious quantities than Type II supernovae.

A number of astronomers are now measuring the distances to remote galaxies by looking for Type Ia supernovae in those galaxies. Because all such supernovae have about the same maximum luminosity, a measurement of a Type Ia supernova's peak apparent brightness tells us (through the inverse-square law) the distance to the supernova, and, therefore, the distance to the supernova's host galaxy. The tremendous luminosity of Type Ia supernovae allows this method to be used for galaxies more than $10^9$ light-years distant. In Chapter 28 we will learn what such studies tell us about the size and evolution of the universe as a whole.

## 22-10　A supernova remnant can be detected at many wavelengths for centuries after the explosion

Astronomers find the debris of supernova explosions, called **supernova remnants,** scattered across the sky. A beautiful example is the Veil Nebula, seen in Figure 22-23. The doomed star's outer layers were blasted into space with such violence that they are still traveling through the interstellar medium at supersonic speeds. As this expanding shell of gas plows through space, it collides with atoms in the interstellar medium, exciting this sparse gas and causing it to glow.

A few nearby supernova remnants cover sizable areas of the sky. The largest is the Gum Nebula, named after astronomer Colin Gum, who first noticed its faint glowing wisps on photographs of the southern sky (Figure 22-24). Its 60° angular diameter is centered on the constellation Vela.

The Gum Nebula looks big simply because it is quite close to us. Its near side is only about 100 parsecs (330 light-years) from the Earth, and the center of the nebulosity is just 460 parsecs (1300 light-years) away. Studies of the nebula's expansion rate suggest that this supernova exploded around 9000 B.C. At maximum brilliance, the exploding star probably was as bright as the Moon at first quarter. Like the first quarter Moon, it would have been visible in the daytime!

Many supernova remnants are virtually invisible at optical wavelengths. However, when the expanding gases collide with the interstellar medium, they radiate energy at a wide range of wavelengths, from X rays through radio waves. For example, Figure 22-25 shows both X-ray and radio images of the supernova remnant Cassiopeia A. Optical photographs of this part of the sky reveal only a few small, faint wisps. In fact, radio searches for supernova remnants are more fruitful than optical searches. Only two dozen supernova remnants have been found on photographic plates, but more than 100 remnants have been discovered by radio astronomers.

**Figure 22-23**　R I **V** U X G

**The Veil Nebula—A Supernova Remnant** This nebulosity is a portion of the Cygnus Loop (see Figure 20-23), which is the remnant of a supernova that exploded about 15,000 years ago. The distance to the nebula is about 800 parsecs (2600 light-years), and the overall diameter of the loop is about 35 parsecs (120 light-years). (Palomar Observatory)

From the expansion rate of the nebulosity in Cassiopeia A, astronomers conclude that this supernova explosion occurred about 300 years ago. Although telescopes were in wide use by the late 1600s, no one saw the outburst (and no one today knows why). In fact, the last supernova seen in our Galaxy, which occurred in 1604, was observed by Johannes Kepler, and in 1572, Tycho Brahe also recorded the sudden appearance of an exceptionally bright star in the sky. To find any other accounts of nearby bright supernovae, we must delve into astronomical records that are almost 1000 years old.

At first glance, this apparent lack of nearby supernovae may seem puzzling. From the frequency with which supernovae occur in distant galaxies, it is reasonable to suppose that a galaxy such as our own should have as many as five supernovae per century. Where have they been?

As we will learn from the study of galaxies in Chapters 25 and 26, the plane of our Galaxy is where massive stars are born and supernovae explode. This region is so rich in interstellar dust, however, that we simply cannot see very far into space in the directions occupied by the Milky Way (see Section 20-2). In other words, supernovae probably do in fact erupt every few decades in remote parts of our Galaxy, but

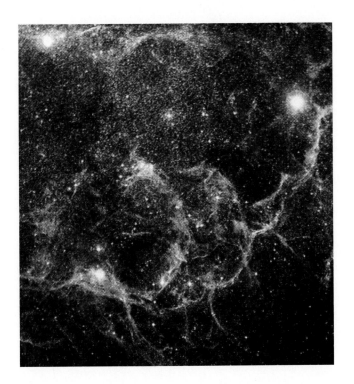

ƒigure 22-24  R I **V** U X G

**The Gum Nebula—A Supernova Remnant**  The Gum Nebula, which spans 60° of the sky, has the largest angular size of any known supernova remnant. Only the central regions of the nebula are shown here. The supernova explosion occurred about 11,000 years ago, and the supernova remnant now has a diameter of about 700 parsecs (3200 light-years). (Royal Observatory, Edinburgh)

their detonations are hidden from our view by intervening interstellar matter.

For some supernovae, a supernova remnant may be all that is left when the explosion is over. But for supernovae of Types II, Ib, and Ic, which are caused by core collapse in a massive star, the core itself may also remain. If there is a relic of the core, it may be either a **neutron star** or a **black hole,** depending on the mass of the core and the conditions within it during the collapse. Neutron stars, as the name suggests, are made primarily of neutrons. Wholly unlike anything we have studied so far, these exotic objects are the subject of Chapter 23. We will study black holes, which are far stranger even than neutron stars, in Chapter 24.

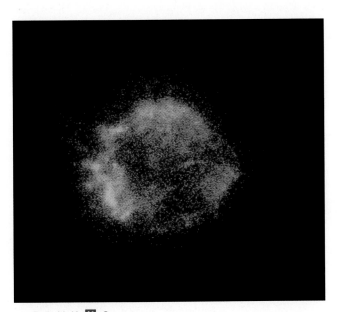

a  R I V U **X** G

b  **R** I V U X G

ƒigure 22-25

**Cassiopeia A—A Supernova Remnant**  Some supernova remnants, such as Cassiopeia A, are strong sources of X rays and radio waves. **(a)** This X-ray image of Cassiopeia A was taken by the Einstein Observatory. As matter streams at high speed away from the site of the supernova explosion (which occurred at the center of the region seen here), it collides with the interstellar medium and heats it to such high temperatures that it emits X rays. **(b)** This radio image of Cassiopeia A was produced by the Very Large Array. The impact of supernova material ionizes the interstellar medium, and the liberated electrons generate radio waves as they move. The supernova explosion that produced this nebula occurred 300 years ago, 3000 parsecs (10,000 light-years) from the Earth. (Smithsonian Institution and the Very Large Array)

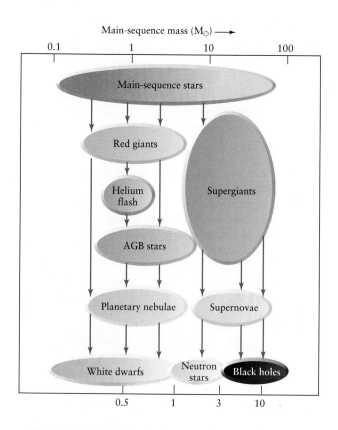

Main-sequence mass ($M_\odot$) ⟶

## figure 22-26

**Pathways of Stellar Evolution** This flow chart shows how the evolution of an isolated star (one that is not part of a close multiple-star system) depends on the star's mass. The scale on the top gives the mass of the star when it is on the main sequence, and the scale on the bottom shows the mass of the resulting stellar corpse. Stars whose original masses are less than about 4 $M_\odot$ can eject enough mass over their lifetimes so that their corpses can become white dwarfs (which requires that the corpse be less massive than the Chandrasekhar limit of 1.4 $M_\odot$). High-mass stars can evolve into supernovae and become neutron stars or black holes. It also possible that a high-mass star can leave no corpse at all. (The supernovae referred to in this chart are Type II. Type Ia supernovae happen only in binary star systems and occur in a very different way.)

 Although neutron stars and black holes can be part of the debris from a supernova explosion, they are *not* called "supernova remnants." That term is applied exclusively to the gas that spreads away from the site of the supernova explosion.

We have seen in this chapter that only the most massive stars end their lives as a supernova of Types II, Ib, or Ic. We learned earlier that mass plays a central role in determining the speed with which a star forms and joins the main sequence (see Figure 20-9), the star's luminosity and surface temperature while on the main sequence (see Figure 19-21 and Figure 19-22), and how long a star can remain on the main sequence (see Table 21-1). Now we see that mass also determines the eventual fate of a star—to end in the whimper of a planetary nebula, as do low-mass stars like the Sun, or to follow the evolutionary path of a high-mass star and end with an almost inconceivably intense bang (Figure 22-26).

## Key Words

## Key Ideas

**Late Evolution of Low-Mass Stars:** A low-mass star becomes a red giant when shell hydrogen burning begins, a horizontal-branch star when core helium burning begins, and an asymptotic giant branch (AGB) star when the helium in the core is exhausted and shell helium burning begins.

• As a low-mass star ages, convection occurs over a larger portion of its volume. This takes heavy elements formed in the star's interior and distributes them throughout the star.

**Planetary Nebulae and White Dwarfs:** Helium shell flashes in an old, low-mass star produce thermal pulses during which more than half the star's mass may be ejected into space. This exposes the hot carbon-oxygen core of the star.

• Ultraviolet radiation from the exposed core ionizes and excites the ejected gases, producing a planetary nebula.

• No further nuclear reactions take place within the exposed core. Instead, it becomes a degenerate, dense sphere about the size of the Earth and called a white dwarf. It glows from thermal radiation; as the sphere cools, it becomes dimmer.

**Late Evolution of High-Mass Stars:** Unlike a low-mass star, a high-mass star undergoes an extended sequence of thermonuclear reactions in its core and shells. These include carbon burning, neon burning, oxygen burning, and silicon burning.

• In the last stages of its life, a high-mass star has an iron-rich core surrounded by concentric shells hosting the various thermonuclear reactions. The sequence of thermonuclear reactions stops here, because the formation of elements heavier than iron requires an input of energy rather than causing energy to be released.

**The Deaths of Massive Stars:** A high-mass star dies in a violent cataclysm in which its core collapses and most of its matter is ejected into space at high speeds. The luminosity of the star increases suddenly by a factor of around $10^8$ during this explosion, producing a supernova.

• More than 99% of the energy from such a supernova is emitted in the form of neutrinos from the collapsing core.

• The matter ejected from the supernova, moving at supersonic speeds through interstellar gases and dust, glows as a nebula called a supernova remnant.

**Other Types of Supernovae:** An accreting white dwarf in a close binary system can also become a supernova when carbon burning ignites explosively throughout such a degenerate star.

• Type Ia supernovae are those produced by accreting white dwarfs in close binaries. Type II supernovae are created by the deaths of massive stars, as are supernovae of Type Ib and Type Ic; these latter types occur when the star has lost a substantial part of its outer layers before exploding.

• Most supernovae occurring in our Galaxy are hidden from our view by interstellar dust and gases.

## Review Questions

1. What is the difference between a red giant and a red supergiant?

2. What is the asymptotic red giant branch? Where is it located on a H-R diagram?

3. How is a planetary nebula formed?

4. What is a white dwarf? Does it produce light in the same way as a star like the Sun?

5. What is the significance of the Chandrasekhar limit?

6. How does the radius of a white dwarf depend on its mass?

7. On a H-R diagram, sketch the evolutionary track that the Sun will follow from when it leaves the main sequence to when it becomes a white dwarf. Approximately how much mass will the Sun have when it becomes a white dwarf? Where will the rest of the mass go?

8. Why do you suppose that all the white dwarfs known to astronomers are relatively close to the Sun?

9. What prevents thermonuclear reactions from occurring at the center of a white dwarf? If no thermonuclear reactions are occurring in its core, why doesn't the star collapse?

10. Why does the mass of a star play such an important role in determining the star's evolution?

11. Why is the temperature in a star's core so important in determining which nuclear reactions can occur there?

12. What is the difference between Type Ia, Type Ib, Type Ic, and Type II supernovae?

13. Why have radio searches for supernova remnants been more fruitful than optical searches?

14. Is our own Sun likely to become a supernova? Why or why not?

## Advanced Questions

*Questions preceded by an asterisk (*) involve topics discussed in the Boxes in Chapter 7 and Chapter 19.*

**Problem-solving tips and tools:**

You may find it useful to review Box 19-5, which discusses stellar radii and their relationship to temperature and luminosity. The formula for gravitational force was given in Section 4-7, and that for escape speed was given in Box 7-2. We discussed the relationship between luminosity, apparent brightness, and distance in Box 19-2. The relationship between absolute magnitude, apparent magnitude, and distance was the topic of Box 19-3.

**15.** The central star in a newly formed planetary nebula has a luminosity of 1000 $L_\odot$ and a surface temperature of 100,000 K. What is the star's radius? Give your answer as a multiple of the Sun's radius.

**16.** You want to determine the age of a planetary nebula. What observations should you make, and how would you use the resulting data?

**17.** The Ring Nebula in the constellation Lyra is expanding at the rate of about 20 km/s. Approximately how long ago did the central star shed its outer layers? Use the data in Table 22-1 and assume that the nebula is 2,700 light-years from the Earth.

**18.** (a) Find the average density of a 1-$M_\odot$ white dwarf having the same diameter as the Earth. (b) What speed is required to eject gas from the white dwarf's surface? (This is also the speed with which interstellar gas falling from a great distance would strike the star's surface.)

**19.** What kinds of stars would you monitor if you wished to observe a supernova explosion from its very beginning? Look up lists of the brightest and nearest stars. Which, if any, of these stars are possible supernova candidates? Explain.

**20.** The shock wave that traveled through the progenitor star of SN 1987A took 3 hours to reach the star's surface. (a) Given the size of a blue supergiant star (see Section 22-7), estimate the speed with which the shock wave traveled through the star's outer layers. (The core of the progenitor star was very small, so you may consider the shock wave to have started at the very center of the star.) Give your answer in meters per second. (b) Compare your answer to the speed of sound waves in our atmosphere, about 340 m/s, and to the speed of light. (c) A shock wave traveling through a gas is a special case of a sound wave. In general, sound waves travel faster through denser, less easily compressed materials. Thus, sound travels faster through water (about 1500 m/s) than through our atmosphere, and faster still through steel (about 5900 m/s). Use this idea to compare the gases within the progenitor star of SN 1987A to the gases in our atmosphere in terms of their average density and how easily they are compressed.

**\*21.** Compared to SN 1987A (Figure 22-15), the supernova SN 1993J (Figure 22-20) had a maximum apparent brightness only $9.1 \times 10^{-4}$ as great. Using the distances from Earth to each of these supernovae, determine the ratio of the maximum luminosity of SN 1993J to that of SN 1987A. Which of the two supernovae had the greater maximum luminosity?

**\*22.** Consider a high-mass star just prior to a supernova explosion, with a core of diameter 20 km and density $4 \times 10^{17}$ kg/m$^3$. (a) Calculate the mass of the core. Give your answer in kilograms and in solar masses. (b) Calculate the force of gravity on a 1-kg object at the surface of the core. How many times larger is this than the gravitational force on such an object at the surface of the Earth, about

10 newtons? (c) Calculate the escape speed from the surface of the star's core. Give your answer in m/s and as a fraction of the speed of light. What does this tell you about how powerful a supernova explosion must be in order to blow material away from the star's core?

**23.** The neutrinos from SN 1987A arrived 3 hours before the visible light. While they were en route to the Earth, what was the distance between the neutrinos and the first photons from SN 1987A? Assume that neutrinos are massless and thus travel at the speed of light. Give your answers in kilometers and in AU.

**\*24.** Suppose that the brightness of a star becoming a supernova increases by 20 magnitudes. Show that this corresponds to an increase of $10^8$ in luminosity.

**\*25.** Suppose that the red supergiant star Betelgeuse, which lies some 1400 light-years from the Earth, becomes a Type II supernova. (a) At the height of the outburst, how bright would it appear in the sky? Give your answer as a fraction of the brightness of the Sun ($b_\odot$). (b) How would it compare with the brightness of Venus (about $10^{-9}$ $b_\odot$)?

**\*26.** In July 1997, a supernova named SN 1997cw explod-ed in the galaxy NGC 105 in the constellation Cetus (the Whale). It reached an apparent magnitude of +16.5 at maximum brilliance, and its spectrum showed an absorption line of ionized silicon. Use this information to find the distance to NGC 105. (*Hint:* Inspect the light curves in Figure 22-22 to find the *absolute* magnitudes of typical supernovae at peak brightness.)

**27.** Consult the World Wide Web or recent issues of *Sky & Telescope* and *Astronomy* to learn about the latest observations of SN 1993J. Has the shape of the supernova's light curve been adequately explained? Has the supernova produced any surprises?

**28.** Search the World Wide Web for information about SN 1994I, a supernova that occurred in the galaxy M51 (NGC 5194). Why was this supernova unusual? Was it bright enough to have been seen by amateur astronomers?

**29.** Figure 22-23 shows a portion of the Veil Nebula in Cygnus. Use the information given in the caption to find the average speed at which material has been moving away from the site of the supernova explosion over the past 15,000 years. Express your answer in km/s and as a fraction of the speed of light.

## Discussion Questions

**30.** Suppose that you discover a small, glowing disk of light while searching the sky with a telescope. How would you decide if this object is a planetary nebula? Could your object be something else? Explain.

**31.** Speculate about the connection between carbon stars and life on the Earth. How might the origin and evolution of life have been affected if the convective zone in AGB stars did *not* reach all the way down into their carbon-rich cores?

**32.** SN 1987A did not agree with the theoretical picture outlined in Section 22-6. Does this mean that the theory was wrong? Discuss.

# Observing Projects

### Observing tips and tools:

While planetary nebulae are rather bright objects, their brightness is spread over a relatively large angular size, which can make seeing them a challenge for the beginning observer. For example, the Helix Nebula (Figure 22-6) has the largest angular size of any planetary nebula but is also one of the most difficult planetary nebulae to see. To improve your view, make your observations on a dark, moonless night from a location well shielded from city lights. Another useful trick, mentioned in Chapters 19 and 20, is to use "averted vision." Once you have the nebula centered in the telescope, you will get a brighter and clearer image if you look at the nebula out of the corner of your eye. The so-called Blinking Planetary in Cygnus (see Table 22-1) affords an excellent demonstration of this effect; the nebula seems to disappear when you look straight at it, but it reappears as soon as you look toward the side of your field of view. Another useful tip is to view the nebula through a green filter (a #58, or O III, filter available from telescope supply houses). Green light is emitted by excited, doubly ionized oxygen atoms, which are common in planetary nebulae but not in most other celestial objects. Using such a filter can make a planetary nebula stand out more distinctly against the sky. As a side benefit, it also helps to block out stray light from street lamps. The same tips also apply to observing supernova remnants.

**33.** Although they represent a fleeting stage at the end of a star's life, planetary nebulae are found all across the sky. Some of the brightest are listed in Table 22-1. Observe as many of these planetary nebulae as you can on a clear, moonless night using the largest telescope at your disposal. Note and compare the various shapes of the different nebulae. In how many cases can you see the central star? The central star in the Eskimo Nebula is supposed to be the "nose" of an Eskimo wearing a parka. Can you see this pattern?

**34.** Two supernova remnants can be seen through modest telescopes, one in the winter sky and the other in the summer sky. Both are quite faint, however, so you should schedule your observations for a moonless night. The winter sky contains the Crab Nebula, which is discussed in detail in Chapter 23. The coordinates are R.A. = $5^h$ $34.5^m$ and Decl. = 22° 00′, which places the object near the star marking the eastern horn of Taurus (the Bull). Whereas the entire Crab Nebula easily fits in the field of view of an eyepiece, the Veil or Cirrus Nebula in the summer sky is so vast that you can see only a small fraction of it at a time. The easiest way to find the Veil Nebula is to aim the telescope at the star 52 Cygni (R.A. = $20^h$ $45.7^m$ and Decl. = +30° 43′), which lies on one of the brightest portions of the nebula. If you then move the telescope slightly north or south until 52 Cygni is just out of the field of view, you should see faint wisps of glowing gas. (52 Cygni is the bright star toward the lower right corner of Figure 20-23, which shows the entire Cygnus Loop.)

**35.** Use a telescope to observe the remarkable triple star 40 Eridani, whose coordinates are R.A. = $4^h$ $15.3^m$ and Decl. = −7° 39′. The primary, a 4.4-magnitude yellowish star like the Sun, has a 9.6-magnitude white dwarf companion, the most easily seen white dwarf in the sky. On a clear, dark night with a moderately large telescope, you should also see that the white dwarf has an 11th-magnitude companion, which completes this most interesting trio.

# Where to Learn More

*Books and magazine articles*

Davidson, K. "From Swords to Supernovae." *Sky & Telescope*, November 1997. This article describes how scientists are trying to duplicate on the Earth the conditions within a supernova—all within a volume no larger than the tip of a pencil.

———. "Crisis at Eta Carinae?" *Sky & Telescope*, January 1998. Eta Carinae is an incredibly luminous, highly evolved blue supergiant star that has displayed complex changes over the past century and a half. This article describes some of the mysterious aspects of this unusual star.

Filippenko, A. V. "A Supernova with an Identity Crisis." *Sky & Telescope,* December 1993. SN 1993J—which proved to be an unusual supernova—is the subject of this article by a supernova specialist. An excellent overview of all the different types of supernovae is also given.

Kirschner, R. P. "Supernova 1987A: The First Ten Years." *Sky & Telescope*, February 1997. This article by a leading researcher summarizes what has been learned in a decade of observing the brightest supernova since 1604.

Mann, A. K. *Shadow of a Star: The Neutrino Story of Supernova 1987A.* W. H. Freeman, 1997. Written by one of the principal scientists who discovered neutrinos from SN 1987A, this book describes the trials and tribulations of carrying out a cutting-edge experiment.

Nather, R. E., and Winget, D. E. "Taking the Pulse of White Dwarfs." *Sky & Telescope*, April 1992. Many white dwarf stars vibrate. By studying these vibrations, astronomers can learn more about the structure of white dwarfs and about the stars that eventually evolved into white dwarfs.

Soker, N., "Planetary Nebulae." *Scientific American*, May 1992. This article describes the structure and evolution of fluorescent clouds of gas that represent the last gasp of dying low-mass stars.

Woosley, S., and Weaver, T. "The Great Supernova of 1987." *Scientific American*, August 1989. Two preeminent authorities on supernovae teamed up to write this superb article, which describes many fascinating details of the discovery and early evolution of SN 1987A.

### W *World Wide Web*

A number of beautiful images of planetary nebulae and supernova remnants can be found at Bill Arnett's "The Web Nebulae" (**http://www.seds.org/billa/twn/**), the SEDS Messier Catalog (**http://www.seds.org/messier/**), and David Malin's site at the Anglo-Australian Observatory (**http://www.aao.gov.au/images.html**).

The Hubble Space Telescope has been used for a number of studies of planetary nebulae and supernovae, and some of the finest images are available at the Space Telescope Science Institute web site (**http://oposite.stsci.edu/pubinfo/ Subject.html**).

The SEDS Messier Catalog also has information about and images of the Large Magellanic Cloud, in which SN 1987A occurred (**http://www.seds.org/messier/xtra/ngc/lmc.html**), as well as the galaxy M81, home of the supernova SN 1993J shown in Figure 22-20 (**http://www.seds.org/messier/ m/m081.html**)

Links to current supernova research and discoveries can be found at "The Supernova Nexus," a web site at the Harvard-Smithsonian Center for Astrophysics (**http:// cfa-www. harvard.edu/cfa/oir/Research/supernova.html**) and at a web site maintained by the Supernovae Research Group of the University of Texas (**http://tycho.as.utexas. edu/SN/**).

Modern computers have made it possible for telescopes to carry out automated searches for supernovae. An excellent web site with a wealth of images of supernovae found in this way can be found at the Mount Stromlo observatory in Australia **http://msowww.anu.edu.au/~reiss/Abell_ SNSearch/**). Unfortunately, many supernovae are so distant that they can be detected only with very large telescopes. A list of supernovae bright enough to be seen with moderately-sized amateur telescopes (30 cm or larger) is maintained by the Astronomy Section of the Rochester Academy of Science (**http://www.ggw.org/freenet/a/asras/ supernova.html**).

For the most up-to-date list of recently discovered supernovae, the best place to look is the International Astronomical Union's Central Bureau for Astronomical Telegrams (**http://cfa-www.harvard.edu/cfa/ps/ Headlines.html**).

# Neutron Stars

**A Neutron Star** This Hubble Space Telescope image is the first direct, visible-light view ever recorded of an isolated neutron star. No more than 28 km (17 mi) in diameter but with a mass comparable to the Sun, this star is more than $10^{13}$ times denser than lead. For such a small object to be visible at a distance of hundreds of light-years with even a powerful telescope, its surface temperature must be about 700,000 K. Neutron stars are thought to be the corpses of massive stars that died in supernova explosions. (Fred Walter, State University of New York at Stony Brook; NASA)

*In this chapter you will find the answers to the following questions:*

23-1 What led scientists to the idea of a neutron star?

23-2 What are pulsars, and how were they discovered?

23-3 How did astronomers determine the connection between pulsars and neutron stars?

23-4 How can a neutron star supply energy to a surrounding nebula?

23-5 What is it like inside a neutron star?

23-6 How are some neutron stars able to spin several hundred times per second?

23-7 Why do some pulsars emit fantastic amounts of X rays?

23-8 Are X-ray bursters and novae similar to supernovae?

23-9 How massive can a neutron star be?

**figure 23-2** R I **V** U X G

**The Crab Nebula** This beautiful nebula in the constellation Taurus, named for the armlike appearance of its filamentary structure, is the remnant of the supernova of A.D. 1054. The distance to the nebula is about 2000 parsecs (6500 light-years), so its present angular dimensions of 4 × 6 arcminutes correspond to linear dimensions of approximately 2 by 3 parsecs (7 by 10 light-years). The Crab Nebula can be seen with even a small telescope. (Palomar Observatory)

are too big and bulky to generate 30 signals per second, because calculations demonstrated that a white dwarf could not rotate that fast without tearing itself apart. Hence, the existence of the Crab pulsar means that the stellar corpse at the center of the Crab Nebula has to be much smaller and more compact than a white dwarf.

This was a truly unsettling realization, because most astronomers in the mid-1960s believed that all stellar corpses are white dwarfs. The number of white dwarfs in the sky seemed to account for all the stars that must have died since our Galaxy was formed. It was generally assumed that all dying stars—even the most massive ones, which produce supernovae—somehow manage to eject enough matter so that their corpses do not exceed the Chandrasekhar limit. But the discovery of the Crab pulsar showed these conservative opinions to be in error.

It was now clear that pulsars had to be something far more exotic than rotating white dwarfs. As Thomas Gold of Cornell University emphasized, astronomers would now have to seriously consider the existence of neutron stars.

## 23-3 Pulsars are rapidly rotating neutron stars with intense magnetic fields

Science is, by its very nature, based on skepticism. If a scientist proposes a new and exotic explanation for an unexplained phenomenon, that explanation will only be accepted if it satisfies two demanding criteria. First, there must be excellent reasons why more conventional explanations cannot work. Second, the new explanation must help scientists understand a range of other phenomena. A good example is the Copernican idea that the Earth orbits the Sun, which gained general acceptance only after Galileo's observations. First, Galileo showed the older geocentric explanation of celestial motions to be untenable, and, second, it became clear that the Copernican model could also explain the motions of Jupiter's moons (see Section 4-5 for details). By contrast, almost all scientists reject the idea that an alien spacecraft crashed near Roswell, New Mexico, in 1947, because this idea fails to satisfy the first criterion. There is a perfectly good, conventional explanation for the much-ballyhooed bits of "spacecraft wreckage" found near Roswell—they are merely remnants of an unmanned balloon.

We have seen how the discovery of the Crab pulsar ruled out the idea that pulsars were white dwarfs. But astronomers soon accepted the more exotic idea that pulsars were actually neutron stars. To understand why, we must assume the skeptical outlook of the scientist and ask some probing questions about this idea. Why should neutron stars emit radiation at all? In particular, why should they emit radio waves? Why should the emissions be pulsed? And how could the pulses occur as rapidly as 30 times per second, as in the Crab pulsar? The answers to all of these questions are intimately related to the small size of neutron stars.

Because neutron stars are very small, they should also rotate rapidly. All stars rotate, but most of them do so in a leisurely fashion. For example, our Sun takes nearly a full month to rotate once about its axis. However, just as an ice skater doing a pirouette speeds up when she pulls in her arms, a collapsing star also speeds up as its size shrinks. (This phenomenon is a direct consequence of a law of physics called the *conservation of angular momentum*, discussed in Box 23-1.) In fact, if our Sun was compressed to the size of a neutron star, it would spin about 1000 times per second! Because neutron stars are so small and dense, they can spin this rapidly without flying apart.

The small size of neutron stars also implies that they have intense magnetic fields. It seems safe to say that every star possesses some magnetic field, but the strength is typically quite low. The magnetic field of a main-sequence star is spread out over millions upon millions of square kilometers of the star's surface. However, if such a star collapses down to a neutron star, its surface area (which is proportional to the square of its radius) shrinks by a factor of about $10^{10}$. The magnetic field, which is bonded to the star's gases, becomes concentrated onto an area $10^{10}$ times smaller than before the collapse, and thus the field strength increases by a factor of $10^{10}$.

The strength of a magnetic field is usually measured in the unit called the *gauss* (G), named after the famous German mathematician and astronomer Karl Friedrich Gauss. As described in Box 18-1, a magnetic field splits spectral lines into two or more lines, whose spacing reveals the field's strength. For example, magnetically split lines in the spectra of main-sequence stars reveal overall magnetic fields with strengths in the range from about 1 G (the value for the Sun) to several thousand gauss. Line splitting in the spectra of certain white dwarfs corresponds to field strengths in excess of $10^6$ G, comparable to the strongest fields ever produced in a laboratory. But the magnetic fields surrounding neutron stars are a million times stronger still, on the order of 1 trillion gauss ($10^{12}$ G).

The magnetic field of a neutron star makes it possible for the star to radiate pulses of energy toward telescopes on the Earth. A number of different models have been proposed for how rotating neutron stars generate radiation; we'll discuss just one leading model. The idea behind this model is that the magnetic axis of a neutron star, the line connecting the north and south magnetic poles, is likely to be inclined at an angle to the rotation axis (Figure 23-3). After all, there is no fundamental reason for these two axes to coincide. (Indeed, these two axes do not coincide for any of the planets of our solar system.) In 1969, Peter Goldreich at Caltech pointed out that the combination of such a powerful magnetic field and rapid rotation would act like a giant electric generator, creating very strong electric fields near the neutron star's surface. Indeed, these fields are so intense that part of their energy is used to create electrons and positrons out of nothingness in a process called **pair production**. The powerful electric fields push these charged particles out from the neutron star's surface and into its curved magnetic field, as sketched in Figure 23-3. As the particles spiral along the curved field, they are accelerated and emit energy in the form of electromagnetic radiation. (A radio transmission antenna works in a similar way. Electrons are accelerated back and forth along the length of the antenna, producing radio waves.) The result is that two narrow beams of radiation pour out of the neutron star's north and south magnetic polar regions.

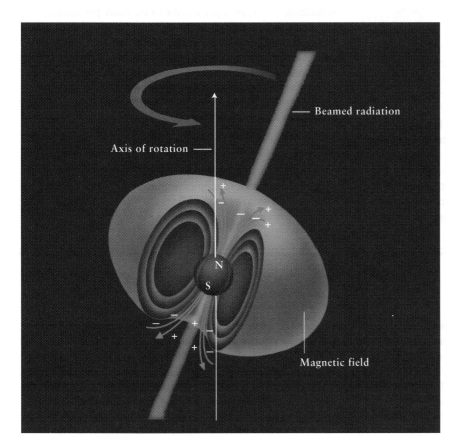

**Figure 23-3**

**A Rotating, Magnetized Neutron Star**
It is reasonable to suppose that a neutron star is rotating rapidly and possesses a powerful magnetic field. Charged particles that have been accelerated near the star's magnetic poles produce two oppositely directed beams of radiation that emerge from the magnetic poles. In general, the magnetic axis of the star (a line that connects the north and south magnetic poles) is tilted at an angle from the star's axis of rotation, as shown. Therefore, as the star rotates, the beams sweep around the sky. If the Earth happens to lie in the path of the beams, we detect radiation that appears to pulse on and off—in other words, a pulsar.

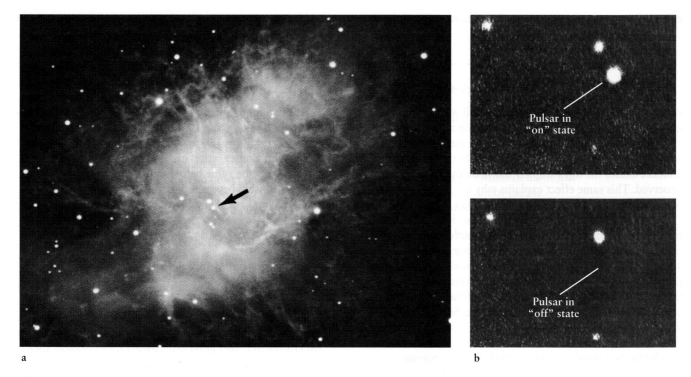

a                                                                                  b

**ƒigure 23-4**   R I **V** U X G

**The Crab Pulsar in Visible Light**   A pulsar is located at the center of the Crab Nebula. **(a)** The
Crab pulsar is identified by the arrow in this visible-light view. **(b)** These two photographs show the
Crab pulsar when its beam is directed toward us (upper image) and when it is not (lower image).
These are called the "on" and "off" states, respectively. Like the radio pulses, these visual flashes
have a period of 0.033 s. (Palomar Observatory; Lick Observatory)

ure 23-4)—exactly the same pulse rate as observed with
radio waves. This observation was strong evidence in favor
of pulsars being rotating, magnetized neutron stars, because
charged particles accelerating in a magnetic field can indeed
radiate strongly over a very wide range of wavelengths. (By
contrast, blackbody radiation from a hot, dense object such
as an ordinary star tends to be concentrated over a relatively
narrow set of wavelengths. Thus, the Sun emits strongly in
the visible part of the spectrum but emits hardly any radio
waves. The Sun's weak radio emissions come not from
blackbody radiation but rather from charged particles spiraling in the solar magnetic field.)

Since those pioneering days, periodic flashes of radiation
have been detected from the Crab pulsar at still other wavelengths. For example, Figure 23-5 shows X-ray pictures
from the Einstein Observatory with the pulsar in its "on"
and "off" states. The light curves of the Crab pulsar at optical, X-ray, and radio wavelengths are displayed in Figure
23-6. The pulsar period is the same at all wavelengths, just
as we would expect if the emissions are coming from a portion of the neutron star which rotates periodically into view
as the star rotates.

The many successes of the neutron star model of pulsars led
to the rapid acceptance of the model. Since 1968 radio astron-

omers have discovered more than 700 pulsars scattered across
the sky, and it is estimated that some $10^5$ more are strewn
around the disk of the Milky Way Galaxy. Each one is presumed to be the neutron star corpse of a massive extinct star.
Radio telescopes have detected pulsars with a wide variety of
pulse periods, from a relatively sluggish 4.308 s to a blindingly
fast 0.0016 s. In each case the pulse period is thought to be the
same as the rotation period of the neutron star.

Several supernovae have been seen throughout history, but
they did not necessarily produce pulsars. For example, those
observed by Tycho Brahe in 1572 and by Johannes Kepler in
1604 were probably Type Ia supernovae (see Section 22-9),
and it is difficult to imagine how an exploding white dwarf
might become a neutron star. Indeed, no pulsar has been
found at the locations of either of those supernovae. The primary sources of pulsars are thought to be Type II supernovae,
in which the cores of massive stars collapse (as described
in Section 22-6). The supernova that left behind the Crab
Nebula and the Crab pulsar is thought to have been of Type
II. The Vela pulsar, which lies at the center of the supernova
remnant called the Gum Nebula (Figure 22-24), also appears
to have been formed in a Type II supernova explosion. Like
the Crab pulsar, the Vela pulsar emits visible and X-ray pulses
as well as radio pulses (Figure 23-7).

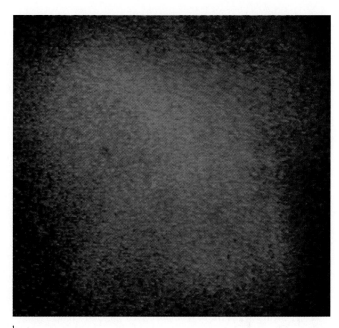

a

b

**figure 23-5**   R I V U ☒ G

**X-Ray Views of the Crab Nebula and Pulsar**   These views of the Crab Nebula were obtained
with an X-ray telescope on the Earth-orbiting Einstein Observatory. The pulsar is shown in its
**(a)** "on" state and **(b)** "off "state. The X-ray pulses have the same period as the visible pulses (shown
in Figure 23-4) and as the radio pulses. (Harvard-Smithsonian Center for Astrophysics)

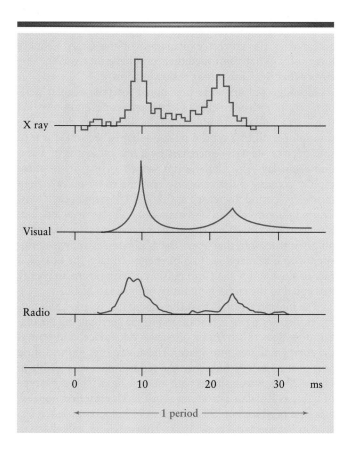

Because the great supernova of 1987 (SN 1987A, described in Section 22-7) was a Type II supernova, astronomers
have been carefully observing its remains for signs of a pulsar.
An important tool in the search is an orbiting observatory
called the Rossi X-ray Timing Explorer, launched in 1995. Its
X-ray sensors are capable of seeing through much of the gas
and dust surrounding the supernova to search for rapid X-ray
variations indicative of a rapidly spinning neutron star. To
date, however, no confirmed observations of pulses have been
made, and the search for a pulsar in SN 1987A goes on.

**figure 23-6**

**Light Curves of the Crab Pulsar**   These three graphs show the
intensity of radiation emitted by the Crab pulsar at X-ray, visible, and
radio wavelengths. One millisecond (ms) equals $10^{-3}$ s. Thus, the
pulsar's period is 33 ms $= 33 \times 10^{-3}$ s $= 0.033$ s. Note that, in addition
to the main pulse, there is a second pulse about halfway through
the pulsar's period. This is a feature of the emission from a number
of pulsars. One interpretation of this effect is that we are seeing
emissions from both of the rotating neutron star's magnetic poles.
In this interpretation, one of the magnetic poles is more directly
aligned toward the Earth than the other, so we detect one strong
pulse and one weak one during each rotation of the neutron star.

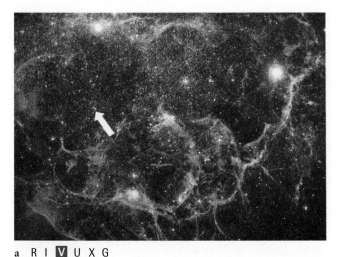

a RI **V** UXG     b RIVU **X** G

## figure 23-7

**The Vela Pulsar in X Rays** The Vela pulsar, named for its location in the southern constellation of Vela, lies within a supernova remnant called the Gum Nebula (see Figure 22-24). Like the Crab pulsar, the Vela pulsar can be detected at X-ray, visible, and radio wavelengths. The Vela pulsar is also quite young. From the size of its supernova remnant and the rate at which it is expanding, astronomers estimate that the Vela supernova occurred roughly 11,000 years ago. Its pulse period is 0.089 s, as compared to 0.033 s for the even younger Crab pulsar (which is only 950 years old). **(a)** The location of the pulsar is identified by an arrow in this visible-light image. **(b)** This X-ray image of the pulsar in its "on" phase was obtained by the orbiting Einstein Observatory. The Vela pulsar is located some 500 parsecs (1600 light-years) from the Earth. (Harvard-Smithsonian Center for Astrophysics)

---

## 23-4 Pulsars gradually slow down as they radiate energy into space

As remarkable as the Crab pulsar is, the Crab Nebula that surrounds it is even more so. This cloud of gases shines with a luminosity 75,000 times that of the Sun (and some $10^8$ times the luminosity of the pulses from the Crab pulsar). It can be spotted in dark skies with even a small telescope, despite its distance of 2000 parsecs (6500 light-years). But why is the Crab Nebula so luminous? Certain other nebulae—the H II regions described in Section 20-2—can also be seen at great distances. But H II regions contain young, hot stars that give off ultraviolet radiation. This radiation ionizes the surrounding gas and provides the energy that the H II region emits into space. The Crab Nebula, however, is different from H II regions—it contains *no* young stars. What, then, is the source of this nebula's prodigious energy output? As we will see, the Crab pulsar provides the energy not only for radio pulses but for the entire Crab Nebula.

In 1966, two years before the discovery of pulsars, John A. Wheeler at Princeton University and Franco Pacini in Italy had already speculated that the ultimate source of the Crab Nebula's immense energy output might be a spinning neutron star. Wheeler and Pacini's speculation quickly came to be accepted as the correct explanation of the Crab Nebula's luminosity. What confirmed their prophetic idea was the discovery that the Crab pulsar is slowing down.

Although pulsars were first noted for their regular periods, careful measurements by radio telescopes soon revealed that many pulsars are indeed gradually slowing down. The rotation period of a typical pulsar—that is, the time for it to spin once on its axis—increases by a few billionths of a second each day. The Crab pulsar, which is one of the most rapidly rotating pulsars, is also slowing more quickly than most—its period increases by $3 \times 10^{-8}$ second each day. Thirty billionths of a second may sound like a trivial amount, but there is so much energy in the rotation of a rapidly spinning neutron star that even the slightest slowdown corresponds to a tremendous loss of energy. In fact, the observed rate of slowing for the Crab pulsar corresponds to an energy loss equal to the entire luminosity of the Crab Nebula.

The interpretation is that the energy lost by the Crab pulsar as it slows down is transferred to the surrounding nebula. But *how* does this transfer of energy take place, and how is the energy radiated into space? The answer lies in the diffuse part of the Crab Nebula (not the reddish filaments visible in Figure 23-2), which shines with an eerie light. This same type of light, as Russian astronomer Iosif Shklovskii first noticed, was first produced on the Earth in 1947 in a particle accelerator in Schenectady, New York. This machine, called a *syn-*

*chrotron,* was built by the General Electric Company to accelerate electrons to nearly the speed of light for experiments in nuclear physics. (The electrons are said to be *relativistic,* because their velocities are near the speed of light and Einstein's theory of relativity must be applied to understand their motions.) As a beam of electrons whirled along a circular path, held in orbit by powerful magnets, scientists noticed that it emitted a strange light. It is now known that this light, called **synchrotron radiation,** is emitted whenever high-speed electrons move along curved paths through a magnetic field. Synchrotron radiation is responsible for part of Jupiter's radio emission, as described in Section 13-6.

The total energy output of the Crab Nebula in synchrotron radiation is $3 \times 10^{31}$ watts, compared to the relatively paltry $4 \times 10^{26}$ watts emitted by the Sun. Thus, the Crab Nebula must contain quite a large number of relativistic electrons spiraling in an extensive magnetic field. As the Crab pulsar slows, the energy of its rotation is transferred to these electrons through the magnetic field and is then emitted by the nebula in the form of synchrotron radiation. This is the reason why the Crab Nebula shines. Recent Hubble Space Telescope images of the Crab pulsar's surroundings have shown that the gases around the pulsar are in a state of great agitation, indicative of the tremendous amount of energy transferred from the star to the nebula every second (Figure 23-8).

Because a spinning neutron star slows down as it radiates away its rotational energy, it follows that an old neutron star should be spinning more slowly than a young one. Two of the fastest pulsars are the Crab pulsar and the Vela pulsar, both of which are at the center of relatively young supernova remnants. Slower pulsars, by contrast, are typically not found within detectable supernova remnants. These pulsars are thought to have been formed in supernova explosions that occurred hundreds of thousands of years ago. Over the ages,

the supernova remnants have been dispersed into the interstellar medium. The pulsars left behind from the exploded stars have slowed to periods of a second or more. (Like all general rules, the one relating a pulsar's age and its period has exceptions. In Section 23-6 we will see how some old pulsars can be reaccelerated to truly dizzying rotation speeds.)

## 23-5 Superfluidity and superconductivity are among the strange properties of neutron stars

The radiation from pulsars gives us information about the surroundings of neutron stars, as well as how they rotate. To deduce what conditions are like *inside* a neutron star, however, astrophysicists must construct theoretical models. We saw examples of such models for the Sun in Section 18-7 and for post–main-sequence stars in Sections 22-1 and 22-5, but the models for neutron stars are very different. For one thing, neutron stars are thought to have a solid crust on their surface. Furthermore, the interior of a neutron star is a sea of densely packed, degenerate neutrons, with properties quite unlike those of ordinary gases or even degenerate electrons. Detailed calculations strongly suggest that degenerate neutron matter can flow without any friction whatsoever, a phenomenon called **superfluidity.**

Superfluidity is observed in the laboratory on the Earth when liquid helium is cooled to temperatures near absolute zero. Because it is frictionless, superfluid liquid helium exhibits such strange properties as being able to creep up the walls of a container in apparent defiance of gravity. Within a neutron star, elongated friction-free whirlpools of superfluid

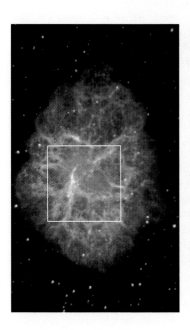

Crab pulsar

**Figure 23-8** R I **V** U X G

**The Dynamic Heart of the Crab Nebula** The pulsar at the center of the Crab Nebula not only powers the synchrotron radiation by which the nebula shines but also stirs up the gases surrounding the pulsar. At left is a visible-light image of the Crab Nebula from Palomar Observatory. The close-up visible-light image on the right, made with the Hubble Space Telescope, shows a region about 2½ arcminutes square centered on the Crab pulsar. (The bright star that appears near the pulsar is located much closer to us than the Crab Nebula. It just happens to be along the line of sight to the nebula.) The bright clumps of gas have been excited by the pulsar, perhaps by jets of material being thrown off the neutron star. The bright clumps move outward from the pulsar at half the speed of light. (Jeff Hester and Paul Scowen, Arizona State University; NASA)

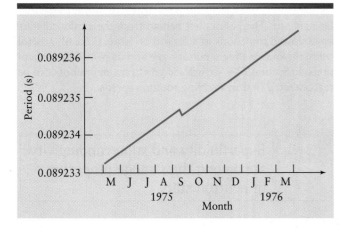

**figure 23-9**

**A Glitch of the Vela Pulsar** This graph shows how the period of the Vela pulsar varied during 1975 and 1976. As the rotational energy of the spinning neutron star was converted into radiation, the star's rotation slowed and the period increased slightly. During September 1975, however, the period *decreased* suddenly, indicating that the neutron star's rotation had undergone a sudden speedup. The speedup, called a glitch, is thought to be caused by an interaction between the neutron star's solid crust and its superfluid interior.

neutrons may form. The interaction of these with the crust above may be the cause of sudden changes in a pulsar's rotation, called *glitches*.

In addition to the general slowing down of pulsars, astronomers have found that they sometimes exhibit a sudden, unexpected speedup, prosaically called a **glitch.** For example, Figure 23-9 shows measurements of the Vela pulsar's period during 1975 and 1976. On this graph, the pulsar's gradual slowdown is shown as a steady increase in its period. In September 1975, however, there was an abrupt speedup, after which the pulsar continued to slow down at its usual rate. This glitch was the third one observed for the Vela pulsar; similar glitches have been observed for the Crab pulsar.

Current opinion among pulsar theorists is that glitches are caused by the superfluid neutrons within the neutron star. As a rotating neutron star radiates energy into space, the rotation of its crust slows down, but the neutron whirlpools in the star's interior continue to rotate with the same speed. Some of these whirlpools cling to the crust, as though they were bungee cords with one end attached to the crust and the other to the star's interior. As the crust slows down relative to the interior, these superfluid "bungee cords" are stretched out. When the tension in them gets too great, they deliver a sharp jolt that makes the crust speed up suddenly. (An older model, in which pulsar glitches were caused by a "starquake" in the neutron star's settling crust, does not appear to agree with the accumulated observational data on glitches.)

Superfluidity is not the only exotic property of neutron star interiors. Models of the internal structure of a neutron

star strongly suggest that the protons in the core can move around without experiencing any electrical resistance whatsoever. This phenomenon, called **superconductivity,** also occurs on the Earth with certain substances at low temperatures. Once you start an electric current moving in a superconducting material, it keeps moving forever.

You may have been surprised to read in the preceding paragraph about *protons* in a *neutron* star. While a neutron star is made up predominantly of neutrons, there must be some protons and electrons scattered throughout the star's interior. Indeed, a pulsar's magnetic field must be anchored to the neutron star by charged particles. Neutrons are electrically neutral, so without the protons and electrons in its interior a neutron star would rapidly lose its magnetic field.

As shown in Figure 23-10, the structure of a neutron star probably consists of a core with superfluid neutrons and superconducting protons, a mantle of superfluid neutrons in which some of the neutrons combine with protons to form nuclei, and a brittle crust less than a kilometer thick. At the center of a neutron star, the density is on the order of $10^{18}$ kg/m$^3$. This is some 10 trillion ($10^{13}$) times greater than the central density of our present-day Sun and approximately twice the density of an atomic nucleus.

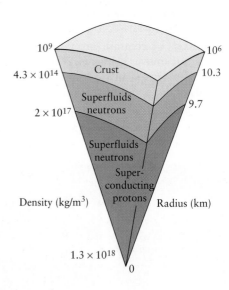

**figure 23-10**

**A Model of a Neutron Star** This theoretical model for a 1.4-M$_\odot$ neutron star has a superconducting, superfluid core 9.7 km in radius. The core is surrounded by a 0.6-km thick mantle containing nuclei, electrons, and superfluid neutrons. The star's crust is only 0.3 km thick (the length of three football fields) and is composed of heavy nuclei (such as iron) and free electrons. The details of such models depend strongly on the relationship between the pressure and density of an ultradense fluid of neutrons. This relationship is only incompletely understood, so there remains a good deal of uncertainty about neutron star interiors.

## 23-6 The fastest pulsars were probably created by mass transfer in close binary systems

In 1982 astronomers at the Arecibo Observatory, shown in Figure 10-5, discovered a pulsar whose period is only 1.558 ms. (One millisecond, or 1 ms, equals a thousandth of a second). This remarkable pulsar, called PSR 1937+21 after its coordinates in the sky, is a neutron star spinning 642 times per second—about 3 times faster than the blades of a kitchen blender! Such incredibly rapid rotation is a clue to this pulsar's unusual history.

As we have seen, pulsars should slow down as they age. The fact that PSR 1937+21 is spinning so rapidly suggests that it has hardly aged at all, so it must be very young. A young pulsar should also be slowing down very rapidly. The faster it rotates, the more rapidly it can transfer rotational energy to its surroundings and the faster the slowdown. But PSR 1937+21 is actually slowing down at a very gradual rate, millions of times more gradually than the Crab pulsar. Such a gradual slowdown is characteristic of a pulsar hundreds of millions of years old. But if PSR 1937+21 is so old, how can it be spinning so rapidly?

PSR 1937+21 is not the only pulsar with unusually rapid rotation. Since 1982, astronomers have discovered more than 30 very fast pulsars, which are now called **millisecond pulsars.** All have periods between 1 and 10 ms, which means that these neutron stars are spinning at rates of 100 to 1000 rotations per second.

An important clue to understanding millisecond pulsars is that the majority are in close binary systems, with only a small separation between the spinning neutron star and its companion. (We know the separation must be small because the orbital periods of these binary systems are short, between 10 and 100 days. Kepler's third law for binary star systems, described in Section 19-9, tells us that the shorter the period, the smaller the distance between the two stars.) This fact suggests a scenario for how millisecond pulsars acquired their very rapid rotation.

Imagine a binary system consisting of a high-mass star and a low-mass star. The high-mass star evolves more rapidly than the low-mass star, and within a few million years becomes a Type II supernova that creates a neutron star. Like most newborn neutron stars, it spins several times per second, and, thus, initially we see a pulsar rather like the Crab or Vela pulsars. Over the next few billion years, the pulsar slows down as it radiates energy into space. Meanwhile, the slowly evolving low-mass star begins to expand as it evolves away from the main sequence to become a red giant. When the red giant gets big enough to fill its Roche lobe, it starts to spill gas over the inner Lagrangian point onto the neutron star. (See Section 21-6 for a review of close binary systems.) The infalling gas strikes the neutron star's surface at high speed and at an angle that causes the star to spin faster. In this way, a slow, aging pulsar is "spun up" by mass transfer from its bloated companion.

What about the few millisecond pulsars, like PSR 1937+21, that are *not* members of close binaries? It may be that solitary millisecond pulsars were once part of close binary systems, but the companion stars have been eroded away by the high-energy particles emitted by the pulsar after it was spun up. The Black Widow Pulsar, discovered in 1988, may be caught in the act of destroying its companion in just such a process (Figure 23-11). This pulsar (a neutron

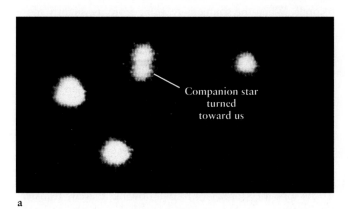

a

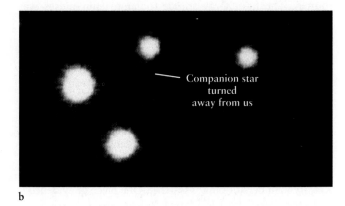

b

**ƒigure 23-11   R I V U X G**

**The Black Widow Pulsar**  The millisecond pulsar PSR 1957+20, called the Black Widow Pulsar, is a rapidly spinning neutron star in a close binary system. The neutron star and its companion orbit each other with an orbital period of 9.16 hours. The plane of the orbit is aligned so that the two stars periodically eclipse each other. Most of the visible light from this system comes from the side of the companion star facing the neutron star, because this side is heated by emissions from the pulsar. That hemisphere is exposed to our view in **(a)** but turned away from us in **(b)**. An unrelated background star less than 1 arcsec from the pulsar shines constantly. Just as the female black widow spider kills and eats its mate, the Black Widow Pulsar is slowly eating away its companion star by blasting it with energetic photons and particles. (Courtesy of Jan van Paradijs)

star spinning 622 times per second) and its companion orbit each other with a period of 9.16 hours. The orbital plane of the system is nearly edge-on to our line of sight, so this system is an eclipsing binary (described in Section 19-11): The companion blocks out the pulsar's radio signals for 45 minutes during each eclipse. Just before and after an eclipse, however, the pulsar's signals are delayed substantially, as if the radio waves were being slowed as they pass through a cloud of ionized gas surrounding the companion star. This circumstellar material is probably the star's outer layers, dislodged by radiation and fast-moving particles emitted from the pulsar. In a few hundred million years or so, the companion of the pulsar will have completely disintegrated, leaving behind only a spun-up, solitary millisecond pulsar.

## 23-7 Pulsating X-ray sources are also neutron stars in close binary systems

Millisecond pulsars are not the only result of having a neutron star in a close binary system. Even more exotic are *pulsating X-ray sources*, in which material from the companion star is drawn onto the magnetic poles of the neutron star, producing intense hot spots that emit breathtaking amounts of X rays. As the neutron star spins and the X-ray beams are directed alternately toward us and away from us, we see the X rays flash on and off. Such pulsating X-ray sources, while similar in many ways to ordinary pulsars, are far more luminous.

**Pulsating X-ray sources** were first discovered in 1971 by the Uhuru spacecraft, the first to give astronomers a comprehensive look at the X-ray sky. (As we saw in Section 6-7, the Earth's atmosphere is opaque to X-rays.) Figure 23-12 is a modern map showing the locations of a number of X-ray sources. The first X-ray source found to emit pulses was Centaurus X-3 (the third X-ray source found in the southern constellation Centaurus). Figure 23-13 shows data from one sweep of Uhuru's detectors across Centaurus X-3. The X-ray pulses have a regular period of 4.84 s. A few months later, similar pulses were discovered coming from a source designated Hercules X-1 (the first X-ray source discovered in the constellation Hercules), which had a period of 1.24 s. Because the periods of these two X-ray sources are so short, astronomers began to suspect that they were actually observing rapidly rotating neutron stars.

It soon became clear that Centaurus X-3 and Hercules X-1 are members of binary systems. One piece of evidence is that every 2.087 days, Centaurus X-3 appears to turn off for almost 12 hours. Apparently, Centaurus X-3 is a neutron star in an eclipsing binary system, and it takes nearly 12 hours for the X-ray source to pass behind its companion star. Hercules X-1, too, appears to turn off periodically, corresponding to a 6-hour eclipse every 1.7 days by its visible companion, the star HZ Herculis. Moreover, careful timing of its X-ray pulses shows a periodic Doppler shifting every 1.7 days, which is direct evidence of orbital motion about a compan-

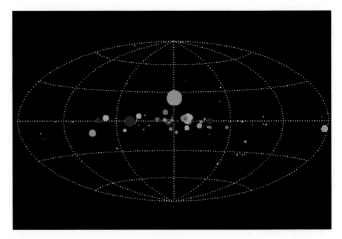

**figure 23-12** R I V U X G

**The X-Ray Sky** Each colored circle on this map shows the location in the sky of an X-ray source. Most of these sources lie within our own Milky Way Galaxy, whose plane is represented by the horizontal line running across the center of the map. The color of each circle indicates the predominant wavelength of the emitted X-ray photons; blue and red represent short and long wavelengths, respectively. The area of the circle for each source is proportional to the source's X-ray brightness. (The exception is Scorpius X-1, which is such a bright source that the area of its circle has been reduced by a factor of 4. Otherwise, the circle for Scorpius X-1 would completely fill the map.) These data are from the Rossi X-Ray Timing Explorer, which was launched in 1995. (D. A. Smith and the All-Sky Monitor Team, Massachusetts Institute of Technology)

ion star. When the X-ray source is approaching us, the pulses from Hercules X-1 are separated by slightly less than 1.24 seconds. When the source is receding from us, slightly more than 1.24 seconds elapse between the pulses.

Putting all the pieces together, astronomers now realize that pulsating X-ray sources like Centaurus X-3 and Hercules X-1 are binary systems in which one of the stars is a neutron star. All these binary systems have very short orbital periods, which means that the distance between the two stars is quite small. The neutron star can therefore capture gases escaping from its ordinary companion. About 20 such exotic binary systems, also called *X-ray binary pulsars*, are known.

In Hercules X-1, the ordinary star is a giant that fills its Roche lobe (Section 21-6) to overflowing, thereby transferring matter onto the neutron star. Although the ordinary star in Centaurus X-3 does not quite fill its Roche lobe, mass transfer still occurs, because stellar winds propel the star's outer layers over the critical surface that marks the boundary of each star's Roche lobe (Figure 23-14). A typical rate of mass transfer from the ordinary star to the neutron star is roughly $10^{-9}$ solar masses per year.

Because of its strong gravity, a neutron star in a pulsating X-ray source easily captures much of the gas escaping from

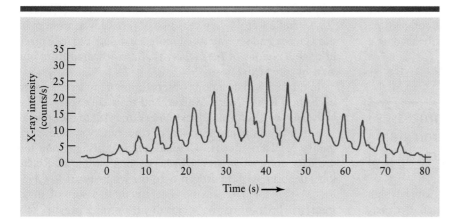

**figure 23-13**

**X-Ray Pulses from Centaurus X-3** This graph shows the intensity of X rays detected by the Uhuru spacecraft as Centaurus X-3 moved across the satellite's field of view. Successive pulses are separated by only 4.84 s. Because this period is so short, Centaurus X-3 is in all probability a rotating neutron star. The gradual variation in the height of the pulses from left to right is the result of the changing orientation of the X-ray detectors toward the source as Uhuru rotated. (Adapted from R. Giacconi and colleagues)

its companion. But like an ordinary pulsar, the neutron star is rotating rapidly and has a powerful magnetic field inclined to the axis of rotation (recall Figure 23-3). As the gas falls toward the neutron star, its magnetic field funnels the incoming matter down onto the star's north and south magnetic polar regions. The neutron star's gravity is so strong that the gas is traveling at nearly half the speed of light by the time it crashes onto the star's surface. This violent impact creates spots at both poles with searing temperatures of about $10^8$ K. These temperatures are so high that the hot spots emit abundant X rays. In fact, their X-ray luminosity is roughly $10^{31}$ watts, nearly 100,000 times greater than the combined luminosity of the Sun at all wavelengths.

As the neutron star rotates, the beams of X rays from its magnetic poles sweep around the sky. If the Earth happens to be in the path of one of the two beams, we can observe a pulsating X-ray source. We detect a pulse once per rotation period, which means that the neutron star in Hercules X-1 spins at the rate of once every 1.24 seconds. If the orbital plane of the binary system is nearly edge-on to our line of sight, as for Centaurus X-3 and Hercules X-1, the pulsating X rays appear to turn off when the neutron star is eclipsed by the companion star. Other pulsating X-ray sources, such as Scorpius X-1, do not undergo this sort of turning off. The orbital planes of these systems are oriented more nearly face-on to our line of sight, so there are no eclipses.

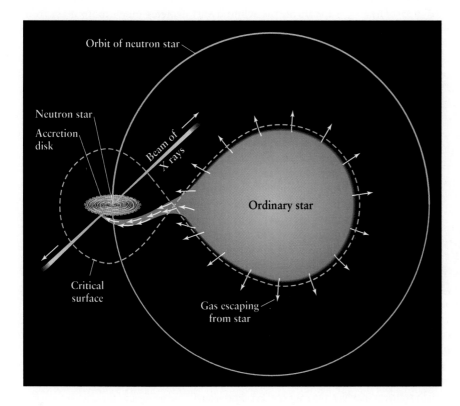

**figure 23-14**

**A Model of a Pulsating X-Ray Source** Pulsating X-ray sources are close binary systems in which one member is a rotating neutron star. Gas from the ordinary star flows across the star's critical surface and is captured by the neutron star. (A similar situation is found in close binary systems made up of two ordinary stars, as shown in Figure 21-16b.) The infalling gas is funneled down onto the neutron star's magnetic poles and strikes the star with enough energy to create two X-ray–emitting hot spots. As the neutron star spins, beams of X rays from the hot spots sweep around the sky. As the beam sweeps into and out of the line of sight to the Earth, we see pulses of X rays.

As time passes, the neutron star in a pulsating X-ray source will accrete more and more mass from its companion. In the process, it will spin up and may eventually become a millisecond pulsar (Section 23-6).

## 23-8 Explosive thermonuclear processes on white dwarfs and neutron stars produce novae and bursters

As seen in the preceding two sections, there can be high drama when a close binary system includes a stellar corpse. The corpse can emit dazzling pulses of X rays and be spun up to truly impressive fast rates of rotation. Such systems can also undergo tremendous explosions, which go by the names of *novae* and *X-ray bursters*.

We have encountered explosions in close binary systems before. As we saw in Section 22-9, one model for a Type Ia supernova is a close binary system with a giant star and a white dwarf. In this model, accretion onto the white dwarf causes a runaway thermonuclear reaction that consumes the entire volume of the white dwarf. Novae and X-ray bursters are less catastrophic but no less remarkable. They are caused by explosive thermonuclear reactions that take place on just the *surface* of a white dwarf or neutron star in a close binary system.

From time to time astronomers observe a faint star suddenly brighten by a factor of $10^4$ to $10^8$ over a few days or hours, reaching a peak luminosity of about $10^5$ L$_\odot$. This phenomenon is called a **nova** (plural **novae**). (By contrast, a *super*nova has a peak luminosity of about $10^9$ L$_\odot$.) The abrupt rise in brightness is followed by a gradual decline that may stretch out over several months or more (see Figures 23-15 and 23-16). Every year two or three novae are seen in our Galaxy, and several dozen more are thought to take place in remote regions of the Galaxy that are obscured from our view by interstellar dust.

In the 1950s, painstaking observations of numerous novae by Robert Kraft, Merle Walker, and their colleagues at the University of California's Lick Observatory led to the conclusion that all novae are members of close binary systems containing a white dwarf. Gradual mass transfer from the ordinary companion star, which presumably fills its Roche lobe, deposits fresh hydrogen onto the white dwarf (see Figure 21-17b for a schematic diagram of this sort of mass transfer). Because of the white dwarf's strong gravity, this hydrogen is compressed into a dense layer covering the hot surface of the white dwarf. As more gas is deposited and compressed, the temperature in the hydrogen layer increases. When the temperature reaches about $10^7$ K, hydrogen burning ignites throughout the gas layer, embroiling the white dwarf's surface in a thermonuclear holocaust that we see as a nova.

CAUTION! It is important to understand the similarities and differences between novae and Type Ia supernovae. Both kinds of celestial explosions are thought to occur in close binary systems where one of the stars is a white dwarf. But, as befits their name, supernovae are much more energetic. A Type Ia supernova explosion radiates $10^{44}$ joules of energy into space, while the corresponding figure for a typical nova is $10^{37}$ joules. (To be fair to novae, this relatively paltry figure is as much energy as our Sun emits in 1000 years.) The difference is thought to be that in a Type Ia supernova, the white dwarf accretes much more mass from its companion. This added mass causes so much compression

a      b

**figure 23-15**   R I  U X G

**Nova Herculis 1934** These two pictures show a nova **(a)** shortly after peak brightness as a star of apparent magnitude +3, bright enough to be seen easily with the naked eye, and **(b)** two months later, when it had faded by a factor of 4000 in brightness to apparent magnitude +12. (See Figure 19-6 for a description of the magnitude scale.) Novae are named after the constellation and year in which they appear. (Lick Observatory)

that nuclear reactions can take place *inside* the white dwarf. Eventually, these reactions blow the white dwarf completely apart. In a nova, by contrast, nuclear reactions occur only within the accreted material. The reaction is more sedate because it takes place only on the white dwarf's surface (perhaps because the accretion rate is less or because the white dwarf had less mass in the first place).

Because the white dwarf itself survives a nova explosion, it is possible for the same star to undergo more than one nova. As an example, the star T Coronae Borealis erupted as a nova in 1866, then put in a repeat performance in 1946. By contrast, a given star can only be a supernova once.

A surface explosion similar to a nova also occurs with neutron stars. Beginning in late 1975, astronomers analyzing data from X-ray satellites realized that their instruments had detected sudden, powerful bursts of X rays from objects in the sky. The record of a typical burst is shown in Figure 23-17. The source emits X rays at a constant low level until suddenly, without warning, there is an abrupt increase in X rays, followed by a gradual decline. An entire burst typically lasts for only 20 s. Unlike pulsating X-ray sources, there is a fairly long interval of hours or days between bursts. Sources that behave in this fashion are known as **X-ray bursters.** Several dozen X-ray bursters have been discovered, most being located toward the center of our Galaxy.

X-ray bursters, like novae, are believed to involve close binaries whose stars are engaged in mass transfer. With a burster, however, the stellar corpse is a neutron star rather than a white dwarf. Gases escaping from the ordinary companion star fall onto the neutron star. The X-ray burster's magnetic field is probably not strong enough to funnel the falling material toward the magnetic poles, so the gases are distributed more evenly over the surface of the neutron star. The energy released as these gases crash down onto the neu-

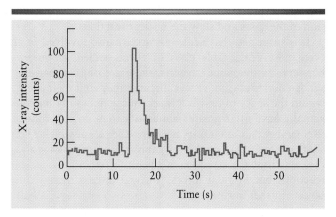

**ƒigure 23-17**

**X Rays from a Burster** A burster emits a constant low intensity of X rays interspersed with occasional powerful X-ray bursts. This particular burst is typical. It was recorded on September 28, 1975, by an Earth-orbiting X-ray telescope pointed toward the globular cluster NGC 6624. About one-third of all known bursters are located in globular clusters. X-ray bursters are thought to occur when gas accretes onto a neutron star from a normal companion star. When the gas layer becomes a few meters thick, it undergoes explosive thermonuclear reactions. These heat the star's surface to such a high temperature that copious X rays are emitted. (Adapted from W. Lewin)

tron star's surface produces the low-level X rays that are continuously emitted by the burster.

Most of the gas falling onto the neutron star is hydrogen, which the star's powerful gravity compresses against its hot surface. In fact, temperatures and pressures in this accreting layer become so high that the arriving hydrogen is converted into helium by hydrogen burning. As a result, the accreted gases develop a layered structure that covers the entire neutron star, with a few tens of centimeters of hydrogen lying atop a similar thickness of helium. The structure is reminiscent of the layers within an evolved giant star (see Figure 22-2), although the layers atop a neutron star are much more compressed, thanks to the star's tremendous surface gravity.

When the helium layer is about 1 m thick, helium burning ignites explosively and heats the neutron star's surface to about $3 \times 10^7$ K. At this temperature the surface predominantly emits X rays, but the emission ceases within a few seconds as the surface cools. Hence, we observe a sudden burst of X rays only a few seconds in duration. New hydrogen then flows onto the neutron star, and the whole process starts over. Indeed, X-ray bursters typically emit a burst every few hours or days.

Whereas explosive *hydrogen* burning on a white dwarf produces a nova, explosive *helium* burning on a neutron star produces a X-ray burster. In both cases, the process is explosive, because the fuel is compressed so tightly against the star's surface that it becomes degenerate, like the star itself. As with the helium flash inside red giants (described in Section 21-2), the ignition of a degenerate thermonuclear fuel involves a sudden thermal runaway. This is because an increase in temperature does not produce a corresponding

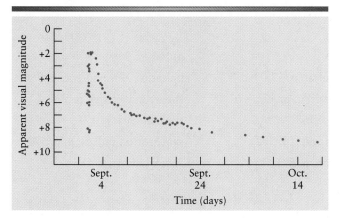

**ƒigure 23-16**

**The Light Curve of Nova Cygni 1975** This graph shows the history of a nova that blazed forth in the constellation Cygnus in September 1975. Its rapid rise and gradual decline in apparent brightness are characteristic of all novae. This nova, also designated V1500 Cyg, was easily visible to the naked eye (that is, was brighter than an apparent magnitude of +6) for nearly a week.

increase in pressure that would otherwise relieve compression of the gases and slow the nuclear reactions.

It is suspected that neutron stars are also involved in an even more remarkable class of events called *gamma-ray* bursters. As the name suggests, these are objects that emit sudden bursts of gamma rays. Because they emit their energy within the space of only a few seconds, gamma-ray bursters actually have a far greater luminosity than do supernovae. (Remember that luminosity measures the *rate* at which a star emits energy.) One model for gamma-ray bursters involves the catastrophic destruction of a neutron star, not by nuclear reactions but by tidal forces. We will discuss these exotic objects further in Chapter 27.

## 23-9 Like a white dwarf, a neutron star has an upper limit on its mass

A white dwarf will collapse if its mass is greater than the Chandrasekhar limit of 1.4 $M_\odot$. At that point, degenerate electron pressure cannot support the overpowering weight of the star's matter, which presses inward from all sides (see Section 22-4). The mass of a neutron star also has an upper limit. However, the pressure within a neutron star is harder to analyze, because it comes from *two* sources. One is the degenerate nature of the neutrons, and the other is the strong nuclear force that acts between the neutrons themselves.

The strong nuclear force is what holds protons and neutrons together in atomic nuclei. Neutrons exert strong nuclear forces on each other only when they are almost touching. This force behaves somewhat like the force that billiard balls exert on each other when they touch. It strongly resists further compression (try squeezing two billiard balls together and see how much success you have), and so it is a major contributor to the star's internal pressure. Unfortunately, there is a good deal of uncertainty about the details of the strong nuclear force. This translates into uncertainties about how much weight the neutron star's internal pressure can support—that is, the neutron star's maximum mass. Theoretical estimates of this maximum mass range from 2 to 3 $M_\odot$.

Before pulsars were discovered, most astronomers believed all dead stars to be white dwarfs. Dying stars were thought to somehow eject enough material so that their corpses could be below the Chandrasekhar limit. The discovery of neutron stars proved this idea incorrect. Inspired by this lesson, astronomers soon began wondering what might happen if a dying massive star failed to eject enough matter to get below the upper limit for a neutron star. For example, what might a 5-$M_\odot$ stellar corpse be like?

The gravity associated with a neutron star is so strong that the escape speed from it is roughly one-half the speed of light. With a stellar corpse greater than 3 $M_\odot$, there is so much matter crushed into such a small volume that the escape speed actually exceeds the speed of light. Because nothing can travel faster than light, nothing—not even light—can leave this dead star. It therefore disappears from the universe, its powerful gravity leaving a hole in the fabric of space and time. Thus the discovery of neutron stars inspired astrophysicists to examine seriously one of the most bizarre and fantastic objects ever predicted by modern science, the black hole. We take up its story in the following chapter.

## Key Words

*Terms preceded by an asterisk (\*) are discussed in Box 23-1.*

*angular momentum, p. 572

*angular velocity, p. 572

*conservation of angular momentum, law of, p. 572

degenerate neutron pressure, p. 568

glitch, p. 578

millisecond pulsar, p. 579

*moment of inertia, p. 572

neutron star, p. 568

nova (*plural* novae), p. 582

pair production, p. 571

pulsar, p. 569

pulsating X-ray source, p. 580

superconductivity, p. 578

superfluidity, p. 577

synchrotron radiation, p. 577

X-ray burster, p. 583

## Key Ideas

**Neutron Stars:** A neutron star is a dense stellar corpse consisting primarily of closely packed degenerate neutrons.

• A neutron star typically has a diameter of about 30 km, a mass less than 3 $M_\odot$, a magnetic field $10^{12}$ times stronger than that of the Sun, and a rotation period of roughly 1 second.

• A neutron star consists of a superfluid, superconducting core surrounded by a superfluid mantle and a thin, brittle crust.

• Energy pours out of the north and south polar regions of a neutron star in intense beams produced by streams of charged particles moving in the star's intense magnetic field.

**Pulsars:** A pulsar is a source of periodic pulses of radio radiation. These pulses are produced as beams of radio waves from a neutron star's polar regions sweep past the Earth.

• The steady slowing of a pulsar's pulse rate presumably reflects the gradual slowing of the rotation rate of the neutron star as it radiates energy into space. Glitches (sudden speedups of the pulse rate) may be caused by interactions between the neutron star's crust and its superfluid interior.

**Millisecond Pulsars:** Mass transfer from an ordinary star to a neutron star in a close binary system can significantly speed up the neutron star's rotation rate, giving rise to a millisecond pulsar.

**Neutron Stars in Close Binary Systems:** If a neutron star is in a close binary system with an ordinary star, tidal forces will draw gas from the ordinary star onto the neutron star.

• Magnetic forces can funnel the gas onto the neutron star's magnetic poles, producing hot spots. These hot spots then radiate intense beams of X rays. As the neutron star rotates, the X-ray beams appear to flash on and off. Such a system is called a pulsating X-ray variable.

**Novae and Bursters:** Material from an ordinary star in a close binary can fall onto the surface of the companion white dwarf or neutron star to produce a surface layer in which thermonuclear reactions can explosively ignite.

• Explosive hydrogen burning may occur in the surface layer of a companion white dwarf, producing the sudden increase in luminosity that we call a nova. The peak luminosity of a nova is only $10^{-4}$ of that observed in a supernova.

• Explosive helium burning may occur in the surface layer of a companion neutron star, producing the sudden increase in X-ray radiation that we call a burster.

# Review Questions

1. What is a neutron star?

2. Why do astronomers think that pulsars are rapidly rotating neutron stars?

3. During the weeks immediately following the discovery of the first pulsar, one suggested explanation was that the pulses might be signals from an extraterrestrial civilization. Why did astronomers soon discard this idea?

4. Compare a white dwarf and a neutron star. Which of these two types of stellar corpses is more common? Explain.

5. How does the law of the conservation of angular momentum help explain why neutron stars rotate so much more rapidly than ordinary stars?

6. What is the difference between superconductivity and superfluidity?

7. How does a glitch affect a pulsar's period? What could be the cause of glitches?

8. How do you think astronomers have deduced that the Vela pulsar is about 11,000 years old?

9. Why do astronomers think that millisecond pulsars are very old?

10. Describe a pulsating X-ray source like Hercules X-1 or Centaurus X-3. What produces the pulsation?

11. What is the connection between pulsating X-ray sources and millisecond pulsars?

12. What is the difference between a nova and a Type Ia supernova?

13. Why is the maximum mass of neutron stars not known as accurately as the Chandrasekhar limit for white dwarfs?

# Advanced Questions

*Questions preceded by an asterisk (*) involve topics discussed in Box 23-1.*

> **Problem-solving tips and tools:**
>
> You will need the formulas for the Doppler effect introduced in Section 5-9 and Box 5-6 (see Figure 5-23). The volume of a sphere of radius $r$ is $\frac{4}{3}\pi r^3$. The small-angle formula is given in Box 1-2. The relationship between the temperature of a blackbody and its wavelength of maximum emission is given in Section 5-4. Section 19-6 gives the formula relating a star's luminosity, surface temperature, and radius (see Box 19-5 for worked examples). The angular diameter of the Moon is given in Section 3-4.

14. Determine the luminosity of the neutron star shown in the image that opens this chapter. Assume a diameter of 28 km and a surface temperature of $7 \times 10^5$ K. Give your answer in terms of the luminosity of the sun.

15. The distance to the Crab Nebula is about 2000 parsecs. In what year did the star actually explode?

*16. When the Sun becomes a red giant, it will expand to 100 times its present radius. Use the law of conservation of angular momentum to determine what the Sun's rotation period will be when this happens. Treat the Sun as a uniform sphere, and assume that the present-day rotation period is 25 days. (This is actually the rotation period at the Sun's equator. As we learned in Section 18-4, different latitudes on the Sun rotate at different rates, but we will ignore this complication here.)

17. How do we know that the Crab pulsar is really embedded in the Crab Nebula and not simply located at a different distance along the same line of sight?

18. The apparent expansion rate of the Crab Nebula is 0.23 arcsec per year. The apparent size of the Crab Nebula

is about 4 by 6 arcmin. If the expansion rate remains constant, calculate how long it will be until the long axis of the Crab Nebula has the same apparent size as the full moon.

**19.** Emission lines in the spectrum of the Crab Nebula exhibit a Doppler shift, which indicates an expansion velocity of about 1200 km/s. (**a**) From this and the data given in the preceding question, calculate the distance to the Crab Nebula and estimate the date on which it exploded. (**b**) Do your answers agree with figures quoted in this chapter? If not, can you point to assumptions you made in your computations that led to the discrepancies? Or do you think your calculations suggest additional physical effects are at work in the Crab Nebula, over and above a constant rate of expansion?

**20.** Consult the World Wide Web and recent issues of such magazines as *Sky & Telescope, Astronomy,* and *Science News* for information about the latest observations of the stellar remnant at the center of SN 1987A. Has a pulsar been detected? If so, how fast is it spinning? Has the supernova's debris thinned out enough to give a clear view of the neutron star?

**21.** To determine accurately the period of a pulsar, astronomers must take into account the Earth's orbital motion about the Sun. (**a**) Explain why. (**b**) Knowing that the Earth's orbital velocity is 30 km/s, calculate the maximum correction to a pulsar's period because of the Earth's motion. Explain why the size of the correction is greatest for pulsars located near the ecliptic.

**22.** The mass of a neutron is about $1.7 \times 10^{-27}$ kg and its radius is about $10^{-15}$ m. (**a**) Compare the density of matter in a neutron with the average density of a neutron star. (**b**) If the neutron star's density is more than that of a neutron, the neutrons within the star are overlapping; if it is less, the neutrons are not overlapping. Which of these seems to be the case? What do you think is happening at the center of the neutron star, where densities are higher than average?

**23.** X-ray pulsars are speeding up but ordinary (radio) pulsars are slowing down. Propose an explanation for this difference.

**24.** If the model for Hercules X-1 discussed in the text is correct, at what orientation of the binary system do we see its maximum optical brightness? Explain your answer.

**25.** In an X-ray burster, the surface of a neutron star 10 km in radius is heated to a temperature of $3 \times 10^7$ K. (**a**) Determine the wavelength of maximum emission of the heated surface (which you may treat as a blackbody). In what part of the electromagnetic spectrum does this lie? (See Figure 5-7.) (**b**) Find the luminosity of the heated neutron star. Give your answer in watts and in terms of the luminosity of the Sun, given in Table 18-1. How does this compare with the peak luminosity of a nova? Of a Type Ia supernova?

## Discussion Questions

**26.** Compare novae and X-ray bursters. What do they have in common? In what ways are they different?

**27.** How might astronomers be able to detect the presence of an accretion disk in a close binary system?

## Observing Projects

**28.** If you did not take the opportunity to observe the Crab Nebula as part of the exercises in Chapter 22, do so now. The Crab Nebula is visible in the night sky from October through March. Its epoch 2000 coordinates are R.A. = $5^h$ $34.5^m$ and Decl. = 22° 00′, which is near the star marking the eastern horn of Taurus (the Bull). Be sure to schedule your observations for a moonless night. The larger the telescope you use, the better, because the Crab Nebula is quite dim.

**29.** Consult such publications as the current issues of *Sky & Telescope* and *Astronomy,* the online versions of these magazines (**http://www.skypub.com/** and **http://www. astronomy.com/** respectively), or recent *IAU Circulars* published by the International Astronomical Union's Central Bureau for Astronomical Telegrams (**http://cfa-www.harvard.edu/cfa/ps/cbat.html**) to see if any novae have been sighted recently. If by good fortune one has been sighted, what is its apparent magnitude? Is it within reach of a telescope at your disposal? If so, arrange to observe it. Draw what you see through the eyepiece, noting the object's brightness in comparison with other stars in the field of view. If possible, observe the same object a few weeks or months later to see how its brightness has changed.

## Where to Learn More

*Books and magazine articles*

Boebinger, G.; Passner, A.; and Bevk, J. "Building World-Record Magnets." *Scientific American,* June 1995. Three leading magnet researchers describe how the world's most powerful magnets are built. The trials and tribulations that they experience in attempting to create fields of $10^6$ G help us better appreciate the fields of neutron stars, which are a million times stronger.

Bell Burnell, J. "The Discovery of Pulsars," in *Serendipitous Discoveries in Radio Astronomy.* National Radio Astronomy Observatory, 1983. The discoverer of the first pulsar describes how it was done.

Cannizzo, J. K., and Kaitchuck, R. H. "Accretion Disks in Interacting Binary Stars." *Scientific American,* January 1992. Two experts in the field describe the wide range of astronomical effects caused by mass transfer between the members of binary star systems.

Feinberg, R. T. "Pulsars, Planets, and Pathos." *Sky & Telescope,* May 1992. The article discusses the controversial discovery of a planet orbiting a millisecond pulsar.

Isles, J. "The Dwarf Nova U Geminorum." *Sky & Telescope,* December 1997. One of the first novae discovered is an easy target for even the smallest telescopes. This article describes where and when to look.

Kaspi, V. M. "Millisecond Pulsars: Timekeepers of the Cosmos." *Sky & Telescope,* April 1995. A leading expert explains how millisecond pulsars provide clues to the environments of these exotic objects and to the internal structure of neutron stars.

Starrfield, S., and Shore, S. N. "Nova Cygni 1992: Nova of the Century." *Sky & Telescope,* February 1994. When a bright nova appeared in the constellation Cygnus in 1992, orbiting telescopes helped make it the best studied nova in history. This article discusses what these observations have taught us.

Van den Heuvel, E. P. J., and van Paradijs, J. "X-ray Binaries." *Scientific American,* November 1993. This article describes in detail how X rays can be emitted from close binary systems.

**W** *World Wide Web*

A number of groups doing research on pulsars and neutron stars have an active presence on the World Wide Web. A list of many of these is maintained by the Pulsar Group at the Swinburne University of Technology in Australia (**http://mania.physics.swin.edu.au/**). Of particular interest is the web site for the Princeton Pulsar Group (**http://pulsar.princeton.edu/**), which has audio recordings made from the radio signals of different pulsars.

Neutron stars and novae are among the targets of the Hubble Space Telescope, and a number of striking images can be found at the Space Telescope Science Institute web site (**http://oposite.stsci.edu/pubinfo/Subject.html**). Much has been learned about neutron stars from other orbiting telescopes designed to observe high-energy radiation. The Laboratory for High Energy Astrophysics at the Goddard Space Flight Center maintains a list of links to all of these spacecraft, past, present, and future (**http://heasarc.gsfc.nasa.gov/docs/heasarc/missions.html**).

An especially good site with data from the Rossi X-ray Timing Explorer is maintained at the Massachusetts Institute of Technology (**http://space.mit.edu/XTE/XTE.html**).

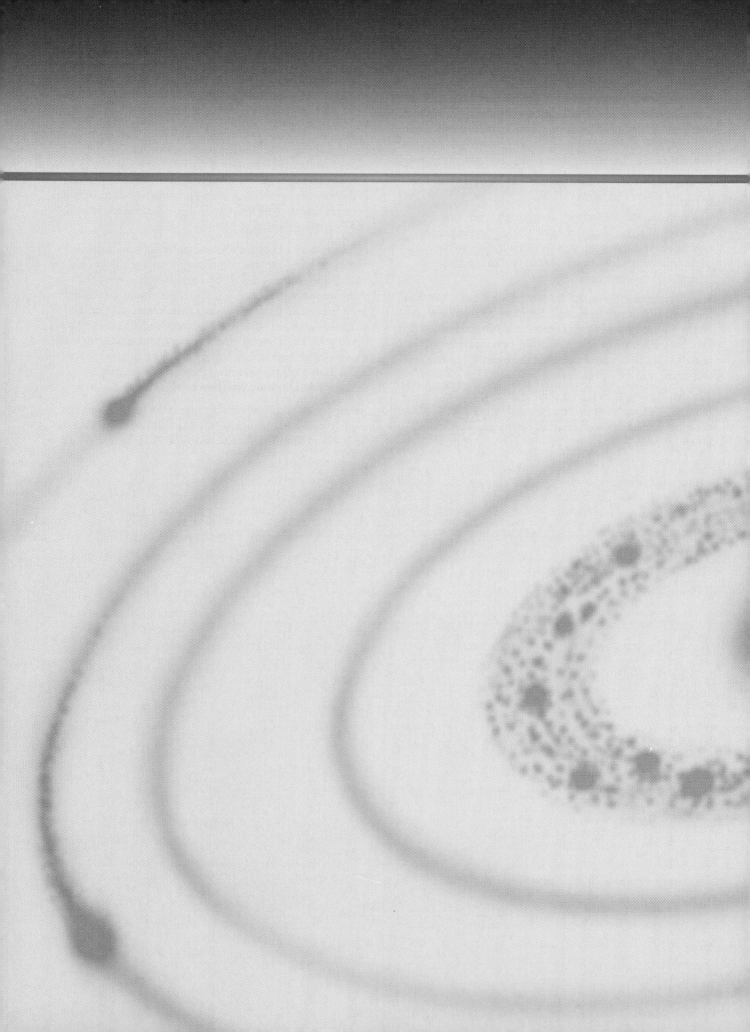

# Black Holes

A Black Hole in a Binary System  This artist's rendition shows the close binary system that contains Cygnus X-1. Cygnus X-1 is a strong source of X rays and is widely believed by astronomers to be a black hole. Gas from the companion star is captured into orbit about the black hole, forming an accretion disk. As gases spiral in toward the black hole, friction heats them to high temperatures. At the inner edge of the accretion disk, the gases are so hot that they emit X rays. (Courtesy of D. Norton, Science Graphics)

*In this chapter you will find the answers to the following questions:*

24-1  What are the two central ideas behind Einstein's special theory of relativity?

24-2  How did Einstein describe gravity, and how is his description different from Newton's?

24-3  How "black" is a black hole? In what sense is it "black"?

24-4  In what way are black holes actually simpler than any other objects in astronomy?

24-5  How do astronomers search for black holes?

to start off with a sizable positive charge, it would vigorously attract vast numbers of negatively charged electrons from the interstellar medium, which would soon neutralize the hole's charge. For this reason, astronomers usually neglect electric charge when discussing real black holes.

Although a black hole might theoretically have a tiny electric charge, it can have no magnetic field of its own whatsoever. When a black hole is created, however, the collapsing star from which it forms may possess an appreciable magnetic field. The star must therefore radiate this magnetic field away before it can settle down inside its event horizon. Theory predicts that the star does this by emitting electromagnetic and gravitational waves. As described in Section 24-2, gravitational waves are ripples in the overall geometry of space. Some physicists are exploring the possibility of observing the creation of black holes by detecting bursts of gravitational radiation emitted by collapsing massive stars (see Box 24-2).

Third, we can detect the effects of a black hole's rotation. Specifically, we can measure a black hole's *angular momentum*. As described in Box 23-1, an object's angular momentum is related to how fast it rotates and how the object's mass is distributed over its volume. As a dead star collapses into a black hole, its rotation naturally speeds up as its mass moves toward the center, just as a figure skater rotates faster when she pulls her arms and legs in. (This same effect explains why neutron stars spin rapidly, as described in Section 23-3.) Hence, we expect a black hole that forms from a rotating star to be spinning very rapidly. Einstein's theory makes the startling prediction that this rotation causes space and time to be dragged around the hole. A spinning black hole is thus surrounded by space that rotates with the hole. In fact, around the event horizon of every rotating black hole, a region exists where this dragging of space and time is so severe that it is impossible to stay in the same place. No matter what you do, you get pulled around the hole, along with the rotating geometry of space and time. This region, where it is impossible to be at rest, is called the **ergosphere** (Figure 24-10).

To measure a black hole's angular momentum, we could hypothetically place two satellites in orbit about the hole. Suppose that one satellite circles the hole in the same direction the hole rotates and the other in the opposite direction. One satellite is thus carried along by the geometry of space and time, but the other is constantly fighting its way "upstream." The two satellites will thus have different orbital periods. By comparing these two periods, astronomers can deduce the total angular momentum of the hole.

These three properties—mass, charge, and angular momentum—are the only ones that a black hole possesses. This simplicity is the essence of the famous **no-hair theorem**, first formulated in the early 1970s: "Black holes have no hair." Any and all additional properties carried by the matter that has fallen into the hole have disappeared from the universe and thus can have no effect on the structure of the hole.

---

| **24-5** | **Certain binary star systems probably contain black holes** |

Are there any black holes out there? Many astronomers think so, but finding black holes in the sky is a difficult business. Because light cannot escape from inside the event hori-

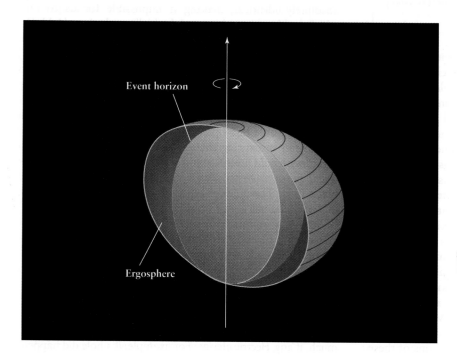

**Figure 24-10**

**The Structure of a Rotating Black Hole** A rotating black hole is surrounded by a region called the ergosphere, where the dragging of space and time around the hole is so severe that nothing can remain at a fixed location. Because the ergosphere (the shaded area in this cross-sectional diagram) is outside the event horizon, this bizarre region is accessible to us, and astronauts or asteroids could travel through it without disappearing into the black hole. According to detailed calculations, objects grazing the ergosphere could be catapulted back out into space at tremendous speeds. In other words, the ejected object could leave the ergosphere with more energy than it had initially, having extracted added energy from the hole's rotation. This is called the *Penrose process*, after Roger Penrose, the British mathematician who proposed it.

Event horizon

Ergosphere

zon, you cannot observe a black hole directly in the way that you can observe a star or a planet. The best you can hope for is to detect the effects of a black hole's powerful gravity.

Close binary star systems offer the best chance of finding black holes in our Galaxy. For example, if the gravitational attraction of a black hole captured gas from its companion star, the fate of this material might reveal the existence of the hole. In fact, since the early 1970s, several good black hole candidates have been discovered in just this way.

Shortly after the launch of the Uhuru X-ray-detecting satellite (described in Section 23-7), astronomers became intrigued with an X-ray source designated Cygnus X-1. Unlike pulsating X-ray sources, which emit regular bursts of X rays every few seconds, the emissions from Cygnus X-1 are highly variable and irregular. Its X-ray emission flickers on time scales that are as short as one-hundredth of a second. One of the fundamental concepts in physics is that nothing can travel faster than the speed of light (recall Section 24-1). Because of this limitation, an object cannot flicker faster than the time required for light to travel across the object. Because light travels 3000 kilometers in a hundredth of a second, Cygnus X-1 must be smaller than the Earth.

Cygnus X-1 occasionally emits radio radiation, and in 1971 radio astronomers used these outbursts to show that the source was at the same location in the sky as the star HDE 226868 (Figure 24-11). Spectroscopic observations revealed that HDE 226868 is a B0 supergiant with a surface temperature of about 31,000 K. Because such stars do not emit significant amounts of X rays, HDE 226868 alone cannot be the X-ray source Cygnus X-1. Because binary stars are very common, astronomers began to suspect that the visible star and the X-ray source are in orbit about each other.

Further spectroscopic observations soon showed that the spectral lines in the spectrum of HDE 226868 shift back and forth with a period of 5.6 days. This behavior is characteristic of a single-line spectroscopic binary (see Section 19-10); the companion of HDE 226868 is just too dim to produce its own set of spectral lines. The clear implication is that HDE 226868 and Cygnus X-1 are the two components of a binary star system.

From the mass-luminosity relation, HDE 226868 is estimated to have a mass of roughly 30 $M_\odot$. As a result, the unseen member of the binary system must have a mass of about 7 $M_\odot$ or more. Otherwise, it would not exert enough gravitational pull to make the B0 star wobble by the amount deduced from the periodic Doppler shifting of its spectral lines. Because 7 solar masses is too large for either a white dwarf or a neutron star, Cygnus X-1 is likely to be a black hole.

The case for Cygnus X-1 being a black hole is not airtight. HDE 226868 might have a low mass for its spectral type, which would imply a somewhat lower mass for Cygnus X-1. In addition, uncertainties in the distance to the binary system could further reduce estimates of the mass of Cygnus X-1. If all these uncertainties combined in just the right way, the estimated mass of Cygnus X-1 could be pushed down to about 3 $M_\odot$. Thus, there is a slim chance

**figure 24-11**   R I **V** U X G

**HDE 226868** This star is the optical companion of the X-ray source Cygnus X-1. The star is a B0 supergiant located 8000 light-years (2500 parsecs) from the Earth. Many astronomers suspect that Cygnus X-1 is a black hole, as shown in Figure 24-12. This photograph was taken with the 5-meter telescope at Palomar. (Courtesy of J. Kristian, Carnegie Observatories)

that Cygnus X-1 might contain the most massive possible neutron star rather than a black hole.

If Cygnus X-1 does contain a black hole, the X rays do not come from the black hole itself. Gas captured from HDE 226868 goes into orbit about the hole, forming an accretion disk about 4 million kilometers in diameter (Figure 24-12). As material in the disk gradually spirals in toward the hole, friction heats the gas to temperatures approaching $2 \times 10^6$ K. In the final 200 kilometers above the hole, these extremely hot gases emit the X rays that our satellites detect. Presumably the X-ray flickering is caused by small hot spots on the rapidly rotating inner edge of the accretion disk. In this way, the black hole's existence is announced by doomed gases just before they plunge to oblivion.

Two other likely black hole candidates are the X-ray sources LMC X-3 (located in a nearby galaxy called the Large Magellanic Cloud) and V404 Cygni (which is within our own Milky Way Galaxy, in the direction of the constellation Cygnus). Like Cygnus X-1, both of these sources orbit

## box 24-4 | Looking Deeper into Astronomy

### *Wormholes and Time Machines*

General relativity is so rich, complex, and fascinating—and so may be the geometry of black holes. For example, in the 1930s, Einstein and his colleague, Nathan Rosen, discovered that a black hole could possibly connect our universe with a second domain of space and time that is separate from ours. The first diagram shows this connection, called an **Einstein-Rosen bridge.** You could think of the upper surface as our universe and the lower surface as a "parallel universe." Alternatively, the upper and lower surfaces could be different regions of our own universe. In that case, an Einstein-Rosen bridge would connect our universe with itself, forming a **wormhole,** shown in the second diagram.

Scientists and science-fiction authors have speculated that a wormhole could be a shortcut to distant places in our universe, but detailed calculations reveal a major obstacle. The powerful gravity of a black hole causes the wormhole to collapse almost as soon as it forms. As a result, to get from one side of a wormhole to the other, you would have to travel faster than the speed of light, which is not possible.

Caltech physicists Kip Thorne, Michael Morris, and Ulvi Yurtsever have proposed a scheme that might get around the difficulty of a collapsing wormhole. According to general relativity, they point out, pressure as well as mass can be a source of gravity. Normally we do not see the

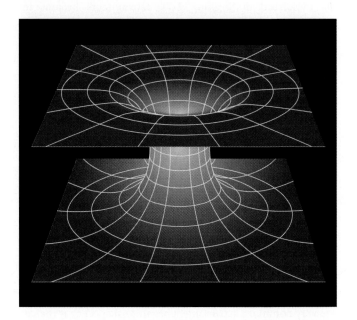

---

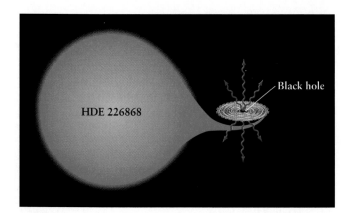

### figure 24-12

**The Cygnus X-1 System** A stellar wind from the blue supergiant HDE 226868 pours matter onto an accretion disk surrounding a possible black hole. The infalling gases from the disk are heated to high temperatures as they spiral in toward the hole. At the inner edge of the disk, just above the black hole, the gases become so hot that they emit vast quantities of X rays. An artist's rendition of this system is shown in the figure that opens this chapter.

ordinary stars. The orbital data suggests that both compact X-ray sources have masses of about 6 $M_\odot$, which is far greater than the probable maximum mass of a neutron star. Once again, however, the case that these are actually black holes is not firm, in part because we can see shifting spectral lines from only the conventional star in each binary system.

By far the most convincing black hole candidate to date is the flickering X-ray source A0620-00 in the constellation Monoceros. (The A refers to the British satellite *Ariel 5* that discovered the source; the numbers refer to its position in the sky). Like Cygnus X-1, LMC X-3, and V404 Cygni, A0620-00 is a member of a spectroscopic binary. The visible companion is an orange K5 main-sequence star called V616 Monocerotis, which orbits the X-ray source every 7.75 hours. Because this star is relatively faint, it is possible to observe the shifting spectral lines from *both* the visible companion and the X-ray source as they orbit each other. With this more complete information about the orbits, astronomers have determined the mass of A0620-00 to be 3.82 $M_\odot$, with an uncertainty of only 0.24 $M_\odot$. Because it is probably impossible for a neutron star to be more massive than 3 $M_\odot$, A0620-00 must almost certainly be a black hole.

gravitational effects of pressure because they are so small. Thorne and his colleagues speculate that a technologically advanced civilization might someday be able to use pressure to produce *antigravity* strong enough to keep the wormhole open.

If a wormhole could be held open, it could also be a "time machine." To see how, imagine you take one end of a wormhole and move it around for a while at speeds very near the speed of light. As discussed in Box 24-1, such motion causes clocks to slow down. Thus, when you stop moving that end of the wormhole, you find that it has not aged as much as the stationary end. In other words, one side of the wormhole has a different time than the other. As a result, you could go into one end of the wormhole at a late time and come out at an early time. For example, you might go in at 10 A.M. and come out at 9 A.M.

Time machines are illogical. If you could get back from a trip an hour before you left, you could meet yourself and tell yourself what a nice journey you had. Then both of you could take the trip. If you and your twin return just before you both left, there would be four of you. And all four of you could take the trip again. And then all eight of you. Then all sixteen of you. . . .

Making copies of yourself is an example of how time machines violate **causality,** the notion that effects must follow their causes. If time machines are possible, they lead to phenomena that are fundamentally illogical, irrational, and unpredictable.

We have never seen a violation of causality. To the contrary, the universe seems remarkably rational. Scientists would therefore like to show that time machines cannot exist. British astrophysicist Stephen Hawking points to one strong bit of observational evidence against time machines: We are not being visited by hordes of tourists from the future. If we could discover why nature precludes time machines, we would have a much deeper understanding of the nature of space and time.

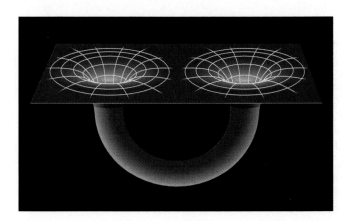

---

The X rays emitted from A0620-00 come not from the black hole but its surrounding accretion disk. What does such a disk look like? Observations cannot give us a direct answer to this question, because accretion disks are too small to be resolved by our best telescopes. Instead, supercomputers have been used to create detailed simulations of an accretion disk around a black hole. One such simulation was carried out in the 1980s by Larry L. Smarr at the University of Illinois and John F. Hawley, now at the University of Virginia. They used a supercomputer to solve the equations that describe how accreting gas is distributed around a black hole. As shown in Figure 24-13a, the simulation begins with gas falling toward the black hole. The infalling gas is endowed with angular momentum, which causes the gas to orbit about the hole. Eventually the orbiting gas settles into a thick disc centered about the black hole, as shown in cross section in Figure 24-13b. The inner edge of an accretion disk occurs where the outward-directed "centrifugal force" on the gas just balances the inward pull of gravity. Along this inner edge, the flow of gas is unstable and hot bubbles can develop (Figure 24-14).

If there really are black holes in close binary systems, how did they get there? One possibility is that one member of the binary explodes as a Type II supernova, leaving a burned-out core whose mass exceeds 3 $M_\odot$. This core undergoes gravitational collapse to form a black hole. But this is *not* the only way a black hole can form. A white dwarf or a neutron star in a binary system can become a black hole if it accretes enough matter from its companion star. This transformation can occur when the companion star becomes a red giant and dumps a significant part of its mass over its Roche lobe.

Another possibility is two dead stars coalescing to form a black hole. For example, imagine a binary system consisting of two neutron stars, such as the binary system discussed in Section 24-2. Because of the emission of gravitational radiation, the two stars gradually spiral in toward each other and eventually merge. If their total mass exceeds 3 $M_\odot$, the entire system may become a black hole.

A black hole is one of the most fantastic concepts ever to emerge from modern physical science. Some physicists have argued that the highly warped geometry of black holes allows for the possibility of wormholes to "other universes" and time machines (see Box 24-4). Although the idea of black holes initially met with skepticism, it is now clear that many

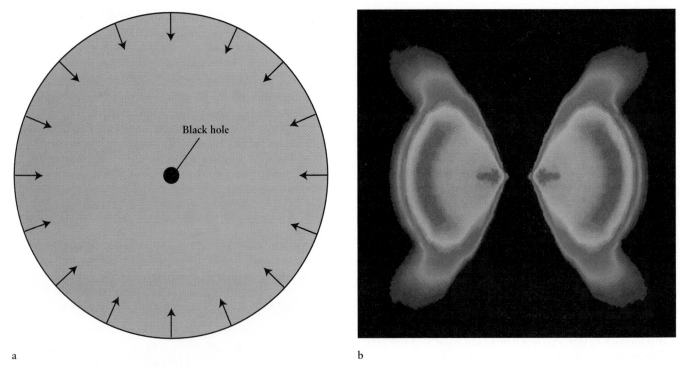

a                                                                    b

## figure 24-13

**An Accretion Disk Around a Black Hole**  A supercomputer can be used to solve the complicated relativistic equations that govern the infalling of matter toward a black hole. This supercomputer simulation begins with matter possessing angular momentum falling toward a black hole, as shown schematically in **(a).** After awhile, the infalling gases become captured into a doughnut-shaped accretion disk shown in cross section in **(b).** False color is used here to display density, from purple and blue for the least dense regions through the colors of the rainbow to red, which indicates the densest regions. (Courtesy of L. L. Smarr and J. F. Hawley)

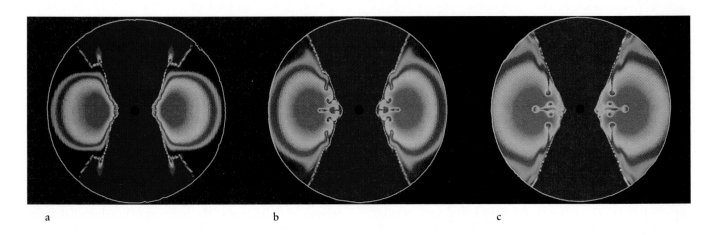

a                                          b                                          c

## figure 24-14

**The Inner Edge of an Accretion Disk**  Three stages in the evolution of the inner edge of an accretion disk are shown here in cross section using false color. The black hole is the dot in the center of each view. View **(a)** shows the undisturbed accretion disk. In **(b),** bubbles and fingers of hot gas develop because of instabilities along the inner edge of the disk. In **(c),** the bubbles have extended into the disk. (Courtesy of L. L. Smarr and J. F. Hawley)

of the stars we see in the sky are doomed to disappear from the universe someday, leaving only black holes behind. Even more astounding is the idea that enormous black holes, containing millions or even billions of solar masses, are located at the centers of many galaxies and quasars. As we will see in Chapter 25, one of these monstrosities is very probably lurking at the center of our Milky Way, only 28,000 light-years from the Earth.

## Key Words

*Terms preceded by an asterisk (\*) are discussed in the Boxes.*

black hole, p. 590

\*causality, p. 605

\*Einstein-Rosen bridge, p. 604

electromagnetic theory, p. 590

ergosphere, p. 602

event horizon, p. 599

general theory of relativity, p. 594

gravitational radiation (gravitational waves), p. 596

gravitational redshift, p. 596

law of cosmic censorship, p. 600

\*length contraction, p. 593

\*Lorentz transformations, p. 592

no-hair theorem, p. 602

principle of equivalence, p. 594

\*proper length (proper distance), p. 593

\*proper time, p. 592

\*Schwarzschild radius ($R_{Sch}$), p. 600

singularity, p. 599

special theory of relativity, p. 591

\*time dilation, p. 593

\*wormhole, p. 604

## Key Ideas

**The Special Theory of Relativity:** This theory, published by Einstein in 1905, is based on the notion that there is no such thing as absolute space.

• The speed of light is the same to all observers, no matter how fast they are moving.

• An observer will note a slowing of clocks and a shortening of rulers that are moving with respect to the observer. This effect becomes significant only if the clock or ruler is moving with a speed near the speed of light.

**The General Theory of Relativity:** Published by Einstein in 1915, this is a theory of gravity. Any massive object causes space to become curved and time to slow down, and these effects manifest themselves as a gravitational force. These distortions of space and time are most noticeable in the vicinity of large masses or compact objects.

• The general theory of relativity is our most accurate description of gravitation. It predicts a number of phenomena, including the bending of light by gravity and the gravitational redshift, whose existence has been confirmed by observation and experiment.

• The general theory of relativity also predicts the existence of gravitational waves, which are ripples in the overall geometry of space and time produced by moving masses. Gravitational waves have been detected indirectly, and specialized antennas are under construction to make direct measurement of the gravitational waves from cosmic cataclysms.

**Black Holes:** If a stellar corpse has a mass greater than about 3 M$_\odot$, gravitational compression will overwhelm any and all forms of internal pressure. The stellar corpse will collapse to such a high density that its escape speed exceeds the speed of light. The corpse then contracts rapidly to a single point called a singularity, where all of the mass is located.

• The singularity is surrounded by a surface called the event horizon, where the escape speed equals the speed of light. Nothing—not even light—can escape from inside the event horizon.

**Properties of Black Holes:** A black hole—a singularity surrounded by an event horizon—has only three physical properties: mass, electric charge, and angular momentum.

• A rotating black hole (one with angular momentum) has a region called the ergosphere around the outside of the event horizon. In the ergosphere, space and time themselves are dragged along with the rotation of the black hole.

**Black Holes in Binary Systems:** Some binary star systems are thought to contain a black hole. In such a system, gases captured from the companion star by the black hole emit detectable X rays.

## Review Questions

1. In Einstein's special theory of relativity, two different observers moving at different speeds will measure the same value of the speed of light. Will these same observers measure the same value of, say, the speed of an airplane? Explain.

**2.** Why does the speed of light represent an ultimate speed limit?

**3.** Why is Einstein's general theory of relativity a better description of gravity than Newton's universal law of gravitation?

**4.** Under what circumstances are degenerate electron pressure and degenerate neutron pressure incapable of preventing the complete gravitational collapse of a dead star?

**5.** In what way is a black hole blacker than black ink or a black piece of paper?

**6.** If the Sun suddenly became a black hole, how would the Earth's orbit be affected? Explain.

**7.** According to the general theory of relativity, why can't some sort of yet-undiscovered degenerate pressure prevent the matter inside a black hole from collapsing all the way down to a singularity?

**8.** What is the law of cosmic censorship?

**9.** What is the no-hair theorem?

**10.** Should we worry about the Earth being pulled into a black hole? Why or why not?

**11.** What kind of black hole is surrounded by an ergosphere? What happens inside the ergosphere?

**12.** Why do you suppose that all the black hole candidates mentioned in the text are members of very short-period binary systems?

**13.** As a binary star system loses energy by emitting gravitational waves, why do its member stars speed up and why does the period become shorter?

## Advanced Questions

*Questions preceded by an asterisk (\*) involve topics discussed in the Boxes.*

> **Problem-solving tips and tools:**
>
> Remember that the time to travel a certain distance is equal to the distance divided by the speed, and that the density of an object is its mass divided by its volume. The volume of a sphere of radius $r$ is $\frac{4}{3}\pi r^3$. Box 4-2 shows how to use Newton's formulation of Kepler's third law, which explicitly includes masses; when using this formula, note that the period $P$ must be expressed in seconds, the semimajor axis $a$ in meters, and the masses in kilograms. Another version, in which period is in years, semimajor axis in AU, and masses in solar masses, appears in Section 19-9.

**\*14.** A clock on board a moving starship shows a total elapsed time of 2 minutes, while an identical clock on the Earth shows a total elapsed time of 10 minutes. What is the speed of the starship relative to the Earth?

**\*15.** How fast should a meter stick be moving in order to appear to be a "centimeter stick"?

**\*16.** An astronaut flies from the Earth to a distant star at 80% of the speed of light. As measured by the astronaut, the one-way trip takes 15 years. (**a**) How long does the trip take as measured by an observer on the Earth? (**b**) What is the distance from the Earth to the star (in light-years) as measured by an Earth observer? As measured by the astronaut?

**17.** In the binary system of two neutron stars discovered by Hulse and Taylor (Section 24-2), one of the neutron stars is a pulsar. The time interval between pulses from the pulsar is not constant: It is greatest when the two stars are closest to each other and least when the two stars are farthest apart. Explain why this is consistent with the gravitational slowing of time, depicted in Figure 24-6a.

**18.** Estimate how long it will be until the two neutron stars that make up the binary system discovered by Hulse and Taylor collide with each other. Assume that the distance between the two stars will continue to decrease at its present rate of 3 mm every 7.75 hours. (You can assume that the two stars are very small, so they will collide when the distance between them is equal to zero.)

**19.** Find the total mass of the neutron star binary system discovered by Hulse and Taylor (Section 24-2), for which the orbital period is 7.75 hours and the average distance between the neutron stars is 2.8 solar radii. Is your result reasonable for a pair of neutron stars? Explain.

**\*20.** Find the Schwarzschild radius for an object having a mass equal to that of the planet Jupiter.

**\*21.** What is the Schwarzschild radius of a black hole whose mass is (**a**) the mass of the Earth, (**b**) the mass of the Sun, and (**c**) the visible mass of the Milky Way Galaxy, which is about $1.1 \times 10^{11}$ $M_\odot$? In each case, also calculate what the density would be if the matter were spread uniformly throughout the volume of the event horizon.

**\*22.** To what density must the matter of a dead 10-$M_\odot$ star be compressed in order for the star to disappear inside its event horizon?

**\*23.** Prove that the density of matter needed to produce a black hole is inversely proportional to the square of the mass of the hole. If you wanted to make a black hole from matter compressed to the density of water (1000 kg/m³), how much mass would you need?

**24.** The orbital period of the binary system containing A0620-00 is 0.32 day, and Doppler shift measurements reveal that the radial velocity of the X-ray source peaks at 457 km/s (about 1 million miles per hour). (**a**) Assuming that the orbit of the X-ray source is a circle, find the radius of its orbit in kilometers. (This is actually an estimate of the semimajor axis of the orbit.) (**b**) By using Newton's form of Kepler's third law, prove that the mass of the X-ray source must be at least 3.1 times the mass of the Sun. (*Hint:* Assume that the mass of the K5V visible star—about 0.5 $M_\odot$ from the mass-luminosity relationship—is negligible compared to that of the invisible companion.)

# Discussion Questions

**25.** The speed of light is the same for all observers, regardless of their motion. Discuss why this requires us to abandon the Newtonian view of space and time.

**26.** Describe the kinds of observations you might make in order to locate and identify black holes.

**27.** Speculate on the effects you might encounter on a trip to the center of a black hole.

# Observing Project

**28.** As you well know, you cannot see a black hole with a telescope. Nevertheless, you might want to observe the visible companion of Cygnus X-1. The epoch 2000 coordinates of this ninth-magnitude star are R.A. = $19^h$ $58.4^m$ and Decl. = +35° 12′, which is quite near the bright star η (eta) Cygni. Compare what you see with the photograph in Figure 24-11.

# Where to Learn More

*Books and magazine articles*

Abramowicz, M. A. "Black Holes and the Centrifugal Force Paradox." *Scientific American,* March 1993. This fascinating article discusses the paradoxical prediction that an object orbiting near a black hole feels a centrifugal force pushing *inward* rather than outward.

Begelman, M., and Rees, M. *Gravity's Fatal Attraction: Black Holes in the Universe.* Scientific American Library, 1996. Two distinguished astrophysicists explain the most exotic objects in the universe in clear, lucid prose.

Davies, P. "Wormholes and Time Machines." *Sky & Telescope,* January 1992. This intriguing article by a noted scientist discusses some of the fanciful aspects of black holes.

Jeffries, A., and others. "Gravitational Wave Observatories." *Scientific American,* June 1987. This article, which includes an excellent description of gravitational radiation, explains how lasers can be used to detect gravitational waves.

Kaufmann, W. *Black Holes and Warped Spacetime.* W. H. Freeman, 1979. This slim book gives a nontechnical overview of black holes and general relativity.

————. *Cosmic Frontiers of General Relativity.* Little, Brown, 1977. This book describes many fascinating aspects of black holes, including views that might greet an astronaut

who plunges through a wormhole connecting our universe with another.

McClintock, J. "Do Black Holes Exist?" *Sky & Telescope,* January 1988. This well-written article presents evidence for black holes in certain X-ray binaries, especially A0620-00.

Taylor, E. F., and Wheeler, J. A. *Spacetime Physics,* 2nd edition. W. H. Freeman, 1992. This classic volume is an excellent introduction to the ideas of the special and general theories of relativity. Only a minimum of mathematics is needed.

Thorne, K. S. *Black Holes and Time Warps: Einstein's Outrageous Legacy.* W. W. Norton, 1994. A well-known expert in general relativity presents a lucid explanation of Einstein's theory and modern observations.

Weisberg, J., Taylor, J., and Fowler, L. "Gravitational Waves from an Orbiting Pulsar." *Scientific American,* October 1981. This article explains how orbital changes in the binary pulsar PSR 1913+16 result from gravitational radiation emitted by the system.

Wheeler, J. A. *A Journey into Gravity and Spacetime.* Scientific American Library, 1990. One of the leading experts in the general theory of relativity explains the subtleties of this remarkable theory of gravitation.

W *World Wide Web*

A number of web sites provide help in visualizing the seemingly paradoxical ideas of the theories of relativity. One of the best has been created by Roberto Salgado at Syracuse University (**http://www.phy.syr.edu/courses/ modules/LIGHTCONE/**). Another site, created by Andrew Hamilton at the University of Colorado, takes an in-depth look at what happens when you fall into a black hole (**http://mensae.colorado.edu/~ajsh/schw.shtml**).

Two spacecraft missions, STEP and Gravity Probe B, are designed to test the predictions of general relativity with unparalleled precision. These missions, scheduled for launch around the turn of the century, are described at a web site at Stanford University (**http://einstein.stanford.edu/**).

There are also web sites for each of the experiments designed to detect gravitational radiation: LIGO (**http://www.ligo. caltech.edu/**), VIRGO (**http://www.pg.infn.it/virgodoc/**), and LISA (**http://www.estec.esa.nl/spdwww/future/html/lisa.htm** and **http://www.lisa.uni-hannover.de/**), as well as a master list of all current and planned experiments (**http://www. roma1.infn.it/rog/others/others.html**).

A spiral galaxy and an elliptical galaxy. (W. Keel and R. White,
University of Alabama, and NASA)

# Galaxies and Cosmology

Just over a century ago, most astronomers thought that the universe was a few thousand light-years across. Beyond the disk of stars we call the Milky Way Galaxy, of which our Sun is a part, they saw only black, empty space.

We now understand that the universe is far more vast than astronomers of the nineteenth century ever realized. The Milky Way is just one of billions of galaxies strewn across the distances of billions of light-years. And by looking at the most remote galaxies, we can see into the distant past of the universe—and gather hints about its distant future.

In the next six chapters we explore the realm of the galaxies. We begin with our home galaxy, the Milky Way, then go on to study both normal and unusual galaxies. And we will investigate cosmology, the science of the universe as a whole. We conclude with a tantalizing and as yet unanswered question: Are there civilizations on other worlds?

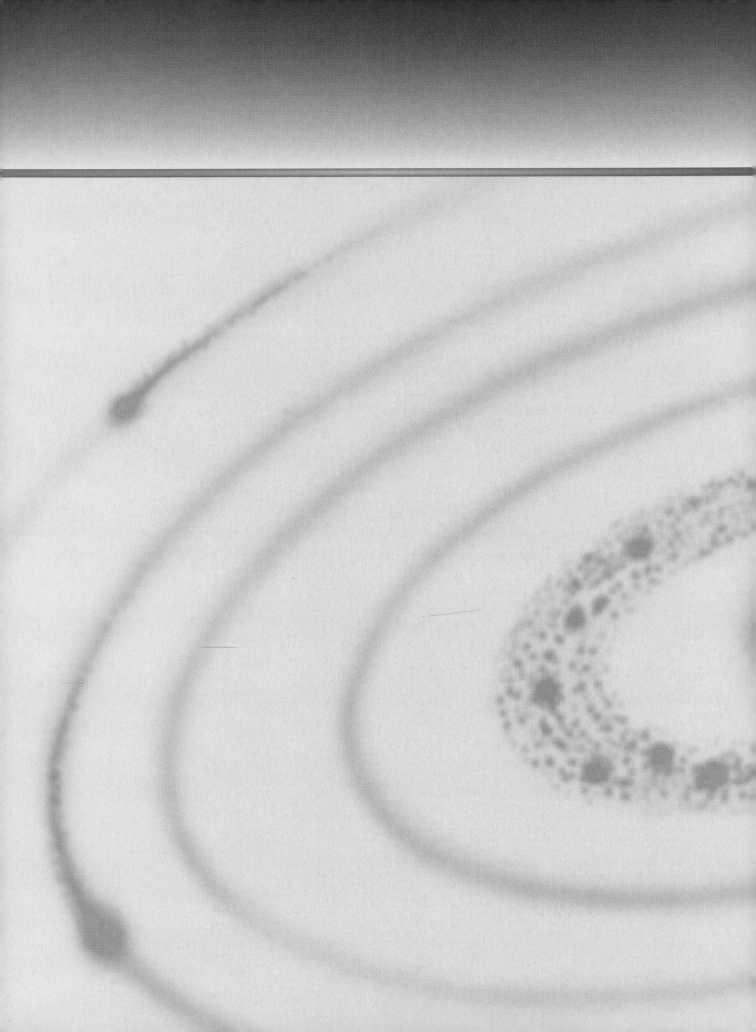

# Our Galaxy

R I **V** U X G

**Our Galaxy** This wide-angle photograph, taken from Australia, spans 180° of the Milky Way, from the constellation Cygnus at the far right to Crux (the Southern Cross) at the left. The center of our Galaxy lies some 8000 parsecs (26,000 light-years) from the Earth in the constellation of Sagittarius, which is in the middle of this photograph. The dark lanes and blotches are caused by interstellar gas and dust obscuring the light from the background stars. This obscuration prevents us from viewing the actual center of the Galaxy with visible-light telescopes. (Courtesy of D. di Cicco)

*In this chapter you will find the answers to the following questions:*

25-1 What is our Galaxy? How do astronomers know where we are located within it?

25-2 What is the shape and size of our Galaxy?

25-3 How do we know that our Galaxy has spiral arms?

25-4 What is most of the Galaxy made of? Is it stars, gas, dust, or something else?

25-5 What is the nature of the spiral arms?

25-6 What lies at the very center of our Galaxy?

*On a clear, moonless night, away from the glare of city lights, you can often see a hazy, luminous band stretching across the sky. This band, which extends all the way around the celestial sphere, is called the Milky Way. Galileo was the first person to view the Milky Way with a telescope. He immediately discovered that it is composed of countless dim stars. Today, we realize that the Milky Way is actually a disk tens of thousands of parsecs wide that contains several hundred billion stars, one of which is our own Sun, as well as vast quantities of gas and dust. It is one among myriads of **galaxies**, or systems of stars and interstellar matter. The disk of the Milky Way, along with a halo of stars that surrounds it, is called the **Milky Way Galaxy.** (It is also called "the Galaxy" or "our Galaxy," with a capital G).*

*Our Galaxy has a rich and fascinating structure, with immense spiral arms that arch gracefully outward from its nucleus. As stars orbit around the galactic center, their motions reveal the presence of vast quantities of matter that emits no measurable radiation. The character of this dark matter, which constitutes the lion's share of the Galaxy's mass, remains mysterious. No less mysterious is the very center of the Galaxy, where stars and gas orbit at high speed around an object with a mass millions of times greater than that of the Sun. This remarkable object is very probably a gigantic black hole.*

*In studying the Milky Way, we explore the universe on a grand scale. Instead of examining individual stars, we look at a system of stars. And instead of focusing on the structure and life of an isolated star, we look at the overall arrangement and history of a huge stellar community of which the Sun is a member. In this way, we gain insights into galaxies in general and prepare ourselves to ask fundamental questions about the cosmos.*

---

## 25-1 The Sun is located in the disk of our Galaxy, about 8,000 parsecs from the galactic center

Eighteenth-century astronomers were the first to suspect that because the Milky Way completely encircles us, all the stars in the sky are part of an enormous disk of stars—the Milky Way Galaxy. We have an edge-on view from inside this disk, which is why the Milky Way appears as a band around the sky (Figure 25-1). But where within this disk is our own Sun? For well over a century, the prevailing opinion was that the Sun and planets lie at the Galaxy's center. We now know that the Milky Way contains not only stars but also copious interstellar dust. This dust hides most of the Galaxy's stars from view and thus obscures the true dimensions of the Galaxy. As we will see, we are actually nowhere near its center.

An early attempt to deduce the Sun's location in the Galaxy was made by the English astronomer William Herschel (the discoverer of Uranus and a pioneering cataloger of binary star systems, as described in <u>Section 16-1 and Section 19-9</u>).

Herschel's approach was to count the number of stars in each of 683 regions of the sky. He reasoned that he should see the greatest number of stars toward the Galaxy's center and a lesser number toward the Galaxy's edge.

Herschel found approximately the same density of stars all along the Milky Way. Therefore, he concluded that we are at the center of our Galaxy (Figure 25-2). In the early 1900s, Dutch astronomer Jacobus Kapteyn came to essentially the same conclusion by analyzing the brightness and proper motion of a large number of stars. According to Kapteyn, the Milky Way is about 17 kpc (17 kiloparsecs = 17,000 parsecs, or 55,000 light-years) in diameter, with the Sun near its center.

Both Herschel and Kapteyn were wrong about the Sun being at the center of our Galaxy. The reason for their mistake was finally discerned in 1930 by Robert J. Trumpler of Lick Observatory. While studying star clusters, Trumpler discovered that the more remote clusters appear unusually dim—more so than would be expected from their distances alone. As a result, Trumpler concluded that interstellar space must not be a perfect vacuum: It must contain dust that absorbs or scatters light from distant stars.

We first encountered interstellar dust in <u>Section 20-2</u>. Like the stars themselves, this material is concentrated in the plane of the Galaxy. As a result, it obscures our view within the plane and makes distant objects appear dim, an effect

## figure 25-1

**Our View of the Milky Way** The Milky Way Galaxy, of which our Sun is a member, is a disk-shaped collection of hundreds of billions of stars (of which our Sun is one). These stars are so numerous that when we look out at the night sky in the plane of the disk, they appear as a continuous band of light that stretches all the way around the sky. When we look in a direction perpendicular to the plane of the Galaxy, we see only those relatively few stars that lie between us and the "top" or "bottom" of the disk.

called **interstellar extinction.** Great patches of interstellar dust are clearly visible in such wide-angle photographs as the one that opens this chapter. Thanks to interstellar extinction, Herschel and Kapteyn were actually seeing only the nearest stars in the Galaxy. Hence, they had no idea of either the enormous size of the Galaxy or of the vast number of stars concentrated around the galactic center.

**ANALOGY** Herschel and Kapteyn faced much the same dilemma as a lost motorist on a foggy night. Unable to see more than a city block in any direction, he would have a hard time deciding what part of town he was in or even if he was in a small town or a large city. If the fog layer were relatively shallow, however, our hapless motorist would be able to see the lights from tall buildings that extend above the fog, and in that way he could determine his location (*Figure 25-3a*). The same principle applies to our Galaxy. While interstellar dust in the plane of our Galaxy hides the sky covered by the Milky Way, we have an almost unobscured view out of the plane (that is, to either side of the Milky Way). To find our location in the Galaxy, we need to locate bright objects that are part of the Galaxy but lie outside its plane in obscured regions of the sky.

Fortunately, bright objects like this do exist. They are a type of star cluster called a **globular cluster,** which are associated with the Galaxy but lie outside of its plane (Figure 25-3b). As discussed in Section 21-3, a typical globular cluster is a spherical distribution of roughly $10^6$ stars packed in a volume only a few hundred light-years across (see Figure 21-7).

However, to use globular clusters to determine our location in the Galaxy, we must first determine the distances from the Earth to these clusters. (Think again of our lost motorist—glimpsing the lights of a skyscraper through the fog is useful to the motorist only if he can tell how far away the skyscraper is.) Variable stars in globular clusters provide the distances, giving astronomers the key to the dimensions of our Galaxy.

In 1912, American astronomer Henrietta Leavitt reported her important discovery of the period-luminosity relation for Cepheid variables. As described in Section 21-5 (recall Figure 21-13), Cepheid variables are pulsating stars that vary periodically in brightness. Leavitt studied numerous Cepheids in the Small Magellanic Cloud (a small galaxy near the Milky Way) and found their periods to be directly related to their average luminosities. As Figure 25-4 shows, the longer a Cepheid's period, the greater its luminosity.

The period-luminosity law is an important tool in astronomy, because it can be used to determine distances. For example, suppose you find a Cepheid variable in the sky. By measuring its period and using a graph like Figure 25-4, you can determine the star's average luminosity. Knowing the

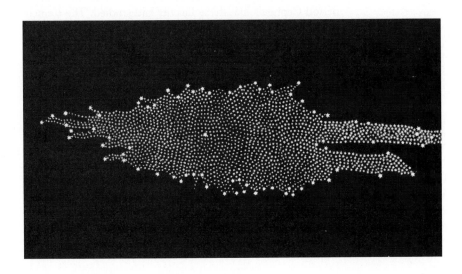

## figure 25-2

**Herschel's Map of Our Galaxy** In the late eighteenth century, English astronomer William Herschel attempted to map the Milky Way Galaxy by counting the numbers of stars in various parts of the sky. Because of interstellar dust that blocked his view of distant stars, Herschel erroneously concluded that the Sun is at the center of the Galaxy. (Yerkes Observatory)

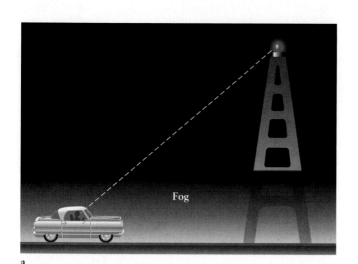

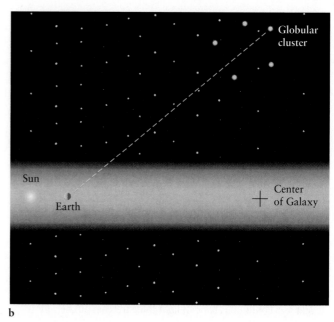

## figure 25-3

**Finding the Center of the Galaxy** **(a)** A motorist lost on a foggy night can determine his location by looking for tall buildings that extend above the fog. **(b)** In the same way, astronomers determine our location in the Galaxy by observing globular clusters that are part of the Galaxy but lie outside the obscuring material in the galactic disk. Observations by Harlow Shapley showed that the globular clusters form a spherical halo centered several thousand parsecs from the Sun in the direction of the constellation Sagittarius. The center of the halo is also the center of the Galaxy.

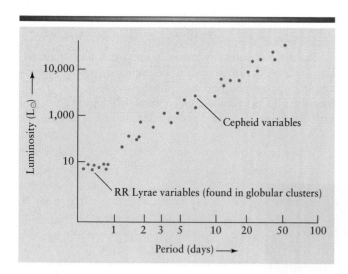

## figure 25-4

**The Period–Luminosity Relation for Cepheids** This graph shows the relationship between the periods and luminosities of Cepheid variables. The brighter the Cepheid, the longer is its pulsation period. Data for RR Lyrae variables, which are similar variable stars found in globular clusters, are also shown. RR Lyrae variables are horizontal branch stars with average luminosities of about 100 $L_\odot$ and pulsation periods of less than a day.

star's average luminosity, you can find out how far away the star must be in order to give the observed brightness. (We explained how this is done in Box 19-2).

Shortly after Leavitt's discovery of the period-luminosity law, Harlow Shapley, a young astronomer at the Mount Wilson Observatory in California, began studying a family of pulsating stars closely related to Cepheid variables called **RR Lyrae variables**. The light curve of an RR Lyrae variable (Figure 25-5) is similar to that of a Cepheid (see Figure 21-13), and Shapley believed RR Lyrae variables to be short-period Cepheids like those Leavitt had studied. The tremendous importance of RR Lyrae variables is that they are commonly found in globular clusters (Figure 25-6). By using the period-luminosity relationship for these stars, Shapley was able to determine the distances to the 93 globular clusters then known. He found that some of them were more than 100,000 light-years from Earth. The large values of these distances immediately suggested that the Galaxy was much larger than Herschel or Kapteyn had thought.

Another striking property of globular clusters is how they are distributed across the sky. Ordinary stars and open clusters of stars (see Section 20-6, especially Figures 20-17 and 20-18) are rather uniformly spread along the Milky Way. However, the majority of the 93 globular clusters that Shapley studied are located in one-half of the sky, widely scattered around the portion of the Milky Way that is in the constella-

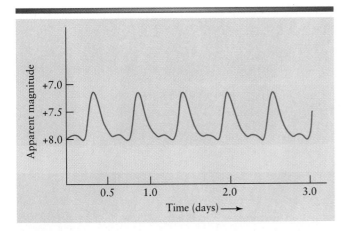

## figure 25-5

**The Light Curve of RR Lyrae** An RR Lyrae variable can be recognized by its characteristic variations in brightness. All RR Lyrae variables have periods of less than a day and brightnesses that vary by about 1 magnitude (that is by a factor about 2.5; see Section 19-3). The example shown here is the curve of RR Lyrae, the prototype of this class of stars.

tion Sagittarius. From the directions to the globular clusters and their distances from us, Shapley mapped out the three-dimensional distribution of these clusters in space. In 1920 he concluded that the globular clusters form a huge spherical distribution centered not on the Earth but rather about a point in

## figure 25-6    R I V U X G

**The Globular Cluster M55** Three RR Lyrae variables are identified with arrows in this globular cluster, located in the constellation Sagittarius. From their average apparent brightness (as seen in this photograph) and average luminosity (known to be roughly 100 L☉), astronomers have deduced that the distance to this cluster is 20,000 light-years. (Harvard Observatory)

the Milky Way several kiloparsecs away in the direction of Sagittarius (Figure 25-3b). This point, reasoned Shapley, must coincide with the center of our Galaxy, because of gravitational forces between the disk of the Galaxy and the "halo" of globular clusters. Therefore, by locating the center of the distribution of globular clusters, Shapley was in effect measuring the location of the galactic center.

Since Shapley's pioneering observations, many astronomers have measured the distance from the Sun to the **galactic nucleus,** the center of our Galaxy. Shapley's estimate of this distance was too large by about a factor of 2, because he did not take into account the effects of interstellar extinction (which were not well understood at the time). Today, the generally accepted distance to the galactic nucleus is about 8 kpc (26,000 light-years), with an uncertainty of about 1 kpc (3300 light-years). Just as Copernicus and Galileo showed that the Earth was not at the center of the solar system, Shapley and his successors showed that the solar system lies nowhere near the center of the Galaxy. We see that the Earth indeed occupies no special position in the universe.

## 25-2   Observations at nonvisible wavelengths reveal the shape of the Galaxy

At visible wavelengths, light suffers so much interstellar extinction that the galactic nucleus is totally obscured from view. But the amount of interstellar extinction is roughly inversely proportional to its wavelength. In other words, the longer the wavelength, the farther radiation can travel through interstellar dust without being scattered or absorbed. As a result, we can see farther into the plane of the Milky Way at infrared wavelengths than at visible wavelengths, and radio waves can traverse the Galaxy freely. For this reason, telescopes sensitive to these nonvisible wavelengths are important tools for studying the structure of our Galaxy.

Infrared light is an important tool for tracing the location of interstellar dust in the Galaxy. Starlight warms the dust grains to temperatures in the range of 10 to 90 K; thus, in accordance with Wien's law (see Section 5-4), the dust emits radiation predominantly at wavelengths from about 30 to 300 μm. These are called **far-infrared** wavelengths, because they lie in the part of the infrared spectrum most different in wavelength from visible light (see Figure 5-7). At these wavelengths, interstellar dust radiates more strongly than stars, so a far-infrared view of the sky is principally a view of where the dust is. In 1983 the Infrared Astronomical Satellite (IRAS) scanned the sky with a 60-cm reflecting telescope at far-infrared wavelengths, giving the panoramic view of the Milky Way's dust shown in Figure 25-7a.

In 1990 an instrument on the Cosmic Background Explorer (COBE) satellite scanned the sky at **near-infrared** wavelengths, that is, relatively short wavelengths closer to the visible spectrum. Figure 25-7b shows the resulting near-infrared view of

a

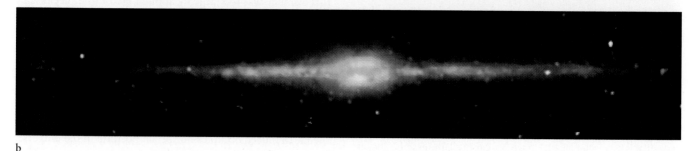

b

## ƒigure 25-7   R ∎ V U X G

**The Infrared Milky Way** **(a)** This far-infrared view was constructed from observations made by the IRAS spacecraft at wavelengths of 25, 60, and 100 μm. Interstellar dust is the principal source of radiation in this wavelength range, so the image shows how this dust is distributed along the Milky Way. Note that the dust is very narrowly confined to the plane of the Galaxy, which runs horizontally across the image. **(b)** In this near-infrared view, constructed from COBE observations at 1.2, 2.2, and 3.4 μm, we see much farther through the dust in the Milky Way than we can at visible wavelengths. The light recorded in this image comes principally from stars in the plane of the Galaxy. Note the bulge at the center of the Galaxy, which is at the center of this image. (NASA)

the plane of the Milky Way. At near-infrared wavelengths, interstellar dust does not emit very much light. Hence, the light sources in Figure 25-7b are stars, which do emit strongly in the near-infrared (especially the cool stars, such as red giants and supergiants). Because interstellar dust causes little interstellar extinction in the near-infrared, many of the stars whose light is recorded in Figure 25-7b lie deep within the Milky Way.

Observations such as those shown in Figure 25-7, along with the known distance to the center of the Galaxy, have helped astronomers to establish the dimensions of the Galaxy. The **disk** of our Galaxy is about 50 kpc (160,000 light-years) in diameter and about 0.6 kpc (2000 light-years) thick, as shown in Figure 25-8. The center of the Galaxy is surrounded by a distribution of stars, called the **central bulge,** that is about 2 kpc (6500 light-years) in diameter. This central bulge is clearly visible in Figure 25-7b. Observations suggest that this bulge is not shaped like a flattened sphere but may be elongated somewhat into a bar or peanut shape. (The potential significance of this elongated shape is discussed in Section 25-5.) The spherical distribution of globular clusters traces the **halo** of the Galaxy. Our Galaxy, seen edge-on from a great distance, would probably look somewhat like the galaxy NGC 4565 shown in Figure 25-9.

Different kinds of stars are found in the various components of our Galaxy. The globular clusters in the halo are composed of old, metal-poor, Population II stars. (We introduced the different populations of stars in Section 21-4.) Although these clusters are conspicuous, they contain only about 1% of the total number of stars in the halo. Most halo stars are single Population II stars in isolation, called **high-velocity stars** because of their high speed relative to the Sun. These ancient stars orbit the Galaxy along paths tilted at random angles to the plane of the Milky Way.

The stars in the disk are mostly young, metal-rich, Population I stars like the Sun. The disk of a galaxy appears bluish, as in Figure 25-9, because its light is dominated by radiation from hot O and B main-sequence stars. Such stars have very short main-sequence lifetimes (see Section 21-1, especially Table 21-1), so they must be quite young by astronomical standards. Hence, their presence shows that there must be active star formation in the galactic disk. By contrast, no O or B stars are present in the halo, which implies that star formation ceased there long ago.

The central bulge contains both Population I and Population II stars, suggesting that some stars there are quite ancient whereas others were created more recently. The cen-

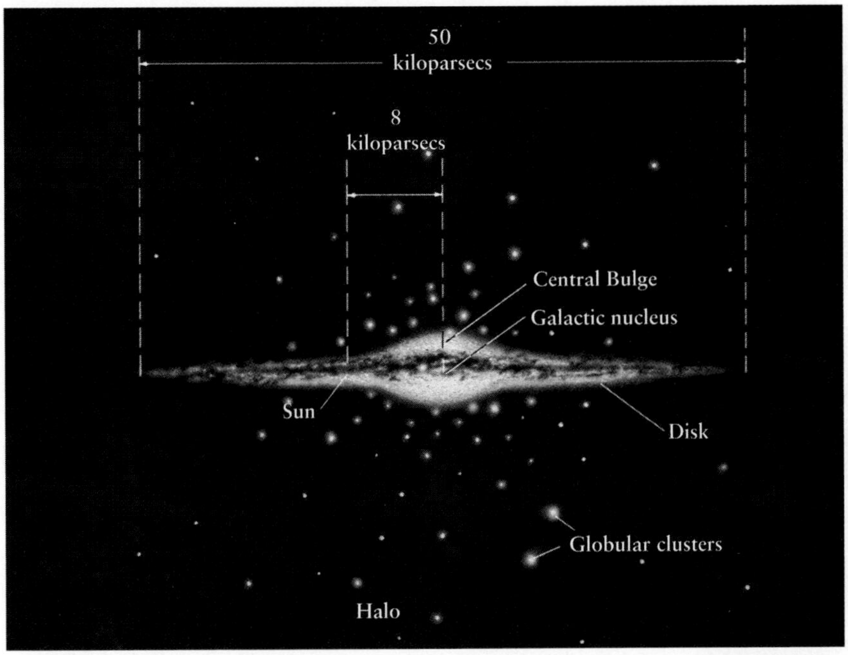


**Figure 25-8**

**Our Galaxy (Schematic Edge-On View)** There are three major components of our Galaxy: a thin disk, a central bulge, and a halo. The disk contains gas and dust along with metal-rich (Population I) stars. The halo is composed almost exclusively of old, metal-poor (Population II) stars. The central bulge is a mixture of Population I and Population II stars.

tral bulge looks reddish because it contains many red giants and red supergiants.

Why are there such different populations of stars in the halo, disk, and central bulge? Why has star formation stopped in some regions of the Galaxy but continues in other regions? The answers to these questions are related to the way that stars, as well as the gas and dust from which stars form, move within the Galaxy.

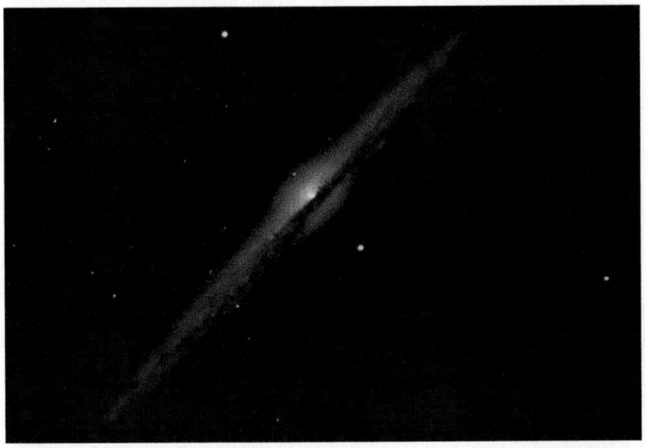

**Figure 25-9** R I **V** U X G

**Edge-On View of the Galaxy NGC 4565** If we could view our Galaxy edge-on from a great distance, it would probably look like this galaxy in the constellation Coma Berenices. A layer of dust and gas is clearly visible in the plane of the galaxy. Also note the reddish color of the bulge that surrounds the galaxy's nucleus. (U.S. Naval Observatory)

## 25-3 Observations of star-forming regions reveal that our Galaxy has spiral arms

Because interstellar dust obscures our visible-light view in the plane of our Galaxy, a detailed understanding of the structure of the galactic disk had to wait until the development of radio astronomy. Because of their long wavelengths, radio waves can penetrate the interstellar medium even more easily than infrared light and can travel without being scattered or absorbed. As we shall see in this section, both radio and optical observations reveal that our Galaxy has **spiral arms,** spiral-shaped concentrations of gas and dust that extend outward from the center in a shape reminiscent of a pinwheel.

Hydrogen is by far the most abundant element in the universe. Hence, by looking for concentrations of hydrogen gas, we should be able to detect important clues about the structure of the disk of our Galaxy. Unfortunately, the major electron transitions in the hydrogen atom produce photons at ultraviolet and visible wavelengths (review Section 5-8 and Box 5-5), which do not penetrate the interstellar medium. Furthermore, to emit even these photons hydrogen atoms must be excited to at least the second Bohr orbit (see Figure 5-22), which is very unlikely in the cold depths of interstellar space. Fortunately, however, hydrogen atoms also emit radio waves, which makes it possible to detect even remote, cold hydrogen clouds. (The hydrogen in such clouds is neutral—that is, not ionized—and is called H I. This distinguishes it from ionized hydrogen, which is

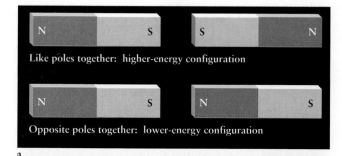

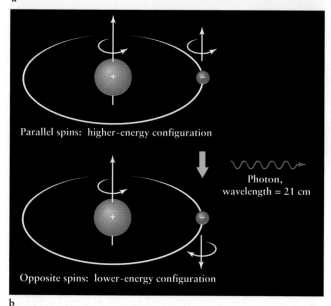

**figure 25-10**

**Magnetic Interactions in the Hydrogen Atom (a)** Two magnets are in a state of low energy when they have opposite poles (one north pole and one south pole) near each other and in a state of high energy when they have like poles (two north poles or two south poles) near each other. **(b)** Thanks to their spin, electrons and protons are both tiny magnets. When the electron and the proton are spinning in the same direction, their energy is higher than when they are spinning in opposite directions. When the electron flips from the higher-energy to the lower-energy configuration, the atom loses a tiny amount of energy. This energy is emitted as a radio photon having a wavelength of 21 cm.

designated H II.) To understand how these radio waves are emitted, we must probe a bit more deeply into the structure of protons and electrons, the particles of which hydrogen atoms are made.

In addition to having mass and charge, particles such as protons and electrons possess a tiny amount of angular momentum commonly called **spin.** Very roughly, you can visualize a proton or electron as a tiny, electrically charged sphere that spins on its axis. Because electric charges in motion generate magnetic fields, this means that a proton or electron is actually a tiny magnet with a north pole and a south pole. If you have ever played with magnets, you know that two magnets attract when the north pole of one magnet is next to the south pole of the other and repel when two like poles (both north or both south) are next to each other (Figure 25-10*a*). In other words, the energy of the two magnets is least when opposite poles are together and highest when like poles are together. Hence, as shown in Figure 25-10*b*, the energy of a hydrogen atom is slightly different depending on whether the spins of the proton and electron are in the same direction or opposite directions. (According to the laws of quantum mechanics, these are the only two possibilities; the spins cannot be at random angles.) If the spin of the electron changes its orientation from the higher-energy configuration to the lower-energy one—called a **spin-flip transition**—a photon is emitted. The energy difference between the two spin configurations is very small, only about $10^{-6}$ as great as those between different electron orbits (shown in Figure 5-22). Therefore, the photon emitted in a spin-flip transition between these configurations has only a small energy, and thus its wavelength is a relatively long 21 cm—a radio wavelength.

The spin-flip transition in neutral hydrogen was first predicted in 1944 by Dutch astronomer Hendrik van de Hulst. His calculations suggested that it should be possible to detect the **21-cm radio emission** from interstellar hydrogen, although a very sensitive radio telescope would be required. In 1951, Harold Ewen and Edward Purcell at Harvard University first succeeded in detecting this faint emission from hydrogen between the stars. Figure 25-11 shows the results of a more recent 21-cm survey of the entire sky. Neutral hydrogen gas (H I) in the plane of the Milky Way stands out promi-

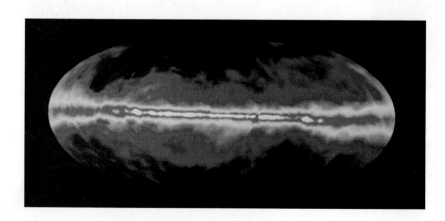

**figure 25-11**   **R** I V U X **G**

**The Entire Sky at 21 Centimeters** This false-color image shows the results of an all-sky survey of 21-cm emission from interstellar hydrogen. The entire sky has been mapped onto an oval, with the plane of the Galaxy extending horizontally across the image. Colors indicate the density of hydrogen gas, ranging from black and blue for the lowest to red and white for the highest. Note how hydrogen gas is concentrated along the plane of the Galaxy. (Courtesy of C. Jones and W. Forman, Harvard-Smithsonian Center for Astrophysics)

| box 25-1 | Looking Deeper into Astronomy |
|---|---|

## *The Local Bubble and the Geminga Pulsar*

As we saw in Section 20-2, interstellar dust and gas (mostly hydrogen) together make up the interstellar medium. This material is not uniformly spread throughout the plane of the Milky Way. Rather, it is quite lumpy and frothy, churned up over billions of years by passing stars and exploding supernovae. Indeed, a number of astronomers think that a recent, nearby supernova may have blown an enormous hole in the interstellar medium in the vicinity of the Sun.

Evidence for such holes comes from observations at wavelengths from 10 to 100 nm, called the extreme-ultraviolet part of the electromagnetic spectrum. Photons of such short wavelength and high energy are able to ionize neutral hydrogen atoms, which means that these photons are readily absorbed by the cold hydrogen in the interstellar medium. As a consequence of this absorption, most astronomers once believed that the Galaxy must be quite opaque to extreme-ultraviolet radiation. This opinion was disproved in 1975, when a team of astronomers led by Stuart Bowyer at the University of California, Berkeley, used a telescope aboard a spacecraft to observe several hot stars at extreme-ultraviolet wavelengths. Their discovery proved there are "holes" in the interstellar gas through which distant objects can be viewed in this part of the spectrum.

Such "holes" are not completely empty, but the hydrogen gas in them is much thinner and hotter than in the rest of the interstellar medium. The high temperature of this gas, about $10^6$ K, means that hydrogen atoms are already ionized and thus do not absorb extreme-ultraviolet photons.

Later space missions indicated that the Sun lies inside just such a region where the interstellar medium is hot and thin. As shown in the illustration, this irregularly shaped region, dubbed the **Local Bubble,** may extend as far as 1000 light-years (300 parsecs) in the direction of the star β (beta) Canis Majoris (the second-brightest star in the constellation Canis Major). In 1992, NASA launched a satellite called the Extreme Ultraviolet Explorer (EUVE) that is now probing the Local Bubble more thoroughly than any previous spacecraft.

Some astronomers think they can identify the supernova that blew the Local Bubble. In 1992 the European ROSAT satellite detected X-ray pulses from a mysterious gamma-ray source in the constellation Gemini that had defied explanation for nearly 20 years. Indeed, the source was so elusive that is was nicknamed Geminga, from a word in an Italian dialect meaning "it is not there" or "does not exist." The ROSAT discovery of pulsed radiation with a period of 0.237 s proved that the source is a pulsar.

In 1993 a team of Italian astronomers led by Giovanni Bignami measured the proper motion of the Geminga pulsar. They found that this pulsar is moving at a very high speed against the background stars, suggesting that it must be quite nearby, probably less than 300 light-years (100 parsecs) from the Earth. Furthermore, the rate at which the pulsar's period is lengthening suggests that the pulsar is quite young, perhaps only 300,000 years old. Calculations demonstrate that a Type II supernova would have had enough energy to evacuate the cavity in the interstellar medium that we now call the Local Bubble. Quite possibly, the Geminga pulsar is the remnant of that supernova.

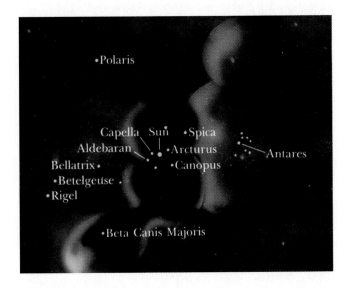

nently as a bright band across the middle of this image. The distribution of gas in the Milky Way is not uniform but is actually quite frothy, as described in Box 25-1. (Remarkably, spin-flip transitions can be used not only to map our Galaxy but also to map out the internal structure of the human body. The application of these transitions to medicine, called magnetic resonance imaging, is discussed in Box 25-2.)

The detection of 21-cm radio radiation was a major breakthrough that permitted astronomers to probe the galactic disk. Figure 25-12 shows how this was done. Suppose that you aim a radio telescope along a particular line of sight across the Galaxy. Your radio receiver, located at *S* (the position of the solar system), picks up 21-cm emission from H I clouds at points 1, 2, 3, and 4. However, the radio waves

## box 25-2 | The Heavens on the Earth

### *Spin-Flip Transitions in Medicine*

Thanks to their spin, protons and electrons are microscopic bar magnets. In a hydrogen atom, the interaction between the magnetism of the electron and that of the proton gives rise to the 21-cm radio emission. But these particles can also interact with outside magnetic fields, such as that produced by a large electromagnet. This is the physical principle behind **magnetic resonance imaging (MRI)**, an important diagnostic tool of modern medicine.

Much of the human body is made of water. Every water molecule has two hydrogen atoms, each of which has a nucleus made of a single proton. If a person's body is placed in a strong magnetic field, the spins of the protons in the hydrogen atoms of their body can either be in the same direction as the field ("aligned") or in the direction opposite to the field ("opposed"). The aligned orientation has lower energy, and therefore the majority of protons end up with their spins in this orientation. But if a radio wave of just the right wavelength is now sent through the person's body, an aligned proton can absorb a radio photon and flip its spin into the higher-energy, opposed orientation. How much of the radio wave is absorbed depends on the number of protons in the body, which in turns depends on how much water (and, thus, how much water-containing tissue) is in the body.

In magnetic resonance imaging, a magnetic field is used whose strength varies from place to place. The difference in energy between the opposed and aligned orientations of a proton depends on the strength of the magnetic field, so radio waves will only be absorbed at places where this energy difference is equal to the energy of a radio photon. (This equality is called *resonance*, which is how magnetic resonance imaging gets its name.) By varying the magnetic field strength over the body and the wavelength of the radio waves, and by measuring how much of the radio wave is

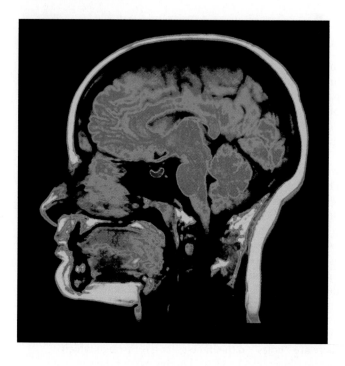

absorbed by different parts of the body, it is possible to map out the body's tissues. A portion of such a map is shown in the accompanying figure.

Unlike X-ray images, which show only the densest parts of the body, such as bones and teeth, magnetic resonance imaging can be used to view less dense but water-containing soft tissue. Just as the 21-cm radio emission has given astronomers a clear view of what were hidden regions of our Galaxy, magnetic resonance imaging allows modern medicine to see otherwise invisible parts of the human body.

---

from these various clouds are Doppler shifted by slightly different amounts, because the clouds are moving at different speeds as they travel with the rotating Galaxy.

It is important to remember that the Doppler shift reveals only motion along the line of sight (<u>review Figure 5-23</u>). In Figure 25-12, cloud 2 has the highest speed along our line of sight, because it is moving directly toward us. Consequently, the radio waves from cloud 2 exhibit a larger Doppler shift than those from the other three clouds along our line of sight. Because clouds 1 and 3 are at the same distance from the galactic center, they have the same orbital speed. The

fraction of their velocity parallel to our line of sight is also the same, so their radio waves exhibit the same Doppler shift, which is less than the Doppler shift of cloud 2. Finally, cloud 4 is the same distance from the galactic center as the Sun. This cloud is thus orbiting the Galaxy at the same speed as the Sun, resulting in no net motion along the line of sight. Radio waves from cloud 4, as well as from hydrogen gas near the Sun, are not Doppler-shifted at all.

These various Doppler shifts cause the 21-cm radiation to be smeared out over a range of wavelengths, that is, radio waves from gases in different parts of the Galaxy

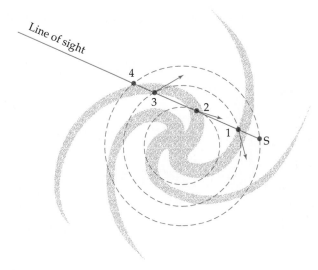

**figure 25-12**

**A Technique for Mapping Our Galaxy** If we look within the plane of our Galaxy from our position at *S*, hydrogen clouds at different locations (shown as 1, 2, 3, and 4) along our line of sight are moving at slightly different speeds. As a result, radio waves from these various gas clouds are subjected to slightly different Doppler shifts. This permits radio astronomers to sort out the gas clouds and thus map the Galaxy.

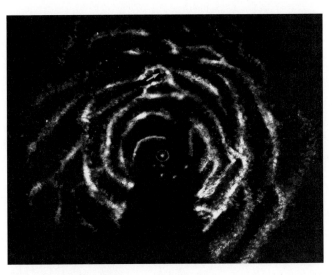

**figure 25-13**  **R** I V U X G

**A Map of Neutral Hydrogen in Our Galaxy** This map, constructed from radio-telescope surveys of 21-cm radiation, shows arched lanes of gas that suggest a spiral structure for our Galaxy. The Sun's location is indicated by a yellow arrow. Details in the blank, wedge-shaped region of the map cannot be determined, because gas in this part of the sky is moving perpendicular to our line of sight and thus does not exhibit a detectable Doppler shift. (Courtesy of G. Westerhout)

arrive at our radio telescopes with slightly different wavelengths. It is therefore possible to sort out the various gas clouds and thus produce a map of the Galaxy like that shown in Figure 25-13.

You can see from Figure 25-13 that neutral hydrogen gas is not spread uniformly around the disk of the Galaxy but is concentrated into numerous arched lanes. Similar features are seen in other galaxies beyond the Milky Way. As an example, the galaxy in Figure 25-14*a* has prominent spiral arms outlined by hot, luminous, blue main-sequence stars and the red emission nebulae (H II regions) found near many such stars. Stars of this sort are very short-lived, so these features indicate that spiral arms are sites of active, ongoing star formation. The 21-cm radio image of this same galaxy, shown in Figure 25-14*b*, shows that spiral arms are also regions where neutral hydrogen gas is concentrated, similar to the structures in our own Galaxy visible in Figure 25-13. This similarity is a strong indication that our Galaxy also has spiral arms.

**CAUTION!** Photographs such as Figure 25-14*a* can lead to the impression that there are very few stars found between the spiral arms of a galaxy. Nothing could be further from the truth! In fact, stars are distributed rather uniformly throughout the disk of a galaxy like the one in Figure 25-14*a*; the density of stars in the spiral arms is only about 5% higher than in the rest of the disk. The spiral arms stand out nonetheless because that is where hot, blue O and B stars are found. One such star is about $10^4$ times more luminous than an average star in the disk, so the light from

O and B stars completely dominates the visible appearance of a spiral galaxy.

Figure 25-14*a* suggests that we can confirm the presence of spiral structure in our own Galaxy by mapping the locations of star-forming regions. Such regions are marked by OB associations, H II regions, and molecular clouds (see Section 20-7). Unfortunately, the first two of these are best observed using visible light, and interstellar extinction limits the range of visual observations in the plane of the Galaxy to less than 3 kpc (10,000 light-years) from the Earth. But there are enough OB associations and H II regions within this range to plot the spiral arms in the vicinity of the Sun. Molecular clouds are easier to observe at great distances, because molecules of carbon monoxide (CO) in these clouds emit radio waves that are relatively unaffected by interstellar extinction. Hence, the positions of molecular clouds have been plotted even in remote regions of the Galaxy (see Figure 20-20).

Taken together, all these observations demonstrate that our Galaxy has four major spiral arms and several short arm segments (Figure 25-15). The Sun is located on a relatively short arm segment called the Orion arm, which includes the Orion Nebula and neighboring sites of vigorous star formation in that constellation. Two major spiral arms border either side of the Sun's position. The Sagittarius arm is on the side toward the galactic center. This is the arm you see during the summer months when you look at the portion of the Milky Way stretching across Scorpius and Sagittarius (see the photograph that opens this chapter). During winter, when our nighttime view is directed away from the galactic center, we see the

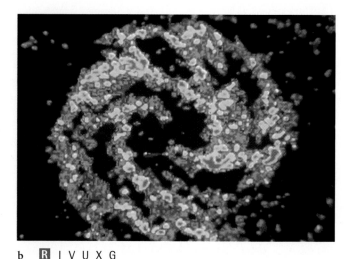

a   R I Ⓥ U X G                    b   Ⓡ I V U X G

## figure 25-14

**A Spiral Galaxy** This galaxy, called M83 or NGC 5236, lies in the southern constellation Hydra about 12 million light-years (4 million parsecs) from the Earth. **(a)** This visible-light photograph clearly shows the spiral arms traced out by the light from hot, young, blue stars and the red glow of H II regions. The presence of these stars and H II regions indicates that star formation takes place in spiral arms. **(b)** This radio view at a wavelength of 21 cm shows the emission from neutral hydrogen gas (H I). Note that essentially the same pattern of spiral arms is traced out in this image as in the visible-light photograph. Hence, spiral arms are also regions where interstellar hydrogen is most densely concentrated. (Anglo-Australian Observatory; VLA, NRAO)

Perseus arm. The remaining two major spiral arms are usually referred to as the Centaurus arm and the Cygnus arm.

Why are the young stars, star-forming regions, and clouds of neutral hydrogen in our Galaxy all found predominantly in the spiral arms? To answer this question, we must understand why spiral arms exist at all. Spiral arms are essentially cosmic "traffic jams," places where matter piles up as it orbits around the center of the Galaxy. This orbital motion, which is essential to grasping the significance of spiral arms, is our next topic as we continue the exploration of the Galaxy.

## figure 25-15

**Our Galaxy (Face-on View)** Our Galaxy has four major spiral arms and several shorter arm segments. The Sun is located on the Orion arm, between two major spiral arms. The Galaxy's diameter is about 50 kiloparsecs (160,000 light-years), and the Sun is about 8 kiloparsecs (26,000 light-years) from the galactic center.

## 25-4 The rotation of our Galaxy reveals the presence of dark matter

The spiral arms in the disk of our Galaxy suggest that the disk rotates. This means that the stars, gas, and dust in our Galaxy are all orbiting the galactic center. Indeed, were this not the case, mutual gravitational attraction would cause the entire Galaxy to collapse into the galactic center. In the same way, the Moon is kept from crashing into the Earth and the planets from crashing into the Sun, because of their motion around their orbits (see Section 4-7). However, measuring the rotation of our Galaxy accurately is a difficult business. But such challenging measurements have been made, as we shall see, and the results lead to a remarkable conclusion. Most of the mass of the Galaxy is in the form of *dark matter*, a mysterious sort of material that emits no light at all.

Radio observations of 21-cm radiation from hydrogen gas provide important clues about our Galaxy's rotation. Doppler shift measurements of this radiation indicate that stars and gas all orbit in the same direction around the galactic center, just as the planets all orbit in the same direction around the Sun. Measurements also show that the orbital speed of stars and gas about the galactic center is fairly constant throughout much of the Galaxy's disk. As a result, stars inside the Sun's orbit complete a trip around the galactic center more quickly than the Sun, because the stars have a shorter distance to travel. Conversely, stars outside the Sun's orbit take longer to go once around the galactic center because they have farther to travel. Consequently, as seen by Earth-based astronomers moving along with the Sun, stars inside the Sun's orbit overtake and pass us, while we overtake and pass stars outside the Sun's orbit (Figure 25-16a).

Figure 25-16 shows that the rotation of the Galaxy is very different from that of a solid disk. When a solid disk rotates, all parts of the disk take the same time to complete one rotation. Moreover, because the outer part of the disk has to travel around a larger circle than the inner part, the speed (distance per time) is greater in the outer part (Figure 25-16b). By contrast, the orbital speed of material in our Galaxy is roughly the same at all distances from the galactic center. This is also quite unlike the orbits of the planets. The

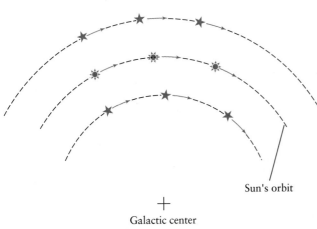

**a** Actual rotation of our Galaxy

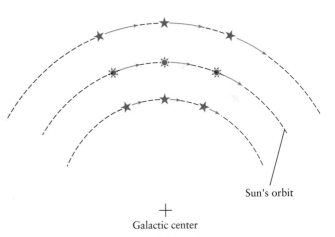

**b** If our Galaxy rotated like a solid disk

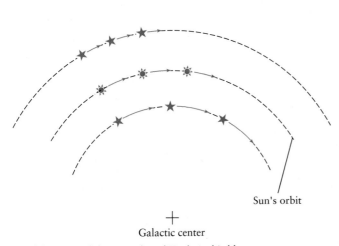

**c** If the Sun and the stars obeyed Kepler's third law

## Figure 25-16

**The Rotation of Our Galaxy** (a) The orbital speed of stars and gas around the galactic center is fairly constant throughout most of our Galaxy. This schematic diagram shows the Sun and two other stars traveling at the same speed but at different distances from the galactic center. Although they start off lined up, the Sun and the stars become increasingly separated as they move along their orbits. As seen by an observer traveling with the Sun, stars inside the Sun's orbit overtake and move ahead of the Sun, while stars far from the galactic center lag behind. **(b)** If the Galaxy rotated like a solid disk, the Sun and stars would always keep the same relative positions. This is not what is observed. **(c)** If the Sun and stars orbited the galactic center in the same way that planets orbit the Sun, the speed would be less for stars farther from the galactic center. This model cannot be correct because it predicts that stars inside the Sun's orbit would overtake us faster than they are observed to do.

larger a planet's orbit, the slower the speed at which it moves around the orbit (Figure 25-16c), as described in Section 4-2.

The 21-cm observations reveal only how fast things are moving relative to the Sun. Of course, the Sun itself is also moving. To get a complete picture of our Galaxy's rotation, we must therefore find out how fast the Sun is traveling. In other words, we need to find a kind of background that is not rotating along with the rest of the Galaxy. Such a background is provided by distant galaxies beyond the Milky Way and by the globular clusters. (Remember that globular clusters lie outside the plane of the Galaxy. Hence, they do not take part in the rotation of the disk.) By measuring the Doppler shifts of these objects and averaging their speeds, astronomers deduce that the Sun's speed along its orbit about the galactic center is about 220 km/s—about 790,000 kilometers per hour or 490,000 miles per hour!

We know that we are about 8000 parsecs from the galactic center, so we can use the Sun's speed $v$ to calculate the time required for one complete trip around the Galaxy, a time called the Sun's orbital period $P$. For the sake of simplicity, we assume that the Sun travels along a circular orbit about the center of the Galaxy. (The orbit is in fact nearly circular.) The distance traveled by the Sun is the circumference of this circle, which is given by $2\pi r$ (with $r$ being the radius of the circle). The time required for one orbit is equal to the distance traveled divided by the Sun's speed:

**Period of the Sun's orbit around the galactic center**

$$P = \frac{2\pi r}{v}$$

$P$ = orbital period of the Sun
$r$ = distance from the Sun to the galactic center
$v$ = orbital speed of the Sun

or

$$P = \frac{2\pi \times 8000 \text{ pc}}{220 \text{ km/s}} \times \frac{3.09 \times 10^{13} \text{ km}}{1 \text{ pc}}$$

$$= 7.1 \times 10^{15} \text{ s} = 2.2 \times 10^8 \text{ years}$$

Traveling at 790,000 kilometers per hour, it takes the Sun about 220 million years to complete one trip around the Galaxy. These results demonstrate the vastness of our Galaxy.

The orbit of the Sun around the center of the Galaxy is roughly circular, just as is the orbit of the Earth around the Sun. There is, however, a key difference between these two orbits. While the most important gravitational force holding the Earth in orbit comes from just one object (the Sun), there is no single object at the center of the Galaxy that is primarily responsible for holding the Sun in its 220-million-year orbit. Instead, what holds the Sun in this orbit is the total gravitational force exerted on the Sun by *all* of the matter (including stars, gas, and dust) that lies inside the Sun's orbit. (It turns out that the gravitational forces from matter *outside* the Sun's orbit have little or no net effect on the Sun's motion.) We can use this fact, along with Newton's

form of Kepler's third law, to determine the total mass of all the matter that lies inside the Sun's orbit. As shown in Box 25-3, such calculations yield $9.0 \times 10^{10} \text{ M}_\odot$. Because the Galaxy extends well beyond the Sun's orbit (Figure 25-8), the mass of the entire Galaxy must be larger than this.

In recent years, astronomers have been astonished to discover how much matter may lie beyond the orbit of the Sun. The clues come from 21-cm radiation emitted by hydrogen in spiral arms that extend far beyond the Sun's orbit. Because we know the true speed of the Sun, we can convert the Doppler shifts of this radiation into actual speeds for the spiral arms. This calculation gives us a **rotation curve**, a graph of the velocity of galactic rotation measured outward from the galactic center (Figure 25-17). Note that the rotation curve keeps rising slowly, even out to a distance of more than 18 kiloparsecs from the galactic nucleus.

According to Kepler's third law, the orbital speeds of stars or gas clouds beyond the confines of most of the Galaxy's mass should decrease with increasing distance from the Galaxy's center, just as the orbital speeds of the planets decrease with increasing distance from the Sun (Figure 25-16c). But the Galaxy's rotation curve is quite flat, indicating uniform orbital speeds well beyond the visible edge of the galactic disk. To explain this constant rotation speed of the

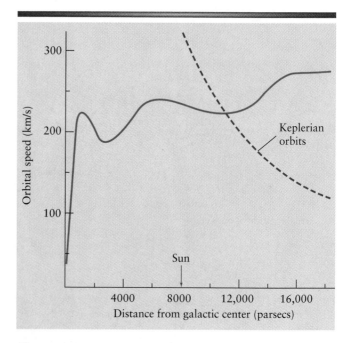

**figure 25-17**

**The Galaxy's Rotation Curve** The blue curve shows the orbital speeds of stars and gas in the disk of the Galaxy out to a distance of 18,000 parsecs from the galactic center. The dashed red curve indicates how this orbital speed should decline beyond the confines of most of the Galaxy's visible mass. Because the data (blue curve) do not show any such decline, there must be an abundance of invisible dark matter that extends to great distances from the galactic center.

## box 25-3 | Tools of the Astronomer's Trade

### *Estimating the Mass Inside the Sun's Orbit*

As mentioned in Section 25-4, the force that keeps the Sun in orbit around the center of the Galaxy is the gravitational pull of all of the matter interior to the Sun's orbit. We can estimate the total mass of all of this matter using Newton's form of Kepler's third law, discussed in Box 4-2:

$$P^2 = \frac{4\pi^2 a^3}{G(M + M_\odot)}$$

In this equation $P$ is the orbital period of the Sun, $a$ is the semimajor axis of the Sun's orbit around the galactic center, $G$ is the universal constant of gravitation, $M$ is the mass of the Galaxy inside the Sun's orbit, and $M_\odot$ is the mass of the Sun.

Because the Sun is only one of more than $10^{11}$ stars in the Galaxy, the Sun's mass is minuscule compared to $M$. Hence, we can safely replace the sum $M + M_\odot$ in the above equation by simply $M$. If we now assume that the Sun's orbit is a circle, the semimajor axis $a$ of the orbit is just the radius of this circle, which we call $r$. From Section 25-4, the period $P$ of the orbit is then equal to $2\pi r/v$, where $v$ is the Sun's orbital speed. You can then show that

$$M = \frac{rv^2}{G}$$

(The derivation is left as an exercise at the end of this chapter.)

Now we can insert known values to obtain the mass within the Sun's orbit. Being careful to express distance in meters and speed in meters per second, we have $v = 220$ km/s $= 2.2 \times 10^5$ m/s, $G = 6.67 \times 10^{-11}$ newton m$^2$/kg$^2$, and

$$r = 26{,}000 \text{ light-years} \times \frac{9.46 \times 10^{12} \text{ km}}{1 \text{ light-year}} \times \frac{10^3 \text{ m}}{1 \text{ km}}$$
$$= 2.5 \times 10^{20} \text{ m}$$

Hence, we find that

$$M = \frac{2.5 \times 10^{20} \times (2.2 \times 10^5)^2}{6.67 \times 10^{-11}} = 1.8 \times 10^{41} \text{ kg} \times \frac{1 \text{ M}_\odot}{1.99 \times 10^{30} \text{ kg}}$$
$$= 9.0 \times 10^{10} \text{ M}_\odot$$

This estimate involves only mass interior to the Sun's orbit. Matter outside the Sun's orbit has no net gravitational effect on the Sun's motion and thus does not enter into Kepler's third law.

---

outer parts of the Galaxy, astronomers suspect that a large amount of matter must lie outside the Sun's orbit. When this matter is included, the total mass of our Galaxy could be $6 \times 10^{11}$ M$_\odot$ or more.

To make matters even more puzzling, this "extra" matter is dark. It does not show up on photographs, nor indeed in images made in any part of the electromagnetic spectrum. We sense its presence only through its gravitational influence on the orbits of stars and gas clouds. It is important to understand what this **dark matter** could be, because it appears to comprise the lion's share of the matter in our Galaxy.

One proposal is that dark matter is composed, at least in part, of dim stars or black holes with masses between 0.01 M$_\odot$ and 1 M$_\odot$. These are called **massive compact halo objects**, or **MACHOs**, because the dark matter extends in a "halo" beyond the part of the Galaxy that emits light. Astronomers have searched for MACHOs by monitoring the light from distant stars. If a MACHO passes between us and the star, its gravity will bend the light coming from the star, and we see the star brighten for a few days. (In Section 24-2 we described how gravity can bend starlight.) MACHOs have indeed been observed in this way, but only in relatively small numbers, which suggests that MACHOs make up no more than 40% of the dark matter.

The remainder of the dark matter is thought to be much more exotic. One candidate is a neutrino with a small amount of mass. If these neutrinos are sufficiently massive, and if enough of them are present in the halo of the Galaxy, they might constitute a reasonable fraction of the dark matter. As we saw in Section 18-9, recent data shows that one type of neutrino can transform into another. These transformations can only take place if neutrinos have a nonzero amount of mass. Thus neutrinos must comprise at least part of the dark matter, though it is not known how much.

Another speculative possibility is a new class of subatomic particle called **weakly interacting massive particles**, or **WIMPs**. These particles, whose existence is suggested by certain theories but has not yet been confirmed experimentally, would not emit or absorb electromagnetic radiation. Physicists are attempting to detect these curious particles, which would have masses 10 to 10,000 times greater than a proton or neutron, by using a large crystal cooled almost to absolute zero. If a WIMP should enter this crystal and collide with one of its atoms, the collision will deposit a tiny but measurable amount of heat in the crystal.

As yet, the true nature of dark matter remains a mystery. Furthermore, this mystery is not confined to our own Galaxy. In Chapter 26 we will find that other galaxies have the same sort of rotation curve as Figure 25-17, indicating that they also contain vast amounts of dark matter. Indeed, dark matter appears to make up most of the mass in our universe. Hence, the quest to understand dark matter is one of the most important in modern astronomy.

## 25-5 Spiral arms are caused by density waves that sweep around the Galaxy

The disk shape of our Galaxy is not difficult to understand. In Section 7-7 we described what happens when a large number of objects are put into orbit around a common center: Over time the objects tend naturally to orbit in the same plane. This is what happened when our solar system formed from the solar nebula. There a giant cloud of material eventually organized itself into planets, all of which orbit in nearly the same plane. The same process shaped the rings of Saturn (Section 15-2). The icy objects that make up the rings lie so nearly in the same plane that the rings all but disappear when viewed edge-on (see Figure 15-3 and the figure that opens Chapter 15). In like fashion, the disk of our Galaxy, which is also made up of a large number of individual objects orbiting a common center, is very flat, as shown in Figure 25-8.

Understanding why our Galaxy has spiral arms presents more of a challenge. One early idea was that the material in the Galaxy somehow condensed into a spiral pattern from the very start. In this view, once stars, gas, and dust had become concentrated within the spiral arms, the pattern would remain fixed. This would only be possible if the Galaxy rotated like a solid disk (Figure 25-16c); the fixed pattern would be like the spokes on a rotating bicycle wheel. But the reality is that the Galaxy is not a solid disk. As we have seen, stars, gas, and dust all orbit the galactic center with approximately the same speed, as shown in Figure 25-16a. Let us see why this makes it impossible for a rigid spiral pattern to persist.

Imagine four stars A, B, C, and D that originally lie on a line extending outward from the galactic center (Figure 25-18a). In a given amount of time, each of the stars travels the same distance around its orbit. But because the innermost star has a smaller orbit than the others, it takes less time to complete one orbit. As a result, a line connecting the four stars is soon bent into a spiral (Figure 25-18b). Moreover, the spiral becomes tighter and tighter with the passage of time (Figure 25-18c and Figure 25-18d). After a few hundred million years this "winding up" of the spiral arms causes the spiral structure to disappear completely.

Figure 25-18 suggests that the Milky Way's spiral arms ought to have disappeared by now. The fact that they have not is called the **winding dilemma**. It shows that the spiral arms cannot simply be assemblages of stars and interstellar matter that travel around the Galaxy together, like a troop of soldiers marching in formation around a flagpole. In other

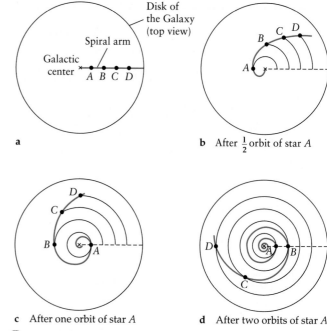

**a**

**b** After $\frac{1}{2}$ orbit of star A

**c** After one orbit of star A

**d** After two orbits of star A

**𝒻igure 25-18**

**The Winding Dilemma** (a) Imagine four stars, A, B, C, and D, that lie along a line that extends out from the galactic center. (b) All four stars have roughly the same orbital speed but travel in orbits of different sizes. Hence, when star A has completed one-half of an orbit, star B has completed less than one-half, star C has completed less than star B, and star D has completed the least of all. The line connecting the four stars has become a spiral. (c) By the time star A has completed one orbit, the spiral has become tighter. (d) The spiral continues to tighten with each successive orbit. This shows that the spiral structure of galaxies, including our own Galaxy, cannot be caused by matter orbiting in spiral formations. Such formations would "wind up" and erase the spiral structure in just a few hundred million years—a very brief time compared to the age of our Galaxy, thought to be roughly $10^{10}$ years.

words, the spiral arms cannot be made of anything *material*. What, then, can the spiral arms be?

In the 1940s, the Swedish astronomer Bertil Lindblad proposed that the spiral arms of a galaxy are actually a pattern that moves through the Galaxy like ripples on water. This idea was greatly enhanced and embellished in the 1960s by the American astronomers Chia Chiao Lin and Frank Shu. In this picture, spiral arms are a kind of wave, like the waves that move across the surface of a pond when you toss a stone into the water. Water molecules pile up at a crest of the wave but spread out again when the crest passes. By analogy, Lindblad, Lin, and Shu pictured a pattern of **density waves** sweeping around the Galaxy. These waves make matter pile up in the spiral arms, which are the crests of the waves. Individual parts of the Galaxy's material are compressed only temporarily when they pass through a spiral arm. The pattern of spiral arms persists, however, just as the waves made by a stone dropped in the water can persist for quite awhile after the stone has sunk.

To understand better how a density wave operates in a galaxy, think again about a water wave in a pond. If one part of the pond is disturbed by dropping a stone into it, the molecules in that part will be displaced a bit. They will nudge the molecules next to them, causing those molecules to be displaced and to nudge the molecules beyond them. In this way the wave disturbance spreads throughout the pond. In a galaxy, stars play the role of water molecules. Although stars and interstellar clouds of gas and dust are separated by vast distances, they can nonetheless exert forces on each other because they are affected by each other's gravity. If a region of above-average density should form, its gravitational attraction will draw nearby material into it. The displacement of this material will change the gravitational force that it exerts on other parts of the galaxy, causing additional displacements. In this way a spiral-shaped density wave can travel around the disk of a galaxy.

A key feature of density waves is that they move more slowly around a galaxy than stars and interstellar gas and dust. To visualize this, imagine workers painting a line down a busy freeway (Figure 25-19). The cars normally cruise along the freeway at 65 miles per hour (105 km/h), but the crew of painters are moving much more slowly. When the cars come up on the painters, they must slow down temporarily to avoid hitting anyone. As seen from the air, there is a noticeable congestion of cars around the painters. An individual car spends only a few moments in the traffic jam before resuming its usual speed, but the traffic jam itself lasts all day, inching its way along the road as the painters advance.

A similar crowding of interstellar matter takes place when it enters a spiral arm. This crowding plays a key role in the formation of stars and the recycling of the interstellar medium. As interstellar gas and dust moves into a spiral arm, it is compressed into new nebulae (Figure 25-20). This compression begins the process by which new stars form, which we described in Section 20-3.

These freshly formed stars continue to orbit around the center of their galaxy, just like the matter from which they formed. The most luminous among these are the hot, massive, blue O and B stars, which may have emission nebulae (H II regions) associated with them. These stars have main-sequence lifetimes of only 3 to 15 million years (see Table 21-1), which is very short compared to the 220 million years required for the Sun to make a complete orbit around the Galaxy. As a result, these luminous blue stars can travel only a fraction of an orbit around a galaxy before dying off. Therefore, these stars, and their associated H II regions, are only seen in or close to the spiral arm in which they formed. (For an example of this in another galaxy, see Figure 25-14.) Less massive stars have much longer main-sequence lifetimes, and thus their orbits are able to carry them all around the galactic disk. These less-luminous stars are found throughout the disk, including between the spiral arms.

The density-wave model of spiral arms explains why the disk of our Galaxy is dominated by metal-rich Population I stars. Because the material left over from the death of ancient stars is enriched in heavy elements, new generations

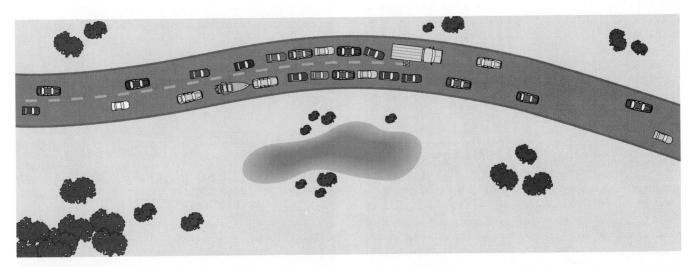

**ƒigure 25-19**

**A Density Wave on the Highway** A slow-moving crew of painters can cause a traffic jam on the highway. As cars slow down in the traffic jam, they form a region where the cars are more densely packed, a region that moves along with the painters. Individual cars, by contrast, enter this region on the left and exit it on the right. In a similar way, a density wave in a spiral galaxy is a slow-moving region where stars, gas, and dust are more densely packed than in the rest of the galaxy. As the material of the galaxy passes through the density wave, it is compressed. This triggers star formation, as shown in Figure 25-20.

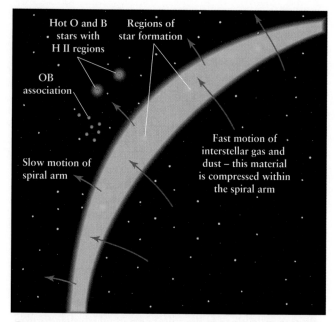

**figure 25-20**

**Star Formation in the Density-Wave Model** In the density-wave model, a spiral arm is a region where the density of material is higher than in the surrounding parts of a galaxy. Interstellar matter moves around the galactic center more rapidly than the spiral arms and is compressed as it passes through the spiral arms. This compression triggers the process of star formation, and new stars appear immediately "downstream" of the densest part of the spiral arms.

of stars formed in spiral arms are likely to be more metal-rich than their ancestors.

The density-wave model is still under development. One problem is finding a driving mechanism that keeps density waves going in spiral galaxies. After all, density waves expend an enormous amount of energy to compress the interstellar gas and dust. Hence, we would expect that density waves should eventually die away, just as do ripples on a pond. The American astronomers Debra and Bruce Elmegreen have suggested that gravity can supply that needed energy. As mentioned in Section 25-2, the nucleus of our Galaxy may be elongated somewhat into a bar shape. The asymmetric gravitational field of such a bar pulls on the stars and interstellar matter of a galaxy to generate density waves. Another factor that may help to generate and sustain spiral arms are gravitational interactions *between* galaxies. We will discuss this in Chapter 26.

Spiral density waves may not be the whole story behind spiral arms in our Galaxy and other galaxies. The reason is that spiral density waves should produce very well-defined spiral arms. We do indeed see many so-called **grand-design spiral galaxies** (Figure 25-21a), with thin, graceful, and well-defined spiral arms. But in some galaxies, called **flocculent spiral galaxies** (Figure 25-21b), the spiral arms are broad, fuzzy, chaotic, and poorly defined. ("Flocculent" means "resembling wool.")

To explain such flocculent spirals, M. W. Mueller and W. David Arnett in 1976 proposed a theory of **self-propagating star formation**. Imagine that star formation begins in a dense

a Grand design spiral galaxy

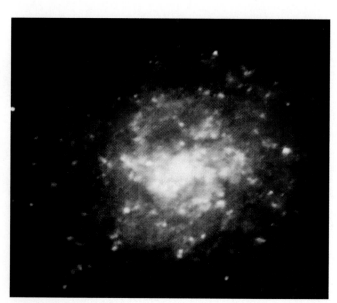

b Flocculent spiral galaxy

**figure 25-21** R I **V** U X G

**Variety in Spiral Arms** The differences from one spiral galaxy to another suggest that more than one process can create spiral arms. **(a)** NGC 628 is a grand-design spiral galaxy with thin, well-defined spiral arms. **(b)** NGC 7793 is a flocculent spiral galaxy with fuzzy, poorly defined spiral arms. (Courtesy of P. Seiden, D. Elmegreen, B. Elmegreen, and A. Mobarak; IBM)

interstellar cloud within the disk of a galaxy that does not yet have spiral arms. As soon as hot, massive stars form, their radiation and stellar winds compress nearby matter, triggering the formation of additional stars in that gas. When massive stars become supernovae, they produce shock waves that further compress the surrounding interstellar medium, thus encouraging still more star formation. Although all parts of this broad, star-forming region have approximately the same orbital speed about the galaxy's center, the inner regions have a shorter distance to travel to complete one orbit than the outer regions. As a result, the inner edges of the star-forming region move ahead of the outer edges as the Galaxy rotates. The bright O and B stars and their nearby glowing nebulae soon become stretched out in the form of a spiral arm. These spiral arms come and go essentially at random across a galaxy. Bits and pieces of spiral arms appear where star formation has recently begun but fade and disappear at other locations where all the massive stars have died off. Self-propagating star formation therefore tends to produce flocculent spiral galaxies that have a chaotic appearance with poorly defined spiral arms, like NGC 7793 in Figure 25-21b.

The two theories presented here are very different in character. In the density-wave model, star formation is caused by the spiral arms; in the self-propagating star formation model, by contrast, the spiral arms are caused by star formation. The correct description of spiral arms in our Galaxy may involve a combination of these models and remains a topic of active research.

## 25-6 Infrared and radio observations are used to probe the galactic nucleus

The nucleus of our Galaxy is an active, crowded place. If you lived on a planet near the galactic center, you could see a million stars as bright as Sirius, the brightest single star in our own night sky. The total intensity of starlight from all those nearby stars would be equivalent to 200 of our full moons. In effect, night would never really fall on a planet near the center of the Milky Way. At the center of this empire of light, however, may lie the darkest of all objects in the universe—a black hole millions of times more massive than the Sun.

Because of the severe interstellar absorption at visual wavelengths, some of our most important information about the galactic center comes from infrared and radio observations. Figure 25-22 shows two views of the center of our Galaxy. Figure 25-22a is a wide-angle visible-light photograph of the Milky Way, centered on the galactic nucleus in Sagittarius. Figure 25-22b is an infrared view of the galactic center made by the IRAS spacecraft (shown in Figure 6-28). It covers the area indicated by the white rectangle in Figure 25-22a. Numerous arches and streamers of dust surround the galactic nucleus. The strongest infrared emission comes from a region at the center of the galaxy called Sagittarius A, which is also a strong source of synchrotron radiation. As we described in Section 23-4, this is a type of radio radiation

a R I V U X G

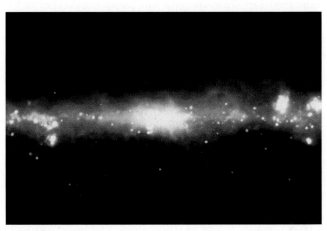

b R I V U X G

figure 25-22

**The Galactic Nucleus in Visible and Infrared Light** **(a)** This wide-angle photograph of the Milky Way at visible wavelengths is centered on the nucleus of our Galaxy in Sagittarius. **(b)** This infrared view from IRAS (the Infrared Astronomical Satellite) covers the area outlined by the white rectangle in (a). Black represents the dimmest regions of infrared emission, with blue the next dimmest, followed by yellow, red, and then white for the strongest emission. The prominent band across this image is interstellar dust in the plane of the Galaxy. The numerous knots and blobs along the layer of dust are interstellar clouds heated by young O and B stars. The center of the Galaxy is in the bright area at the center of this image. (Courtesy of D. di Cicco; NASA)

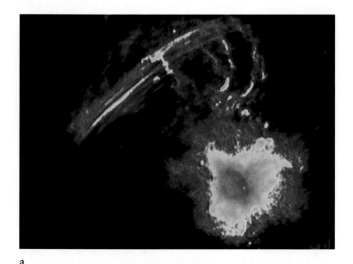

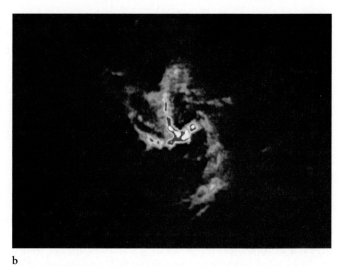

a                                        b

## ƒigure 25-23   R I V U X G

**Two Radio Views of the Galactic Center** These two false-color radio images show the appearance of the center of our Galaxy at radio wavelengths. The strongest radio emission is shown in red; weaker emission is colored green through blue. **(a)** This view covers an area of the sky of about the same angular size as the full moon, corresponding to a true size of 60 parsecs across. The parallel filaments may be associated with a magnetic field. The galactic nucleus is toward the lower right, at the center of the strongest emission. **(b)** This high-resolution view shows an even smaller area around the galactic center. The area shown is some 7 parsecs across, with an angular size of about 3 arcminutes. The pinwheel-like structure is centered on Sagittarius A*, thought to be the very center of the Galaxy. (VLA, NRAO; Courtesy of K. Y. Lo and N. Killeen, VLA, NCSA)

produced by high-speed electrons spiraling around a magnetic field. Despite its compact size, Sagittarius A is one of the brightest radio sources in the entire sky.

Still more detailed views of the galactic center have been provided by radio telescopes such as the Very Large Array (VLA) (see Figure 6-25). Figure 25-23a is a wide-angle radio image of Sagittarius A covering an area 60 parsecs (200 light-years) across. Huge filaments of gas extend perpendicular to the plane of the Galaxy. They stretch for 20 parsecs (65 light-years) northward of the galactic center, then abruptly arch southward (straight down in Figure 25-23a) toward Sagittarius A, at the core of the strongest emission. The orderly arrangement of these filaments is reminiscent of prominences on the Sun (see Section 18-5, especially Figure 18-25). This suggests that as on the Sun, a magnetic field may be controlling the flow of ionized gas at the galactic center.

The inner core of Sagittarius A, shown in Figure 25-23b, covers an area about 7 parsecs (20 light-years) across. Here, a pinwheel-like feature is centered on a radio source called Sagittarius A* ("Sagittarius A star"). One of the arms of this pinwheel is part of a ring of gas and dust orbiting the galactic center. Radio radiation from hydrogen cyanide (HCN) molecules in the ring surrounding Sagittarius A* yields the view in Figure 25-24. Inclined by about 70° to our line of sight, this ring is quite lumpy and consists of turbulent clouds of gas. The total amount of gas, both atomic and molecular, orbiting the galactic nucleus along with this ring-like structure is estimated at $3 \times 10^4$ $M_\odot$. All the material

less than about 2 parsecs from the galactic nucleus seems to have been ionized by a source that coincides with Sagittarius A*. (This ionized gas does not emit radio waves at the same wavelength as HCN molecules. It thus appears as a black area at the center of the ring in Figure 25-24.)

Sagittarius A* is thought to be the galactic nucleus, the very center of the Milky Way Galaxy. Astronomers think this because Sagittarius A* is the brightest radio source within several degrees of the galactic center. Furthermore, its position, pinpointed with simultaneous observations by radio telescopes scattered around the world, seems to be very near the dynamical and gravitational center of the Galaxy.

But what *is* Sagittarius A*? It is neither a star nor a pulsar, based on its high luminosity and radio spectrum. It also cannot be a supernova remnant because it is not expanding rapidly. Instead, Sagittarius A* is almost certainly an extremely massive black hole. The strongest evidence for this comes from recent infrared observations of the motions of stars in the vicinity of Sagittarius A*. Two research groups—Andreas Eckart and Reinhard Genzel of the Max Planck Institute for Extraterrestrial Physics in Garching, Germany, and Andrea Ghez, Mark Morris, and Eric Becklin of UCLA—independently observed stars orbiting around Sagittarius A* at speeds in excess of 1500 km/s (Figure 25-25). A source of gravity must be keeping these stars in orbit about the galactic center. Using Newton's form of Kepler's third law, astronomers calculate that this source of gravity must have a mass of $2.6 \times 10^6$ $M_\odot$. It now appears

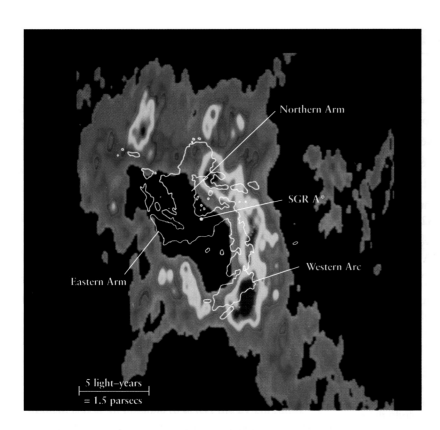

**figure 25-24** ℝ I V U X G

**The Molecular Ring Around the Galactic Center** This false-color, high-resolution map of HCN (hydrogen cyanide) emission reveals a ring of neutral atomic and molecular gas surrounding the center of the Galaxy. The ring has an inner radius of 2 parsecs and an outer radius of 8 parsecs. It consists of gas at a temperature of about 300 K, much warmer than the dust (80 K) in that region. The pinwheel feature in Figure 25-23b is outlined in white. Note that one of the pinwheel's arms (the western arc) coincides with the ring, but the other two arms do not. As in Figure 25-23, the strongest radio emission is shown in red, and weaker emission is colored green through blue. (Courtesy of R. Güsten and M. C. H. Wright)

conclusive that an object this massive and this compact could only be a black hole. Because of its enormous mass, it is called a **supermassive black hole.** As we will see in Chapter 27, astronomers have also found extraordinary activity in the nuclei of many other galaxies, which indicates the possibility of supermassive black holes at their centers. Thus, the supermassive black hole at the center of our own Galaxy is probably not unique.

Astronomers are still groping for a better understanding of the galactic center. With future developments in very-long-baseline interferometry (described in Section 6-6), it may be possible to actually obtain a picture of the supermassive black hole thought to lurk there. During the coming years, observations from Earth-orbiting satellites as well as from radio and infrared telescopes on the ground will certainly add to our knowledge of the core of the Milky Way.

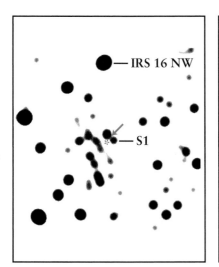

**April 1994**

**April 1996**

**figure 25-25** R I V U X G

**Stars Orbiting the Center of the Galaxy** These infrared images reveal the presence of a supermassive black hole at the center of the Galaxy. Each image shows an area just 3 arcseconds wide centered on Sagittarius A* (shown by the blue asterisk). At a distance of 8 kpc, the region shown is 24,000 AU (0.12 parsec, or 0.38 light-year) across. The images are shown as negatives, so stars appear black. Between April 1994 and April 1996, when the two images were made, the star marked S1 moved quite noticeably. This star, and theothers in these images, appear to be orbiting around Sagittarius A*. Their motions imply that Sagittarius A* contains about 2.6 million solar masses within a very small volume less than 1 AU across. The only plausible explanation seems to be that Sagittarius A* is a supermassive black hole. (Courtesy of A. Eckart, R. Genzel, T. Ott, and F. Eisenhauer, Max Planck Institute for Extraterrestrial Physics)

**20.** The disk of the Galaxy is about 50 kpc in diameter and 600 parsecs thick. **(a)** Find the volume of the disk in cubic parsecs. **(b)** Find the volume (in cubic parsecs) of a sphere 300 parsecs in radius centered on the Sun. **(c)** If supernovae occur randomly throughout the volume of the Galaxy, what is the probability that a given supernova will occur within 300 parsecs of the Sun? If there are about three supernovae each century in our Galaxy, how often, on average, should we expect to see one within 300 parsecs of the Sun?

**\*21.** Show that the form of Kepler's third law stated in Box 25-3, $P^2 = 4\pi^2 a^3 / G(M + M_\odot)$, is equivalent to $M = rv^2 / G$, provided the orbit is a circle. (*Hint:* The mass of the Sun [$M_\odot$] is much less than the mass of the Galaxy inside the Sun's orbit [$M$].)

**22.** The accompanying photographs show the spiral galaxy M74, located about 55 million light-years from the Earth in the constellation Pisces (the Fish). On the top is an optical image taken at visible wavelengths. On the bottom is an ultraviolet image made by NASA's Ultraviolet Imaging Telescope, which was carried into orbit by the space shuttle *Columbia* during the *Astro-1* mission in 1990. Compare these two images and, based on what you know about stellar evolution and spiral structure, explain the differences you see.

R I **V** U X G

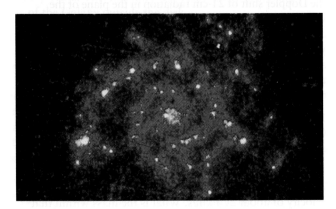

R I V **U** X G

**23.** **(a)** Calculate the Schwarzschild radius of a supermassive black hole of mass $2.6 \times 10^6$ M$_\odot$, the estimated mass of the black hole at the galactic center. Give your answer in both kilometers and in astronomical units. **(b)** What is the angular diameter of such a black hole as seen at a distance of 8 kpc, the distance from the Earth to the galactic center? Give your answer in arcseconds. Observing an object with such a small angular size will be a challenge indeed!

**24.** The star S1 shown in Figure 25-25 lies about 1000 AU from Sagittarius A\*. Assume that its orbit around Sagittarius A\* is a circle. **(a)** At an orbital speed of about 1500 km/s, what is the orbital period of S1? Give your answer in years. **(b)** Determine the sum of the masses of Sagittarius A\* and S1. Give your answer in solar masses. (Your answer is an estimate of the mass of Sagittarius A\*, because the mass of S1 is negligibly small by comparison.)

## Discussion Questions

**25.** From what you know about stellar evolution, the interstellar medium, and the density-wave theory, explain the appearance and structure of the spiral arms of grand-design spiral galaxies.

**26.** What observations would you make to determine the nature of the hidden mass in our Galaxy's halo?

## Observing Project

**27.** Examine the Milky Way on a clear, moonless night away from city lights. Use a high-quality pair of binoculars or a telescope with a wide field of view. During the summer months, look toward the galactic center in Sagittarius and follow the Milky Way arching across the sky through the constellations of Aquila and Cygnus. In the winter, follow the Milky Way as it sweeps across such recognizable constellations as Canis Major, Auriga, and Perseus. Take your time to note details in the structure of the Milky Way. For example, toward the galactic center you should be able to see a few H II regions along with extensive star clouds. Other stretches of the Milky Way clearly display dark patches and mottling where clouds of interstellar gas and dust have blocked the light from background stars. Enjoy the spectacle!

## Where to Learn More

*Magazine articles*

Binney, J., "The Evolution of Our Galaxy." *Sky & Telescope*, March 1995. This article describes current thinking on how the Galaxy was formed and how it continues to evolve.

Chien, P. "EUVE Probes the Local Bubble." *Sky & Telescope*, February 1992. This article explains how the

Extreme Ultraviolet Explorer is giving us views of the heavens at wavelengths once thought to be unobservable.

Jayawardhana, R. "Destination: Galactic Center." *Sky & Telescope,* June 1995. Written as a travelogue of a journey to the center of the Galaxy, this entertaining and informative article describes some of the evidence that a supermassive black hole lurks there.

Lomberg, J. "A Portrait of Our Galaxy." *Sky & Telescope,* December 1993. A well-known astronomical artist describes the challenges in creating an accurate painting of our Galaxy as viewed from outside.

Mateo, M. "Searching for Dark Matter." *Sky & Telescope,* January 1994. One of the scientists involved in hunting for dark matter describes how the search is carried out.

Palmer, E. S. "Unveiling the Hidden Milky Way." *Astronomy,* November 1989. Beautiful illustrations grace this article, which explains how radio astronomers trace the spiral structure of our Galaxy.

Trimble, V., and Parker, S. "Meet the Milky Way." *Sky & Telescope,* January 1995. First in a four-part series on the Milky Way, this article gives a guided tour of the structure of our home galaxy as it is presently understood.

**W** *World Wide Web*

Maps of the Milky Way in eight wavelength bands—infrared, near-infrared, visible, X-ray, gamma-ray, radio emissions from hydrogen atoms and from free electrons, and the microwave emissions from carbon monoxide molecules—are available on the World Wide Web courtesy of NASA's Astrophysics Data Facility (**http://adc.gsfc.nasa.gov/mw/milkyway.html**).

A number of web sites describe the search for MACHOs (**http://wwwmacho.mcmaster.ca/**) and for WIMPs (**http://physics7.berkeley.edu/group/directdet/gen.html**).

Among the spacecraft that have contributed to our understanding of the Galaxy are IRAS (**http://www. gsfc.nasa.gov/astro/iras/iras_home.html**), the Extreme Ultraviolet Explorer (**http://www.cea. berkeley.edu/**), and ROSAT (**http://www.rosat.mpe-garching.mpg.de/**).

The latest research news about the galactic center can be found in the on-line newsletter *GCNews* (**http://www. astro.umd.edu/~gcnews/**).

A web site at the Rochester Institute of Technology (**http://www.cis.rit.edu/htbooks/mri/**) gives an in-depth introduction to magnetic resonance imaging.

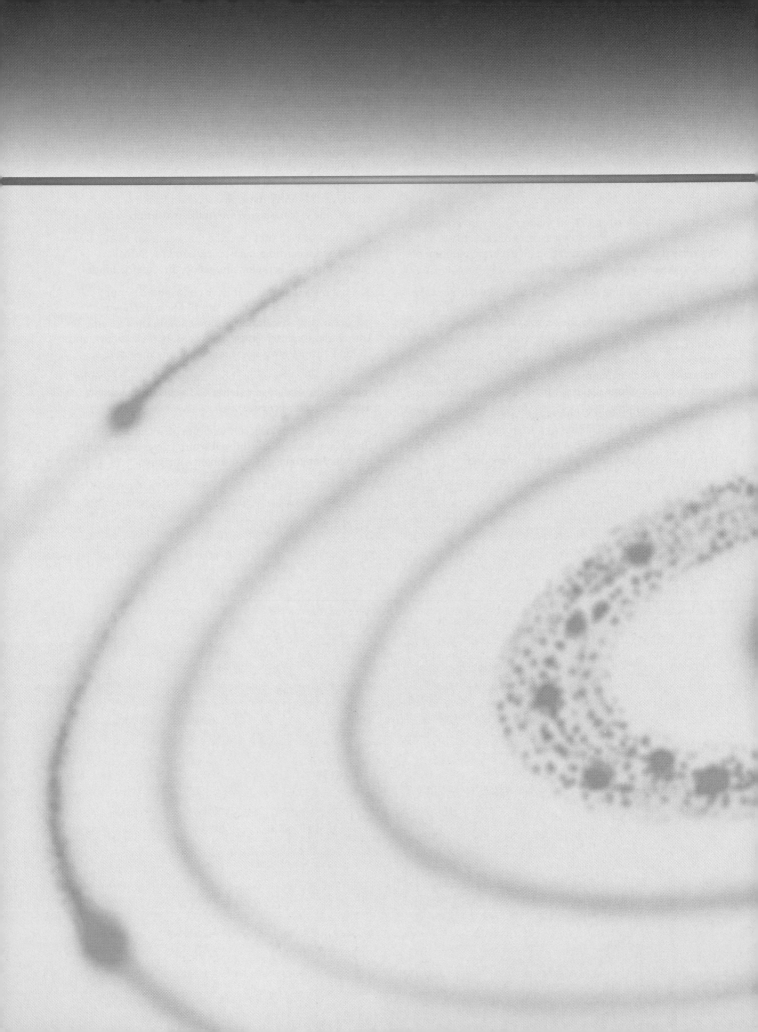

# Galaxies

R I **V** U X G

**A Spiral Galaxy** This galaxy, called NGC 1365, is the largest and most impressive member of a cluster of galaxies about 60 million light-years from the Earth. NGC 1365 is classified as a barred spiral because of the bar crossing its nucleus. From the ends of the bar two distinct and quite open spiral arms branch off. The yellow color of the nucleus and the bar shows that old, relatively cool stars occupy these parts of the galaxy. Light from young, hot stars causes the blue color of the arms. (European Southern Observatory)

*In this chapter you will find the answers to the following questions:*

26-1  How did astronomers first discover galaxies?

26-2  How did astronomers first determine the distances to galaxies?

26-3  Do all galaxies have spiral arms, like the Milky Way?

26-4  How do modern astronomers tell how far away galaxies are?

26-5  How do the spectra of galaxies tell astronomers that the universe is expanding?

26-6  Are galaxies isolated in space, or are they found near other galaxies?

26-7  What happens when galaxies collide with each other?

26-8  Is dark matter found in galaxies beyond the Milky Way?

26-9  How do astronomers think galaxies formed?

*One of the greatest scientific discoveries of the twentieth century was the true size and scope of our universe. At the beginning of the century, many astronomers felt that the Milky Way and its retinue of stars comprised the entire universe. We now realize that the Milky Way is just one of billions of galaxies spread across the observable universe.*

*Galaxies come in many shapes and sizes. Some are disk shaped, like our own Milky Way, with arching spiral arms that are active sites of star formation. Others are featureless, ellipse-shaped agglomerations of stars, virtually devoid of interstellar gas and dust. Some galaxies are only one one-hundredth the size and one ten–thousandth the mass of the Milky Way. Others are giants, typically 5 times the size and 50 times the mass of the Milky Way. Just as most stars are grouped together into galaxies, most galaxies are located in groups and clusters. These clusters of galaxies stretch across the universe, forming huge, lacy patterns.*

*Like planets within the solar system or stars within a galaxy, galaxies are in ceaseless motion. As they move through space, galaxies occasionally collide with each other. These titanic collisions produce spectacular results, including violent bursts of star formation and enormous trails of stars and gas cast outward into intergalactic space. More important still, remote galaxies are receding from us with speeds that are proportional to their distances from our Galaxy. This relationship between distance and recessional velocity, called the Hubble law, reveals that we live in an expanding universe.*

## 26-1 When galaxies were first discovered, it was not clear that they lie far beyond the Milky Way

William Parsons was the third Earl of Rosse in Ireland. He was rich, he liked machines, and he was fascinated with astronomy. Accordingly, he set about building gigantic telescopes. In February 1845 his *pièce de résistance* was finished. The telescope's massive mirror measured 6 feet in diameter and was mounted at one end of a 60-foot tube controlled by cables, straps, pulleys, and cranes. For many years, this triumph of nineteenth-century engineering was the largest telescope in the world.

With this new telescope, Lord Rosse examined many of the nebulae that had been discovered and cataloged by William Herschel. Lord Rosse observed that some of these nebulae have a distinct spiral structure. One of the best examples is M51, also called NGC 5194.

Lacking photographic equipment, the earl had to make drawings of what he saw. Figure 26-1 shows a drawing he made of M51, and Figure 26-2 shows a modern photograph of M51. Views such as this inspired Lord Rosse to echo the famous German philosopher Immanuel Kant, who in 1755 had suggested that these objects might be "island universes," that is, vast collections of stars far beyond the confines of the Milky Way.

## ƒigure 26-1

**Lord Rosse's Sketch of M51** Using a large telescope of his own design, the nineteenth-century astronomer Lord Rosse was able to distinguish spiral structure in this "spiral nebula." This galaxy, whose angular size is 8 × 11 arcminutes (about a third the angular size of the full moon), is today called the Whirlpool Galaxy because of its distinctive appearance. (Courtesy of Lund Humphries)

**figure 26-2** R I **V** U X G

**A Modern View of the Spiral Galaxy M51** This spiral galaxy, also called NGC 5194, lies in the constellation Canes Venatici (the Hunting Dogs). It is some 8.5 million parsecs (28 million light-years) from the Earth and is about 20 kiloparsecs (65,000 light-years) in diameter. The "blob" at the end of one spiral arm is the companion galaxy NGC 5195. (Courtesy of D. Elmegreen, B. Elmegreen, and P. Seiden; IBM)

## 26-2 Hubble proved that the spiral nebulae are far beyond the Milky Way

After completing his studies, Edwin Hubble joined the staff of the Mount Wilson Observatory in Pasadena, California. On October 6, 1923, he took a historic photograph of the Andromeda "Nebula," one of the spiral nebulae around which controversy raged. A modern photograph of this object appears in Figure 26-3. Hubble carefully examined his photographic plate and discovered what he at first thought to be a nova. Referring to previous plates of that region, he soon realized that the object was actually a Cepheid variable star. Further scrutiny of additional plates over the next several months revealed several more Cepheids, two of which can be seen in Figure 26-4.

As we saw in Section 21-5, Cepheid variables help astronomers determine distances. An astronomer begins by carefully measuring the variations in apparent brightness of a Cepheid variable, then recording the results in the form of a light curve (a plot of brightness versus time). This graph gives the variable star's period and average brightness. Given the

Many astronomers did not subscribe to this notion of island universes. A considerable number of nebulae are in fact scattered throughout the Milky Way. (Examples include the nebulae in the figure that opens Chapter 20, as well as those shown in Figures 20-2 and 20-3.) It therefore seemed reasonable that "spiral nebulae," even though they are very different in shape from other sorts of nebulae, could also be components of our Galaxy.

The astronomical community became increasingly divided over the nature of the spiral nebulae. In April 1920, a debate was held at the National Academy of Sciences in Washington, D.C. On one side was Harlow Shapley, a young, brilliant astronomer renowned for his recent determination of the size of the Milky Way Galaxy (see Section 25-1). Shapley believed the spiral nebulae to be relatively small, nearby objects scattered around our Galaxy like the globular clusters he had studied. Opposing Shapley was Heber D. Curtis of Lick Observatory near San Jose, California. Curtis championed the island universe theory, arguing that each of these spiral nebulae is a rotating system of stars much like our own Galaxy.

The Shapley-Curtis debate generated much heat but little light. Nothing was decided, because no one could present conclusive evidence to demonstrate exactly how far away are the spiral nebulae. Astronomy desperately needed a definitive determination of the distance to a spiral nebula. Such a measurement became the first great achievement of a young man who studied astronomy at the Yerkes Observatory, near Chicago. His name was Edwin Hubble.

**figure 26-3** R I **V** U X G

**The Andromeda Galaxy** This nearby galaxy (also called M31 or NGC 224) covers an area of the sky roughly 5 times as large as the full moon. Under good observing conditions, the galaxy's bright central bulge can be glimpsed with the naked eye in the constellation of Andromeda. The distance to this galaxy is about 900 kiloparsecs (2.9 million light-years). The white rectangle outlines the area shown in Figure 26-4. (Palomar Observatory)

## box 26-1 | Looking Deeper into Astronomy

### *Cepheids and Supernovae as Indicators of Distance*

Because their periods are directly linked to their luminosities, Cepheid variables are one of the most reliable tools astronomers have for determining the distances to galaxies. To this day, astronomers use this link—much as Hubble did back in the 1920s—to measure intergalactic distances. More recently, they have begun to use Type Ia supernovae, which are far more luminous and can thus can be seen much farther away, to determine the distances to very remote galaxies.

In 1992, for example, a team of astronomers used Cepheid variables in a galaxy called IC 4182 to deduce this galaxy's distance from the Earth. The team of astronomers used the Hubble Space Telescope on 20 separate occasions to record images of the stars in IC 4182. By comparing these images, the astronomers could pick out which stars vary in brightness. One of the 27 Cepheids that they discovered in IC 4182 is shown in the two images below (indicated in each by an arrow). From these images, the astronomers estimated the brightnesses of the Cepheid variables and plotted their light curves. The Hubble Space Telescope is particularly well-suited for studies of this kind, because its extraordinary angular resolution makes it possible to pick out individual stars at great distances.

R I **V** U X G

R I **V** U X G

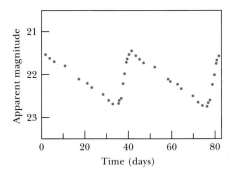

R I **V** U X G

### figure 26-4  R I **V** U X G

**Cepheid Variables in the Andromeda Galaxy**
The arrows in this view indicate two Cepheid variables on the outskirts of the Andromeda Galaxy. Because these stars appear so faint, even though they are intrinsically luminous, Edwin Hubble successfully demonstrated that the Andromeda "Nebula" is extremely far away. (The Carnegie Observatories)

The accompanying graph shows the light curve of one such Cepheid, whose period is 42.0 days. Note that this curve is plotted in terms of apparent magnitude, which is a common way in which astronomers express brightness (see Box 19-3). This Cepheid has an average apparent magnitude ($m$) of +22.0. (By comparison, the dimmest star you can see with the naked eye has $m = +6$; this Cepheid in IC 4182 appears less than one ten–millionth as bright.) According to the period-luminosity relation shown in Figure 21-14, a Cepheid variable with a period of 42.0 days has an average luminosity of 33,000 $L_\odot$. (Note that Figure 21-14 has period-luminosity relationships for two types of Cepheids, Type I and Type II. The Cepheids in IC 4182 are of Type I, which means that they are metal-rich Population I stars.) This can expressed by saying that this Cepheid variable has an average absolute magnitude ($M$) of –6.5 (compared to $M = +4.8$ for the Sun). Hence, the difference between the Cepheid's apparent and absolute magnitudes, called its distance modulus, is

$$m - M = (22.0) - (-6.5) = 22.0 + 6.5 = 28.5$$

In Box 19-3, we saw that the distance modulus of a star is related to its distance in parsecs ($d$) by

$$m - M = 5 \log d - 5$$

This can be rewritten as

$$d = 10^{(m - M + 5)/5} \text{ parsecs}$$

Inserting the value for the distance modulus in this equation, we get the distance to the Cepheid variable

and, hence, the distance to the galaxy of which the star is part:

$$d = 10^{(28.5 + 5)/5} \text{ parsecs} = 10^{6.7} \text{ parsecs} = 5 \times 10^6 \text{ parsecs}$$

The galaxy is therefore 5 Mpc (1 Mpc = 1 megaparsec = $10^6$ parsecs), or 16 million ($1.6 \times 10^7$) light-years, from Earth.

Astronomers are interested in IC 4182 because a Type Ia supernova was observed there in 1937. All Type Ia supernovae are exploding white dwarfs that reach nearly the same maximum brightness at the peak of their outburst (see Section 22-9). Once astronomers know the peak absolute magnitude of Type Ia supernovae, they can use these supernovae as distance indicators. Because the distance to IC 4182 is known from its Cepheids, the 1937 observations of the supernova in that galaxy allow us to calibrate Type Ia supernovae as distance indicators.

At maximum brightness, the 1937 supernova reached an apparent magnitude of $m = +8.6$. Since the distance modulus of the galaxy ($m - M$) is 28.5, we see that when a Type Ia supernova is at maximum brightness, its absolute magnitude is

$$M = m - (m - M) = 8.6 - 28.5 = -19.9$$

Whenever astronomers find a Type Ia supernova in a remote galaxy, they can combine this absolute magnitude with the observed maximum apparent magnitude to get the galaxy's distance modulus, from which the galaxy's distance can be easily calculated (just as we did above for the Cepheids in IC 4182). This technique has been used to determine the distances to galaxies hundreds of millions of parsecs away.

star's period, the astronomer then uses the period-luminosity relation (shown in Figure 21-14) to find the average luminosity of the Cepheid variable. Knowing both the apparent brightness and luminosity of the Cepheid, the astronomer can then use the inverse-square law to calculate the distance to the star (see Box 19-2). Box 26-1 presents an example of how this is done.

Cepheid variables are intrinsically quite luminous, with average luminosities that can exceed $10^4$ $L_\odot$. Hubble realized that for these luminous stars to appear as dim as they were on his photographs of the Andromeda "Nebula," they must be extremely far away. Straightforward calculations using modern data reveal that M31 is some 900 kiloparsecs (2.9 million light-years) from the Earth. Based on its angular size, M31 has a diameter of 70 kiloparsecs—larger than the diameter of our own Milky Way Galaxy! These results prove that the Andromeda "Nebula" is not a nearby nebula but an

enormous stellar system, far beyond the confines of the Milky Way. Today, this system is properly called the Andromeda Galaxy.

Hubble's results, which were presented at a meeting of the American Astronomical Society on December 30, 1924, settled the Shapley-Curtis debate once and for all. The universe was recognized to be far larger and populated with far bigger objects than anyone had thus seriously imagined. Hubble had discovered the realm of the galaxies.

CAUTION! In everyday language, the words "galaxy" and "universe" are often used interchangeably. It is true that before Hubble's discoveries our Milky Way Galaxy was thought to comprise essentially the entire universe. But we now know that the universe contains literally billions of galaxies. A single galaxy, vast though it may be, is just a tiny part of the entire observable universe.

SO

SBO

**Figure 26-11** R I **V** U X G

**Lenticular Galaxies** A lenticular galaxy has a disk but no spiral arms. A SB0 galaxy seems to have a barlike structure across its nucleus, whereas an S0 galaxy does not. (Courtesy of J. D. Wray; McDonald Observatory)

Galaxies that do not fit into the scheme of spirals, barred spirals, and ellipticals are usually referred to as **irregular galaxies.** Hubble defined two types of irregulars. Irr I galaxies, like underdeveloped spiral galaxies, contain many OB associations and H II regions. Irr II galaxies have asymmetrical, distorted shapes that seem to have been caused by collisions with other galaxies or by violent activity in their nuclei.

The best-known examples of Irr I galaxies are the Large Magellanic Cloud (LMC) and the Small Magellanic Cloud (SMC), which are nearby companions of our Milky Way

that can be seen with the naked eye from southern latitudes. Telescopic views of the Magellanic clouds can easily resolve individual stars (Figures 26-12 and 26-13). The SMC does not exhibit any of the geometric symmetry characteristic of spirals or ellipticals, but the LMC does have a faint barlike structure. Both these galaxies contain substantial amounts of interstellar gas. Tidal forces exerted on these irregular galaxies by the Milky Way help to compress the gas, which is why both the LMC and SMC are sites of active star formation.

**Figure 26-12** R I **V** U X G

**The Large Magellanic Cloud (LMC)** At a distance of only 50 kiloparsecs (160,000 light-years), this galaxy is the second closest companion of our Milky Way Galaxy. (The Milky Way's closest companion, the Sagittarius Dwarf, is shown in Figure 26-19a.) About 19 kiloparsecs (62,000 light-years) across, the LMC spans 22° across the sky, or about fifty times the size of the full Moon. Note the huge H II region (called the Tarantula Nebula, or 30 Doradus) toward the left side of the photograph. Its diameter of 250 parsecs (800 light-years) and mass of $5 \times 10^6$ $M_\odot$ make it the largest known H II region. (Anglo-Australian Observatory)

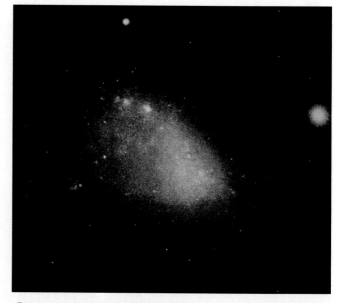

**Figure 26-13** R I **V** U X G

**The Small Magellanic Cloud (SMC)** The SMC is 63 kiloparsecs (200,000 light-years) away from us and just 8 kiloparsecs (26,000 light-years) across. Because of its sprawling, asymmetrical shape, the SMC is classified as an irregular galaxy. Note that the SMC is rich in young, blue stars. (Anglo-Australian Observatory)

## 26-4 Astronomers use standard candles to determine the distance to remote galaxies

One of the key questions that astronomers ask about galaxies is "How far away are they?" To determine the distance to a remote galaxy, they look for a **standard candle**—an object, such as a star, that lies within that galaxy and for which we know the luminosity (or, equivalently, the absolute magnitude, described in Section 19-3). By measuring how bright the standard candle appears, astronomers can calculate its distance—and hence the distance to the galaxy of which it is part—using the inverse-square law. The challenge is to find standard candles that are luminous enough to be seen across the tremendous distances to galaxies.

For nearby galaxies, Cepheid variables make fairly reliable standard candles. These variables can be seen out to about 60 Mpc (200 million light-years) using the Hubble Space Telescope, and their luminosity can be determined from their period through the period-luminosity relation (Figure 21-14). Box 26-1 gives an example of using Cepheid variables to determine distances.

Beyond about 60 Mpc even the brightest Cepheid variables, which have luminosities of about $2 \times 10^4 \ L_\odot$, fade from view. Astronomers then turn to even more luminous stars to serve as standard candles. The brightest red supergiants have luminosities of about $10^5 \ L_\odot$; the brightest blue supergiants are about twice again as luminous. These two types of stars can be seen out to distances of 150 Mpc (500 million light-years) and 250 Mpc (800 million light-years), respectively.

At distances so great that individual stars are no longer discernible, astronomers use entire star clusters and nebulae as standard candles. The brightest globular clusters, which have a total luminosity of about $10^6 \ L_\odot$, can be seen out to 400 Mpc (1.3 billion light-years) from the Earth. The brightest H II regions have luminosities of $6 \times 10^6 \ L_\odot$ and can be detected out to 900 Mpc (3 billion light-years). From the measured apparent brightness of these clusters and nebulae, distances to remote galaxies can be estimated.

To get beyond 900 Mpc, astronomers must wait for supernova explosions. Type Ia supernovae (described in Section 22-9) are fairly reliable as standard candles, because they all reach about the same maximum luminosity of about $3 \times 10^9 \ L_\odot$ at the peak of their outbursts (Figure 26-14). If a Type Ia supernova is seen in a distant galaxy and its maximum apparent brightness measured, the inverse-square law can be used to find the galaxy's distance (see Box 26-1). Such supernova explosions have been seen at distances of more than 1000 Mpc (3 billion light-years) from the Earth and, in theory, should be visible at even greater distances.

Cepheid variables, the most luminous supergiants, globular clusters, H II regions, and Type Ia supernovae are good standard candles for four reasons. First, they are bright, so we can see them out to great distances. Second, we are fairly certain about their luminosities, so we can be equally certain of any distance calculated from a standard candle's apparent brightness and luminosity. Third, they are easily identifiable—for example, by the shape of their light curve or their unique color. Fourth, they are relatively common, so that astronomers can use them to determine the distances to many different galaxies. As you might suspect, astronomers go to great lengths to check the accuracy and reliability of their standard candles.

As astronomers study increasingly remote galaxies, their distance determinations become more and more uncertain. This is because the farther we look into space, the fewer standard candles we have. For example, the distance to a nearby galaxy can be cross-checked in many ways. The distance computed by applying the period-luminosity relation to that galaxy's Cepheids can be compared to the distance determined from the apparent brightness of its most luminous supergiants. These results can then be compared with distances found from the apparent brightness of the galaxy's globular clusters and H II regions. The results from all these methods can then be averaged to obtain a distance to the galaxy in which astronomers have confidence. As we turn to more distant galaxies, however, we can see fewer and fewer

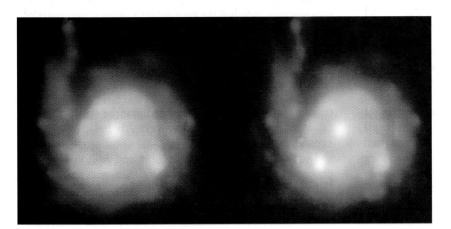

**figure 26-14** R I V U X G

**A Supernova in a Spiral Galaxy** In 1991 a supernova erupted in the spiral galaxy NGC 3310. The left view shows the galaxy before the outburst. The right view shows the supernova near maximum brightness in one of the spiral arms of the galaxy. Type Ia supernovae, which can be seen even in extremely remote galaxies, are important standard candles used to determine the distances to these faraway galaxies. (NRAO)

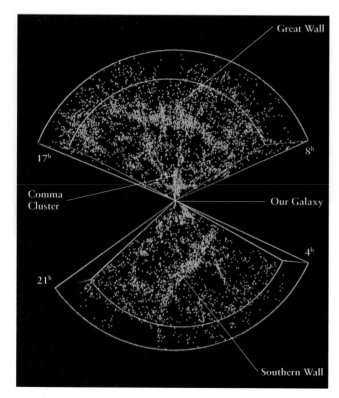

**figure 26-23**

**The Large-Scale Distribution of Galaxies** This map shows the distribution of 9325 galaxies in two wedges extending out to a redshift of $z = 0.04$. The Earth is located at the apex of the wedge-shaped slices. The upper wedge includes galaxies in the right ascension range $8^h$ to $17^h$ and declination range $8.5°$ to $44.5°$; the lower wedge covers right ascension from $4^h$ to $21^h$ and declination from $2.5°$ to $-40°$. If you squint while looking at this map, you can see the Great Wall in the upper wedge and the Southern Wall in the bottom wedge. The Coma cluster (see Figure 26-17) forms part of the Great Wall. Note the prominent voids along the inner edge of the Great Wall and on either side of the Southern Wall. (Courtesy of M. J. Geller, Harvard-Smithsonian Center for Astrophysics)

wall covers an area at least 80 by 230 Mpc (about 250 by 750 million light-years). A similar structure called the Southern Wall, visible in the southern part of the map, contains more than a thousand galaxies and extends across approximately 100 Mpc.

In recent years, astronomers have time and again been startled to find vast structures whose sizes seem to be limited only by the area of the sky covered in their surveys. Indeed, when larger areas are explored, larger features are found. Furthermore, some of these features exhibit significant movement, indicating large-scale motions in the universe (see Alan Dressler's essay "The Great Attractor" at the end of this chapter).

At present, astrophysicists are working to explain how these enormous structures came into existence. Ongoing redshift surveys to explore even larger volumes of the universe may well uncover even more surprises during the next few years.

## 26-7 Colliding galaxies produce starbursts, spiral arms, and other spectacular phenomena

Galaxies in a cluster are not fixed in space but are in continuous motion. As a result, occasionally two galaxies in a cluster collide. Collisions that have happened in the past have hurled vast numbers of stars into intergalactic space. In some cases, we can even observe a collision in progress, a cosmic catastrophe that gives birth to new stars. And we can predict collisions that will not take place for billions of years, such as the collision that is fated to occur between the galaxy M31 and our own Milky Way Galaxy.

When two galaxies collide at high speed, the huge clouds of interstellar gas and dust in the galaxies slam into each other and can be completely stopped in their tracks. In this way, two colliding galaxies can be stripped of their interstellar gas and dust. The best evidence that such collisions take place is that many rich clusters of galaxies are strong sources of X rays (Figure 26-24). This emission reveals the presence of substantial amounts of hot **intracluster gas** (that is, gas within the cluster) at temperatures between $10^7$ and $10^8$ K. The only way that such large amounts of gas could be heated to such extremely high temperatures is in violent collisions between galaxies.

**figure 26-24** R I V U X G

**X-ray Emission from a Cluster of Galaxies** This view from the Einstein Observatory, an X-ray telescope placed in orbit in 1978, shows the rich cluster Abell 1367. (The name denotes that it is the 1367th in a list of rich clusters cataloged by UCLA astronomer George O. Abell.) The X-ray emission shown here comes from gas between the galaxies at a temperature of $10^7$ to $10^8$ K. (Harvard-Smithsonian Center for Astrophysics)

Although galaxies can and do collide, it is highly unlikely that the *stars* from two colliding galaxies actually run into each other. The reason is that the stars within a galaxy are very widely separated from each other, with a tremendous amount of empty space between them.

In a less violent collision or a near-miss between two galaxies, the compressed interstellar gas may have more time to cool, allowing many protostars to form. Such collisions may account for **starburst galaxies,** which blaze with the light of numerous newborn stars. These galaxies have bright centers surrounded by clouds of warm interstellar dust, indicating recent, vigorous star birth (Figure 26-25). Because the warm dust is so abundant, starburst galaxies are among the most luminous objects in the universe at infrared wavelengths.

The starburst galaxy M82 shown in Figure 26-25 is one member of a nearby cluster that includes the beautiful spiral galaxy M81 and a fainter elliptical companion called NGC 3077 (Figure 26-26a). Radio surveys of that region of the sky reveal enormous streams of hydrogen gas connecting the three galaxies (Figure 26-26b). The loops and twists in these streamers suggest that the three galaxies have had several close encounters over the ages. A similar stream of hydrogen gas connects our Galaxy with its second nearest neighbor, the Large Magellanic Cloud (LMC), suggesting a history of close encounters between our Galaxy and the LMC.

**Figure 26-25**   R I V U X G

**A Starburst Galaxy**   Prolific star formation is occurring at the center of this irregular galaxy, called M82 or NGC 3034. This activity was probably triggered by gravitational interactions with neighboring galaxies. Note the turbulent appearance of the interstellar gas and dust around the galaxy's center. M82 lies in the constellation Ursa Major at a distance of some 3.5 Mpc (12 million light-years) and has an angular width of 9 arcminutes. (Lick Observatory)

a   R I V U X G

b   R I V U X G

**Figure 26-26**

**The M81-M82-NGC 3077 Cluster**   The starburst galaxy M82 is in a nearby cluster whose three members are connected by streamers of hydrogen gas. **(a)** This wide-angle photograph shows the three galaxies at visible wavelengths. M82 and NGC 3077 are separated in the sky by 1°, twice the apparent size of the full moon. **(b)** This radio image of the same region (actually a mosaic of 13 fields observed with radio telescopes of the Very Large Array) shows the streams of hydrogen gas that connect the three galaxies. (Palomar Sky Survey; M. S. Yun, VLA, and Harvard)

Tidal forces between colliding galaxies can deform the galaxies from their original shapes, just as the tidal forces of the Moon on the Earth deform the oceans and help give rise to the tides (see Section 9-4). The deformation is so great that thousands of stars can be hurled into intergalactic space along huge, arching streams. Supercomputer simulations of such collisions show that while some of the stars are flung far and wide, other stars slow down and the galaxies may merge. Figure 26-27 shows one such simulation by Joshua Barnes, now at the Institute for Astronomy at the University of Hawaii. As the two galaxies pass through each other, they are severely distorted by gravitational interactions and throw out a pair of extended tails. The interaction also prevents the galaxies from continuing on their original paths. Instead, they fall back together for a second encounter (at 625 million years), when they look remarkably similar to the actual colliding galaxies in Figure 26-28. The simulated galaxies merge soon thereafter, leaving a single object.

Our own Milky Way Galaxy is expected to undergo a galactic collision like that shown in Figure 26-28. The Milky Way and the Andromeda Galaxy (Figure 26-3) are actually approaching each other and should collide in another 6 bil-

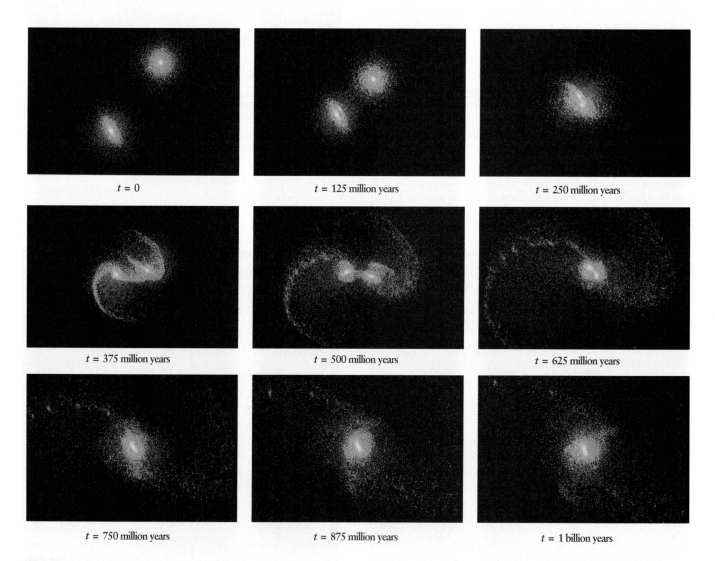

$t = 0$      $t = 125$ million years      $t = 250$ million years

$t = 375$ million years      $t = 500$ million years      $t = 625$ million years

$t = 750$ million years      $t = 875$ million years      $t = 1$ billion years

## ƒigure 26-27

**A Simulated Collision Between Two Galaxies** These frames from a supercomputer simulation show the collision and merger of two disk-shaped galaxies. Stars in the disk of each galaxy are colored blue, while stars in their central bulges are yellow. Red indicates dark matter that surrounds each galaxy. The pictures display progress at 125-million-year intervals. Compare the final frames with the image of NGC 4038 and NGC 4039 in Figure 26-28. (Courtesy of J. Barnes, Canadian Institute for Theoretical Astrophysics)

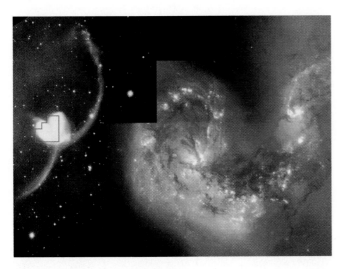

**figure 26-28**   R I **V** U X G

**Colliding Galaxies with "Antennae"**  Many pairs of colliding galaxies exhibit long tendrils of stars ejected by the collision. The left-hand image shows these for the colliding spiral galaxies NGC 4038 and NGC 4039, called "the Antennae" for their shape. The area outlined in green is shown in the right-hand image from the Hubble Space Telescope. The collision of the two galaxies has compressed vast amounts of interstellar gas and dust, triggering a firestorm of star formation. This is shown by the numerous red H II regions and blue clusters of hot, young stars. The Antennae are in the southern constellation Corvus (the Crow) and are about 20 Mpc (63 million light-years) from the Earth. The ends of the two "antennae" are separated in angle by 15 arcminutes. (Brad Whitmore, Space Telescope Science Institute; NASA)

propagating star formation, discussed in Section 25-5). Supercomputer simulations clearly demonstrate that spiral arms can be created during a collision, either by drawing out long streamers of stars or by compressing clouds of interstellar gas. For example, the spiral arms of M51 (examine Figure 26-2) may have been produced by a close encounter with a second galaxy. The disruptive galaxy is now located at the end of one of the spiral arms created by the collision. Some astronomers argue that the spiral arms of our Milky Way Galaxy were similarly produced by a close encounter with the Large Magellanic Cloud. If this is so, then the chain of events that led to the formation of our Sun, our solar system, and ourselves was initiated by this long-ago interaction between two galaxies.

## 26-8   Most of the matter in the universe has yet to be discovered

A cluster of galaxies must be held together by gravity. In other words, there must be enough matter in the cluster to prevent the galaxies from wandering away. Nevertheless, careful examination of a rich cluster, like the Coma cluster, reveals that the mass of the visually luminous matter (principally the stars in the galaxies) is not at all sufficient to bind the cluster gravitationally. The observed line-of-sight speeds of the galaxies, measured by Doppler shifts, are so large that the cluster should have broken apart long ago. Considerably more mass than is visible is needed to keep the galaxies bound in orbit about the center of the cluster.

We encountered a similar situation in studying our own Milky Way Galaxy in Section 25-4: The total mass of our Galaxy is more than the amount of visible mass. As for our Galaxy, we conclude that clusters of galaxies must contain significant amounts of nonluminous **dark matter.** If this dark matter were not there, the galaxies would have long ago dispersed in random directions and the cluster would no longer exist today. Analyses demonstrate that the total mass needed to bind a typical rich cluster is about 10 times greater than the mass of material that shows up on photographs.

As for our Galaxy, the problem is to determine what form the invisible mass takes. A partial solution to this **dark-matter problem,** which dates from the 1930s, was solved by the discovery in the late 1970s of hot, X-ray emitting gas within clusters of galaxies (see Figure 26-24). By measuring the amount of X-ray emission, astronomers find that the total mass of intracluster gas in a typical rich cluster is comparable to or greater than the combined mass of all the stars in all the cluster's galaxies. This is sufficient to account for only about 10% of the invisible mass, however. The remainder is dark matter of unknown composition.

Although we do not know what dark matter is made of, it is possible to investigate how dark matter is distributed in galaxies and clusters of galaxies. It appears that dark matter lies within and immediately surrounding galaxies, not in the

lion years or so. (Recall that our solar system is only 4.6 billion years old.) When this happens, the sky will light up with a plethora of newly formed stars, followed in rapid succession by a string of supernovae as the most massive of these stars complete their lifespans. Any inhabitants of either galaxy will see a night sky far more dramatic and tempestuous than ours.

When two galaxies merge, the result is a bigger galaxy. If this new galaxy is located in a rich cluster, it may capture and devour additional galaxies, growing to enormous dimensions by **galactic cannibalism.** Cannibalism differs from mergers in that the galaxy that does the devouring is bigger than its "meal," whereas merging galaxies are about the same size.

Many astronomers suspect that galactic cannibalism is the reason that giant ellipticals are so huge. As we have seen, giant galaxies typically occupy the centers of rich clusters. In many cases, smaller galaxies are located around these giants (examine Figure 26-8). As they pass through the extended halo of a giant elliptical, these smaller galaxies slow down and are eventually devoured by the larger galaxy.

Close encounters between galaxies provide a third way of forming spiral arms (in addition to density waves and self-

vast spaces between galaxies. The evidence for this comes principally from observations of the rotation curves of galaxies and of the gravitational bending of light by clusters of galaxies.

As we saw in Section 25-4, a rotation curve is a graph that shows how fast stars in a galaxy are moving at different distances from that galaxy's center. For example, Figure 25-17 is the rotation curve for our Galaxy. Many other galaxies have rotation curves similar to that of our Galaxy, as shown in Figure 26-29. These curves for four spiral galaxies all remain remarkably flat out to surprisingly great distances from each galaxy's center. In other words, the orbital speed of the stars remains roughly constant out to the visible edges of these galaxies. This tells us that we still have not detected the *true* edges of these galaxies (and many similar ones). Near the true edge of a galaxy we should see a decline in orbital speed, in accordance with Kepler's third law (see Figure 25-17). Because this decline has not been observed, astronomers conclude that there must be a considerable amount of dark material within a galaxy's disk that extends well beyond the visible portion of the disk.

Further evidence about how dark matter is distributed comes from the gravitational bending of light rays, which we described in Section 24-2. As Figure 24-5 shows, the gravity of a single star like the Sun can deflect light by only a few arcseconds. But a more massive object such as a galaxy can produce much greater deflections, and the amount of this deflection can be used to determine the galaxy's mass. For example, suppose that the Earth, a massive galaxy, and a background light source (such as a more distant galaxy) are in nearly perfect alignment, as sketched in Figure 26-30. Because of the warped space around the massive galaxy, light from the background source curves around the galaxy as it heads toward us. As a result, there are two paths along which light rays can travel from the background source to us here on the Earth. Thus, we should see two images of the background source.

A powerful source of gravity that distorts background images is called a **gravitational lens.** For gravitational lensing

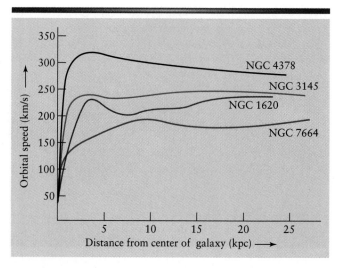

**figure 26-29**

**The Rotation Curves of Four Spiral Galaxies** This graph shows how the orbital speed of material in the disks of four spiral galaxies varies with the distance from the center of each galaxy. If most of each galaxy's mass were concentrated near the center of the galaxy, these curves would fall off at large distances. But these and many other galaxies have flat rotation curves that do not fall off. This indicates the presence of extended halos of dark matter. (Adapted from V. Rubin and K. Ford)

to work, the alignment between the Earth, the massive galaxy, and a remote background light source must be almost perfect. Without nearly perfect alignment, the second image of the background star is too faint to be noticeable.

A number of gravitational lenses have been discovered, involving images of remote, luminous objects called quasars. As we will see in Chapter 27, a quasar is a very bright, distant object. Typical quasars shine as brightly as hundreds of galax-

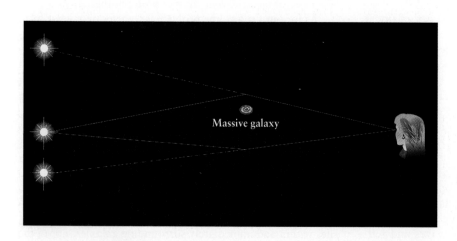

**figure 26-30**

**A Gravitational Lens** A massive object such as a galaxy can deflect light rays from a distant light source so that an observer sees more than one image of the distant source. Depending on the alignment of observer, massive galaxy, and distant source, the observer can see several images of the light source (as in Figure 26-32).

ies and are located billions of parsecs from the Earth. Several thousand quasars have been discovered across the sky.

In 1979, Dennis Walsh, Robert Carswell, and Ray Weymann discovered two faint quasars separated by only 6 arcsec. They took spectra of both quasars and discovered that they were nearly identical. They thus concluded that they were looking at two images of the same quasar. Further observations revealed a galaxy between the quasar images (Figure 26-31). The gravitational field of this galaxy bends the light from the remote quasar and thus acts like a gravitational lens.

The Hubble Space Telescope has found many cases in which galaxies and clusters of galaxies act as gravitational lenses. Figure 26-32 is an HST image of an ordinary-looking rich cluster of yellowish elliptical and spiral galaxies—but with a number of curious blue arcs. Reconstruction of the light paths through the cluster show that all of these blue arcs are actually distorted images of the same galaxy, which lies billions of light-years beyond the cluster. By measuring the distortion of the images of such background galaxies, J. Anthony Tyson of Bell Laboratories and his colleagues have determined that dark matter, which constitutes about 90% of the cluster's mass, is distributed much like the visible matter in the cluster. In other words, the overall

**Figure 26-32**   R I **V** U X G

**Gravitational Lensing by a Cluster of Galaxies** The Hubble Space Telescope image of the rich cluster 0024 + 1654 (named for its celestial coordinates) records not only the cluster's galaxies, which appear as yellowish blobs, but also a number of blue arcs. These blue arcs are distorted multiple images of a single more distant galaxy, the result of gravitational lensing by the matter in 0024 + 1654. The cluster is about 1600 Mpc (5 billion light-years) away in the constellation Pisces (the Fish); the blue galaxy is about twice as far away. The blue color of the remote galaxy suggests that it is very young and is actively forming stars. (W. N. Colley and E. Turner, Princeton University; J. A. Tyson, Bell Labs, Lucent Technologies; NASA)

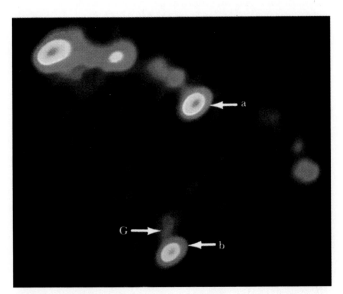

**Figure 26-31**   **R** I V U X G

**A "Double" Quasar** Two images of the same quasar, labeled **a** and **b,** are seen in this radio view made by the Very Large Array. Light from the distant quasar is deflected to either side of a massive galaxy located between us and the quasar. A faint image of the deflecting galaxy **G** is seen directly above image **b.** The jetlike feature protruding to the left and right of the upper image **a** does not appear alongside the lower image because the jet is too far away from the required quasar-galaxy-Earth alignment. (VLA, NRAO)

arrangement of visible galaxies seems to trace the location of dark matter.

Many proposals have been made to explain the nature of dark matter. One reasonable suggestion was that clusters might contain a large number of faint, red, low-mass (0.2 $M_\odot$ or less) stars. These faint stars could be located in extended halos surrounding individual galaxies or scattered throughout the spaces between the galaxies of a cluster. They would have escaped detection because their luminosity and hence apparent brightness would be very low. (Recall the mass-luminosity relation for main-sequence stars, discussed in Section 19-9.) Such faint red stars would, however, be detectable by the Hubble Space Telescope (HST), and searches for these stars around other galaxies as well as around the Milky Way have been carried out using HST. None have yet been detected, so it is thought that faint stars are unlikely to constitute much of the dark matter in the universe. As we described in Section 25-4, more exotic dark matter candidates include massive neutrinos, subatomic particles called WIMPs (weakly interacting massive particles), and MACHOs (massive compact halo objects, such as small black holes or brown dwarfs). To date, however, the true nature of dark matter remains unknown.

## 26-9 Galaxies may have formed from the merger of smaller objects

How do galaxies form? How do they evolve? Why are some galaxies spirals, other ellipticals, and still others irregulars? Unfortunately, the birth and evolution of galaxies happens so slowly that we can cannot simply watch one over time and expect to learn much about how it changes. But astronomers can gain important clues about galactic evolution by simply looking deep into space. The more distant a galaxy is, the longer its light takes to reach us. Consequently, as we examine galaxies that are at increasing distances from the Earth, we are actually looking farther and farther back in time. By looking into the past in this way, we can see galaxies in the earliest stages of their lives.

By observing remote galaxies, astronomers have discovered that galaxies were bluer in the past than they are today. This trait was noticed in 1978 by astronomers Harvey Butcher and Augustus Oemler, who drew attention to large numbers of blue galaxies in the remote, rich clusters they were studying. Spectroscopic studies of such galaxies in the 1980s by James Gunn and Alan Dressler demonstrated that most owe their blue color to vigorous star formation, often occurring in intense, episodic bursts. The hot, luminous, and short-lived O and B stars produced in these bursts of star formation give blue galaxies their characteristic color. (The arcs in Figure 26-32 are gravitational-lens images of such a distant blue galaxy.)

In 1992, Dressler, Oemler, Gunn, and Butcher teamed up to use the Hubble Space Telescope to examine two remote, rich clusters of galaxies with redshifts $z = 0.4$. This redshift corresponds to looking back about 4 billion years into the past. Careful examination of the HST images revealed a surprisingly large number of spiral galaxies. About 30% of the galaxies in these remote clusters are spirals, whereas only 5% of the galaxies in nearby rich clusters are spirals. The astronomers also found that most of the blue, star-forming galaxies are spirals, many of which show signs of collisions or mergers.

If spiral galaxies were more common in the distant past, where have they gone? Galactic collisions and mergers are probably responsible for depleting rich clusters of their spiral galaxies. During a collision, interstellar gas in the colliding galaxies is vigorously compressed, which triggers a burst of star formation. A succession of collisions produces a series of star-forming episodes that create numerous bright, hot O and B stars that become dispersed along arching spiral arms by the galaxy's rotation. Eventually, however, the gas is used up; star formation then ceases and the spiral arms become less visible.

Although collisions and mergers deplete the population of spiral galaxies, these events do not produce elliptical or S0 galaxies. The HST images show that elliptical and S0 galaxies were already well developed 4 billion years ago. Indeed, the preponderance of old Population II stars in elliptical galaxies, as well as the lack of gas and dust in these galaxies, demonstrates that their stars formed in a burst of activity 10 to 15 billion years ago. In contrast, spiral galaxies have been forming stars continually over the past 10 to 15 billion years, although at a gradually decreasing rate as their interstellar gas gets used up. Figure 26-33 compares the rates at which spiral and elliptical galaxies form stars.

One of the challenges that faces astronomers today is to understand how galaxies were created. One theory, popularized in the 1960s, argues that galaxies formed from the gravitational contraction of huge clouds of primordial gas. The rate of star formation in a contracting gas cloud determines whether it becomes a spiral or an elliptical galaxy. If the rate of star formation is low, then the gas has plenty of time to settle in a flattened disk. A flattened disk is the natural consequence of the overall rotation of the original gas cloud. Star formation continues in the disk because it contains an abundance of hydrogen, and thus a spiral galaxy is created. But if the stellar birthrate is high, then virtually all of the gas is used up in the formation of stars before a disk has time to form. In this case, an elliptical galaxy is created.

An alternative theory, proposed in 1977, suggests that galaxies formed by the merging of several gas clouds rather than from the gravitational contraction of huge, isolated clouds. A third theory also contends that galaxies formed from mergers of gas clouds, except that the ancestral fragments were rather small and quite numerous. These three main theories of galaxy formation are illustrated in Figure 26-34.

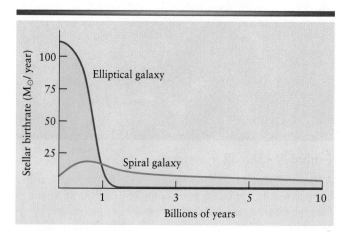

**figure 26-33**

**The Stellar Birthrate in Galaxies** Most of the stars in an elliptical galaxy are created in a brief burst of star formation when the galaxy is very young. In spiral galaxies, star formation occurs at a more leisurely pace and may extend over billions of years. (Adapted from J. Silk)

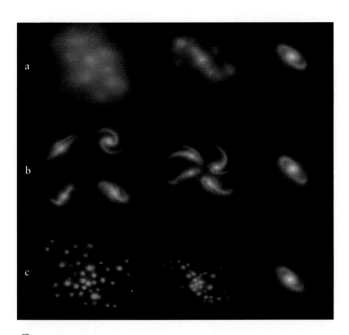

## figure 26-34

**Theories of Galaxy Formation** Theories of galaxy formation fall into three categories. **(a)** A single huge primordial gas cloud contracts to form a galaxy. **(b)** Several moderate-sized gas clouds merge to form a galaxy. **(c)** Numerous small gas clouds coalesce to form a galaxy. (Adapted from van den Bergh and Hesser)

Observations from the Hubble Space Telescope favor the idea that galaxies formed from the merging of gas clouds. Figure 26-35 shows a number of galaxylike objects some 11 billion light-years (3400 Mpc) away and are thus seen as they were 11 billion years ago. These objects are smaller than even the smallest galaxies we see in the present-day universe and have unusual, irregular shapes. Furthermore, these objects are scattered over an area only 600 kpc (2 million light-years) across—less than the distance between the Milky Way Galaxy and M31—making it quite probable that they would collide and merge with each other. These collisions would be aided by the dark matter associated with each subgalactic object, which increases the object's mass and, hence, the gravitational forces pulling the objects together. Such mergers would eventually give rise to a normal-sized galaxy.

Galactic evolution is a difficult and controversial subject full of unresolved questions. Where did the subgalactic clouds of gas come from? And what happened in the early universe to cause the primordial hydrogen and helium to clump up in clouds destined to evolve into galaxies, instead of becoming objects a million times bigger or smaller? Even more troublesome is the issue of the dark matter. We know that the observable stars, gas, and dust in a galaxy constitute only 10% of the mass associated with the galaxy. We have no idea what the remaining 90% looks like or what it is made of. These questions are among the most challenging that face astronomers today.

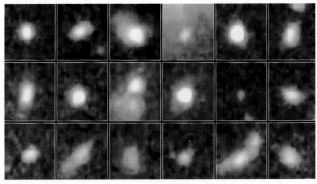

## figure 26-35   R I **V** U X G

**The Building Blocks of Galaxies** **(a)** This Hubble Space Telescope image shows a portion of the sky $2\frac{1}{2}$ arcminutes across in the constellation Hercules. The objects outlined by boxes lie about 3400 Mpc (11 billion light-years) from the Earth and are only 600 to 900 pc (2,000 to 3,000 light-years) across—larger than a star cluster but smaller than even dwarf elliptical galaxies like those in Figure 26-19. **(b)** Close-up images of these curious subgalactic objects show their irregular shapes and blue color, indicating the presence of young stars. It is thought that gravitational attraction would make these objects coalesce rapidly. In so doing, they would form a full-sized galaxy, such as we see in the nearby universe today. (Rogier Windhorst and Sam Pascarelle, Arizona State University; NASA)

## Key Words

## Key Ideas

**The Hubble Classification:** Galaxies can be grouped into four major categories: spirals, barred spirals, ellipticals, and irregulars.

• The disks of spiral and barred spiral galaxies are sites of active star formation.

• Elliptical galaxies are nearly devoid of interstellar gas and dust, which severely inhibits star formation.

• Lenticular (S0) galaxies are intermediate between spiral and elliptical galaxies.

• Irregular galaxies have ill-defined, asymmetrical shapes. They are often found associated with other galaxies.

**Distance to Galaxies:** Standard candles, such as Cepheid variables and the most luminous supergiants, globular clusters, H II regions, and supernovae in a galaxy, are used in estimating intergalactic distances.

• The Tully-Fisher relation, which correlates the width of the 21-cm line of hydrogen in a spiral galaxy with its luminosity, can also be used for determining distance.

**The Hubble Law:** There is a simple linear relationship between the distance from the Earth to a remote galaxy and the redshift of that galaxy (which is a measure of the speed with which it is receding from us). This relationship is the Hubble law, $v = H_0 d$.

• Because of difficulties in measuring the distances to galaxies, the value of the Hubble constant, $H_0$, is not known with certainty but is probably close to 75 km/s/ Mpc. As a result of this uncertainty, distances to remote galaxies calculated from the Hubble law are uncertain by about 20%.

**Clusters and Superclusters:** Galaxies are grouped into clusters rather than being scattered randomly throughout the universe.

• A rich cluster contains hundreds or even thousands of galaxies; a poor cluster, often called a group, may contain only a few dozen.

• A regular cluster has a nearly spherical shape with a central concentration of galaxies; in an irregular cluster, galaxies are distributed asymmetrically.

• Our Galaxy is a member of a poor, irregular cluster called the Local Group.

• Rich, regular clusters contain mostly elliptical and S0 galaxies; irregular clusters contain spiral and irregular galaxies along with ellipticals.

• Giant elliptical galaxies are often found near the centers of rich clusters.

**Galactic Collisions and Mergers:** When two galaxies collide, their stars pass each other, but their interstellar media collide violently, either stripping the gas and dust from the galaxies or triggering prolific star formation.

• The gravitational effects during a galactic collision can throw stars out of their galaxies into intergalactic space.

• Galactic mergers may occur; a large galaxy in a rich cluster may tend to grow steadily through galactic cannibalism, perhaps producing in the process a giant elliptical galaxy.

**The Dark-Matter Problem:** The luminous mass of a cluster of galaxies is not large enough to account for the observed motions of the galaxies; a large amount of unobserved mass must be present between the galaxies. This situation is called the dark-matter problem.

• Hot intergalactic gases emit X rays in rich clusters; massive extended halos probably surround all the galaxies.

• Gravitational lensing of remote galaxies by a foreground cluster enables astronomers to glean information about the distribution of dark matter in the foreground cluster.

**Formation and Evolution of Galaxies:** Observations of remote clusters of galaxies reveal that spiral galaxies were more common billions of years ago than they are today. This excess of spiral galaxies probably resulted from collisions and mergers when galaxies had an abundance of interstellar gas with which to form stars.

• One theory states that galaxies formed by the gravitational contraction of huge, individual clouds of gas; others explain that galaxies arose from mergers of several smaller gas clouds. Hubble Space Telescope observations suggest that the latter model explains at least some of the galaxies we see today.

## Review Questions

**1.** What was the Shapley-Curtis debate all about? Was a winner declared at the end of the debate? Whose ideas turned out to be correct?

**2.** How did Edwin Hubble prove that the Andromeda "Nebula" is not a nebula in our Milky Way Galaxy?

**3.** What is the Hubble classification scheme? Which category includes the largest galaxies? Which includes the smallest? Which type of galaxy is the most common?

**4.** What types of galaxies are most likely to have new stars forming? Describe the observational evidence that supports your answer.

**5.** Are there any galaxies besides our own that can be seen with the naked eye? If so, which one(s)?

**6.** How is it possible that galaxies in our Local Group still remain to be discovered? In what part of the sky would these galaxies be located? What sorts of observations might reveal these galaxies?

**7.** What is the Hubble law?

**8.** Why do you suppose there are many discordant determinations of $H_0$?

**9.** Some galaxies in the Local Group exhibit blueshifted spectral lines. Why aren't these blueshifts violations of the Hubble law?

**10.** What kinds of stars would you expect to find populating space between galaxies in a cluster?

**11.** What evidence is there for the existence of dark matter in clusters of galaxies?

**12.** On what grounds do astronomers believe that spiral galaxies were more numerous in the past than they are today? What could account for this excess of spiral galaxies in the distant past?

**13.** What observations suggest that present-day galaxies formed from smaller assemblages of matter?

## Advanced Questions

*Questions preceded by an asterisk (*) involve topics discussed in the Boxes.*

> **Problem-solving tips and tools:**
>
> The relationship between apparent magnitude, absolute magnitude, and distance is discussed in Box 19-3. As explained in Box 25-3, a useful form of Kepler's third law is $M = rv^2/G$, where $M$ is the mass within an orbit of radius $r$, $v$ is the orbital speed, and $G$ is the gravitational constant. Another form of Kepler's third law, particularly useful for two stars or two galaxies orbiting each other, is given in Section 19-9. The volume of a sphere of radius $r$ is $\frac{4}{3}\pi r^3$. The mass of the hydrogen atom ($^1$H) is given in Appendix 7.

**14.** As Figure 21-14 shows, there are two types of Cepheid variables. Type I Cepheids are metal-rich stars of Population I, while Type II Cepheids are metal-poor stars of Population II. (a) Which type of Cepheid variables would you expect to be found in globular clusters? Which type would you expect to be found in the disk of a galaxy, as in Figure 26-4? Explain your reasoning. (b) When Hubble discovered Cepheid variables in M31, the distinction between Type I and Type II Cepheids was not yet known. Hence, Hubble thought that the Cepheids seen in the disk of M31 were identical to those seen in globular clusters in our own Galaxy. As a result, his calculations of the distance to M31 were in error. Using Figure 21-14, explain whether Hubble's calculated distance was too small or too large.

*15. Astronomers often state the distance to remote galaxies in terms of their distance modulus, which is the difference between the apparent magnitude $m$ and the absolute magnitude $M$ (see Box 19-3). (a) By measuring the brightness of supernova 1994I in the galaxy M51 (Figure 26-2), the distance modulus for this galaxy was determined to be $m - M = 29.2$. Find the distance to M51 in megaparsecs (Mpc). (b) A separate distance determination, which involved measuring the brightnesses of planetary nebulae in M51, found $m - M = 29.6$. What is the distance to M51 that you calculate from this information? (c) What is the difference between your answers to parts (a) and (b)? Compare this difference to the distance from the Earth to M31, the Andromeda Galaxy (Figure 26-3). The difference between your answers illustrates the uncertainties involved in determining the distances to galaxies!

*16. Suppose you discover a Type Ia supernova in a distant galaxy. At maximum brilliance, the supernova reaches an apparent magnitude of +10. How far away is the galaxy? (*Hint:* See Box 26-1.)

**17.** A certain galaxy is observed to be receding from the Sun at a rate of 75,000 km/s. The distance to this galaxy is measured independently and found to be $1.4 \times 10^9$ pc. Based on these data, what is the value of the Hubble constant?

**18.** The average radial velocity of galaxies in the Hercules cluster pictured in Figure 26-21 is 10,800 km/s. (**a**) Using $H_0 = 75$ km/s/Mpc, find the distance to this cluster. Give your answer in megaparsecs and in light-years. (**b**) How would your answer to (a) differ if the Hubble constant had a smaller value? A larger value? Explain.

**19.** It is estimated that the Coma cluster (Figure 26-17) contains about $10^{13}$ $M_\odot$ of intracluster gas. (**a**) Assuming that this gas is made of hydrogen atoms, calculate the total number of intracluster gas atoms in the Coma cluster. (**b**) The Coma cluster is roughly spherical in shape, with a radius of about 3 Mpc. Calculate the number of intracluster gas atoms per cubic centimeter in the Coma cluster. Assume that the gas fills the cluster uniformly. (**c**) Compare the intracluster gas in the Coma cluster with the gas in our atmosphere ($3 \times 10^{19}$ molecules per cubic centimeter, temperature 300 K); a typical gas cloud within our own Galaxy (a few hundred molecules per cubic centimeter, temperature 50 K or less); and the corona of the Sun ($10^5$ atoms per cubic centimeter, temperature $10^6$ K).

**\*20.** Two galaxies separated by 600 kpc are orbiting each other with a period of 40 billion years. What is the total mass of the two galaxies?

**\*21.** The rotation curve of the Sa galaxy NGC 4378 is shown in Figure 26-29. Using data from that graph, calculate the orbital period of stars 20 kpc from the galaxy's center. How much mass lies within 20 kpc from the center of NGC 4378?

**22.** How might you determine what part of a galaxy's redshift is caused by the galaxy's orbital motion about the center of mass of its cluster?

# Discussion Questions

**23.** The Earth is composed principally of heavy elements, such as silicon, nickel, and iron. Would you be likely to find such planets orbiting stars in the disk of a spiral galaxy? In the nucleus of a spiral galaxy? In an elliptical galaxy? In an irregular galaxy? Explain your answers.

**24.** Discuss whether the various Hubble types of galaxies actually represent some sort of evolutionary sequence.

**25.** Discuss the advantages and disadvantages of using the various standard candle distance indicators to obtain extragalactic distances.

**26.** How would you distinguish star images from unresolved images of remote galaxies on a photographic plate or CCD?

**27.** When galaxies pass close to each other, as should happen frequently in a rich cluster, tidal forces between the galaxies can strip away their outlying stars. The result should be a loosely dispersed sea of "intergalactic stars" populating the space between galaxies in a cluster. Search the World Wide Web and magazines such as

*Sky & Telescope* and *Astronomy* for information about intergalactic stars. Have they been observed? Where are they found? What would our nighttime sky look like if our Sun were an intergalactic star?

**28.** Explain why the dark matter in galaxy clusters could not be neutral hydrogen.

**29.** The Hubble Space Telescope (HST) has made extensive observations of very distant galaxies. Visit the HST web site (**http://oposite.stsci.edu/pubinfo/Subject.html**) to learn about these investigations. How far back in time has HST been able to look? What sorts of early galaxies are observed? What is the current thinking about how galaxies formed and evolved?

# Observing Projects

**30.** Using a telescope with an aperture of at least 30 cm (12 in.), observe as many of the spiral galaxies listed in the following table as you can. Many of these galaxies are members of the Virgo cluster, which can best be seen during the spring. Because all galaxies are quite faint, be sure to schedule your observations for a moonless night. The best view is obtained when a galaxy is near the meridian. While at the eyepiece, make a sketch of what you see. Can you distinguish any spiral structure? After completing your observations, compare your sketches with photographs found in such popular books as *Galaxies* by Timothy Ferris (Sierra Club Books, 1980) and *The Color Atlas of the Galaxies* by James Wray (Cambridge University Press, 1988).

| Spiral galaxy | Right ascension | Declination | Hubble type |
|---|---|---|---|
| M31 (NGC 224) | $0^h$ $42.7^m$ | +41° 16′ | Sb |
| M58 (NGC 4579) | 12 37.7 | +11 49 | Sb |
| M61 (NGC 4303) | 12 21.9 | +4 28 | Sc |
| M63 (NGC 5055) | 13 15.8 | +42 02 | Sb |
| M64 (NGC 4826) | 12 56.7 | +21 41 | Sb |
| M74 (NGC 628) | 1 36.7 | +15 47 | Sc |
| M83 (NGC 5236) | 13 37.0 | −29 52 | Sc |
| M88 (NGC 4501) | 12 32.0 | +14 25 | Sb |
| M90 (NGC 4569) | 12 36.8 | +13 10 | Sb |
| M91 (NGC 4548) | 12 35.4 | +14 30 | SBb |
| M94 (NGC 4736) | 12 50.9 | +41 07 | Sb |
| M98 (NGC 4192) | 12 13.8 | +14 54 | Sb |
| M99 (NGC 4254) | 12 18.8 | +14 25 | Sc |
| M100 (NGC 4321) | 12 22.9 | +15 49 | Sc |
| M101 (NGC 5457) | 14 03.2 | +54 21 | Sc |
| M104 (NGC 4594) | 12 40.0 | −11 37 | Sa |
| M108 (NGC 3556) | 11 11.5 | +55 40 | Sc |

*Note: The right ascensions and declinations are given for epoch 2000.*

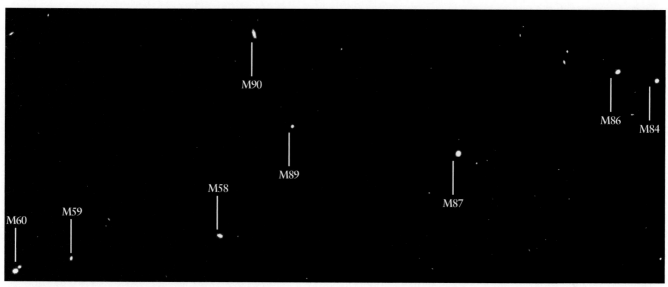

R I [V] U X G

**31.** Using a telescope with an aperture of at least 30 cm (12 in.), observe as many of the following elliptical galaxies as you can. Six of these galaxies are in the Virgo cluster, which is conveniently located in the evening sky from March through June. The photograph above, which covers a 2° × 5° swath of the cluster, may help you find some of these galaxies. As in the previous exercise, be sure to schedule your observations for a moonless night, when the galaxies you wish to observe will be near the meridian. Do these elliptical galaxies differ in appearance from spiral galaxies?

| Elliptical galaxy | Right ascension | Declination | Hubble type |
|---|---|---|---|
| M58 (NGC 4472) | $12^h$ $29.8^m$ | +8° 00′ | E4 |
| M59 (NGC 4621) | 12 42.0 | +11 39 | E3 |
| M60 (NGC 4649) | 12 43.7 | +11 33 | E1 |
| M84 (NGC 4374) | 12 25.1 | +12 53 | E1 |
| M86 (NGC 4406) | 12 26.2 | +12 57 | E3 |
| M89 (NGC 4552) | 12 35.7 | +12 33 | E0 |
| M110 (NGC 205) | 00 40.4 | +41 41 | E6 |

*Note: The right ascensions and declinations are given for epoch 2000.*

**32.** Using a telescope with an aperture of at least 30 cm (12 in.), observe as many of the following interacting galaxies as you can. As in the previous exercises, be sure to schedule your observations for a moonless night, when the galaxies you wish to observe will be near the meridian. While at the eyepiece, make a sketch of what you see. Can you distinguish hints of interplay among the galaxies? After completing your observations, compare your sketches with photographs found in such popular books as *Galaxies* by Timothy Ferris (Sierra Club Books, 1980).

| Interacting galaxies | Right ascension | Declination |
|---|---|---|
| M51 (NGC 5194) | $13^h$ $29.9^m$ | +47° 12′ |
| NGC 5195 | 13 30.0 | +47 16 |
| M65 (NGC 3623) | 11 18.9 | +13 05 |
| M66 (NGC 3627) | 11 20.2 | +12 59 |
| NGC 3628 | 11 20.3 | +13 36 |
| M81 (NGC 3031) | 9 55.6 | +69 04 |
| M82 (NGC 3034) | 9 55.8 | +69 41 |
| M95 (NGC 3351) | 10 44.0 | +11 42 |
| M96 (NGC 3368) | 10 46.8 | +11 49 |
| M105 (NGC 3379) | 10 47.8 | +12 35 |

*Note: The right ascensions and declinations are given for epoch 2000.*

## Where to Learn More

*Books and magazine articles*

Bothun, G. D. "The Ghostliest Galaxies." *Scientific American*, February 1997. This article describes some of the largest galaxies known in the universe–and why they are so difficult to find.

Dalcanton, J. "The Overlooked Galaxies." *Sky & Telescope*, April 1998. This article describes a new class of galaxies, called low-surface-brightness galaxies, that astronomers have only recently begun to discover.

Dressler, A. "Galaxies Far Away and Long Ago." *Sky & Telescope,* April 1993. This article discusses recent observations by the Hubble Space Telescope of remote spiral-rich clusters of galaxies.

Elmegreen, D. M., and Elmegreen, B. "What Puts the Spiral in Spiral Galaxies?" *Astronomy,* September 1993. This excellent overview of spiral structure includes an assortment of beautiful computer-enhanced images of galaxies.

Jayawardhana, R. "Our Galaxy's Nearest Neighbor." *Sky & Telescope,* May 1998. The galaxy nearest to our own, the Sagittarius Dwarf (Figure 26-19a), is slowly dissolving into the halo of the Milky Way. This article explains how astronomers made this remarkable discovery.

Lake, G. "Cosmology of the Local Group." *Sky & Telescope,* December 1992. This article explains how astronomers can learn about galaxy dynamics and dark matter by studying the Local Group.

———. "Understanding the Hubble Sequence." *Sky & Telescope,* May 1992. This article describes current thinking about how galaxies get their shapes.

Leverington, D. *A History of Astronomy: From 1890 to the Present.* Springer, 1995. This very readable history of modern astronomy has an excellent chapter on how our understanding of galaxies has evolved.

Macchetto, F. D., and Dickinson, M. "Galaxies in the Young Universe." *Scientific American,* May 1997. The Hubble Space Telescope has revolutionized our knowledge of distant galaxies. This article describes what has been learned about galactic formation and evolution.

Osterbrock, D.; Brashear, R.; and Gwinn, J. "Young Edwin Hubble." *Mercury,* January/February 1990. This fascinating article chronicles the adolescence and early adulthood of one of the greatest astronomers of all time.

Roth, J. "When Galaxies Collide." *Sky & Telescope,* March 1998. This beautifully illustrated article describes what the Hubble Space Telescope is telling us about interactions between galaxies.

van den Bergh, S., and Hesser, J. E. "How the Milky Way Formed." *Scientific American,* January 1993. Although this article focuses on the Milky Way, it contains an excellent summary of theories of galaxy formation in general.

West, M. "Galaxy Clusters: Urbanization of the Cosmos." *Sky & Telescope,* January 1997. An expert on clusters of galaxies describes how these gigantic structures tell us about the history of the universe.

**W** *World Wide Web*
The Great Debate between Shapley and Curtis is the subject of an insightful article by Michael Hoskin, available on-line (**http://antwrp.gsfc.nasa.gov/htmltest/gifcity/cs_real.html**).

Several beautiful images of galaxies are available from the Anglo-Australian Observatory (**http://www.aao.gov.au/images.html**). A number of the more prominent galaxies are part of the Messier Catalog of bright deep-sky objects. Images and descriptions of these can be found at the on-line version of the catalog (**http://www.seds.org/messier/**). Up-to-date information about the Local Group is available within the same web site (**http://www.seds.org/messier/more/local.html**). Still more galaxy images are available at Princeton University (**http://astro.princeton.edu/~frei/Gcat/galaxy_form_0.html**) and from the Hubble Space Telescope (**http://oposite.stsci.edu/pubinfo/Subject.html**).

The web site for the Las Campanas Redshift Survey provides a wealth of information about how matter is distributed in the universe and includes some remarkable animations (**http://manaslu.astro.utoronto.ca/~lin/lcrs.html**).

A number of still images and video clips of simulated galaxy collisions are available at Joshua Barnes's web site (**http://www.ifa.hawaii.edu/~barnes/transform.html**). (Some of his work is shown in Figure 26-27.) The forthcoming collision between the Milky Way and M31 is the topic of a web site at the Canadian Institute for Theoretical Astrophysics (**http://www.cita.utoronto.ca/~dubinski/merger/bigmerger.html**).

# ALAN DRESSLER

## The Great Attractor

Alan Dressler is an astronomer at the Observatories of the Carnegie Institution in Pasadena, California. As a teenager he polished mirrors for 4- and 8-inch telescopes; he later studied physics at the University of California, Berkeley and received his Ph.D. in astronomy from the University of California, Santa Cruz.

In addition to his research into large-scale structure of the universe, Dressler has focused his studies on the nature of galaxies—their present-day structure and stellar populations and how these have evolved. Dressler and his colleague James E. Gunn of Princeton University have pursued the difficult task of obtaining electronic pictures and spectra of very faint, high-redshift galaxies. Looking far into space is a view to earlier cosmic times, so Dressler and Gunn have been able to study the evolution of galaxies directly. They find that many galaxies formed stars at a much higher rate 5 to 10 billion years ago.

Dressler was one of the first to find observational evidence for massive black holes in the centers of galaxies. He found that the speeds of stars increase rapidly as they cross the very center of the Andromeda Galaxy (M31), indicating a central mass concentration of tens of millions of solar masses in a region only a few light-years across. He is now leading a team of astronomers who are taking pictures of high-redshift galaxies with the Hubble Space Telescope to see for the first time whether galaxies looked noticeably different 5 billion years ago.

O ur gaze into a starry sky scarcely reveals the magnificent tapestry that nature has woven with galaxies. To see these features we must look deep into space, to the billions of galaxies beyond the Milky Way.

To gain a true three-dimensional view, we need to know how these galaxies are distributed in space. In other words, we need to place each galaxy at its approximate distance. We can do that by measuring its redshift and applying Hubble's relation: Because of the uniform expansion of the universe, a galaxy's distance is proportional to its redshift. Advances in telescopes have made it much easier to measure galaxy redshifts—nearly 100,000 have now been measured—and a clearer picture is emerging of how galaxies fill the vastness of space. More like sculpture than spatter, galaxies are found strewn in long chains and spread in giant, cavernous walls that surround the nearly empty regions we call voids.

We now believe that the Big Bang itself carved these patterns, faintly, into the map of primeval matter. As cosmic time passed, gravity amplified the patterns by drawing more and more matter into denser regions. Recovering a map of these patterns engraved by the Big Bang may be our best opportunity to test theories of particle physics that describe the nature of matter in its ultimate, high-density, high-temperature form.

However, our theories predict the distribution of *all* matter, and most of the matter in the universe appears to be not in galaxies but around them and in between them in a form that gives off no light. Observational evidence for "dark matter" has built steadily since the 1930s, and we now believe that it overwhelms the matter in galaxies by at least 10 to 1. It probably takes a very different form than protons, neutrons, and electrons—the "ordinary matter" of which we and our world are made. A popular model suggests that dark matter is made of elementary particles that interact only through the relatively weak force of gravity. Immune to electromagnetism, nature's strongest force, dark matter would not clump together to form such things as stars, planets, or puppy dogs. On the contrary, the mysterious particles would pass right through them!

To test these ideas, we need an accurate map of not just galaxies but also the dark matter as well. It may seem like a great problem to map something you cannot see, but the galaxies themselves provide the solution—they are attracted by gravity to the concentrations of dark matter. Just by measuring how galaxies are moving—how fast and in which direction—we can map the dark matter. Of course, there's a hitch. Most of a galaxy's motion is a result of the Hubble expansion, whereas we want to measure only the additional "peculiar velocity" due to dark matter. To do that, we need to measure a galaxy's distance, so that we can separate the two contributions.

In 1986, I and six of my colleagues made the most ambitious study of galactic distances ever. From measurements on hundreds of galaxies, we discovered that our Milky Way Galaxy and its neighbors are streaming at speeds of many hundreds of miles per second toward the same distant point. That point, in the general direction marked by the constellation Centaurus, lies roughly 150 million light years away. We suggested that the cause of galaxy streaming is the gravitational pull of a vast continent of dark matter and galaxies, which we named the "Great Attractor."

Although we discovered the Great Attractor from the *motions* of galaxies, we later found that the *number* of galaxies in this region is also far enhanced. There is also at least 10 times more dark matter in the Great Attractor than the "ordinary" matter of visible galaxies. The total mass in the Great Attractor is approximately $10^{17}$ times the mass of our Sun, equivalent to the most massive *superclusters* of galaxies, and it extends about 500 million light-years.

Our discovery of the Great Attractor supports the popular conclusion that the universe is very lumpy, including piles of dark matter as well as highly clustered galaxies. If so, there must be many structures like the Great Attractor that we have yet to find. Further exploration of the large-scale structure of the universe promises to teach us more about how the universe was born and how our destinies were forged in the first moments of creation.

# Quasars, Blazars, and Active Galaxies

*In this chapter you will find the answers to the following questions:*

27-1 Why are quasars unusual? How did astronomers discover that they are extraordinarily distant and luminous?

27-2 How does the light from quasars differ from starlight?

27-3 How are Seyfert galaxies and radio galaxies related to quasars?

27-4 How can material ejected from quasars and blazars appear to travel faster than light?

27-5 What could power the incredible energy output from active galaxies?

27-6 Why do many active galaxies emit ultrafast jets of material?

**The Core of a Radio Galaxy** This artist's rendition shows a scenario that many astronomers believe is responsible for the enormous energy output from the centers of active galaxies. A supermassive black hole at the center of the galaxy is surrounded by an accretion disk. In the inner regions of the accretion disk, matter crowding toward the hole is diverted outward along two oppositely directed beams, which deposit energy into two huge, radio-emitting lobes on either side of the galaxy. (*Astronomy*)

As we learned in Chapter 26, galaxies are vast, majestic collections of stars. As a general rule, the larger the galaxy, the greater its luminosity and the more light-emitting stars it contains. Since the 1960s, however, astronomers have discovered thousands of galaxies whose luminosity is far too great to come from starlight alone. These objects—called quasars, Seyfert galaxies, radio galaxies, and blazars—are also called **active galaxies**, because their substantial luminosity can actually fluctuate up and down over a period of months, weeks, or even days. Many active galaxies also have jets of particles streaming outward from their centers at nearly the speed of light.

What makes these active galaxies so tremendously energetic? Unlike an ordinary galaxy, whose luminosity comes from all the galaxy's stars, much of an active galaxy's luminosity emanates from a very small region at its center less than a light-year across. Many astronomers think that a supermassive black hole lies at the center of these galaxies. As the surrounding matter accretes onto the black hole, it releases fantastic amounts of energy. Recent observations suggest that supermassive black holes also reside at the centers of many "normal" galaxies, including the Milky Way.

## 27-1  Quasars look like stars but have huge redshifts

Our picture of active galaxies developed over many years, starting with the development of radio astronomy in the late 1940s. Before then, the distant universe was known only from visual observations. By looking at wavelengths far removed from visible light, radio telescopes provided an unexpected glimpse of a previously invisible universe. Among the strangest new discoveries were remote, dazzlingly luminous objects called *quasi-stellar radio sources* or *quasars*.

As we saw in Section 6-6, Grote Reber, an amateur astronomer, built the first radio telescope in 1936 in his backyard in Illinois. By 1944 he had detected strong radio emissions from the constellations Sagittarius, Cassiopeia, and Cygnus. The first two of these sources, nicknamed Sgr A and Cas A, happen to lie in our own Galaxy; they are the galactic nucleus (Section 25-6) and a supernova remnant (see Section 22-10, especially Figure 22-25). The location of the third source, called Cygnus A (Cyg A), was finally established in 1951, using a newly constructed radio interferometer (Section 6-6). Walter Baade and Rudolph Minkowski then used the 200-inch visible-light telescope on Palomar Mountain to discover a strange-looking galaxy at this same location. Figure 27-1 shows a photograph of the visible counterpart of Cygnus A.

When Baade and Minkowski photographed the spectrum of Cygnus A, they were surprised to find a number of bright *emission* lines. By contrast, a normal galaxy has *absorption* lines in its spectrum (see Section 26-3 and Figure 26-15). This absorption takes place in the atmospheres of the stars that make up the galaxy, as described in Section 19-5. Further-

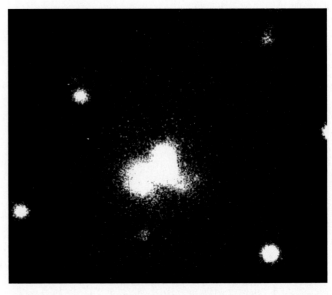

**Figure 27-1**  R I **V** U X G

**Cygnus A (3C 405)**  This strange-looking galaxy was discovered at the location of the radio source Cygnus A. This galaxy has a substantial redshift, which means that it must be extremely far away (about 220 Mpc, or 720 million light-years, from Earth). Because Cygnus A is one of the brightest radio sources in the sky, the great distance of this remote galaxy means that its energy output must be enormous. (Palomar Observatory)

more, the wavelengths of Cygnus A's emission lines are all shifted by 5.7% toward the red end of the spectrum. As described in Section 26-5, astronomers use the letter $z$ to denote redshift, and thus this object has $z = 0.057$. If Cygnus A

**Figure 27-2** R I **V** U X G

**The Quasar 3C 48** For several years astronomers erroneously believed that this object in the constellation Triangulum was simply a peculiar nearby star that happened to emit radio waves. Actually, the redshift of this starlike object is so great ($z = 0.367$) that, according to the Hubble law, it must be approximately 1200 Mpc (4 billion light-years) away. (Palomar Observatory)

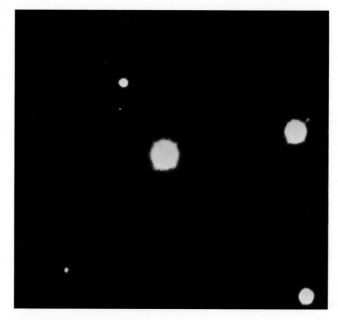

**Figure 27-3** R I **V** U X G

**The Quasar 3C 273** This greatly enlarged view of a portion of the constellation Virgo shows the starlike object associated with the radio source 3C 273. The spectrum of this object has a redshift $z = 0.158$, which according to the Hubble law implies that 3C 273 is about 600 Mpc (2 billion light-years) from the Earth. The luminosity of 3C 273 at visible wavelengths is so great that, despite its great distance, this object can be seen with even a 15-cm (6-in.) telescope. Note the luminous jet projecting from the core of 3C 273 and pointing toward the lower right. The angular size of this jet is about 18 arcsec, corresponding to an actual length of 52 kpc (170,000 light-years)—comparable in size to the entire Milky Way Galaxy. (NOAO)

participates in the same Hubble flow as clusters of galaxies (described in Section 26-5), then this substantial redshift corresponds to a substantial distance from the Earth. If the Hubble constant equals 75 km/s/Mpc, the redshift of Cygnus A translates to a distance of about 220 Mpc (720 million light-years) from the Earth. This is farther away than any galaxy known to astronomers at that time.

Despite its tremendous distance from the Earth, radio waves from Cygnus A can be picked up by amateur astronomers with backyard equipment. This means that Cygnus A must be one of the most luminous radio sources in the sky. In fact, its radio luminosity is $10^7$ times as great as that of an ordinary galaxy like the Milky Way. Obviously, the object that comprises Cygnus A must be something quite extraordinary.

Cygnus A was not the only unusual radio source to draw the attention of astronomers. In 1960, Allan Sandage used the 200-inch telescope to discover a "star" at the location of a radio source designated 3C 48 (Figure 27-2). Because ordinary stars are not strong sources of radio emission, astronomers knew 3C 48 must be something unusual. Like Cygnus A, its spectrum showed a series of emission lines, but astronomers were unable to identify the chemical elements that produced these lines. Although 3C 48 was clearly an oddball, many astronomers thought it was just a strange star in our own Galaxy. This "star" proved not to be unique. Two years later, astronomers discovered a similar starlike optical counterpart to the radio source 3C 273. As Figure 27-3

shows, this "star" is even more unusual, with a luminous jet protruding from one side. Like 3C 48, its visible spectrum contains a series of unexplained bright emission lines.

What prevented astronomers from deciphering the spectra of 3C 48 and 3C 273 was the preconceived idea that these objects are nearby stars. All stars in our Galaxy have comparatively small Doppler shifts, because any star with an extremely high speed would have long ago escaped from the Galaxy. Astronomers, therefore, erroneously assumed that the mysterious spectral lines exhibited by 3C 48 and 3C 273 were not shifted far from their normal wavelengths.

A breakthrough occurred in 1963, when Maarten Schmidt at the California Institute of Technology took another look at the spectrum of 3C 273. He realized that 3C 273 actually exhibits an enormous redshift. Schmidt identified four of its brightest spectral lines as Balmer lines of hydrogen that have suffered a redshift $z = 0.158$, corresponding to a recessional velocity of 45,000 km/s (15% of the speed of light). According to the Hubble law, this implies that the distance to 3C 273 is about 600 Mpc (2 billion light-years). To be detected at such distances, 3C 273 cannot be a star but must be a very powerful source of both visible light and radio radiation.

Upon learning how Schmidt deciphered the spectrum of 3C 273, two other Caltech astronomers, Jesse Greenstein and Thomas Matthews, found they could identify the spectral lines of 3C 48 as having suffered an even larger redshift of $z = 0.367$. That shift corresponds to a speed three-tenths that of light, which places 3C 48 twice as far away as 3C 273, or approximately 1200 Mpc (4 billion light-years) from Earth.

Because of their strong radio emission and starlike appearance, 3C 48 and 3C 273 were dubbed **quasi-stellar radio sources**, a term soon shortened to **quasars.** After the first quasars were discovered by their radio emission, many similar, high-redshift, starlike objects were found that emit little or no radio radiation. These "radio-quiet" quasars were originally called **quasi-stellar objects,** or **QSOs,** to distinguish them from radio emitters. Today, however, the term "quasar" is often used to include both types. Only about 10% of quasars are "radio-loud" like 3C 48 and 3C 273.

Figure 27-4 shows the visible spectrum of 3C 273 as obtained with a charge-coupled device (CCD). As we saw in Section 6-5 (review Figure 6-23), such a spectrum is a graph of intensity versus wavelength on which emission lines appear as peaks that rise above the overall background intensity. Emission lines are caused by excited atoms that emit radiation at specific wavelengths. Spectra of ordinary galaxies (recall Figure 26-15) are dominated by dark absorption lines. Like Cygnus A, most quasars and many peculiar galaxies exhibit strong emission lines in their spectra. Because ordinary galaxies have only absorption lines in their visible spectra, this is a sign that something unusual is going on.

More than 10,000 quasars are now known. They all look rather like stars (see Figure 1-9), and all have large redshifts, ranging from 0.06 up to almost 5. Most quasars have redshifts of 0.3 or more, which implies that they are more than 1000 Mpc (3 billion light-years) from the Earth. For example, the quasar PC 1247+3406 has a redshift of $z = 4.897$, which corresponds to a speed of more than 94% of the speed of light and a distance of roughly 3600 Mpc (10 billion light-years) from the Earth.

**CAUTION** A value of $z$ greater than 1 does *not* mean that a quasar is receding from us faster than the speed of light. At high speeds, the relationship between redshift and recessional velocity must be modified by the special theory of relativity, as explained in Box 27-1. As the speed of a receding source approaches the speed of light, its redshift can become very much greater than 1. An infinite redshift ($z = \infty$) corresponds to a recessional velocity equal to the speed of light.

Astronomers in the 1960s used the large redshifts of quasars to show that these exotic objects are very distant. Astronomers in the 1990s used the same logic when they discovered large redshifts for *gamma-ray bursters*, mysterious objects that emit intense bursts of high-energy radiation. They concluded that gamma-ray bursters must, like quasars, be at tremendous distances from the Earth. (Except for their great distance, gamma-ray bursters seem to be quite unrelated to quasars.) Box 27-2 describes gamma-ray bursters in more detail.

For very remote objects, the relationship between redshift and distance from the Earth depends on how the universe evolved. As we will see in Chapter 28, the Hubble law reveals that the universe is expanding. In other words, if you could watch the motions of widely separated clusters of galaxies over millions of years, you would see them gradually moving away from each other. Box 27-3 explains how expansion

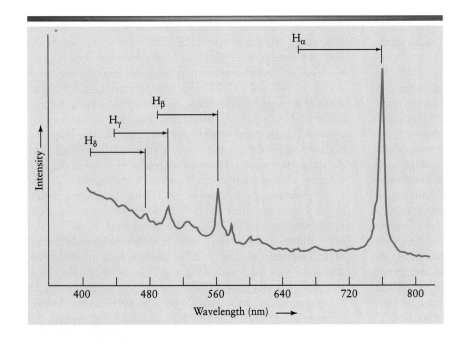

**Figure 27-4**

**The Spectrum of 3C 273** This graph shows the spectrum of the quasar 3C 273 in visible light and in the near infrared. The spectrum is dominated by four bright emission lines of hydrogen (see Section 5-8). The arrows indicate how far these spectral lines are redshifted from their normal wavelengths. For example, the wavelength of $H_\beta$ is shifted from 486 nm (blue-green) to 563 nm (yellow).

# box 27-1 | Tools of the Astronomer's Trade

## *The Relativistic Redshift*

 s we saw in Section 26-5, the redshift ($z$) of an object moving toward or away from us is defined as

**The redshift formula**

$$z = \frac{\lambda - \lambda_0}{\lambda_0}$$

$\lambda$ = observed wavelength of a spectral line in the spectrum of an object (for example, a star, galaxy, or quasar)

$\lambda_0$ = normal (unshifted) wavelength of that same spectral line, as deduced from laboratory experiments

$z$ = redshift of the object

If the object is moving away from us, the observed wavelength is greater than the normal wavelength and $z$ is a positive number.

**EXAMPLE:** The strongest emission line in the spectrum of the quasar PKS 2000-330 is the Lyman-alpha line of hydrogen, observed at a wavelength of 582.5 nm. From laboratory measurements, we know that this spectral line is normally seen in the ultraviolet at a wavelength of 121.6 nm. Thus, this quasar has a redshift of

$$z = \frac{582.5 - 121.6}{121.6} = 3.79$$

The average redshift for this quasar, determined from measurements of a number of spectral lines, is $z = 3.78$.

For low speeds, we ignore the effects of relativity and use the following equation for the Doppler shift:

$$z = \frac{v}{c}$$

where $c$ is the speed of light. For example, a 5% shift in wavelength ($z = 0.05$) corresponds to a speed of 5% of the speed of light ($v = 0.05c$). For speeds greater than about $0.1c$ and redshifts greater than about 0.1, however, we must use the relativistic equation for the Doppler shift:

$$z = \sqrt{\frac{c + v}{c - v}} - 1$$

We can also write this relationship in the useful form

**Relativistic relationship between redshift and speed**

$$\frac{v}{c} = \frac{(z + 1)^2 - 1}{(z + 1)^2 + 1}$$

$z$ = redshift of an object

$v$ = apparent speed of that object away from the Earth

$c$ = speed of light

This relativistic relationship between $z$ and $v$ is displayed in the accompanying graph. Note that $z$ approaches infinity as $v$ approaches the speed of light.

**EXAMPLE:** As stated earlier, quasar PKS 2000-330 has a redshift of $z = 3.78$. Using this value and applying the full, relativistic equation to find the radial velocity for the quasar, we obtain

$$\frac{v}{c} = \frac{(4.78)^2 - 1}{(4.78)^2 + 1} = \frac{21.85}{23.85} = 0.92$$

In other words, this quasar appears to be receding from us at 92% of the speed of light.

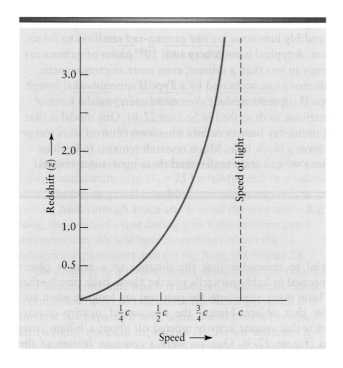

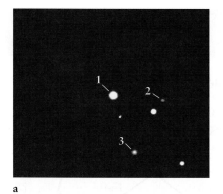
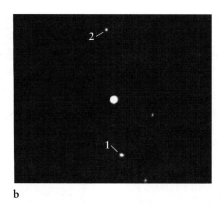
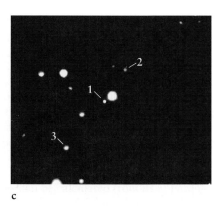

a　　　　　　　　　　b　　　　　　　　　　c

**figure 27-8**　R I **V** U X G

**Quasars in Remote Clusters of Galaxies**  A quasar is located at the center of each of these photographs. Very distant galaxies, identified by the numbers, are faintly visible near each of the quasars. In each case, the redshift of the quasar is virtually the same as the redshifts of the galaxies surrounding it. These observations reinforce the idea that quasars really are very distant objects. (Courtesy of A. Stockton)

of these quasars is embedded in a galaxy. In each case, the absorption lines of the stars have the same redshift as the quasar's emission lines, further supporting the belief that quasars are at distances indicated by their redshifts and the Hubble law.

It is very difficult to observe the "host galaxy" in which a quasar is located because the quasar's light overwhelms light from the galaxy's stars. Nevertheless, painstaking observations have revealed some basic properties of these host galaxies (*Figure 27-9a*). *Radio-quiet quasars tend to be located in spiral galaxies, whereas radio-loud quasars tend to be located in ellipticals. However, a large percentage of these host galaxies have distorted shapes or are otherwise peculiar. Many have nearby companion galaxies, suggesting*

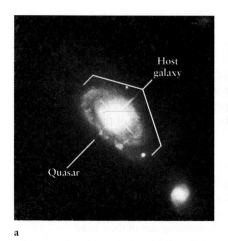

a　　　　　　　　　　b　　　　　　　　　　c

**figure 27-9**　R I **V** U X G

**Quasars and Their "Host Galaxies"**  These Hubble Space Telescope images illustrates the connections between quasars and galaxies. **(a)** Quasar PG 0052+251 is located at the center of an apparently normal spiral galaxy some 430 Mpc (1.4 billion light-years) from the Earth. Other quasars are found at the centers of ordinary-looking elliptical galaxies. **(b)** The galaxy that hosts quasar PG 1012+008 is in the process of merging with a second luminous galaxy. The wispy material surrounding the quasar has presumably been pulled out of the galaxies by tidal forces. (Similar effects can be seen in Figure 26-27 and Figure 26-28.) The two merging galaxies are 490 Mpc (1.6 billion light-years) from the Earth but just 9.5 kpc (31,000 light-years) from each other. Another small galaxy to the left of the quasar may also be merging with the others. **(c)** A long tail of gas and dust extends upward in this image from quasar 0316-346, located 670 Mpc (2.2 billion light-years) from the Earth. This tail was presumably formed when the quasar's host galaxy collided with a second galaxy that lies beyond this image. (J. Bahcall, Institute for Advanced Study; M. Disney, University of Wales; NASA)

*a link between collisions or mergers and the quasar itself (Figure 27-9b and Figure 27-9c).* As we will see in Section 27-6, this link can help to explain why quasars are found only at great distances from Earth—and hence only in the distant past of the universe.

## 27-3 Seyfert galaxies and radio galaxies bridge the gap between normal galaxies and quasars

We have seen that some scientists in the 1970s preferred to challenge the Hubble law rather than accept the existence of highly luminous objects. One reason for their skepticism was the huge gap in energy output between normal galaxies and quasars. The gap was soon bridged, however, when astronomers realized that the centers of certain *peculiar galaxies* look like low-luminosity quasars.

Examples of peculiar galaxies had been known since the early 1900s. It was not until 1943, however, that Carl Seyfert at the Mount Wilson Observatory made the first systematic study of these strange objects. Seyfert focused his attention on spiral galaxies with bright, compact nuclei that seem to show signs of intense and violent activity. An example is shown in **Figure 27-10**. The nuclei of these galaxies, which are intense sources of quasarlike radiation, have strong emis-

a  R I **V** U X G

b  R I V U **X** G

## figure 27-11

**The Seyfert Galaxy NGC 1275** This Seyfert galaxy, also called 3C 84, is the largest and most luminous member of the Perseus cluster of galaxies located about 75 Mpc (250 million light-years) from the Earth. Spectroscopic studies indicate that NGC 1275 is actually two galaxies in collision. **(a)** This Hubble Space Telescope image at visible wavelengths shows about 50 small, bright, blue objects surrounding NGC 1275. These are thought to be young globular clusters formed by the collision. The small white spot is the nucleus of the galaxy. (J. Holtzman, Lowell Observatory; S. Faber, University of California, Santa Cruz; NASA) **(b)** NGC 1275 is also a strong source of X rays and radio radiation. This X-ray image from the Einstein Observatory shows that most of the galaxy's X-ray emission comes from its nucleus. (Harvard-Smithsonian Center for Astrophysics)

## figure 27-10  R I **V** U X G

**A Seyfert Galaxy** This Sc spiral galaxy, called NGC 1566, is a Seyfert galaxy that lies some 16 Mpc (50 million light-years) from the Earth in the southern constellation Dorado (the Goldfish). The nucleus of this galaxy is a strong source of radiation whose spectrum shows emission lines of highly ionized atoms. (Anglo-Australian Observatory)

sion lines in their spectra. These galaxies are now referred to as **Seyfert galaxies.**

About 10% of the most luminous spiral galaxies are Seyfert galaxies. More than 700 Seyferts are known and more continue to be found. Seyfert galaxies range in luminosity from about $10^{36}$ to $10^{38}$ watts, which makes the brightest Seyferts as luminous as faint quasars. Like radio-quiet quasars, they tend to have only weak radio emissions. Some Seyferts are members of interacting pairs or exhibit the vestiges of mergers and collisions. For example, the Seyfert galaxy NGC 1275 (Figure 27-11) is actually two colliding galaxies.

a R I **V** U X G   b R **I** V U X G

## figure 27-12

**The Radio Galaxy M87** This giant elliptical galaxy is in the Virgo cluster, about 15 Mpc (50 million light-years) from the Earth. **(a)** This visible-light photograph shows the full extent of the galaxy, which measures about 100 kpc (300,000 light-years) across. The numerous fuzzy dots that surround the galaxy are globular clusters. (Anglo-Australian Observatory) **(b)** This Hubble Space Telescope image, made with infrared light at a wavelength of 890 nm, shows the inner portion of the galaxy. From the galaxy's tiny, bright nucleus—less than 2 pc (6.5 light-years) in diameter—a jet extends outward some 1500 pc (5000 light-years). (T. Lauer, NOAO; S. Faber, University of California, Santa Cruz; NASA)

While Seyfert galaxies resemble dim, radio-quiet quasars, certain elliptical galaxies, called **radio galaxies** because of their strong radio emission, are like dim, radio-loud quasars. The first of these peculiar galaxies was discovered in 1918 by Heber D. Curtis. His short-exposure photograph of the giant elliptical galaxy M87 revealed a bright, starlike nucleus with a protruding jet. *Figure 27-12a shows an overall view of the entire galaxy. Figure 27-12b*, from the Hubble Space Telescope, shows the jet extending outward 1500 pc (5000 light-years) from the nucleus of M87. Light from the galaxy's central regions is thermal radiation, suggesting a profusion of stars crowded around the galaxy's nucleus. Light from the jet is synchrotron radiation, signifying that relativistic particles are being ejected from the nucleus.

During the 1960s and 1970s, radio astronomers systematically examined extragalactic radio sources. They found that a radio galaxy generally consists of two **radio lobes** on either side of a parent galaxy. The radio lobes generally span a distance that is 5 to 10 times the size of the parent galaxy, although smaller and much larger examples are known. The parent galaxy—almost always a giant elliptical—sits midway between the radio lobes. Because of their overall shape, radio galaxies (and many similar quasars) are sometimes called **double radio sources**. Cygnus A, shown in Figure 27-13, is a fine example.

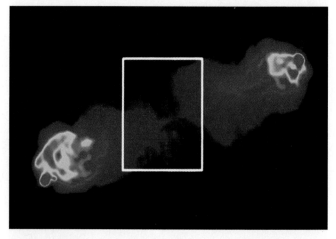

## figure 27-13   **R** I V U X G

**A Radio Image of Cygnus A** This false-color radio image of Cygnus A shows regions of strong radio emission in red and weak emission in blue. Most of the radio emission comes from the radio lobes located on either side of the visible peculiar galaxy. These two radio lobes are each about 50 kpc (160,000 light-years) from the optical galaxy. Each lobe contains a brilliant, condensed region of radio emission. The white rectangle indicates the area shown in Figure 27-1. (VLA, NRAO)

The spectrum and polarization of the radiation from a radio galaxy bear all the characteristics of synchrotron emission, which suggests that radio galaxies should have jets of relativistic particles. In fact, some radio galaxies appear to have a "head" of concentrated radio emission, with a weaker "tail" trailing behind it. A good example of these **head-tail sources** is the active elliptical galaxy NGC 1265 in the Perseus cluster of galaxies. This galaxy is known to be moving at a high speed (2500 km/s) relative to the cluster as a whole. Figure 27-14 is a radio image of NGC 1265. Note that its radio emission has a distinctly windswept appearance. Just as smoke pouring from a steam locomotive trails behind a rapidly moving train, particles ejected along two jets from this galaxy are deflected by the galaxy's passage through the sparse intergalactic medium.

Like most giant ellipticals, many radio galaxies are found near the centers of rich clusters of galaxies and thus are probably subjected to collisions and mergers (see Section 26-7). A spectacular example is Centaurus A, one of the brightest radio sources in the southern sky, which lies some 13 million light-years from Earth. The peculiar parent galaxy of Centaurus A, shown in Figure 27-15a, has a broad dust lane studded with young, hot, massive stars. This galaxy, called NGC 5128, is thought to be an elliptical galaxy that has collided with a spiral galaxy, seen roughly edge-on by us. As Figure 27-15b shows, radio waves pour from two lobes on either side of the galaxy. A second set of radio lobes farther from the galaxy spans a volume 2 million

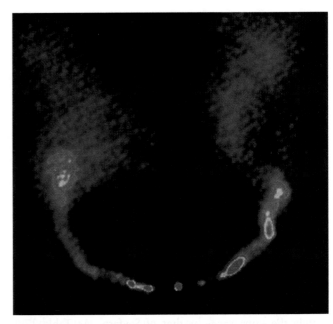

**figure 27-14**   R I V U X G

**The Head-Tail Source NGC 1265** The elliptical galaxy NGC 1265 would probably be an ordinary double radio source except that this galaxy is moving at a high speed through the intergalactic medium. Because of this motion, its two jets trail behind the galaxy, giving this radio source its distinctly windswept appearance. Like NGC 1275 (Figure 27-11), this galaxy is a member of the Perseus cluster of galaxies. (NRAO)

a  R I V U X G

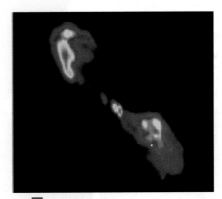

b  R I V U X G

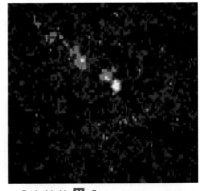

c  R I V U X G

**figure 27-15**

**The Radio Galaxy NGC 5128** This extraordinary galaxy, also called Cen A, is located in the southern constellation Centaurus roughly 4 Mpc (13 million light-years) from Earth. NGC 5128 is so large that despite its great distance, its angular diameter is half that of the full moon. These three views show visible, radio, and X-ray images of this galaxy. **(a)** This photograph at visible wavelengths shows a broad dust lane across the face of the galaxy. A bluish glow along the edges of the dust lane disclose the presence of numerous young stars. (Cerro Tololo Inter-American Observatory, NOAO) **(b)** As this false-color radio image shows, vast quantities of radio radiation pour from extended regions of the sky on either side of the dust lane. Oppositely directed radio jets emanating from the core of NGC 5128 are nearly perpendicular to the galaxy's dust lane. The structure shown here is usually referred to as the inner lobes, to differentiate them from the much larger lobes that extend much farther out from the galaxy. (VLA, NRAO) **(c)** This X-ray image from the Einstein Observatory shows that NGC 5128 has a bright X-ray nucleus. An X-ray jet protrudes from this nucleus along a direction perpendicular to the galaxy's dust lane. Diffuse X-ray emission comes from the regions surrounding the galaxy's center. (Harvard-Smithsonian Center for Astrophysics)

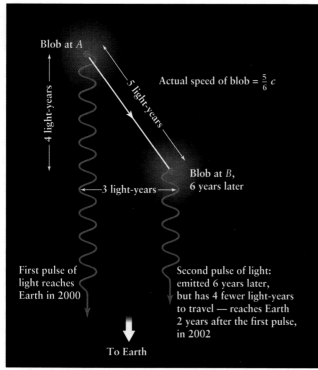

Blob at *A*

Actual speed of blob = $\frac{5}{6}$ *c*

4 light-years

5 light-years

3 light-years

Blob at *B*,
6 years later

First pulse of
light reaches
Earth in 2000

Second pulse of light:
emitted 6 years later,
but has 4 fewer light-years
to travel — reaches Earth
2 years after the first pulse,
in 2002

To Earth

a   View from above

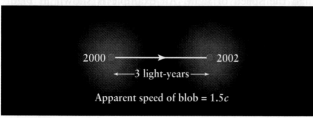

2000 ———→ 2002

3 light-years

Apparent speed of blob = 1.5*c*

b   View from Earth

ƒigure 27-19

**An Explanation of Superluminal Motion** These diagrams show
how an object moving almost directly toward the Earth at a high
speed (but slower than light) can appear to be going faster than
light. **(a)** If a blob of material ejected from a quasar moves at 5/6 the
speed of light, it covers the 5 light-years from point *A* to point *B* in
6 years. In the case shown here, it moves 4 light-years toward the
Earth and 3 light-years in a transverse direction. The light emitted by
the blob at *A* reaches us in year 2000. The light emitted by the blob
at *B* reaches us just two years later, in 2002. The light left the blob six
years later than the light from *A* but had four fewer light-years to
travel to reach us. **(b)** From Earth we can only see the blob's transverse motion across the sky, as in Figure 27-18. It appears that the
blob has traveled 3 light-years in just 2 years, so its apparent speed is
3/2 (1.5) times *c*.

An important clue about the energy sources that power
active galactic nuclei comes from their variability. Some quasars vary in brightness over a few weeks or months. Some
blazars actually fluctuate from night to night. X-ray observa-

tions reveal that some blazars vary in brightness over as few
as 3 hours.

These fluctuations in brightness allow astronomers to
place strict limits on the maximum size of a light source,
because an object cannot vary in brightness faster than light
can travel across that object. For example, an object that is
one light-year in diameter cannot vary significantly in brightness over a period of less than one year.

To understand this limitation, imagine an object that measures one light-year across, as shown in Figure 27-20. Suppose the entire object emits a brief flash of light. Photons from
that part of the object nearest the Earth arrive at our telescopes first. Photons from the middle of the object arrive at
Earth six months later. Finally, light from the far side of the
object arrives a year after the first photons. Although the
object emitted a sudden flash of light, we observe a gradual
increase in brightness that lasts a full year. In other words, the
flash is stretched out over an interval equal to the difference in
the light travel time between the nearest and most remote
observable regions of the object.

The rapid flickering exhibited by active galactic nuclei
means that they emit their energy from a small volume, possibly less than one light-day across. In other words, a region
no larger than our solar system emits as much energy per
second as a thousand galaxies! Astrophysicists therefore face
the challenge of explaining how so much energy can be produced in such a very small volume.

| **27-5** | **Supermassive black holes may be the "central engines" that power active galaxies** |

As long ago as 1968, Donald Lynden-Bell pointed out that a
black hole lurking at the center of a galaxy could be the "central engine" that powers an active galactic nucleus. Lynden-
Bell, a British astronomer working at Caltech, theorized that
as gases fall onto a black hole, their gravitational energy
would be converted into radiation. (As we saw in Section
24-5, a similar process produces radiation from black holes in
close binary star systems.) To produce as much radiation as is
seen from active galactic nuclei, the black hole would have
to be very massive. But even a gigantic black hole would occupy a volume much smaller than our solar system—exactly
what is needed to explain how active galactic nuclei can vary
so rapidly.

How large a black hole would be needed to power an active galactic nucleus? You might think that what really matters is not the size of the black hole, but rather the amount of
gas that falls onto it and releases energy. In fact, there is a
natural limit to the luminosity that can be radiated by accretion onto a compact object like a black hole. This is called
the **Eddington limit**, after British astrophysicist Sir Arthur
Eddington. If the luminosity exceeds the Eddington limit,
there is so much *radiation pressure*—the pressure produced

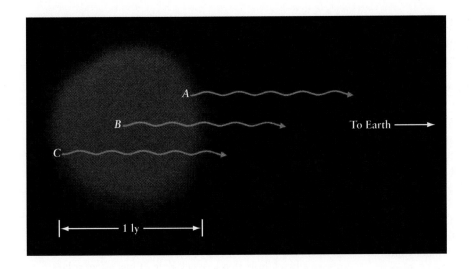

A

B

To Earth →

C

|← 1 ly →|

ƒigure 27-20

**A Limit on the Speed of Variations in Brightness** The rapidity with which the brightness of an object can vary significantly is limited by the time it takes light to travel across the object. Even if the object, shown here to be one light-year in size, emits a sudden flash of light, photons from point *A* arrive at Earth one year before photons from point *C*. Thus, as seen from Earth, the sudden flash is observed over a full year.

by photons streaming outward from the infalling material (see Section 17-7)—that the surrounding gas is pushed outward rather than falling inward onto the black hole. Without a source of gas to provide energy, the luminosity naturally decreases to below the Eddington limit, at which point gas can again fall inward. This limit allows us to calculate the minimum mass of an active galactic nucleus.

Numerically, the Eddington limit is:

**Eddington limit**

$$L_{\text{Edd}} = 30{,}000 \left( \frac{M}{M_\odot} \right) L_\odot$$

$L_{\text{Edd}}$ = maximum luminosity that can be radiated by accretion onto a compact object

$M$ = mass of the compact object

$M_\odot$ = mass of the Sun

$L_\odot$ = luminosity of the Sun

The luminosity of an active galactic nucleus must be less than its Eddington limit, so this limit must be very high indeed. Hence, the mass of the black hole must also be quite large. For example, consider the quasar 3C 273, whose luminosity is about $3 \times 10^{13}$ $L_\odot$. To calculate the minimum mass of a black hole that could continue to attract gas to power the quasar, assume that the quasar's luminosity equals the Eddington limit. Inserting $L_{\text{Edd}} = 3 \times 10^{13}$ $L_\odot$ into the above equation, we find that $M = 10^9$ $M_\odot$. Therefore, if a black hole is responsible for the energy output of 3C 273, its mass must be greater than a billion Suns!

Astronomers use the term **supermassive black hole** when referring to a black hole that contains millions or billions of solar masses. As Box 27-4 explains, extremely massive black holes are not as far-fetched as they might seem at first. Unlike stellar-mass black holes, which may require a supernova to produce them (see Section 23-9), supermassive black holes can be produced without extreme conditions of pressure and density. Furthermore, there is now a good deal of observational evidence that supermassive black holes really exist. As

discussed in Section 25-6, a black hole of 2.6 million solar masses almost certainly lies at the center of our own Milky Way Galaxy. Evidence for supermassive black holes has also been found at the centers of several nearby normal galaxies.

One galaxy that probably has a black hole at its center is the Andromeda Galaxy (M31), shown in Figure 26-3. At a distance of only 900 kiloparsecs (2.9 million light-years) from the Earth, M31 is so close to us that details in its core as small as 1 parsec across can be resolved under the best seeing conditions. Figure 27-21 is a photograph of the central bulge of M31.

In the mid-1980s, several astronomers detected a rapidly moving disk of stars surrounding the core of M31—clear evidence of an enormous mass holding them in orbit. This discovery was made by made high-resolution spectroscopic observations of M31's core. By measuring the Doppler shifts of spectral lines at various locations in the core, we can determine the orbital speeds of the stars about the galaxy's nucleus. Figure 27-22 plots the results for the innermost 350 parsecs (1100 light-years) of the galaxy. (At M31's distance, 350 pc equals 80 arcsec.) Note that the rotation curve in the galaxy's nucleus does not follow the trend set in the outer core. Rather, there are sharp peaks—one on the approaching side of the galaxy and the other on the receding side—within 5 arcsec of the galaxy's center.

The most straightforward interpretation is that the peaks are caused by the rotation of a disk of stars orbiting M31's center. One side of the disk is approaching us while the other side is receding from us. The highest observed radial velocity (at 1.1 arcsec from the galaxy's center; see Figure 27-22) is 110 km/s. This is surely an underestimate, because unsteadiness of the Earth's atmosphere prohibits the detection of features smaller than about 0.5 arcsec across.

The high-speed stars orbiting close to M31's center indicate the presence of a massive central object. Calculations using Newton's form of Kepler's third law (described in Section 4-7 and Box 4-2) show that there must be about 10 million solar masses within 5 pc (16 light-years) of the galaxy's center. That much matter confined to such a small

## box 27-4 | Looking Deeper into Astronomy

### *How Plausible Are Extremely Massive Black Holes?*

Despite their exotic properties, massive black holes do not necessarily require exotic circumstances for their creation. To see why, we examine the density of material needed to produce a massive black hole, but first a word of caution. As we saw in Section 24-3, space around a black hole is highly curved. Simple geometric equations (such as a volume $V = \frac{4}{3}\pi R^3$ for a sphere of radius $R$) are thus not exactly correct. Nevertheless, they are accurate enough to ensure that the following arguments remain valid.

We want to know how tightly we must compress a mass $M$ inside a sphere of radius $R$ in order to create a black hole. Just before the creation of the hole, the average density ($\rho$) of the compressed matter is the mass $M$ divided by the volume $\frac{4}{3}\pi R^3$:

$$\rho = \frac{3M}{4\pi R^3}$$

As explained in Box 24-3, however, the radius of a black hole is related to its mass by Schwarzschild's equation:

$$R = \frac{2GM}{c^2}$$

where $c$ is the speed of light and $G$ is the gravitational constant. Substituting this expression for $R$ into the previous equation, we obtain

$$\rho = \frac{3c^6}{32\pi G^3 M^2}$$

This formula tells us that the density required to create a black hole is inversely proportional to the square of the mass of the hole. In other words, as the mass increases, the density needed to make a black hole drops dramatically. Given enough mass, it is much easier to form a supermassive black hole than a black hole of just a few solar masses.

As an example, the equation tells us that to make a 1-$M_\odot$ black hole, we must compress this mass to a density of roughly $10^{19}$ kg/m$^3$. This is about 20 times the typical density inside a neutron star.

To make a $10^9$-$M_\odot$ black hole, however, the required average density is lowered by a factor of $10^{18}$. Thus, we need to squeeze the matter to a density of only 10 kg/m$^3$. This is only one-hundredth the density of water, or just 10 times the density of the air you breathe. Put another way, this is the same density to which gasoline and air are compressed by the pistons in an automobile engine. Once $10^9$ $M_\odot$ of matter collects together, this matter can quite readily become a black hole.

**Figure 27-21** R I **V** U X G

**The Central Bulge of M31** The central bulge of the Andromeda Galaxy can be seen with the naked eye on a clear, moonless night. A photograph of the entire galaxy is shown in Figure 26-3. Spectroscopic observations suggest that a supermassive black hole with a mass of roughly $10^7$ $M_\odot$ is located at the center of this large, nearby galaxy. (NOAO)

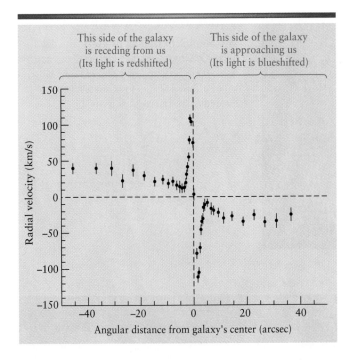

**figure 27-22**

**Rotation Curve of the Core of M31** The radial velocity of matter in the core of M31 is plotted against the angular distance from the galaxy's center. Note the sharp peaks, one blueshifted and one redshifted, within 5 arcsec of the galaxy's center. At the distance of M31, 1 arcsec corresponds to 4.36 pc (14.2 ly). (Adapted from J. Kormendy)

at exceptionally high speeds. These orbital motions suggest the presence of a black hole of about 3 million solar masses.

In addition, an image made by the Hubble Space Telescope (Figure 27-23b) shows a remarkable concentration of stars at the core of M32. The density of stars there is more than a hundred million times greater than the density of stars in the Sun's neighborhood. This strong concentration of stars further suggests the presence of a supermassive black hole whose powerful gravity causes the stars to crowd around it.

Astronomers have uncovered more evidence of supermassive black holes in distant galaxies. For example, high-resolution spectroscopy of M104 (Figure 27-24) reveals high-speed orbital motions around a bright, starlike nucleus. These motions suggest that $10^9$ solar masses lie within 3.5 arcsec of the galaxy's center. With an estimated distance for M104 of 15 Mpc (50 million light-years) from the Earth, all this material must be jammed into a region only 500 pc (1700 light-years) across. Once again, the observations imply the presence of a supermassive black hole.

volume strongly suggests the presence of a supermassive black hole.

Located near M31 is a small satellite galaxy called M32, an elliptical with a bright, starlike nucleus (Figure 27-23a). High-resolution spectroscopy of this galaxy also indicates that stars quite close to M32's center are orbiting its nucleus

a

b

**figure 27-23**   R I **V** U X G

**The Elliptical Galaxy M32** **(a)** This small galaxy is a satellite of M31, a portion of which is seen at the left of this wide-angle photograph. Both galaxies are about 900 kpc (2.9 million light-years) from the Earth. **(b)** This high-resolution image from the Hubble Space Telescope shows the center of M32. Note the concentration of stars at the nucleus of the galaxy. The area covered in this view is 70 pc (230 light-years) across. (Palomar Observatory, NASA, ESA)

<figure 27-24 R I **V** U X G

**The Sombrero Galaxy (M104)** This spiral galaxy in Virgo is nearly edge-on to our Earth-based view. Spectroscopic observations suggest that a billion-solar-mass black hole is located at the galaxy's center. (ESO)

## 27-6 A unified model attempts to explain active galaxies of all types and their energetic jets

Astrophysicists have tried to construct models in which super-massive black holes can generate the energy output of active galactic nuclei and account for their basic properties. The scenario that has emerged is called the *unified model*, because it attempts to explain different types of active galaxies as just being different views of one type of object. Each is a super-massive black hole surrounded by an **accretion disk**, a disk of matter captured by the hole's gravity and spiraling into it.

Imagine a billion-solar-mass black hole sitting at the center of a galaxy, surrounded by a rotating accretion disk. According to Kepler's third law, the inner regions of this accretion disk would orbit the hole more rapidly than would the outer parts. Thus, the rapidly spinning inner regions would constantly rub against the slower moving gases in the outer regions. This friction would heat up the gases and cause them to spiral inward toward the hole.

Supercomputer simulations can help us visualize what happens in such an accretion disk. These simulations combine general relativity with equations describing gas flow. At first matter is accelerated to supersonic speeds as it spirals inward toward a black hole. But because the matter is rotating around the hole as well as moving toward it, this inward rush stops abruptly near the hole. The reason is the law of conservation of angular momentum (described in [Section 7-7 and Box 23-1](#)). Thanks to this law, rotating objects of all kinds tend to expand (like a spinning lump of dough that expands to make a pizza) rather than contract. The inward motion of gases stops where this tendency to expand outward balances the pull of the black hole's gravity. As a result, not all of the infalling gas reaches the black hole. Instead, part of the gas can become concentrated in high-speed orbits quite close to the hole. The sudden halt in the supersonic inflow creates a shock wave, which defines the inner edge of the accretion disk (Figure 27-25).

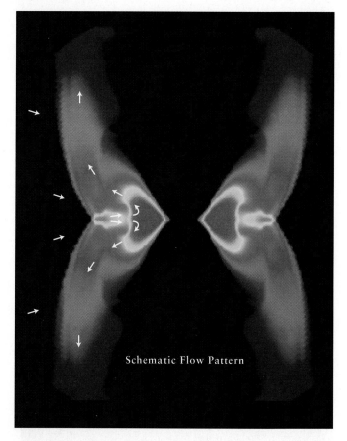

Schematic Flow Pattern

figure 27-25

**Flow Patterns along the Inner Edge of an Accretion Disk** This computer-generated picture shows the flow pattern of gas in an accretion disk surrounding a black hole. The black hole itself is located at the center of this cross-sectional diagram, where the broad-angled inner edges of the accretion disk point. The colors display gas pressure, from red for high pressure across the spectrum to blue for low pressure. The flow pattern, indicated by white arrows, shows how the infalling gas is channeled into two oppositely directed jets. (Courtesy of J. F. Hawley and L. L. Smarr)

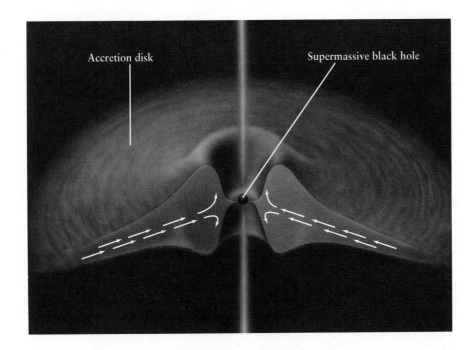

**figure 27-26**

**A Supermassive Black Hole as the "Central Engine"** The energy output of an active galaxy may involve a supermassive black hole that captures matter from its surroundings. In the scenario depicted here, the inflow of matter through an accretion disk is redirected to produce powerful jets of particles traveling near the speed of light. (See also the artist's rendition in the figure that open this chapter.)

Because of the constant inward crowding of hot gases, pressures climb rapidly in the inner accretion disk. To relieve this congestion, matter is expelled from the hole along the inner edges of the empty funnel-shaped cavity. The result is two beams of relativistic particles, oppositely directed at right angles to the accretion disk, like the axle through a wheel. The scenario is sketched in Figure 27-26, and an artist's rendition is seen in the figure that opens this chapter. The magnetic field of the hot, ionized accretion disk helps to focus the ejected matter into narrow jets. (The steering of fast-moving ionized gases by magnetic fields also takes place in the Sun's atmosphere, as described in Section 18-5. The fields around an accretion disk can be much stronger than those in the Sun, however.)

This model—a supermassive black hole surrounded by an accretion disk with oppositely directed high-speed jets—offers a single explanation for the wide variety of active galaxies. Many astronomers suspect that the main difference between double radio galaxies, quasars, and blazars is only the angle at which the black hole "central engine" is viewed. As Figure 27-27 shows, an observer sees a double radio source when the accretion disk is viewed nearly edge-on, so that the jets are nearly in the plane of the sky. At a steeper angle, the observer sees a quasar if the luminosity is high and a radio galaxy if the luminosity is low. If one of the jets is aimed almost directly at the Earth, a blazar is observed.

The unified model can also explain the two types of spectra seen in Seyfert galaxies. Type I Seyfert galaxies have both

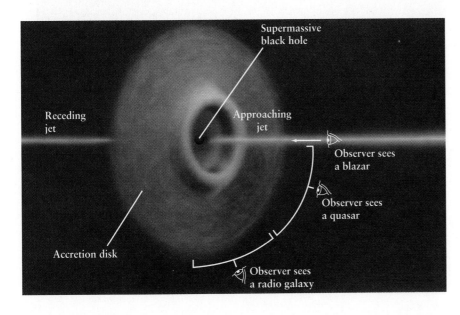

**figure 27-27**

**The Unified Model of Active Galaxies** Double radio sources, quasars, and blazars may be the same type of object viewed at different angles. If one of the jets is aimed almost exactly at the Earth, we see a blazar. If the jet is somewhat tilted to our line of sight, we see a quasar. If the jets are nearly perpendicular to our line of sight, we see a double radio source.

broad and narrow emission lines, usually with a narrow, weak emission peak superimposed on each of the broad lines. The spectra of Type II Seyfert galaxies have only narrow emission lines. These spectral differences may simply be the result of viewing the "central engine" from different angles. A Type II Seyfert galaxy may be a mostly edge-on view of matter spiraling in toward a supermassive black hole, whereas a Type I Seyfert provides a more pole-on view, allowing us to see the turbulent region around the black hole. Because gases move at many different velocities in this region, we see spectral lines that have been broadened by the Doppler effect.

The unified model also helps to explain why there are no nearby quasars (Figure 27-5). Over time, most of the available gas and dust surrounding a quasar's "central engine" is accreted onto the black hole. The "central engine" becomes starved of new infalling matter to act as "fuel," and the quasar becomes much less active. Hence, with the passage of time, a quasar might either turn off altogether or become a less luminous Seyfert galaxy. As a result, there are essentially no quasars in the present-day universe. It may be that galaxies such as our own Milky Way, which have supermassive black holes at their centers but are not presently active, once led much more dramatic lives as quasars or radio galaxies.

Finally, the unified model explains what happens when active galaxies collide or merge with other galaxies. Such collisions and mergers can transfer gas and dust from one galaxy to another, providing more "fuel" for a supermassive black hole. This helps to power and sustain that galaxy's "central engine." Thus, it is not surprising that many of the most luminous active galaxies are also those that are undergoing collisions.

Recent observations of active galactic nuclei provide additional evidence for the unified model. In 1992 astronomers used the Hubble Space Telescope to examine the nucleus of the radio galaxy NGC 4261 in the Virgo cluster (Figure 27-28a). Their image of the galaxy's center, shown in Figure 27-28b, reveals a cold, dark, dusty disk that could represent the outer regions of an accretion disk surrounding a suspected supermassive black hole.

The probable existence of supermassive black holes in active galaxies is one of the most exciting topics in modern astronomy. In the coming years, instruments like the Hubble Space Telescope and the twin Keck telescopes will be used to further probe the nuclei of galaxies with unprecedented resolution. These observations will undoubtedly give us a much better understanding of the powerful engines at the cores of active galaxies and quasars.

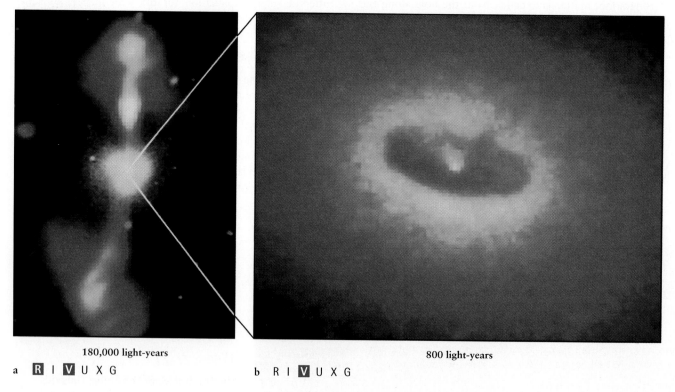

180,000 light-years

a R I **V** U X G

800 light-years

b R I **V** U X G

## figure 27-28

**The Core of an Active Galaxy** The giant elliptical galaxy NGC 4261 is a double radio source located in the Virgo cluster, about 30 Mpc (100 million light-years) from Earth. **(a)** An optical photograph of the galaxy (white) is combined with a radio image (orange) to show both the visible galaxy (which does not emit appreciable amounts of radio energy) and its jets (which do). **(b)** This Hubble Space Telescope image of the nucleus of NGC 4261 shows a disk of gas and dust about 250 pc (800 light-years) in diameter. (NASA, ESA)

# Key Words

*Terms preceded by an asterisk (\*) are discussed in the Boxes.*

accretion disk, p. 690

active galactic nucleus, p. 685

active galaxy, p. 672

blazar, p. 684

\*deceleration parameter, p. 677

double radio source, p. 682

Eddington limit, p. 686

\*gamma-ray bursters, p. 676

head-tail source, p. 683

nonthermal radiation, p. 678

polarized radiation, p. 678

quasar, p. 674

quasi-stellar object (QSO), p. 674

quasi-stellar radio source, p. 674

radio galaxy, p. 682

radio lobes, p. 682

Seyfert galaxy, p. 681

superluminal motion, p. 685

supermassive black hole, p. 687

thermal radiation, p. 678

# Key Ideas

**Quasars:** A quasar is an object in the sky that looks like a star but has a huge redshift. This corresponds to an extreme distance of several hundred megaparsecs or more from the Earth, according to the Hubble law.

• To be seen at such large distances, quasars must be very luminous, typically about 1000 times brighter than an ordinary galaxy.

• About 10% of all quasars are strong sources of radio emission and are therefore called "radio-loud"; the remaining 90% are radio-quiet, because they are weak radio sources at best.

• Some of the energy emitted by quasars is synchrotron radiation produced by high-speed particles traveling in a strong magnetic field.

**Seyfert Galaxies:** Seyfert galaxies are spiral galaxies with bright nuclei that are strong sources of radiation. Seyfert galaxies seem to be nearby, low-luminosity, radio-quiet quasars.

**Radio Galaxies:** Radio galaxies are elliptical galaxies located midway between the lobes of a double radio source. Radio galaxies seem to be nearby, low-luminosity, radio-loud quasars.

• Relativistic particles are ejected from the nucleus of a radio galaxy along two oppositely directed beams.

**Blazars:** Blazars are bright, starlike objects that can vary rapidly in their luminosity. They are probably radio galaxies seen end-on, with a jet of relativistic particles aimed toward the Earth.

• Rapid fluctuations in the brightness of blazars indicate that the energy-producing region is quite small.

**Active Galaxies:** Quasars, blazars, and Seyfert and radio galaxies are examples of active galaxies. The energy source at the center of an active galaxy is called an active galactic nucleus.

**Black Holes and Active Galactic Nuclei:** The preponderance of evidence suggests that an active galactic nucleus consists of a supermassive black hole onto which matter accretes.

• As gases spiral in toward the supermassive black hole, some of the gas may be redirected to become two jets of high-speed particles that are aligned perpendicularly to the accretion disk.

• An observer sees a double radio source when the accretion disk is viewed nearly edge-on, so that the jets are nearly in the plane of the sky. At a steeper angle, the observer sees a quasar. If one of the jets is aimed almost directly at the Earth, a blazar is observed.

# Review Questions

1. Suppose you saw an object in the sky that you suspected might be a quasar. What sort of observations might you perform to find out if it was indeed a quasar?

2. Explain why astronomers cannot use any of the standard candles described in Section 26-4 to determine the distances to quasars.

3. How would you distinguish between thermal and nonthermal radiation?

4. It was suggested in the 1960s that quasars might be compact objects ejected at high speeds from the centers of nearby ordinary galaxies. Why does the absence of blueshifted quasars disprove this hypothesis?

5. Why do you suppose there are no quasars relatively near our Galaxy?

6. What is a Seyfert galaxy? Why do astronomers think that Seyfert galaxies may be related to radio-quiet quasars?

7. Explain why the blue color of the globular clusters in NGC 1275 (Figure 27-11*a*) indicates that these clusters formed relatively recently.

8. What is a radio galaxy? What is a double radio source? Why do astronomers think these objects may be related to radio-loud quasars?

9. How could a supermassive black hole, from which nothing—not even light—can escape, be responsible for the extraordinary luminosity of a quasar?

10. Why do most astronomers conclude that the centers of certain nearby galaxies contain supermassive black holes?

11. Explain how the unified model of active galaxies accounts for the basic properties of quasars, blazars, and double radio sources.

## Advanced Questions

*Questions preceded by an asterisk (\*) involve topics discussed in the Boxes.*

> **Problem-solving tips and tools:**
>
> Relativistic redshift is discussed in Box 27-1. The density of matter needed to form a black hole is given in Box 27-4. Time dilation in the special theory of relativity is discussed in Box 24-1. You may find it useful to recall from Section 1-7 that 1 light-year = 63,240 AU. A speed of 1 km/s is the same as 0.211 AU/yr. Newton's form of Kepler's law is discussed in Box 4-2, and the formula for the Schwarzschild radius is given in Box 24-3.

*12. What redshift would be observed for an object receding from the Sun at a rate of 99% of the speed of light?

*13. The quasar HS 1946+7658 is the most luminous known object in the universe. Its redshift is $z = 3.02$. At what speed does this quasar seem to be receding from us? Give your answer in km/s and as a fraction of the speed of light $c$.

*14. Suppose that someday an astronomer discovers a quasar with a redshift of 6.0. With what speed would this quasar seem to be receding from us? Give your answer in km/s and as a fraction of the speed of light.

*15. Imagine driving down a street toward a traffic light. How fast would you have to go so that the red light ($\lambda = 700$ nm) would appear green ($\lambda = 500$ nm)?

*16. Contrast gamma-ray bursters with X-ray bursters (discussed in Section 23-8). Based on our models of what causes these energetic phenomena, explain why X-ray bursters emit repeated pulses but gamma-ray bursters apparently emit just once.

17. Explain how the existence of gravitational lenses involving quasars (review Section 26-8) constitutes evidence that quasars are located at the great distances from the Earth inferred from the Hubble law.

*18. Suppose a blazar at $z = 1.00$ goes through a fluctuation in brightness that lasts one week (168 hours) as seen from the Earth. (a) At what speed does the blazar seem to be moving away from us? (b) Using the idea of time dilation, determine how long this fluctuation lasted as measured by an astronomer within the blazar's host galaxy. (c) What is the maximum size (in AU) of the region from which this blazar emits energy?

19. (a) Calculate the maximum luminosity that could be generated by accretion onto a $2.6 \times 10^6$-$M_\odot$ black hole. (This is the size of the black hole found at the center of the Milky Way, as described in Section 25-6.) Compare this to the total luminosity of the Milky Way (see Table 27-2). (b) Speculate on what we might see if the center of our Galaxy became an active galactic nucleus with the luminosity you calculated in (a).

20. It is stated in the text that the observed orbital velocity (110 km/s) at 1.1 arcsec from the center of M31 demonstrates the presence of more than 10 million solar masses within 15.6 light-years of the galaxy's center. Verify this by direct calculation using Newton's form of Kepler's third law.

*21. What density is required to make a black hole whose mass is equal to that of the Earth?

*22. How much water at a density of 1 g/cm$^3$ would it take to make a black hole?

*23. Calculate the Schwarzschild radius of a $10^9$-solar-mass black hole. How does your answer compare with the size of our solar system (given by the diameter of Pluto's orbit)?

## Discussion Questions

24. The accompanying image from the VLA shows the radio galaxy 3C 75 in the constellation Cetus. This galaxy has several radio-emitting jets. High-resolution optical photographs reveal that the galaxy has two nuclei, which are the two reddish spots near the center of the VLA image. Propose a scenario that might explain the appearance of 3C 75.

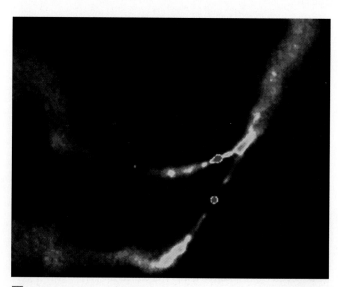

R I V U X G

**25.** Some quasars show several sets of absorption lines whose redshifts are less than the redshifts of their emission lines. For example, the quasar PKS 0237-23 has five sets of absorption lines with redshifts in the range of 1.364 to 2.202, whereas the quasar's emission lines have a redshift of 2.223. Propose an explanation for these sets of absorption lines.

**26.** Search the World Wide Web and magazines such as *Sky & Telescope* and *Astronomy* for information about "microquasars." These are objects that are found *within* the Milky Way Galaxy. How are they detected? What are the similarities and differences between these objects and true quasars? Are they long-lasting or short-lived?

## Observing Projects

**27.** Use a telescope with an aperture of at least 20 cm (8 in.) to observe the Seyfert galaxy NGC 1068. Located in the constellation Cetus (the Whale), this galaxy is most easily seen during autumn and winter (September through January). The epoch 2000 coordinates are: R.A. = $2^h 2.7^m$ and Decl. = $-0° 01'$. Sketch what you see. Is the galaxy's nucleus diffuse or starlike? How does this compare with other galaxies you have observed?

f at least 20 cm (8 in.)
Andromeda Galaxy—
iptical galaxies, but
ipermassive black
es of the Andromeda
September through
are:

| | n | Declination |
|---|---|---|
| | | +40° 52′ |
| | | +41 41 |

ou see any obvious
galaxies? How does
u know about these

with an aperture of
finitely observe the
l in Virgo and can
and early summer
)0 coordinates are
′ Make a sketch
ne? How does the
enters of other

with an aperture of at
l compare it with the
84 and M86, that
dominate the central regions of the Virgo cluster. The epoch 2000 coordinates are:

| Galaxy | Right ascension | Declination |
|---|---|---|
| M84 (NGC 4374) | $12^h 25.1^m$ | +12°53′ |
| M86 (NGC 4406) | 12 26.2 | +12 57 |
| M87 (NGC 4486) | 12 30.8 | +12 24 |

The photograph from the Palomar Sky Survey shown below may be helpful in identifying the galaxies. This photograph covers an area 3° × 3°; north is at the top.

R I **V** U X G

**31.** If you have access to a telescope with an aperture of at least 40 cm (16 in.), you might try to observe the brightest-appearing quasar, 3C 273, which has an apparent magnitude of nearly +13. It is located in Virgo at coordinates R.A. = $12^h 29^m 07^s$ and Decl. = +2° 03′ 07″. You may find it helpful to consult J. Mood's article in *Astronomy*, listed under "Where to Learn More."

## Where to Learn More

*Books and magazine articles*

Arp, H. *Quasars, Redshifts and Controversies*. Interstellar Media, 1987. In this book, which has both technical and nontechnical sections, Arp sets out his controversial theories about quasars with considerable conviction.

Bechtold, J. "Shadows of Creation: Quasar Absorption Lines and the Genesis of Galaxies." *Sky & Telescope*, September 1997. As light from distant quasars traverses the universe, certain wavelengths of this light are absorbed by clouds of gas. This article describes how such absorption tells astronomers about the early history of our universe.

Courvoisier, T. J.-L., and Robson, E. I. "The Quasar 3C 273." *Scientific American*, June 1991. This article takes a close look at the brightest-appearing quasar in the sky and shows how its energy output could be derived from a supermassive black hole.

Finkbeiner, A. "Active Galactic Nuclei: Sorting Out the Mess." *Sky & Telescope,* August 1992. This article describes the unified model, which explains how quasars and active galactic nuclei may derive their energy from an accretion disk surrounding a supermassive black hole.

Ford, H., and Tsvetanov, Z. I. "Massive Black Holes in the Hearts of Galaxies." *Sky & Telescope,* June 1996. Two leading astronomers weigh the evidence that black holes with masses between $10^6$ and $10^9$ $M_\odot$ lurk at the centers of many galaxies.

Leonard. P. J. T., and Bonnell, J. T. "Gamma Rays of Doom." *Sky & Telescope,* February 1998. This intriguing article describes what is known about the mysterious gamma-ray bursters—and the devastation that would follow if one went off within a few thousand light-years from Earth.

Miley, G. K., and Chambers, K. C. "The Most Distant Radio Galaxies." *Scientific American,* June 1993. This fascinating article explains how observations of remote radio galaxies provide a glimpse of the early evolution of giant galaxies.

Mood, J. "Star Hopping to a Quasar." *Astronomy,* May 1987. If you have access to a fairly large telescope and want to hunt for the brightest-appearing quasar in the sky, you might consult this article on 3C 273.

Rees, M. J. "Black Holes in Galactic Centers." *Scientific American,* November 1990. This article explores the idea that supermassive black holes are located at the centers of active galaxies and quasars.

Veilleux, S., Cecil, G., and Bland-Hawthorn, J. "Colossal Galactic Explosions." *Scientific American,* February 1996. This clearly written article describes recent research into the nature of active galaxies.

W *World Wide Web*

Some of the most exciting recent observations of quasars and other active galaxies, as well as of galaxies that may harbor supermassive black holes, have been made with the Hubble Space Telescope (**http://oposite.stsci.edu/pubinfo/ Subject.html**), the National Radio Astronomy Observatory (**http://www.nrao.edu/**), and the Australia Telescope National Facility (**http://wwwatnf.atnf.csiro.au/**).

An interactive map of the sky at radio wavelengths (**http://wwwpks.atnf.csiro.au/databases/surveys/aitoff/ aitoff.html**) is a great help in visualizing the distribution of quasars.

The search for high-energy radiation from gamma-ray bursters and other cosmic exotica is the subject of a web site maintained by NASA's Laboratory for High Energy Astrophysics (**http://heasarc.gsfc.nasa.gov/**).

Maximilian Ruffert of the Max Planck Institute for Astrophysics has created supercomputer simulation "movies" of colliding neutron stars—a leading model for gamma-ray bursters (**http://www.mpa-garching.mpg.de/ ~mor/nsgrb.html**).

# Cosmology: The Creation and Fate of the Universe

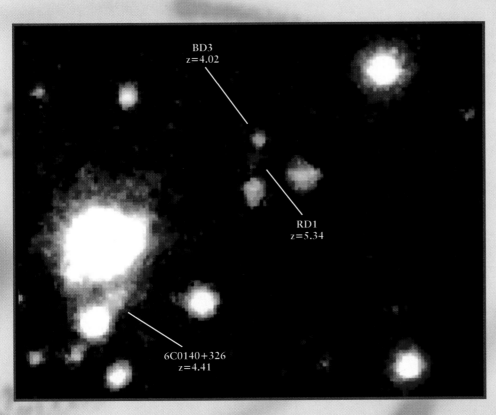

BD3
z=4.02

RD1
z=5.34

6C0140+326
z=4.41

R **I** V U X G

**A Glimpse of the Early Universe**
The red smudge labeled RD1 is, as of this writing, the most distant galaxy known. Its light has suffered such a large redshift ($z = 5.34$) due to the expansion of the universe that it is best seen in infrared light. When the light used to make this image left RD1, the universe was less than a billion years old. By studying such remote galaxies, astronomers can learn not only about the early universe but also whether the universe will continue to expand forever. (A. Dey, Johns Hopkins U.; H. Spinrad, D. Stern and J. R. Graham, U. of California, Berkeley; F. H. Chaffee, W. M. Keck Observatory)

*In this chapter you will find the answers to the following questions:*

28-1 What does the darkness of the night sky tell us about the nature of the universe?

28-2 As the universe expands, what, if anything, is it expanding into?

28-3 Where did the Big Bang take place?

28-4 How do we know that the Big Bang was hot?

28-5 What was the early universe like?

28-6 What determines whether the universe will keep expanding forever?

28-7 How can we tell if the expansion of the universe is slowing down?

28-8 Is it possible to measure how the universe is curved?

28-9 What would happen if the universe stopped expanding and began to contract?

28-10 What will happen if the universe keeps expanding forever?

Imagine a catalog of the universe. It might start with microscopically small objects, such as nuclei, atoms, and molecules. It would extend to the incomprehensibly large, from galaxies to clusters of galaxies and galactic superclusters. In between would be stars, planets, and moons. You have already learned about all these in this text, along with how astronomers know about them. But now we turn our focus beyond what we find in the universe to the nature of the universe itself. How large is the universe? What is its structure? How long has it existed, how has it changed over time, and how will it continue to change in the distant future? These questions lead us to **cosmology**, the science concerned with the structure and evolution of the universe as a whole.

In this chapter we will see that the universe is expanding, and that this expansion began with an explosion at the beginning of time called the Big Bang. We look first at direct evidence of the Big Bang, microwave photons that fill the universe today. These photons are ghostly relics of a primordial fireball that filled all space shortly after the Big Bang. Although very feeble now, this radiation dominated the universe for nearly a million years.

The universe may or may not expand forever. One possibility is that it may instead collapse in a Big Crunch. The ultimate fate of the universe depends in part on the density of matter around us. We will find clues to this ultimate fate by studying the motions of the most distant galaxies.

## 28-1 The darkness of the night sky tells us about the nature of the universe

One of the most profound and basic questions in cosmology may at first seem foolish: *Why is the sky dark at night?* This question, which haunted Johannes Kepler as long ago as 1610, was brought to public attention in the early 1800s by the German amateur astronomer Heinrich Olbers.

Olbers and his contemporaries pictured a universe of stars scattered more or less randomly throughout infinite space. Isaac Newton himself thought that no other model made sense. The gravitational forces between any *finite* number of stars, he argued, would in time cause them all to fall together, and the universe would soon be a compact blob. Obviously, this has not happened. Newton concluded that we must be living amid a static, infinite expanse of stars. In this model, the universe is infinitely old, and it will exist forever without major changes in its structure. Olbers noticed, however, that a static, infinite universe presents a major puzzle.

Imagine looking out into space. If space goes on forever, with stars scattered throughout it, then any line of sight must eventually hit a star. Thus, no matter where you look, you should ultimately see a star. The entire sky should be as bright as an average star, so even at night, the sky should blaze like the surface of the Sun. **Olbers's paradox** is that the night sky is actually *dark*.

Olbers's paradox suggests that something is wrong with Newton's infinite, static universe. According to the classical, Newtonian picture of reality, space is like a gigantic flat sheet of inflexible, rectangular graph paper. (Space is actually three-dimensional, but it is easier to visualize just two of its three dimensions. In a similar way, an ordinary map represents the three-dimensional surface of the Earth, with its hills and valleys, as a flat, two-dimensional surface.) This rigid, flat space stretches on and on, totally independent of stars or galaxies or anything else. The same is true of time in Newton's view of the universe; a Newtonian clock ticks steadily and monotonously forever, never slowing down or speeding up. Furthermore, Newtonian space and time are unrelated, in that a clock runs at the same rate no matter where in the universe it is located.

Albert Einstein overturned this view of space and time. His special theory of relativity (recall Section 24-1 and Box 24-1) shows that measurements with clocks and rulers depend on the motion of the observer. What is more, Einstein's general theory of relativity (Section 24-2) tells us that gravity curves the fabric of space. As a result, the matter that occupies the universe influences the overall shape of space throughout the universe. If we represent the universe as a sheet of graph paper, the sheet is not perfectly flat but has a dip wherever there is a concentration of mass, such as a person, a planet, or a star (see Figure 24-3). Because of gravitational effects, clocks run at different rates depending on whether they are close to or far from a massive object, as shown in Figure 24-6a.

What does the general theory of relativity, with its many differences from the Newtonian picture, have to say about the structure of the universe as a whole? Einstein attacked this problem shortly after formulating his general theory in 1915. At that time, the prevailing view was that the universe was static, just as Newton had thought. Einstein was therefore dismayed to find that his calculations could not produce a truly static universe; according to general relativity, the universe must be either expanding or contracting. In a desperate move to force his theory to predict a static universe, he added to the equations of general relativity a term called the **cosmological constant** (denoted by $\Lambda$, the capital Greek letter lambda). This cosmological constant was supposed to represent a pressure that tends to make the universe as a whole expand. Einstein's idea was that this pressure would just exactly balance gravitational attraction, so that the universe would be static and not collapse.

**ANALOGY** Einstein's cosmological constant is analogous to the pressure of gas inside a bicycle tire. This pressure exactly balances the inward force exerted by the stretched rubber of the tire itself, so the tire maintains the same size.

Unlike other aspects of Einstein's theories, the cosmological constant did not have a firm basis in physics. He just added it to make the general theory of relativity agree with the prejudice that the universe is static. Because he doubted his original equations, he missed an incredible opportunity: He could have postulated that *we live in an expanding universe.* Indeed, Einstein has been quoted in his later years as calling the cosmological constant "the greatest blunder of my life." As it was, the first hint that the universe is not static at all but is actually expanding came more than a decade later from the observations of Edwin Hubble. As we will see in Section 28-3, Hubble's discovery provides the resolution of Olbers's paradox.

---

## 28-2 We live in an expanding universe

Hubble is usually credited with discovering that we live in an expanding universe. He found a simple linear relationship between the distances to remote galaxies and the redshifts of the spectral lines of those galaxies (review Section 26-5, especially Figures 26-15 and 26-16). This relationship, now called the *Hubble law*, states that the greater the distance to a galaxy, the greater is the galaxy's redshift. Thus, remote galaxies are moving away from us with speeds proportional to their distances. Specifically, the recessional velocity $v$ of a galaxy is related to its distance $d$ from the Earth by the equation

$$v = H_0 d$$

where $H_0$ is the Hubble constant. Because clusters of galaxies are getting farther and farther apart as time goes on, astronomers say that the universe is expanding.

**ANALOGY** What does it actually mean to say that the universe is expanding? According to general relativity, space itself is not rigid. The amount of space between widely separated locations changes over time. A good analogy is that of a person blowing up a balloon, as sketched in Figure 28-1. Small coins, each representing a galaxy, are glued onto the surface of the balloon. As the balloon expands, the amount of space between the coins gets larger and larger. In the same way, as the universe expands, the amount of space between widely separated galaxies increases. The expansion of the universe *is* the expansion of space.

An expanding balloon illustrates why distant galaxies obey the Hubble law. Imagine that you are stationed on one of the coins in Figure 28-1. As the balloon expands, you see all the other coins moving away from you. The nearby coins appear to be moving away from you slowly, whereas more distant coins are moving away more rapidly. This relationship is just like the Hubble law, which says that the greater the distance to a galaxy, the greater its recessional velocity. Box 28-1 explains in more detail the connection between the Hubble law and the expansion of the universe.

**CAUTION!** It may seem that if the universe is expanding, and if we see all the distant galaxies rushing away from us, then we must be at the center of the universe. The analogy in Figure 28-1 shows that this is not the case, however. No matter which coin you decided to call home, you would see all the other coins moving away from you in the same way. No one coin lies at the center of the balloon; indeed, the surface of the balloon does not actually have a center. In the same way, no matter which galaxy you call home, you will see all the other distant galaxies receding from you in accordance with the same Hubble law that we observe from the Earth. (Box 28-1 explains why this is so.)

**Figure 28-1**

**The Expanding Balloon Analogy** The expanding universe can be compared to the expanding surface of an inflating balloon. All the coins on the balloon recede from one another as the balloon expands, just as all the galaxies recede from one another as the universe expands. (Adapted from C. Misner, K. Thorne, and J. Wheeler)

## box 28-1 | Tools of the Astronomer's Trade

### *The Hubble Law and the Expansion of the Universe*

The Hubble law states that the recessional velocity of a galaxy is directly proportional to its distance from the Earth; a galaxy twice as far away is receding from us twice as fast. This is exactly what we expect in an expanding universe. To see why this is so, imagine a grid of parallel lines (like on a piece of graph paper) crisscrossing the universe. Part *a* of the accompanying figure shows a series of such gridlines 100 Mpc apart, along with five galaxies labeled A, B, C, D, and E that happen to lie where gridlines cross. As the universe expands in all directions, the gridlines and the attached galaxies spread apart. (This is just what would happen if the universe were a two-dimensional rubber sheet that was being pulled equally on all sides.) Part *b* of the figure shows the universe at a later time, when the gridlines are 50% farther apart (150 Mpc) and all the distances between galaxies are 50% greater than in part *a*.

Imagine that A represents our galaxy, the Milky Way. The table shows by how much each of the other galaxies have moved away from us during the expansion: Galaxies A and B and galaxies A and E were originally 100 Mpc apart, and have moved away from each other by an additional 50 Mpc; A and C, which were originally 200 Mpc apart, have increased their separation by an additional 100 Mpc; and the distance between A and D, originally 300 Mpc, has increased by an additional 150 Mpc. In other words, the increase in distance between any pair of galaxies is in direct proportion to the original distance; if the original distance is twice as great, the increase in distance is also twice as great.

To see what this tells us about the recessional velocities of galaxies, remember that velocity is equal to the distance moved divided by the elapsed time. (For example, if you traveled in a straight line for 300 kilometers in 4 hours, your velocity was 300/4 = 75 kilometers per hour). Because the distance that each galaxy moves away from A during the expansion is directly proportional to its original distance from A, it follows that the velocity at which each galaxy moves away from A is also directly proportional to the original distance. Using the symbol $v$ for velocity and $d$ for distance, this proportionality can be written as

$$v = H_0 d$$

Here $H_0$ is a constant that has the same value for all galaxies; this is just the Hubble constant, and the above equation is just the Hubble law.

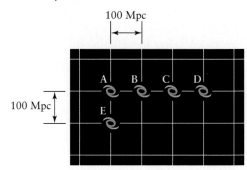

a

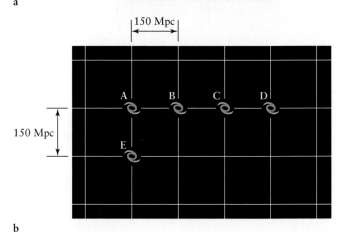

b

We have concentrated on how things look from the perspective of galaxy A, but the expansion of the universe looks the same from *any* vantage point. For example, as seen from galaxy D, the initial distances to galaxies A, B, and C are 300 Mpc, 200 Mpc, and 100 Mpc respectively. Between parts *a* and *b* of the figure, these distances increase by 150 Mpc, 100 Mpc, and 50 Mpc respectively. So, as seen from D as well, the recessional velocity increases in direct proportion with the distance, and in the same proportion as seen from A. In other words, the Hubble constant has the same value as measured from any galaxy in the universe.

|  | Original distance (Mpc) | Later distance (Mpc) | Change in distance (Mpc) |
|---|---|---|---|
| A–B | 100 | 150 | 50 |
| A–C | 200 | 300 | 100 |
| A–D | 300 | 450 | 150 |
| A–E | 100 | 150 | 50 |

Because every point in the universe appears to be at the center, it must be that, like the surface of the balloon, our universe has no center at all.

CAUTION If the universe is expanding, what is it expanding into? This commonly asked question arises only if we take our balloon analogy too literally. In Figure 28-1, the two-dimensional surface of the balloon (representing the universe) expands in three-dimensional space into the surrounding air. But the real universe is three-dimensional, and it is expanding in three dimensions. There is nothing "beyond" it, because there is no "beyond." Asking "What lies beyond the universe?" is as meaningless as asking "Where on the Earth is north of the north pole?"

The ongoing expansion of space explains why the light from remote galaxies is redshifted. Imagine a photon coming toward us from a distant galaxy. As the photon travels through space, the space is expanding, so the photon's wavelength becomes stretched. When the photon reaches our eyes, we see a longer wavelength than usual, which is the redshift. The longer the photon's journey, the more its wavelength will have been stretched. Thus, photons from distant galaxies have larger redshifts than those of photons from nearby galaxies, as expressed by the Hubble law.

A redshift caused by the expansion of the universe is properly called a **cosmological redshift**. It is *not* the same as a Doppler shift. Doppler shifts are caused by an object's *motion through space*, whereas a cosmological redshift is caused by the *expansion of space*.

We can calculate the factor by which the universe has expanded since some ancient time from the redshift of light emitted by objects at that time. As we saw in Section 26-5, redshift ($z$) is defined as

$$z = \frac{\lambda - \lambda_0}{\lambda_0}$$

where $\lambda_0$ is the unshifted wavelength of a photon and $\lambda$ is the wavelength we observe. For example, $\lambda_0$ could be the wavelength of a particular spectral line in the spectrum of light leaving the surface of a remote quasar. As the quasar's light travels through space, its wavelength is stretched by the expansion of the universe. Thus, at our telescopes we observe the spectral line to have a wavelength $\lambda$. The ratio $\lambda/\lambda_0$ is a measure of the amount of stretching. By rearranging terms in the preceding equation, we can solve for this ratio to obtain

$$\lambda/\lambda_0 = 1 + z$$

For example, consider a quasar with a redshift of $z = 3$. Since the time that light left that quasar, the universe has expanded by a factor of $1 + z = 1 + 3 = 4$. In other words, when the light left that quasar, representative distances between widely separated galaxies were only one-quarter as large as they are today. A representative volume of space,

which is proportional to the cube of its dimensions, was only $1/4^3$, or $\frac{1}{64}$ as large as it is today. Thus, the density of matter in such a volume was 64 times greater than it is today.

CAUTION It is important to realize that the expansion of space occurs primarily in the intergalactic space that separates the various clusters of galaxies. Just as the coins in Figure 28-1 do not expand as the balloon inflates, galaxies themselves do not expand. Einstein and others have established that an object that is held together by its own gravity, such as a galaxy, is always contained within a patch of nonexpanding space. A galaxy's gravitational field produces this nonexpanding region, which is indistinguishable from the rigid space described by Newton. Thus, the Earth and your body, for example, are not getting any bigger. Only the distance between widely separated galaxies increases with time.

These ideas about the expanding universe demonstrate the central philosophy of cosmology. In cosmology, unlike other sciences, we cannot carry out controlled experiments or even make comparisons: There is only one observable universe. To make progress in cosmology, we must accept certain philosophical assumptions or abandon hope of understanding the nature of the universe. The Hubble law provides a classic example. It could be interpreted to mean that we are at the center of the universe (see Box 28-1). We reject this interpretation, however, because it violates a cosmological extension of Copernicus's belief that we do not occupy a special location in space.

When Einstein began applying his general theory of relativity to cosmology, he made a daring assumption: Over very large distances the universe is **homogeneous** (that is, every region is the same as every other region) and **isotropic** (that is, the universe looks the same in every direction). In other words, if you could stand back and look at a very large region of space, any one part of the universe would look basically the same as any other part, with the same kinds of galaxies distributed through space in the same way. The assumption that the universe is homogeneous and isotropic constitutes the **cosmological principle**. It gives precise meaning to the idea that we do not occupy a special location in space.

Models of the universe based on the cosmological principle have proven remarkably successful in describing the structure and evolution of the universe and in interpreting observational data. All the discussion about the universe in this chapter and the next assumes that the universe is homogeneous and isotropic on the largest scale.

## 28-3 The expanding universe probably emerged from a cataclysmic event called the Big Bang

The Hubble flow shows that the universe has been expanding for billions of years. This means that in the past the matter in the universe must have been closer together and therefore denser than it is today. If we look far enough into the very distant past, there must have been a time when the density of matter was almost inconceivably high. Presumably some sort of colossal explosion of this ultradense matter initiated the expansion of the universe. This explosion, commonly called the **Big Bang,** marks the creation of the universe.

CAUTION! Although we can think of the Big Bang as an explosion, it was not at all like an exploding bomb. When a bomb explodes, pieces of debris fly off *into space* from a central location. If you could trace all the pieces back to their origin, you could find out exactly where the bomb had been. This process is not possible with the universe, however, because the universe itself always has and always will consist of all space. There is genuinely nothing—not even space—beyond the "edge" of the universe. As we have seen, the universe logically cannot have an edge (see the discussion of the expanding balloon in Section 28-2).

How long ago did the Big Bang take place? To estimate an answer, imagine watching a movie of any two galaxies that today are separated by a distance $d$ and receding from each other with a velocity $v$. Now run the film backward, and observe the two galaxies approaching each other as time runs in reverse. Then we can calculate the time $T_0$ it will take for the galaxies to collide by using the simple equation

$$T_0 = \frac{d}{v}$$

This says that the time to travel a distance $d$ at velocity $v$ is equal to the ratio $d/v$. (As an example, to travel a distance of 360 km at a velocity of 90 km/h takes 360/90 = 4 hours.) If we use the Hubble law, $v = H_0 d$, to replace the velocity $v$ in this equation, we get

$$T_0 = \frac{d}{H_0 d} = \frac{1}{H_0}$$

Note that the distance of separation, $d$, has canceled out and does not appear in the final expression. This means that $T_0$ is the same for *all* galaxies. This is the time in the past when all galaxies were crushed together, the time back to the Big Bang. In other words, the reciprocal of the Hubble constant $H_0$ gives us an estimate of the age of the universe, which is one reason why $H_0$ is such an important quantity in cosmology.

If we use our standard value of $H_0 = 75$ km/s/Mpc, our estimate for the age of the universe is

$$T_0 = \frac{1}{75 \text{ km/s/Mpc}}$$

To convert this into units of time, we simply need to remember that 1 Mpc equals $3.09 \times 10^{19}$ km and 1 year equals $3.156 \times 10^7$ s:

$$T_0 = \frac{1}{75} \frac{\text{Mpc-s}}{\text{km}} \times \frac{3.09 \times 10^{19} \text{ km}}{1 \text{ Mpc}} \times \frac{1 \text{ year}}{3.156 \times 10^7 \text{ s}}$$

$$= 1.3 \times 10^{10} \text{ years} = 13 \text{ billion years}$$

This is about 3 times the age of the solar system, which is estimated to be about 4.5 billion years. Thus, the formation of our home planet is a relatively recent event in the history of the cosmos.

As we saw in Section 26-5, the value of the Hubble constant is uncertain by approximately 20 percent. Some astronomers think that $H_0$ is as low as 60 km/s/Mpc, whereas others think it may be as high as 90 km/s/Mpc. This range of $H_0$ corresponds to a range in $T_0$ from about 11 to 16 billion years.

Of course, the formula $T_0 = 1/H_0$ is only an *estimate* of the age of the universe. In deriving it, we assumed that the universe expands at a constant rate, and that may not be so. If the expansion of the universe was faster in the past than it is today, then the time it has taken the universe to expand is less, and thus the age of the universe is less than $1/H_0$. If the expansion was slower, the age of the universe is greater than $1/H_0$. (In Sections 28-6 and 28-7 we describe what may actually have happened.)

Whatever the true age of the universe, it must be at least as old as the oldest stars. The oldest stars that we can observe readily lie in the Milky Way's globular clusters (see Section 21-3 and Section 25-1). Most calculations based on the theory of stellar evolution indicate that these stars are 16 to 18 billion years old. Recently, however, a team led by Neill Reid of the California Institute of Technology has used the high-precision data from the Hipparcos satellite (Section 19-1) to show that the globular clusters are farther away than had been thought previously. In order to account for how bright these clusters appear to us, the stars in the globular clusters must be more luminous, have consumed their nuclear fuel more rapidly, and be younger than previous calculations had suggested. This leads to an age estimate of 11 to 13 billion years for these stars. Encouragingly enough, this is not too different from our rough estimate of the age of the universe obtained from $T_0 = 1/H_0$. As a compromise, in this book we use a figure of 15 billion years for the age of the universe.

The Big Bang helps resolve Olbers's paradox. We know that the universe had a definite beginning, and thus its age is finite (as opposed to infinite). If the universe is about 15 billion years old, then light from stars more than about 15 billion light-years away has just not had enough time to get here. As a result, we cannot see any objects that are more than about 15 billion light-years away. This is true even if the universe is infinite, with galaxies scattered throughout its limitless expanse.

You can think of the Earth as being at the center of an enormous sphere having a radius of approximately 15 billion light-years (Figure 28-2). The surface of this sphere is called the **cosmic particle horizon.** Our entire **observable universe** is located inside this sphere. We cannot see anything beyond the cosmic particle horizon because the time required for light to reach us from these incredibly remote distances is greater than the age of the universe. (As Figure 28-2 shows, the size of this sphere increases as the universe ages. When the universe was 5 billion years old, for example, our observable universe was only 5 billion light-years in radius.) Throughout the observable universe, galaxies are distributed sparsely enough that

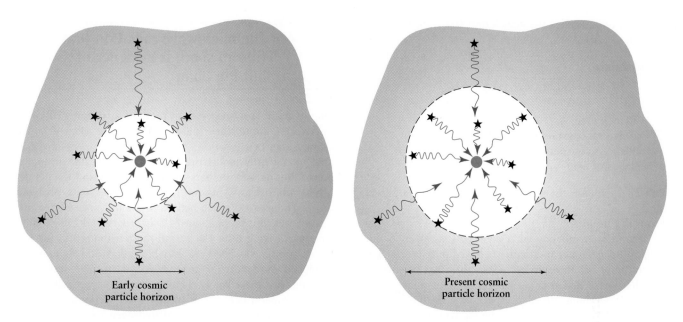

## figure 28-2

**The Observable Universe**  The observable universe (that part of the universe that we can observe) lies within a sphere called the cosmic particle horizon. The radius of this horizon is equal to the distance that light has traveled since the Big Bang. This event occurred about 15 billion years ago, so at present the cosmic particle horizon is about 15 billion light-years away. We cannot see objects beyond the cosmic particle horizon, because their light has not had enough time to reach us. As the universe ages, the radius of the observable universe increases; early in the history of the universe, as shown in the left-hand figure, this radius was smaller. The illustration also indicates that the wavelength of light reaching us from distant objects has been stretched (that is, redshifted) by the expansion of the universe.

there are no stars along most of our lines of sight, which helps explain why the night sky is dark.

Besides the finite age of the universe, a second effect also contributes significantly to the darkness of the night sky—the redshift. According to the Hubble law, the greater the distance to a galaxy, the greater the redshift. When a photon is redshifted, its wavelength becomes longer and its energy—which is inversely proportional to its wavelength (see Section 5-5)—decreases. Consequently, even though there are many galaxies far from the Earth, they have large redshifts and their light does not carry much energy. A galaxy nearly at the cosmic particle horizon has a nearly infinite redshift, meaning that the light we receive from that galaxy carries practically no energy at all. This decrease in photon energy because of the expansion of the universe decreases the brilliance of remote galaxies, helping to make the night sky dark.

The concept of a Big Bang origin for the universe is a straightforward, logical consequence of having an expanding universe. If you can just imagine far enough back into the past, you can arrive at a time approximately 15 billion years ago, when the density throughout the universe was infinite. The entire universe was in effect like the center of a black hole.

Comparing the Big Bang to the center of a black hole can help us appreciate certain aspects of the creation of the universe. As we saw in Section 24-3, matter at the center of a black hole is crushed to infinite density. At this location, called the singularity, the curvature of space and time is infinite, and the very distinction between space and time becomes muddled. Without a clear background of space and time, such concepts as "past," "future," "here," and "now" cease to have meaning. For this reason, a better name for the Big Bang is the **cosmic singularity.**

At the moment of the Big Bang, a state of infinite density filled the universe. Throughout the universe, space and time were completely jumbled up in a condition of infinite curvature like that at the center of a black hole. Thus, we cannot use the known laws of physics to tell us exactly what happened at the moment of the Big Bang. And we certainly cannot use these laws to tell us what existed before the Big Bang. These things are fundamentally unknowable. The phrases "*before* the Big Bang" or "at the *moment* of the Big Bang" are meaningless, because time did not really exist until *after* that moment.

A very short time after the Big Bang, space and time did begin to behave in the way we think of them today. This short time interval, called the **Planck time** ($t_P$), is given by the expression

**The Planck time**

$$t_P = \sqrt{\frac{Gh}{c^5}} = 1.35 \times 10^{-43} \text{ s}$$

$t_P$ = Planck time
$G$ = universal constant of gravitation
$h$ = Planck's constant
$c$ = speed of light

Box 28-2 explains the significance of this minute interval of time.

From the beginning of the Big Bang to the Planck time $10^{-43}$ second later, all known science fails us. We do not know how space, time, and matter behaved in that brief but important interval. Nevertheless, some physicists have speculated that space and time as we know them today burst forth from a seething, foamlike, space-time mishmash during the Planck time. If we think of the Big Bang as an explosion of space at the beginning of time, then the universe that we see today is the debris from that cosmic explosion.

### 28-4 The microwave radiation that fills all space is evidence of a hot Big Bang

One of the major successes in modern astronomy involves discoveries about the origin of the heavy elements. We know today that essentially all the heavy elements are created in the infernos at the centers of stars and in supernovae. But as astronomers began to understand the details of thermonuclear synthesis in the 1960s, a new problem arose: There is too much helium in the universe. For example, the Sun consists of about 74% hydrogen and 25% helium by mass, leaving only 1% for all the remaining heavier elements combined. This 1% can be understood as material produced inside earlier generations of massive stars that long ago cast these heavy elements out into space when they became supernovae. Some freshly made helium, produced by the thermonuclear fusion of hydrogen within the stars, certainly accompanied these heavy elements. But calculations showed that the amount of helium produced in this

---

## $box$ 28-2 | Looking Deeper into Astronomy

### *The Planck Time: A Limit of Knowledge*

In the general theory of relativity, space and time are treated together as a continuum. Because of quantum effects, however, this smooth picture of space and time breaks down at extremely small distances or short time intervals. To see why this breakdown occurs, we must first examine some basic ideas in quantum mechanics.

Quantum concepts were introduced around 1900 to explain fundamental properties of light, particularly details of blackbody radiation and the photoelectric effect. As we saw in Section 5-5, this new understanding revealed that light has *both* wavelike and particlelike properties. If you perform an experiment to study the wavelike properties of light (for example, Young's double-slit experiment, shown in Figure 5-5), you find that light behaves like waves. If you perform an experiment like the photoelectric effect (see Section 5-5) to study the particlelike properties of light, you find that light behaves like particles. This chameleonlike behavior is called **wave-particle duality.**

As experiment and theory continued to progress through the 1920s, physicists discovered that particles, such as electrons, can behave like waves. In other words, wave-particle duality applies to particles as well as photons. This

was first shown in experiments performed by the American physicists Clinton Davisson and Lester Germer, who sent a beam of electrons through a nickel crystal and found interference phenomena—just as would be expected if they had sent waves through the crystal.

Because massive particles have wavelike properties, their positions are somewhat uncertain. A particle typically cannot be precisely located within a distance $\lambda_C$ given by

$$\lambda_C = \frac{h}{mc}$$

where $h$ is Planck's constant, $m$ is the particle's mass, $c$ is the speed of light, and $\lambda_C$ is the **Compton wavelength** (after the American physicist Arthur Compton, who first realized the significance of this quantity). For example, an electron has a Compton wavelength of $2.43 \times 10^{-3}$ nm, which is roughly a thousandth the size of an atom.

The Compton wavelength is a measure of how fuzzy the physical world is at the quantum level. Quantum effects limit our ability to locate a particle on scales smaller than its Compton wavelength, which is a characteristic distance over which the particle exists.

way was not nearly enough to account for one-quarter of the Sun's mass. Because it was thought that the universe originally contained only hydrogen—the simplest of all the chemical elements—the presence of so much helium posed a major dilemma.

Shortly after World War II, Ralph Alpher and Robert Hermann proposed that the universe immediately following the Big Bang must have been so incredibly hot that thermonuclear reactions occurred everywhere throughout space. Following up this idea in 1960, Princeton physicists Robert Dicke and P. J. E. Peebles discovered that they could indeed account for today's high abundance of helium by assuming that the early universe had been at least as hot as the Sun's center, where helium is currently being produced. The hot early universe must therefore have been filled with many high-energy, short-wavelength photons, which formed a radiation field with a temperature that can be given by Planck's blackbody law (review Figure 5-10).

The universe has expanded so much since those ancient times that all those short-wavelength photons have had their wavelengths stretched by a tremendous factor. As a result, they have become low-energy, long-wavelength photons. The temperature of this cosmic radiation field is now quite low, only a few degrees above absolute zero. By Wien's law, radiation at such a low temperature should have its peak intensity at microwave wavelengths of approximately 1 millimeter. In the early 1960s, Dicke and his colleagues began designing an antenna to detect this microwave radiation.

Meanwhile, just a few miles from Princeton University, Arno Penzias and Robert Wilson of Bell Telephone Laboratories were working on a new microwave horn antenna designed to relay telephone calls to Earth-orbiting communications satellites (Figure 28-3). Penzias and Wilson were deeply puzzled when, no matter where in the sky they pointed their antenna, they detected faint background noise. Thanks to a colleague, they happened to learn about the work of Dicke and Peebles and came to realize that they had discovered the cooled-down cosmic background radiation left over from the hot Big Bang.

You can actually detect cosmic background radiation with an ordinary television set. This radiation is responsible for about 1% of the random noise or "hash" that appears on the screen when you tune a TV to a station that is off the air. Using far more sophisticated detectors than TV sets, scientists

---

General relativity, the second great revolution in physics to emerge in the early twentieth century, tells us that events that occur within the Schwarzschild radius of an object of mass $m$ are hidden from the outside world (recall Figure 24-9). This limiting distance ($D$) is approximately

$$D = \frac{Gm}{c^2}$$

where $G$ is the constant of gravitation.

To see why general relativity breaks down, imagine a black hole whose mass is so tiny that its Schwarzschild radius is smaller than its Compton wavelength. Because of quantum-mechanical uncertainty, we cannot be sure that the singularity is inside the event horizon. Indeed, there is a probability that the singularity spends some time outside the event horizon.

The limiting case of the smallest black hole that does not suffer from this breakdown is obtained by equating $D$ and $\lambda_C$. If we solve the resulting equation for the mass, the result is called the **Planck mass** ($m_P$):

$$m_P = \sqrt{\frac{hc}{G}} = 5.46 \times 10^{-8} \text{ kg}$$

(Be careful not to confuse the Planck mass with the mass of a proton, which is also symbolized by $m_P$ but which is

$10^{19}$ times smaller.) A black hole whose mass is less than the Planck mass is not correctly described by general relativity.

The limiting distance corresponding to $m = m_P$ is called the **Planck length** ($\lambda_P$):

$$\lambda_P = \sqrt{\frac{Gh}{c^3}} = 4.05 \times 10^{-35} \text{ m}$$

This is the size of the smallest black hole that can be described by general relativity. Note that the size of a proton (about $10^{-15}$ m) is enormous compared to the Planck length.

The time it takes light to travel across one Planck length is called the Planck time ($t_P$):

$$t_P = \frac{\lambda_C}{c} = \sqrt{\frac{Gh}{c^5}} = 1.35 \times 10^{-43} \text{ s}$$

At one Planck time after the Big Bang (that is, at the moment $t = t_P$), the cosmic particle horizon around each point in space was of the order of 1 Planck length in radius, and the mass contained within each such horizon was of the order of the Planck mass. Hence $t = t_P$ was the earliest time at which general relativity is valid. Because general relativity breaks down for times earlier than the Planck time, no one knows how to describe the universe between $t = 0$ and $t = t_P$.

**figure 28-3**

**The Bell Labs Horn Antenna** Using this microwave horn antenna, actually built for communications purposes, Arno Penzias and Robert Wilson of the Bell Telephone Laboratories in New Jersey detected a signal that seemed to come from all parts of the sky. After carefully removing all potential sources of electronic "noise" (including bird droppings inside the antenna) that could create a false signal, Penzias and Wilson realized that they were actually detecting radiation from space. This radiation is the afterglow of the Big Bang. (Bell Labs)

have made many measurements of the intensity of the background radiation at a variety of wavelengths. Unfortunately, the Earth's atmosphere is almost totally opaque to wavelengths between about 10 μm and 1 cm (see Figure 6-27), which is just the wavelength range in which the background radiation is most intense. As a result, scientists have had to place detectors either on high-altitude balloons (which can fly above the majority of the obscuring atmosphere) or, even better, on board orbiting spacecraft. The most accurate measurements to date have come from the Cosmic Background Explorer (COBE) satellite, which was placed in orbit about the Earth in 1989 (Figure 28-4). Data from COBE's spectrometer, shown in Figure 28-5, demonstrate that this ancient radi-

ation has the spectrum of a blackbody with a temperature of 2.726 K. This radiation field, which fills all of space, is commonly called the **cosmic microwave background.**

An important feature of the microwave background is that its intensity is almost perfectly isotropic, that is, the same in all directions. In other words, we detect nearly the same background intensity from all parts of the sky. This is a striking confirmation of Einstein's assumption that the universe is isotropic (Section 28-2). However, extremely accurate measurements first made from high-flying airplanes, and more recently from high-altitude balloons and from COBE, reveal a very slight variation in temperature across the sky. The microwave background appears slightly warmer than

**figure 28-4**

**The Cosmic Background Explorer (COBE)** This satellite, launched in 1989, measured |the spectrum and angular distribution of the cosmic microwave background over a wavelength range from 1 μm to 1 cm. COBE (pronounced CO-bee) was designed to detect deviations from a perfect blackbody spectrum and from perfect isotropy. (Courtesy of J. Mather; NASA)

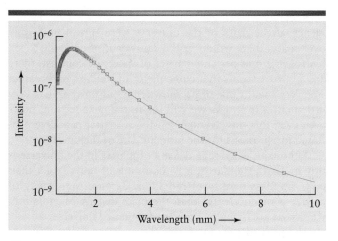

## figure 28-5

**The Spectrum of the Cosmic Microwave Background** The little squares on this graph are COBE's measurements of the brightness of the cosmic microwave background plotted against wavelength. The data fall along a blackbody curve for 2.726 K to a remarkably high degree of accuracy. The peak of the curve is at a wavelength of 1.1 mm, in accordance with Wien's law (Section 5-4). (Courtesy of E. Cheng; NASA COBE Science Team)

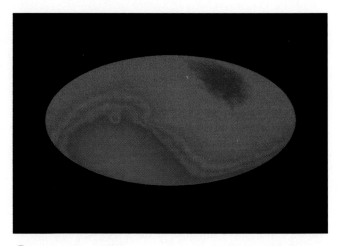

## figure 28-6   R I V U X G

**The Microwave Sky** This map of microwave radiation from the entire sky was produced from data taken by instruments on board COBE. The galactic center is in the middle of the map, and the plane of the Milky Way runs horizontally across the map. Color indicates temperature—red is warm and blue is cool. The temperature variation across the sky is caused by the Earth's motion through the microwave background. The variation is quite small, only 0.0033 K above or below the average radiation temperature of 2.726 K. (NASA)

average toward the constellation of Leo and slightly cooler than average in the opposite direction toward Aquarius. Between the warm spot in Leo and the cool spot in Aquarius, the background temperature declines smoothly across the sky. A map of the microwave sky showing this variation is seen in Figure 28-6.

This apparent variation in temperature can be explained as a result of the Earth's overall motion through the cosmos. If we were at rest with respect to the microwave background, the radiation would be even more isotropic. Because we are moving through this radiation field, however, we see a Doppler shift. Specifically, we see shorter-than-average wavelengths in the direction toward which we are moving, as sketched in Figure 28-7. A decrease in wavelength corresponds to an increase in photon energy and thus an increase in the temperature of the radiation. The slight temperature excess observed, about 0.0033 K, corresponds to a speed of 370 km/s. Conversely, we see longer-than-average wavelengths in that part of the sky from which we are receding. An increase in wavelength corresponds to a decline in photon energy and hence a decline in radiation temperature.

Our solar system is thus traveling away from Aquarius and toward Leo at a speed of 370 km/s. Taking into account the known velocity of the Sun around the center of our Galaxy, we find that the entire Milky Way Galaxy is moving at 600 km/s toward the Hydra-Centaurus supercluster. Observations show that thousands of other galaxies are being carried in this direction, as is the Hydra-Centaurus supercluster itself. This tremendous flow of matter may be

caused by the gravitational pull of an enormous mass, dubbed the **Great Attractor,** lying in that direction. Alan Dressler's essay at the end of Chapter 26 discusses the Great Attractor in more detail.

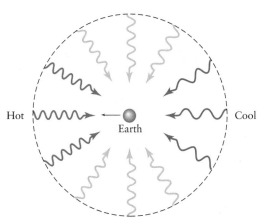

## figure 28-7

**Our Motion Through the Microwave Background** Because of the Doppler effect, we detect shorter wavelengths in the microwave background and a higher temperature of radiation in that part of the sky toward which we are moving. This part of the sky is the area shown in red in Figure 28-6. In the opposite part of the sky, shown in blue in Figure 28-6, the microwave radiation has longer wavelengths and a colder temperature.

## 28-5 The universe was a hot, opaque plasma during its first 300,000 years

Everything in the universe falls into one of two categories—energy or matter. The energy in the universe consists of radiation, that is, photons. There are many photons of starlight traveling across space, but the vast majority of photons in the universe belong to the cosmic microwave background. The matter in the universe is contained in such luminous objects as stars, planets, and galaxies, as well as in nonluminous dark matter. A natural question to ask is this: Which plays a more important role in the universe, radiation or matter? As we will see, the answer to this question was different for the early universe than for our universe today.

To make a comparison between radiation and matter, recall Einstein's famous equation $E = mc^2$ (Section 18-6). We can think of the photon energy in the universe ($E$) as being equal to a quantity of mass ($m$) multiplied by the square of the speed of light ($c$). The amount of this equivalent mass in a volume $V$, divided by that volume, is the **mass density of radiation** ($\rho_{rad}$; say "rho sub rad"). We can combine $E = mc^2$ with the Stefan-Boltzmann law (Section 5-4) to give the following formula:

**Mass density of radiation**

$$\rho_{rad} = \frac{4\sigma T^4}{c^3}$$

$\rho_{rad}$ = mass density of radiation
$T$ = temperature of radiation
$\sigma$ = Stefan-Boltzmann constant
    = $5.67 \times 10^{-8}$ W m$^{-2}$ K$^{-4}$
$c$ = speed of light = $3 \times 10^8$ m/s

For the present-day temperature of the cosmic background radiation, $T = 2.726$ K, this equation yields

$$\rho_{rad} = 4.6 \times 10^{-31} \text{ kg/m}^3$$

The **average density of matter** ($\rho_m$; say "rho sub em") in the universe is harder to determine. To find this density, we look at a large volume ($V$) of space, determine the total mass ($M$) of all the stars, galaxies, and dark matter in that volume, and divide the volume into the mass: $\rho_m = M/V$. (We emphasize that this is the *average* density of matter. It would be the actual density if all the matter in the universe was spread out uniformly rather than being clumped into galaxies and clusters of galaxies.) Determining how much mass is in a large volume of space is a challenging task, especially because the amount of dark matter remains unknown (see Section 25-4 and Section 26-8). The present-day average density of matter in the universe is thought to be in the range

$$\rho_m = 2 - 11 \times 10^{-27} \text{ kg/m}^3$$

The mass of a single hydrogen atom is $1.7 \times 10^{-27}$ kg/m$^3$. Hence, if the mass of the universe were spread uniformly over space, there would be the equivalent of only one to six hydrogen atoms per cubic meter of space. By contrast, there are $5 \times 10^{25}$ atoms in a cubic meter of the air you breathe! The very small value of $\rho_m$ shows that our universe has very little matter in it. (As we will see in the next section, however, even this tiny average density of matter can have profound implications for the ultimate fate of the universe.)

Although the average density of matter in the universe is tiny by Earth standards, it is thousands of times larger than $\rho_{rad}$, the mass density of radiation. Because the density of matter is much greater than the mass density of radiation, we say we are living in a **matter-dominated universe.**

Although matter dominates the universe today, this was not always the case. Matter prevails over radiation today only because the energy now carried by microwave photons is so small. Nevertheless, the number of photons in the microwave background is astounding. From the physics of blackbody radiation it can be demonstrated that there are today 550 million photons in every cubic meter of space. In other words, the photons in space outnumber atoms by roughly a billion to one. In terms of total number of particles, the universe thus consists almost entirely of microwave photons. This radiation field no longer has much "clout," however, because its photons have been redshifted to long wavelengths and low energies after approximately 15 billion years of being stretched by the expansion of the universe.

In contrast, think back toward the Big Bang. The universe becomes increasingly compressed, and thus the density of matter increases as we go back in time. The photons in the background radiation also become more crowded together as we go back in time. But in addition the photons become less redshifted and thus have shorter wavelengths and higher energy than they do today. Because of this added energy, the mass density of radiation ($\rho_{rad}$) increases more quickly as we go back in time than the average density of matter ($\rho_m$). In fact, as Figure 28-8 shows, there was a time in the ancient past when $\rho_{rad}$ equaled $\rho_m$. Before this time, $\rho_{rad}$ was greater than $\rho_m$, and radiation thus held sway over matter. Astronomers call this state a **radiation-dominated universe.**

This transition from a radiation-dominated universe to a matter-dominated universe occurred about 2500 years after the Big Bang, at a time that corresponds to a redshift of about $z = 25,000$. In other words, since that time the wavelengths of photons have been stretched by a factor of 25,000. Today, these microwave photons typically have wavelengths of about 1 mm, but when the universe was about 2500 years old, they had wavelengths of about 40 nm (in the ultraviolet part of the spectrum).

To calculate the temperature of the cosmic background radiation at the time of this transition from a radiation-dominated universe to a matter-dominated one, we use Wien's law (Section 5-4). This law says that the wavelength of maximum emission ($\lambda_{max}$) of a blackbody is inversely proportional to its temperature ($T$): a *decrease* of $\lambda_{max}$ by a factor of 2 corresponds to an *increase* of $T$ by a factor of 2. The

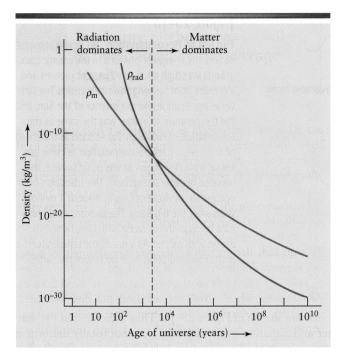

**figure 28-8**

**The Evolution of Density** During approximately 2500 years after the Big Bang, the mass density of radiation ($\rho_{rad}$, shown in red) exceeded the matter density ($\rho_m$, shown in blue), and the universe was radiation-dominated. Later, however, continued expansion of the universe caused $\rho_{rad}$ to become less than $\rho_m$, at which point the universe became matter-dominated.

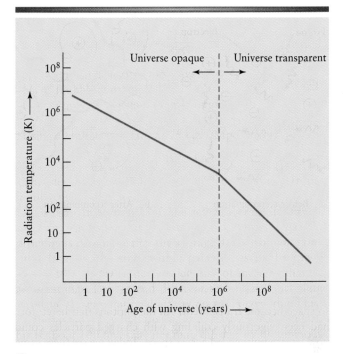

**figure 28-9**

**The Evolution of Radiation Temperature** As the universe expanded, the photons in the radiation background became increasingly redshifted and the temperature of the radiation fell. Approximately 300,000 years after the Big Bang, when the temperature fell below 3000 K, hydrogen atoms formed and the radiation field "decoupled" from the matter in the universe. After that point, the temperature of matter in the universe was not the same as the temperature of radiation.

present-day peak wavelength of the cosmic background radiation is roughly 1 mm, corresponding to a blackbody temperature of about 3 K; hence, a peak wavelength 25,000 times smaller (40 nm) corresponds to $T = 25,000 \times 3$ K = 75,000 K. In other words, the temperature of the radiation background is in direct proportion to the factor $1 + z$ and has been declining over the ages, as shown in Figure 28-9.

The nature of the universe changed again in a fundamental way some 300,000 years after the Big Bang, when $z$ was roughly 1000 and the temperature of the radiation background was about $1000 \times 3$ K = 3000 K. To see the significance of this moment in cosmic history, recall that hydrogen is by far the most abundant element in the universe—hydrogen atoms outnumber helium atoms by about 12 to 1. A hydrogen atom consists of a single proton orbited by a single electron, and it takes relatively little energy to knock the proton and electron apart. In fact, ultraviolet radiation warmer than about 3000 K easily ionizes hydrogen. Thus, hydrogen atoms could not survive in the universe that existed before $z$ = 1000. That is, in the first 300,000 years after the Big Bang (Figure 28-10a), the background photons had energies great enough to prevent electrons and protons from binding to

form hydrogen atoms. Only since $t$ = 300,000 years have the energies of these photons been low enough to permit hydrogen atoms to exist (Figure 28-10b).

The epoch when atoms first formed at $t$ = 300,000 years is called the **era of recombination**. This refers to electrons "recombining" to form atoms. (The name is a bit misleading, because the electrons and protons had never before combined into atoms.)

Prior to $t$ = 300,000 years, the universe was completely filled with a shimmering expanse of high-energy photons colliding vigorously with protons and electrons. This state of matter, called a **plasma,** is opaque, just like the glowing gases inside a discharge tube (like a neon advertising sign). The interior of the Sun is also a hot, glowing, opaque plasma (Section 18-7). Princeton University physicist P. J. E. Peebles coined the term **primordial fireball** to describe the universe during its first 300,000 years of existence.

After $t$ = 300,000 years, the photons no longer had enough energy to keep the protons and electrons apart. As soon as the temperature of the radiation field fell below about 3000 K, protons and electrons began combining to form hydrogen atoms. These atoms do not absorb low-energy photons, and

## box 28-3 | Looking Deeper into Astronomy

### *Cosmology with a Nonzero Cosmological Constant*

Some astronomers have suggested recently that the cosmological constant is not zero. If true, the evolution of the universe can be relatively complex.

The cosmological constant behaves like a constant energy density, present even if the universe is totally devoid of matter and radiation. $\Lambda$ can therefore be thought of as the energy density of the vacuum. Einstein originally introduced this energy density to provide a cosmological repulsion that can balance the gravity of the universe, thereby making the universe static. The critical value of the cosmological constant for which this balance occurs is called $\Lambda_c$ (say "lambda sub cee").

If $\Lambda$ is only slightly larger than $\Lambda_c$, the universe starts with a Big Bang but then has a long "quasi-stationary" phase with gravity and cosmological repulsion almost in balance before cosmological repulsion wins and the expansion continues. This scenario is sometimes called a Lemaître model, after French cosmologist Georges Lemaître, who discovered it. If $\Lambda$ is much larger than $\Lambda_c$, then the universe will expand forever.

If $\Lambda$ is less than $\Lambda_c$, there are two possibilities: a bounce model or an oscillating model. A **bounce universe** contracts from an infinitely extended state (that is, there is no Big Bang) until cosmological repulsion halts the contraction and the universe rebounds out to infinity again. An **oscillating universe** expands from a Big Bang, reaches a maximum size, and then collapses, only to erupt again in another Big Bang. Whether the universe bounces or oscillates depends on factors such as the overall curvature of space. If the cosmological constant is negative ($\Lambda < 0$), then only

oscillating models are possible. These various scenarios are sketched in the accompanying graph.

Astronomers who have attempted to estimate $\Lambda$ typically suggest that its value lies somewhere between $2 \times 10^{-51}$ m$^2$ and $-2 \times 10^{-51}$ m$^2$, which is a range extremely close to zero. Thus, most researchers find it convenient to assume that $\Lambda = 0$; the actual value remains an open question in cosmology.

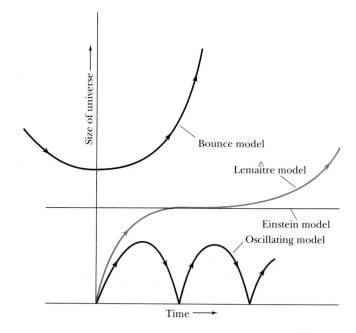

---

### 28-7 The rate of deceleration of the universe can be determined by observing extremely distant galaxies

Another way of determining the future of the universe is to measure the rate at which the cosmological expansion has been gradually slowing down. This slowing can be measured because it causes the relationship between redshift and distance for extremely remote galaxies to deviate from the direct proportion specified by the Hubble law.

If the universe has always expanded at the same rate, the recessional velocity of any galaxy, no matter how remote, will be proportional to the distance to that galaxy. A graph of recessional velocity versus distance, like Figure 26-16, will be a straight line. The graph in Figure 26-16 is indeed a

very straight line. However, this graph was based on measurements of galaxies no farther than 500 Mpc (1.6 billion light-years) from the Earth, which means we are looking only 1.6 billion years into the past. The straightness of the line in Figure 26-16 means only that the expansion of the universe has been relatively constant over the past 1.6 billion years—a very brief interval on the cosmic scale.

Now suppose that you were to measure the redshifts of galaxies *several* billion light-years from the Earth. The light from these galaxies has taken billions of years to arrive at your telescope, so what your measurements will reveal is how fast the universe was expanding billions of years ago. If the universe was expanding faster then than it is today, your data will deviate slightly from the straight-line Hubble law.

The curves in Figure 28-12a display the deviation from the straight-line Hubble law for different rates of deceleration of the universe. Astronomers denote the amount of

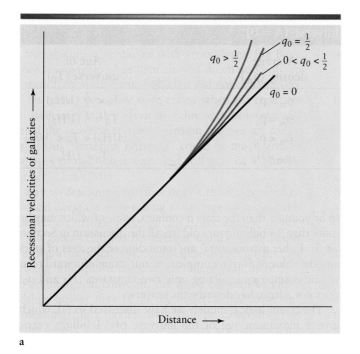

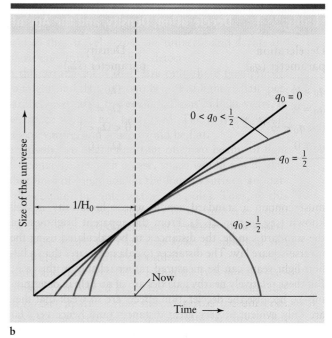

a                                                            b

## figure 28-12

**Deceleration and the Hubble Diagram** These two graphs compare the appearance of the Hubble diagram with the evolution of the universe. **(a)** This diagram shows the relationship between recessional velocity and distance for various values of $q_0$. If the expansion of the universe is decelerating ($q_0 > 0$), then the expansion must have been more rapid in the distant past. In this case the recessional velocities that we see for very distant galaxies (which we see as they were in the distant past) are greater than if there were no deceleration ($q_0 = 0$). **(b)** This diagram shows the size of the universe for various values of $q_0$. The case $q_0 = 0$ is an empty universe ($\rho = 0$) that expands at a constant rate. The case $q_0 = \frac{1}{2}$ is a marginally bounded universe ($\rho = \rho_c$). If $0 < q_0 < \frac{1}{2}$, then the universe is unbounded and will expand forever. If $q_0 > \frac{1}{2}$, the universe is bounded and will someday collapse.

deceleration with the **deceleration parameter** ($q_0$). Appropriately, $q_0 = 0$ corresponds to no deceleration at all. This is possible if the universe is completely empty and thus has no gravity to slow down the expansion. As sketched in Figure 28-12b, a universe with $q_0 = 0$ expands forever at a constant rate. If there were no deceleration, the present age of the universe would be exactly $1/H_0$, as explained in Section 28-3.

The case $q_0 = \frac{1}{2}$ corresponds to a marginally bounded universe. Such a universe just barely manages to continue expanding forever. If the cosmological constant, $\Lambda$, is zero (so there is no pressure propelling the expansion of the universe), then the universe contains matter at the critical density $\rho_c$. In this case, the present age of the universe equals $\frac{2}{3}(1/H_0)$. For simplicity, we continue to assume that $\Lambda = 0$.

If $q_0$ is between 0 and ½, the universe is unbounded and will continue to expand forever. Such a universe contains matter at less than the critical density and has a present age between $\frac{2}{3}(1/H_0)$ and $1/H_0$.

If $q_0$ is greater than ½, the universe is bounded and is filled with matter having a density greater than $\rho_c$. The present age

of such a universe is less than $\frac{2}{3}(1/H_0)$. The gravitational pull of all matter through all space is strong enough to eventually halt the expansion of such a universe.

Since the deceleration parameter $q_0$ and the density parameter $\Omega_0$ are both related to the average density of the universe, it is natural that they are related to each other. If $\Lambda = 0$, then $\Omega_0 = 2q_0$. For example, the marginally bounded case, $q_0 = \frac{1}{2}$, corresponds to $\Omega_0 = 1$, so that the universe is filled with matter at the critical density. Table 28-1 shows how the deceleration parameter, the density parameter, and the age of the universe are related.

In principle, it should be possible to determine the nature of the universe by measuring the redshifts and distances of many remote galaxies and then plotting the data on a Hubble diagram like the one in Figure 28-12a. If the data points fall above the $q_0 = \frac{1}{2}$ line, the universe is bounded. If the data points fall between the $q_0 = 0$ and $q_0 = \frac{1}{2}$ lines, the universe is unbounded.

Unfortunately, making these observations of remote galaxies is extremely difficult. To measure the distance, the galaxy

holes. The theorems predict that the black holes will contain singularities that will be an end of time for anyone unfortunate or foolhardy enough to fall in.

Einstein's general theory of relativity is probably one of the two greatest intellectual achievements of the twentieth century. It is, however, incomplete because it is what is called a classical theory; that is, it does not incorporate the uncertainty principle of the other great discovery of this century, quantum mechanics. The uncertainty principle states that certain pairs of quantities, such as the position and velocity of a particle, cannot be predicted simultaneously with an arbitrarily high degree of accuracy. The more accurately one predicts the position of the particle, the less accurately one will be able to predict its velocity, and vice versa.

Quantum mechanics was developed in the early years of this century to describe the behavior of very small systems such as atoms or individual elementary particles. In particular, there was a problem with the structure of the atom, which was supposed to consist of a number of electrically charged particles called electrons orbiting around a central nucleus, as the planets orbit the Sun. The previous classical theory predicted that electrons would radiate light waves because of their motion. The waves would carry away energy and so would cause the electrons to spiral inwards until they collided with the nucleus. However, such behavior is not allowed by quantum mechanics because it would violate the uncertainty principle: If an electron were to sit on the nucleus, it would have both a definite position and a definite velocity. Instead, quantum mechanics predicts that the electron does not have a definite position; rather, the probability of finding it is spread out over some region around the nucleus, remaining finite even at the nucleus.

The prediction of classical theory of an infinite probability of finding the electron at the nucleus is rather similar to the prediction of classical general relativity that there should be a Big Bang singularity of infinite density. Thus one might hope that if one were able to combine general relativity and quantum mechanics into a theory of quantum gravity, one would find that the singularities of gravitational collapse or expansion were smeared out as in the case of the collapse of an atom.

The first indication that singularities might be smeared out came with the discovery that black holes, formed by the collapse of localized regions such as stars, would not be completely black if one took into account the uncertainty principle of quantum mechanics. Instead, a black hole would radiate particles and radiation like a hot body. The radiation would carry away energy and so reduce the mass of the black hole. This in turn would increase the rate of emissions. Eventually the black hole would disappear completely in a tremendous burst of emissions. All the matter that collapsed to form the black hole—and any astronaut who was unlucky enough to fall into the black hole—would disappear, at least from our region of the universe. However, the energy that corresponded to this mass (by Einstein's famous equation $E = mc^2$) would be emitted by the black hole in the form of radiation. Thus the astronaut's mass-energy would be recycled to the universe. However, this would be rather a poor sort of immortality, because the astronaut's subjective concept of time would almost certainly come to an end, and the particles out of which he had been composed would not in general be the same as the particles that would be re-emitted by the black hole. Still, black hole evaporation did indicate that gravitational collapse might not lead to a complete end of time.

The real problem with spacetime having an edge or boundary at a singularity is that the laws of science do not determine the initial state of the universe at the singularity but only how it evolves thereafter. This problem would remain even if there were no singularity and time continued back indefinitely: The laws of science would not fix what the state of the universe was in the infinite past. In order to pick out one particular state for the universe from among all possible states that are allowed by the laws of science, one has to supplement the laws by boundary conditions that say what the state of the universe was at an initial singularity or in the infinite past. Many scientists are embarrassed at talking about the boundary conditions of the universe

A spiral galaxy.
(Anglo-Australian Telescope board)

because they feel that it verges on metaphysics or religion. After all, they might say, the universe could have started off in a completely arbitrary state. That may be so, but in that case it could also have evolved in a completely arbitrary manner. Yet all the evidence that we have suggests it evolves in a well-determined way according to certain laws. It is therefore not unreasonable to suppose that there may also be simple laws that govern the boundary conditions and determine the state of the universe.

In the classical general theory of relativity, which does not incorporate the uncertainty principle, the initial state of the universe is a point of infinite density. It is very difficult to define where the boundary conditions of the universe should be at such a singularity. However, when quantum mechanics is taken into account, there is the possibility that the singularity may be smeared out and that space and time together may form a closed four-dimensional surface without boundary or edge, like the surface of the Earth but with two extra dimensions. This would mean that the universe is completely self-contained and does not require boundary conditions. One would not have to specify the state in the infinite past, and there would not be any singularities at which the laws of physics break down. One could say that the boundary conditions of the universe are that it has no boundary.

It should be emphasized that this is simply a *proposal* for the boundary conditions of the universe. One cannot deduce them from some other principle: One can merely pick a reasonable set of boundary conditions, calculate what they predict for the present state of the universe, and see if they agree with observations. The calculations are very difficult and have been carried out so far only in simple models with a high degree of symmetry. However, the results are very encouraging. They predict that the universe must have started out in a fairly smooth and uniform state. It would have undergone a period of what is called exponential or "inflationary" expansion, during which its size would have increased by a very large factor but the density would have remained the same. The

universe would then have become very hot and would have expanded to the state that we see today, cooling as it expanded. It would be uniform and the same in every direction on very large scales but would contain local irregularities that would develop into stars and galaxies.

What happened at the beginning of the expansion of the universe? Did spacetime have an edge at the Big Bang? The answer is that if the boundary conditions of the universe are that it has no boundary, time ceases to be well-defined in the very early universe just as the direction north ceases to be well defined at the North Pole of the Earth. Asking what happened before the Big Bang is like asking for a point one mile north of the North Pole. The quantity that we measure as time had a beginning, but that does not mean spacetime has an edge, just as the surface of the Earth does not have an edge at the North Pole, or at least, so I am told: I have not been there myself.

If spacetime is indeed finite but without boundary or edge, this would have important philosophical implications. It would mean that we could describe the universe by a mathematical model that was completely determined by the laws of science alone; they would not have to be supplemented by boundary conditions. We do not yet know the precise form of the laws: At the moment we have a number of partial laws governing the behavior of the universe under all but the most extreme conditions. However, it seems likely that these laws are all part of some unified theory that we have yet to discover. We are marking progress and there is a reasonable chance that we will discover it by the end of the century.

At first sight it might appear that such a theory would enable us to predict everything in the universe. However, our powers of prediction would be severely limited, first by the uncertainty principle that states that certain quantities cannot be exactly predicted but only their probability dis-tribution, and second, and even more importantly, by the complexity of the equations, which makes them impossible to solve in any but very simple situations. Thus we would still be a long way from omniscience.

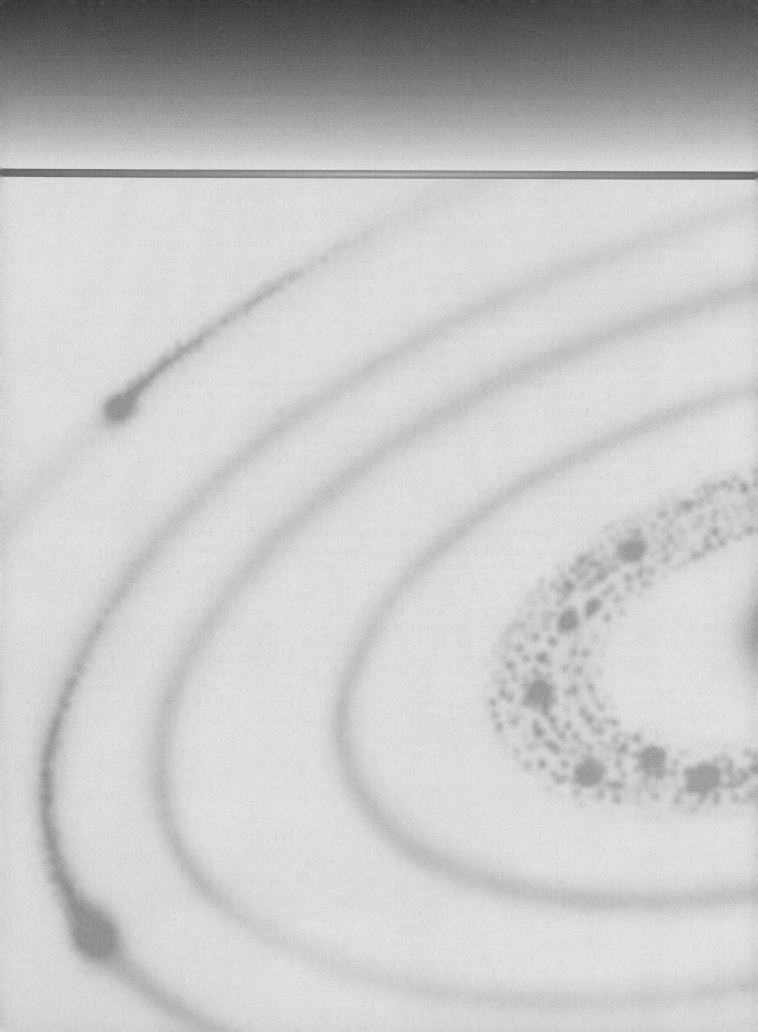

# Exploring the Early Universe

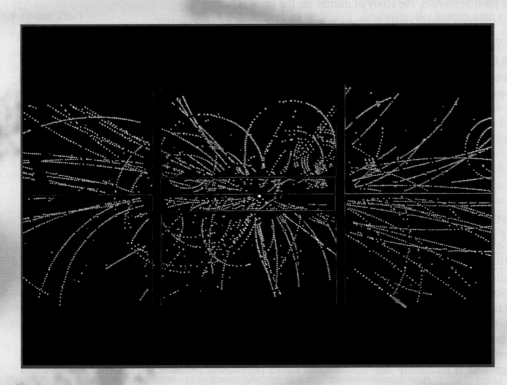

**The Creation of Matter from Energy** This image shows the result of a head-on collision between a proton and its anti-matter equivalent, an antiproton, both of which were traveling at nearly the speed of light. When they collided at the center of this image, the proton and antiproton annihilated each other. Their energy was transformed into a host of new particles, which made the tracks shown in this image. Similar collisions took place in the very early universe during the first ten-thousandth of a second after the Big Bang. (Courtesy of CERN, the European Laboratory for Particle Physics)

*In this chapter you will find the answers to the following questions:*

29-1   Has the universe always expanded as it does today, or might it have suddenly "inflated?"

29-2   What is antimatter? How can it be created, and how is it destroyed?

29-3   Why is antimatter so rare today?

29-4   What materials in today's universe are remnants of nuclear reactions in the hot early universe?

29-5   How did the first galaxies form?

29-6   How were the fundamental forces of nature different during the first second after the Big Bang?

29-7   What exotic relics could have been left behind by a brief period of inflation?

29-8   Are scientists close to developing an all-encompassing "theory of everything?"

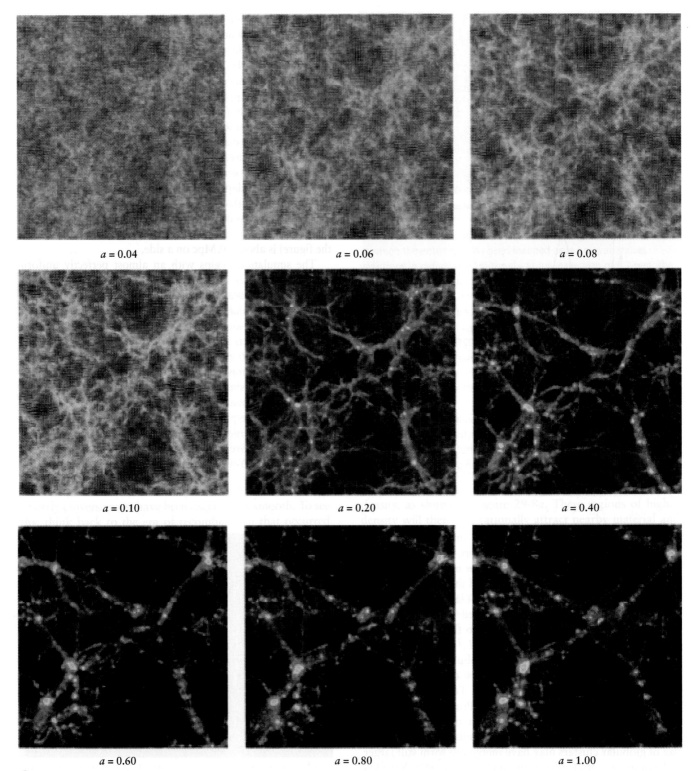

*a* = 0.04          *a* = 0.06          *a* = 0.08

*a* = 0.10          *a* = 0.20          *a* = 0.40

*a* = 0.60          *a* = 0.80          *a* = 1.00

ƒigure 29-10

**A Cold Dark Matter Simulation** These nine views show slices through a large, box-shaped volume of space as the universe expands. The "expansion factor" *a* indicates the relative size of the box at the time when the slice was taken. For example, the actual dimensions of the box labeled *a* = 0.10 are one-tenth those of the box labeled *a* = 1.00. The first four views represent early epochs within the first few billion years after the Big Bang. The last five views are from much later times, with the *a* = 0.60 view possibly representing the present time. Galaxy formation probably occurs around the time of the *a* = 0.20 slice. (Courtesy of J. M. Gelb and E. Bertschinger; Massachusetts Institute of Technology)

and filaments. Ordinary matter, which follows the course set by the cold dark matter, accumulates along density enhancements and forms galaxies and clusters of galaxies. The slice $a = 0.20$ seems to represent the era during which our Galaxy formed.

Similar simulations have been carried out using hot dark matter in the form of neutrinos. Today there should be, on the average, about $10^8$ neutrinos per cubic meter of space. If each neutrino has a mass about $2 \times 10^{-5}$ that of an electron, then the cosmic neutrino background alone has an average density very nearly equal to the critical density $\rho_c$.

If a neutrino has no mass, it will always travel at the speed of light, just as a photon does. If, however, it has a little mass, it must slow down as the universe expands and cools. In this case, slow-moving neutrinos would accumulate over time within density fluctuations. The gravitational pull of these neutrinos on surrounding matter would eventually lead to the formation of clusters of galaxies.

A primary difference between simulations based on cold and hot dark matter is the way in which galaxies form. In calculations based on cold dark matter, galaxies form "from the bottom up." First, small clumps of matter appear, which then group together into galaxies; clusters and superclusters appear later. But in calculations based on hot dark matter, galaxies form "from the top down." Huge supercluster-sized sheets of matter form first and then fragment into galaxies. Recent observations of remote galaxies favor the "bottom up" scenario (see Figure 26-35), which argues that dark matter is cold. However, cold dark matter simulations produce too few superclusters and voids, which seems to favor the hot dark matter hypothesis. It may be that a mixture of cold and hot dark matter is required to explain the observed large-scale structure of the universe.

As is often the case in science, we need better observations to determine the true nature of dark matter. To this end, a new generation of experiments will measure fluctuations in the cosmic background radiation with much finer detail than COBE. These experiments make use of both spacecraft (NASA's Microwave Anisotropy Probe and the European Space Agency's Planck) and high-altitude balloons (BOOMERanG, a joint project of scientists in the United States, Italy, and the United Kingdom). Armed with better data, astronomers may soon be able to pin down the true nature of dark matter, and thus better understand how large-scale structure developed in our universe.

## 29-6 In grand unified theories, all the forces had the same strength immediately after the Big Bang

We have pushed our description of the early universe back to the first few minutes after the Big Bang, when neutrinos and helium nuclei formed. Remarkably, scientists can now describe events during the first *second* after the Big Bang, when the universe was a hot, dense sea of particles and antiparticles colliding with each other at high speeds. To discover what happened during that brief moment of time, we must understand how particles interact at very high energies.

Just *four* physical forces—gravity, electromagnetism, and the strong and weak nuclear forces—explain the interactions of everything in the universe. Of these forces, gravity is the most familiar. It is a long-range force that dominates the universe over astronomical distances. The electromagnetic force is also a long-range force, but it is intrinsically much stronger than the gravitational force. For example, the electromagnetic force between an electron and a proton is about $10^{39}$ times stronger than the gravitational force between those two particles. That is why the electromagnetic force, not the gravitational force, is responsible for holding electrons in orbit about the nuclei in atoms. We do not generally observe longer-range effects of the electromagnetic force, because there is usually a negative electric charge for every positive charge and a south magnetic pole for every north magnetic pole. Thus, over great volumes of space the effects of electromagnetism effectively cancel out. No similar canceling occurs with gravity because there is no equivalent "negative mass." That is why the force that holds the Moon in orbit about the Earth is gravitational, not electromagnetic.

Both the strong and the weak nuclear forces are said to be *short-range forces*. Their influence extends only over distances less than about $10^{-15}$ m, about the diameter of a proton. The **strong nuclear force** holds protons and neutrons together inside the nuclei of atoms. Without the strong nuclear force, nuclei would disintegrate because of the electromagnetic repulsion of the positively charged protons. In fact, the strong nuclear force overpowers the electromagnetic forces inside nuclei.

The weak nuclear force is so weak that it cannot hold anything together. Instead, the **weak nuclear force** is at work in certain kinds of radioactive decay. An example is the transformation of a neutron (n) into a proton (p), in which an electron ($e^-$) and an antineutrino ($\bar{v}$) are released:

$$n \rightarrow p + e^- + \bar{v}$$

Numerous experiments in nuclear physics strongly suggest that protons and neutrons are composed of more basic particles called **quarks**, the most common varieties being "up" (u) quarks and "down" (d) quarks. A proton is composed of two up quarks and one down quark, a neutron of two down quarks and one up quark.

In the 1970s the concept of quarks led to a breakthrough in our understanding of the strong and weak nuclear forces. The strong nuclear force holds quarks together, while the weak nuclear force is at work whenever a quark changes from one variety to another. For example, when a neutron decays into a proton, one of the neutron's down quarks changes into an up quark. Thus, the weak nuclear force is responsible for transformations such as

$$d \rightarrow u + e^- + \bar{v}$$

In the 1940s, physicists Richard P. Feynman and Julian S. Schwinger (working independently in the United States) and

spontaneous symmetry breaking separated the electro-magnetic force from the weak nuclear force; from that moment on, the universe behaved as it does today.

**The Frontier of Knowledge:** Many speculative aspects of the early universe remain the subject of active research.

• Cosmic strings—long, massive remnants of the spontaneous symmetry breaking that separated the strong nuclear force from the electromagnetic and weak forces—may yet be found to exist.

• The search for a theory that unifies gravity with the other physical forces suggests that the universe actually has 11 dimensions (ten of space and one of time), seven of which are folded on themselves so that we cannot see them.

## Review Questions

1. What aspects of the present-day universe can be explained by the idea of inflation?

2. What is the Heisenberg uncertainty principle? How does it lead to the idea that all space is filled with virtual particle-antiparticle pairs?

3. What is the difference between an electron and a positron?

4. Explain why antimatter was present in copious amounts in the early universe but is very rare today.

5. Explain the connection between particles and anti-particles in the early universe and the cosmic microwave background that we observe today.

6. What is meant by the threshold temperature of a particle?

7. Why is it reasonable to suppose that all space is filled with a neutrino background analogous to the cosmic microwave background?

8. What is the deuterium bottleneck? Why was it important during the formation of nuclei in the early universe?

9. Why were only the four lightest chemical elements produced in the early universe?

10. Describe an example of each of the four basic types of interactions in the physical universe. Do you think it possible that a fifth force might be discovered someday? Explain your answer.

11. Describe the observational evidence for (a) the Big Bang, (b) the inflationary epoch, (c) the era of recombination, and (d) the confinement of quarks.

12. Describe the large-scale structure of the universe as revealed by the distribution of clusters and superclusters of galaxies.

## Advanced Questions

*Questions preceded by an asterisk (\*) involve topics discussed in Box 29-1.*

> **Problem-solving tips and tools:**
>
> The mass of a proton is $1.67 \times 10^{-27}$ kg, the mass of an electron is $9.11 \times 10^{-31}$ kg, and the mass of the Sun is $1.99 \times 10^{30}$ kg. As explained in Section 28-6, the critical density $\rho_c$ is equal to $3H_0^2/8\pi G$; that is, $\rho_c$ is proportional to the square of the Hubble constant $H_0$. If $H_0 = 75$ km/s/Mpc, the critical density is equal to about $1.1 \times 10^{-26}$ kg/m³. It is also useful to know that 1 m³ $= 10^6$ cm³, that 1 light-year $= 9.46 \times 10^{15}$ m, and that 1 GeV $= 10^3$ MeV $= 10^9$ eV.

13. An electron has a lifetime of $1.0 \times 10^{-8}$ s in a given energy state before it makes a transition to a lower state. What is the uncertainty in the energy of the photon emitted in this process?

14. How long can a proton-antiproton pair exist without violating the principle of the conservation of mass?

*15. The mass of the intermediate vector boson $W^+$ (and of its antiparticle, the $W^-$) is 85.6 times the mass of the proton. The weak nuclear force involves the exchange of the $W^+$ and the $W^-$. (a) Find the rest energy of the $W^+$. Give your answer in GeV. (b) Find the threshold temperature for the $W^+$ and $W^-$. (c) Based on Figure 29-12, how long after the Big Bang did $W^+$ and $W^-$ particles begin to disappear from the universe? Explain.

16. Using the physical conditions present in the universe during the era of recombination ($T = 3000$ K and $\rho_m = 10^{-18}$ kg/m³), show by calculation that the Jeans length for the universe at that time was about 100 ly and that the total mass contained in a sphere with this diameter was about $4 \times 10^5$ M$_\odot$.

17. Suppose that the average density of neutrinos throughout space is 100 neutrinos per cubic centimeter and that these neutrinos are responsible for giving the universe a density equal to the critical density. What range of neutrino masses corresponds to $H_0$ in the range from 65 to 85 km/s/Mpc? Give your answers in kilograms and as a fraction of the mass of the electron.

18. A typical dark nebula (see Figure 20-3) has a temperature of 30 K and a density of about $10^{-12}$ kg/m³. (a) Calculate the Jeans length for such a dark nebula, assuming that the nebula is mostly composed of hydrogen. Express your answer in meters and in light-years. (b) A typical dark nebula is several light-years across. Is it likely that density fluctuations within such nebulae will grow with time? (c) Explain how your answer to (b) relates to the idea that protostars form within dark nebulae (see Section 20-3).

## Discussion Questions

**19.** Some GUTs predict that the proton is unstable, although with a half-life far longer than the present age of the universe. What would it be like to live at a time when protons were decaying in large numbers?

**20.** If you hold an iron rod next to a strong magnet, the rod will become magnetized; one end will be a north pole and the other a south pole. But if you heat the iron rod to 1043 K (770°C = 1418°F) or higher, it will lose its magnetization. This demagnetization is said to be an example of *restoring* a spontaneously broken symmetry. Explain why.

## Where to Learn More

*Books and magazine articles*

Barrow, J., and Silk, J. *The Left Hand of Creation: Origin and Evolution of the Universe.* Oxford University Press, 1993. This authoritative book tells the exciting story of how space, time, and matter may have been created.

Carrigan, R., and Trower, W. *Particle Physics in the Cosmos* and *Particles and Forces: At the Heart of Matter.* W. H. Freeman, 1989 and 1990, respectively. These two superb collections of articles from *Scientific American* cover many exciting discoveries and developments in particle physics and cosmology during the 1980s.

Duff, M. J. "The Theory Formerly Known as Strings." *Scientific American*, February 1998. This challenging but rewarding article describes the latest efforts toward a "theory of everything," in which all particles and interactions are described in terms of membranes in an 11-dimensional space-time.

Feynman, R. P. *QED: The Strange Theory of Light and Matter.* Princeton University Press, 1985. Richard Feynman was both an innovative physicist and a superb teacher. In this slim book, he explains the theory of quantum electrodynamics in easy-to-understand terms, using simple diagrams and no equations at all.

Guth, A. *The Inflationary Universe: The Quest for a New Theory of Cosmic Origins.* Addison-Wesley, 1997. The originator of the theory of cosmic inflation describes how the theory came to be and the search for observations to confirm it.

Halliwell, J. J. "Quantum Cosmology and the Creation of the Universe." *Scientific American,* December 1991. This article describes recent efforts in applying quantum mechanics to the universe as a whole.

Lederman, L. M., and Schramm, D. N. *From Quarks to the Cosmos.* Scientific American Library, 1995. Two distinguished scientists, one a particle physicist and the other an astrophysicist, explore the connection between their subjects and what this connection tells us about the early universe.

Linde, A. "The Self-Reproducing Inflationary Universe." *Scientific American,* November 1994. A pioneer in inflationary theory describes recent speculative developments that may change our picture of the Big Bang.

Riordan, M., and Schramm, D. N. *The Shadows of Creation: Dark Matter and the Structure of the Universe.* W. H. Freeman, 1991. The role of dark matter in the formation of our present-day universe is the subject of this very accessible book.

Silk, J., Szalay, A., and Zeldovich, Ya. B. "The Large-Scale Structure of the Universe." *Scientific American,* October 1983. The article shows how the large-scale structure of the universe may have resulted from perturbations in the matter density that emerged from the Big Bang.

**W** *World Wide Web*

Excellent introductions to particle physics and the four physical forces can be found on the World Wide Web, courtesy of the University of California's Lawrence Berkeley Laboratory (**http://pdg.lbl.gov/cpep/adventure.html**) and the Fermi National Accelerator Center (**http://www.fnal.gov/pub/hep_overview.html**).

The next generation of experiments to probe the cosmic background radiation is described at the web sites for the Microwave Anisotropy Probe (**http://map.gsfc.nasa.gov/**), for the Planck spacecraft (**http://astro.estec.esa.nl/SA-general/Projects/Cobras/cobras.html**), and for the BOOMERanG project (**http://astro.caltech.edu/~lgg/boom/boom.html**).

Supercomputer simulations of galaxy formation, including movies and animations, can be seen at the "Cosmos in a Computer" web site (**http://www.ncsa.uiuc.edu/Cyberia/Cosmos/CosmosCompHome.html**) from the National Center for Supercomputing Applications and at the web site for the Galaxy Formation Group of the Max Planck Institute for Astrophysics in Germany (**http://www.mpa-garching.mpg.de/Galaxien/Galaxy.html**).

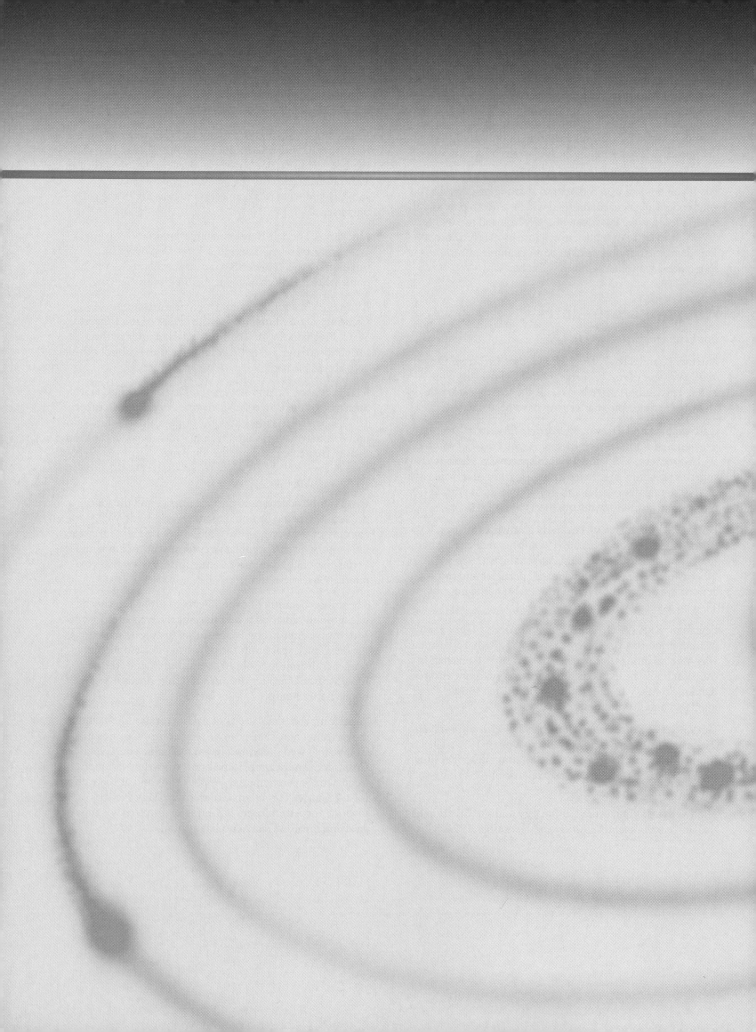

# The Search for Extraterrestrial Intelligence

**The Prevalence of Life** This painting, entitled *DNA Embraces the Planets*, artistically expresses the suspicion of many scientists that carbon-based life may be a common phenomenon in the universe. Other scientists argue, however, that we may be unique and no intelligent alien civilizations exist. In either case, humanity has the clear mandate to preserve and protect the abundance of life forms with which we share our planet. (Courtesy of J. Lomberg)

*In this chapter you will find the answers to the following questions:*

30-1 Do we have solid evidence of life, or its chemical building blocks, on other planets?

30-2 How likely is it that other civilizations exist in our Galaxy?

30-3 How do astronomers search for evidence of civilizations on planets orbiting other stars?

30-4 Will it ever be possible to see Earthlike planets orbiting other stars?

The heavens inspire us to contemplate profound questions, from the creation of the universe to the nature of the stars. Of all the fascinating subjects we might explore, perhaps none is as compelling as intelligent life on other worlds. Are we alone? Does life exist elsewhere in the universe? What are the chances that we might someday make contact with an alien civilization?

There are no certain answers to questions about extraterrestrial life. Earth is the only planet on which life is known to exist, and our spacecraft have so far failed to detect life elsewhere in the solar system. Reports of UFO "sightings" and of humans being abducted by aliens make compelling science fiction, but none of these reports has ever been verified. Some evidence from a single meteorite suggests that life may have existed on Mars in the distant past, but this conclusion is very controversial (see Section 12-8). The presence of abundant water on Jupiter's moon Europa (Section 14-6) holds the tantalizing prospect that life may once have existed there (and may yet exist). But even the high-resolution cameras of the Galileo spacecraft cannot confirm or deny the presence of life on Europa.

Given the lack of hard data, why should we think that life exists beyond the Earth? The answer to this question comes from one of the great lessons of modern astronomy (one that you have seen throughout this book), namely, that our circumstances are quite ordinary. Contrary to the beliefs of our ancestors, our Earth does not occupy a special location and is by no means the "center of the universe." We now know that we inhabit one of nine planets orbiting an unremarkable star, a star that is just one of billions in an undistinguished galaxy. Is it possible that life such as ours is also commonplace? The answer to this question is sought by scientists involved in the **search for extraterrestrial intelligence,** or **SETI.**

---

## 30-1 The chemical building blocks of life are found throughout space

All terrestrial life is based on the unique properties of the carbon atom. Although we cannot rule out the possibility that other forms of biochemistry exist, most scientists confine their search to life as we now know it. Among all the elements, carbon has the most versatile chemistry, since carbon atoms can form chemical bonds to create especially long and complex molecules. Among these carbon-based compounds, called **organic molecules,** are the molecules of which living organisms are made.

Organic molecules can be linked together to form elaborate structures, such as chains, lattices, and fibers. Some of these structures are capable of complex, self-regulating chemical reactions. Furthermore, the primary constituents of organic molecules—carbon, hydrogen, nitrogen, oxygen, sulfur, and phosphorus—are among the most abundant elements in the universe. The versatility and abundance of carbon suggest that extraterrestrial biology, also called **exobiology,** may also be based on organic chemistry.

Organic molecules are liberally scattered throughout the Galaxy. Giant molecular clouds in interstellar space (Section 20-7) contain substantial amounts of carbon atoms that have combined with oxygen atoms to make the simple molecule carbon monoxide ($CO$). But carbon atoms have also combined with other elements to produce an impressive variety of organic compounds. Since the 1960s, radio astronomers have detected telltale microwave emission lines from interstellar clouds that help identify dozens of these carbon-based chemicals. Examples include ethyl alcohol ($CH_3CH_2OH$), formaldehyde ($H_2CO$), methyl cyanoacetylene ($CH_3C_3N$), and acetaldehyde ($CH_3CHO$), to name a few.

Further evidence of extraterrestrial organic molecules comes from newly fallen meteorites called **carbonaceous chondrites** (Figure 30-1), which are often found to contain a variety of organic substances. As we saw in Section 17-6, carbonaceous chondrites are ancient meteorites dating from the formation of the solar system. Therefore, it seems reasonable to conclude that, even from their earliest days, the planets have been continually bombarded with organic compounds.

Interstellar space is not the only source of organic material. In a classic experiment performed in 1952, American

**figure 30-1**

**A Carbonaceous Chondrite** Carbonaceous chondrites are ancient meteorites that date back to the formation of the solar system. Chemical analyses of newly fallen specimens disclose that they are rich in organic molecules, many of which are the chemical building blocks of life. This sample is a piece of the Allende meteorite, a large carbonaceous chondrite that fell in Mexico in 1969. (From the collection of Ronald A. Oriti)

suppose that life could have originated as the result of chemical processes. Furthermore, because the molecules that combine to form these compounds are rather common, it seems equally reasonable that life could have originated in the same way on other planets.

An abundance of organic building blocks does not guarantee that life is commonplace throughout the universe. If a planet's environment is hostile, life may never get started or may quickly become extinct. But we now have evidence that there are planets orbiting other stars (see Section 7-9 and the essay "Alien Planets" at the end of Chapter 7), as well as evidence for additional planetary systems forming around young stars (see Section 7-7, especially Figure 7-15, and

chemists Stanley Miller and Harold Urey demonstrated that simple chemicals can combine to form prebiological compounds under conditions that are thought to have prevailed on the primitive Earth. In a closed container, they prepared a mixture of hydrogen, ammonia, methane, and water vapor. These are the most common molecules in the solar system, and are among the primary constituents of the present-day atmosphere of Jupiter (see Section 13-1). Miller and Urey then exposed this mixture of gases to an electric arc (to simulate lightning bolts) for a week. At the end of this period, the inside of the container had become coated with a reddish-brown substance rich in amino acids and other compounds essential to life.

Since Miller and Urey did their original experiment, most scientists have come to the conclusion that Earth's primordial atmosphere was composed of carbon dioxide, nitrogen, and water vapor outgassed from volcanoes, along with some hydrogen (see Section 12-5). Modern versions of the Miller-Urey experiment (Figure 30-2) using these common gases have also succeeded in producing a wide variety of organic compounds. These experiments suggest that the chemical building blocks of life were available in substantial quantities on the young Earth.

**CAUTION!** It is important to emphasize that scientists have *not* created life in a test tube. While organic molecules may have been available on the ancient Earth, biologists have yet to figure out how these molecules gathered themselves into cells and developed systems for self-replication. Nevertheless, because so many chemical components of life are so easily synthesized under conditions that simulate the primordial Earth, it seems reasonable to

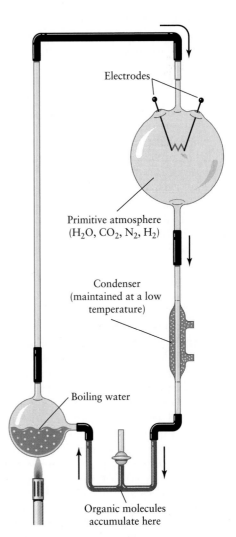

**figure 30-2**

**The Miller–Urey Experiment (Updated)** Modern versions of this classic experiment prove that numerous organic compounds important to life can be synthesized from gases that were present in the Earth's primordial atmosphere. This experiment supports the hypothesis that life on the Earth arose as a result of ordinary chemical reactions.

Section 20-5). It seems probable that there are terrestrial-type planets orbiting other stars, and conditions on some of these worlds may be suitable for life as we know it.

---

## 30-2 The Drake equation helps scientists estimate how many civilizations may inhabit our Galaxy

The development of life on the Earth seems to suggest that extraterrestrial life, including intelligent species, might evolve on terrestrial planets around other stars, given sufficient time and hospitable conditions. How might we ascertain whether such worlds exist, given the tremendous distances that separate us from them?

With our present technology, sending even a small unmanned spacecraft to another star requires a flight time of tens of thousands of years. Speculative design studies have been made for unmanned probes that could reach other stars within a century or less, but these are prohibitively expensive. (Building one such design would require the economic resources of the entire Earth for 20 years!) Instead, many astronomers hope to learn about extraterrestrial civilizations by detecting radio transmissions from them. Radio waves can travel immense distances without being significantly degraded by the gas and dust through which they pass (see Section 25-2). Because of this ability to penetrate the interstellar medium, radio waves are a logical choice for interstellar communication.

Over the past several decades, astronomers have proposed various ways to search for alien radio transmissions, and several limited searches have been undertaken. In 1960 Frank Drake used a radio telescope at the National Radio Astronomy Observatory in West Virginia to listen to two Sunlike stars, Tau Ceti and Epsilon Eridani, without success. About 40 similar unsuccessful searches using radio telescopes have taken place since then in both the United States and Russia. In 1973, for example, astronomers listened to 600 nearby solar-type stars for half an hour each, but no unusual signals were detected. From 1985 to 1994, a radio telescope belonging to Harvard University was used along with a sophisticated computer program to scan a wide range of frequencies over a large portion of the sky. Since 1995 the independent SETI Institute has been using advanced radio techniques to listen for transmissions from some 1,000 solar-type stars. So far, no confirmed, repeating signals have turned up.

Should we be discouraged by this lack of success? What are the chances that a radio astronomer might someday detect radio signals from an extraterrestrial civilization? The first person to tackle this issue was Frank Drake, now at the University of California, Santa Cruz. Drake proposed that the number of technologically advanced civilizations in the Galaxy could be estimated by a simple equation, now called the **Drake equation:**

**Drake equation**

$$N = R_* f_p\, n_e\, f_l\, f_i\, f_c\, L$$

$N$ = the number of technologically advanced civilizations in the Galaxy whose messages we might be able to detect

$R_*$ = the rate at which solar-type stars form in the Galaxy

$f_p$ = the fraction of stars that have planets

$n_e$ = the number of planets per solar system that are Earthlike (i.e., suitable for life)

$f_l$ = the fraction of those Earthlike planets on which life actually arises

$f_i$ = the fraction of those life forms that evolve into intelligent species

$f_c$ = the fraction of those species that develop adequate technology and then choose to send messages out into space

$L$ = the lifetime of that technologically advanced civilization

The Drake equation is enlightening because it expresses the number of extraterrestrial civilizations in a simple series of terms. We can estimate some of these terms from what we know about stars and stellar evolution. For example, the first two factors, $R_*$ and $f_p$, can be determined by observation. In estimating $R_*$, we should probably exclude stars with masses larger than about 1.5 solar masses, because they have main-sequence lifetimes shorter than the time it took to develop intelligent life here on the Earth. (We described the connection between a star's mass and its main-sequence lifetime in Section 21-1 and Box 21-2.) Life on the Earth originated some 3.5 to $4.0 \times 10^9$ years ago. If that is typical of the time needed to evolve higher life forms, then a star of 1.5 $M_\odot$ or more probably becomes a red giant or a supernova before intelligent creatures can appear on any of its planets.

Although stars less massive than the Sun have much longer main-sequence lifetimes, they, too, seem unsuited for life because they are so dim. (See Section 19-9 for a discussion of how a main-sequence star's mass determines its luminosity.) Only planets very near a low-mass star would be sufficiently warm for life as we know it, and a planet that close can become tidally coupled to the star, with one side continually facing the star, while the other is in perpetual, frigid darkness. This leaves us with main-sequence stars not too different from the Sun, those with spectral types between F5 and M0. (Like Goldilocks sampling the three bears' porridge, we must have a star that is not too hot and not too cold, but just right.) Based on statistical studies of star formation in the Milky Way, some astronomers estimate that roughly one of these Sunlike stars forms in the Galaxy each year, thus setting $R_*$ at 1 per year.

As we saw in Sections 7-7 and 7-8, the planets in our solar system formed as a natural consequence of the birth of the Sun. We have also seen evidence suggesting that planetary formation may be commonplace around single stars.

Many astronomers believe that most Sunlike stars probably have planets, and so they give $f_p$ a value of 1.

Unfortunately, the rest of the terms in the Drake equation are very uncertain. Let's play with some hypothetical values. The chances that a planetary system has an Earthlike world suitable for life are not known. Were we to consider our own solar system as representative, we could put $n_e$ at 1. Let's be more conservative, however, and suppose that one in ten solar-type stars is orbited by a habitable planet, making $n_e = 0.1$. From what we know about the evolution of life on the Earth, we might assume that, given appropriate conditions, the development of life is a certainty, which would make $f_l = 1$. This is an area of intense interest to biologists.

For the sake of argument, we might also assume that evolution might naturally lead to the development of intelligence (a conjecture that is hotly debated) and also make $f_i = 1$. It's anyone's guess as to whether these intelligent extraterrestrial beings would attempt communication with other civilizations in the Galaxy, but were we to assume they would, $f_c$ would be put at 1 also.

The last variable, $L$, involving the longevity of a civilization, is the most uncertain of all, and certainly cannot be subjected to direct testing! Looking at our own example, we see a planet whose atmosphere and oceans are increasingly polluted by creatures that possess nuclear weapons. If we are typical, perhaps $L$ is as short as 100 years. Putting all these numbers together, we arrive at

$$N = \frac{1}{yr} \times 1 \times 0.1 \times 1 \times 1 \times 1 \times 100 \text{ yr} = 10$$

In other words, out of the hundreds of billions of stars in the Galaxy, we would estimate that there are only ten technologically advanced civilizations from which we might receive communications.

A wide range of values has been proposed for the terms in the Drake equation, and these various guesses produce vastly different estimates of $N$. Some scientists argue that there is exactly one advanced civilization in the Galaxy and that we are it. Others speculate that there may be hundreds or thousands of planets inhabited by intelligent creatures.

## 30-3 Radio searches for alien civilizations are under way

If there are other technological civilizations in our Galaxy, and if they are trying to communicate with us using radio waves, what frequency are they using? This is an important question, because if we fail to tune our radio telescopes to the right frequency, we might never know whether the aliens are out there. A reasonable choice would be a frequency that is fairly free of interference from extraneous sources. SETI pioneer Bernard Oliver was the first to draw attention to a range of relatively noise-free frequencies in the neighborhood of the microwave emission lines of hydrogen (H) and hydroxide (OH) (Figure 30-3). This region of the microwave spectrum is called the **water hole**, because H and OH together make $H_2O$, or water. A specific possibility is that such beings would

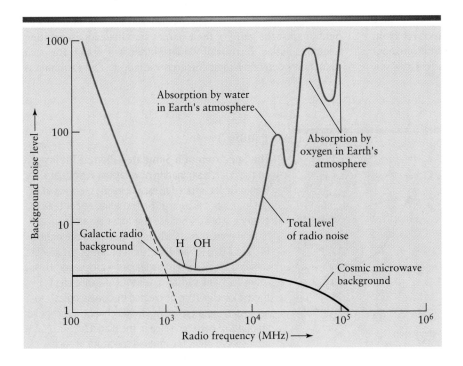

**Figure 30-3**

**The Water Hole** This graph shows the background noise level from the sky at various radio and microwave frequencies. The so-called water hole is a range of radio frequencies—from about 1000 to 10,000 megahertz (Mhz)—that happen to have relatively little cosmic noise. At lower frequencies there is substantial emission from interstellar gas (labeled "Galactic radio background"), and at higher frequencies radio waves tend to be absorbed by the Earth's atmosphere. Some scientists suggest that this noise-free region would be well suited for interstellar communication. To put this graph in perspective, a frequency of 100 MHz corresponds to "100" on an FM radio, and $10^3$ MHz is a frequency used for various types of radar. (Adapted from C. Sagan and F. Drake)

choose to transmit at a wavelength of 21 cm (corresponding to a frequency of $1.4 \times 10^3$ MHz, within the water hole), because astronomers studying the distribution of hydrogen around the Galaxy would already have their radio telescopes tuned to that wavelength (recall Figure 25-11).

Even if there are only a few alien civilizations scattered across the Galaxy, we have the technology to detect radio transmissions from them. One of the most ambitious projects was NASA's High Resolution Microwave Survey (HRMS). This would have scanned the entire sky at frequencies spanning the water hole, from $10^3$ to $10^4$ MHz. HRMS would also have observed more than 800 nearby solar-type stars over a narrower frequency range in the hope of detecting signals that were either pulsed (like Morse code) or continuous (like the carrier wave for a TV or radio broadcast). The sophisticated signal-processing technology of HRMS would have been able to sift through tens of millions of individual frequency channels simultaneously. It would even have been able to detect the minute Doppler shifts in a signal coming from an alien planet as that planet spun on its axis and moved around its star.

Sadly, just one year after HRMS began operation in 1992, the U.S. Congress imposed a mandate requiring that NASA no longer support HRMS or any other radio searches for extraterrestrial intelligence. This decision was entirely political and made only to give the legislators the appearance of saving tax dollars. In fact, the total budgetary savings was only a few million dollars, entirely negligible compared to the total NASA budget. Paradoxically, the senator who spearheaded this move was from the state of Nevada, where tax dollars have been spent to signpost a remote desert road as "The Extraterrestrial Highway"!

Even though NASA funding is no longer available, several teams of scientists remain actively involved in SETI programs. At the Big Ear Radio Observatory in Ohio, a huge radio telescope has been upgraded with a new receiver from Stanford University capable of monitoring 10 million separate microwave frequencies in an all-sky search. In 1997 sci-entists from the University of California, Berkeley attached a new 168-million channel receiver, dubbed SERENDIP IV, to the 1000-ft Arecibo antenna in Puerto Rico. (The acronym stands for *Search for Extraterrestrial Radio Emissions from Nearby, Developed, Intelligent Populations.*) This instrument "piggybacks" on any other observations being made at Arecibo and thus is able to search for signals on an essentially full-time basis. At Harvard University, BETA (the *Billion-channel ExtraTerrestrial Assay*) is scanning the sky at even more individual frequencies. And a group of astronomers from the SETI Institute in California has been using new techniques at radio observatories in Australia and West Virginia to eliminate almost all "false alarm" signals. Funding for these projects has come from nongovernmental organizations such as The Planetary Society and from private individuals.

## 30-4 Infrared telescopes in space will soon begin searching for Earthlike planets

Although no longer involved in SETI, NASA is planning a major effort to search for planets suitable for the evolution of an advanced civilization. In 2002 and after, NASA's Origins program will place a new generation of telescopes into space in the hope of answering some of the grandest questions in astronomy, such as "How did the first galaxies form?" and "How do planetary systems develop?" One important goal will be to search for Earthlike planets that show evidence of biological activity.

Such a search poses a major challenge. Planets the size of the Earth are too small to be detected by the indirect methods described in Section 7-9, and are too dim to be seen in visible light against the glare of their parent star. Instead, an orbiting telescope called Terrestrial Planet Finder will search for such planets by detecting their infrared radiation. The rationale is

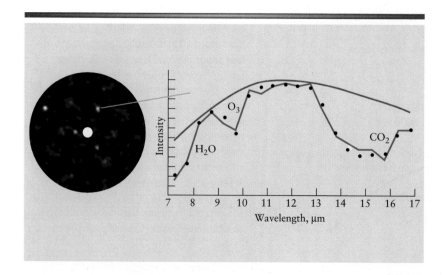

**Figure 30-4**

**The Spectrum of a Simulated Planet** The image on the left is a simulation of what the Terrestrial Planet Finder infrared telescope might see when it is put into space around 2010. The white dot at the center is a nearby Sunlike star, and the smaller dots around it are planets orbiting the star. On the right is the simulated infrared spectrum of one of the planets, showing broad absorption lines of water vapor ($H_2O$), ozone ($O_3$), and carbon dioxide ($CO_2$). While all of these molecules can be created by nonbiological processes, the presence of life will change the relative amounts of each molecule in the planet's atmosphere. Thus, the infrared spectrum of such planets will make it possible to identify worlds on which life may have evolved. (Jet Propulsion Laboratory)

that stars like the Sun emit much less infrared radiation than visible light, while planets are relatively strong emitters of infrared. Hence, observing in the infrared makes it less difficult (although still technically challenging) to detect any planets that might be orbiting a star. Over its planned six-year lifetime, Terrestrial Planet Finder will search for planets around the brightest 1000 stars within 13 pc (40 ly) of the Sun. It will also analyze the infrared spectra of any planets that it finds, in the hope of seeing the characteristic absorption of atmospheric gases such as ozone, carbon dioxide, and water vapor (Figure 30-4). The relative amounts of these gases, as determined from a planet's spectrum, can reveal whether life is present on that planet.

To achieve enough resolution to detect individual planets, Terrestrial Planet Finder will need to make use of interferometry. We discussed this technique for improving the resolution of telescopes in Section 6-6. By combining the light from four widely spaced 8-m dishes, Terrestrial Planet Finder will make the sharpest infrared images of any telescope in history.

A more speculative project using interferometry is Planet Imager. If funded, this will be an infrared telescope with sufficient resolution that some detail would be visible in the image of an extrasolar planet. One concept for such a mission would consist of five Terrestrial Planet Finder-type telescopes flying in a geometrical formation some 6000 km across (equal to the radius of the Earth). All five telescopes would collect light from the same extrasolar planet, then reflect it onto a single 8-m mirror. The combined light would go to another spacecraft, which would carry the light detectors (Figure 30-5).

Sometime in the first few decades of the twenty-first century, Terrestrial Planet Finder and Planet Imager may answer the question "Are there worlds like Earth orbiting other stars?" If the answer is yes, radio searches for intelligent signals will have even more impetus. The potential rewards from such searches is great, because detecting a message from an alien civilization would be one of the greatest events in human history. Such a message could dramatically change the course of civilization through the sharing of scientific

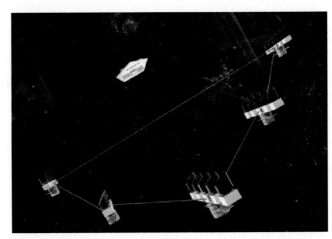

**figure 30-5**

**Planet Imager** This illustration shows the concept for Planet Imager, an infrared telescope array designed to create images of Earthlike planets orbiting other stars. Five Terrestrial Planet Finder-type telescopes—each of which has four 8-m mirrors—collect infrared light from a distant planet and reflect it onto a single 8-m mirror, which in turn reflects the combined light onto a spacecraft with the infrared-sensitive detectors. The combined light-gathering area is 1000 square meters, equivalent to a single 33-m mirror. The key to achieving the necessary resolution is to have the individual telescopes be very widely spaced. The entire Planet Imager array is about 6000 km across, very much larger than shown here. The technology to build Planet Imager does not yet exist but may be available after 2015. (Jet Propulsion Laboratory)

information or an awakening of social or humanistic enlightenment. In only a few years our technology, industry, and social structure might advance the equivalent of centuries into the future. Such changes would touch every person on the Earth. Mindful of these profound implications, scientists push ahead with the search for extraterrestrial intelligence.

## Key Words

carbonaceous chondrite, p. 752

Drake equation, p. 754

exobiology, p. 752

organic molecules, p. 752

search for extraterrestrial intelligence (SETI), p. 752

water hole, p. 755

## Questions

**Problem-solving tips and tools:**

The small-angle formula, discussed in Box 1-2, will be useful. Section 6-3 gives the relationship between the angular resolution of a telescope, the telescope diameter, and the wavelength used.

**1. (a)** Of the visually brightest stars in the sky listed in Appendix 5, which might be candidates for having Earthlike planets on which intelligent civilizations have evolved? Explain your selection criteria. **(b)** Repeat part (a) for the nearest stars, listed in Appendix 4.

**2.** Assume that all of the terms in the Drake equation have the values given in the text, except for $N$ and $L$. **(a)** If there are 1000 civilizations in the Galaxy today, what must be the average lifetime of a technological civilization? **(b)** What if there are a million such civilizations?

**3.** What do you think will set the limit on the lifetime of our technological civilization? Explain your reasoning.

**4.** Explain why Terrestrial Planet Finder and Planet Imager, both of which are planned to be infrared telescopes, need to be placed in space.

**5.** Suppose that Planet Imager has an effective diameter of 6000 km and uses infrared radiation with a wavelength of 10 μm. If it is used to observe an Earthlike planet orbiting the star Epsilon Eridani, 10.7 light-years from Earth, what is the size of the smallest detail that *Planet Imager* will be able to resolve on the face of that planet? Give your answer in kilometers.

**6.** How do you think our society would respond to the discovery of intelligent messages coming from a civilization on a planet orbiting another star? Explain your reasoning.

## Where to Learn More

*Books and magazine articles*

Gould, S. J. "The Evolution of Life on the Earth." *Scientific American*, October 1994. A renowned scientist and writer describes the complicated, seemingly unpredictable course of biological evolution. This article is one of 11 that appeared in a special issue of *Scientific American* on the topic of life in the universe. The entire issue is available in book form (*Life in the Universe*, W. H. Freeman, 1995).

Naeye, R. "SETI at the Crossroads." *Sky & Telescope*, November 1992. This article eloquently describes ongoing SETI programs.

Orgel, L. E. "The Origin of Life on the Earth." *Scientific American*, October 1994. A distinguished biologist describes current thinking about how living creatures first arose from inanimate matter.

Pendleton, Y. J., and Famer, J. D. "Life: A Cosmic Imperative?" *Sky & Telescope*, July 1997. The authors explore the notion that life may have originated elsewhere in the solar system but was only able to survive in the hospitable climate of the Earth.

Purves, W. K., Orians, G. H., Heller, H. C., and Sadava, D. *Life: The Science of Biology*, 5th ed. W. H. Freeman,

1998. This clearly written introduction to biology includes an extensive discussion of the origin of life on the Earth and its subsequent evolution.

Sagan, C. *Contact*. Simon and Schuster, 1985. What would happen if we actually made radio contact with an alien civilization? How would our society respond? This excellent novel by a distinguished scientist explores these questions. (The film of the same name, which was based on this novel, is also very worthwhile.)

Sagan, C., and Drake, F. "The Search for Extraterrestrial Intelligence." *Scientific American*, May 1975. This classic article explores the possibility and methodology of communicating with extraterrestrials.

Shklovskii, I. S., and Sagan, C. *Intelligent Life in the Universe*, Holden-Day, 1966. This somewhat dated but nevertheless classic book was one of the first to examine the question of extraterrestrial life in light of our modern understanding of astronomy and biology.

Zuckerman, B., and Hart, M. eds. *Extraterrestrials— Where Are They?*, 2nd ed. Cambridge University Press, 1995. This thought-provoking volume, written for nonscientists, examines the implications of our failure to observe extraterrestrials.

Ⓦ *World Wide Web*

The ongoing search for radio transmissions from planets orbiting other stars is described at the web sites for the SETI Institute (**http://www.seti-inst.edu/**), the Harvard SETI project (**http://mc.harvard.edu/seti/**), the SERENDIP project at the University of California, Berkeley (**http://sag-www.ssl.berkeley.edu/serendip/**), and the Big Ear Radio Observatory (**http://www.bigear.org/**).

NASA's plans to search for extrasolar planets on which life might exist are described at the web site for the Origins program (**http://origins.jpl.nasa.gov/**).

Like other popular media, the World Wide Web is full of claims of the existence of "extraterrestrial intelligence"— namely, UFO sightings and alien abductions. One of the few sites that analyzes these claims in a rational, scientific manner is maintained by the Committee for the Scientific Investigation of Claims of the Paranormal, or CSICOP (**http://www.csicop.org/**). One particularly worthwhile article is "An Astronomer's Personal Statement on UFOs" by Alan Hale, codiscoverer of Comet Hale-Bopp (**http://www.csicop.org/si/9703/ufo.html**).

# Appendices

## Appendix 1 | The Planets: Orbital Data

| Planet | Semimajor axis (AU) | Semimajor axis ($10^6$ km) | Sidereal period (years) | Sidereal period (days) | Synodic period (days) | Average orbital speed (km/s) | Orbital eccentricity | Inclination of orbit to ecliptic (°) |
|---|---|---|---|---|---|---|---|---|
| Mercury | 0.387 | 57.9 | 0.241 | 87.97 | 115.88 | 47.9 | 0.206 | 7.00 |
| Venus | 0.723 | 108.2 | 0.615 | 224.70 | 583.92 | 35.0 | 0.007 | 3.39 |
| Earth | 1.000 | 149.6 | 1.000 | 365.26 | — | 29.8 | 0.017 | 0.00 |
| Mars | 1.524 | 227.9 | 1.881 | 686.98 | 779.94 | 24.1 | 0.093 | 1.85 |
| Jupiter | 5.203 | 778.3 | 11.862 | | 398.9 | 13.1 | 0.048 | 1.30 |
| Saturn | 9.554 | 1429.0 | 29.42 | | 378.1 | 9.7 | 0.056 | 2.49 |
| Uranus | 19.22 | 2875 | 83.75 | | 369.7 | 6.8 | 0.046 | 0.77 |
| Neptune | 30.11 | 4504 | 163.7 | | 367.5 | 5.5 | 0.009 | 1.77 |
| Pluto | 39.54 | 5916 | 248.0 | | 366.7 | 4.7 | 0.249 | 17.14 |

## Appendix 2 | The Planets: Physical Data

| Planet | Equatorial diameter (km) | Equatorial diameter (Earth = 1) | Mass (kg) | Mass (Earth = 1) | Average density (kg/m$^3$) | Rotation period* (days) | Inclination of equator to orbit (°) | Surface gravity (Earth = 1) | Albedo | Escape speed (km/s) |
|---|---|---|---|---|---|---|---|---|---|---|
| Mercury | 4,879 | 0.383 | $3.302 \times 10^{23}$ | 0.055 | 5430 | 58.646 | 2 (?) | 0.39 | 0.106 | 4.3 |
| Venus | 12,104 | 0.949 | $4.869 \times 10^{24}$ | 0.815 | 5240 | $243.01^R$ | 177.3 | 0.91 | 0.65 | 10.4 |
| Earth | 12,756 | 1.000 | $5.974 \times 10^{24}$ | 1.000 | 5515 | 1.000 | 23.45 | 1.000 | 0.37 | 11.2 |
| Mars | 6,794 | 0.533 | $6.419 \times 10^{23}$ | 0.107 | 3940 | 1.026 | 25.19 | 0.38 | 0.15 | 5.0 |
| Jupiter | 142,984 | 11.209 | $1.899 \times 10^{27}$ | 317.83 | 1330 | 0.414 | 3.12 | 2.5 | 0.52 | 59.5 |
| Saturn | 120,536 | 9.449 | $5.685 \times 10^{26}$ | 95.16 | 700 | 0.444 | 26.73 | 1.1 | 0.47 | 35.5 |
| Uranus | 51,118 | 4.007 | $8.662 \times 10^{25}$ | 14.50 | 1300 | $0.718^R$ | 97.86 | 0.90 | 0.50 | 21.3 |
| Neptune | 49,528 | 3.883 | $1.028 \times 10^{26}$ | 17.204 | 1640 | 0.671 | 29.56 | 1.1 | 0.5 | 23.5 |
| Pluto | 2,300 | 0.18 | $1.3 \times 10^{22}$ | 0.002 | 2030 | $6.387^R$ | 118 | 0.07 | 0.5 | 1.3 |

* For Jupiter, Saturn, Uranus, and Neptune, the internal rotation period is given. A superscript R means that the rotation is retrograde (opposite the planet's orbital motion).

## Appendix 3 | Satellites of the Planets

| Planet | Satellite | Discoverers | Average distance from center of planet (km) | Orbital (sidereal) period* (days) | Orbital eccentricity | Diameter of satellite (km) | Mass (kg) |
|---|---|---|---|---|---|---|---|
| EARTH | Moon | — | 384,400 | 27.322 | 0.0549 | 3476 | $7.348 \times 10^{22}$ |
| MARS | Phobos | Hall (1877) | 9,380 | 0.319 | 0.01 | $28 \times 23 \times 20$ | $1.1 \times 10^{16}$ |
| | Deimos | Hall (1877) | 23,460 | 1.263 | 0.00 | $16 \times 12 \times 10$ | $1.8 \times 10^{15}$ |
| JUPITER | Metis | Synott (1979) | 127,960 | 0.2948 | 0.00 | 40 | $1 \times 10^{17}$ |
| | Adrastea | Jewitt et al. (1979) | 128,980 | 0.2983 | 0 (?) | $24 \times 16 \times 20$ | $1.9 \times 10^{16}$ |
| | Amalthea | Barnard (1892) | 181,300 | 0.4981 | 0.00 | $270 \times 200 \times 155$ | $7.2 \times 10^{18}$ |
| | Thebe | Synott (1979) | 221,900 | 0.6745 | 0.01 | 100 | $8 \times 10^{17}$ |
| | Io | Galileo (1610) | 421,600 | 1.769 | 0.00 | 3630 | $8.94 \times 10^{22}$ |
| | Europa | Galileo (1610) | 670,900 | 3.551 | 0.01 | 3138 | $4.92 \times 10^{22}$ |
| | Ganymede | Galileo (1610) | 1,070,000 | 7.155 | 0.00 | 5262 | $1.48 \times 10^{23}$ |
| | Callisto | Galileo (1610) | 1,883,000 | 16.689 | 0.01 | 4800 | $1.08 \times 10^{23}$ |
| | Leda | Kowal (1974) | 11,094,000 | 238.72 | 0.15 | 16 | $6 \times 10^{15}$ |
| | Himalia | Perrine (1904) | 11,480,000 | 250.57 | 0.16 | 180 | $1 \times 10^{19}$ |
| | Lysithea | Nicholson (1938) | 11,720,000 | 259.22 | 0.11 | 40 | $8 \times 10^{16}$ |
| | Elara | Perrine (1905) | 11,737,000 | 259.65 | 0.21 | 80 | $8 \times 10^{17}$ |
| | Ananke | Nicholson (1951) | 21,200,000 | $631^R$ | 0.17 | 30 | $4 \times 10^{16}$ |
| | Carme | Nicholson (1938) | 22,600,000 | $692^R$ | 0.21 | 44 | $1 \times 10^{17}$ |
| | Pasiphae | Melotte (1908) | 23,500,000 | $735^R$ | 0.38 | 70 | $2 \times 10^{17}$ |
| | Sinope | Nicholson (1914) | 23,700,000 | $758^R$ | 0.28 | 40 | $8 \times 10^{16}$ |
| SATURN | Pan | Showlater (1990) | 133,570 | 0.573 | 0.00 | 20 | (?) |
| | Atlas | Terrile (1980) | 137,640 | 0.602 | 0 (?) | $40 \times 30 \times 30$ | (?) |
| | Prometheus | Collins et al. (1980) | 139,350 | 0.613 | 0.00 | $140 \times 80 \times 100$ | $3 \times 10^{17}$ |
| | Pandora | Collins et al. (1980) | 141,700 | 0.629 | 0.00 | $110 \times 70 \times 100$ | $2 \times 10^{17}$ |
| | Epimetheus | Walker (1966) | 151,422 | 0.694 | 0.01 | $140 \times 100 \times 100$ | $6 \times 10^{17}$ |
| | Janus | Dolfuss (1966) | 151,472 | 0.695 | 0.01 | $220 \times 160 \times 200$ | $2 \times 10^{18}$ |
| | Mimas | Herschel (1789) | 185,520 | 0.942 | 0.02 | 390 | $3.8 \times 10^{19}$ |
| | Enceladus | Herschel (1789) | 238,020 | 1.370 | 0.00 | 500 | $7.3 \times 10^{19}$ |
| | Tethys | Cassini (1684) | 294,660 | 1.888 | 0.00 | 1050 | $6.2 \times 10^{20}$ |
| | Calypso | Smith et al. (1980) | 294,660 | 1.888 | 0 (?) | $30 \times 20 \times 25$ | (?) |
| | Telesto | Smith et al. (1980) | 294,660 | 1.888 | 0 (?) | 24 | (?) |
| | Dione | Cassini (1684) | 377,400 | 2.737 | 0.00 | 1120 | $1.1 \times 10^{21}$ |
| | Helene | Laques et al. (1980) | 377,400 | 2.737 | 0.01 | $40 \times 30 \times 30$ | (?) |
| | Rhea | Cassini (1672) | 527,040 | 4.518 | 0.00 | 1530 | $2.3 \times 10^{21}$ |
| | Titan | Huygens (1655) | 1,221,850 | 15.945 | 0.03 | 5150 | $1.4 \times 10^{23}$ |
| | Hyperion | Bond (1848) | 1,481,000 | 21.277 | 0.10 | $410 \times 260 \times 220$ | $2 \times 10^{19}$ |
| | Iapetus | Cassini (1671) | 3,561,300 | 79.331 | 0.03 | 1440 | $1.6 \times 10^{21}$ |
| | Phoebe | Pickering (1898) | 12,952,000 | $550.48^R$ | 0.16 | 220 | $4 \times 10^{18}$ |

## Appendix 3 (continued) | Satellites of the Planets

| Planet | Satellite | Discoverers | Average distance from center of planet (km) | Orbital (sidereal) period* (days) | Orbital eccentricity | Diameter of satellite (km) | Mass (kg) |
|---|---|---|---|---|---|---|---|
| URANUS | Cordelia | *Voyager* 2 (1986) | 49,750 | 0.335 | 0 (?) | 40 | (?) |
| | Ophelia | *Voyager* 2 (1986) | 53,760 | 0.376 | 0 (?) | 30 | (?) |
| | Bianca | *Voyager* 2 (1986) | 59,160 | 0.435 | 0 (?) | 40 | (?) |
| | Cressida | *Voyager* 2 (1986) | 61,770 | 0.464 | 0 (?) | 70 | (?) |
| | Desdemona | *Voyager* 2 (1986) | 62,660 | 0.474 | 0 (?) | 60 | (?) |
| | Juliet | *Voyager* 2 (1986) | 64,360 | 0.493 | 0 (?) | 80 | (?) |
| | Portia | *Voyager* 2 (1986) | 66,100 | 0.513 | 0 (?) | 110 | (?) |
| | Rosalind | *Voyager* 2 (1986) | 69,930 | 0.558 | 0 (?) | 60 | (?) |
| | Belinda | *Voyager* 2 (1986) | 75,260 | 0.624 | 0 (?) | 70 | (?) |
| | Puck | *Voyager* 2 (1986) | 86,010 | 0.762 | 0 (?) | 150 | (?) |
| | Miranda | Kuiper (1948) | 129,780 | 1.414 | 0.00 | 470 | $6.6 \times 10^{19}$ |
| | Ariel | Lassell (1851) | 191,240 | 2.520 | 0.00 | 1160 | $1.4 \times 10^{21}$ |
| | Umbriel | Lassell (1851) | 265,790 | 4.144 | 0.00 | 1170 | $1.2 \times 10^{21}$ |
| | Titania | Herschel (1787) | 435,840 | 8.706 | 0.00 | 1580 | $3.5 \times 10^{21}$ |
| | Oberon | Herschel (1787) | 582,600 | 13.463 | 0.00 | 1520 | $3.0 \times 10^{21}$ |
| NEPTUNE | Naiad | *Voyager* 2 (1989) | 48,230 | 0.296 | 0 (?) | 60 | (?) |
| | Thalassa | *Voyager* 2 (1989) | 50,070 | 0.312 | 0 (?) | 80 | (?) |
| | Despina | *Voyager* 2 (1989) | 52,530 | 0.333 | 0 (?) | 150 | (?) |
| | Galatea | *Voyager* 2 (1989) | 61,950 | 0.429 | 0 (?) | 180 | (?) |
| | Larissa | *Voyager* 2 (1989) | 73,550 | 0.554 | 0 (?) | 190 | (?) |
| | Proteus | *Voyager* 2 (1989) | 117,640 | 1.121 | 0 (?) | 415 | (?) |
| | Triton | Lassell (1846) | 354,800 | 5.877$^R$ | 0.00 | 2700 | $2.15 \times 10^{22}$ |
| | Nereid | Kuiper (1949) | 5,513,400 | 359.2 | 0.75 | 340 | (?) |
| PLUTO | Charon | Christy (1978) | 19,640 | 6.387 | 0.00 | 1190 | $1.9 \times 10^{21}$ |

* A superscript R means that the satellite orbits in a retrograde direction (opposite to the planet's rotation).

## Appendix 4 | The Nearest Stars

| Name* | Parallax (arcsec) | Distance (light-years) | Spectral type | Radial velocity** (km/s) | Proper motion (arcsec/year) | Apparent visual magnitude | Absolute visual magnitude | Luminosity (Sun = 1) |
|---|---|---|---|---|---|---|---|---|
| Sun | | | G2 V | | | −26.7 | +4.83 | 1.00 |
| Proxima Centauri | 0.772 | 4.22 | M5.5 V | −22 | 3.853 | +11.01 | +15.45 | $8.2 \times 10^{-4}$ |
| Alpha Centauri A | 0.742 | 4.40 | G2 V | −25 | 3.710 | −0.01 | +4.34 | 1.77 |
| Alpha Centauri B | 0.742 | 4.40 | K0 V | −21 | 3.724 | +1.35 | +5.70 | 0.55 |
| Barnard's Star | 0.549 | 5.94 | M4 V | −111 | 10.358 | +9.54 | +13.24 | $3.6 \times 10^{-3}$ |
| Wolf 359 | 0.418 | 7.80 | M6 V | +13 | 4.689 | +13.45 | +16.6 | $3.5 \times 10^{-4}$ |
| Lalande 21185 | 0.392 | 8.32 | M2 V | −84 | 4.802 | +7.49 | +10.46 | 0.023 |
| L 726-8 A | 0.381 | 8.56 | M5.5 V | +29 | 3.366 | +12.41 | +15.3 | $9.4 \times 10^{-4}$ |
| L 726-8 B | 0.381 | 8.56 | M6 V | +32 | 3.366 | +13.2 | +16.1 | $5.6 \times 10^{-4}$ |
| Sirius A | 0.379 | 8.61 | A1 V | −9 | 1.339 | −1.44 | +1.45 | 26.1 |
| Sirius B | 0.379 | 8.61 | white dwarf | −9 | 1.339 | +8.44 | +11.3 | $2.4 \times 10^{-3}$ |
| Ross 154 | 0.336 | 9.71 | M3.5 V | −12 | 0.666 | +10.37 | +13.00 | $4.1 \times 10^{-3}$ |
| Ross 248 | 0.316 | 10.32 | M5.5 V | −78 | 1.626 | +12.29 | +14.8 | $1.5 \times 10^{-3}$ |
| Epsilon Eridani | 0.311 | 10.49 | K2 V | +17 | 0.977 | +3.72 | +6.18 | 0.40 |
| Lacaille 9352 | 0.304 | 10.73 | M1.5 V | +10 | 6.896 | +7.35 | +9.76 | 0.051 |
| Ross 128 | 0.300 | 10.87 | M4 V | −31 | 1.361 | +11.12 | +13.50 | $2.9 \times 10^{-3}$ |
| L 789-6 | 0.294 | 11.09 | M5 V | −60 | 3.259 | +12.33 | +14.7 | $1.3 \times 10^{-3}$ |
| 61 Cygni A | 0.287 | 11.36 | K5 V | −65 | 5.281 | +5.20 | +7.49 | 0.16 |
| 61 Cygni B | 0.285 | 11.44 | K7 V | −64 | 5.172 | +6.05 | +8.33 | 0.095 |
| Procyon A | 0.286 | 11.40 | F5 IV–V | −4 | 1.259 | +0.40 | +2.68 | 7.73 |
| Procyon B | 0.286 | 11.40 | white dwarf | −4 | 1.259 | +10.7 | +13.0 | $5.5 \times 10^{-4}$ |
| BD +59° 1915 A | 0.281 | 11.61 | M3 V | −1 | 2.238 | +8.94 | +11.18 | 0.020 |
| BD +59° 1915 B | 0.281 | 11.61 | M3.5 V | +1 | 2.313 | +9.70 | +11.97 | 0.010 |
| Groombridge 34 A | 0.280 | 11.65 | M1.5 V | +12 | 2.918 | +8.09 | +10.33 | 0.030 |
| Groombridge 34 B | 0.280 | 11.65 | M3.5 V | +11 | 2.918 | +11.07 | +13.3 | $3.1 \times 10^{-3}$ |
| Epsilon Indi | 0.276 | 11.82 | K5 V | −40 | 4.704 | +4.69 | +6.89 | 0.27 |
| GJ 1111 | 0.276 | 11.82 | M6.5 V | −5 | 1.288 | +14.79 | +17.0 | $2.7 \times 10^{-4}$ |
| Tau Ceti | 0.274 | 11.90 | G8 V | −17 | 1.922 | +3.49 | +5.68 | 0.62 |
| GJ 1061 | 0.270 | 12.08 | M5.5 V | −20 | 0.836 | +13.03 | +15.2 | $1.0 \times 10^{-3}$ |
| L 725-32 | 0.269 | 12.12 | M4.5 V | +28 | 1.372 | +12.10 | +14.25 | $1.7 \times 10^{-3}$ |
| BD +05° 1668 | 0.263 | 12.40 | M3.5 V | +18 | 3.738 | +9.84 | +11.94 | 0.011 |
| Kapteyn's star | 0.255 | 12.79 | M0 V | +246 | 8.670 | +8.86 | +10.89 | 0.013 |
| Lacaille 8760 | 0.253 | 12.89 | M0 V | +28 | 3.455 | +6.69 | +8.71 | 0.094 |
| Krüger 60 A | 0.250 | 13.05 | M3 V | −33 | 0.990 | +9.85 | +11.9 | 0.010 |
| Krüger 60 B | 0.250 | 13.05 | M4 V | −32 | 0.990 | +11.3 | +13.3 | $3.4 \times 10^{-3}$ |

*Stars that are components of binary systems are labeled A and B.*
**A positive radial velocity means the star is receding; a negative radial velocity means the star is approaching.*

*Compiled from the* Hipparcos General Catalogue *and from data reported by the Research Consortium on Nearby Stars. The table lists all known stars within 4.00 parsecs (13.05 light-years).*

# Appendix 5 | The Visually Brightest Stars

| Name | Designation | Distance (light-years) | Spectral type | Radial velocity* (km/s) | Proper motion (arcsec/year) | Apparent visual magnitude | Apparent visual brightness** (Sirius = 1) | Absolute visual magnitude | Luminosity (Sun = 1) |
|---|---|---|---|---|---|---|---|---|---|
| Sirius A | α CMa A | 8.61 | A1 V | −9 | 1.339 | −1.44 | 1.000 | +1.45 | 26.1 |
| Canopus | α Car | 313 | F0 I | +21 | 0.031 | −0.62 | 0.470 | −5.53 | $1.4 \times 10^4$ |
| Arcturus | α Boo | 36.7 | K2 III | −5 | 2.279 | −0.05 | 0.278 | −0.31 | 190 |
| Alpha Centauri A | α Cen A | 4.4 | G2 V | −25 | 3.71 | −0.01 | 0.268 | +4.34 | 1.77 |
| Vega | α Lyr | 25.3 | A0 V | −14 | 0.035 | +0.03 | 0.258 | +0.58 | 61.9 |
| Capella | α Aur | 42.2 | G8 III | +30 | 0.434 | +0.08 | 0.247 | −0.48 | 180 |
| Rigel | β Ori A | 773 | B8 Ia | +21 | 0.002 | +0.18 | 0.225 | −6.69 | $7.0 \times 10^5$ |
| Procyon | α CMi A | 11.4 | F5 IV–V | −4 | 1.259 | +0.4 | 0.184 | +2.68 | 7.73 |
| Achernar | α Eri | 144 | B3 IV | +19 | 0.097 | +0.45 | 0.175 | −2.77 | 5250 |
| Betelgeuse | α Ori | 427 | M2 Iab | +21 | 0.029 | +0.45 | 0.175 | −5.14 | $4.1 \times 10^4$ |
| Hadar | β Cen | 525 | B1 II | −12 | 0.042 | +0.61 | 0.151 | −5.42 | $8.6 \times 10^4$ |
| Altair | α Aql | 16.8 | A7 IV–V | −26 | 0.661 | +0.76 | 0.132 | +2.2 | 11.8 |
| Aldebaran | α Tau A | 65.1 | K5 III | +54 | 0.199 | +0.87 | 0.119 | −0.63 | 370 |
| Spica | α Vir | 262 | B1 V | +1 | 0.053 | +0.98 | 0.108 | −3.55 | $2.5 \times 10^4$ |
| Antares | α Sco A | 604 | M1 Ib | −3 | 0.025 | +1.06 | 0.100 | −5.28 | $3.7 \times 10^4$ |
| Pollux | β Gem | 33.7 | K0 III | +3 | 0.627 | +1.16 | 0.091 | +1.09 | 46.6 |
| Fomalhaut | α PsA | 25.1 | A3 V | +7 | 0.368 | +1.17 | 0.090 | +1.74 | 18.9 |
| Deneb | α Cyg | 3230 | A2 Ia | −5 | 0.002 | +1.25 | 0.084 | −8.73 | $3.2 \times 10^5$ |
| Becrux | β Cru | 353 | B0.5 III | +20 | 0.05 | +1.25 | 0.084 | −3.92 | $3.4 \times 10^4$ |
| Regulus | α Leo A | 77.5 | B7 V | +4 | 0.249 | +1.36 | 0.076 | −0.52 | 331 |

*Data in this table was compiled from the* Hipparcos General Catalogue.

* *A positive radial velocity means the star is receding; a negative radial velocity means the star is approaching.*

** *This is the ratio of the star's apparent brightness to that of Sirius, the brightest star in the night sky.*

Note: *Acrux, or α Cru (the brightest star in Crux, the Southern Cross), appears to the naked eye as a star of apparent magnitude +0.87, the same as Aldebaran, but it does not appear in this table because Acrux is actually a binary star system. The blue-white component stars of this binary system have apparent magnitudes of +1.4 and +1.9, and thus they are dimmer than any of the stars listed here.*

## Appendix 6 | Some Important Astronomical Quantities

| | | |
|---|---|---|
| Astronomical unit | 1 AU | $= 1.4960 \times 10^{11}$ m |
| | | $= 1.4960 \times 10^{8}$ km |
| Light-year | 1 ly | $= 9.4605 \times 10^{15}$ m |
| | | $= 63{,}240$ AU |
| Parsec | 1 pc | $= 3.0857 \times 10^{16}$ m |
| | | $= 3.2616$ ly |
| Solar mass | $1\ M_\odot$ | $= 1.989 \times 10^{30}$ kg |
| Solar radius | $1\ R_\odot$ | $= 6.9599 \times 10^{8}$ m |
| Solar luminosity | $1\ L_\odot$ | $= 3.90 \times 10^{26}$ W |

## Appendix 8 | Some Useful Mathematics

| | |
|---|---|
| Area of a rectangle of sides $a$ and $b$ | $A = ab$ |
| Volume of a rectangular solid of sides $a$, $b$, and $c$ | $V = abc$ |
| Hypotenuse of a right triangle whose other sides are $a$ and $b$ | $c = \sqrt{a^2 + b^2}$ |
| Circumference of a circle of radius $r$ | $C = 2\pi r$ |
| Area of a circle of radius $r$ | $A = \pi r^2$ |
| Surface area of a sphere of radius $r$ | $A = 4\pi r^2$ |
| Volume of sphere of radius $r$ | $V = \frac{4}{3}\pi r^3$ |
| Value of $\pi$ | $\pi = 3.1415926536$ |

## Appendix 7 | Some Important Physical Constants

| | | |
|---|---|---|
| Speed of light | $c$ | $= 2.9979 \times 10^{8}$ m/s |
| Gravitational constant | $G$ | $= 6.6726 \times 10^{-11}$ N m$^2$/kg$^2$ |
| Planck's constant | $h$ | $= 6.6261 \times 10^{-34}$ J s |
| | | $= 4.1357 \times 10^{-15}$ eV s |
| Boltzmann constant | $k$ | $= 1.3807 \times 10^{-23}$ J/K |
| | | $= 8.6174 \times 10^{-5}$ eV/K |
| Stefan-Boltzmann constant | $\sigma$ | $= 5.6705 \times 10^{-8}$ W m$^{-2}$ K$^{-4}$ |
| Mass of electron | $m_{\mathrm{e}}$ | $= 9.1094 \times 10^{-31}$ kg |
| Mass of proton | $m_{\mathrm{p}}$ | $= 1.6726 \times 10^{-27}$ kg |
| Mass of neutron | $m_{\mathrm{n}}$ | $= 1.6749 \times 10^{-27}$ kg |
| Mass of hydrogen atom | $m_{\mathrm{H}}$ | $= 1.6735 \times 10^{-27}$ kg |
| Rydberg constant | $R$ | $= 1.0968 \times 10^{7}$ m$^{-1}$ |
| Electron volt | 1 eV | $= 1.6022 \times 10^{-19}$ J |

# Glossary

**A ring** One of three prominent rings encircling Saturn.

**absolute magnitude** The apparent magnitude that a star would have if it were at a distance of 10 parsecs.

**absolute zero** A temperature of −273%C (or 0 K), at which all molecular motion stops; the lowest possible temperature.

**absorption line spectrum** Dark lines superimposed on a continuous spectrum.

**acceleration** The rate at which an object's velocity changes due to a change in speed, a change in direction, or both.

**accretion** The gradual accumulation of matter in one location, typically due to the action of gravity.

**accretion disk** A disk of gas orbiting a star or black hole.

**active galactic nucleus (AGN)** The center of an active galaxy.

**active galaxy** A galaxy that is emitting exceptionally large amounts of energy; a Seyfert galaxy or a quasar.

**adaptive optics** A technique for improving a telescopic image by altering the telescope's optics in a way that compensates for distortion caused by the Earth's atmosphere.

**aerosol** Tiny droplets of liquid dispersed in a gas.

**AGB star** See *asymptotic giant branch star*.

**AGN** See *active galactic nucleus*.

**albedo** The fraction of sunlight that a planet, asteroid, or satellite reflects.

**alpha particle** The nucleus of a helium atom, consisting of two protons and two neutrons.

**amino acids** The chemical building blocks of proteins.

**angle** The opening between two lines that meet at a point.

**angular diameter** The angle subtended by the diameter of an object.

**angular measure** The size of an angle, usually expressed in degrees, arcminutes, and arcseconds.

**angular momentum** A measure of the momentum associated with rotation.

**angular resolution** The angular size of the smallest feature that can be distinguished with a telescope.

**angular size** See *angular diameter*.

**angular velocity** The speed with which an object revolves about an axis.

**anisotropic** Possessing unequal properties in different directions; not isotropic.

**annihilation** The process by which the masses of a particle and antiparticle are converted into energy.

**annular eclipse** An eclipse of the Sun in which the Moon is too distant to cover the Sun completely, so that a ring of sunlight is seen around the Moon at mid-eclipse.

**anorthosite** Rock commonly found in ancient, cratered highlands on the Moon.

**Antarctic Circle** A circle of latitude 23½° north of the Earth's South Pole.

**antimatter** Matter consisting of antiparticles, such as antiprotons, antineutrons, and positrons.

**antiparticle** A particle with the same mass as an ordinary particle but with other properties reversed, such as electric charge.

**antiproton** A particle with the same mass as a proton but with a negative electric charge.

**aphelion** The point in its orbit where a planet is farthest from the Sun.

**apogee** The point in its orbit where a satellite or the Moon is farthest from the Earth.

**apparent brightness** The flux of a star's light arriving at the Earth.

**apparent magnitude** A measure of the brightness of light from a star or other object as measured from the Earth.

**apparent solar day** The interval between two successive transits of the Sun's center across the local meridian.

**apparent solar time** Time reckoned by the position of the Sun in the sky.

**arcminute** One-sixtieth (1/60) of a degree, designated by the symbol ′

**arcsecond** One-sixtieth (1/60) of an arcminute or 1/3600 of a degree, designated by the symbol ″.

**Arctic Circle** A circle of latitude 23½° south of the Earth's north pole.

**asteroid** One of tens of thousands of small, rocky, planetlike objects in orbit about the Sun; also called a minor planet.

**asteroid belt** A region between the orbits of Mars and Jupiter that encompasses the orbits of many asteroids.

**asthenosphere** A warm, plastic layer of the mantle beneath the lithosphere of the Earth.

**astrometric method** A technique for detecting extrasolar planets by looking for stars that "wobble" periodically.

**astronomical unit (AU)** The semimajor axis of the Earth's orbit; the average distance between the Earth and the Sun.

**asymptotic giant branch** A region on the Hertzsprung-Russell diagram occupied by stars in their second and final red-giant phase.

**asymptotic giant branch star (AGB star)** A red supergiant star that has ascended the giant branch for the second and final time.

**atmosphere (atm)** A unit of atmospheric pressure.

**atmospheric pressure** The force per unit area exerted by a planet's atmosphere.

**atom** The smallest particle of an element that has the properties characterizing that element.

**atomic number** The number of protons in the nucleus of an atom of a particular element.

**AU** See *astronomical unit*.

**aurora** (plural **aurorae**) Light radiated by atoms and ions in the Earth's upper atmosphere, mostly in the polar regions.

**aurora australis** Aurorae seen from southern latitudes; the southern lights.

**aurora borealis** Aurorae seen from northern latitudes; the northern lights.

**autumnal equinox** The intersection of the ecliptic and the celestial equator where the Sun crosses the equator from north to south; also used to refer to the date on which the Sun passes through this intersection.

**average density** The mass of an object divided by its volume.

**average density of matter** In cosmology, the average density of all the material substance in the universe.

**B ring** One of three prominent rings encircling Saturn.

**Balmer line** An emission or absorption line in the spectrum of hydrogen caused by an electron transition between the second and higher energy levels.

**Balmer series** The entire pattern of Balmer lines.

**bar** A unit of atmospheric pressure.

**Barnard object** One of a class of dark nebulae discovered by E. E. Barnard.

**barred spiral galaxy** A spiral galaxy in which the spiral arms begin from the ends of a "bar" running through the nucleus rather than from the nucleus itself.

**baseline** In interferometry, the distance between two telescopes whose signals are combined to give a higher-resolution image.

**belt** A dark band in Jupiter's atmosphere.

**belt asteroid** An asteroid whose orbit lies in the asteroid belt, between the orbits of Mars and Jupiter.

**Big Bang** An explosion of all space that took place roughly 15 billion years ago and from which the universe emerged.

**Big Crunch** The fate of the universe if it is bounded and ultimately collapses in upon itself.

**binary star** (also **binary star system** or **binary**) Two stars orbiting about each other.

**biosphere** The layer of soil, water, and air surrounding the Earth in which living organisms thrive.

**bipolar outflow** Oppositely directed jets of gas expelled from a young star.

**bipolar sunspot group** A sunspot group that has roughly comparable areas covered by north and south magnetic polarity.

**black hole** An object whose gravity is so strong that the escape speed exceeds the speed of light.

**black-hole evaporation** The process by which black holes emit particles.

**blackbody** A hypothetical perfect radiator that absorbs and reemits all radiation falling upon it.

**blackbody curve** The intensity of radiation emitted by a blackbody plotted as a function of wavelength or frequency.

**blackbody radiation** The radiation emitted by a perfect blackbody.

**blazar** A type of active galaxy whose nucleus has a featureless spectrum without spectral lines.

**blueshift** A decrease in the wavelength of photons emitted by an approaching source of light.

**Bohr orbits** In the model of the atom described by Niels Bohr, the only orbits in which electrons are allowed to move about the nucleus.

**Bok globule** A small, roundish, dark nebula.

**bounce universe** A universe that contracts from an infinitely extended state, then expands out to infinity again.

**bounded universe** A universe throughout which the average density of matter exceeds the critical density; a closed universe.

**bright terrain (Ganymede)** Young, reflective, relatively crater-free terrain on the surface of Ganymede.

**brown dwarf** A starlike object that is not massive enough to ignite hydrogen burning in its core.

**brown ovals** Elongated, brownish features usually seen in Jupiter's northern hemisphere.

**C ring** One of three prominent rings encircling Saturn.

**capture theory** The hypothesis that the Moon was gravitationally captured by the Earth.

**carbon burning** The thermonuclear fusion of carbon to produce heavier nuclei.

**carbon star** A peculiar red giant star whose spectrum shows strong absorption lines of carbon and carbon compounds.

**carbonaceous chondrite** A type of meteorite that has a high abundance of carbon and volatile compounds.

**Cassegrain focus** An optical arrangement in a reflecting telescope in which light rays are reflected by a secondary mirror to a focus behind the primary mirror.

**Cassini division** An apparent gap between Saturn's A and B rings.

**causality** The notion that effects follow their causes.

**CCD** See *charge-coupled device*.

**celestial equator** A great circle on the celestial sphere 90° from the celestial poles.

**celestial poles** Points about which the celestial sphere appears to rotate.

**celestial sphere** An imaginary sphere of very large radius centered on an observer; the apparent sphere of the sky.

**center of mass** The point between a star and a planet or between two stars around which both objects orbit.

**central bulge** A spherical distribution of stars around the nucleus of a spiral galaxy.

**centrifugal force** A fictitious force in a rotating system that seems to push objects outward from the axis of rotation.

**Cepheid variable** A type of yellow, supergiant, pulsating star.

**Cerenkov radiation** The radiation emitted by a particle traveling through a substance at a speed greater than the speed of light in that substance.

**Chandrasekhar limit** The maximum mass of a white dwarf.

**charge-coupled device (CCD)** A type of solid-state silicon wafer designed to detect photons.

**chemical differentiation** The process by which the heavier elements in a planet sink toward its center while lighter elements rise toward its surface.

**chemical element** A chemical that cannot be broken down into more basic chemicals.

**chromatic aberration** An optical defect whereby different colors of light passing through a lens are focused at different locations.

**chromosphere** A layer in the atmosphere of the Sun between the photosphere and the corona.

**circumstellar accretion disk** An accretion disk that surrounds a protostar.

**close binary** A binary star system in which the stars are separated by a distance roughly comparable to their diameters.

**closed universe** See *bounded universe*.

**cluster of galaxies** A collection of galaxies containing a few to several thousand member galaxies.

**cluster of stars** A group of stars that formed together and that have remained together due to their mutual gravitational attraction.

**CNO cycle** A series of nuclear reactions in which carbon is used as a catalyst to transform hydrogen into helium.

**co-creation theory** The hypothesis that the Earth and the Moon formed at the same time from the same material.

**cocoon nebula** The nebulosity surrounding a protostar.

**cold dark matter** Slowly moving, weakly interacting particles presumed to constitute the bulk of matter in the universe.

**collisional ejection theory** The hypothesis that the Moon formed from material ejected from the Earth by the impact of a large asteroid.

**color-magnitude diagram** A plot of the apparent magnitudes (that is, apparent brightnesses) of stars in a cluster against their color indices.

**color ratio** The ratio of the apparent brightness of a star measured in one spectral region to its brightness measured in a different region.

**coma (of a comet)** The diffuse gaseous component of the head of a comet.

**coma (optical)** The distortion of off-axis images formed by a parabolic mirror.

**comet** A small body of ice and dust in orbit about the Sun. While passing near the Sun, a comet's vaporized ices give rise to a coma and tail.

**Compton wavelength** A characteristic length associated with the size of an elementary particle.

**condensation temperature** The temperature at which a particular substance in a low-pressure gas condenses into a solid.

**conduction** The transfer of heat by directly passing energy from atom to atom.

**confinement** The permanent bonding of quarks to form particles like protons and neutrons.

**conic section** The curve of intersection between a circular cone and a plane; this curve can be a circle, ellipse, parabola, or hyperbola.

**conjunction** The geometric arrangement of a planet in the same part of the sky as the Sun; a planet with an elongation of 0°.

**conservation of angular momentum** A law of physics stating that the total amount of angular momentum in an isolated system remains constant.

**constellation** A configuration of stars in the same region of the sky.

**contact binary** A binary star system in which both members fill their Roche lobes.

**continental drift** The gradual movement of the continents over the surface of the Earth due to plate tectonics.

**continuous spectrum** A spectrum of light over a range of wavelengths without any spectral lines.

**continuum** The continuous part of a spectrum (which may also include emission or absorption lines).

**convection** The transfer of energy by moving currents of fluid or gas containing that energy.

**convection cell** A circulating loop of gas or liquid that transports heat from a warm region to a cool region.

**convection current** The pattern of motion in a gas or liquid in which convection is taking place.

**convective envelope** The outer layers of a star like the Sun where energy is transported by convection.

**convective zone** The region in a star where convection is the dominant means of energy transport.

**core helium burning** The thermonuclear fusion of helium at the center of a star.

**core hydrogen burning** The thermonuclear fusion of hydrogen at the center of a star.

**corona (of the Sun)** The Sun's outer atmosphere, which has a high temperature and a low density.

**corona (Venus) (plural coronae)** A large oval or circular area where the surface of Venus has been pushed upward by the flow of magma from a "hot spot" in the mantle.

**coronagraph** An instrument for photographing the solar corona in which a disk inside the telescope produces an artificial eclipse.

**coronal hole** A region in the Sun's corona that is deficient in hot gases.

**coronal mass ejection** An event in which billions of tons of gas from the Sun's corona is suddenly blasted into space at high speed.

**cosmic censorship, law of** See *law of cosmic censorship*.

**cosmic microwave background** An isotropic radiation field with a blackbody temperature of about 2.7 K and permeates the entire universe.

**cosmic particle horizon** An imaginary sphere, centered on the Earth, whose radius equals the distance light has traveled since the Big Bang.

**cosmic singularity** The Big Bang.

**cosmic string** A hypothetical long, thin, wirelike concentration of matter left over from the Big Bang.

**cosmogony** A theory of the Earth's place in the universe and of the motions of the Sun and planets.

**cosmological constant Λ** In the equations of general relativity, a quantity signifying a pressure throughout all space that helps the universe expand.

**cosmological principle** The assumption that the universe is homogeneous and isotropic on the largest scale.

**cosmological redshift** A redshift that is caused by the expansion of the universe.

**cosmology** The study of the structure and evolution of the universe.

**coudé focus** An optical arrangement with a reflecting telescope. A series of mirrors is used to direct light to a remote focus away from the moving parts of the telescope.

**crater** A circular depression on a planet or satellite caused by the impact of a meteoroid.

**critical density** The average density of matter throughout the universe at which space would be flat and galaxies would just barely continue receding from each other infinitely far into the future.

**critical surface** The surface of the Roche lobes in a binary star system.

**crust (of a planet)** The surface layer of a terrestrial planet.

**crystal** A mineral sample in which the atoms are arranged in orderly rows, so that the sample has a symmetrical shape.

**current sheet** A broad, flat region in Jupiter's magnetosphere that contains an abundance of charged particles.

**D ring** One of several faint rings encircling Saturn.

**dark matter** Nonluminous matter that appears to be quite abundant in galaxies and throughout the universe.

**dark-matter problem** The enigma that most of the matter in the universe seems not to emit radiation of any kind and is detectable by its gravity alone.

**dark nebula** A cloud of interstellar gas and dust that obscures the light of more distant stars.

**dark terrain (Ganymede)** Older, heavily cratered, dark-colored terrain on the surface of Ganymede.

**decametric radiation** Radiation from Jupiter whose wavelength is about 10 meters.

**deceleration parameter ($q0$)** A number whose value indicates the rate at which the expansion of the universe is slowing down.

**decimetric radiation** Radiation from Jupiter whose wavelength is about one-tenth of a meter.

**declination** Angular distance of a celestial object north or south of the celestial equator.

**deferent** A stationary circle in the Ptolemaic system along which another circle (an epicycle) moves, carrying a planet, the Sun, or the Moon.

**degeneracy** The phenomenon, due to quantum mechanical effects, whereby the pressure exerted by a gas does not depend on its temperature.

**degenerate electron (neutron) pressure** The pressure exerted by degenerate electrons (neutrons).

**degenerate gas** A gas in which all the allowed states for particles (electrons or neutrons) have been filled, thereby causing the gas to behave differently from ordinary gases.

**degree** A basic unit of angular measure, designated by the symbol °.

**degree Celsius** A basic unit of temperature, designated by the symbol °C and used on a scale where water freezes at 0° and boils at 100°.

**degree Fahrenheit** A basic unit of temperature, designated by the symbol %F and used on a scale where water freezes at 32° and boils at 212°.

**density** The ratio of the mass of an object to its volume.

**density fluctuation** A variation of density from one place to another.

**density parameter ($\Omega_0$)** The ratio of the average density of matter in the universe to the critical density.

**density wave** In a spiral galaxy, a localized region in which matter piles up as it orbits the center of the galaxy; this is a proposed explanation of spiral structure.

**detached binary** A binary star system in which neither star fills its Roche lobe.

**deuterium** An isotope of hydrogen whose nucleus contains one proton and one neutron; also called heavy hydrogen.

**deuterium bottleneck** The situation during the first 3 minutes after the Big Bang when deuterium inhibited the formation of heavier elements.

**differential rotation** The rotation of a nonrigid object in which parts adjacent to each other at a given time do not always stay close together.

**differentiated asteroid** An asteroid in which chemical differentiation has taken place, so that denser material is toward the asteroid's center.

**differentiation** See *chemical differentiation*.

**diffraction** The spreading out of light passing through an aperture or opening in an opaque object.

**diffraction grating** An optical device, consisting of thousands of closely spaced lines etched in glass or metal, that disperses light into a spectrum.

**direct motion** The apparent eastward movement of a planet seen against the background stars.

**disk (of a galaxy)** The disk-shaped distribution of Population I stars that dominates the appearance of a spiral galaxy.

**distance modulus** The difference between the apparent and absolute magnitudes of an object.

**diurnal motion** Any apparent motion in the sky that repeats on a daily basis, such as the rising and setting of stars.

**Doppler effect** The apparent change in wavelength of radiation due to relative motion between the source and the observer along the line of sight.

**double-line spectroscopic binary** A spectroscopic binary in which spectral lines of both stars can be seen.

**double radio source** An extragalactic radio source characterized by two large regions of radio emission, typically located on either side of an active galaxy.

**double star** A pair of stars located at nearly the same position in the night sky. Some, but not all, double stars are binary stars.

**Drake equation** An equation used to estimate the number of intelligent civilizations in the Galaxy with whom we might communicate.

**dredge-up** The transporting of matter from deep within a star to its surface by convection.

**dust grains** Microscopic particles found in interplanetary and interstellar space with a typical size of 0.01 mm.

**dust tail** The tail of a comet that is composed primarily of dust particles.

**dwarf elliptical galaxy** A low-mass elliptical galaxy that contains only a few million stars.

**dynamo effect** The creation of a magnetic field by a moving electric current.

**E ring** A very broad, faint ring encircling Saturn.

**earthquake** A sudden vibratory motion of the Earth's surface.

**eccentricity** A number between 0 and 1 that describes the shape of an ellipse.

**eclipse** The cutting off of part or all the light from one celestial object by another.

**eclipse path** The track of the tip of the Moon's shadow along the Earth's surface during a total or annular solar eclipse.

**eclipse year** The interval between successive passages of the Sun through the same node of the Moon's orbit.

**eclipsing binary** A binary star system in which, as seen from Earth, the stars periodically pass in front of each other.

**ecliptic** The apparent annual path of the Sun on the celestial sphere.

**Eddington limit** A constraint on the rate at which a black hole can accrete matter.

**Einstein-Rosen bridge** A speculative, topological feature of a black hole that connects our universe with another universe.

**electromagnetic radiation** Radiation consisting of oscillating electric and magnetic fields. Examples include gamma rays, X rays, visible light, ultraviolet and infrared radiation, radio waves, and microwaves.

**electromagnetic spectrum** The entire array of electromagnetic radiation.

**electromagnetism** Electric and magnetic phenomena, including electromagnetic radiation.

**electromagnetic theory** The branch of physics that deals with electricity, magnetism, and electromagnetic radiation.

**electron** A subatomic particle with a negative charge and a small mass, usually found in orbits about the nuclei of atoms.

**electron volt (eV)** The energy acquired by an electron accelerated through an electric potential of 1 volt.

**element** See *chemical element*.

**ellipse** A conic section obtained by cutting completely through a circular cone with a plane.

**elliptical galaxy** A galaxy with an elliptical shape and no conspicuous interstellar material.

**elongation** The angular distance between a planet and the Sun as viewed from Earth.

**emission line spectrum** A spectrum that contains bright emission lines.

**emission nebula** A glowing gaseous nebula whose spectrum has bright emission lines.

**Encke gap** A narrow gap in Saturn's A ring.

**energy flux** The rate of energy flow, usually measured in joules per square meter per second.

**energy level** In an atom, a particular amount of energy possessed by an atom above the atom's least energetic state.

**energy-level diagram** A diagram showing the arrangement of an atom's energy levels.

**epicenter** The location on the Earth's surface directly over the focus of an earthquake.

**epicycle** A moving circle in the Ptolemaic system about which a planet revolves.

**epoch** The date used to define the coordinate system for objects on the sky.

**equal areas, law of** See *Kepler's second law*.

**equations of stellar structure** The equations used by astrophysicists to calculate the structure, properties, and evolution of a star.

**equinox** One of the intersections of the ecliptic and the celestial equator; also used to refer to the date on which the Sun passes through such an intersection.

**equivalent width** A measure of the amount of light subtracted from the continuous spectrum by an absorption line.

**era of recombination** The moment, approximately 300,000 years after the Big Bang, when the universe had cooled sufficiently to permit the formation of hydrogen atoms.

**ergosphere** The region of space immediately outside the event horizon of a rotating black hole where it is impossible to remain at rest.

**escape speed** The speed needed by an object (such as a spaceship) to leave a second object (such as a planet or star) permanently and to escape into interplanetary space.

**eV** See *electron volt*.

**event horizon** The location around a black hole where the escape speed equals the speed of light; the surface of a black hole.

**evolutionary track** The path on a Hertzsprung-Russell diagram followed by a star as it evolves.

**excited state** A state of an atom, ion, or molecule with a higher energy than the ground state.

**exobiology** The biology of life on other planets.

**exponent** A number placed above and after another number to denote the power to which the latter is to be raised, as 4 in 10⁴.

**extinction (interstellar)** See *interstellar extinction*.

**extrasolar planet** A planet orbiting a star other the Sun.

**eyepiece lens** A magnifying lens used to view the image produced at the focus of a telescope.

**F ring** A thin, faint ring encircling Saturn just beyond the A ring.

**false vacuum** Empty space with an abnormally high energy density that possibly filled the universe immediately after the Big Bang.

**far-infrared** The part of the infrared spectrum most different in wavelength from visible light.

**far side (of the Moon)** The side of the Moon that faces perpetually away from the Earth.

**favorable opposition** An opposition of Mars that affords good Earth-based views of the planet.

**filament** A portion of the Sun's chromosphere that arches to high altitudes.

**first quarter moon** The phase of the Moon that occurs when the Moon is 90° east of the Sun.

**fission theory** The hypothesis that the Moon was pulled out of a rapidly rotating proto-Earth.

**flare** See *solar flare*.

**flat space** Space that is not curved; space with zero curvature.

**flatness problem** The dilemma posed by the fact that the average density of matter throughout the universe is very nearly equal to the critical density.

**flocculent spiral galaxy** A spiral galaxy with fuzzy, poorly defined spiral arms.

**fluorescence** A process in which high-energy ultraviolet photons are absorbed and the absorbed energy is radiated as lower-energy photons of visible light.

**focal length** The distance from a lens or mirror to the point where converging light rays meet.

**focal plane** The plane in which a lens or mirror forms an image of a distant object.

**focal point** The point at which a lens or mirror forms an image of a distant point of light.

**focus (of an ellipse) (plural foci)** One of two points inside an ellipse such that the combined distance from the two foci to any point on the ellipse is a constant.

**focus (of a lens or mirror)** The point to which a light ray converges after passing through a lens or being reflected from a mirror.

**force** A push or pull that acts on an object.

**frequency** The number of crests or troughs of a wave that cross a given point per unit time; also, the number of vibrations per unit time.

**full moon** A phase of the Moon during which its full daylight hemisphere can be seen from the Earth.

**fusion crust** The coating on a stony meteorite caused by the heating of the meteorite as it descended through the Earth's atmosphere.

**G ring** A thin, faint ring encircling Saturn.

**galactic cannibalism** A collision between two galaxies of unequal mass and size in which the smaller galaxy seems to be absorbed by the larger galaxy.

**galactic cluster** See *open cluster*.

**galactic nucleus** The center of a galaxy.

**galaxy** A large assemblage of stars, nebulae, and interstellar gas and dust.

**Galilean satellites** The four large moons of Jupiter.

**gamma-ray bursters** Objects found in all parts of the sky that emit intense bursts of high-energy radiation.

**gamma rays** The most energetic form of electromagnetic radiation.

**gauss** A unit of magnetic field strength.

**general theory of relativity** A description of gravity formulated by Albert Einstein, which explains that gravity affects the geometry of space and the flow of time.

**geocentric cosmogony** An Earth-centered theory of the universe.

**giant** A star whose diameter is typically 10 to 100 times that of the Sun and whose luminosity is roughly that of 100 Suns.

**giant elliptical galaxy** A large, massive, elliptical galaxy containing many billions of stars.

**giant molecular cloud** A large cloud of interstellar gas and dust in which temperatures are low enough and densities are high enough for atoms to form into molecules.

**glitch** A sudden speedup in the period of a pulsar.

**globular cluster** A large spherical cluster of stars, typically found in the outlying regions of a galaxy.

**gluon** A particle that is exchanged between quarks.

**grand-design spiral galaxy** A galaxy with well-defined spiral arms.

**grand unified field theory (GUT)** A theory that describes and explains the four physical forces.

**granulation** The rice-grain–like structure found in the solar photosphere.

**granule** A convective cell in the solar photosphere.

**grating** See *diffraction grating*.

**gravitational force** See *gravity*.

**gravitational lens** A massive object that deflects light rays from a remote source, forming an image much as an ordinary lens does.

**gravitational radiation (gravitational waves)** Oscillations of space produced by changes in the distribution of matter.

**gravitational redshift** The increase in the wavelength of a photon as it climbs upward in a gravitational field.

**graviton** The particle that is responsible for the gravitational force.

**gravity** The force with which all matter attracts all other matter.

**Great Attractor** A huge concentration of matter toward which many galaxies, including the Milky Way, appear to be moving.

**Great Dark Spot** A prominent high-pressure system in Neptune's southern hemisphere.

**Great Red Spot** A prominent high-pressure system in Jupiter's southern hemisphere.

**greatest eastern elongation** The configuration of an inferior planet at its greatest angular distance east of the Sun.

**greatest western elongation** The configuration of an inferior planet at its greatest angular distance west of the Sun.

**greenhouse effect** The trapping of infrared radiation near a planet's surface by the planet's atmosphere.

**ground state** The state of an atom, ion, or molecule with the least possible energy.

**group (of galaxies)** A poor cluster of galaxies.

**GUT** See *grand unified theory*.

**H II region** A region of ionized hydrogen in interstellar space.

**half-life** The time required for one-half of the radioactive nuclei of an isotope to disintegrate.

**halo (of a galaxy)** A spherical distribution of globular clusters and Population II stars that surround a spiral galaxy.

**head-tail source** A radio source consisting of a bright "head"' and a long, dim "tail."

**heavy hydrogen** See *deuterium*.

**Heisenberg uncertainty principle** A principle of quantum mechanics that places limits on the precision of simultaneous measurements.

**heliocentric cosmogony** A Sun-centered theory of the universe.

**helioseismology** The study of the vibrations of the Sun as a whole.

**helium burning** The thermonuclear fusion of helium to form carbon and oxygen.

**helium flash** The nearly explosive beginning of helium burning in the dense core of a red giant star.

**helium shell flash** A brief thermal runaway that occurs in the helium-burning shell of a red supergiant.

**Herbig-Haro object** A small, luminous nebula associated with the end point of a jet emanating from a young star.

**Hertzsprung-Russell (H-R) diagram** A plot of the luminosity (or absolute magnitude) of stars against their surface temperature (or spectral type).

**high-velocity star** A star traveling at an exceptionally high speed relative to the Sun.

**highlands (on Mars)** Older, cratered terrain in the Martian southern hemisphere.

**highlands (on the Moon)** See *lunar highlands*.

**Hirayama family** A group of asteroids that have nearly identical orbits about the Sun.

**homogeneous** Having the same property in one region as in every other region.

**horizon problem** See *isotropy problem*.

**horizontal branch** A group of stars on the color-magnitude diagram of a typical globular cluster that lie near the main sequence and having roughly constant luminosity.

**horizontal-branch star** A low-mass, post–helium flash star on the horizontal branch.

**hot dark matter** Dark matter consisting of particles moving at high speeds.

**hot-spot volcanism** Volcanic activity that occurs over a hot region buried deep within a planet.

**H-R diagram** See *Hertzsprung-Russell diagram*.

**Hubble classification scheme** A method of classifying galaxies as spirals, barred spirals, ellipticals, or irregulars, according to their appearance.

**Hubble constant ($H_0$)** In the Hubble law, the constant of proportionality between the recessional velocities of remote galaxies and their distances.

**Hubble flow** The recessional motions of remote galaxies caused by the expansion of the universe.

**Hubble law** The empirical relationship stating that the redshifts of remote galaxies are directly proportional to their distances from the Earth.

**hydrocarbon** Any one of a variety of chemical compounds composed of hydrogen and carbon.

**hydrogen burning** The thermonuclear conversion of hydrogen into helium.

**hydrogen envelope** A huge, tenuous sphere of gas surrounding the head of a comet.

**hydrostatic equilibrium** A balance between the weight of a layer in a star and the pressure that supports it.

**hyperbola** A conic section formed by cutting a circular cone with a plane at an angle steeper than the sides of the cone.

**hyperbolic space** Space with negative curvature.

**hypothesis** An idea or collection of ideas that seems to explain a specified phenomenon; a conjecture.

**ice rafts (Europa)** Segments of Europa's icy crust that have been moved by tectonic disturbances.

**ices** Solid materials with low condensation temperatures, including ices of water, methane, and ammonia.

**ideal gas** A gas in which the pressure is directly proportional to both the density and the temperature of the gas; an idealization of a real gas.

**igneous rock** A rock that formed from the solidification of molten lava or magma.

**imaging** The process of recording the image made by a telescope of a distant object.

**impact crater** A crater formed by the impact of a meteoroid.

**inertia, law of** See *Newton's first law*.

**inferior conjunction** The configuration when an inferior planet is between the Sun and the Earth.

**inferior planet** A planet that is closer to the Sun than the Earth is.

**inflation** A sudden expansion of space.

**inflationary epoch** A brief period shortly after the Big Bang during which the scale of the universe increased very rapidly.

**infrared radiation** Electromagnetic radiation of wavelength longer than visible light but shorter than radio waves.

**inner core (of the Earth)** The solid innermost portion of the Earth's iron core

**inner Lagrangian point** The point between the two stars comprising a binary system where their Roche lobes touch; the point across which mass transfer can occur.

**instability strip** A region of the H-R diagram occupied by pulsating stars.

**interferometry** A technique of combining the observations of two or more telescopes to produce images better than one telescope alone could make.

**intermediate vector boson** The particle that is responsible for the weak nuclear force.

**internal rotation period** The period with which the core of a Jovian planet rotates.

**interstellar extinction** The dimming of starlight as it passes through the interstellar medium.

**interstellar medium** Gas and dust in interstellar space.

**interstellar reddening** The reddening of starlight passing through the interstellar medium as a result of blue light being scattered more than red light.

**intracluster gas** Gas found between the galaxies that make up a cluster of galaxies.

**inverse-square law** The statement that the apparent brightness of a light source varies inversely with the square of the distance from the source.

**Io torus** A doughnut-shaped ring of gas circling Jupiter at the distance of Io's orbit.

**ion** An atom that has become electrically charged due to the addition or loss of one or more electrons.

**ion tail** The relatively straight tail of a comet produced by the solar wind acting on ions.

**ionization** The process by which an atom loses electrons.

**ionopause** The boundary around a planet like Venus where the pressure exerted by ions counterbalances the pressure of the solar wind.

**ionosphere** A layer in the Earth's upper atmosphere in which many of the atoms are ionized.

**iron meteorite** A meteorite composed primarily of iron.

**irregular cluster (of galaxies)** A sprawling collection of galaxies whose overall distribution in space does not exhibit any noticeable spherical symmetry.

**irregular galaxy** An asymmetrical galaxy having neither spiral arms nor an elliptical shape.

**isotope** Any of several forms for the same chemical element whose nuclei all have the same number of protons but different numbers of neutrons.

**isotropic** Having the same property in all directions.

**isotropy problem** The dilemma posed by the fact that the cosmic microwave background is isotropic; also called the horizon problem.

**Jeans length** The smallest scale over which a density fluctuation in a medium will contract to form a gravitationally bound object.

**joule (J)** A unit of energy.

**Jovian planet** Any of the four largest planets—Jupiter, Saturn, Uranus, and Neptune.

**Kaluza-Klein theory** A theory that attempts to describe the four physical forces in terms of the geometry of space-time with more than four dimensions.

**kelvin (K)** A unit of temperature on the Kelvin temperature scale and equivalent to a degree Celsius.

**Kelvin-Helmholtz contraction** The contraction of a gaseous body, such as a star or nebula, during which gravitational energy is transformed into thermal energy.

**Kepler's first law** The statement that each planet moves around the Sun in an elliptical orbit with the Sun at one focus of the ellipse.

**Kepler's second law** The statement that a planet sweeps out equal areas in equal times as it orbits the Sun; also called the law of equal areas.

**Kepler's third law** A relationship between the period of an orbiting object and the semimajor axis of its elliptical orbit.

**kiloparsec (kpc)** One thousand parsecs; about 3260 light-years.

**kinetic energy** The energy possessed by an object because of its motion.

**Kirchhoff's laws** Three statements about circumstances that produce absorption lines, emission lines, and continuous spectra.

**Kirkwood gaps** Gaps in the spacing of asteroid orbits, discovered by Daniel Kirkwood.

**Kuiper belt** A region that extends from around the orbit of Pluto to about 500 AU from the Sun and where many icy objects orbit the Sun.

**Lagrangian points** Five points in the orbital plane of two bodies revolving about each other in circular orbits where a third object of negligible mass can remain in equilibrium.

**Lamb shift** A tiny shift in the spectral lines of hydrogen caused by the presence of virtual particles.

**Large Magellanic Cloud** See *Magellanic clouds*.

**last quarter moon** The phase of the Moon that occurs when the Moon is 90% west of the Sun.

**lava** Molten rock flowing on the surface of a planet.

**law of cosmic censorship** The hypothesis that all singularities must be surrounded by an event horizon.

**law of equal areas** See *Kepler's second law*.

**law of inertia** See *Newton's first law*.

**laws of physics** A set of physical principles with which we can understand natural phenomena and the nature of the universe.

**length contraction** In the special theory of relativity, the shrinking of an object's length along its direction of motion.

**lenticular galaxy** A galaxy with a central bulge and a disk but no spiral structure; an S0 galaxy.

**libration** An apparent rocking of the Moon whereby an Earth-based observer can, over time, see slightly more than one-half the Moon's surface.

**light curve** A graph that displays how the brightness of a star or other astronomical object varies over time.

**light-gathering power** A measure of the amount of radiation brought to a focus by a telescope.

**light pollution** Light from cities and towns that degrades telescope images.

**light scattering** The process by which light bounces off particles in its path.

**light-year (ly)** The distance light travels in a vacuum in one year.

**limb (of the Sun)** The apparent edge of the Sun as seen in the sky.

**limb darkening** The phenomenon whereby the Sun is darker near its limb than near the center of its disk.

**line of apsides** The major axis (that is, long axis) of an elliptical orbit.

**line of nodes** The line where the plane of the Earth's orbit intersects the plane of the Moon's orbit.

**line profile** The shape of a spectral line.

**liquid metallic hydrogen** Hydrogen compressed to such a density that it behaves like a liquid metal.

**lithosphere** The solid, upper layer of the Earth; essentially the Earth's crust.

**LMC** See *Magellanic clouds*.

**Local Bubble** A large cavity in the interstellar medium within which the Sun and nearby stars are located.

**Local Group** The cluster of galaxies of which our Galaxy is a member.

**local meridian** See *meridian*.

**long-period comet** A comet that takes hundreds of thousands of years to complete one orbit of the Sun.

**long-period variable** A variable star with a period longer than about 100 days.

**Lorentz transformations** Equations that relate the measurements of different observers who are moving relative to each other at high speeds.

**lower meridian** The half of the meridian that lies below the horizon.

**lowlands (on Mars)** Relatively young and crater-free terrain in the Martian northern hemisphere.

**luminosity** The rate at which electromagnetic radiation is emitted from a star or other object.

**luminosity class** A classification of a star of a given spectral type according to its luminosity.

**luminosity function** The numbers of stars of differing brightness per cubic parsec.

**lunar eclipse** An eclipse of the Moon by the Earth; a passage of the Moon through the Earth's shadow.

**lunar highlands** Ancient, high-elevation, heavily cratered terrain on the Moon.

**lunar month** See *synodic month*.

**lunar phase** The appearance of the illuminated area of the Moon as seen from Earth.

**Lyman series** A series of spectral lines of hydrogen produced by electron transitions to and from the lowest energy state of the hydrogen atom.

**M-theory** A theory that attempts to describe fundamental particles as 11-dimensional membranes.

**MACHO** See *massive compact halo object.*

**Magellanic clouds** Two nearby galaxies, the Large Magellanic Cloud (LMC) and Small Magellanic Cloud (SMC), visible to the naked eye from southern latitudes.

**magma** Molten rock beneath a planet's surface.

**magnetic axis** A line connecting the north and south magnetic poles of a planet or star possessing a magnetic field.

**magnetic-dynamo model** A theory that explains the solar cycle as a result of the Sun's differential rotation acting on the Sun's magnetic field.

**magnetic resonance imaging (MRI)** A technique for viewing the interior of the human body that makes use of how protons respond to magnetic fields.

**magnetogram** An artificial picture of the Sun that shows regions of different magnetic polarity.

**magnetopause** That region of a planet's magnetosphere where the magnetic field counterbalances the pressure from the solar wind.

**magnetosphere** The region around a planet occupied by its magnetic field.

**magnification** The factor by which the apparent angular size of an object is increased when viewed through a telescope.

**magnitude scale** A system for denoting the brightnesses of astronomical objects.

**main sequence** A grouping of stars on a Hertzsprung-Russell diagram extending diagonally across the graph from hot, luminous stars to cool, dim stars.

**main-sequence lifetime** The total time that a star spends burning hydrogen in its core, and thus the total time that it will spend as a main-sequence star.

**main-sequence star** A star whose luminosity and surface temperature place it on the main sequence on a Hertzsprung-Russell diagram; a star that derives its energy from core hydrogen burning.

**major axis (of an ellipse)** The longest diameter of an ellipse.

**mantle (of a planet)** That portion of a terrestrial planet located between its crust and core.

**mare (plural maria)** Latin for "sea"; a large, relatively crater-free plain on the Moon.

**mare basalt** A type of lunar rock commonly found in the mare basins.

**marginally bounded universe** A universe throughout which the average density of matter equals the critical density.

**mass** A measure of the total amount of material in an object.

**mass density of radiation** The energy possessed by a radiation field per unit volume divided by the square of the speed of light.

**mass loss** A process by which a star gently loses matter.

**mass-luminosity relation** A relationship between the masses and luminosities of main-sequence stars.

**mass-radius relation** A relationship between the masses and radii of white dwarf stars.

**mass transfer** The flow of gases from one star in a binary system to the other.

**massive compact halo object (MACHO)** A dim star or low-mass black hole that may comprise part of the unseen dark matter.

**matter-dominated universe** A universe in which the average density of matter exceeds the mass density of radiation.

**Maunder minimum** An interval of about 70 years, starting around 1645, during which very few sunspots were seen.

**mean solar day** The interval between successive meridian passages of the mean Sun; the average length of a solar day.

**mean Sun** A fictitious object that moves eastward at a constant speed along the celestial equator, completing one circuit of the sky with respect to the vernal equinox in one tropical year.

**medium (plural media)** A material through which light travels.

**medium, interstellar:** See *interstellar medium.*

**megaparsec (Mpc)** One million parsecs.

**melting point** The temperature at which a substance changes from solid to liquid.

**meridian (or local meridian)** The great circle on the celestial sphere that passes through an observer's zenith and the north and south celestial poles.

**meridian transit** The crossing of the meridian by any astronomical object.

**mesosphere** A layer in a planet's atmosphere above the stratosphere.

**metal** In reference to stars and galaxies, any element other than hydrogen and helium; in reference to planets, a material such as iron, silver, and aluminum that is a good conductor of electricity and of heat.

**metal-poor star** A star that, compared to the Sun, is deficient in elements heavier than helium.

**metal-rich star** A star whose abundance of heavy elements is roughly comparable to that of the Sun.

**metamorphic rock** A rock whose properties and appearance have been transformed by the action of pressure and heat beneath the Earth's surface.

**meteor** The luminous phenomenon seen when a meteoroid enters the Earth's atmosphere; a shooting star.

**meteor shower** Many meteors that seem to radiate from a common point in the sky.

**meteorite** A fragment of a meteoroid that has survived passage through the Earth's atmosphere.

**meteoritic swarm** A collection of meteoroids moving together along an orbit about the Sun.

**meteoroid** A small rock in interplanetary space.

**microwaves** Short-wavelength radio waves.

**Milky Way** Our Galaxy; the band of faint stars seen from the Earth in the plane of our Galaxy's disk.

**millibar** A unit of atmospheric pressure; one-thousandth of a bar.

**millisecond pulsar** A pulsar with a period of roughly 1 to 10 milliseconds.

**mineral** A naturally occurring solid composed of a single element or chemical combination of elements, often in the form of crystals.

**minor planet** See *asteroid*.

**minute of arc** See *arcminute*.

**model** A hypothesis that has withstood experimental or observational tests.

**model (of the Sun or a star)** The results of a theoretical calculation that gives the values of temperature, pressure, density, and so forth throughout the Sun or a star.

**molecule** A combination of two or more atoms.

**moment of inertia** A quantity that expresses both the amount of mass in an object and how that mass is distributed.

**moonquake** Sudden, vibratory motion of the Moon's surface.

**MRI** See *magnetic resonance imaging*.

**neap tide** An ocean tide that occurs when the Moon is near first-quarter or third-quarter phase.

**near-Earth object (NEO)** An asteroid whose orbit lies wholly or partly within the orbit of Mars.

**near-infrared** The part of the infrared spectrum closest in wavelength to visible light.

**nebula (plural nebulae)** A cloud of interstellar gas and dust.

**nebulosity** See *nebula*.

**negative curvature** The curvature of a surface or space in which parallel lines diverge and the sum of the angles of a triangle is less than 180°.

**negative hydrogen ion** A hydrogen atom that has acquired a second electron.

**NEO** See *near-Earth object*.

**neon burning** The thermonuclear fusion of neon to produce heavier nuclei.

**neutrino** A subatomic particle with no electric charge and little or no mass, but one that is important in many nuclear reactions.

**neutrino flux** The number of neutrinos from the Sun arriving at the Earth per square meter per second.

**neutron** A subatomic particle with no electric charge and with a mass nearly equal to that of the proton.

**neutron capture** The buildup of neutrons inside nuclei.

**neutron star** A very compact, dense star composed almost entirely of neutrons.

**new moon** The phase of the Moon when the dark hemisphere of the Moon faces the Earth.

**Newtonian mechanics** The branch of physics based on Newton's laws of motion.

**Newtonian reflector** A reflecting telescope that uses a small mirror to deflect the image to one side of the telescope tube.

**Newton's first law of motion** The statement that a body remains at rest, or moves in a straight line at a constant speed, unless acted upon by an net outside force; the law of inertia.

**Newton's form of Kepler's third law** A relationship between the period of two objects orbiting each other, the semimajor axis of their orbit, and the masses of the objects.

**Newton's second law of motion** A relationship between the acceleration of an object, the object's mass, and the net outside force acting on the mass.

**Newton's third law of motion** The statement that whenever one body exerts a force on a second body, the second body exerts an equal and opposite force on the first body.

**no-hair theorem** A statement of the simplicity of black holes.

**node** See *line of nodes*.

**nonthermal radiation** Radiation other than that emitted by a heated body.

**north celestial pole** The point directly above the Earth's north pole where the Earth's axis of rotation, if extended, would intersect the celestial sphere.

**northern lights** See *aurora borealis*.

**nova (plural novae)** A star that experiences a sudden outburst of radiant energy, temporarily increasing its luminosity approximately a thousandfold.

**nuclear density** The density of matter in an atomic nucleus; about $4 \times 10^{17}$ kg/m3.

**nuclear fusion** The combining of lighter nuclei to make heavier ones.

**nucleosynthesis** The process of building up nuclei such as deuterium and helium from protons and neutrons.

**nucleus (of an atom)** The massive part of an atom, composed of protons and neutrons, about which electrons revolve.

**nucleus (of a comet)** A collection of ices and dust that constitute the solid part of a comet.

**nucleus (of a galaxy)** See *galactic nucleus*.

**OB association** A grouping of hot, young, massive stars, predominantly of spectral types O and B.

**OBAFGKM** The temperature sequence (from hot to cold) of spectral types.

**objective lens** The principal lens of a refracting telescope.

**objective mirror** The principal mirror of a reflecting telescope.

**oblate** Flattened at the poles.

**oblateness** A measure of how much a flattened sphere (or spheroid) differs from a perfect sphere.

**observable universe** That portion of the universe inside the cosmic particle horizon.

**Occam's razor** The notion that a straightforward explanation of a phenomenon is more likely to be correct than a convoluted one.

**occultation** The eclipsing of an astronomical object by the Moon or a planet.

**oceanic rift** A crack in the ocean floor that exudes lava.

**Olbers's paradox** The dilemma associated with the fact that the night sky is dark.

**1-to-1 spin-orbit coupling** See *synchronous rotation*.

**Oort cloud** A presumed accumulation of comets and cometary material surrounding the Sun at distances of about 50,000 to 100,000 AU.

**open cluster** A loose association of young stars in the disk of our Galaxy; a galactic cluster.

**open universe** See *unbounded universe*.

**opposition** The configuration of a planet when it is at an elongation of 180° and thus appears opposite the Sun in the sky.

**optical double star** Two stars that lie along nearly the same line of sight but are actually at very different distances from us.

**optical telescope** A telescope designed to detect visible light.

**optical window** The range of visible wavelengths to which the Earth's atmosphere is transparent.

**organic molecules** Molecules containing carbon, some of which are the molecules that make up living organisms.

**oscillating universe** A universe that expands from a Big Bang, reaches a maximum size, and then collapses, only to erupt again in another Big Bang.

**outer core (of the Earth)** The outer, molten portion of the Earth's core.

**overcontact binary** A close binary system in which the two stars share a common atmosphere.

**oxygen burning** The thermonuclear fusion of oxygen to produce heavier nuclei.

**ozone** A type of oxygen whose molecules contain three oxygen atoms.

**ozone hole** A region of the Earth's atmosphere over Antarctica where the concentration of ozone is abnormally low.

**ozone layer** A layer in the Earth's upper atmosphere where the concentration of ozone is high enough to prevent much ultraviolet light from reaching the surface.

**P wave** One of three kinds of seismic waves produced by an earthquake; a primary wave.

**pair production** The creation of a particle and its antiparticle from energy.

**parabola** A conic section formed by cutting a circular cone at an angle parallel to one of the sides of the cone.

**parallax** The apparent displacement of an object due to the motion of the observer.

**parsec (pc)** A unit of distance; 3.26 light-years.

**partial lunar eclipse** A lunar eclipse in which the Moon does not appear completely covered.

**partial solar eclipse** A solar eclipse in which the Sun does not appear completely covered.

**Paschen series** A series of spectral lines of hydrogen produced by electron transitions to and from the third ($n = 3$) energy level of the hydrogen atom.

**Pauli exclusion principle** A principle of quantum mechanics stating that two identical particles cannot have the same position and momentum.

**penumbra (of a shadow)** (plural **penumbrae**) The portion of a shadow in which only part of the light source is covered by an opaque body.

**penumbra (of a sunspot)** The grayish region that usually surrounds the umbra of a sunspot.

**penumbral eclipse** A lunar eclipse in which the Moon passes only through the Earth's penumbra.

**perigee** The point in its orbit where a satellite or the Moon is nearest the Earth.

**perihelion** The point in its orbit where a planet or comet is nearest the Sun.

**period (of a planet)** The interval of time between successive geometric arrangements of a planet and an astronomical object, such as the Sun.

**period-luminosity relation** A relationship between the period and average density of a pulsating star.

**periodic table** A listing of the chemical elements according to their properties, invented by Dmitri Mendeleev.

**permafrost** Frozen soil that lies just underneath the surface.

**photodisintegration** The breakup of nuclei by high-energy gamma rays.

**photoelectric effect** The phenomenon whereby certain metals emit electrons when exposed to short-wavelength light.

**photometry** The measurement of light intensities.

**photon** A discrete unit of electromagnetic energy.

**photosphere** The region in the solar atmosphere from which most of the visible light escapes into space.

**physics, laws of** See *laws of physics*.

**pixel** A picture element.

**plage** A bright region in the solar atmosphere as observed in the monochromatic light of a spectral line.

**Planck length** A fundamental length defined by basic physical constants.

**Planck mass** A fundamental mass defined by basic physical constants.

**Planck time** A fundamental interval of time defined by basic physical constants.

**Planck's law** The relationship between the energy of a photon and its wavelength or frequency; $E = h\nu$.

**planetary nebula** A luminous shell of gas ejected from an old, low-mass star.

**planetesimal** A small body of primordial dust and ice from which the planets formed.

**plasma** A hot ionized gas.

**plastic** The attribute of being nearly solid yet able to flow.

**plate** A large section of the Earth's lithosphere that moves as a single unit.

**plate tectonics** The motions of large segments (plates) of the Earth's surface over the underlying mantle.

**polarized radiation** Electromagnetic waves whose electric fields are aligned in a particular direction.

**polymer** A long molecule consisting of many smaller molecules joined together.

**poor cluster (of galaxies)** A cluster of galaxies with very few members; a group of galaxies.

**Population I star** A star whose spectrum exhibits spectral lines of many elements heavier than helium; a metal-rich star.

**Population II star** A star whose spectrum exhibits comparatively few spectral lines of elements heavier than helium; a metal-poor star.

**positional astronomy** The study of the apparent positions of the planets and stars and how those positions change.

**positive curvature** The curvature of a surface or space in which parallel lines converge and the sum of the angles of a triangle is greater than 180°.

**positron** An electron with a positive rather than negative electric charge; the antiparticle of the electron.

**power of ten** The exponent $n$ in $10n$.

**powers-of-ten notation** A shorthand method of writing numbers, involving 10 followed by an exponent.

**precession (of the Earth)** A slow, conical motion of the Earth's axis of rotation caused by the gravitational pull of the Moon and Sun on the Earth's equatorial bulge.

**precession of the equinoxes** The slow westward motion of the equinoxes along the ecliptic due to precession of the Earth.

**primary mirror** See *objective mirror*.

**prime focus** The point in a telescope where the objective focuses light.

**primitive asteroid** See *undifferentiated asteroid*.

**primordial fireball** The extremely hot gas that filled the universe immediately following the Big Bang.

**principle of equivalence** A principle of general relativity stating that in a small volume it is impossible to distinguish between the effects of gravitation and acceleration.

**progenitor star** A star that later explodes into a supernova.

**prograde orbit** An orbit of a satellite around a planet that is in the same direction as the rotation of the planet.

**prograde rotation** A situation in which an object (such as a planet) rotates in the same direction that it orbits around another object (such as the Sun).

**prominence** Flamelike protrusions seen near the limb of the Sun and extending into the solar corona.

**proper distance** See *proper length*.

**proper length** A length measured by a ruler at rest with respect to an observer.

**proper motion** The angular rate of change in the location of a star on the celestial sphere, usually expressed in arcseconds per year.

**proper time** A time interval measured with a clock at rest with respect to an observer.

**proplyd** See *protoplanetary disk*.

**proton** A heavy, positively charged subatomic particle that is one of two principal constituents of atomic nuclei.

**proton-proton chain** A sequence of thermonuclear reactions by which hydrogen nuclei are built up into helium nuclei.

**protoplanet** An object intermediate in size between a planetesimal and a planet; an intermediate stage in the formation of a planet.

**protoplanetary disk (proplyd)** A disk of material encircling a protostar or a newborn star.

**protostar** A star in its earliest stages of formation.

**protosun** The part of the solar nebula that eventually developed into the Sun.

**Ptolemaic system** The definitive version of the geocentric cosmogony of ancient Greece.

**pulsar** A pulsating radio source believed to be associated with a rapidly rotating neutron star.

**pulsating variable star** A star that pulsates in size and luminosity.

**pulsating X-ray source** An object that emits pulses of X rays at regular intervals of a few seconds.

**QSO** See *quasi-stellar object*.

**quantum electrodynamics** A theory that describes details of how charged particles interact by exchanging photons.

**quantum mechanics** The branch of physics dealing with the structure and behavior of atoms and their interaction with light.

**quark** One of several hypothetical particles thought to be the internal constituents of certain heavy subatomic particles such as protons and neutrons.

**quasar** A starlike object with a very large redshift; a quasi-stellar object or quasi-stellar radio source.

**quasi-stellar object (QSO)** A quasar that emits little or no radio radiation.

**quasi-stellar radio source** A quasar that emits detectable radio radiation.

**radial velocity** That portion of an object's velocity parallel to the line of sight.

**radial velocity curve** A plot showing the variation of radial velocity with time for a binary star or variable star.

**radial velocity method** A technique used to detect extrasolar planets.

**radiant (of a meteor shower)** The point in the sky from which meteors of a particular shower seem to originate.

**radiation darkening** The darkening of methane ice by electron impacts.

**radiation-dominated universe** A universe in which the mass density of radiation exceeds the average density of matter.

**radiation pressure** Pressure exerted on an object by radiation falling on the object.

**radiative diffusion** The random migration of photons from a star's center toward its surface.

**radiative zone** A region within a star where radiative diffusion is the dominant mode of energy transport.

**radio galaxy** A galaxy that emits an unusually large amount of radio waves.

**radio lobe** A region near an active galaxy from which significant radio radiation emanates.

**radio telescope** A telescope designed to detect radio waves.

**radio waves** The longest-wavelength electromagnetic radiation.

**radio window** The range of radio wavelengths to which the Earth's atmosphere is transparent.

**recombination** The process in which an electron combines with a positively charged ion.

**red giant** A large, cool star of high luminosity.

**red giant branch** The region of a Hertzsprung-Russell diagram occupied by stars in their first red giant phase.

**reddening (interstellar)** See *interstellar reddening*.

**redshift** The shifting to longer wavelengths of the light from remote galaxies and quasars; the Doppler shift of light from a receding source.

**reflecting telescope** A telescope in which the principal optical component is a concave mirror.

**reflection** The return of light rays by a surface.

**reflection nebula** A comparatively dense cloud of dust in interstellar space that is illuminated by a star.

**reflector** A reflecting telescope.

**refracting telescope** A telescope in which the principal optical component is a lens.

**refraction** The bending of light rays when they pass from one transparent medium to another.

**refractor** A refracting telescope.

**refractory element** An element with high melting and boiling points.

**regolith** The layer of rock fragments covering the surface of the Moon.

**regression of the line of nodes** The slow motion of the line of nodes of the Moon due to the gravitational pull of the Sun.

**regular cluster (of galaxies)** A spherical cluster of galaxies.

**relativistic cosmology** A cosmology based on the general theory of relativity.

**residual polar cap** An ice-covered polar region on Mars that does not completely evaporate during the Martian summer.

**rest energy** The rest mass of a particle multiplied by the square of the speed of light.

**retrograde motion** The apparent westward motion of a planet with respect to background stars.

**retrograde orbit** An orbit of a satellite around a planet that is in the direction opposite to which the planet rotates.

**retrograde rotation** A situation in which an object (such as a planet) rotates in the direction opposite to which it orbits around another object (such as the Sun).

**rich cluster (of galaxies)** A cluster of galaxies containing many members.

**rift valley** A feature created when a planet's crust breaks apart along a line.

**right ascension** A coordinate for measuring the east-west positions of objects on the celestial sphere.

**ring particles** Small particles that constitute a planetary ring.

**ringlet** One of many narrow bands of particles of which Saturn's ring system is composed.

**Roche limit** The smallest distance from a planet or other object at which a second object can be held together by purely gravitational forces.

**Roche lobe** A teardrop-shaped volume surrounding a star in a binary inside which gases are gravitationally bound to that star.

**rock** A mineral or combination of minerals.

**rotation curve** A plot of the orbital speeds of stars and nebulae outward from the center of a galaxy.

**rotation of the Moon's orbit** The slow motion of the major axis of the Moon's orbit due to the gravitational pull of the Sun.

**RR Lyrae variable** A type of pulsating star with a period of less than one day.

**runaway greenhouse effect** A greenhouse effect in which the temperature continues to increase.

**S wave** One of three kinds of seismic waves produced by an earthquake; a secondary wave.

**saros** A particular cycle of similar eclipses that recur about every 18 years.

**scarp** A line of cliffs formed by the faulting or fracturing of a planet's surface.

**scattering of light** See *light scattering*.

**Schwarzschild radius ($RSch$)** The distance from the singularity to the event horizon in a nonrotating black hole.

**scientific method** The basic procedure used by scientists to investigate phenomena.

**seafloor spreading** The separation of plates under the ocean due to lava emerging in an oceanic rift.

**search for extraterrestrial intelligence (SETI)** The scientific search for evidence of intelligent life on other planets.

**second of arc** See *arcsecond*.

**sedimentary rock** A rock that is formed from material deposited on land by rain or winds, or on the ocean floor.

**seeing disk** The angular diameter of a star's image.

**seismic wave** A vibration traveling through a terrestrial planet, usually associated with earthquakelike phenomena.

**seismograph** A device used to record and measure seismic waves, such as those produced by earthquakes.

**self-propagating star formation** The process by which the formation of stars in one location in a galaxy stimulates the formation of stars in a neighboring location.

**semidetached binary** A binary star system in which one star fills its Roche lobe.

**semimajor axis** One-half of the major axis of an ellipse.

**SETI** See *search for extraterrestrial intelligence*.

**Seyfert galaxy** A spiral galaxy with a bright nucleus whose spectrum exhibits emission lines.

**shell helium burning** The thermonuclear fusion of helium in a shell surrounding a star's core.

**shell hydrogen burning** The thermonuclear fusion of hydrogen in a shell surrounding a star's core.

**shepherd satellite** A satellite whose gravity restricts the motions of particles in a planetary ring, preventing them from dispersing.

**shield volcano** A volcano with long, gently sloping sides.

**shock wave** An abrupt, localized region of compressed gas caused by an object traveling through the gas at a speed greater than the speed of sound.

**shooting star** See *meteor*.

**short-period comet** A comet that orbits the Sun with a period of less than about 200 years.

**SI units** The International System of Units, based on the meter (m), the second (s), and the kilogram (kg).

**sidereal clock** A clock that measures sidereal time.

**sidereal day** The interval between successive meridian passages of the vernal equinox.

**sidereal month** The period of the Moon's revolution about the Earth with respect to the stars.

**sidereal period** The orbital period of one object about another as measured with respect to the stars.

**sidereal time** Time reckoned by the location of the vernal equinox.

**sidereal year** The orbital period of the Earth about the Sun with respect to the stars.

**silicon burning** The thermonuclear fusion of silicon to produce heavier elements, especially iron.

**single-line spectroscopic binary** A spectroscopic binary in which the spectral lines of only one star can be detected.

**singularity** A place of infinite space-time curvature; the center of a black hole.

**small-angle formula** A relationship between the angular and linear sizes of a distant object.

**Small Magellanic Cloud** See *Magellanic clouds*.

**SMC** See *Magellanic clouds*.

**SNC meteorite** A meteorite that came to Earth from Mars.

**solar atmosphere** The outer layers of the Sun, consisting of the photosphere, chromosphere, and corona.

**solar constant** The average amount of energy received from the Sun per square meter per second, measured just above the Earth's atmosphere.

**solar corona** Hot, faintly glowing gases seen around the Sun during a total solar eclipse; the uppermost regions of the solar atmosphere.

**solar cycle** See *22-year solar cycle*.

**solar eclipse** An eclipse of the Sun by the Moon; a passage of the Earth through the Moon's shadow.

**solar flare** A sudden, temporary outburst of light from an extended region of the solar surface.

**solar interior** Everything below the solar atmosphere; the inside of the Sun.

**solar nebula** The cloud of gas and dust from which the Sun and solar system formed.

**solar neutrino** A neutrino emitted from the core of the Sun.

**solar system** The Sun, planets and their satellites, asteroids, comets, and related objects that orbit the Sun.

**solar transit** The passage of an object in front of the Sun.

**solar wind** An outward flow of particles (mostly electrons and protons) from the Sun.

**south celestial pole** The point directly above the Earth's south pole where the Earth's axis of rotation, if extended, would intersect the celestial sphere.

**southern lights** See *aurora australis*.

**space-time** A four-dimensional combination of the three dimensions of space and time.

**space velocity** How fast and in what direction a star moves through space.

**special theory of relativity** A description of mechanics and electromagnetic theory formulated by Albert Einstein, which explains that measurements of distance, time, and mass are affected by the observer's motion.

**spectral analysis** The identification of chemical substances from the patterns of lines in their spectra.

**spectral class** A classification of stars according to the appearance of their spectra.

**spectral line** In a spectrum, an absorption or emission feature that is at a particular wavelength.

**spectral type** A subdivision of a spectral class.

**spectrograph** An instrument for photographing a spectrum.

**spectroscopic binary** A binary star system whose binary nature is deduced from the periodic Doppler shifting of lines in its spectrum.

**spectroscopic parallax** The distance to a star derived by comparing its apparent brightness to a luminosity inferred from the star's spectrum.

**spectroscopy** The study of spectra and spectral lines.

**spectrum** (plural **spectra**) The result of dispersing a beam of electromagnetic radiation so that components with different wavelengths are separated in space.

**spectrum binary** A binary star whose binary nature is deduced from the presence of two sets of incongruous spectral lines.

**spherical aberration** The distortion of an image formed by a telescope due to differing focal lengths of the optical system.

**spherical space** Space with positive curvature.

**spicule** A narrow jet of rising gas in the solar chromosphere.

**spin** The intrinsic angular momentum possessed by certain particles.

**spin-flip transition** A transition in the ground state of the hydrogen atom, which occurs when the orientation of the electron's spin changes.

**spin-orbit coupling** See *synchronous rotation* and *3-to-2 spin-orbit coupling*.

**spiral arms** Lanes of interstellar gas, dust, and young stars that wind outward in a plane from the central regions of a galaxy.

**spiral galaxy** A flattened, rotating galaxy with pinwheel-like spiral arms winding outward from the galaxy's nucleus.

**spontaneous symmetry breaking** A symmetry breaking that occurs suddenly.

**spring tide** An ocean tide that occurs at new moon and full moon phases.

**standard candle** An astronomical object of known intrinsic brightness that can be used to determine the distances to other galaxies.

**star cluster** See *cluster of stars*.

**starburst galaxy** A galaxy that is experiencing an exceptionally high rate of star formation.

**stationary absorption line** An absorption line in the spectrum of a binary star that does not show the same Doppler shift as other lines, indicating that it originates in the interstellar medium.

**Stefan-Boltzmann law** A relationship between the temperature of a blackbody and the rate at which it radiates energy.

**stellar association** A loose grouping of young stars.

**stellar evolution** The changes in size, luminosity, temperature, and so forth that occur as a star ages.

**stellar model** The result of theoretical calculations that give details of physical conditions inside a star.

**stellar parallax** The apparent displacement of a star due to the Earth's motion around the Sun.

**stony iron meteorite** A meteorite composed of both stone and iron.

**stony meteorite** A meteorite composed of stone.

**stratosphere** A layer in the atmosphere of a planet directly above the troposphere.

**strong nuclear force** The force that binds protons and neutrons together in nuclei.

**subduction zone** A location where colliding tectonic plates cause the Earth's crust to be pulled down into the mantle.

**subtend** To extend over an angle.

**summer solstice** The point on the ecliptic where the Sun is farthest north of the celestial equator; also used to refer to the date on which the Sun passes through this point.

**sunspot** A temporary cool region in the solar photosphere.

**sunspot cycle** The semiregular 11-year period with which the number of sunspots fluctuates.

**sunspot group** A cluster of sunspots.

**sunspot maximum/minimum** That time during the sunspot cycle when the number of sunspots is highest/lowest.

**supercluster** A collection of clusters of galaxies.

**superconductivity** The phenomenon whereby a flowing electric current does not experience any electrical resistance.

**superfluidity** The phenomenon whereby a fluid flows without experiencing any viscosity.

**supergiant** A very large, extremely luminous star; stars of luminosity class I.

**supergrand unified theory** A complete description of all forces and particles, as well as the structure of space and time; a "theory of everything" (TOE).

**supergranule** A large convective feature in the solar atmosphere, usually outlined by spicules.

**superior conjunction** The configuration of a planet being behind the Sun as viewed from the Earth.

**superior planet** A planet that is more distant from the Sun than the Earth is.

**superluminal motion** Motion that appears to involve speeds greater than the speed of light.

**supermassive black hole** A black hole with a mass of a million or more solar masses.

**supernova** (plural **supernovae**) A stellar outburst during which a star suddenly increases its brightness about a millionfold.

**supernova remnant** The gases ejected by a supernova.

**supersonic** Faster than the speed of sound.

**symmetry** Any situation in which certain properties of particles or forces are equivalent.

**symmetry breaking** A process by which certain symmetries in the mathematics of particle physics are altered to produce new particles and forces.

**synchronous rotation** The rotation of a body with a period equal to its orbital period; also called 1-to-1 spin-orbit coupling.

**synchrotron radiation** A type of nonthermal radiation emitted by charged particles moving through a magnetic field.

**synodic month** The period of revolution of the Moon with respect to the Sun or the length of one cycle of lunar phases; also called the lunar month.

**synodic period** The interval between successive occurrences of the same configuration of a planet.

**S0 galaxy** See *lenticular galaxy*.

**T Tauri stars** Young variable stars associated with interstellar matter that show erratic changes in luminosity.

**T Tauri wind** A flow of particles away from a T Tauri star.

**tail (of a comet)** Gas and dust particles from a comet's nucleus that have been swept away from the comet's head by the radiation pressure of sunlight and the solar wind.

**tangential velocity** That portion of an object's velocity perpendicular to the line of sight.

**temperature** See *degree Celsius, degree Fahrenheit,* and *kelvin*.

**terminator** The line dividing day and night on the surface of the Moon or a planet; the line of sunset or sunrise.

**terrae** Cratered lunar highlands.

**terrestrial planets** Mercury, Venus, Earth, and Mars; the classification sometimes also includes the Galilean satellites and Pluto.

**theory** A hypothesis that has withstood experimental or observational tests.

**theory of everything (TOE)** See *supergrand unified theory*.

**thermal equilibrium** A balance between the input and outflow of heat in a system.

**thermal pulses** Brief bursts in energy output from the helium-burning shell of an aging low-mass star.

**thermal radiation** The radiation naturally emitted by any object that is not at absolute zero. Blackbody radiation is an idealized case of thermal radiation.

**thermonuclear fusion** The combining of nuclei under conditions of high temperature in a process that releases substantial energy.

**thermonuclear reaction** A reaction resulting from the high-speed collision of nuclear particles that are moving rapidly because they are at a high temperature.

**thermosphere** A region in the Earth's atmosphere between the mesosphere and the exosphere.

**3-to-2 spin-orbit coupling** The rotation of Mercury, which makes three complete rotations on its axis for every two complete orbits around the Sun.

**threshold temperature** The temperature above which photons spontaneously produce particles and antiparticles of a particular type.

**tidal force** A gravitational force whose strength and/or direction varies over a body and thus tends to deform the body.

**tidal heating** The heating of the interior of a satellite by continually varying tidal stresses.

**time dilation** The slowing of time due to relativistic motion.

**time zone** A region on the Earth where, by agreement, all clocks have the same time.

**Titius-Bode Law** A series of numbers that corresponds roughly to the sizes of the orbits of the planets.

**TOE** See *supergrand unified theory*.

**total lunar eclipse** A lunar eclipse during which the Moon is completely immersed in the Earth's umbra.

**total solar eclipse** A solar eclipse during which the Sun is completely hidden by the Moon.

**totality** The period during a total solar eclipse when the disk of the Sun is completely hidden.

**transit** See *meridian transit, solar transit*.

**triple alpha process** A sequence of two thermonuclear reactions in which three helium nuclei combine to form one carbon nucleus.

**triple bands (Europa)** Linear features on Europa's surface in which a bright stripe is flanked by two dark stripes.

**Trojan asteroid** One of several asteroids that share Jupiter's orbit about the Sun.

**Tropic of Cancer** A circle of latitude 23½° north of the Earth's equator.

**Tropic of Capricorn** A circle of latitude 23½° south of the Earth's equator.

**tropical year** The period of revolution of the Earth about the Sun with respect to the vernal equinox.

**troposphere** The lowest level in the Earth's atmosphere.

**Tully-Fisher relation** A correlation between the width of the 21-cm line of a spiral galaxy and the total luminosity of that galaxy.

**tuning fork diagram** A diagram that summarizes Edwin Hubble's classification scheme for spiral, barred spiral, and elliptical galaxies.

**turnoff point** The point on a Hertzsprung-Russell diagram where the stars in a cluster are leaving the main sequence.

**21-cm radio emission** Radio radiation emitted by neutral hydrogen atoms in interstellar space.

**22-year solar cycle** The semiregular 22-year interval between successive appearances of sunspots at the same latitude and with the same magnetic polarity.

**Type I Cepheid** A metal-rich Cepheid variable.

**Type Ia supernova** A supernova whose spectrum lacks hydrogen lines but has a strong absorption line of ionized silicon.

**Type Ib supernova** A supernova whose spectrum lacks hydrogen and silicon lines but has a strong helium absorption line.

**Type Ic supernova** A supernova whose spectrum is almost devoid of emission or absorption lines.

**Type II Cepheid** A metal-poor Cepheid variable.

**Type II supernova** A supernova with hydrogen emission lines in its spectrum, caused by the explosion of a massive star.

**UBV filters** Three colored filters that are transparent to ultraviolet (U), blue (B), and visible (V) light.

**UBV photometry** A system for determining the surface temperature of a star by measuring the star's brightness in the ultraviolet, blue, and visible spectral regions.

**ultraviolet radiation** Electromagnetic radiation of wavelengths shorter than those of visible light but longer than those of X rays.

**umbra (of a shadow)** (plural **umbrae**) The central, completely dark portion of a shadow.

**umbra (of a sunspot)** The dark, central region of a sunspot.

**unbounded universe** A universe throughout which the average density of matter is less than the critical density; an open universe.

**undifferentiated asteroid** An asteroid within which chemical differentiation did not occur.

**universal constant of gravitation** (*G*) The constant of proportionality in Isaac Newton's law of gravitation.

**universal law of gravitation** A formula deduced by Isaac Newton that expresses the strength of the force of gravity that two masses exert on each other.

**upper meridian** The half of the meridian that lies above the horizon.

**Van Allen belts** Two doughnut-shaped regions around the Earth where many charged particles (protons and electrons) are trapped by the Earth's magnetic field.

**velocity** The speed and direction with which an object moves.

**vernal equinox** The point on the ecliptic where the Sun crosses the celestial equator from south to north; also used to refer to the date on which the Sun passes through this intersection.

**very-long-baseline interferometry (VLBI)** A method of connecting widely separated radio telescopes to make observations of very high resolution.

**virtual pair** A particle and antiparticle that exist for such a brief interval that they cannot be observed.

**visible light** Photons detectable by the human eye.

**visual binary** A binary star in which the two components can be resolved through a telescope.

**VLBI** See *very-long-baseline interferometry*.

**void** A large volume of space, typically 100 to 400 million light-years in diameter, that contains very few galaxies.

**volatile element** An element with low melting and boiling points.

**waning crescent moon** The phase of the Moon that occurs between third quarter and new moon.

**waning gibbous moon** The phase of the Moon that occurs between full moon and third quarter.

**water hole** A region in the microwave spectrum suitable for interstellar radio communication.

**wave-particle duality** The notion from quantum mechanics that particles have wavelike properties and radiation has particlelike properties.

**wavelength** The distance between two successive wave crests.

**wavelength of maximum emission** The wavelength at which a heated object emits the greatest intensity of radiation.

**waxing crescent moon** The phase of the Moon that occurs between new moon and first quarter.

**waxing gibbous moon** The phase of the Moon that occurs between first quarter and full moon.

**weak nuclear force** The short-range force that is responsible for transforming certain particles into other particles, such as the decay of a neutron into a proton.

**weakly interacting massive particle (WIMP)** A hypothetical massive particle that may comprise part of the unseen dark matter.

**weight** The force with which gravity acts on a body.

**white dwarf** A low-mass star that has exhausted all its thermonuclear fuel and contracted to a size roughly equal to the size of the Earth.

**white hole** A black hole from which matter and radiation emerge.

**white ovals** Round, whitish features usually seen in Jupiter's southern hemisphere.

**Widmanstätten patterns** Crystalline structure seen in certain types of meteorites.

**Wien's law** A relationship between the temperature of a blackbody and the wavelength at which it emits the greatest intensity of radiation.

**WIMP** See *weakly interacting massive particle*.

**winding dilemma** The problem of why the spiral arms of a galaxy like the Milky Way do not disappear.

**winter solstice** The point on the ecliptic where the Sun reaches its greatest distance south of the celestial equator; also used to refer to the date on which the Sun passes through this point.

**wormhole** A speculative, topological feature of a black hole that connects our universe with another universe.

**X-ray burster** A nonperiodic X-ray source that emits powerful bursts of X rays.

**X rays** Electromagnetic radiation whose wavelength is between that of ultraviolet light and gamma rays.

**ZAMS** See *zero-age main sequence*.

**Zeeman effect** A splitting or broadening of spectral lines due to a magnetic field.

**zenith** The point on the celestial sphere directly overhead an observer.

**zero-age main sequence** The main sequence of young stars that have just begun to burn hydrogen at their cores.

**zero-age main sequence star** A newly formed star that has just arrived on the main sequence.

**zero curvature** The curvature of a surface or space in which parallel lines remain parallel and the sum of the angles of a triangle is exactly $180°$.

**zodiac** A band of 12 constellations around the sky centered on the ecliptic.

**zonal winds** The pattern of alternating eastward and westward winds found in the atmospheres of Jupiter and Saturn.

**zone** A light-colored band in Jupiter's atmosphere.

# Answers to Selected Questions

*Only mathematical answers are given, not answers which require interpretation or discussion. Your instructor will expect you to show the steps required to get each mathematical answer.*

## Chapter 1
11. (a) $1.08 \times 10^{14}$ km (b) 11.4 yr   12. $4.99 \times 10^2$ s
13. About $3 \times 10^{36}$ times larger   14. $8.7 \times 10^3$ km
15. $2.9 \times 10^7$ Suns   16. $4.3 \times 10^9$ km   17. $4.7 \times 10^{17}$ s
18. $8.94 \times 10^{56}$ hydrogen atoms   19. $5.669 \times 10^{-2}$ arcmin
20. $3.4 \times 10^3$ km   21. (a) 1.1 m (b) 69 m (c) 4100 m
22. 6.9 m   23. 3.7 km

## Chapter 2
20. (b) About 10 hours   24. 9:30 P. M.   26. (a) 6:00 P. M.
sidereal time (b) September 21 30. 26.5°   33. R. A. = $0^h 0^m 0^s$
34. R. A. = $18^h 0^m 0^s$, Decl. = $-23° 27'$   36. 80°

## Chapter 3
20. One   25. (a) March 9, 2016; Indian Ocean, East Indies, Pacific (b) April 1, 2052   28. 49 arcsec

## Chapter 4
10. $4.9$ AU$^2$ in 2002, $24.5$ AU$^2$ in five years   11. Average distance = 100 AU, maximum distance = 200 AU   12.
Semimajor axis = 2.0 AU, period = 2.8 yr   15. 4 N, 8 m/s$^2$
16. $1.98 \times 10^{20}$ N; the force between the Sun and the Earth is about 180 times greater   17. The Sun's gravitational pull would be 1/25 as strong   22. 224.7 days   25. (a) 0.86 AU
(b) 146 days   26. (a) 24 hours (b) 42,300 km   27. You would weigh only 1/9 as much as on Earth   28. You would weigh only 0.377 as much as on Earth   29. The forces are approximately the same; the acceleration of the Earth is 100 times greater   30. The star is 81 times more massive than the Sun

## Chapter 5
4. $3.64 \times 10^{14}$ Hz   5. 3.34 m   7. 333 nm   8. 2400 K
16. 9980°F   17. About 10 μm, or 10,000 nm   18. 9.4 μm, or 9400 nm   19. 2.65 times as much   20. $\lambda_{max} = 1.0$ μm, or 1000 nm; this star would be 16 times dimmer than the Sun
21. 2.9 nm   22. 10,000 K   23. The other star would be 6.25 times brighter   25. $2.43 \times 10^{-3}$ nm   26. 434.1 nm
27. Possible transitions are 1 eV ($\lambda = 1240$ nm), 2 eV ($\lambda = 620$ nm), and 3 eV ($\lambda = 414$ nm)   28. Coming toward us at 13.0 km/s   29. 656.9 nm   30. 8,600 km/s

## Chapter 6
17. $1/25 = 0.04$   18. The Keck telescope has 17.4 times the light–gathering power of HST   19. (a) 222× (b) 100× (c) 36×; diffraction–limited angular resolution at $\lambda = 600$ nm is

0.6 arcsec   21. HST can see features as small as 300 km on Jupiter's moons; the human eye can see features as small as 110 km on our Moon   22. HST can see features as small as 2900 km   23. (a) 0.18 m (b) $5.9 \times 10^{-4}$ arcsec

## Chapter 7
17. (a) $3.3 \times 10^{21}$ J (b) Equivalent to 40 million Hiroshima–type weapons   18. 12 km/s   19. 617 km/s
20. (a) 2.02 km/s (b) 19.4 km/s   22. Mass = $6.4 \times 10^{23}$ kg, average density = 3900 kg/m$^3$   23. (a) The Earth's mass would be about 50 times greater   26. $2.03 \times 10^{30}$ kg, or 1.02 times the Sun's mass   27. (a) 28 m/s (b) 0.25 arcsec (c) 0.0016 arcsec

## Chapter 8
17. The present–day separation is 6600 km; they were together $2.2 \times 10^8$ years ago   18. $1.6 \times 10^7$ years
20. 17% for the core, 82% for the mantle, and 1% for the crust   21. Only 0.020 (2%) of the Earth's mass lies in the inner core   22. The average density of the core is about 15,000 kg/m$^3$   24. 4 km

## Chapter 9
18. 130 N on the Moon, 780 N on Earth   19. In 100 years, probability is $8.3 \times 10^{-8}$, or 1 chance in 12 million; in $10^6$ years, probability is $8.3 \times 10^{-4}$, or 1 chance in 1200
20. $5.5 \times 10^5$ km

## Chapter 10
14. 12.6 arcsec   15. Features larger than about 730 km can be resolved   17. For a temperature of 700 K, $\lambda_{max} = 4.1$ mm
19. (a) 3.03 m/s (b) 1.26 nm   20. (a) The minimum molecular weight is about 30   21. (a) 310 N on Mercury, 130 N on the Moon, 780 N on Earth   23. 0.615 AU, or $9.19 \times 10^7$ km

## Chapter 11
18. $\lambda_{max} = 3.9$ mm   19. $6 \times 10^{-6}$ nm   24. 920 m

## Chapter 12
16. 270 km with a 1 arcsec telescope, 27 km with HST
17. The Earth and Moon would be separated by 350 arcsec
18. $6.41 \times 10^{23}$ kg   25. Angular size of Phobos = 16 arcmin; angular size of Deimos = 3 arcmin (as compared to 30 arcmin for the Moon as seen from the Earth)

## Chapter 13
14. 124 K   15. (a) About 160 K = $-113°$C = $-171°$F (b) About 6 atm (c) About 50 m   17. 7770 N   18. 59.5 km/s, or 5.3 times escape speed for the Earth   19. Roughly 300 km/h

## Chapter 14

21. 900 km   22. 0.046 arcsec   25. $2.8 \times 10^{11}$ (280 billion) years   26. About 8.1 minutes

## Chapter 15

18. (a) 14.4 hours for outer edge of A ring, 7.9 hours for inner edge of B ring (b) opposite   20. Escape speed = 2.6 km/s; mass of molecule = $2.0 \times 10^{-26}$ kg, corresponding to a molecular weight of 12   23. 0.08 arcsec

## Chapter 16

16. 2 hours, 5 minutes   17. Sun–Uranus force = $1.39 \times 10^{21}$ N, Neptune–Uranus force = $2.24 \times 10^{17}$ N; Neptune reduces the sunward gravitational pull on Uranus by $1.61 \times 10^{-4}$, or 0.0161%   18. As seen from Pluto, angular diameter of Charon = 3.47° and angular diameter of Sun = 24.3 arcsec   20. (a) At perihelion, $1.13 \times 10^{-3}$ as bright as on Earth (b) At aphelion, $4.10 \times 10^{-4}$ as bright as on Earth (c) 2.77 times brighter   23. About 7200 km

## Chapter 17

19. Orbital period = 678 years, angular speed = 0.53 degrees/year   20. 50 km   21. 2.82 AU   23. (a) $3.6 \times 10^{12}$ kg (b) escape speed = 0.83 m/s   24. The diameter of the iron-rich core of a typical parent asteroid is about 1/3 of the asteroid's overall diameter   25. (a) Orbital period = 350 yr, lifetime = $3.5 \times 10^{4}$ yr (b) Orbital period = $1.1 \times 10^{4}$ yr, lifetime = $1.1 \times 10^{6}$ yr (c) Orbital period = $3.5 \times 10^{5}$ yr, lifetime = $3.5 \times 10^{7}$ yr (d) Orbital period = $1.1 \times 10^{7}$ yr, lifetime = $1.1 \times 10^{9}$ yr

## Chapter 18

19. (a) 1700 nm   21. For the photosphere, 500 nm; for the chromosphere, = 58 nm; for the corona, 1.9 nm   22. For the umbra, 670 nm; for the penumbra, 580 nm   23. (a) (flux from patch of penumbra)/(flux from patch of photosphere) = 0.55 (b) (Flux from patch of penumbra)/(flux from patch of umbra) = 1.8   25. (a) $1.8 \times 10^{-9}$ J (b) $9.0 \times 10^{16}$ J (c) $5.4 \times 10^{41}$ J   26. (a) $4.6 \times 10^{-36}$ s (b) $2.3 \times 10^{-10}$ s (c) $1.4 \times 10^{15}$ s = $4.4 \times 10^{7}$ yr   27. $1.4 \times 10^{13}$ kilograms per second   29. 0.048 (4.8%) of the Sun's mass will be converted from hydrogen into helium; chemical composition of Sun (by mass) will be 69% hydrogen, 30% helium   31. $1.9 \times 10^{-7}$ nm

## Chapter 19

17. Average Sun–Pluto distance = $1.92 \times 10^{-4}$ pc; Proxima Centauri is 6780 times farther from the Sun   18. (a) 9.7 pc (b) 0.010 arcsec   19. 3.88 pc   20. (a) 159 km/s (b) 293 km/s   21. Distance = 132 pc; tangential velocity would have to be about 6500 km/s   22. 80 km/s   23. (a) +50 km/s (c) 486.21 nm   25. As seen from Saturn, the Sun is 91.0 times dimmer than as seen from Earth   26. Star $B$ is 9 times more luminous than star $B$   27. Star C appears 16 times brighter than star $D$   28. 37.0 AU   29. The farther star is 100 times more luminous than the closer star   30. 6.8 $L_\odot$   31. 0.19 pc = $4.0 \times 10^{4}$ AU   32. (a) +13.3 (b) (Luminosity of Krüger 60 B)/(luminosity of Sun) = $1.3 \times 10^{-9}$   33. +17   34. $1.6 \times 10^{4}$ pc   35. 28 $R_\odot$

36. Radius increases by a factor of 2, luminosity increases by a factor of 64   38. Surface temperature = 5000 K, diameter = $2.4 \times 10^{7}$ km (17 times the diameter of the Sun)   39. Using a luminosity of 35 $L_\odot$, distance = 14 pc   40. (a) 5.0 pc (b) 22.5 AU (c) 1.5 $M_\odot$   41. (a) 40 $M_\odot$ (b) 8 $M_\odot$ and 32 $M_\odot$   42. *Hipparcos* could measure stars out to 500 pc; the volume covered by these observations is 125 times greater than with Earth–based observations

## Chapter 20

17. 61%   18. $3.4 \times 10^{4}$ atoms/cm$^3$   20. 1100 $R_\odot$ = $7.4 \times 10^{8}$ km = 4.5 AU   22. (a) 250 AU = $3.7 \times 10^{10}$ km (b) 4000 yr (c) 1000 AU; 24 years   23. $1.3 \times 10^{31}$ m$^3$   24. (a) 2.9° (b) 1300 km/s (c) 180 yr

## Chapter 21

19. 657,000 km   21. $1.9 \times 10^{21}$ kg, or 13% of the original mass of hydrogen   22. $3.4 \times 10^{7}$ yr   23. (a) Nearly 180° (it will cover most of the sky) (b) about 2000 K (1700°C, or 3200°F)   27. (a) $1.65 \times 10^{6}$ km   28. About 680 pc   29. About 100 pc

## Chapter 22

15. 0.11 $R_\odot$   17. 7000 years ago   18. (a) $1.8 \times 10^{9}$ kg/m$^3$ (b) 6400 km/s   20. (a) $6.5 \times 10^{5}$ m/s   21. The maximum luminosity of SN 1993J was 4.5 times greater than that of SN 1987A   22. (a) $1.7 \times 10^{30}$ kg = 0.84 $M_\odot$ (b) $1.1 \times 10^{12}$ N (c) $1.5 \times 10^{8}$ m/s, or 0.5 of the speed of light   23. $3.24 \times 10^{9}$ km = 22.7 AU   25. (a) $1.3 \times 10^{-7}$ $b_\odot$ (b) It would be 130 times brighter than Venus   26. $1.3 \times 10^{8}$ pc (130 Mpc)   29. 1100 km/s, or $3.8 \times 10^{-3}$ of the speed of light

## Chapter 23

14. 0.086 $L_\odot$   15. About 5400 BC   16. 680 years   18. It will take about 3100 years, until about 5100 AD (note that "expansion rate" is the rate at which the surface of the nebula moves)   19. (a) Distance = 1100 pc = 3600 ly; explosion occurred around 2500 BC   21. The maximum correction is $10^{-4}$ of the pulsar period   22. (a) Density of matter in a neutron = $4 \times 10^{17}$ kg/m$^3$   25. (a) 0.097 nm (b) Luminosity of neutron star = $5.8 \times 10^{31}$ W = $1.5 \times 10^{5}$ $L_\odot$

## Chapter 24

14. 0.98 of the speed of light   15. 0.99995 of the speed of light   16. (a) 25 yr (b) 20 light–years as measured by an Earth observer, 12 light–years as measured by the astronaut   18. $5.7 \times 10^{8}$ yr   19. 2.8 $M_\odot$   20. 2.8 m   21. (a) $R_{Sch}$ = 8.86 mm, density = $2.05 \times 10^{30}$ kg/m$^3$ (b) $R_{Sch}$ = 2.95 km, density = $1.85 \times 10^{19}$ kg/m$^3$ (c) $R_{Sch}$ = $3.24 \times 10^{11}$ km = 2170 AU, density = $1.53 \times 10^{-3}$ kg/m$^3$   22. $1.9 \times 10^{17}$ kg/m$^3$   23. $2.7 \times 10^{38}$ kg = $1.4 \times 10^{8}$ $M_\odot$   24. (a) $2.01 \times 10^{6}$ km

## Chapter 25

14. $9.5 \times 10^{-25}$ J; it takes $3.2 \times 10^{5}$ such photons to equal the energy of one $H_a$ photon   15. (a) $3.1 \times 10^{8}$ yr (b) $7.4 \times 10^{11}$ $M_\odot$   16. A 10% error in radius results in a 10% error in mass; a 10% error in velocity results in a 20% error in mass   17. (b) 5700 pc   19. $2.7 \times 10^{11}$ $M_\odot$   20. (a) $1.2 \times 10^{12}$ pc$^3$

**(b)** $1.1 \times 10^8 \text{ pc}^3$ **(c)** Probability $= 9.6 \times 10^{-5}$; we can expect to see a supernova within 300 pc once every 350,000 years **23. (a)** $7.7 \times 10^6 \text{ km} = 0.051 \text{ AU}$ **(b)** $6.4 \times 10^{-6}$ arcsec **24. (a)** 20 years **(b)** $2.5 \times 10^6 \text{ M}_\odot$

## Chapter 26

**15. (a)** 6.9 Mpc **(b)** 8.3 Mpc **(c)** 1.4 Mpc = 1400 kpc, greater than the 900–kpc distance to M31 **16.** 9.6 Mpc **17.** 54 km/s/Mpc **18. (a)** 144 Mpc $= 4.69 \times 10^8$ light–years **(b)** Distance would be larger if $H_0$ were smaller, and vice versa **19. (a)** $1.2 \times 10^{70}$ **(b)** $3.6 \times 10^{-6}$ atoms/cm³ **20.** $1.2 \times 10^{12}$ $\text{M}_\odot$ **21.** Orbital period $= 4.4 \times 10^8$ yr; mass within 20 kpc $= 3.6 \times 10^{11} \text{ M}_\odot$

## Chapter 27

**12.** 13.1 **13.** $2.65 \times 10^5 \text{ km/s} = 0.883$ of the speed of light **14.** $2.9 \times 10^5 \text{ km/s} = 0.96$ of the speed of light **15.** $9.7 \times 10^4$ km/s $= 3.5 \times 10^8$ km/h $= 0.32$ of the speed of light **18. (a)** $1.80 \times 10^5 \text{ km/s} = 0.600c$ **(b)** 134 hours **(c)** 970 AU **19. (a)** $7.8 \times 10^{10} \text{ L}_\odot = 3.0 \times 10^{37}$ watts **21.** $2.05 \times 10^{30} \text{ kg/m}^3$ **22.** $2.7 \times 10^{38} \text{ kg} = 1.4 \times 10^8 \text{ M}_\odot$ **23.** $2.9 \times 10^9 \text{ km} = 20 \text{ AU}$

## Chapter 28

**13.** 771 nm **14. (a)** 20 billion years **(b)** 13 billion years **(c)** 9.7 billion years **15.** $1.6 \times 10^8$ km/s/Mpc **16.** 65 times denser **17.** 10.3 K **18. (a)** $4.6 \times 10^{-27} \text{ kg/m}^3$ **19. (a)** $\rho_{rad} = 9.5 \times 10^{-18} \text{ kg/m}^3$ **(b)** $\rho_{rad} = 4.8 \times 10^{-4} \text{ kg/m}^3$ **(c)** $\rho_{rad} = 1.3 \times 10^{-7} \text{ kg/m}^3$ **20. (a)** $4.7 \times 10^{-27} \text{ kg/m}^3$ **(b)** $1.9 \times 10^{-26} \text{ kg/m}^3$ **21. (a)** 79 Mpc **(b)** $\rho_m = 5 \times 10^{-25}$ kg/m³ **(c)** temperature $= 17.3$ K, $\rho_{rad} = 7.5 \times 10^{-28}$ kg/m³ **22.** $2.2 \times 10^{-42}$ m **25.** $-7.4 \times 10^{-53} \text{ m}^2$

## Chapter 29

**13.** $1.1 \times 10^{-26}$ J **14.** $3.2 \times 10^{-25}$ s **15. (a)** 80.5 GeV **(b)** $9.34 \times 10^{14}$ K **17.** For $H_0 = 65$ km/s/Mpc, mass $= 7.9 \times 10^{-35} \text{ kg} = 8.7 \times 10^{-5}$ the mass of the electron; for $H_0 = 85$ km/s/Mpc, mass $= 1.4 \times 10^{-34}$ kg $= 1.5 \times 10^{-4}$ the mass of the electron **18.** $1.1 \times 10^{14}$ m $= 0.011$ light–year

## Chapter 30

**2. (a)** 10,000 years **(b)** 10 million years **5.** 200 km

# Index

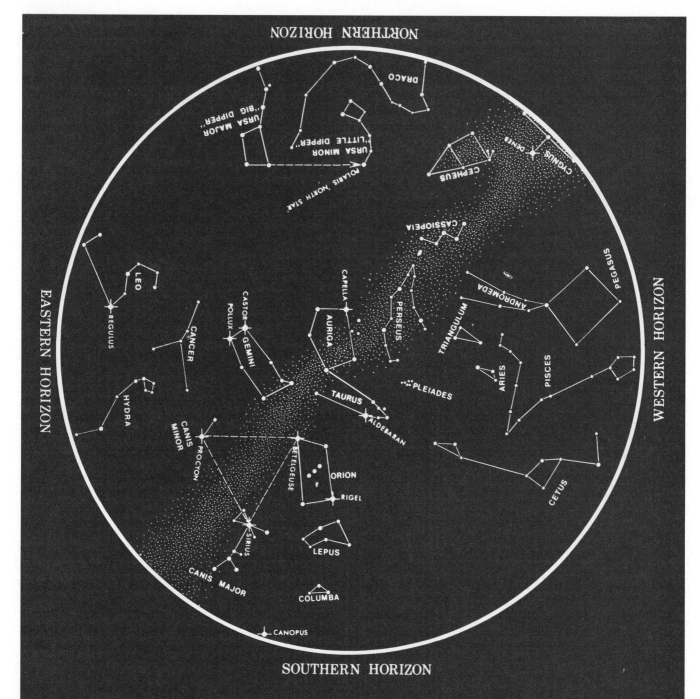

# THE NIGHT SKY IN JANUARY

Chart time (Local Standard Time):

10 pm...First of January
9 pm...Middle of January
8 pm...Last of January

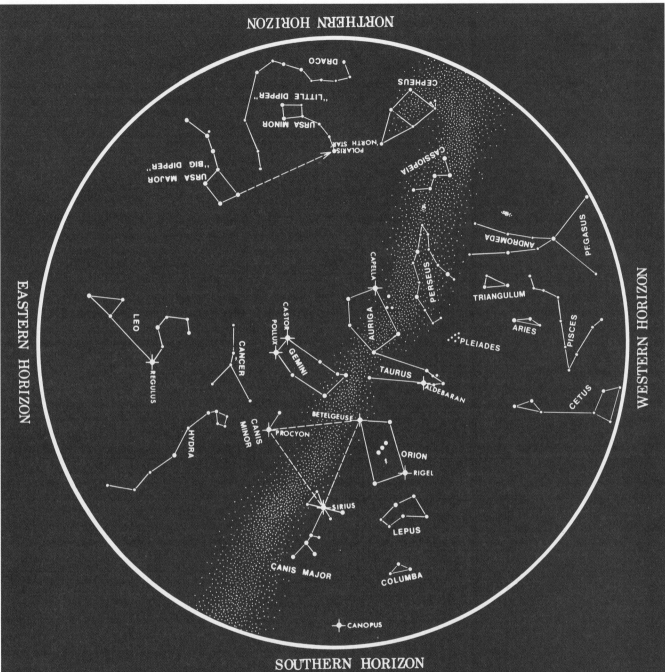

EASTERN HORIZON

WESTERN HORIZON

SOUTHERN HORIZON

# THE NIGHT SKY IN FEBRUARY

Chart time (Local Standard Time):

10 pm...First of February
9 pm...Middle of February
8 pm...Last of February

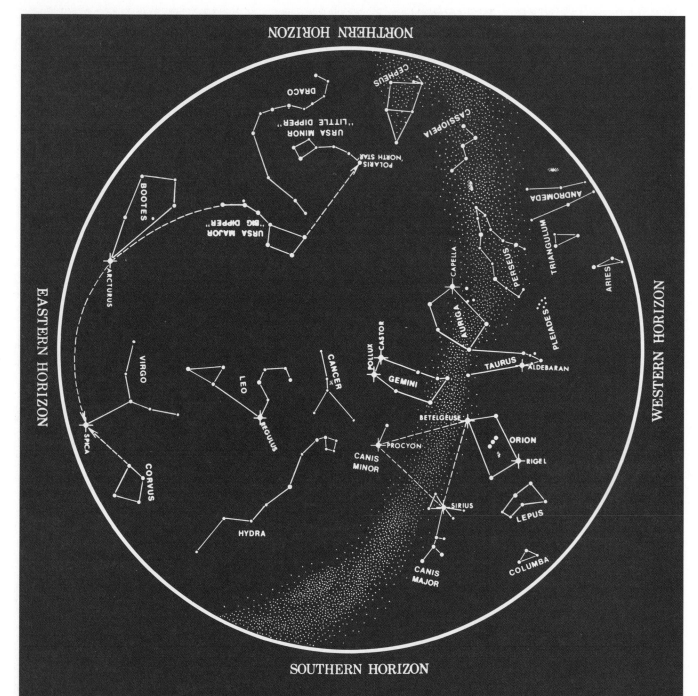

## THE NIGHT SKY IN MARCH

Chart time (Local Standard Time):

10 pm...First of March
9 pm...Middle of March
8 pm...Last of March

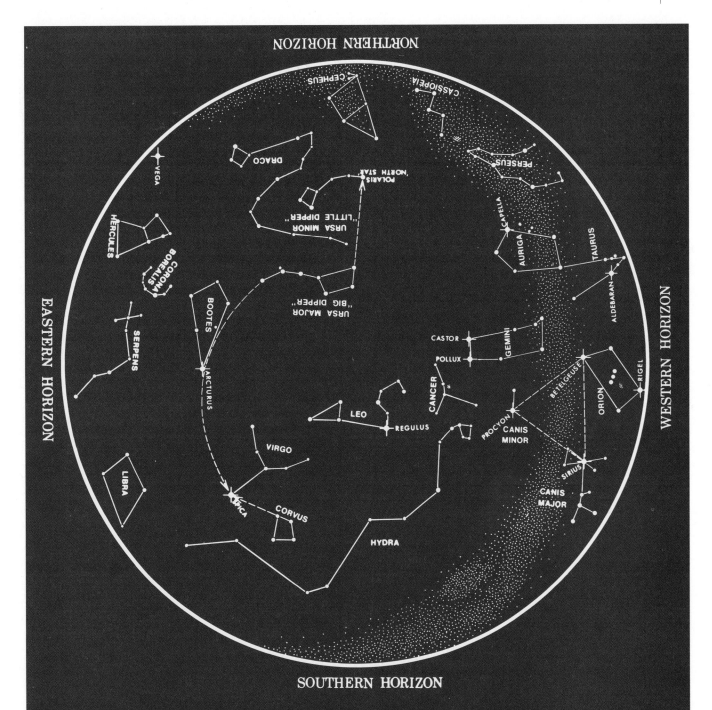

SOUTHERN HORIZON

# THE NIGHT SKY IN APRIL

Chart time (Daylight Savings Time):

11 pm...First of April
10 pm...Middle of April
9 pm...Last of April

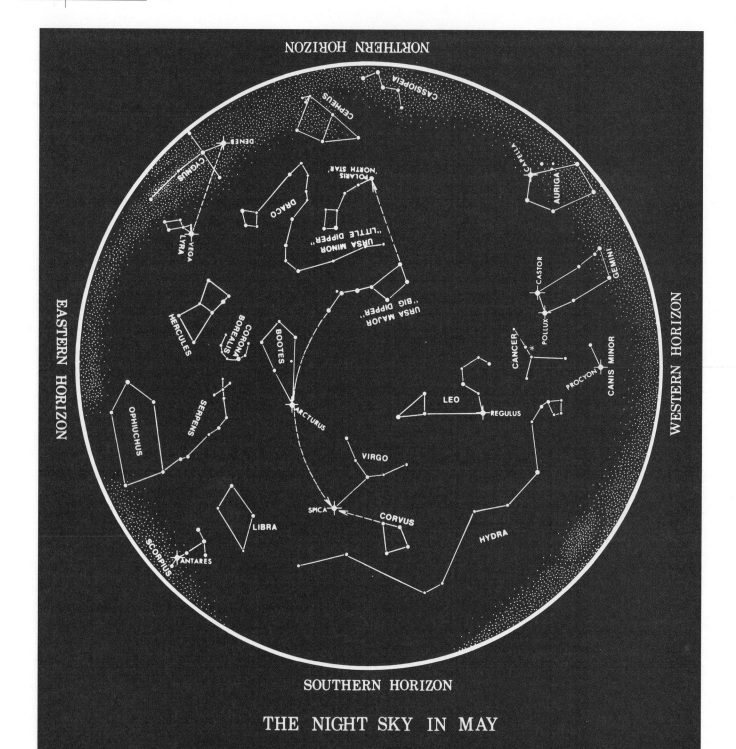

## THE NIGHT SKY IN MAY

Chart time (Daylight Savings Time):

11 pm...First of May
10 pm...Middle of May
9 pm...Last of May

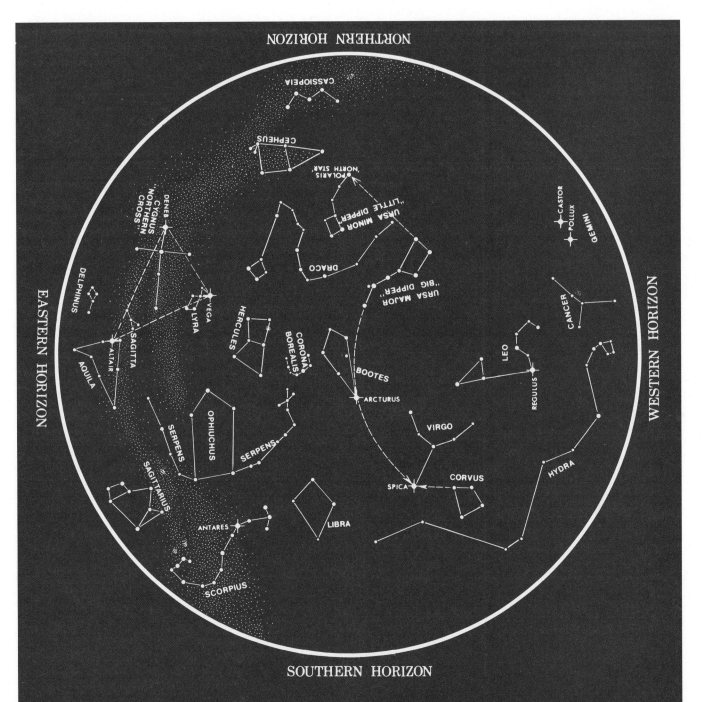

# THE NIGHT SKY IN JUNE

Chart time (Daylight Savings Time):

11 pm...First of June
10 pm...Middle of June
9 pm...Last of June

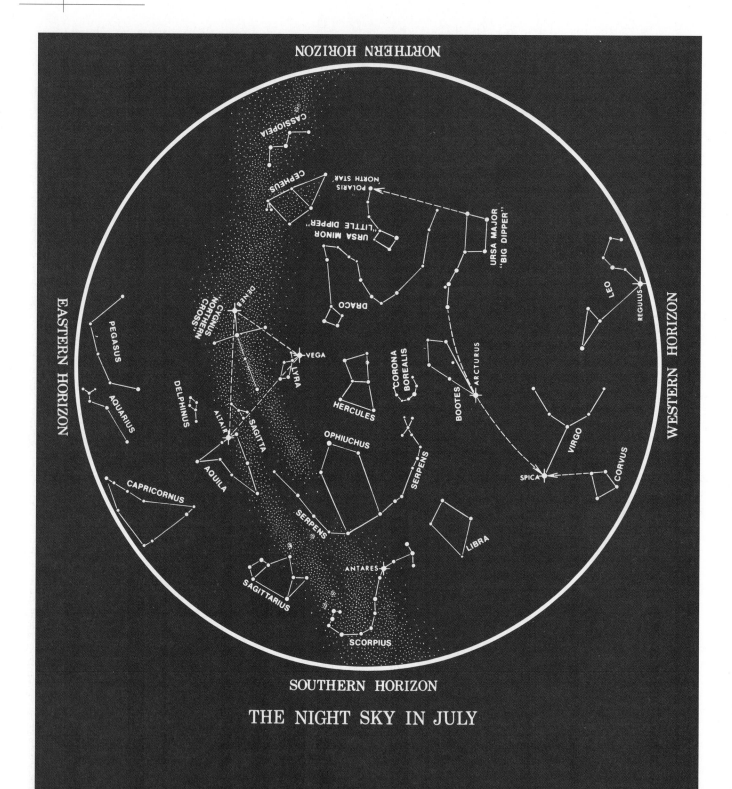

## THE NIGHT SKY IN JULY

SOUTHERN HORIZON

Chart time (Daylight Savings Time):

11 pm...First of July
10 pm...Middle of July
9 pm...Last of July

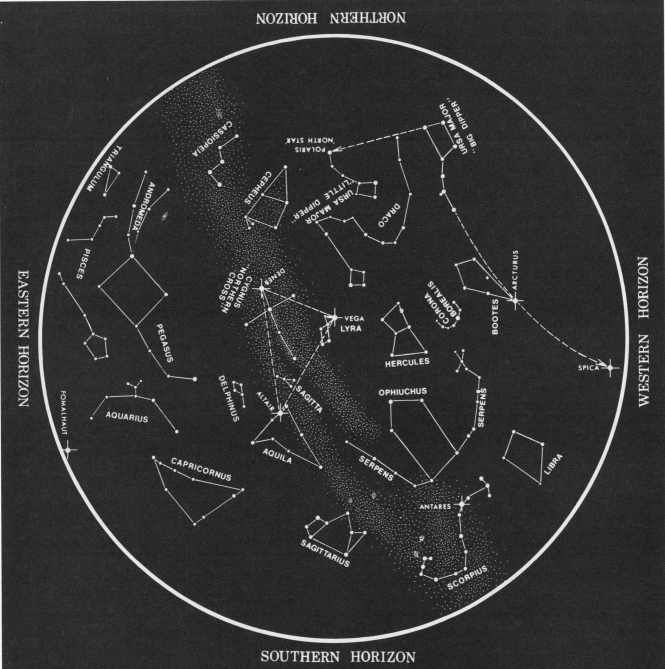

## THE NIGHT SKY IN AUGUST

Chart time (Daylight Savings Time):

11 pm...First of August
10 pm...Middle of August
9 pm...Last of August

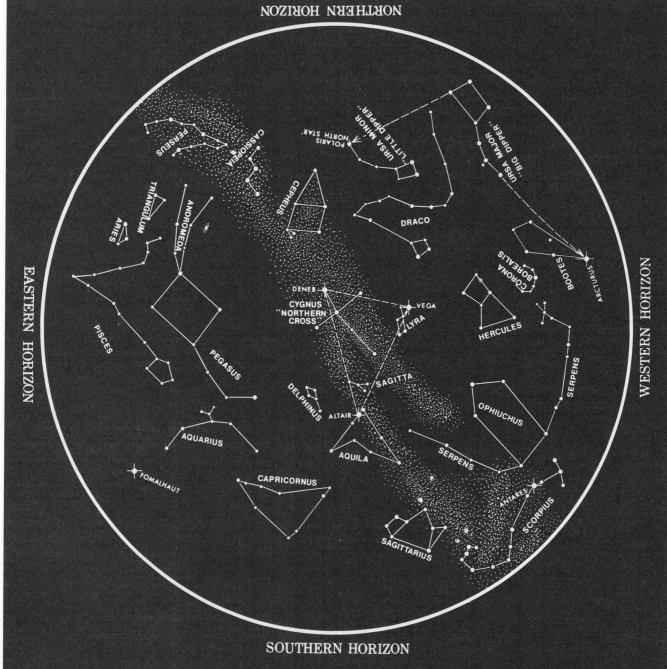

SOUTHERN HORIZON

# THE NIGHT SKY IN SEPTEMBER

Chart time (Daylight Savings Time):

11 pm...First of September
10 pm...Middle of September
9 pm...Last of September

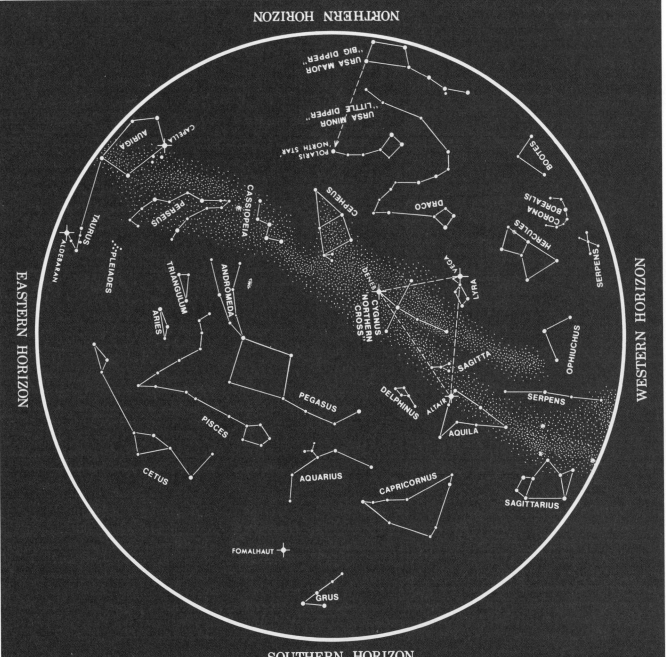

# THE NIGHT SKY IN OCTOBER

Chart time (Daylight Savings Time):

11 pm...First of October
10 pm...Middle of October
9 pm...Last of October

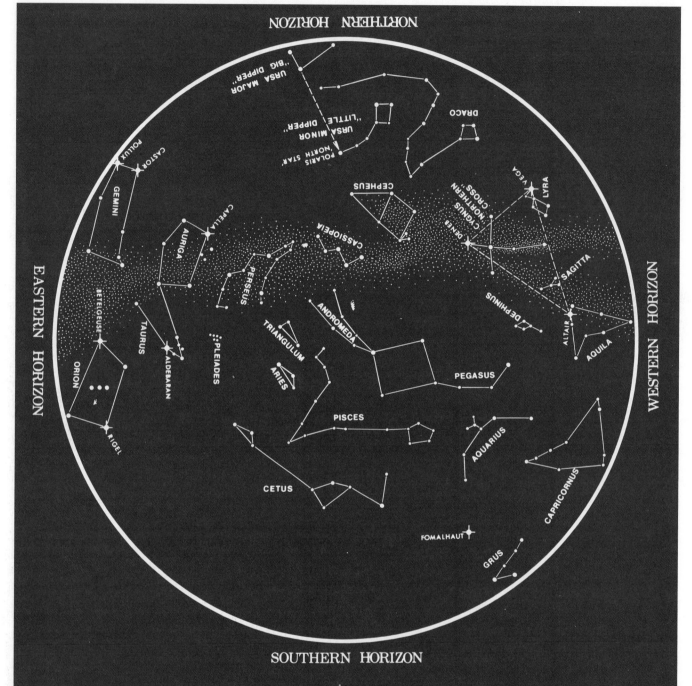

THE NIGHT SKY IN NOVEMBER

Chart time (Local Standard Time):

10 pm...First of November
9 pm...Middle of November
8 pm...Last of November

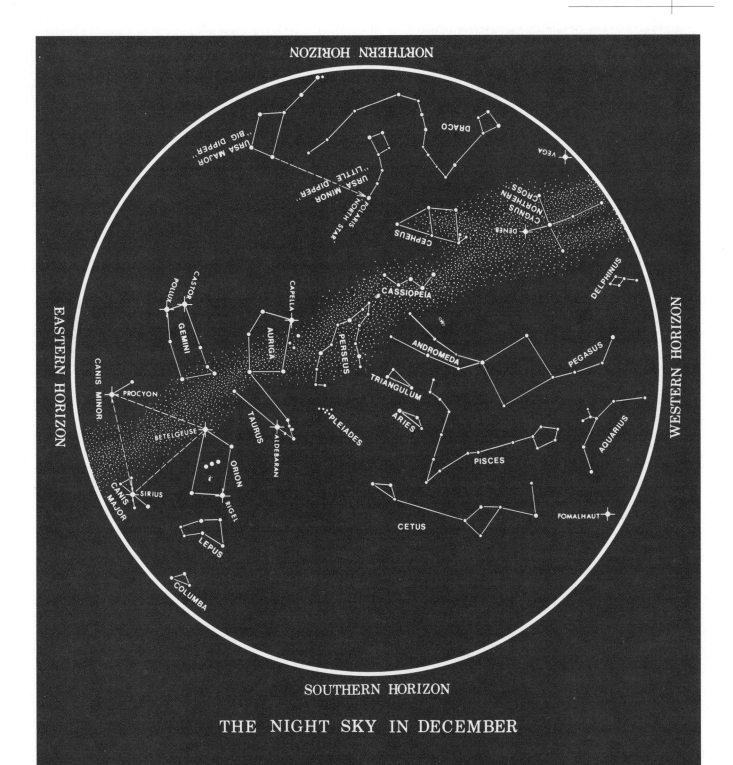

THE NIGHT SKY IN DECEMBER

Chart time (Local Standard Time):

10 pm...First of December
9 pm...Middle of December
8 pm...Last of December